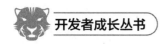
开发者成长丛书

Spark原理深入与编程实战
微课视频版

辛立伟　张　帆　张会娟◎编著

清华大学出版社
北京

内 容 简 介

本书系统讲述 Apache Spark 大数据计算平台的原理,以及如何将 Apache Spark 应用于大数据的实时流处理、批处理、图计算等各个场景。通过深入学习原理和对实践示例、案例的综合应用,使读者了解并掌握 Apache Spark 大数据计算平台的基本原理和技能,接近理论与实践的距离。

全书共分为 13 章,主要内容包括 Spark 架构原理与集群搭建、开发和部署 Spark 应用程序、Spark 核心编程、Spark SQL、Spark SQL 高级分析、Spark Streaming 流处理、Spark 结构化流、Spark 结构化流高级处理、Spark 图处理库 GraphFrame、下一代大数据技术(Delta Lake 数据湖、Iceberg 数据湖和 Hudi 数据湖)、Spark 大数据处理综合案例。本书源码全部在 Apache Spark 3.1.2 上调试成功,所有示例和案例均基于 Scala 语言。

为降低读者学习大数据技术的门槛,本书除了提供丰富的上机实践操作和范例程序详解外,还为购买和使用本书的读者提供了搭建好的 Hadoop、Hive 数据仓库和 Spark 大数据开发及学习环境。读者既可参照本书的讲解自行搭建 Hadoop 和 Spark 环境,也可直接使用作者提供的开发和学习环境,快速开始大数据和 Spark、数据湖的学习。

本书适合大数据学习爱好者、想要入门 Apache Spark 的读者作为入门和提高的技术参考书,也适合用作高等院校大数据专业相关的学生和老师的教材或教学参考书。

本书封面贴有清华大学出版社防伪标签,无标签者不得销售。
版权所有,侵权必究。举报: 010-62782989, beiqinquan@tup.tsinghua.edu.cn。

图书在版编目(CIP)数据

Spark 原理深入与编程实战:微课视频版/辛立伟,张帆,张会娟编著.—北京:清华大学出版社,2023.7
(开发者成长丛书)
ISBN 978-7-302-62886-6

Ⅰ.①S… Ⅱ.①辛… ②张… ③张… Ⅲ.①数据处理软件 Ⅳ.①TP274

中国国家版本馆 CIP 数据核字(2023)第 037787 号

责任编辑:赵佳霓
封面设计:刘　键
责任校对:时翠兰
责任印制:宋　林

出版发行:清华大学出版社
　　网　　址:http://www.tup.com.cn, http://www.wqbook.com
　　地　　址:北京清华大学学研大厦 A 座　　邮　编:100084
　　社 总 机:010-83470000　　邮　购:010-62786544
　　投稿与读者服务:010-62776969, c-service@tup.tsinghua.edu.cn
　　质 量 反 馈:010-62772015, zhiliang@tup.tsinghua.edu.cn
　　课 件 下 载:http://www.tup.com.cn, 010-83470236
印 装 者:大厂回族自治县彩虹印刷有限公司
经　　销:全国新华书店
开　　本:186mm×240mm　　印　张:49.25　　字　数:1107 千字
版　　次:2023 年 7 月第 1 版　　印　次:2023 年 7 月第 1 次印刷
印　　数:1~2000
定　　价:189.00 元

产品编号:097803-01

前言
PREFACE

大数据分析一直是热门话题，需要大数据分析的场景也越来越多。Apache Spark 是一个用于快速、通用、大规模数据处理的开源项目。现在，Apache Spark 已经成为一个统一的大数据处理平台，拥有快速的统一分析引擎，可用于大数据的批处理、实时流处理、机器学习和图计算。

2009 年，Spark 诞生于伯克利大学 AMP 实验室，最初属于伯克利大学的研究性项目。它于 2010 年被正式开源，于 2013 年被转交给 Apache 软件基金会，并于 2014 年成为 Apache 基金的顶级项目，整个过程不到五年时间。Apache Spark 诞生以后，迅速发展成为大数据处理技术中的佼佼者，目前已经成为大数据处理领域炙手可热的技术，其发展势头非常强劲。

自 2010 年首次发布以来，Apache Spark 已经成为最活跃的大数据开源项目之一。如今，Apache Spark 实际上已经是大数据处理、数据科学、机器学习和数据分析工作负载的统一引擎，是从业人员及希望进入大数据行业人员必须学习和掌握的大数据技术之一，但是作为大数据的初学者，在学习 Spark 时通常会遇到以下几个难题：

（1）缺少面向零基础读者的 Spark 入门教程。
（2）缺少系统化的 Spark 大数据教程。
（3）现有的 Spark 资料、教程或图书陈旧或者碎片化。
（4）官方全英文文档难以阅读和理解。
（5）缺少必要的数据集、可运行的实验案例及学习平台。

特别是 Spark 3 发布以后，性能得到了极大提升，并且增加了对数据湖等下一代大数据技术的支持。为此，编写本书一方面是为了笔者自己能更系统、更及时地跟进 Spark 的演进和迭代；另一方面也是为了降低面向零基础读者学习 Spark（及其他大数据技术）的入门难度。本书具有以下特点：

（1）面向零基础读者，知识点深浅适当，代码完整易懂。
（2）内容全面系统，包括架构原理、开发环境及程序部署、流和批计算、图计算等，并特别包含了 Delta Lake、Iceberg、Hudi 等数据湖内容。
（3）版本先进，所有代码均基于 Spark 3.1.2。特别值得一提的是，图计算部分抛弃了性能欠佳的 GraphX，而是引入了下一代 Spark 图计算库 GraphFrame 进行讲解和演示。
（4）全书包含大量的示例代码讲解和完整项目案例。

本书特别适合想要入门并深入掌握 Apache Spark 大数据开发和大数据分析、大数据 OLAP 引擎、流计算的读者，希望拥有大数据系统参考教材的教师，以及想要了解最新 Spark 技术应用的从业人员。

本书第 1~3 章由郑州财经学院张帆副教授编写，第 4 章和第 5 章由华北水利水电大学公共管理学院张会娟老师编写，其余部分由辛立伟编写。由于编者水平所限，行文及内容难免存在疏漏之处，请读者见谅并予以反馈，笔者会在后续的版本重构中不断提升质量。

编　者

2023 年 1 月

目录
CONTENTS

本书源代码

教学课件(PPT)

第1章　Spark 架构原理与集群搭建 (▶93min) ··· 1
  1.1　Spark 简介 ·· 1
  1.2　Spark 技术栈 ··· 3
    1.2.1　Spark Core ··· 4
    1.2.2　Spark SQL ·· 4
    1.2.3　Spark Streaming 和 Structured Streaming ·· 5
    1.2.4　Spark MLlib ·· 5
    1.2.5　Spark GraphX ·· 6
    1.2.6　SparkR ·· 6
  1.3　Spark 架构原理 ·· 7
    1.3.1　Spark 集群和资源管理系统 ·· 7
    1.3.2　Spark 应用程序 ·· 7
    1.3.3　Spark Driver 和 Executor ··· 9
  1.4　Spark 程序部署模式 ··· 10
  1.5　安装和配置 Spark 集群 ··· 11
    1.5.1　安装 Spark ·· 11
    1.5.2　了解 Spark 目录结构 ·· 12
    1.5.3　配置 Spark 集群 ··· 13
    1.5.4　验证 Spark 安装 ··· 14
  1.6　配置 Spark 历史服务器 ··· 15
    1.6.1　历史服务器配置 ··· 16
    1.6.2　启动 Spark 历史服务器 ·· 17
  1.7　使用 spark-shell 进行交互式分析 ··· 18
    1.7.1　运行模式--master ·· 18
    1.7.2　启动和退出 spark-shell ··· 19
    1.7.3　spark-shell 常用命令 ·· 21
    1.7.4　SparkContext 和 SparkSession ·· 22
    1.7.5　Spark Web UI ·· 23
  1.8　使用 spark-submit 提交 Spark 应用程序 ··· 25
    1.8.1　spark-submit 指令的各种参数说明 ··· 25
    1.8.2　提交 SparkPi 程序，计算圆周率 π 值 ·· 29
    1.8.3　将 SparkPi 程序提交到 YARN 集群上执行 ······································ 30

第 2 章 开发和部署 Spark 应用程序 (▶ 86min) ········· 33

- 2.1 使用 IntelliJ IDEA 开发 Spark SBT 应用程序 ········· 33
  - 2.1.1 安装 IntelliJ IDEA ········· 34
  - 2.1.2 配置 IntelliJ IDEA Scala 环境 ········· 37
  - 2.1.3 创建 IntelliJ IDEA SBT 项目 ········· 39
  - 2.1.4 配置 SBT 构建文件 ········· 42
  - 2.1.5 准备数据文件 ········· 42
  - 2.1.6 创建 Spark 应用程序 ········· 43
  - 2.1.7 部署分布式 Spark 应用程序 ········· 47
  - 2.1.8 远程调试 Spark 程序 ········· 49
- 2.2 使用 IntelliJ IDEA 开发 Spark Maven 应用程序 ········· 51
  - 2.2.1 创建 IntelliJ IDEA Maven 项目 ········· 51
  - 2.2.2 验证 SDK 安装和配置 ········· 53
  - 2.2.3 项目依赖和配置管理 ········· 55
  - 2.2.4 测试 Spark 程序 ········· 58
  - 2.2.5 项目编译和打包 ········· 58
- 2.3 使用 Java 开发 Spark 应用程序 ········· 59
  - 2.3.1 创建一个新的 IntelliJ 项目 ········· 59
  - 2.3.2 验证 SDK 安装和配置 ········· 61
  - 2.3.3 安装和配置 Maven ········· 63
  - 2.3.4 创建 Spark 应用程序 ········· 64
  - 2.3.5 部署 Spark 应用程序 ········· 66
  - 2.3.6 远程调试 Spark 应用程序 ········· 67
- 2.4 使用 Zeppelin 进行交互式分析 ········· 69
  - 2.4.1 下载 Zeppelin 安装包 ········· 70
  - 2.4.2 安装和配置 Zeppelin ········· 70
  - 2.4.3 配置 Spark 解释器 ········· 71
  - 2.4.4 创建和执行 Notebook 程序 ········· 72

第 3 章 Spark 核心编程 (▶ 252min) ········· 75

- 3.1 理解数据抽象 RDD ········· 75
  - 3.1.1 RDD 结构 ········· 75
  - 3.1.2 RDD 容错 ········· 76
- 3.2 RDD 编程模型 ········· 77
  - 3.2.1 单词计数应用程序 ········· 77
  - 3.2.2 理解 SparkSession ········· 79
  - 3.2.3 理解 SparkContext ········· 80
- 3.3 创建 RDD ········· 81
  - 3.3.1 将现有的集合并行化以创建 RDD ········· 81
  - 3.3.2 从存储系统读取数据集以创建 RDD ········· 82
  - 3.3.3 从已有的 RDD 转换得到新的 RDD ········· 83
  - 3.3.4 创建 RDD 时指定分区数量 ········· 83

3.4 操作 RDD ··· 84
　3.4.1 RDD 上的 Transformation 和 Action ·· 85
　3.4.2 RDD Transformation 操作 ·· 87
　3.4.3 RDD Action 操作 ·· 92
　3.4.4 RDD 上的描述性统计操作 ·· 95
3.5 Pair RDD ··· 96
　3.5.1 创建 Pair RDD ·· 97
　3.5.2 操作 Pair RDD ·· 98
　3.5.3 关于 reduceByKey() 操作 ·· 101
　3.5.4 关于 aggregateByKey() 操作 ·· 103
　3.5.5 关于 combineByKey() 操作 ·· 106
3.6 持久化 RDD ·· 109
　3.6.1 缓存 RDD ··· 109
　3.6.2 RDD 缓存策略 ··· 112
　3.6.3 检查点 RDD ·· 113
3.7 RDD 数据分区 ··· 113
　3.7.1 获取和指定 RDD 分区数 ··· 114
　3.7.2 调整 RDD 分区数 ·· 114
　3.7.3 内置数据分区器 ·· 116
　3.7.4 自定义数据分区器 ··· 118
　3.7.5 避免不必要的 shuffling ··· 120
　3.7.6 基于数据分区的操作 ·· 122
3.8 深入理解 RDD 执行过程 ·· 125
　3.8.1 Spark RDD 调度过程 ·· 125
　3.8.2 Spark 执行模型 ··· 126
3.9 Spark 资源管理 ·· 131
　3.9.1 CPU 资源分配策略 ··· 131
　3.9.2 Spark 内存管理 ··· 132
3.10 使用共享变量 ··· 134
　3.10.1 广播变量 ·· 134
　3.10.2 累加器 ··· 139
3.11 Spark RDD 编程案例 ··· 143
　3.11.1 合并小文件 ··· 143
　3.11.2 二次排序实现 ·· 145
　3.11.3 Top N 实现 ··· 146
　3.11.4 酒店数据预处理 ··· 150

第 4 章　Spark SQL（▶ 202min）··· 154
4.1 Spark SQL 数据抽象 ··· 155
4.2 Spark SQL 架构组成 ··· 156
4.3 Spark SQL 编程模型 ··· 157
4.4 程序入口 SparkSession ·· 160

- 4.5 Spark SQL 支持的数据类型 ... 162
  - 4.5.1 Spark SQL 基本数据类型 ... 162
  - 4.5.2 Spark SQL 复杂数据类型 ... 162
  - 4.5.3 模式 ... 163
  - 4.5.4 列对象和行对象 ... 163
- 4.6 创建 DataFrame ... 164
  - 4.6.1 简单创建单列和多列 DataFrame ... 165
  - 4.6.2 从 RDD 创建 DataFrame ... 168
  - 4.6.3 读取外部数据源创建 DataFrame ... 172
- 4.7 操作 DataFrame ... 191
  - 4.7.1 列的多种引用方式 ... 192
  - 4.7.2 对 DataFrame 执行 Transformation 转换操作 ... 193
  - 4.7.3 对 DataFrame 执行 Action 操作 ... 205
  - 4.7.4 对 DataFrame 执行描述性统计操作 ... 206
  - 4.7.5 取 DataFrame Row 中特定字段 ... 209
  - 4.7.6 操作 DataFrame 示例 ... 211
- 4.8 存储 DataFrame ... 212
  - 4.8.1 写出 DataFrame ... 212
  - 4.8.2 存储模式 ... 215
  - 4.8.3 控制 DataFrame 的输出文件数量 ... 216
  - 4.8.4 控制 DataFrame 实现分区存储 ... 220
- 4.9 使用类型化的 DataSet ... 221
  - 4.9.1 了解 DataSet ... 221
  - 4.9.2 创建 DataSet ... 222
  - 4.9.3 操作 DataSet ... 229
  - 4.9.4 类型安全检查 ... 244
  - 4.9.5 编码器 ... 246
- 4.10 临时视图与 SQL 查询 ... 249
  - 4.10.1 在 Spark 程序中执行 SQL 语句 ... 249
  - 4.10.2 注册临时视图并执行 SQL 查询 ... 250
  - 4.10.3 使用全局临时视图 ... 252
  - 4.10.4 直接使用数据源注册临时视图 ... 254
  - 4.10.5 查看和管理表目录 ... 255
- 4.11 缓存 DataFrame/DataSet ... 256
  - 4.11.1 缓存方法 ... 256
  - 4.11.2 缓存策略 ... 257
  - 4.11.3 缓存表 ... 258
- 4.12 Spark SQL 编程案例 ... 259
  - 4.12.1 实现单词计数 ... 259
  - 4.12.2 用户数据集分析 ... 261
  - 4.12.3 电商用户评论数据集分析 ... 264
  - 4.12.4 航空公司航班数据集分析 ... 266
  - 4.12.5 数据增量抽取和全量抽取 ... 274

第 5 章　Spark SQL（高级）(▷ 190min) ··· 276

- 5.1　Spark SQL 函数 ··· 276
- 5.2　内置标量函数 ··· 276
  - 5.2.1　日期时间函数 ··· 277
  - 5.2.2　字符串函数 ··· 280
  - 5.2.3　数学计算函数 ··· 284
  - 5.2.4　集合元素处理函数 ··· 285
  - 5.2.5　其他函数 ·· 288
  - 5.2.6　函数应用示例 ··· 291
  - 5.2.7　Spark 3 数组函数 ·· 294
- 5.3　聚合与透视函数 ··· 301
  - 5.3.1　聚合函数 ·· 301
  - 5.3.2　分组聚合 ·· 307
  - 5.3.3　数据透视 ·· 311
  - 5.3.4　谓词子查询 ··· 312
- 5.4　高级分析函数 ··· 313
  - 5.4.1　使用多维聚合函数 ··· 313
  - 5.4.2　使用时间窗口聚合 ··· 316
  - 5.4.3　使用窗口分析函数 ··· 321
- 5.5　用户自定义函数（UDF） ··· 330
- 5.6　数据集的 join 连接 ··· 332
  - 5.6.1　join 表达式和 join 类型 ·· 332
  - 5.6.2　执行 join 连接 ·· 333
  - 5.6.3　处理重复列名 ··· 340
  - 5.6.4　join 连接策略 ··· 342
- 5.7　读写 Hive 表 ·· 344
  - 5.7.1　Spark SQL 的 Hive 配置 ·· 345
  - 5.7.2　Spark Maven 项目的 Hive 配置 ·· 346
  - 5.7.3　Spark SQL 读写 Hive 表 ·· 347
  - 5.7.4　分桶、分区和排序 ··· 352
- 5.8　查询优化器 Catalyst ··· 359
  - 5.8.1　窄转换和宽转换 ·· 360
  - 5.8.2　Spark 执行模型 ·· 361
  - 5.8.3　Catalyst 实践 ·· 363
  - 5.8.4　可视化 Spark 程序执行 ··· 366
- 5.9　项目 Tungsten ·· 372
- 5.10　Spark 性能调优 ··· 373
- 5.11　Spark SQL 编程案例 ··· 375
  - 5.11.1　电影数据集分析 ·· 375
  - 5.11.2　电商数据集分析 ·· 379
- 5.12　Spark SQL 分析案例 ··· 387
  - 5.12.1　用户行为数据集说明 ··· 387

  5.12.2 分析需求说明 388
  5.12.3 数据探索和预处理 390
  5.12.4 平台流量分析 394
  5.12.5 用户行为分析 397
  5.12.6 转化漏斗分析 405
  5.12.7 用户 RFM 价值分析 408
  5.12.8 推荐效果分析 412
  5.12.9 项目分析总结 417

第 6 章　Spark Streaming 流处理 (▶ 73min) 418

6.1　Spark DStream 418
6.2　Spark 流处理示例 419
  6.2.1 Spark Streaming 编程模型 420
  6.2.2 实时股票交易分析 423
  6.2.3 使用外部数据源 Kafka 440

第 7 章　Spark 结构化流 (▶ 162min) 446

7.1　结构化流简介 446
7.2　结构化流编程模型 448
7.3　结构化流核心概念 451
  7.3.1 数据源 451
  7.3.2 输出模式 452
  7.3.3 触发器类型 452
  7.3.4 数据接收器 453
  7.3.5 水印 454
7.4　使用各种流数据源 454
  7.4.1 使用 Socket 数据源 454
  7.4.2 使用 Rate 数据源 456
  7.4.3 使用 File 数据源 458
  7.4.4 使用 Kafka 数据源 461
7.5　流 DataFrame 操作 467
  7.5.1 选择、投影和聚合操作 468
  7.5.2 执行 join 连接操作 471
7.6　使用数据接收器 474
  7.6.1 使用 File Data Sink 474
  7.6.2 使用 Kafka Data Sink 475
  7.6.3 使用 Foreach Data Sink 478
  7.6.4 使用 Console Data Sink 480
  7.6.5 使用 Memory Data Sink 482
  7.6.6 Data Sink 与输出模式 483
7.7　深入研究输出模式 483
  7.7.1 无状态流查询 483

		7.7.2 有状态流查询	484
7.8	深入研究触发器		489
	7.8.1	固定间隔触发器	490
	7.8.2	一次性的触发器	492
	7.8.3	连续性的触发器	492

第 8 章 Spark 结构化流（高级）(▶ 72min) — 496

8.1	事件时间和窗口聚合	496
	8.1.1 固定窗口聚合	496
	8.1.2 滑动窗口聚合	500
8.2	水印	504
	8.2.1 限制维护的聚合状态数量	504
	8.2.2 处理迟到的数据	507
8.3	任意状态处理	512
	8.3.1 结构化流的任意状态处理	513
	8.3.2 处理状态超时	514
	8.3.3 任意状态处理实战	515
8.4	处理重复数据	524
8.5	容错	526
8.6	流查询度量指标和容错	528
	8.6.1 流查询指标	528
	8.6.2 流监控指标	530
8.7	结构化流案例：运输公司车辆超速实时监测	531
8.8	结构化流案例：实时订单分析	536
	8.8.1 数据集说明和数据源	537
	8.8.2 计算每 10s 的销售和购买订单数量	538
	8.8.3 根据购买或售出的总金额统计前 5 个客户	541
	8.8.4 找出过去一小时内前 5 个交易量最多的股票	543

第 9 章 Spark 图处理库 GraphFrame (▶ 80min) — 546

9.1	图基本概念	546
9.2	GraphFrame 图处理库简介	548
9.3	GraphFrame 的基本使用	549
	9.3.1 添加 GraphFrame 依赖	549
	9.3.2 构造图模型	550
	9.3.3 简单图查询	551
	9.3.4 示例：简单航班数据分析	556
9.4	应用 motif 模式查询	559
	9.4.1 简单 motif 查询	560
	9.4.2 状态查询	562
9.5	构建子图	563

9.6 内置图算法 565
　9.6.1 广度优先搜索（BFS）算法 565
　9.6.2 连通分量算法 567
　9.6.3 强连通分量算法 568
　9.6.4 标签传播算法 569
　9.6.5 PageRank 算法 570
　9.6.6 最短路径算法 573
　9.6.7 三角计数算法 574
9.7 保存和加载 GraphFrame 575
9.8 深入理解 GraphFrame 576
9.9 案例：亚马逊产品联购分析 578
　9.9.1 基本图查询和操作 579
　9.9.2 联购商品分析 580
　9.9.3 处理子图 584
　9.9.4 应用图算法进行分析 584

第10章 Delta Lake 数据湖 (▶ 61min) 586

10.1 从数据仓库到数据湖 586
10.2 解耦存储层和分析层 588
10.3 Delta Lake 介绍 591
10.4 Delta Lake 架构 593
10.5 Delta Lake 使用 595
　10.5.1 安装 Delta Lake 595
　10.5.2 表批处理读写 596
　10.5.3 表流处理读写 602
　10.5.4 文件移除 605
　10.5.5 压缩小文件 612
　10.5.6 增量更新与时间旅行 615
　10.5.7 合并更新（upsert） 621

第11章 Iceberg 数据湖 (▶ 90min) 630

11.1 Apache Iceberg 简介 630
11.2 配置和使用 Catalog 631
　11.2.1 配置 Catalog 632
　11.2.2 使用 Catalog 633
　11.2.3 替换 Session Catalog 633
　11.2.4 运行时配置 634
11.3 管理 Catalog 中的数据库 635
11.4 管理 Iceberg 表 635
　11.4.1 基本的 CRUD 操作 635
　11.4.2 创建和删除表 640

- 11.4.3 使用分区表和分桶表 ... 642
- 11.4.4 数据覆盖 ... 652
- 11.4.5 修改表结构 ... 654
- 11.5 探索 Iceberg 表 ... 663
  - 11.5.1 History 历史表 ... 663
  - 11.5.2 Snapshots 快照表 ... 664
  - 11.5.3 Files 数据文件表 ... 664
  - 11.5.4 Manifests 文件清单表 ... 665
- 11.6 Apache Iceberg 架构 ... 666
  - 11.6.1 Iceberg Catalog ... 666
  - 11.6.2 元数据文件 ... 667
  - 11.6.3 清单列表（Manifest List） ... 670
  - 11.6.4 清单文件（Manifest File） ... 673
- 11.7 CRUD 操作的底层实现 ... 676
- 11.8 增量更新与合并更新 ... 679
- 11.9 时间旅行 ... 685
- 11.10 隐藏分区和分区演变 ... 688
  - 11.10.1 分区概念 ... 688
  - 11.10.2 分区演变示例 ... 691
- 11.11 使用存储过程维护表 ... 698
- 11.12 整合 Spark 结构化流 ... 701
  - 11.12.1 流读取 ... 701
  - 11.12.2 流写入 ... 701
  - 11.12.3 维护流表 ... 704

第 12 章 Hudi 数据湖（▶19min） ... 707

- 12.1 Apache Hudi 特性 ... 707
  - 12.1.1 Hudi Timeline ... 707
  - 12.1.2 Hudi 文件布局 ... 709
  - 12.1.3 Hudi 表类型 ... 709
  - 12.1.4 Hudi 查询类型 ... 712
- 12.2 在 Spark 3 中使用 Hudi ... 713
  - 12.2.1 配置 Hudi ... 713
  - 12.2.2 初始设置 ... 714
  - 12.2.3 插入数据 ... 715
  - 12.2.4 查询数据 ... 716
  - 12.2.5 更新数据 ... 718
  - 12.2.6 增量查询 ... 718
  - 12.2.7 时间点查询 ... 720
  - 12.2.8 删除数据 ... 720
  - 12.2.9 插入覆盖 ... 722

第13章 Spark 大数据处理综合案例 (▶15min) ……724

- 13.1 项目需求说明 ……724
- 13.2 项目架构设计 ……725
- 13.3 项目实现：数据采集 ……726
- 13.4 项目实现：数据集成 ……729
  - 13.4.1 Flume 简介 ……729
  - 13.4.2 安装和配置 Flume ……730
  - 13.4.3 实现数据集成 ……731
- 13.5 项目实现：数据 ETL ……732
- 13.6 项目实现：数据清洗与整理 ……735
- 13.7 项目实现：数据分析 ……739
- 13.8 项目实现：分析结果导出 ……747
- 13.9 项目实现：数据可视化 ……748
  - 13.9.1 Spring MVC 框架简介 ……749
  - 13.9.2 ECharts 图表库介绍 ……750
  - 13.9.3 Spring MVC Web 程序开发 ……751
  - 13.9.4 前端 ECharts 组件开发 ……763
- 13.10 项目部署和测试 ……768

第 1 章 Spark 架构原理与集群搭建
CHAPTER 1

Apache Spark 是一个用于快速、通用、大规模数据处理的开源项目。它类似于 Hadoop 的 MapReduce，但对于批处理执行来讲速度更快、更高效。Apache Spark 可以部署在大量廉价的硬件设备上，以创建大数据并行计算集群。

Apache Spark 作为一个用于大数据处理的内存并行计算框架，利用内存缓存和优化执行来获得更高的性能，并且支持以任何格式读取/写入 Hadoop 数据，同时保证了高容错性和可扩展性。现在，Apache Spark 已经成为一个统一的大数据处理平台，拥有一个快速的统一分析引擎，可用于大数据的批处理、实时流处理、机器学习和图计算。

自 2010 年首次发布以来，Apache Spark 已经成为最活跃的大数据开源项目之一。如今，Apache Spark 实际上已经是大数据处理、数据科学、机器学习和数据分析工作负载的统一引擎。

1.1 Spark 简介

2009 年，Spark 诞生于伯克利大学 AMP 实验室，最初属于伯克利大学的研究性项目。它于 2010 年被正式开源，于 2013 年被转交给 Apache 软件基金会，并于 2014 年成为 Apache 基金的顶级项目，整个过程不到五年时间。Apache Spark 诞生以后，迅速发展成为大数据处理技术中的佼佼者，目前已经成为大数据处理领域炙手可热的技术，其发展势头非常强劲。

Spark 的内存计算模型如图 1-1 所示。

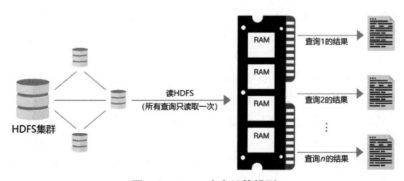

图 1-1　Spark 内存计算模型

在图 1-1 中，Spark 一次性从 HDFS 中读取所有的数据，并以分布式的方式缓存在计算机集群各节点的内存中。

Spark 用于迭代算法的内存数据共享表示如图 1-2 所示。

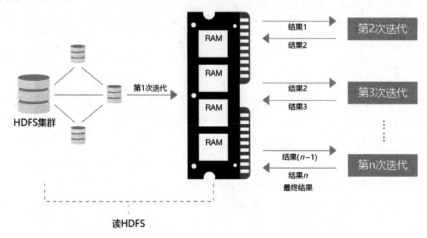

图 1-2　Spark 用于迭代算法的内存数据共享表示

Spark 与其他分布式计算平台相比有许多独特的优势：

（1）它是用于迭代机器学习和交互式数据分析的更快的执行平台。

（2）它是用于批处理、SQL 查询、实时流处理、图处理和复杂数据分析的单一技术栈。

（3）它通过隐藏分布式编程的复杂性，提供高级 API 来供用户开发各种分布式应用程序。

（4）它提供对各种数据源的无缝支持，如 RDBMS、HBase、Cassandra、Parquet、MongoDB、HDFS、Amazon S3 等。

Spark 隐藏了编写核心 MapReduce 作业的复杂性，并通过简单的函数调用提供了大部分功能。由于它的简单性，受到了用户的广泛应用和认同，例如数据科学家、数据工程师、统计学家，以及 R /Python/Scala/Java 开发人员。由于 Spark 采用了内存计算，并采用函数式编程，提供了大量高阶函数和算子，因此它具有以下 3 个显著特性：快速性、易用性和灵活性。

在 2014 年，Spark 赢得了 Daytona GraySort 竞赛，该竞赛是对 100 TB 数据进行排序的行业基准（1 万亿条记录）。来自 Databricks 的提交声称 Spark 能够以比之前的 Hadoop MapReduce 所创造的世界纪录快 3 倍的速度对 100 TB 的数据进行排序，并且使用的资源只为原来的十分之一，如图 1-3 所示。

Spark 可以连接到许多不同的数据源，包括文件（CSV、JSON、Parquet、Avro）、MySQL、MongoDB、HBase 和 Cassandra。此外，它还可以连接到特殊用途的引擎和数据源，如

Elasticsearch、Apache Kafka 和 Redis。这些引擎支持 Spark 应用程序中的特定功能，如搜索、流、缓存等。Spark 提供了 DataSource API 以支持各种数据源（包括自定义数据源）的 Spark 连接，如图 1-4 所示。

图 1-3　与 Hadoop MapReduce 相比，Spark 计算速度更快，使用资源更少

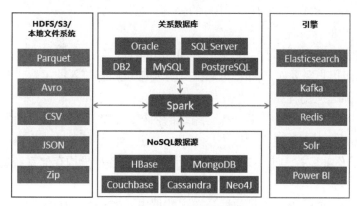

图 1-4　Spark 支持各种数据源

Apache Spark 为用户提供了 4 种编程语言接口，分别是 Java、Scala、Python 和 R。因为 Apache Spark 本身是用 Scala 构建的，所以 Scala 是首选语言。由于 Spark 内置了对 Scala、Java、R 和 Python 的支持，因此大多数的开发人员和数据工程师能够利用整个 Spark 栈应用于不同的应用场景。

1.2　Spark 技术栈

Apache Spark 提供了一个统一的数据处理引擎，称为 Spark 栈。Spark 栈的基础是其核心模块（称为 Spark Core）。Spark Core 提供了管理和运行分布式应用程序的所有必要功能，如调度、协调和容错。此外，它还为数据处理提供了强大的通用编程抽象，称为弹性分布式数据集（Resilient Distributed Datasets，RDD）。

在 Spark Core 之上是一个组件集合，其中每个组件都是为特定的数据处理工作而设计的，它们建立在 Spark Core 的强大基础引擎之上。Spark 技术栈如图 1-5 所示。

图 1-5　Spark 技术栈

下面分别讲解 Spark Core 引擎和各个功能组件。

1.2.1　Spark Core

Spark Core 由两部分组成：分布式计算基础设施和 RDD 编程抽象，其中分布式计算基础设施的职责如下。

（1）负责集群中多节点上的计算任务的分发、协调和调度。

（2）处理失败的计算任务。

（3）高效地跨节点传输数据（数据传输 shuffling）。

Spark 的高级用户需要对 Spark 分布式计算基础设施有深入的了解，从而能够有效地设计高性能的 Spark 应用程序。

Spark Core 在某种程度上类似于操作系统的内核。它是通用的执行引擎，它既快速又容错。整个 Spark 生态系统建立在这个核心引擎之上。它主要用于工作调度、任务分配和跨 worker 节点的作业监控。此外它还负责内存管理，与各种异构存储系统交互，以及各种其他操作。

Spark Core 的主要编程抽象是弹性分布式数据集（RDD），RDD 是一个不可变的、容错的对象集合，可以在一个集群中进行分区，因此可以并行操作。本质上，RDD 为 Spark 应用程序开发人员提供了一组编程 API，开发人员能够轻松高效地执行大规模的数据处理，而不必担心数据驻留在集群上的什么位置或处理机器故障。

Spark 可以从各种数据源创建 RDD，如 HDFS、本地文件系统、Amazon S3、其他 RDD、NoSQL 数据存储等。RDD 适应性很强，会在失败时自动重建。RDD 是通过惰性并行转换构建的，它们可能被缓存和分区，可能会也可能不会被具体化。

1.2.2　Spark SQL

Spark SQL 是构建在 Spark Core 之上的组件，被用来在结构化数据上执行查询、分析操作。因为 Spark SQL 的灵活性、易用性和良好性能，现在它是 Spark 技术栈中最受欢迎、应用最多的组件。

Spark SQL 提供了一种名为 DataFrame 的分布式编程抽象。DataFrame 是分布式二维表集合，类似于 SQL 表或 Python 的 Pandas 库中的 DataFrame。可以从各种数据源构造

DataFrame，如 Hive、Parquet、JSON、关系型数据库（如 MySQL 等）及 Spark RDD。这些数据源可以具有各种模式。

Spark SQL 可以用于不同格式的 ETL 处理，然后进行即时查询分析。Spark SQL 附带一个名为 Catalyst 的优化器框架，它能解析 SQL 查询并自动进行优化以提高效率。Spark SQL 利用 Catalyst 优化器来执行许多分析数据库引擎中常见的优化类型。Spark SQL 的座右铭是"write less code, read less data, and let the optimizer do the hard work"。

1.2.3 Spark Streaming 和 Structured Streaming

为了解决企业的数据实时处理需求，Spark 提供了流处理组件，它具有容错能力和可扩展性。Spark 支持实时数据流的实时数据分析。因为具有统一的 Spark 技术栈，所以在 Spark 中可以很容易地将批处理和交互式查询及流处理结合起来。

目前的 Spark 流处理模块实际上包含两代流处理引擎，分别是第 1 代的 Spark Streaming 和第 2 代的 Spark Structured Streaming（结构化流），其中 Spark Streaming 的数据抽象是基于 RDD 的，而 Spark Structured Streaming 的数据抽象是基于 DataFrame 的。

Spark Streaming 和 Spark Structured Streaming 模块能够以高吞吐量和容错的方式处理来自各种数据源的实时流数据。数据可以从像 Kafka、Flume、Kinesis、Twitter、HDFS 或 TCP 套接字这样的资源中摄取。

在第 1 代 Spark Streaming 处理引擎中，主要的数据抽象是离散化流，简称 DStream。它通过将输入数据分割成小批量数据（基于时间间隔）实现增量流处理模型，该模型可以定期地组合当前的处理状态以产生新的结果。换句话说，一旦传入的数据被分成微批，每批数据都将被视为一个 RDD，并将其复制到集群中，这样它们就可以被作为基本的 RDD 进行处理。通过在 DStream 上应用一些更高级别的操作，可以产生其他的 DStream。Spark 流的最终结果可以被写回 Spark 所支持的各种数据存储，或者可以被推送到任何仪表盘进行可视化。

从 Spark 2.1 开始，Spark 引入了一个新的可扩展和容错的流处理引擎，称为结构化流（Structured Streaming）。结构化流构建在 Spark SQL 引擎之上，它进一步简化了流处理应用程序开发，处理流计算就像在静态数据上处理批计算一样。随着新的流数据的持续到来，结构化流引擎将自动地、增量地、持续地执行流处理逻辑。结构化流提供的一个新的重要特性是基于事件时间（Event Time）处理输入流数据的能力。在结构化流引擎中还支持端到端的、精确一次性保证。

1.2.4 Spark MLlib

MLlib 是 Spark 栈中内置的机器学习库，它的目标是使机器学习变得可扩展并且更容易。MLlib 提供了执行各种统计分析的必要功能，如相关性、抽样、假设检验等。该组件还开箱即用地提供了常用的机器学习算法实现，如分类、回归、聚类和协同过滤。

Spark 机器学习库实际上包含两种，分别是基于 RDD 的第 1 代机器学习库（Spark 0.8 引入）和基于 DataFrame 的第 2 代机器学习库（Spark 2.0 引入）。目前基于 RDD 的机器学习库已经处于维护模式，因此本书的机器学习部分基于第 2 代机器学习库进行讲解，它受益于 Spark SQL 引擎中的 Catalyst 优化器和 Tungsten 项目，以及这些组件所提供的许多优化。

机器学习工作流程包括收集和预处理数据、构建和部署模型、评估结果和改进模型。在现实世界中，预处理步骤需要付出很大的努力。这些都是典型的多阶段工作流，涉及昂贵的中间读/写操作。通常，这些处理步骤可以在一段时间内多次执行。Spark 机器学习库引入了一个名为机器学习管道的新概念，以简化这些预处理步骤。管道是一个转换序列，其中一个阶段的输出是另一个阶段的输入，从而形成工作流链。

除了提供超过 50 种常见的机器学习算法外，Spark MLlib 库还提供了一些功能抽象，用于管理和简化许多机器学习模型构建任务，如特征化，用于构建、评估和调优模型的管道，以及模型的持久性（以帮助将模型从开发转移到生产环境）。

1.2.5　Spark GraphX

GraphX 是 Spark 的统一图分析框架。它被设计成一个通用的分布式数据流框架，取代了专门的图处理框架。它具有容错特性，并且利用内存进行计算。

GraphX 是一种嵌入式图处理 API，用于操纵图（例如，社交网络）和执行图并行计算（例如，谷歌的 Pregel）。它结合了 Spark 栈上的图并行和数据并行系统的优点，以统一探索性数据分析、迭代图计算和 ETL 处理。它扩展了 RDD 抽象来引入弹性分布式图（RDG，Resilient Distributed Graph），这是一个有向图，具有与每个顶点和边相关联的属性。

GraphX 组件包括一组通用图处理算法，包括 PageRank、K-Core、三角计数、LDA、连接组件、最短路径等。

目前的 Spark GraphX 组件是基于 RDD 的，而社区正在构建基于 DataFrame（及其底层的 Catalyst 优化器和 Tungsten 项目）的图计算库版本，称为 GraphFrames。GraphFrames 目前还没有集成到 Spark 发行版中，但已经得到了广泛的应用。本书第 9 章将会详细讲解 GraphFrames 的安装和使用。

1.2.6　SparkR

SparkR 项目将 R 的统计分析和机器学习能力与 Spark 的可扩展性集成在一起。它解决了 R 的局限性，即它处理单个机器内存中所需要的大量数据的能力。R 程序现在可以通过 SparkR 在分布式环境中进行扩展。

SparkR 实际上是一个 R 包，它提供了一个 R Shell 来利用 Spark 的分布式计算引擎。有了 R 丰富的用于数据分析的内置包，数据科学家可以交互式地分析大型数据集。

注意：本书不涉及 SparkR 的内容。

1.3 Spark 架构原理

在深入了解 Spark 的架构之前，一定要对 Spark 的核心概念和各种核心组件有一个深入的理解。这些核心概念和组件如下：

（1）Spark 集群。
（2）资源管理系统。
（3）Spark 应用程序。
（4）Spark Driver。
（5）Spark Executor。

1.3.1 Spark 集群和资源管理系统

Spark 本质上是一个分布式系统，设计的目的是用来高效、快速地处理海量数据。这个分布式系统通常部署在一个计算机集群上，称为 Spark 集群。为了高效和智能地管理这个集群，通常依赖于一个资源管理系统，如 Apache YARN 或 Apache Mesos。

资源管理系统内部有两个主要组件：集群管理器（Cluster Manager）和工作节点（Worker）。它有点像主从架构，其中集群管理器充当主节点，工作节点充当集群中的从节点。集群管理器跟踪与工作节点及其当前状态相关的所有信息。集群管理器负责维护的信息如下：

（1）Worker 节点的状态（busy/available）。
（2）Worker 节点的位置。
（3）Worker 节点的内存。
（4）Worker 节点的总 CPU 核数。

集群管理器知道 Worker 节点的位置，其内存大小，以及每个 Worker 的 CPU 核数量。集群管理器的主要职责之一是管理 Worker 节点并根据 Worker 节点的可用性和容量为它们分配任务（Task）。每个 Worker 节点都向集群管理器提供自己可用的资源（内存、CPU 等），并负责执行集群管理器分配的任务，如图 1-6 所示。

1.3.2 Spark 应用程序

Spark 应用程序也采用了主从架构，其中 Spark Driver 是 Master，Spark Executors 是 Slave。每个组件都作为一个独立的 JVM 进程运行在 Spark 集群上。Spark 应用程序由一个且只有一个 Spark Driver 和一个或多个 Spark Executors 组成，如图 1-7 所示。

Spark 应用程序由两部分组成，分别是：

（1）应用程序数据处理逻辑，使用 Spark API 表示。
（2）Spark 驱动程序（Spark Driver）。

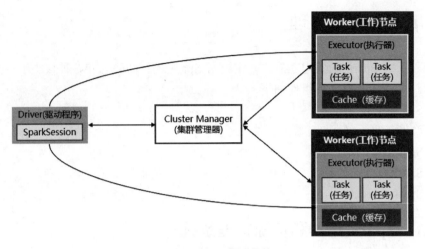

图 1-6　Spark 集群架构

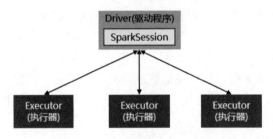

图 1-7　Spark 程序组成

应用程序数据处理逻辑(Task）是用 Java、Scala、Python 或 R 语言编写的数据处理逻辑代码。它可以简单到用几行代码来执行一些数据处理操作，也可以复杂到训练一个大型机器学习模型（这个模型需要多次迭代，可能要运行很多小时才能完成）。

Spark 驱动程序用于运行应用程序 main()函数并创建 SparkSession 的进程。它是 Spark 应用程序的主控制器，负责组织和监控一个 Spark 应用程序的执行。它与集群管理器进行交互，以确定哪台机器来运行数据处理逻辑。Driver 及其子组件（Spark Session 和 Scheduler）负责的职责如下：

（1）向集群管理器请求内存和 CPU 资源。

（2）将应用程序逻辑分解为阶段（Stage）和任务（Task）。

（3）请求集群管理器启动名为 Executor 的进程（在运行 Task 的节点上）。

（4）向 Executor 发送 Tasks（应用程序数据处理逻辑），每个 Task 都在一个单独的 CPU Core 上执行。

（5）与每个 Executor 协调以收集计算结果并将它们合并在一起。

Spark 应用程序的入口是通过一个名为 SparkSession 的类实现的。一旦 Driver 程序被启动之后,它就会启动并配置 SparkSession 的一个实例。SparkSession 是访问 Spark 运行时的主要接口。SparkSession 对象连接到一个集群管理器,并提供了设置配置的工具,以及用于表示数据处理逻辑的 API。

除此之外,还需要一个客户端组件。客户端进程负责启动 Driver 程序。客户端进程可以是一个用于运行程序的 spark-submit 脚本,也可以是一个 spark-shell 脚本或一个使用 Spark API 的自定义应用程序。客户端进程为 Spark 程序准备 class path 和所有配置选项,并将应用程序参数(如果有)传递给运行在 Driver 中的程序。

1.3.3 Spark Driver 和 Executor

每个 Spark 应用程序都有一个 Driver 进程。Spark Driver 包含多个组件,负责将用户代码转换为在集群上执行的实际作业,如图 1-8 所示。

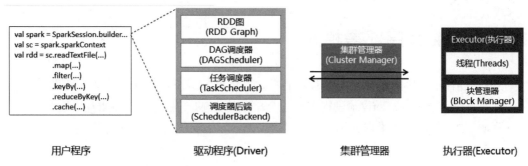

图 1-8 Spark 程序从代码到执行的转换过程

Spark Driver 中各个组件的功能如下。

(1)SparkContext:表示到 Spark 集群的连接,可用于在该集群上创建 RDD、累加器和广播变量。

(2)DAGScheduler:计算每个作业的 Stages 的 DAG,并将它们提交给 TaskScheduler,确定任务的首选位置(基于缓存状态或 shuffle 文件位置),并找到运行作业的最优调度。

(3)TaskScheduler:负责将任务(Tasks)发送到集群,然后运行它们,在出现故障时重试,并减少掉队的情况。

(4)SchedulerBackend:用于调度系统的后端接口,允许插入不同的实现(Mesos、YARN、单机、本地)。

(5)BlockManager:提供用于在本地和远程将 block 块放入和检索到各种存储(内存、磁盘和非堆)中的接口。

每个 Spark 应用程序都有一组 Executor 进程。每个 Executor 都是一个 JVM 进程,扮演 Slave 角色,专门分配给特定的 Spark 应用程序,以便执行命令,并以任务的形式执行数据处理逻辑。每个任务在一个单独的 CPU 核上执行。

Executors 驻留在 Worker 节点上，一旦集群管理器建立连接，就可以直接与 Driver 通信，接受来自 Driver 的任务，然后执行这些任务，并将结果返给 Driver。每个 Executor 都有几个并行运行任务的任务槽。可以将任务槽的数量设置为 CPU 核心数量的 2 倍或 3 倍。尽管这些任务槽通常被称为 Spark 中的 CPU Cores，但它们是作为线程实现的，并且不需要与机器上的物理 CPU Cores 数量相对应。另外，每个 Spark Executor 都由一个 Block Manager 组件组成，Block Manager 负责管理数据块。这些块可以缓存 RDD 数据、中间处理的数据或广播数据。当可用内存不足时，它会自动地将一些数据块移动到磁盘。Block Manager 还有一个职责，即执行跨节点的数据复制。

在启动一个 Spark 应用程序时，可以向资源管理器请求该应用程序所需的 Executor 数量，以及每个 Executor 应该拥有的内存大小和 CPU 核数。要计算出适当数量的 Executor、内存大小和 CPU 数量，需要了解将要处理的数据量、数据处理逻辑的复杂性及 Spark 应用程序完成处理逻辑所需的持续时间。

1.4 Spark 程序部署模式

Spark Driver 程序的运行有两种基本的模式：集群部署模式和客户端部署模式。

在集群部署模式下，Driver 进程作为一个单独的 JVM 进程运行在集群中，集群负责管理其资源（主要是 JVM 堆内存），如图 1-9 所示。

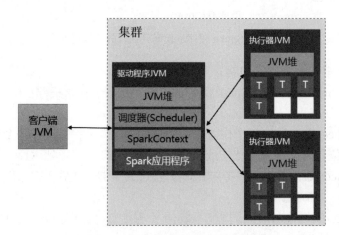

图 1-9　Spark 程序集群部署模式

在客户端部署模式下，Driver 进程运行在客户端的 JVM 进程中，并与受集群管理的 Executors 进行通信，如图 1-10 所示。

选择不同的部署模式将影响如何配置 Spark 和客户端 JVM 的资源需求。通常使用客户端部署模式，在这种模式下，用户可以在客户端获取并显示作业执行情况。

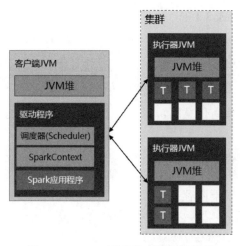

图 1-10　Spark 程序客户端部署模式

1.5　安装和配置 Spark 集群

为了学习 Spark，最好在用户自己的计算机上安装 Spark。通过这种方式，用户可以轻松地尝试 Spark 的特性或使用小型数据集测试数据处理逻辑。

Apache Spark 是用 Scala 编程语言编写的，而 Scala 需要运行在 JVM 上，因此，在安装 Spark 之前，需确保已经在自己的计算机上安装了 Java（JDK 8）。

1.5.1　安装 Spark

要在自己的计算机上安装 Spark，建议按以下步骤操作。

（1）将预先打包的二进制文件下载到 ~/software 目录下，它包含运行 Spark 所需的 JAR 文件。下载网址为 http://spark.apache.org/downloads.html，下载界面如图 1-11 所示。

图 1-11　选择要下载的 Spark 版本

本书使用的版本为 Spark 3.1.2。

（2）将下载的压缩包解压缩到~/bigdata/目录下，并重命名为 spark-3.1.2，命令如下：

```
$ cd ~/bigdata
$ tar -zxvf ~/software/spark-3.1.2-bin-hadoop3.2.tgz
$ mv spark-3.1.2-bin-hadoop3.2 spark-3.1.2
```

（3）配置环境变量。使用任意编辑器（本书使用的编辑器是 nano）打开/etc/profile 文件，命令如下：

```
$ cd
$ sudo nano /etc/profile
```

在文件的最后，添加以下内容：

```
export SPARK_HOME=/home/hduser/bigdata/spark-3.1.2
export PATH=$SPARK_HOME/bin:$PATH
```

保存文件并关闭。

（4）执行/etc/profile 文件使配置生效，命令如下：

```
$ source /etc/profile
```

1.5.2 了解 Spark 目录结构

查看解压缩后的 Spark 安装目录，会发现其中包含多个目录，如图 1-12 所示。

名称	修改日期	类型	大小
bin	2021-08-05 16:44	文件夹	
conf	2021-08-05 16:44	文件夹	
data	2021-08-05 16:44	文件夹	
examples	2021-08-05 16:44	文件夹	
jars	2021-08-05 16:44	文件夹	
kubernetes	2021-08-05 16:44	文件夹	
licenses	2021-08-05 16:44	文件夹	
python	2021-08-05 16:44	文件夹	
R	2021-08-05 16:44	文件夹	
sbin	2021-08-05 16:44	文件夹	
yarn	2021-08-05 16:44	文件夹	
LICENSE	2021-05-24 12:45	文件	23 KB
NOTICE	2021-05-24 12:45	文件	57 KB
README.md	2021-05-24 12:45	MD 文件	5 KB
RELEASE	2021-05-24 12:45	文件	1 KB

图 1-12　Spark 安装目录

其中几个主要目录的作用见表 1-1。

表 1-1　Spark 安装目录说明

目　　录	描　　述
bin	包含各种可执行文件，以启动 Scala 或 Python 中的 Spark Shell、提交 Spark 应用程序和运行 Spark 示例
conf	包含用于 Spark 的各种配置文件
data	包含用于各种 Spark 示例的小示例数据文件
examples	包含所有 Spark 示例的源代码和二进制文件
jars	包含运行 Spark 所需的二进制文件
sbin	包含管理 Spark 集群的可执行文件

1.5.3　配置 Spark 集群

Spark 的配置文件位于 conf 目录下。conf 目录下默认存放着几个 Spark 的配置示例模板文件，见表 1-2。

表 1-2　Spark conf 目录下的模板文件说明

文 件 名	说　　明
fairscheduler.xml.template	Hadoop 公平调度配置模板文件
log4j.properties.template	Spark Driver 节点的日志配置模板文件
metrics.properties.template	Metrics 系统性能监控工具的配置模板文件
spark-defaults.conf.template	Spark 运行时的属性配置模板文件
spark-env.sh.template	Spark 环境变量配置模板文件
workers.template	Spark 集群的 Worker 节点配置模板文件

这些模板文件均不会被 Spark 直接读取，需要将.template 后缀去除，这样 Spark 才会读取这些文件。在这些配置文件中主要需要关注的是 spark-env.sh、spark-defaults.conf 和 workers 这 4 个配置文件。

接下来，对 Spark 进行配置，包括其运行环境和集群配置参数。建议按以下步骤执行。

（1）从模板文件 spark-env.sh.template 复制一份，并重命名为 spark-env.sh，命令如下：

```
$ cd ~/spark-3.1.2/conf/
$ cp spark-env.sh.template spark-env.sh
```

（2）使用编辑器打开 spark-env.sh 文件进行编辑，命令如下：

```
$ nano spark-env.sh
```

加入以下内容，并保存：

```
export JAVA_HOME=/usr/local/jdk1.8.0_251
export HADOOP_CONF_DIR=/home/hduser/bigdata/hadoop-3.2.2/etc/hadoop
export YARN_CONF_DIR=/home/hduser/bigdata/hadoop-3.2.2/etc/hadoop
export SPARK_HOME=/home/hduser/bigdata/spark-3.1.2
export SPARK_DIST_CLASSPATH=$(/home/hduser/bigdata/hadoop-
```

3.2.2/bin/hadoop classpath)

注意，应将 JDK 和 Hadoop 修改为读者自己安装的版本和路径，然后保存并关闭文件。

（3）复制一份模板文件 workers.template，并重命名为 workers，命令如下：

```
$ cp workers.template workers
$ nano workers
```

将其中的 localhost 删除，将集群中所有 Worker 节点的机器名或 IP 地址填写进去，一个一行。例如，笔者的机器名是 xueai8，因此 workers 文件的内容修改如下：

```
xueai8
```

（4）将配置好的 Spark 目录复制到集群中其他的节点，放在相同的路径下。

1.5.4 验证 Spark 安装

Spark 配置完成后就可以直接使用了，不需要像 Hadoop 那样运行启动命令。下面通过运行 Spark 自带的蒙特卡洛求圆周率 π 值示例，以验证 Spark 是否安装成功。

Spark 支持以本地模式运行 Spark 程序，或者以集群模式运行 Spark 程序。在本地模式下，直接使用 spark-submit 命令来提交示例程序 JAR 包运行即可，命令如下：

```
$ ./bin/spark-submit \
  --class org.apache.spark.examples.SparkPi \
  --master local[*] \
  ./examples/jars/spark-examples_2.12-3.1.2.jar
```

执行过程如图 1-13 所示。

图 1-13 以本地模式运行 Spark 自带的蒙特卡洛求圆周率 π 值程序

执行结果如图 1-14 中所示。

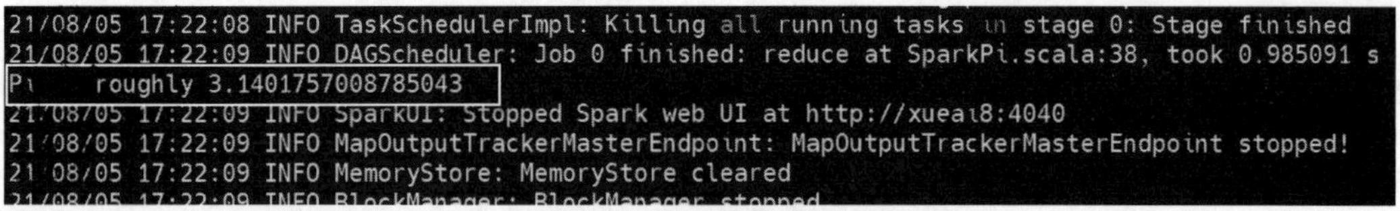

图 1-14 计算出的 π 值

或者，也可以 standalone 模式（需要先执行 ./sbin/start-all.sh 文件启动 Spark 集群），命令如下：

```
$ cd ~/bigdata/spark-3.1.2
$ ./sbin/start-all.sh
```

```
$ ./bin/spark-submit \
--class org.apache.spark.examples.SparkPi \
--master spark://xueai8:7077 \
./examples/jars/spark-examples_2.12-3.1.2.jar
```

执行过程如图 1-15 所示。

图 1-15　以 standalone 模式运行 Spark 自带的蒙特卡洛求圆周率 π 值程序

执行结果如图 1-16 中所示。

图 1-16　蒙特卡洛求圆周率 π 值程序执行结果

1.6　配置 Spark 历史服务器

当用户提交一个 Spark 应用程序时，会创建一个 SparkContext，它提供了 Spark Web UI 来监视应用程序的执行。监控包括以下内容。

（1）Spark 使用的配置。
（2）Spark jobs、stages 和 tasks 细节。
（3）DAG 执行。
（4）Driver 和 Executor 资源利用率。
（5）应用程序日志等。

当应用程序完成处理后，SparkContext 将终止，因此 Web UI 也将终止。如果用户还想看到已经完成的应用程序的监控信息，就必须配置一个单独的 Spark 历史记录服务器。

Spark History Server（历史记录服务器）是一个用户界面服务器，用于监控已完成的 Spark 应用程序的指标和性能。它是 Spark 的 Web UI 的扩展，保存了所有已完成的应用程序的历史（事件日志）及其运行时信息，允许用户稍后检查度量并及时监控应用程序。当用户试图改进应用程序的性能时，历史度量非常有用，用户可以将以前的运行度量与最近的运行度量进行比较。

Spark History Server 可以保存事件日志的历史信息，用于如下操作：
(1) 所有通过 spark-submit 提交的应用程序。
(2) 通过 REST API 提交的。
(3) 运行的每个 spark-shell。
(4) 通过 Notebook 提交的作业。

1.6.1 历史服务器配置

为了存储所有提交的应用程序的事件日志，首先，Spark 需要在应用程序运行时收集信息。默认情况下，Spark 不收集事件日志信息。用户可以通过在 spark-defaults.conf 文件中设置下面的配置来启用它：

(1) 将配置项 spark.eventLog.enabled 设置为 true，以此来启用事件日志功能。
(2) 使用 spark.history.fs.logDirectory 和 spark.eventLog.dir 指定存储事件日志历史的位置。默认位置为 file://tmp/spark-events。需要提前创建该目录。

建议按以下步骤操作。
(1) 在 Spark 安装目录下，创建存储事件日志历史的文件夹，命令如下：

```
$ cd ~/bigdata/spark-3.1.2
$ mkdir spark-events
```

(2) 在 spark-defaults.conf 文件中启用事件日志记录功能。首先从模板文件复制一份，并去掉.template 后缀，得到 spark-defaults.conf 文件，命令如下：

```
$ cd ~/bigdata/spark-3.1.2/conf
$ cp spark-defaults.conf.template spark-defaults.conf

#编辑 spark-defaults.conf 文件
$ nano spark-defaults.conf
```

将下面的配置项添加到文件的末尾：

```
#启用存储事件日志
spark.eventLog.enabled true

#存储事件日志的位置
spark.eventLog.dir file://home/hduser/bigdata/spark-3.1.2/spark-events

#历史服务器读取事件日志的位置
spark.history.fs.logDirectory file://home/hduser/bigdata/spark-3.1.2/spark-events

#日志记录周期
spark.history.fs.update.interval 10s

#历史服务器端口号
spark.history.ui.port 18080
```

Spark 通过为每个应用程序创建一个子目录来保存运行的每个应用程序的历史，并在该目录中记录与该应用程序相关的事件。

还可以设置如 HDFS 目录这样的位置，以便历史文件可以被历史服务器读取，代码如下：

```
spark.eventLog.dir hdfs://xueai8:8020/user/spark/spark-events
```

如果想要为 org.apache.spark.deploy.history 的日志记录器（logger）启用 INFO 日志记录级别，则可以将下面这行配置添加到 conf/log4j.properties 文件中：

```
log4j.logger.org.apache.spark.deploy.history=INFO
```

当启用事件日志记录时，默认的行为是保存所有日志，这会导致存储空间随时间增大。要启用自动清理功能，应编辑 spark-defaults.conf 文件并编辑以下选项：

```
#设置日志清除周期
spark.history.fs.cleaner.enabled true
spark.history.fs.cleaner.interval 1d
spark.history.fs.cleaner.maxAge 7d
```

对于这些设置，将启用自动清理，每天执行清理，并删除超过 7 天的日志。

1.6.2 启动 Spark 历史服务器

要启动 Spark 历史服务器，需要打开一个终端窗口，执行的命令如下：

```
$ SPARK_HOME/sbin/start-history-server.sh
```

如果未明确指定，则 start-history-server.sh 文件会使用默认配置文件 spark-defaults.conf。另外，它也可以接受 --properties-file [propertiesFile] 命令行选项，该选项用于指定带有自定义 Spark 属性的属性文件，命令如下：

```
$ SPARK_HOME/sbin/start-history-server.sh --properties-file
history.properties
```

如果使用更显式的 spark-class 方法来启动 Spark History Server，则可以更容易地跟踪执行，因为可以看到日志被打印到标准输出（直接输出到终端），命令如下：

```
$ SPARK_HOME/bin/spark-class
org.apache.spark.deploy.history.HistoryServer
```

如果要在 Windows 系统下运行 Spark，则启动历史记录服务器的命令如下：

```
$ SPARK_HOME/bin/spark-class.cmd
org.apache.spark.deploy.history.HistoryServer
```

默认情况下，历史记录服务器监听 18080 端口，可以使用 http://localhost:18080/ 从浏览器访问它，如图 1-17 所示。

在每个 App ID 上单击，可以得到该 Spark 应用程序的 job、stage、task、executor 的详细环境信息。

如果要停止 Spark 历史服务器，则需要在终端窗口中执行的命令如下：

```
$ $SPARK_HOME/sbin/stop-history-server.sh
```

使用 Spark 历史服务器，用户可以跟踪所有已完成的应用程序，因此需要启用此功能以保持历史记录。在进行性能调优时，这些指标会派上用场。

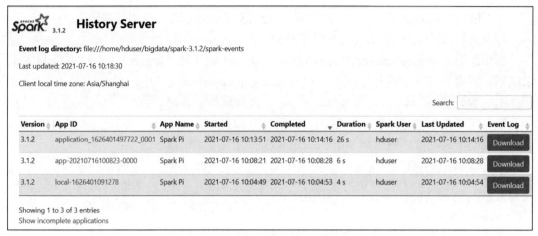

图 1-17 监控 Spark 应用程序

1.7 使用 spark-shell 进行交互式分析

在进行数据分析时，通常需要进行交互式数据探索和数据分析。为此，Spark 提供了一个交互式的工具 spark-shell。通过 spark-shell，用户可以和 Spark 进行实时交互，以进行数据探索、数据清洗和整理及交互式数据分析等工作。

使用 spark-shell 命令，格式如下：

```
$ ./bin/spark-shell [options]
```

要查看完整的参数选项列表，可以执行 spark-shell --help 命令，命令如下：

```
$ spark-shell -help
```

1.7.1 运行模式--master

Spark 的运行模式取决于传递给 SparkContext 的 Master URL 的值。参数选项--master 表示当前的 Spark Shell 要连接到哪个 Master(告诉 Spark 使用哪种集群类型)。

如果是 local[*]，就是使用本地模式启动 spark-shell，其中，中括号内的星号（*）表示需要使用几个 CPU 核，也就是启动几个线程模拟 Spark 集群。如果不指定，则默认为 local。

当运行 spark-shell 命令时，可以像下面这样定义参数：

```
$ spark-shell --master <master_connection_url>
```

在上面的命令中，<master_connection_url>根据所使用的集群的类型而变化。当本地部署时，Master URL(--master 参数）的值见表 1-3。

当集群部署时，Master URL(--master 参数）的值见表 1-4。

表 1-3　Spark 本地运行时的 --master 参数值

部署模式	Master URL	说　　明
本地部署	local	使用一个 Worker 线程本地化运行 Spark（完全不并行），相当于 local[1]
	local[N]	使用 N 个 Worker 线程本地化运行 Spark（最好 N 等于机器的 CPU 核数）
	local[*]	使用逻辑 CPU 个数数量的线程来本地化运行 Spark

表 1-4　Spark 集群运行时的 --master 参数值

部署模式	集群管理器	Master URL	说　　明	示　　例
集群部署	standalone	spark://HOST:PORT	连接到指定的 Spark Master	spark://master:7077
	on YARN	yarn	连接到 YARN 集群	yarn
	on Mesos	mesos://HOST:PORT	连接到指定的 Mesos 集群	mesos://master:5050

1.7.2　启动和退出 spark-shell

以下操作均在终端窗口中进行。

（1）启动 spark-shell 方式一：local 模式。以 local 模式启动 spark-shell，命令如下：

```
$ cd ~/bigdata/spark-3.1.2
$ ./bin/spark-shell
```

启动过程如图 1-18 所示。

图 1-18　以 local 模式启动 spark-shell 过程

从图 1-18 中可以看出，spark-shell 在启动时，已经帮用户创建好了 SparkContext 对象的实例 sc 和 SparkSession 对象的实例 spark，用户可以在 spark-shell 中直接使用 sc 和 spark 这两个对象。另外，默认情况下，启动的 spark-shell 采用 local 部署模式。

在创建 SparkContext 对象的实例 sc 后，它将等待资源。一旦资源可用，sc 将设置内部服务并建立到 Spark 执行环境的连接。

退出 spark-shell，命令如下：

```
scala> :quit
```

（2）启动 spark-shell 方式二：standalone 模式。

首先要确保启动了 Spark 集群。启动 Spark 集群的命令如下：

```
$ cd ~/bigdata/spark-3.1.2
$ ./sbin/start-all.sh
```

使用 jps 命令查看启动的进程。如果看到了 Master 和 Worker 进程，则说明 Spark 集群已经正常启动，如图 1-19 所示。

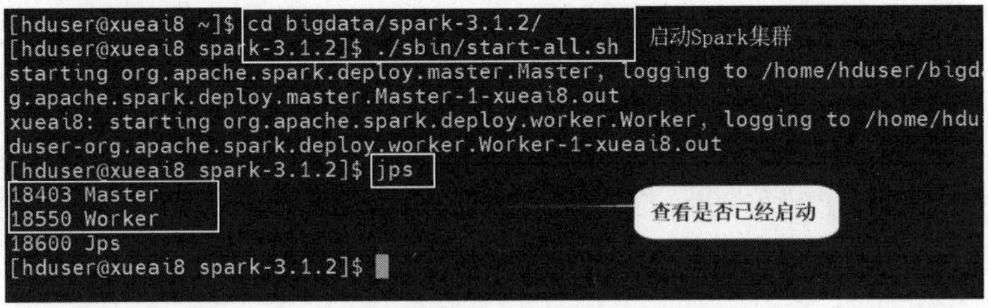

图 1-19　使用 jps 命令查看 Spark 启动的进程

然后启动 spark-shell，并指定 --master spark://xueai8:7077 参数，以 standalone 模式运行，命令如下：

```
$ ./bin/spark-shell --master spark://xueai8:7077
```

在 Master URL 中指定的 xueai8 是笔者当前的机器名，读者需要修改为自己的机器名。spark-shell 的启动过程如图 1-20 所示。

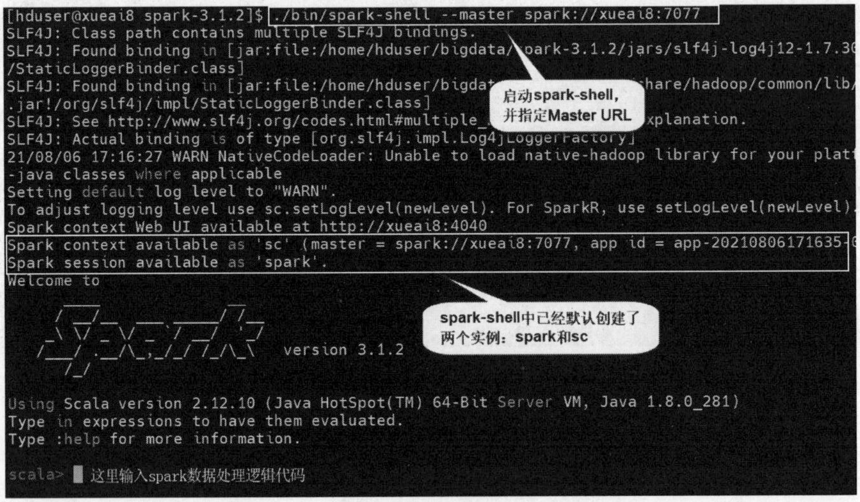

图 1-20　以集群模式启动 spark-shell 过程

1.7.3　spark-shell 常用命令

可以在 spark-shell 中输入以下命令，查看 spark-shell 常用的命令：

```
scala> :help
```

执行以上命令，会在终端窗口中显示帮助信息，如图 1-21 所示。

```
scala> :help
All commands can be abbreviated, e.g., :he instead of :help.
:edit <id>|<line>         edit history
:help [command]           print this summary or command-specific help
:history [num]            show the history (optional num is commands to show)
:h? <string>              search the history
:imports [name name ...]  show import history, identifying sources of names
:implicits [-v]           show the implicits in scope
:javap <path|class>       disassemble a file or class name
:line <id>|<line>         place line(s) at the end of history
:load <path>              interpret lines in a file
:paste [-raw] [path]      enter paste mode or paste a file
:power                    enable power user mode
:quit                     exit the interpreter
:replay [options]         reset the repl and replay all previous commands
:require <path>           add a jar to the classpath
:reset [options]          reset the repl to its initial state, forgetting all session entries
:save <path>              save replayable session to a file
:sh <command line>        run a shell command (result is implicitly => List[String])
:settings <options>       update compiler options, if possible; see reset
:silent                   disable/enable automatic printing of results
:type [-v] <expr>         display the type of an expression without evaluating it
:kind [-v] <expr>         display the kind of expression's type
:warnings                 show the suppressed warnings from the most recent line which had any
```

图 1-21　查看 spark-shell 帮助信息

例如，可以使用“:history”命令查看历史操作记录，使用“:quit”命令退出 Shell 界面，显示内容如图 1-22 所示。

```
scala> :history    查看历史操作信息
 52  ssc.start()
 53  ssc.stop()
 54  :q
 55  spark.sql("use cda")
 56  spark.sql("use ota")
 57  spark.sql("show tables")
 58  spark.sql("show tables").show
 59  spark.table("hotelorder").show(5)
 60  :q
 61  spark.sql("show databases")
 62  spark.sql("show databases").show
 63  spark.sql("use ota")
 64  spark.sql("show tables")
 65  spark.sql("show tables").show
 66  spark.sql("select * from order_num_bydate").show
 67  :q
 68  3.14 * 5 * 5
 69  println("Hello World!")
 70  :help
 71  :history

scala> :quit     退出shell
[hduser@cda spark-2.4.5]$
```

图 1-22　退出 spark-shell

可以在 spark-shell 中输入 Scala 代码进行调试，如图 1-23 所示。

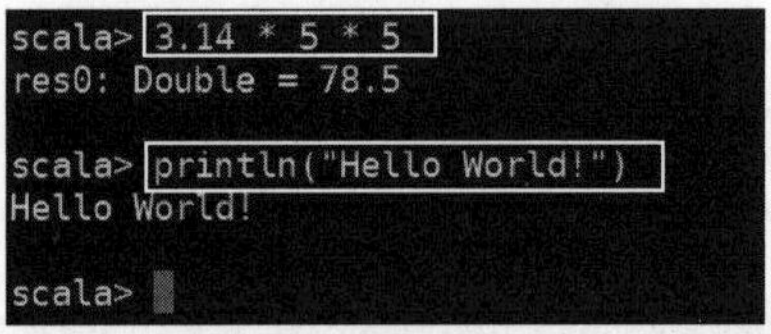

图 1-23　在 spark-shell 中交互执行 Scala 代码

1.7.4　SparkContext 和 SparkSession

在 Spark 2.0 中引入了 SparkSession 类，以提供与底层 Spark 功能交互的单一入口点。这个类具有用于从非结构化文本文件及各种格式的结构化数据和二进制数据文件读取数据的 API，包括 JSON、CSV、Parquet、ORC 等。此外，SparkSession 还提供了检索和设置与 Spark 相关的配置的功能。

SparkContext 在 Spark 2.0 中成为 SparkSession 的一个属性对象。

一旦成功启动一个 spark-shell，它就会初始化一个 SparkSession 类的实例（名为 spark），以及一个 SparkContext 类的实例（名为 sc）。这个 spark 变量和 sc 变量可以在 spark-shell 中直接使用。可以使用 :type 命令来验证这一点，代码如下：

```
scala> :type spark
scala> :type sc
```

执行过程如图 1-24 所示。

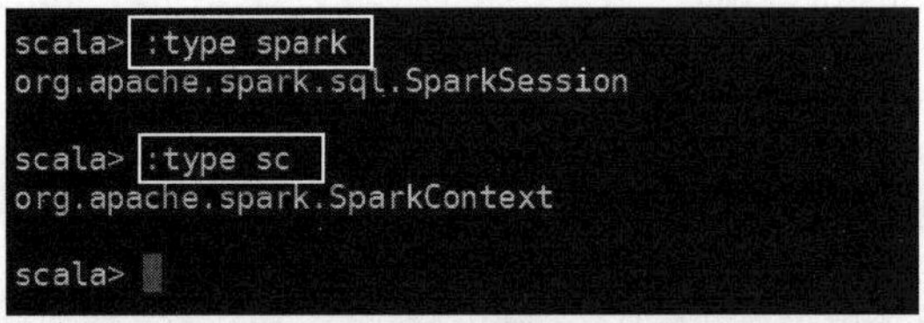

图 1-24　验证 SparkSession 实例和 SparkContext 实例

如果想要查看当前使用的 Spark 版本号，则可执行的命令如下：

```
scala> spark.version
scala> sc.version
```

执行过程如图 1-25 所示。

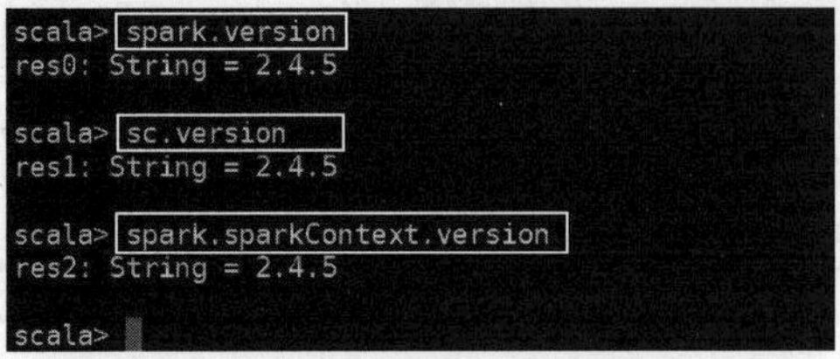

图 1-25　查看当前使用的 Spark 版本号

如果要查看在 spark-shell 中的默认配置，则可以访问 spark 的 conf 变量。例如，显示 spark-shell 中默认的配置信息，命令如下：

```
scala> spark.conf.getAll.foreach(println)
```

执行过程如图 1-26 所示。

图 1-26　查看在 Spark-shell 中的默认配置

1.7.5　Spark Web UI

每次初始化 SparkSession 对象时，Spark 都会启动一个 Web UI，提供关于 Spark 环境和作业执行统计信息的信息。Web UI 的默认端口是 4040，但是如果这个端口已经被占用（例如，被另一个 Spark Web UI 占用），则 Spark 会增加该端口号值，直到找到一个空闲的端口号为止。

在启动一个 spark-shell 时，将看到与此类似的输出行（除非关闭了 INFO log 消息）：

```
Spark context Web UI available at http://xueai8:4040
```

启动 spark-shell 时的输出行信息如图 1-27 所示。

图 1-27　初始化 SparkSession 对象时，Spark 都会启动一个 Web UI

注意：可以通过将 spark.ui.enabled 配置参数设为 false 来禁用 Spark Web UI。可以用 spark.ui.port 参数来改变它的端口。

查看 Spark Web UI 欢迎页面的一个示例，如图 1-28 所示。

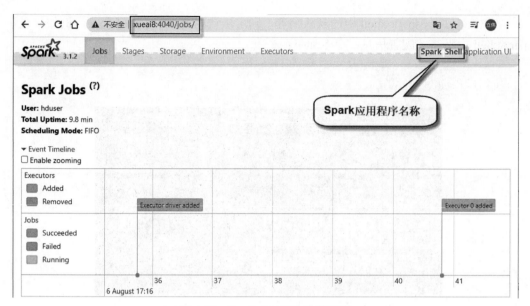

图 1-28 Spark Web UI 欢迎页面

这个 Web UI 是从一个 PySpark Shell 启动的,所以它的名字被设置为 Spark Shell,如图 1-28 右上角所示。

在运行 spark-submit 命令时,也可以使用--conf spark.app.name=<new_name>在命令行上设置程序名称,但不能在启动 Spark Shell 时更改应用程序名称。在这种情况下,它总是默认为 Spark Shell。

在 Spark Web UI 的 Environment 页面,可以查看影响 Spark 应用程序的配置参数的完整列表,如图 1-29 所示。

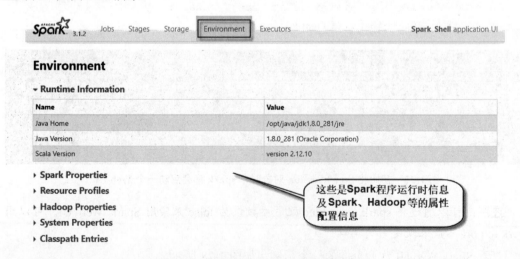

图 1-29 影响 Spark 应用程序的配置参数

1.8 使用 spark-submit 提交 Spark 应用程序

9min

对于公司大数据的批量处理或周期性数据分析/处理任务，通常采用编写好的 Spark 程序，并通过 spark-submit 指令的方式提交给 Spark 集群进行具体的任务计算，spark-submit 指令可以指定一些向集群申请资源的参数。

Spark 安装包附带有 spark-submit.sh 脚本文件（适用于 Linux、Mac）和 spark-submit.cmd 命令文件（适用于 Windows）。这些脚本可以在$SPARK_HOME/bin 目录下找到。

实际上，spark-submit 命令是一个实用程序，通过指定选项和配置向集群中运行或提交 Spark 或 PySpark 应用程序（或作业-job），提交的应用程序可以用 Scala、Java 或 Python 编写。spark-submit 命令支持以下功能：

（1）在 YARN、Kubernetes、Mesos、Standalone 等不同的集群管理器上提交 Spark 应用。

（2）在客户端部署模式或集群部署模式下提交 Spark 应用程序。

在 spark-submit 命令内部使用 org.apache.spark.deploy.SparkSubmit 类和用户指定的选项和命令行参数。使用 spark-submit 命令时最常用的选项如下：

```
./bin/spark-submit \
  --master <master-url> \
  --deploy-mode <deploy-mode> \
  --conf <key<=<value> \
  --driver-memory <value>g \
  --executor-memory <value>g \
  --executor-cores <number of cores>  \
  --jars  <comma separated dependencies>
  --class <main-class> \
  <application-jar> \
  [application-arguments]
```

也可以不使用 spark-submit，而是像下面这样提交应用程序，命令如下：

```
./bin/spark-class org.apache.spark.deploy.SparkSubmit <options & arguments>
```

1.8.1 spark-submit 指令的各种参数说明

在 Linux 环境下，可通过 spark-submit --help 命令来了解 spark-submit 指令的各种参数说明，命令如下：

```
$ cd ~/bigdata/spark-3.1.2
$ ./bin/spark-submit -help
```

使用 spark-submit 命令的完整语法如下：

```
$ ./bin/spark-submit [options] <app jar | python file> [app options]
```

其中 options 的主要标志参数说明如下。

（1）--master：指定使用哪个集群管理器来运行应用程序。Spark 目前支持 YARN、Mesos、Kubernetes、Standalone 和 local。

（2）--deploy-mode：是否要在客户端（选项值为 client）启动驱动程序，或者在集群中（选项值为 cluster）的一台 Worker 机器上启动驱动程序。在 client 模式下，驱动程序在调用 spark-submit 的机器上本地运行。客户端模式主要用于交互和调试。在 cluster 模式下，驱动程序会被发送到集群的一个 Worker 节点上运行，该节点在应用程序的 Spark Web UI 上显示为 driver。集群模式用于运行生产作业。默认为 client 模式。

（3）--class：应用程序的主类（带有 main()方法的类），如果运行 Java 或 Scala 程序。

（4）--name：应用程序易读的名称，这将显示在 Spark 的 Web UI 上。

（5）--jars：一系列 JAR 文件的列表，会被上传并设置到应用程序的 classpath 上。如果应用程序依赖于少量的第三方 JAR 包，则可以将它们加到这里（逗号分隔）。

（6）--files：使用逗号分隔的文件。通常，这些文件可以来自 resource 文件夹。使用此选项，Spark 将所有这些文件提交到集群。这个标志参数可被用于想要分布到每个节点上的数据文件（注意，使用--files 指定的文件会被上传到集群）。

（7）--py-files：一系列文件的列表，会被添加到应用程序的 PYTHONPATH。这可以包括.ipynb、.egg 或.zip 文件。

（8）--executor-memory：executor 使用的内存数量，以字节为单位。可以指定不同的后缀，如 512m 或 15g。

（9）--driver-memory：driver 进程所使用的内存数量，以字节为单位。可以指定不同的后缀，如 512m 或 15g。

（10）--verbose：显示详细信息。例如，将 Spark 应用程序使用的所有配置写入日志文件。

（11）--config：用于指定应用程序配置、shuffle 参数、运行时配置。

关于 driver 和 executor 资源（CPU 核和内存）配置，需要深入了解一下。在提交应用程序时，用户可以指定需要为 driver 和 executor 提供多少内存和核数。与这些资源相关的选项说明见表 1-5。

表 1-5 Spark 程序提交时的资源选项

选 项	说 明
--driver-memory	Spark driver 需要使用的内存
--driver-cores	Spark driver 需要使用的 CPU 内核数
--num-executors	要使用的执行器 executors 总数
--executor-memory	executor 进程使用的内存量
--executor-cores	executor 进程使用的 CPU 核数
--total-executor-cores	要使用的执行器 executor 内核总数

下面这个示例将 Spark 应用程序运行在 Standalone 集群上，采用 cluster 集群部署模式，指定每个 executor 分配 5GB 内存和 8 个核，命令如下：

```
$ ./bin/spark-submit \
    --class org.apache.spark.examples.SparkPi \
```

```
  --master spark://192.168.231.132:7077 \
  --deploy-mode cluster \
  --executor-memory 5g \
  --executor-cores 8 \
  $SPARK_HOME/examples/jars/spark-examples_3.1.2.jar 80
```

下面这个示例将 Spark 应用程序运行在 YARN 集群上，采用 cluster 集群部署模式，指定 driver 进程分配 8GB 内存，每个 executor 分配 16GB 内存和 2 个核，命令如下：

```
$ ./bin/spark-submit \
  --master yarn \
  --deploy-mode cluster \
  --driver-memory 8g \
  --executor-memory 16g \
  --executor-cores 2 \
  --class org.apache.spark.examples.SparkPi \
  $SPARK_HOME/examples/jars/spark-examples_3.1.2.jar 80
```

下面的示例使用集群部署模式将应用程序提交给 YARN 集群管理器，并指定 8GB driver 内存，指定每个 executor 有 16GB 内存和 2 个内核，命令如下：

```
$ ./bin/spark-submit \
  --verbose
  --master yarn \
  --deploy-mode cluster \
  --driver-memory 8g \
  --executor-memory 16g \
  --executor-cores 2 \
  --files /path/log4j.properties,/path/file2.conf,/path/file3.json
  --class org.apache.spark.examples.SparkPi \
  $SPARK_HOME/examples/jars/spark-examples_3.1.2.jar 80
```

命令 spark-submit 使用--conf 支持几种配置，这些配置用于指定应用程序配置、shuffle 参数、运行时配置。这些配置对于用 Java、Scala 和 Python 编写的 Spark 应用程序（PySpark）来讲是相同的。几种常用的配置 key 及其说明见表 1-6。

表 1-6　spark-submit 使用--conf 支持几种配置

选项	说明
spark.sql.shuffle.partitions	为宽 shuffle 转换（join 连接和聚合）创建的分区数
spark.executor.memoryOverhead	在集群模式下为每个 executor 进程分配的额外内存量，这通常是用于 JVM 开销的内存（PySpark 不支持）
spark.serializer	org.apache.spark.serializer. JavaSerializer (default) org.apache.spark.serializer.KryoSerializer
spark.sql.files.maxPartitionBytes	读取文件时为每个分区使用的最大字节数，默认为 128 MB
spark.dynamicAllocation.enabled	指定是否根据工作负载动态增加或减少 executor 的数量，默认值为 true
spark.dynamicAllocation.minExecutors	启用动态分配时使用的最小 executor 数量
spark.dynamicAllocation.maxExecutors	启用动态分配时使用的最大 executor 数量
spark.extraJavaOptions	指定 JVM 选项

当使用spark-submit提交Spark应用程序时,使用--conf指定选项配置参数的代码如下:

```
$ ./bin/spark-submit \
    --master yarn \
    --deploy-mode cluster \
    --conf "spark.sql.shuffle.partitions=20000" \
    --conf "spark.executor.memoryOverhead=5244" \
    --conf "spark.memory.fraction=0.8" \
    --conf "spark.memory.storageFraction=0.2" \
    --conf "spark.serializer=org.apache.spark.serializer.KryoSerializer" \
    --conf "spark.sql.files.maxPartitionBytes=168435456" \
    --conf "spark.dynamicAllocation.minExecutors=1" \
    --conf "spark.dynamicAllocation.maxExecutors=200" \
    --conf "spark.dynamicAllocation.enabled=true" \
    --conf "spark.executor.extraJavaOptions=-XX:+PrintGCDetails -XX:+PrintGCTimeStamps" \
    --files /path/log4j.properties,/path/file2.conf,/path/file3.json \
    --class org.apache.spark.examples.SparkPi \
    $SPARK_HOME/examples/jars/spark-examples_repace-spark-3.1.2.jar 80
```

也可以在$SPARK_HOME/conf/spark-defaults.conf文件中将这些配置设置为全局的,以应用于每个Spark应用程序。

也可以通过编程方式使用SparkConf进行设置,代码如下:

```
val config = new SparkConf()
config.set("spark.sql.shuffle.partitions","300")
val spark = SparkSession.builder().config(config).master("local[3]")
    .appName("SparkExamples")
    .getOrCreate();

val arrayConfig = spark.sparkContext.getConf.getAll
for (conf <- arrayConfig)
println(conf._1 +", "+ conf._2)

//使用SparkConf的get()方法获取特定配置项的值
print("spark.sql.shuffle.partitions ==> " +
spark.sparkContext.getConf.get("spark.sql.shuffle.partitions"))
//显示如下的值:spark.sql.shuffle.partitions ==> 300
```

这几个配置的优先顺序是,首先选择代码中的SparkConf,然后选择命令行spark-submit --conf选项,最后选择spark-defaults.conf文件中提到的配置。

无论使用哪种语言,大多数选项是相同的,但是也有少数选项是特定用于某种语言的。

1. 用于Scala或Java程序的参数

例如,要运行用Scala或Java语言编写的Spark应用程序,需要使用的额外选项见表1-7。

2. 用于PySpark(Python)程序的参数

当想要使用spark-submit命令提交一个PySpark应用程序时,需要指定想要运行的.py文件，并为依赖库指定.egg 文件或.zip 文件。这时，除了可以使用上面提到的大多数选项和配置外，一些特定于 Spark 应用程序的选项和配置见表 1-8。

表 1-7　用于运行 Scala 或 Java 语言编写的 Spark 应用程序的额外选项

选　　项	说　　明
--jars	如果在一个文件夹中有所有的依赖 JAR，则可以使用 spark-submit --jars 选项传递所有这些 JAR。所有的 JAR 文件都应该用逗号分隔。例如，--jars jar1.jar, jar2.jar, jar3.jar
--packages	用此命令时将处理所有传递依赖项
--class	指定想运行的 Scala 或 Java 类。这应该是带有包名的完全限定名，例如 org.apache.spark.examples.SparkPi

表 1-8　用于运行 Python 语言编写的 Spark 应用程序的额外选项

PySpark 专用配置	说　　明
--py-files	使用--py-files 添加.py、.zip 或.egg 文件
--config spark.executor.pyspark.memory	PySpark 为每个 executor 进程使用的内存量
--config spark.pyspark.driver.python	用于 PySpark driver 的 Python 二进制可执行文件
--config spark.pyspark.python	用于 PySpark driver 和 executor 的 Python 二进制可执行文件

注意：使用--py-files 指定的文件将在集群运行应用程序之前被上传到集群。还可以提前上传这些文件，并在 PySpark 应用程序中引用它们。

例如，提交 PySpark 应用程序，命令如下：

```
$ ./bin/spark-submit \
   --master yarn \
   --deploy-mode cluster \
   wordcount.py
```

当使用其他 Python 文件作为依赖项时，命令如下：

```
$ ./bin/spark-submit \
   --master yarn \
   --deploy-mode cluster \
   --py-files file1.py,file2.py,file3.zip
   wordcount.py
```

1.8.2　提交 SparkPi 程序，计算圆周率 π 值

Spark 安装包中自带了一个使用蒙特卡洛方法求圆周率 π 值的程序。下面通过 spark-submit 将其提交到 Spark 集群上以 standalone 模式运行，以掌握 spark-submit 提交 Spark 程序的方法。

建议按以下步骤操作。

（1）确保已经启动了 Spark 集群（standalone）模式（启动方式见 1.5.4 节）。

（2）打开终端窗口，进入 Spark 主目录下，执行的命令如下：

```
$ cd ~/bigdata/spark-3.1.2
$ ./bin/spark-submit --master spark://xueai8:7077 --class
org.apache.spark.examples.SparkPi
  examples/jars/spark-examples_2.12-3.1.2.jar 10
```

说明：

（1）--master 参数用于指定要连接的集群管理器，这里是 standalone 模式。

（2）--class 参数用于指定要执行的主类名称（带包名的全限定名称）。

（3）接下来的一个参数是所提交的 JAR 包。

（4）最后一个参数是需要传入应用程序内的外部参数。本例中的 10 指的是 Spark 程序创建的分区的数量，也就是计算的并行度。计算 Pi 的任务被划分为 10 个任务（Pi 值是通过迭代算法计算的）。

运行结果如图 1-30 所示。

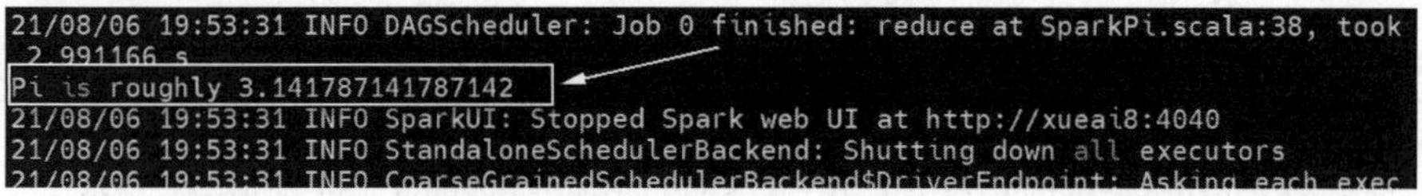

图 1-30 运行蒙特卡洛方法求圆周率 π 值的 Spark 程序计算结果

1.8.3 将 SparkPi 程序提交到 YARN 集群上执行

也可以将 Spark 程序运行在 YARN 集群上，由 YARN 来管理集群资源。下面使用 spark-submit 将 SparkPi 程序提交到 Spark 集群上以 YARN 模式运行。

建议按以下步骤执行。

（1）打开终端窗口。

（2）不需要启动 Spark 集群，但是需要启动 YARN 集群，命令如下：

```
$ start-dfs.sh
$ start-yarn.sh
```

执行过程如图 1-31 所示。

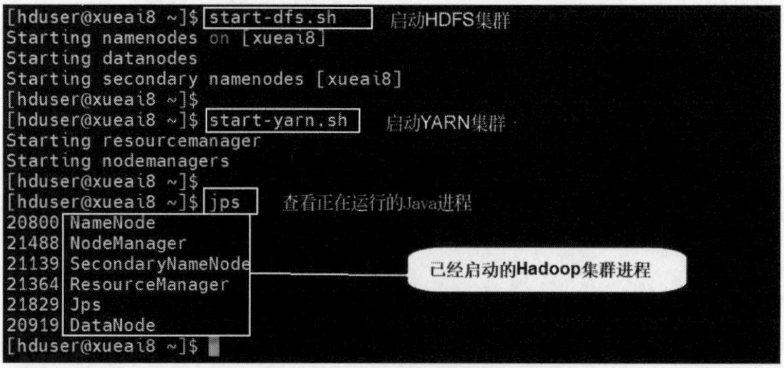

图 1-31 启动 HDFS 和 YARN 集群

（3）进入 Spark 主目录下，执行的命令如下：

```
$ cd ~/bigdata/spark-3.1.2
$ ./bin/spark-submit \
--class org.apache.spark.examples.SparkPi \
--master yarn \
./examples/jars/spark-examples_2.12-3.1.2.jar 10
```

执行过程如图 1-32 所示。

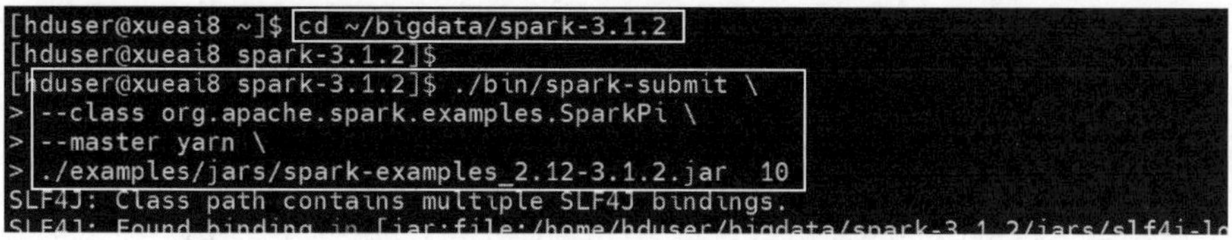

图 1-32　Spark on YARN 执行过程

执行结果如图 1-33 所示。

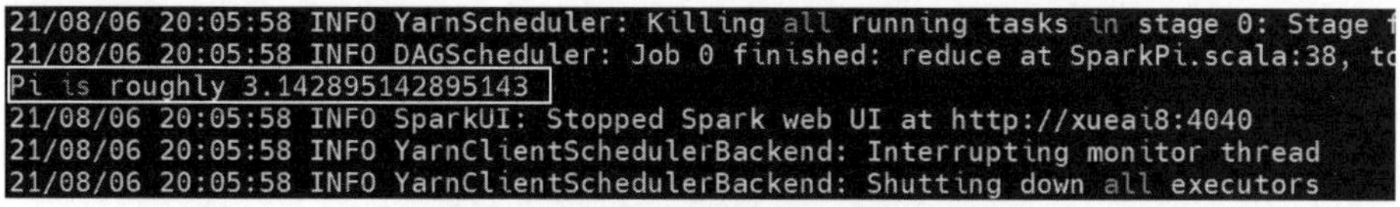

图 1-33　Spark on YARN 执行结果

当在 YARN 上部署 Spark 时，有可能会出现异常信息，异常内容如下：

```
...
20/05/26 14:12:55 ERROR spark.SparkContext: Error initializing
SparkContext.
java.lang.IllegalStateException: Spark context stopped while waiting for
backend
...
Caused by: java.nio.channels.ClosedChannelException
java.lang.IllegalStateException: Spark context stopped while waiting for
backend
...
```

这个问题的原因是由于 Java 8 的内存过量分配问题（虚拟内存不足），导致与 YARN 集群的关联可能会丢失。可以通过在 yarn-site.xml 文件中设置以下属性来强制 YARN 忽略它，配置如下：

```
<property>
     <name>yarn.nodemanager.pmem-check-enabled</name>
     <value>false</value>
</property>

<property>
     <name>yarn.nodemanager.vmem-check-enabled</name>
     <value>false</value>
</property>
```

（1）yarn.nodemanager.pmem-check-enabled：是否启动一个线程检查每个任务正使用的物理内存量，如果任务超出分配值，则直接将其杀掉，默认为 true。

（2）yarn.nodemanager.vmem-check-enabled：是否启动一个线程检查每个任务正使用的虚拟内存量，如果任务超出分配值，则直接将其杀掉，默认为 true。

也可以修改 yarn-site.xml 配置文件，设置虚拟内存与物理内存的比例（默认为 2.1，这里提高到 4），配置如下：

```
<property>
    <name>yarn.nodemanager.vmem-pmem-ratio</name>
    <value>4</value>
    <description>在为容器设置内存限制时，虚拟内存与物理内存的比率</description>
</property>
```

这个参数表示每单位的物理内存总量对应的虚拟内存量，默认为 2.1，表示每使用 1MB 的物理内存，最多可以使用 2.1MB 的虚拟内存总量。

经过以上修改，然后执行 Spark on YARN，异常消失，程序正常执行。

第 2 章　开发和部署 Spark 应用程序

CHAPTER 2

要开发 Spark 应用程序，业界普遍采用 IntelliJ IDEA 集成开发环境。在使用集成开发环境工具开发 Spark 应用程序时，需要很多 Hadoop 和 Spark 等的依赖库，目前企业中普遍采用一些项目管理和构建工具（如 Maven、SBT 等）来管理依赖，以及项目的编译和打包等。Spark 官方推荐使用 SBT（Scala Build Tool，Scala 构建工具）来构建 Spark 项目、管理依赖、编译和打包项目。

2.1　使用 IntelliJ IDEA 开发 Spark SBT 应用程序

IntelliJ IDEA，一般简称 IDEA，是 Java 语言开发的集成环境。IntelliJ 在业界被公认为最好的 Java 开发工具之一，尤其在智能代码助手、代码自动提示、重构、J2EE 支持、Ant、JUnit、CVS 整合、代码审查、创新的 GUI 设计等方面的功能可以说是超常的。IDEA 是 JetBrains 公司的产品，这家公司的总部位于捷克共和国的首都布拉格，开发人员以严谨著称的东欧程序员为主。IntelliJ IDEA 的下载网址为 https://www.jetbrains.com/idea/download。

IntelliJ IDEA 提供 Community（指社区版）和 Ultimate（指企业版）两个版本，其中 Community 是完全免费的，而 Ultimate 版本可以免费使用 30 天，过了这段时间后需要收费。开发 Spark 应用程序，下载一个最新的 Community 版本即可，如图 2-1 所示。

图 2-1　下载 IntelliJ IDEA 社区版

2.1.1 安装 IntelliJ IDEA

首先在官网下载 IntelliJ IDEA 社区版安装包，例如 ideaIU-2018.2.5.exe（版本号可能和本书有所不同，读者下载最新版即可），然后建议按以下步骤执行安装过程。

（1）双击 ideaIU-2018.2.5.exe 安装文件，打开安装向导，如图 2-2 所示。

图 2-2　IntelliJ IDEA 安装向导

（2）单击 Next 按钮，选择安装路径（需要记住这个路径，后面会用到），如图 2-3 所示。

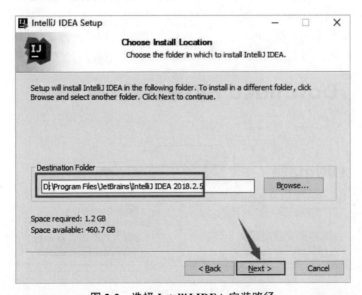

图 2-3　选择 IntelliJ IDEA 安装路径

（3）单击 Next 按钮，继续按向导提示配置安装选项，如图 2-4 所示。

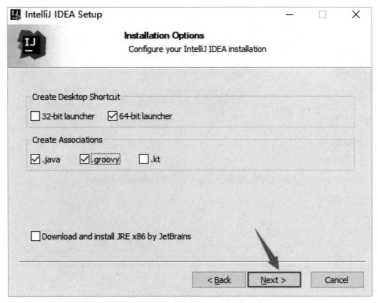

图 2-4　配置 IntelliJ IDEA 安装选项

（4）单击 Next 按钮，向导会提示选择将 IntelliJ IDEA 添加到操作系统的开始菜单中。单击 Install 按钮进行安装，如图 2-5 所示。

图 2-5　选择开始菜单文件夹

（5）在接下来的安装向导界面上，单击 Next 按钮，开始安装。这可能需要一点时间，需耐心等待，如图 2-6 所示。

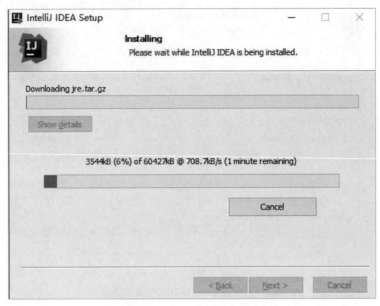

图 2-6　IntelliJ IDEA 安装过程

（6）当 IntelliJ IDEA 安装完成时，会出现提示界面，如图 2-7 所示。

图 2-7　IntelliJ IDEA 安装完成

此时单击 Finish 按钮，结束安装。

IntelliJ IDEA 安装完成后，就可以启动 IntelliJ IDEA 了。可以通过以下两种方式启动：

（1）到 IntelliJ IDEA 安装目录下，进入 bin 目录后双击 idea.sh 文件便可启动 IntelliJ IDEA。

（2）在命令行终端中，进入$IDEA_HOME/bin 目录，输入 ./idea.sh 进行启动。

2.1.2　配置 IntelliJ IDEA Scala 环境

在 IntelliJ IDEA 中开发 Scala 程序（及 Spark 程序）需要安装 Scala 插件。IntelliJ IDEA 默认情况下并没有安装 Scala 插件，需要手动进行安装。安装过程并不复杂，按下面的演示步骤操作即可。

（1）在 IntelliJ IDEA 启动界面上选择 Configure→Plugins 菜单项（或者在项目界面，选择 File→Settings→Plugins 菜单项），然后会弹出插件管理界面，在该界面上列出了所有安装好的插件。由于 Scala 插件没有安装，需要单击 Install JetBrains plugin 进行安装，如图 2-8 所示。

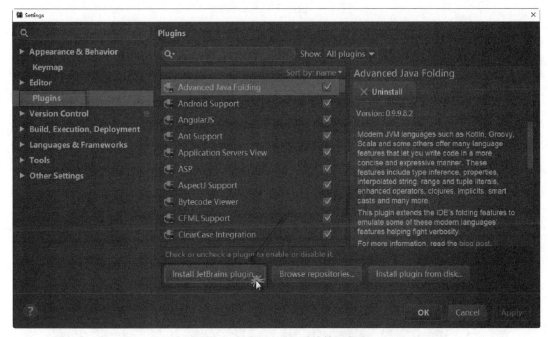

图 2-8　IntelliJ IDEA 插件管理界面

（2）待安装的插件很多，可以通过查询或者字母顺序找到 Scala 插件，选择插件后在界面的右侧会出现该插件的详细信息，在绿色按钮 Install 上单击便可安装插件，如图 2-9 所示。

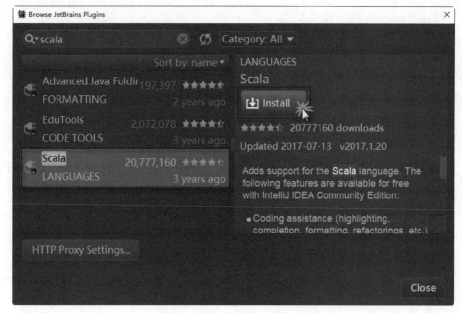

图 2-9　选择安装 Scala 插件

（3）安装过程中将出现安装进度界面，通过该界面可了解插件安装进度，如图 2-10 所示。

图 2-10　Scala 插件安装过程

（4）安装完成后，将看到一个按钮 Restart IntelliJ IDEA，用于重新启动 IntelliJ IDEA。单击它来重启 IntelliJ IDEA，如图 2-11 所示。

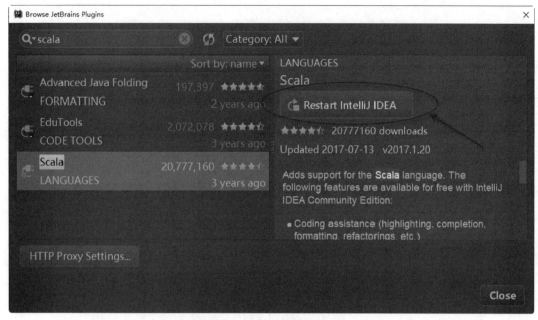

图 2-11　Scala 插件安装完成

（5）这时会出现一个提示框，询问是否确定要重启 IntelliJ IDEA。单击 Restart 按钮，重启 IntelliJ IDEA，如图 2-12 所示。

图 2-12　重启 IntelliJ IDEA

现在已经安装好了 IntelliJ IDEA、Scala 插件和 SBT，可以开始构建 Spark 程序了。

2.1.3　创建 IntelliJ IDEA SBT 项目

SBT 之于 Scala，就像 Maven 之于 Java，用于管理项目依赖和构建项目。使用 IntelliJ IDEA 开发 Spark 应用程序，使用 SBT 作为构建管理器，这是官方推荐的开发方式（安装 Scala 插件时，该 Scala 插件已经自带 SBT 工具）。

建议按以下步骤使用 IntelliJ IDEA 创建一个新的 Spark SBT 项目。

（1）启动 IntelliJ IDEA，在开始界面中选择 Create New Project，创建一个新项目，如图 2-13 所示。

图 2-13 在开始界面选择创建新的项目

（2）在打开的项目引导界面上，依次选择 Scala→SBT 选项，然后单击 Next 按钮，如图 2-14 所示。

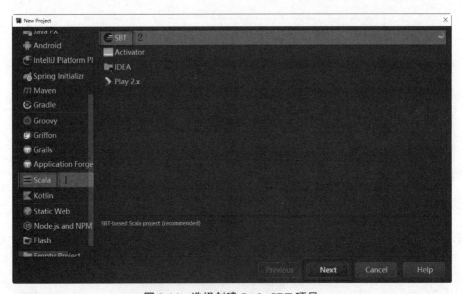

图 2-14 选择创建 Scala SBT 项目

（3）在接下来的向导窗口中，将项目命名为 HelloSpark，指定项目存放的位置，并选择合适的 SBT 和 Scala 版本，如图 2-15 所示。

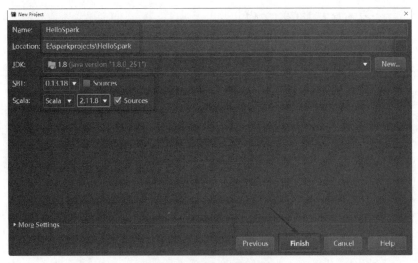

图 2-15 填写项目相关信息

（4）单击 Finish 按钮继续安装过程。IntelliJ IDEA 应该会创建一个具有默认目录结构的新项目。生成所需的所有文件夹可能需要一两分钟，最终的项目目录结构如图 2-16 所示。

图 2-16 生成的项目目录结构

对上面生成的项目目录结构的说明如下。

① .idea：IntelliJ 配置文件。

② project：编译期间使用的文件。例如，在 build.properties 中指定编译项目时使用的 SBT 版本。

③ src：源代码。大多数代码应该放在 main/scala 目录下。测试脚本放在 test/scala 文件夹下。

④ target：当编译项目时，会生成这个文件夹。

⑤ build：SBT 配置文件。可使用该文件导入第三方库和文档。

2.1.4 配置 SBT 构建文件

在开始编写 Spark 应用程序之前，需要将 Spark 库和文档导入 IntelliJ IDEA，这需要在 build.sbt 文件中进行配置。编辑 build.sbt 文件，创建工程时生成的初始内容如下：

```
name := "HelloSpark"

version := "1.0"

scalaVersion := "2.12.14"
```

向其中添加 Spark Core 和 Spark SQL 依赖，内容如下：

```
libraryDependencies ++= Seq(
  "org.apache.spark" %% "spark-core" % "3.1.2",
  "org.apache.spark" %% "spark-sql" % "3.1.2"
)
```

注意：可以到 https://mvnrepository.com/artifact/org.apache.spark/spark-core 查询 SBT 依赖库的内容和格式。

保存文件后，IntelliJ IDEA 将自动导入运行 Spark 所需的库和文档，因此要确保用户的计算机是可以联网的，结果如图 2-17 所示。

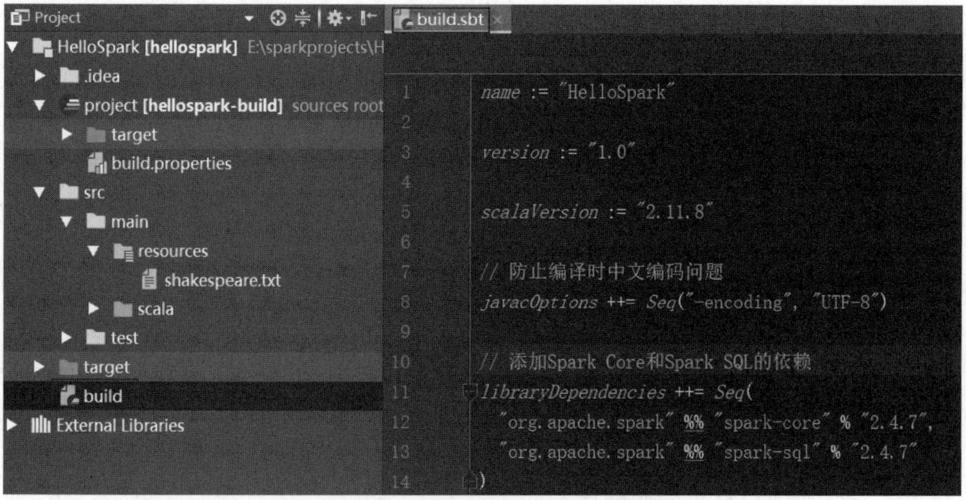

图 2-17　build.sbt 项目构建文件配置

2.1.5 准备数据文件

接下来，将构建一个简单的 Spark 应用程序，用来对莎士比亚文集（shakespeare.txt）执行单词计数任务。

需要在两个地方保存 shakespeare.txt 数据集。一个在项目中用于本地系统测试，另一个在 HDFS（Hadoop 分布式文件系统）中用于集群测试。

将 PBDP 大数据平台中的/home/hduser/data/spark/shakespeare.txt 文件上传到 HDFS，步骤如下。

（1）确保已经启动了 HDFS。

（2）在终端窗口中，执行 HDFS Shell 命令，将文件上传到 HDFS 上，命令如下：

```
$ cd /home/hduser/data/spark
$ hdfs dfs -put shakespeare.txt /data/spark/
```

2.1.6　创建 Spark 应用程序

现在，准备开始编写 Spark 应用程序。建议按以下步骤编写。

（1）回到项目中，在 src/main 文件夹下创建一个名为 resources 的文件夹（如果它不存在），并将 shakespear .txt 文件复制到该文件夹下，如图 2-18 所示。

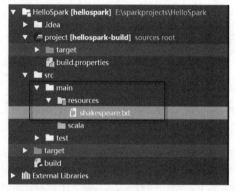

图 2-18　将数据文件复制到项目的 resources 目录中

（2）在 src/main/scala 文件夹下创建一个新的 Scala 类。在项目的 scala 目录上右击，然后选择 New→Scala Class 菜单项，如图 2-19 所示。

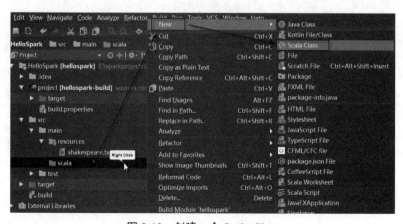

图 2-19　创建一个 Scala Class

（3）IntelliJ IDEA 会询问是创建一个 class、object 还是 trait？此处选择 object，将该类命名为 HelloWorld，如图 2-20 所示。

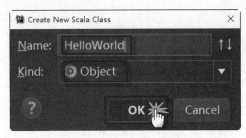

图 2-20　填写类名和类型

（4）在图 2-20 中单击 OK 按钮，会打开创建的 HelloWorld.scala 文件。编辑该源文件，代码如下：

```
object HelloWorld {
  def main(args: Array[String]): Unit ={
    println("Hello World!")
  }
}
```

（5）代码编写完成后，就可以运行程序了。在源文件的任何空白位置右击，在弹出的环境菜单中选择 Run 'HelloWorld' 菜单项，如图 2-21 所示。

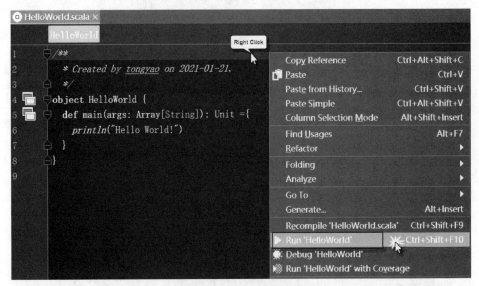

图 2-21　运行 Scala 程序

（6）如果一切正确，IntelliJ IDEA 应该在其下方的控制台窗口输出字符串"Hello World!"，如图 2-22 所示。

图 2-22　查看 Scala 程序运行结果

（7）现在环境已经正确设置了，接下来用以下代码替换文件中原来的内容：

```scala
import org.apache.spark.{SparkConf, SparkContext}
import org.apache.spark.sql.SparkSession

object HelloWorld {

  def main(args: Array[String]) {

    //在 Windows 系统下开发时设置
    System.setProperty("HADOOP_USER_NAME", "hduser")

    //创建一个 SparkContext 来初始化 Spark
    //Spark 2.0 以前的用法
    //val conf = new SparkConf().setMaster("local").setAppName("Word Count")
    //val sc = new SparkContext(conf)

    //Spark 2.0 以后的用法
    val spark = SparkSession
            .builder()
            .master("local[*]")
            .appName("Word Count")
            .getOrCreate()
    val sc = spark.sparkContext

    //将文本加载到 Spark RDD 中，它是文本中每行的分布式表示
    val file = "src/main/resources/shakespeare.txt"
    val textFile = sc.textFile(file)

    //transformation 转换
    val counts = textFile.flatMap(line => line.split(" "))
            .map(word => (word, 1))
            .reduceByKey(_ + _)

    counts.collect.foreach(println)
    System.out.println("全部单词: " + counts.count());

    //将单词计数结果保存到指定文件中
    val output = "tmp/shakespeareWordCount"
```

```
        counts.saveAsTextFile(output)
    }

}
```

（8）同样，在文件上的任意位置右击，并选择 Run 'HelloScala'菜单项来运行程序。这将运行 Spark 作业并打印莎士比亚作品中出现的每个单词的次数，如果一切正确，输出结果如图 2-23 所示。

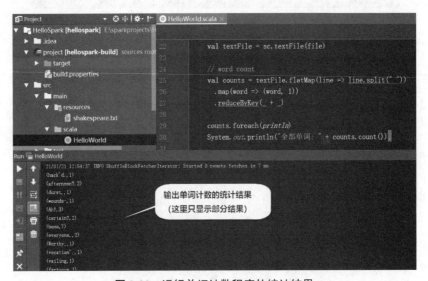

图 2-23　运行单词计数程序的统计结果

（9）此外，浏览下面代码中指定的目录：

```
counts.saveAsTextFile("tmp/shakespeareWordCount");
```

将找到程序运行所输出的内容，如图 2-24 所示。

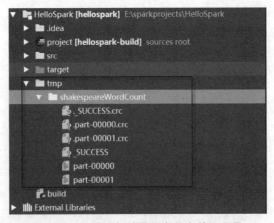

图 2-24　程序运行结果保存的文件和目录

注意在代码中设置了下面这一行代码:

```
.master("local[*]")
```

这是用来告诉 Spark 使用这台计算机在本地运行程序,而不是在分布式模式下运行。要在多台机器上运行 Spark,需要更改此值,稍后将讲解如何更改。

2.1.7 部署分布式 Spark 应用程序

现在已经了解了如何在 IntelliJ IDEA 中直接部署应用程序。这是一种快速构建和测试应用程序的好方法,但是在生产环境中,Spark 通常会处理存储在 HDFS 等分布式文件系统中的数据。Spark 通常也以集群模式运行(分布在许多计算机上)。

接下来,修改代码,使其能部署到 Spark 分布式集群上运行。建议按以下步骤操作。

(1)修改源代码,修改后的内容如下:

```
import org.apache.spark.{SparkConf, SparkContext}

object HelloScala {

  def main(args: Array[String]) {

    //创建一个 SparkContext 来初始化 Spark
    val spark = SparkSession.builder().appName("Word Count").getOrCreate()
    val sc = spark.sparkContext

    //将文本加载到 Spark RDD 中,它是文本中每行的分布式表示
    val file = "hdfs://localhost:8020/data/spark/shakespeare.txt"
    val textFile = sc.textFile(file)

    //word count
    val counts = textFile.flatMap(line => line.split(" "))
                    .map(word => (word, 1))
                    .reduceByKey(_ + _)

    counts.collect.foreach(println)
    System.out.println("全部单词: " + counts.count());

    //将单词计数结果保存到指定输出目录中
    val output = "hdfs://localhost:8020/data/spark/shakespeareWordCount"
    counts.saveAsTextFile(output)
  }

}
```

这将告诉 Spark 读写 HDFS 分布式文件系统,而不是读写本地 Linux 文件系统。

(2)将项目打包成 JAR 文件。

将这些已编译的代码打包到一个 JAR 文件中,该文件可以部署在 Spark 集群上。在 IntelliJ IDEA 菜单栏选择 Tools→Start SBT Shell 菜单项,在编辑窗口下方打开 SBT Shell

交互窗口，然后就可以应用 SBT 的 clean、compile、package 等命令进行操作。例如，执行打包操作，命令如下：

```
> package
```

打包过程如图 2-25 所示。

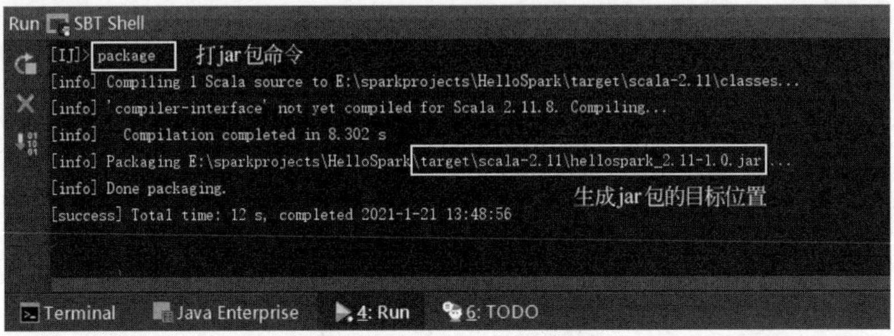

图 2-25 将 SBT 项目打包为 JAR 文件

这将在项目的 target/scala-2.11 目录下创建一个名为 hellospark_2.11-1.0.jar 的编译过的 JAR 文件。

（3）将该 hellospark_2.11-1.0.jar 包提交到 Spark 集群上执行。使用 spark-submit 命令，提交任务时需要指定主类、要运行的 JAR 包和运行模式（本地或集群），命令如下：

```
$ cd ~/bigdata/spark-3.1.2
$ ./bin/spark-submit
  --class HelloWorld
  --master local[*]
  ./hellospark_2.11-1.0.jar
```

（4）查看输出结果。控制台应该打印莎士比亚作品中出现的每个单词的次数，部分内容如下：

```
...
(GENTLEWOMAN,4)
(honors,10)
(returnest,1)
(topp'd?,1)
(compass?,1)
(toothache?,1)
(miserably,1)
(hen?,1)
(luck?,2)
(call'd,162)
(lecherous,2)
......
```

（5）此外，也可通过 HDFS 或 Web UI 查看输出文件的内容。例如，应用 HDFS Shell 命令查看 HDFS 上输出文件的内容，命令如下：

```
$ hdfs dfs -cat /data/spark_demo/shakespeareWordCount/part-00000
```

2.1.8 远程调试 Spark 程序

在本节中，将学习如何将正在运行的 Spark 程序连接到调试器。通过调试器，可以设置断点并逐行执行代码。在直接从 IDEA 运行时，调试 Spark 和调试其他任何程序一样，但是调试远程集群需要一些配置。建议按以下步骤操作。

（1）在计划提交 Spark 作业的机器上，打开一个终端，执行的命令如下：

```
$ export SPARK_SUBMIT_OPTS=-agentlib:jdwp=transport=dt_socket,server=y,suspend=y,address=5005
```

（2）提交 Spark job 作业，使用的命令如下：

```
$ spark-submit --class HelloWorld --master local[*] --driver-java-options bigdata/hellospark_2.11-1.0.jar
```

（3）运行时会出现程序挂起，这表明正在监听端口中，如图 2-26 所示。

```
[hduser@localhost ~]$ spark-submit --class HelloScala  --master local bigdata/helloscala_2.11-1.0.jar
Listening for transport dt_socket at address: 5005
```

图 2-26　运行时程序挂起

（4）在 IntelliJ IDEA 中配置 remote Debug（远程调试）。在 IntelliJ IDEA 中选择菜单项 Run→Edit Configurations，编辑运行配置信息，如图 2-27 所示。

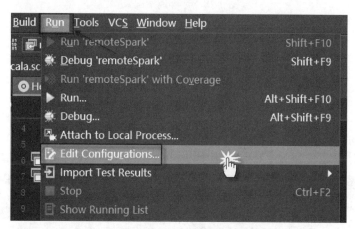

图 2-27　在 IntelliJ IDEA 中配置 remote Debug

（5）单击左上角的+按钮，选择面板左侧的 Remote 项，增加一个新的远程配置。在面板右侧，使用主机 IP 地址填写 Host（Port 字段默认为 5005，不需要修改），这将允许在端口 5005 关联调试器。需要确保端口 5005 能够接收入站连接，如图 2-28 所示。

（6）从 IntelliJ IDEA 中 Debug（调试）此调试配置，调试器将附加此调试配置，如图 2-29 所示。

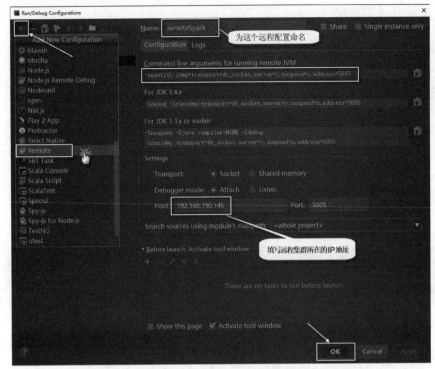

图 2-28　远程调试配置

图 2-29　选择远程调试

（7）远程 Spark 程序将在断点处停止，还可以检查程序中活动变量的值。在试图确定代码中的 Bug 时，这是非常宝贵的，这个过程如图 2-30 所示。

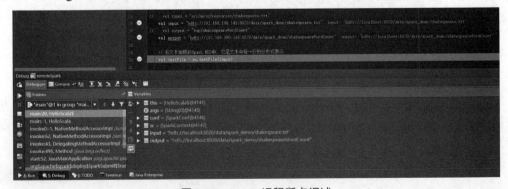

图 2-30　Spark 远程断点调试

2.2 使用 IntelliJ IDEA 开发 Spark Maven 应用程序

虽然 Spark 官方推荐使用 SBT 来构建 Spark 项目，但仍然有很多用户习惯使用 Maven。接下演示如何使用 IntelliJ IDEA 开发 Spark Maven 应用程序。

注意：集成开发环境 IntelliJ IDEA 的安装，可参考 2.1.1 节中的安装步骤。需要确保已经安装了 JDK 8 和 Scala 插件。

2.2.1 创建 IntelliJ IDEA Maven 项目

首先启动 IntelliJ IDEA，依次选择 File→New→Project→Maven，从模板创建 Scala Maven 项目，如图 2-31 所示。

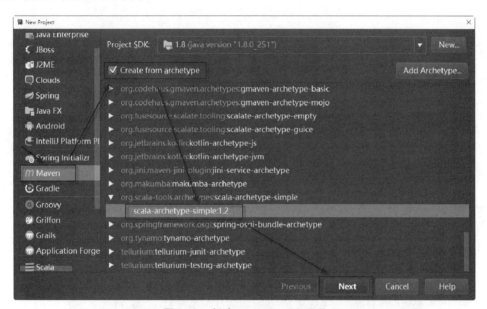

图 2-31 创建 Scala Maven 项目

在接下来的向导窗口中，对项目命名，内容如下：

```
GroupId: com.xueai8
ArtifactId: sparkexamples
Version: 1.0-SNAPSHOT
```

如图 2-32 所示。

单击 Next 按钮继续。接下来，设置 Maven 的 settings 文件和 repository 位置，如图 2-33 所示，可选择默认设置。

最后，选择项目名称和位置。这些字段应该自动填充，所以使用默认值就好了，然后单击 Finish 按钮，开始生成项目，如图 2-34 所示。

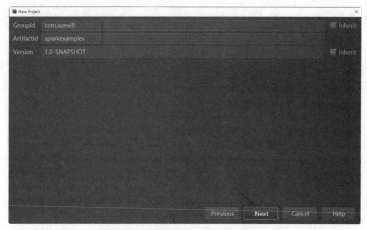

图 2-32　命名 Scala Maven 项目

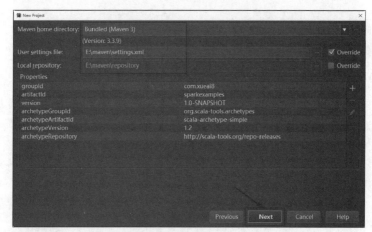

图 2-33　设置 Maven 的 settings 文件和 repository 位置

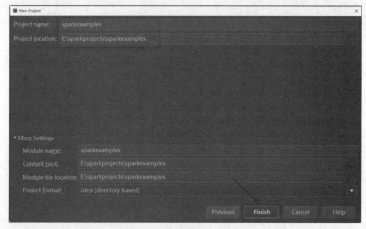

图 2-34　选择项目名称和位置

生成所有文件夹可能需要一两分钟，然后 IntelliJ IDEA 应该创建一个具有默认目录结构的新项目，如图 2-35 所示。

图 2-35 Maven 项目目录结构

对新创建的 Maven 项目的结构说明如下。

（1）.idea：IntelliJ 配置文件。

（2）src/main/scala：源代码。源代码应该位于此目录下，而 test 文件夹应该保留，用于测试脚本。

（3）target：当对项目编译时会生成此目录。

（4）pom.xml：Maven 配置文件。可以使用这个文件导入第三方库和文档。

2.2.2 验证 SDK 安装和配置

在继续之前，需要先验证几个 IntelliJ IDEA 依赖库的设置，确保项目所依赖的 JDK 和 SDK 都被正确设置。建议按以下步骤进行验证操作。

（1）验证导入 Maven 项目是否可自动打开。

选择 IntelliJ IDEA 的 File→Settings 菜单项，打开项目设置面板。在设置面板左侧，依次选择 Build,Execution,Deployment→Build Tools→Maven→Importing 选项，然后查看面板右侧的 Import Maven projects automatically 复选框是否勾选。确保该复选框处于被勾选状态，如图 2-36 所示。

（2）验证项目的 Project SDK 和 Project language level 已设置为 Java 8 版本。

选择 IntelliJ IDEA 的 File→Project Structure 菜单项，在打开的项目结构设置面板左侧，选择 Project 项，然后在面板右侧分别选择 Project SDK 和 Project language level 为 Java 8 版本，如图 2-37 所示。

（3）验证项目中模块所依赖的 Java 语言版本。

选择 IntelliJ IDEA 的 File→Project Structure 菜单项，在打开的项目结构设置面板左侧，选择 Modules 项，然后在面板右侧选择 Language level 为 Java 8 版本，如图 2-38 所示。

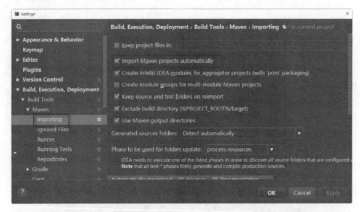

图 2-36　设置导入的 Maven 项目可自动打开

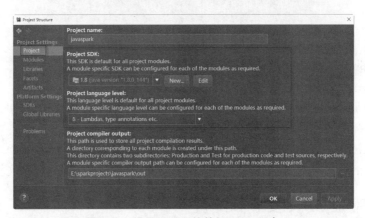

图 2-37　设置项目所依赖的 SDK 版本

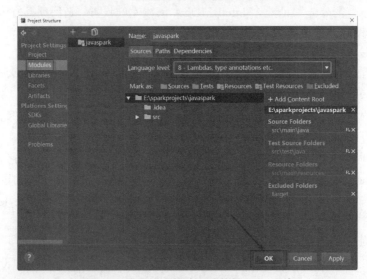

图 2-38　设置项目模块所依赖的 Java 版本

（4）验证全局库是否已经配置了 Scala SDK 依赖。

选择 IntelliJ IDEA 的 File→Project Structure 菜单项，在打开的项目结构设置面板左侧，选择 Global Libraries 项，然后在面板右侧，单击绿色十字按钮，选择本机安装的 Scala SDK 目录（如果本机没有安装 Scala SDK，则需要先下载并解压一个 Scala SDK），如图 2-39 所示。

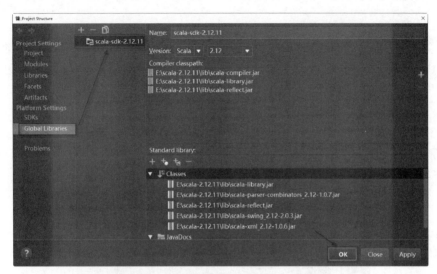

图 2-39　为 IDEA 配置全局库 Scala SDK

确保以上几项验证都通过，然后继续后面的内容。

2.2.3　项目依赖和配置管理

在开始编写 Spark 应用程序之前，需要将 Spark 库和文档导入 IntelliJ IDEA。Spark 库的导入需要使用依赖管理器 Maven，可通过项目中的 pom.xml 文件对 Maven 进行配置。在项目中打开 pom.xml 配置文件，按以下步骤编辑。

（1）首先，将 Scala 版本修改为最新的版本，本书编写时使用的是 2.12.11。修改 Scala 的版本如下：

```
<properties>
    <scala.version>2.12.11</scala.version>
</properties>
```

（2）从 pom.xml 文件中删除如下部分：

```
<plugin>
  <groupId>org.scala-tools</groupId>
  <artifactId>maven-scala-plugin</artifactId>
  <executions>
    <execution>
      <goals>
        <goal>compile</goal>
```

```xml
          <goal>testCompile</goal>
        </goals>
      </execution>
    </executions>
    <configuration>
      <scalaVersion>${scala.version}</scalaVersion>
      <args>
        <arg>-target:JVM-1.8</arg>
      </args>
    </configuration>
</plugin>
```

(3) 删除不必要的文件。

从项目结构中，删除如下部分：

① 删除 src/test。

② 删除 src/main/scala/org.xueai8.App。

操作如图 2-40 所示。

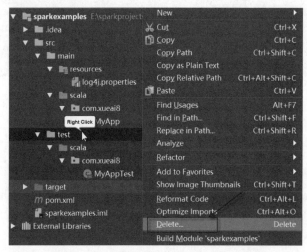

图 2-40　从项目结构中，删除不必要的目录

(4) 将 Spark 3.1.2 依赖添加到 Maven 中。在 pom.xml 文件中增加以下内容：

```xml
<dependency>
  <groupId>org.apache.spark</groupId>
  <artifactId>spark-core_2.12</artifactId>
  <version>3.1.2</version>
  <scope>compile</scope>
</dependency>

<dependency>
  <groupId>org.apache.spark</groupId>
  <artifactId>spark-sql_2.12</artifactId>
  <version>3.1.2</version>
```

```xml
      <scope>compile</scope>
</dependency>
```

（5）经过以上修改，最终的 pom.xml 文件如下：

```xml
<project xmlns="http://maven.apache.org/POM/4.0.0"
         xmlns:xsi="http://www.w3.org/2001/XMLSchema-instance"
         xsi:schemaLocation="http://maven.apache.org/POM/4.0.0
http://maven.apache.org/maven-v4_0_0.xsd">
  <modelVersion>4.0.0</modelVersion>

    <groupId>com.xueai8</groupId>
    <artifactId>sparkexamples</artifactId>
    <version>1.0-SNAPSHOT</version>

    <properties>
        <scala.version>2.12.11</scala.version>
    </properties>

  <dependencies>
    <!--scala-->
    <dependency>
      <groupId>org.scala-lang</groupId>
      <artifactId>scala-library</artifactId>
      <version>${scala.version}</version>
    </dependency>

    <!--spark-->
    <dependency>
      <groupId>org.apache.spark</groupId>
      <artifactId>spark-core_2.12</artifactId>
      <version>3.1.2</version>
    </dependency>
    <dependency>
      <groupId>org.apache.spark</groupId>
      <artifactId>spark-sql_2.12</artifactId>
      <version>3.1.2</version>
    </dependency>
  </dependencies>
  <build>
    <sourceDirectory>src/main/scala</sourceDirectory>
    <testSourceDirectory>src/test/scala</testSourceDirectory>
    <plugins>
        <plugin>
            <groupId>org.apache.maven.plugins</groupId>
            <artifactId>maven-compiler-plugin</artifactId>
            <configuration>
                <source>1.8</source>
                <target>1.8</target>
            </configuration>
        </plugin>
    </plugins>
  </build>
</project>
```

保存文件后，IntelliJ IDEA 将自动导入运行 Spark 所需的库和文档。这个过程有可能会持续较长时间。

2.2.4 测试 Spark 程序

接下来就可以开发基于 Maven 的 Spark 程序了。

首先创建源程序。在项目的 src/main/scala 目录上右击，创建一个 Scala Object，并命名为 HelloWorld。在 HelloWorld.scala 文件中编辑的代码如下：

```
object HelloWorld {
  def main(args: Array[String]): Unit = {
    println("Hello World!")
  }
}
```

注意：有些时候，pom.xml 文件中的依赖项可能不会自动加载，因此，需要重新导入依赖项或重启 IntelliJ IDEA。

接下来，在文件的任意空白位置右击，在弹出的菜单中选择 Run 'MyApp'项，开始运行程序。如果一切正常，则应该可以得到如下的输出结果：

```
Hello World!
```

2.2.5 项目编译和打包

对于 Maven 项目，可以简单地在 IntelliJ IDEA 的终端（Terminal）窗口运行打包命令，这样就会自动编译并打 JAR 包。打包命令如下：

```
$ mvn clean package
```

操作过程如图 2-41 所示。

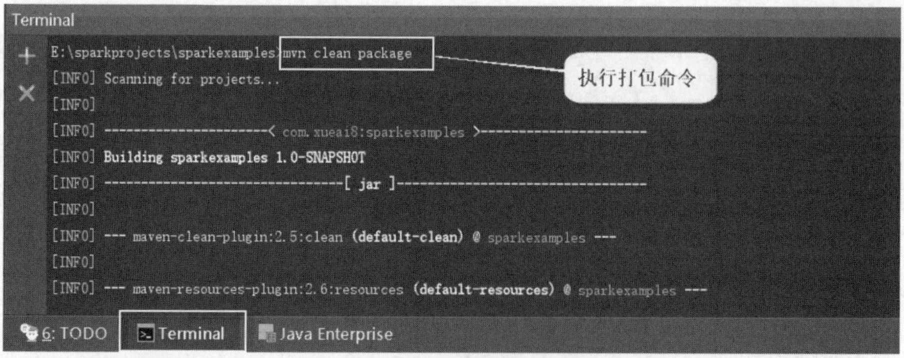

图 2-41　执行项目打包命令

如果一切顺利，则会在项目中生成一个 target 目录，打好的 JAR 包就位于此目录下，如图 2-42 所示。

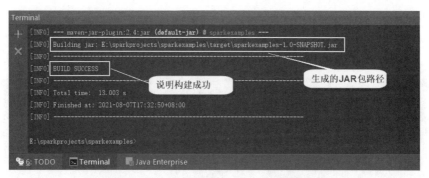

图 2-42　打包成功信息和 JAR 包位置提示

2.3　使用 Java 开发 Spark 应用程序

除了提供 Scala 和 Python API，Spark 还支持使用 Java 语言来开发 Spark 应用程序。本节将演示如何使用 IntelliJ IDEA 作为开发工具，使用 Java 作为开发语言，并使用 Maven 作为构建管理器。在本节中，将了解如何设置 IntelliJ IDEA、如何使用 Maven 管理依赖项、如何将 Spark 应用程序打包和部署到集群，以及如何将实时程序连接到调试器。

2.3.1　创建一个新的 IntelliJ 项目

建议按以下步骤使用 IntelliJ IDEA 创建一个新的 Java Maven 项目。

（1）启动 IntelliJ IDEA，在开始界面中，选择 create new project，或者在打开的编辑界面选择 File→New→Project 菜单项，打开项目创建向导，然后在左侧面板单击选中 Maven 这一项，如图 2-43 所示。

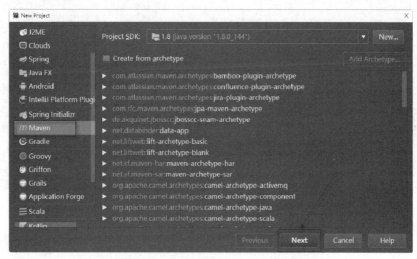

图 2-43　创建 Java Maven 项目界面

（2）在图 2-43 中单击 Next 按钮，在打开的新向导窗口中设置项目的 GroupId、ArtifactId 和 Version 等唯一标识信息，如图 2-44 所示。

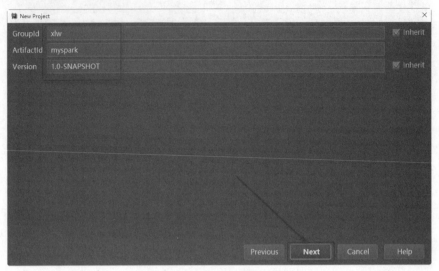

图 2-44　设置项目的唯一标识

（3）单击 Next 按钮，进入项目名称和位置选择的向导界面。项目名称和项目位置字段值应该会自动填充，保持默认值就可以了，如图 2-45 所示。

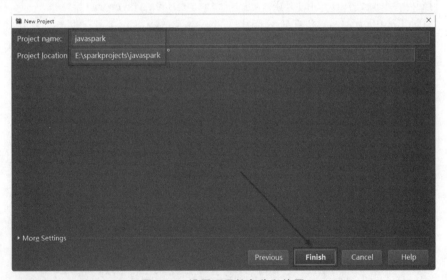

图 2-45　设置项目的名称和位置

（4）单击 Finish 按钮，开始创建项目，IntelliJ 应该创建一个具有默认目录结构的新项目。这个过程会生成所有项目文件夹，可能需要一两分钟。最终生成的项目结构如图 2-46 所示。

项目结构中各部分的含义说明如下。

（1）.idea：IntelliJ 的配置文件。

（2）src：源代码目录。大多数的源代码文件应该位于 src/main/java 目录下，而 test 文件夹主要用于保留测试脚本。

（3）target：当对项目进行编译时才会生成此目录。

（4）pom.xml：这是 Maven 的配置文件。Maven 项目可使用这个文件导入第三方库和文档。

图 2-46　最终生成的项目结构

2.3.2　验证 SDK 安装和配置

在继续之前，需要先验证几个 IntelliJ 依赖库的设置，确保项目所依赖的 JDK 和 SDK 都被正确设置。建议按以下步骤进行验证操作。

（1）验证导入 Maven 项目是否可自动打开。

选择 IntelliJ IDEA 的 File→Settings 菜单项，打开项目设置面板。在设置面板左侧，依次选择 Build, Execution, Deployment→Build Tools→Maven→Importing 选项，然后查看面板右侧的 Import Maven projects automatically 复选框是否勾选。确保该复选框处于被勾选状态，如图 2-47 所示。

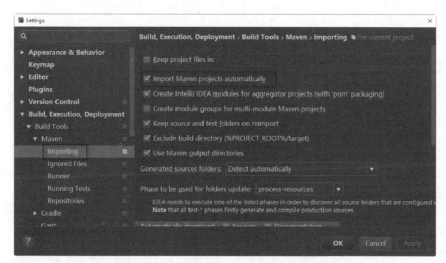

图 2-47　设置导入的 Maven 项目可自动打开

（2）验证项目的 Project SDK 和 Project language level 设置为 Java 8 版本。

选择 IntelliJ IDEA 的 File→Project Structure 菜单项，在打开的项目结构设置面板左侧，选择 Project 项，然后在面板右侧分别选择 Project SDK 和 Project language level 为 Java 8 版本，如图 2-48 所示。

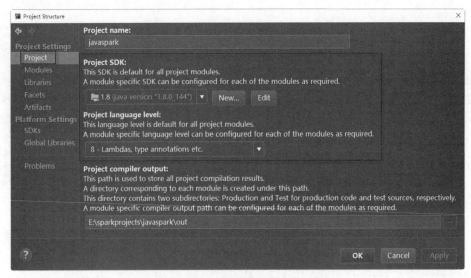

图 2-48　设置项目所依赖的 SDK 版本

（3）验证项目中模块所依赖的 Java 语言版本。

选择 IntelliJ IDEA 的 File→Project Structure 菜单项，在打开的项目结构设置面板左侧，选择 Modules 项，然后在面板右侧选择 Language level 为 Java 8 版本，如图 2-49 所示。

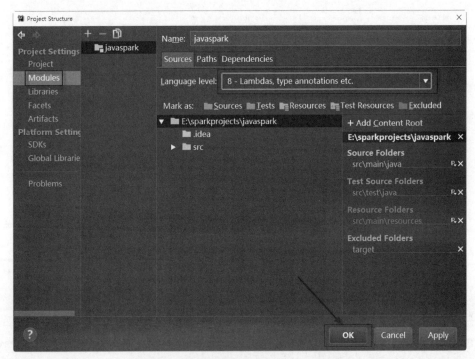

图 2-49　设置项目模块所依赖的 Java 版本

（4）验证全局库是否已经配置了 Scala SDK 依赖。

选择 IntelliJ IDEA 的 File→Project Structure 菜单项，在打开的项目结构设置面板左侧，选择 Global Libraries 项，然后在面板右侧，单击绿色十字按钮，选择本机安装的 Scala SDK 目录（如果本机没有安装 Scala SDK，则需要先下载并解压一个 Scala SDK），如图 2-50 所示。

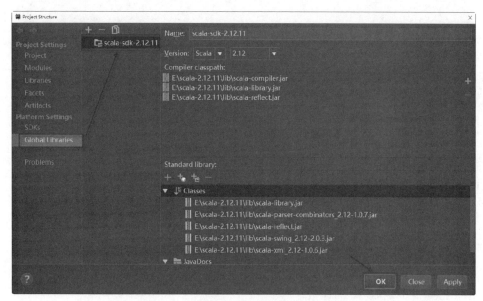

图 2-50　为 IDEA 配置全局库 Scala SDK

确保以上几项验证都通过，然后继续后面的内容。

2.3.3　安装和配置 Maven

在开始编写 Spark 应用程序之前，需要使用 Maven 将 Spark 库和文档导入 IntelliJ IDEA。如果想让 IntelliJ 识别 Spark 代码，则导入是必要的。向 pom.xml 文件中添加以下几行内容：

```
<?xml version="1.0" encoding="UTF-8"?>
<project xmlns="http://maven.apache.org/POM/4.0.0"
       xmlns:xsi="http://www.w3.org/2001/XMLSchema-instance"
       xsi:schemaLocation="http://maven.apache.org/POM/4.0.0
http://maven.apache.org/xsd/maven-4.0.0.xsd">
    <modelVersion>4.0.0</modelVersion>

    <groupId>xlw</groupId>
    <artifactId>myspark</artifactId>
    <version>1.0-SNAPSHOT</version>

    <build>
       <plugins>
          <plugin>
             <groupId>org.apache.maven.plugins</groupId>
```

```xml
                <artifactId>maven-compiler-plugin</artifactId>
                <configuration>
                    <source>1.8</source>
                    <target>1.8</target>
                </configuration>
            </plugin>
        </plugins>
    </build>

    <dependencies>
        <dependency>
            <groupId>org.apache.spark</groupId>
            <artifactId>spark-core_2.11</artifactId>
            <version>2.3.2</version>
        </dependency>
    </dependencies>
</project>
```

保存文件后，IntelliJ IDEA 将自动导入运行 Spark 所需的库和文档。这个过程有可能会持续较长时间。

2.3.4 创建 Spark 应用程序

接下来要创建第 1 个 Spark 应用程序，这是一个简单的程序，来对收集的莎士比亚作品（shakespear .txt）执行单词计数。

在后面的部署阶段，将希望 Spark 从 HDFS（Hadoop 分布式文件系统）读取这个文件，所以现在先把它上传到 HDFS 中去。在本书配置的源码包中可以找到这个数据文件，在笔者的环境中，这个 shakespeare.txt 文件位于 Linux 系统本地的/home/hduser/data/spark/目录下。建议按以下步骤将这个文件上传到 HDFS 的指定位置。

(1) 确保已经启动了 HDFS；如果还未启动，则可执行命令来启动它，命令如下：

```
$ start-dfs.sh
```

(2) 将文件复制到 HDFS 的指定目录下。在终端窗口中，执行的命令如下：

```
$ cd /home/hduser/data/spark
$ hdfs dfs -put shakespeare.txt /data/spark_demo/
```

现在准备创建 Spark 应用程序。在 IntelliJ IDEA 中打开文件夹 src/main/resources（这个文件夹在创建项目时应该已经自动生成了）。将 shakespeare.txt 文件复制到该文件夹下。

右击文件夹 src/main/java，在弹出的环境菜单中选择 New→Java Class，创建一个名为 Main.java 的 Java 类。编辑 Main.java 的源代码如下：

```
public class Main {
    public static void main(String[] args) {
        System.out.println("Hello World");
    }
}
```

现在转到 IntelliJ IDEA 顶部的 Run 下拉菜单并选择 Run 菜单项，然后选择 Main。如果一切设置正确，则在 IntelliJ IDEA 的控制台中应该打印"Hello World"。这证明到目前为止我们创建的项目一切正常。

用以下代码替换 Main.java 文件原来的内容：

```java
//第2章/javaspark/src/Main.java
import org.apache.spark.api.java.JavaPairRDD;
import org.apache.spark.api.java.JavaSparkContext;
import org.apache.spark.api.java.JavaRDD;
import org.apache.spark.SparkConf;
import scala.Tuple2;

import java.util.Arrays;

public class Main {

    public static void main(String[] args){

        //创建要初始化的SparkContext
        SparkConf conf = new SparkConf().setMaster("local").setAppName("Word Count");

        //创建一个Java版本的Spark Context
        JavaSparkContext sc = new JavaSparkContext(conf);

        //将文件加载到一个Spark RDD，这是每行文本的分布式表示
        JavaRDD<String> textFile =
                sc.textFile("./src/main/resources/shakespeare.txt");

        //单词计数
        JavaPairRDD<String, Integer> counts = textFile
            .flatMap(s -> Arrays.asList(s.split("[ ,]")).iterator())
            .mapToPair(word -> new Tuple2<>(word, 1))
            .reduceByKey((a, b) -> a + b);
        counts.foreach(p -> System.out.println(p));
        System.out.println("全部单词: " + counts.count());

        //将统计结果保存到指定文件
        counts.saveAsTextFile("./shakespeareWordCount");
    }
}
```

如前所述，单击 Run→Run 菜单以运行该程序文件。这应该执行 Spark 作业并打印莎士比亚作品中出现的每个单词的次数。程序创建的输出文件位于代码中指定的目录下，在示例中为./shakespeareWordCount 文件夹。

到目前为止，已经了解了如何在 IntelliJ IDEA 中直接部署 Spark 应用程序。这是一种快速构建和测试应用程序的好方法，但是与实际生产环境运行是不一样的，因为 Spark 只

在一台机器上运行。在生产环境中，Spark 通常会处理存储在 HDFS 等分布式文件系统中的数据。Spark 通常也以集群模式运行(分布在许多机器上)。

在接下来的部分中，将学习如何部署分布式 Spark 应用程序。

2.3.5 部署 Spark 应用程序

在本节中，将针对 Spark 集群进行程序部署。尽管仍然在一台机器上运行 Spark，但将使用 HDFS 和 Spark Standalone 集群资源管理器。

将 2.3.4 节中的源代码修改一下，注意粗体字部分的改变。修改后的内容如下：

```java
//第 2 章/javaspark/src/Main.java
import org.apache.spark.api.java.JavaPairRDD;
import org.apache.spark.api.java.JavaSparkContext;
import org.apache.spark.api.java.JavaRDD;
import org.apache.spark.SparkConf;
import scala.Tuple2;

import java.util.Arrays;

public class Main {

    public static void main(String[] args){

        //创建要初始化的 SparkContext
        SparkConf conf = new SparkConf().setMaster("local").setAppName("Word Count");

        //创建一个 Java 版本的 Spark Context
        JavaSparkContext sc = new JavaSparkContext(conf);

        //将文件加载到一个 Spark RDD，这是每行文本的分布式表示
        JavaRDD<String> textFile =
                sc.textFile("hdfs://localhost:8020/data/shakespeare.txt");

        //单词计数
        JavaPairRDD<String, Integer> counts = textFile
            .flatMap(s -> Arrays.asList(s.split("[ ,]")).iterator())
            .mapToPair(word -> new Tuple2<>(word, 1))
            .reduceByKey((a, b) -> a + b);
        counts.foreach(p -> System.out.println(p));
        System.out.println("全部单词: " + counts.count());

        //将统计结果保存到指定文件
        val target = "hdfs://localhost:8020/data/shakespeareWordCount";
        counts.saveAsTextFile(target);
    }
}
```

粗体字部分告诉 Spark 应读写 HDFS 上的数据文件，而不是读写本地数据文件。需要确保保存了该源代码文件。

接下来，将把这些代码打包到一个编译后的 JAR 文件中，该文件可以部署在集群上。为了简化工作，将创建一个 assembly jar：一个包含代码和代码所依赖的所有 JAR 包的单个 JAR 文件。通过将代码打包为程序集，可以确保在代码运行时包含所有依赖 JAR 包（这些依赖是在 pom.xml 文件中定义的）。

打开一个 IntelliJ IDEA 的 Terminal 终端窗口，运行 mvn package 命令便可打包（可参考 2.2.5 节的打包步骤说明），这将会在 target 目录下创建一个名为 myspark-1.0-SNAPSHOT.jar 的 JAR 包。

将 myspark-1.0-SNAPSHOT.jar 包复制到虚拟机的用户主目录下，然后打开一个终端窗口，使用 spark-submit 运行代码。这时需要指定主类、要运行的 JAR 包和运行模式（本地或集群），命令如下：

```
$ cd ~/bigdata/spark-3.1.2
$ spark-submit --class "xlw.myspark.Main" --master local
 ./myspark-1.0-SNAPSHOT.jar
```

注意观察，在终端窗口中应该可以看到输出莎士比亚作品中出现的每个单词的次数，部分结果如下：

```
...
(comutual,1)
(ban-dogs,1)
(rut-time,1)
(ORLANDO],4)
(Deceitful,1)
(commits,3)
(GENTLEWOMAN,4)
(honors,10)
(returnest,1)
(topp'd?,1)
(compass?,1)
(toothache?,1)
(miserably,1)
(hen?,1)
(luck?,2)
(call'd,162)
(lecherous,2)
...
```

2.3.6　远程调试 Spark 应用程序

在 Spark 程序的开发过程中，经常需要对代码进行调试。当直接在 IntelliJ IDEA 中运行时，因为代码是在本地运行的，所以调试 Spark 程序和其他任何程序都是一样的，但是要对远程集群上运行的 Spark 程序进行调试，需要一些配置，可将正在运行的 Spark 程序连接到调试器，通过调试器设置断点并逐行调试代码。

建议按以下步骤操作。

（1）在计划提交 Spark 作业的机器上，从终端运行以下命令：

```
$ export SPARK_SUBMIT_OPTS=-agentlib:jdwp=transport=dt_socket,server=y,suspend=y,address=5005
```

（2）提交 Spark 作业程序，命令如下：

```
$ spark-submit --class xlw.myspark.Main --master local bigdata/javaspark.jar
```

运行时会出现程序挂起现象，表示处于监听端口中，如图 2-51 所示。

图 2-51　程序挂起，处于监听端口中

（3）在 IntelliJ IDEA 中配置 remote Debug。在 IntelliJ IDEA 中依次选择菜单 Run→Edit Configurations，编辑运行配置信息，如图 2-52 所示。

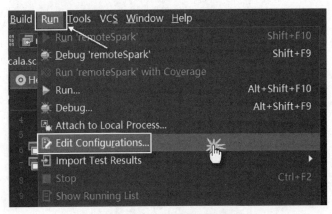

图 2-52　编辑运行时配置

（4）在打开的 Run/Debug Configurations 配置界面中，单击面板左侧的 Remote 项，然后单击左上角的绿色+（加号）按钮，增加一个新的远程配置。在面板右侧，使用主机 IP 地址填写 Host（Port 字段默认为 5005，不需要修改），这将允许在端口 5005 上关联调试器。需要确保端口 5005 能够接收入站连接，如图 2-28 所示。

（5）从 IntelliJ IDEA 中以 Debug 模式运行 Spark 程序，如图 2-53 所示。

（6）调试器将附加此调试配置，并且远程 Spark 程序将在断点处停止。还可以检查程序中活动变量的值，这在试图确定代码中的 Bug 时，非常有价值，如图 2-54 所示。

图 2-53　执行 Debug 调试

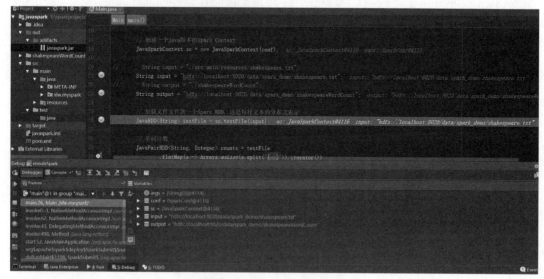

图 2-54　Spark 程序断点调试

2.4　使用 Zeppelin 进行交互式分析

Apache Zeppelin 是一款基于 Web 的 Notebook，支持交互式数据分析，专注于企业级应用。使用 Zeppelin，可以使用丰富的预构建语言后端（或解释器）制作精美的数据驱动、交互式和协作文档。目前，Apache Zeppelin 支持 Apache Spark、Python、JDBC、Markdown 和 Shell 等多种解释器。Zeppelin Notebook 可以满足企业用户的以下需求：

（1）数据摄取。

（2）数据发现。

（3）数据分析。

（4）数据可视化与协作。

特别是，Apache Zeppelin 提供了内置的 Apache Spark 集成。不需要为它构建单独的模块、插件或库。Apache Zeppelin 与 Spark 集成，提供了如下功能：

（1）自动注入 SparkContext 和 SQLContext。

（2）从本地文件系统或 Maven 存储库加载运行时 JAR 依赖项。

（3）取消作业并显示进度。

在接下来的部分将学习如何安装 Zeppelin 和配置 Zeppelin 解释器，并演示如何使用 Zeppelin Notebook 作为 Spark 的交互式数据分析工具进行大数据的分析和数据可视化。

2.4.1 下载 Zeppelin 安装包

Apache Zeppelin 的下载网址为 http://zeppelin.apache.org/download.html。选择合适的版本进行下载，如图 2-55 所示。

图 2-55 下载 Zeppelin

将下载的安装包复制到 Linux 的~/software 目录下。

2.4.2 安装和配置 Zeppelin

建议按以下步骤安装和配置 Zeppelin。

（1）将下载的安装包解压缩到~/bigdata 目录下，并改名为 zeppelin-0.9.0，命令如下：

```
$ cd ~/bigdata
$ tar xvf ~/software/zeppelin-0.9.0-bin-netinst.tgz
$ mv zeppelin-0.9.0-bin-netinst zeppelin-0.9.0
```

（2）配置环境变量。用任意编辑器（例如 nano）打开 etc/profile 文件，命令如下：

```
$ cd
$ sudo nano /etc/profile
```

在 etc/profile 文件的末尾添加以下内容：

```
export ZEPPELIN_HOME=/home/hduser/bigdata/zeppelin-0.9.0
export PATH=$PATH:$ZEPPELIN_HOME/bin
```

保存文件并关闭。

执行/etc/profile 文件使配置生效，命令如下：

```
$ source /etc/profile
```

(3) 打开 conf/zeppelin-env.sh 文件（如果默认没有，则应从模板复制一份）：

```
$ cd ~/bigdata/zeppelin-0.9.0/conf
$ cp zeppelin-env.sh.template zeppelin-env.sh
$ nano zeppelin-env.sh
```

在文件的最后添加如下两行内容：

```
export JAVA_HOME=/usr/local/jdk1.8.0_281
export SPARK_HOME=/home/hduser/bigdata/spark-3.1.2
```

(4) 打开 zeppelin-site.xml 文件（如果默认没有，则应从模板复制一份）：

```
$ cd ~/bigdata/zeppelin-0.9.0/conf
$ cp zeppelin-site.xml.template zeppelin-site.xml
$ gedit zeppelin-site.xml
```

修改以下两个属性，设置新的端口号，以避免与 Spark Web UI 默认端口发生冲突，内容如下：

```
<property>
  <name>zeppelin.server.port</name>
  <value>9090</value>
  <description>Server port.</description>
</property>

<property>
  <name>zeppelin.server.ssl.port</name>
  <value>9443</value>
  <description>Server ssl port. (used when ssl property is set to true)</description>
</property>
```

(5) 启动 Zeppelin 服务。

在终端窗口中启动 Zeppelin 服务，执行的命令如下：

```
$ zeppelin-daemon.sh start
```

执行过程如图 2-56 所示。

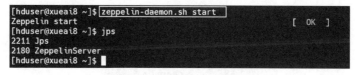

图 2-56　启动 Zeppelin 服务

(6) 关闭 Zeppelin 服务。

在终端窗口中停止 Zeppelin 服务，执行的命令如下：

```
$ zeppelin-daemon.sh stop
```

2.4.3　配置 Spark 解释器

注意：如果使用 Spark standalone 模式，需要配置 Spark 解释器。如果使用 Spark local 模式，则可省略这些步骤。

首先启动浏览器，在浏览器网址栏输入地址 http://xueai8:9090，打开访问界面，如图 2-57 所示。

图 2-57　打开 Zeppelin 解释器配置界面

单击右上角的小三角按钮，打开下拉菜单，单击 Interpreter 菜单项，打开解释器配置界面。打开的解释器配置界面如图 2-58 所示。

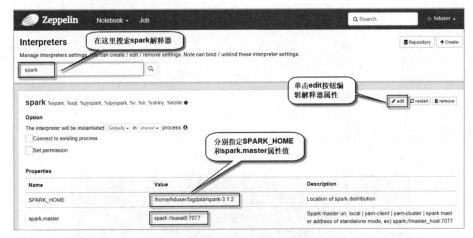

图 2-58　配置 Zeppelin 解释器

按图 2-59 中所示找到 spark 解释器，然后将 master 属性值修改为 spark://xueai8:7077（这实际上是连接到的集群管理器，笔者这里使用的是 Spark standalone 模式。这相当于启动 Spark Shell 时指定--master 参数），然后单击 Save 按钮保存配置。

2.4.4　创建和执行 Notebook 程序

回到浏览器 Zeppelin 首页，单击 Create new note 按钮，创建一个新的 Zeppelin Notebook 文档，如图 2-59 所示。

图 2-59 创建一个新的 Notebook

在弹出的创建窗口填写文档路径和名称信息，解释器保持默认的 spark 即可，然后单击 Create 按钮，如图 2-60 所示。

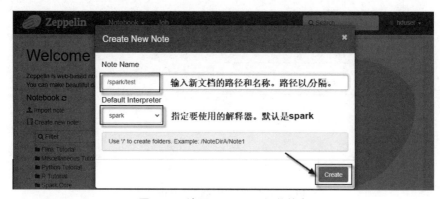

图 2-60 填写 Notebook 相关信息

1. 执行 Spark 交互式操作

在新打开的 Zeppelin Notebook 文档界面的单元格内输入并执行 Spark 代码，如图 2-61 所示。

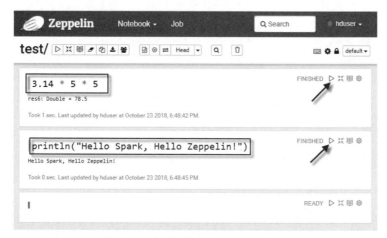

图 2-61 在单元格内执行 Scala 代码

2. 执行 PySpark 交互式操作

另外新创建一个 Notebook 文档，然后在新打开的 Notebook 文档的单元格中执行 Python 代码。在执行 Python 代码时，需要在每个单元格的第 1 行输入"%pyspark"，以告诉 Zeppelin 使用 PySpark 解释器来解释执行代码，如图 2-62 所示。

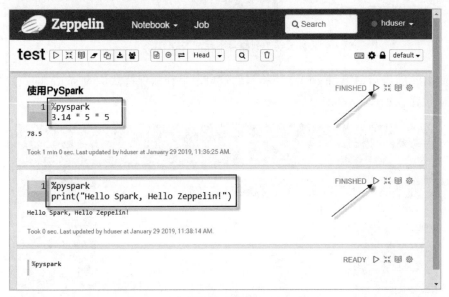

图 2-62　在单元格内执行 Python 代码

可以看到，对于大数据分析来讲，使用 Zeppelin Notebook 可以非常方便地进行交互式的数据探索和分析任务。

第 3 章 Spark 核心编程

CHAPTER 3

Spark Core 模块包含 Spark 的基本功能,例如包括任务调度组件、内存管理、故障恢复、与存储系统交互等。在 Spark Core 模块中,核心的数据抽象被称为弹性分布式数据集（Resilient Distributed Dataset,RDD）。RDD 是 Spark Core 的用户级 API,要真正理解 Spark 的工作原理,就必须理解 RDD 的本质。

Spark 为 Scala、Java、R 和 Python 编程语言提供了 APIs。Spark 本身是用 Scala 编写的,但 Spark 通过 PySpark 支持 Python。PySpark 构建在 Spark 的 Java API 之上（使用 Py4J）。通过 Spark（PySpark）上的交互式 Shell 可以对大数据进行交互式数据分析。数据科学家大多选择 Scala 或 Python 进行 Spark 程序开发和数据分析。

3.1 理解数据抽象 RDD

10min

在 Spark 的编程接口中,每个数据集都被表示为一个对象,称为 RDD。RDD 是一个只读的（不可变的）、分区的（分布式的）、容错的、延迟计算的、类型推断的和可缓存的记录集合。

RDD 提供了一组丰富的常用数据处理操作,包括执行数据转换、过滤、分组、连接、聚合、排序和计数的能力。关于这些操作需要注意的一点是,它们在粗粒度级别上进行操作,这意味着相同的操作应用于许多行,而不是任何特定的行。

3.1.1 RDD 结构

弹性分布式数据集具有以下特点。
（1）Resilient：不可变的、容错的。
（2）Distributed：数据分散在不同节点（机器,进程）。
（3）Dataset：一个由多个分区组成的数据集。

RDD 是分布式内存的一个抽象概念,它提供了一种高度受限的共享内存模型。通常 RDD 很大,会被分成很多个分区,分别保存在不同的节点上,如图 3-1 所示。

综上所述,RDD 只是一个逻辑概念,它可能并不对应磁盘或内存中的物理数据。根据 Spark 官方描述,RDD 由以下五部分组成：

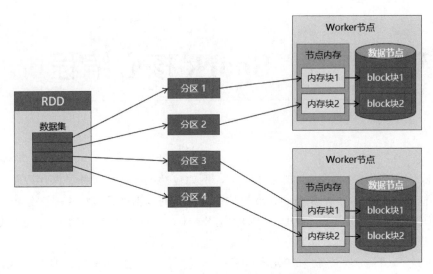

图 3-1　Spark RDD 是弹性分布式数据集

（1）一组 partition（分区），即组成整个数据集的块。
（2）每个 partition（分区）的计算函数（用于计算数据集中所有行的函数）。
（3）所依赖的 RDD 列表(父 RDD 列表)。
（4）（可选的）对于 key-value 类型的 RDD，则包含一个分区器（分区程序，默认为 HashPartitioner）。
（5）（可选的）每个分区数据驻留在集群中的位置；如果数据存放在 HDFS 上，则它就是块所在的位置。

Spark 运行时使用这 5 条信息来调度和执行通过 RDD 操作的用户数据处理逻辑。前 3 条信息组成"血统（lineage）"信息，Spark 将其用于两个目的。第 1 个是确定 RDD 的执行顺序，第 2 个是用于故障恢复。

3.1.2　RDD 容错

RDD 是不可变的、容错的、并行的数据结构。RDD 被设计成不可变的，这意味着不能具体地修改数据集中由 RDD 表示的特定行。如果调用一个 RDD 操作来操纵 RDD 中的行，则该操作将返回一个新的 RDD。原 RDD 保持不变，新的 RDD 将以用户希望的方式包含数据。RDD 的不变性本质上要求 RDD 携带"血统"信息，Spark 利用这些信息有效地提供容错能力。

Spark 通过"血统"信息重建失败的部分，自动地代表其用户处理故障。每个 RDD 或 RDD 分区都知道如何在出现故障时重新创建自己。它有转换的日志，或者血统，可依据此从稳定存储或另一个 RDD 中重新创建自己，因此，任何使用 Spark 的程序都可以确保内置的容错能力，而不用考虑底层数据源和 RDD 类型。

3.2 RDD 编程模型

在 Spark 中，使用 RDD 对数据进行处理，通常遵循的模型如下：
（1）首先，将待处理的数据构造为 RDD。
（2）对 RDD 进行一系列操作，包括 Transformation 和 Action 两种类型操作。
（3）最后，输出或保存计算结果。
这个处理流程如图 3-2 所示。

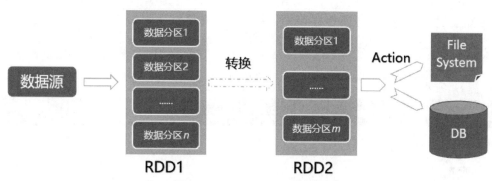

图 3-2　Spark RDD 处理流程

下面通过一个具体的示例来掌握 Spark RDD 编程的一般流程。

3.2.1　单词计数应用程序

使用 Spark RDD 实现经典的单词计数应用程序。这里使用 Zeppelin 作为开发工具，读者可以根据自己的喜好选择任意其他工具。建议按以下步骤执行。
（1）首先启动 HDFS 集群和 Spark 集群，命令如下：

```
$ start-dfs.sh

$ cd ~/bigdata/spark-3.1.2
$ ./sbin/start-all.sh
```

（2）准备一个文本文件 word.txt，内容如下：

```
good good study
day day up
```

（3）将该文本文件上传到 HDFS 的/data/spark_demo/目录下，命令如下：

```
$ hdfs dfs -put wc.txt /data/spark_demo/
```

（4）在 Zeppelin 中新建一个 Notebook。在 Notebook 的单元格中执行代码。
读取数据源文件，构造一个 RDD，代码如下：

```
val source = "hdfs://localhost:8020/data/wc.txt"
var textFile = sc.textFile(source)
```

将每行数据按空格拆分成单词，使用 flatMap() 转换，代码如下：
```
val words = textFile.flatMap(line => line.split(" "))
```
将各个单词加上计数值 1，使用 map() 转换，代码如下：
```
val wordPairs = words.map(word => (word,1))
```
对所有相同的单词进行聚合相加求各单词的总数，使用 reduceByKey() 转换，代码如下：
```
val WordCounts = wordPairs.reduceByKey((a,b) => a + b)
```
将结果返给 Driver 程序，这一步会触发 RDD 开始实际的计算，代码如下：
```
wordCounts.collect()
```
或者，也可以将计算结果保存到文件中，代码如下：
```
val sink = "hdfs://localhost:8020/data/result"
wordCounts.saveAsTextFile(sink)
```
（5）在 Zeppelin Notebook 中交互式数据处理过程如图 3-3 所示。

图 3-3　Spark RDD 单词计数程序

（6）对以上代码也可以进行精简，代码如下：

```
val source = "hdfs://localhost:8020/data/wc.txt"
sc.textFile(source)
  .flatMap(_.split(" ")).
  .map((_,1)).
  .reduceByKey(_+_).
  .collect.
  .foreach(println)
```

（7）如果在 IntelliJ IDEA 或 Eclipse 中开发，则完整的代码如下：

```scala
//第3章/WordCount.scala

import org.apache.spark.sql.SparkSession

object WordCount {
  def main(args: Array[String]): Unit = {

    //创建 SparkSession 实例，入口
    val spark = SparkSession
        .builder
        .master("local[*]")
        .appName("HelloWorld")
        .getOrCreate

    //加载数据源，构造 RDD
    val textFiles = spark.sparkContext.textFile("input/word.txt")

    //对每行数据按单词进行拆分，并扁平化
    val words = textFiles.flatMap(line => line.split(" "))

    //将每个单词转换为元组形式（单词,1）
    val wordTuples = words.map(word => (word, 1))

    //对同一个 key(单词)执行 reduce 操作
    val wordCounts = wordTuples.reduceByKey(_ + _)

    //将结果输出
    wordCounts.collect.foreach(println)

    //将结果保存到文件中
    wordCounts.saveAsTextFile("output/wordcount")
  }
}
```

3.2.2 理解 SparkSession

从 Spark 2.0 开始，SparkSession 已经成为 Spark 使用 RDD、DataFrame 和 DataSet 开发和应用的入口点。在 Spark 2.0 之前，SparkContext 曾经是一个入口点。在这里，将主要通过定义和描述如何创建 SparkSession 和使用 spark-shell 默认的 SparkSession 变量来解释什么是 SparkSession。

1. 什么是 SparkSession

SparkSession 的全限定名称为 org.apache.spark.sql.SparkSession，它是在 Spark 2.0 版本中引入的，是 Spark 底层功能的入口点，用于编程创建 Spark RDD、DataFrame 和 DataSet。

正因为 SparkSession 是 Spark 的一个入口点，创建 SparkSession 实例将是使用 RDD、DataFrame 和 DataSet 编写程序的第 1 个语句。SparkSession 将使用 SparkSession.builder() 构建器模式以编程方式创建。

虽然 SparkContext 是 2.0 之前的一个入口点，但它并没有被 SparkSession 完全取代，SparkContext 的许多特性在 Spark 2.0 中仍然可用，并在 Spark 2.0 和以后的版本中使用。SparkSession 内部使用 SparkSession 提供的配置创建 SparkConfig 和 SparkContext。

2. 交互开发中的 SparkSession

SparkSession 的实例对象 spark 在 spark-shell 中是默认可用的。当调用 spark-shell 时，默认已经创建并提供了 spark，它是 SparkSession 类的一个实例。同样，在 Zeppelin Notebook 中创建程序时也会默认创建并提供 SparkSession 类的实例 spark，因此，在使用 spark-shell 或 Zeppelin Notebook 进行开发时，可以直接调用 spark。

3. 在 Scala 程序中创建 SparkSession

当使用 IntelliJ IDEA 集成开发环境开发 Apache Spark 应用程序时，需要手动创建 SparkSession 对象实例，通过构建器模式方法 builder() 并调用 getOrCreate() 方法。如果 SparkSession 已经存在，则它会返回存在的对象，否则会创建新的 SparkSession。创建 SparkSession 的模板代码如下：

```
val spark = SparkSession.builder()
    .master("local[*]")
    .appName("SparkExamples")
    .getOrCreate();
```

3.2.3 理解 SparkContext

SparkContext 从 Spark 1.x 引入的，在 Spark 2.0 引入 SparkSession 之前，用来作为 Spark 和 PySpark 程序的入口点。SparkContext 是在 org.apache.spark 包中定义的，它用于在集群中通过编程方式创建 Spark RDD、累加器和广播变量。使用 RDD 编程时，连接到 Spark 集群的第 1 步就是创建一个 SparkContext。

注意：每个 JVM 只能创建一个 SparkContext。

在任何给定时间，每个 JVM 应该只有一个 SparkContext 实例是活动的。如果想创建另一个新的 SparkContext，则应该在创建一个新的 SparkContext 之前停止现有的 SparkContext（使用 stop() 方法）。

在使用 spark-shell 工具时，spark-shell 会默认提供一个 sc 对象，该对象是 SparkContext

类的一个实例。在 spark-shell 中可以直接使用。同样，在 Zeppelin Notebook 中也默认提供了一个 sc 对象，同样也可以直接使用。

当使用 Scala（或者 Python/Java）编写 Spark 程序时，首先需要创建一个 SparkConf 实例，并分配应用名称和设置 master（分别使用 SparkConf 的静态方法 setAppName()和 setMaster()），然后将 SparkConf 对象作为参数传递给 SparkContext 构造器来创建 SparkContext，代码如下：

```
//先创建一个SparkConf实例
val sparkConf = new SparkConf()
    .setAppName("sparkexamples")
    .setMaster("local[*]")

//使用sparkConf构造一个SparkContext实例sc
val sc = new SparkContext(sparkConf)
```

一旦创建了 Spark Context 对象，就可以使用它来创建 Spark RDD 了。

从 Spark 2.0 开始，Spark 程序的入口点统一到了 SparkSession，SparkContext 中的大多数方法也存在于 SparkSession 中，并且 SparkSession 内部创建了 SparkContext 并公开了一个 sparkContext 变量以供使用，代码如下：

```
val sc = spark.sparkContext
sc.version
```

3.3 创建 RDD

在对数据进行任何 Transformation 或 Action 操作之前，必须先将这些数据构造为一个 RDD。Spark 提供了创建 RDD 的 3 种方法，分别如下：

（1）第 1 种方法是将现有的集合并行化。
（2）第 2 种方法是加载外部存储系统中的数据集，如文件系统。
（3）第 3 种方法是在现有 RDD 上进行转换，以此来得到新的 RDD。

3.3.1 将现有的集合并行化以创建 RDD

创建 RDD 的第 1 种方法是将对象集合并行化，这意味着将其转换为可以并行操作的分布式数据集。这种方法最简单，是开始学习 Spark 的好方法，因为它不需要任何数据文件。这种方法通常用于快速尝试一个特性或在 Spark 中做一些试验。

对象集合的并行化是通过调用 SparkContext 类的 parallelize()方法实现的。例如，对数组或列表调用 parallelize()方法构造 RDD，代码如下：

```
//可以从列表中创建
val list1 = List(1,2,3,4,5,6,7,8,9,10)
val rdd1 = sc.parallelize(list1)
```

```
rdd1.collect

//通过并行集合(range)创建RDD
val list2 = List.range(1,11)
val rdd2 = sc.parallelize(list2)
rdd2.collect

//通过并行集合(数组)创建RDD
val arr = Array(1,2,3,4,5,6,7,8,9,10)
val rdd3 = sc.parallelize(arr)
rdd3.collect

//通过并行集合(数组)创建RDD
val strList = Array("明月几时有","把酒问青天","不知天上宫阙","今夕是何年")
val strRDD = sc.parallelize(strList)
strRDD.collect
```

3.3.2 从存储系统读取数据集以创建 RDD

创建 RDD 的第 2 种方法是从存储系统读取数据集，存储系统可以是本地计算机文件系统、HDFS、Cassandra、Amazon S3 等。Spark 可以从 Hadoop 支持的任何数据源创建 RDD，包括本地文件系统、HDFS、Cassandra、HBase、Amazon S3 等。Spark 支持 Hadoop InputFormat 支持的任何格式。

例如，要从文件系统中将一个文本文件中的内容加载到一个 RDD 中，代码如下：

```
//从文件系统中加载数据创建RDD
val file = "/data/spark_demo/rdd/wc.txt"      //HDFS上的文件路径
val rdd1 = sc.textFile(file)
```

SparkContext 类的 textFile()方法假设每个文件都是一个文本文件，并且每行由一个新行分隔。这个 textFile()方法返回一个 RDD，它表示所有文件中的所有行。需要注意的重要一点是，textFile()方法是延迟计算的，这意味着如果指定了错误的文件或路径，或者错误地拼写了目录名，则在采取其中一项 Action 操作之前，这个问题不会出现（因此也不会被发现）。

例如，有一个员工信息文件 people.json，以 JSON 格式存储，内容如下：

```
{"name":"张三"}
{"name":"李四", "age":30}
{"name":"王老五", "age":19}
```

要将该 people.json 文件的内容加载到一个 RDD 中，用到了 scala.util.parsing.json.JSON 类，代码如下：

```
//第3章/ReadJsonDemo.scala

import org.apache.spark.sql.SparkSession

import scala.util.parsing.json.JSON
object ReadJsonDemo {
```

```
  def main(args: Array[String]): Unit = {
    //创建 SparkSession 实例，入口
    val spark = SparkSession.builder
          .master("local[*]")
          .appName("HelloWorld")
          .getOrCreate

    //读取文件，构造 RDD
    val jsonRDD = spark.sparkContext.textFile("input/json/people.json")

    //解析
    val result = jsonRDD.map(line => JSON.parseFull(line))

    //输出，使用模式匹配
    result.collect.foreach(r => r match{
      case Some(map) => println(map)
      case None      => println("解析失败")
      case other     => println("未知数据结构: " + other)
    })
  }
}
```

执行以上代码，输出的结果如下：

```
Map(name -> 张三)
Map(name -> 李四, age -> 30.0)
Map(name -> 王老五, age -> 19.0)
```

3.3.3 从已有的 RDD 转换得到新的 RDD

创建 RDD 的第 3 种方法是在现有 RDD 上调用一个转换操作，从而得到一个新的 RDD（因为 RDD 是不可变的）。例如，下面的代码通过对 rdd4 的转换得到一个新的 RDD，即 rdd5：

```
//将字符转换为大写，得到一个新的 RDD
val rdd5 = rdd4.map(line => line.toUpperCase)
rdd5.collect
```

注意：关于 map()函数，在稍后部分讲解。

3.3.4 创建 RDD 时指定分区数量

Spark 在集群的每个分区上运行一个任务，因此必须谨慎地决定优化计算工作。尽管 Spark 会根据集群自动设置分区数量，但可以将其作为第 2 个参数传递给并行化函数。例如，使用 parallelize()方法创建一个 RDD，并将分区数指定为 3，代码如下：

```
sc.parallelize(data,3)        //指定 3 个分区
```

为了进一步理解 RDD 的分区概念，参看下面这个示例。假设现在要创建一个 RDD，包含 14 条记录（或元组），并将该 RDD 的分区数指定为 3，即这 14 条记录分布在 3 个节点上，则该 RDD 的数据分区和对应的执行任务分布如图 3-4 所示。

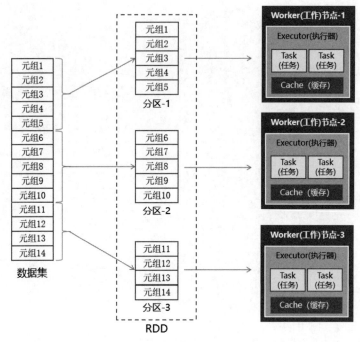

图 3-4 Spark RDD 分区

3.4 操作 RDD

创建了 RDD 之后，就可以编写 Spark 程序对 RDD 进行操作了。RDD 操作分为两种类型：转换（Transformation）和动作（Action）。转换是用来创建 RDD 的方法，而动作是使用 RDD 的方法，如图 3-5 所示。

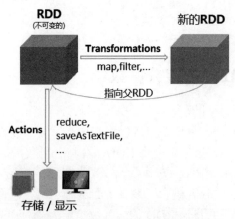

图 3-5 Spark RDD 两种操作类型

3.4.1 RDD 上的 Transformation 和 Action

如前文所述，RDD 支持两种类型的操作：Transformation 和 Action。

Transformation 是定义如何构建 RDD 的延迟操作。大多数 Transformation 转换接受单个函数作为输入参数。每当在 RDD 上执行 Transformation 转换时，都会生成一个新的 RDD，如图 3-6 所示。

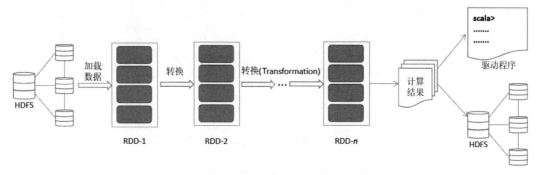

图 3-6　Spark RDD 上的 Transformation 转换操作

RDD 操作作用在粗粒度级别上。数据集中的每行都表示为 Java 对象，这个 Java 对象的结构对于 Spark 来讲是不透明的。RDD 的用户可以完全控制如何操作这个 Java 对象。

因为 RDD 是不可变的（只读的）数据结构，因此任何转换都会产生新的 RDD。转换操作被延迟计算，称为"惰性转换"，这意味着 Spark 将延迟对被调用的操作的执行，直到采取 Action 操作。换句话说，Transformation 转换操作仅仅记录指定的转换逻辑，并没有立即应用这些转换，而是在稍后才应用它们。调用 Action 操作将触发对它之前的所有转换的求值，然后向驱动程序返回一些结果，或者将数据写入存储系统，如 HDFS 或本地文件系统。基于延迟计算概念的一个重要优化技术是在执行期间将类似的转换折叠或组合为单个操作，即优化转换步骤。例如，如果 Action 操作要求返回第 1 行，则 Spark 就只计算单个分区，然后跳过其余部分。

简而言之，RDD 是不可变的，RDD 的 Transformation 转换是延迟计算的，而 RDD 的 Action 操作是即时计算的，并触发数据处理逻辑的计算，在 RDD 的内部实现机制中，底层接口则是基于迭代器的，从而使数据访问变得更高效，也避免了大量中间结果对内存的消耗。

通过应用程序操作 RDD 与操作数据的本地集合类似。以一个简单应用为例进行说明，代码如下：

```
val lines = sc.textFile("hdfs://path/to/the/file")
val filteredLines = lines.filter(line => line.contains("spark")).cache()
val result = filteredLines.count()
```

上面这段代码的意思是，从 HDFS 上加载指定的日志文件，找出包含单词 spark 的行数。其在内存中的计算和转换过程如下。

（1）一个 300MB 的日志文件，分布式存储在 HDFS 上，如图 3-7 所示。

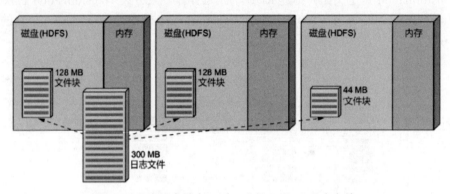

图 3-7　分布式存储在 HDFS 上的 300MB 日志文件

（2）将该日志文件加载到分布式的内存中，代码如下：

```
val lines = sc.textFile("hdfs://path/to/the/file")
```

分布式内存数据集合 RDD 如图 3-8 所示。

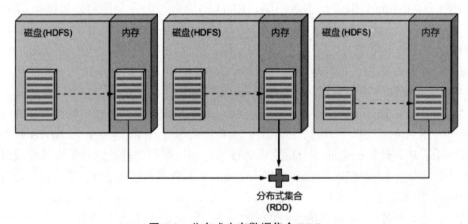

图 3-8　分布式内存数据集合 RDD

（3）过滤满足条件的行(只包含单词 spark 的行)，这是原始数据集的一个子集，并将这个中间结果缓存到内存中，代码如下：

```
val filteredLines = lines.filter(line => line.contains("spark")).cache()
```

在 RDD 上执行过滤操作，如图 3-9 所示。

（4）统计过滤后的行数，返给驱动程序 Driver，代码如下：

```
val result = filteredLines.count()
```

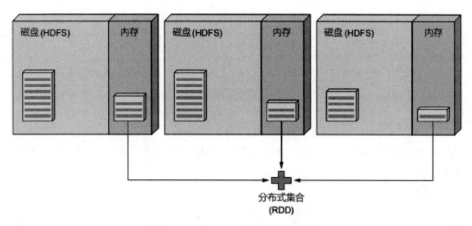

图 3-9　在 RDD 上执行过滤操作

3.4.2　RDD Transformation 操作

Transformation 是操作 RDD 并返回一个新的 RDD，如 map()和 filter()方法，而 Action 是一个将结果返给驱动程序或将结果写入存储的操作，并开始一个计算，如 count()和 first()方法。

PySpark 对于 Transformation RDD 是延迟计算的，只在遇到 Action 时才真正进行计算。许多转换是作用于元素范围内的，也就是一次作用于一个元素。

现在假设有一个 RDD，包含的元素为{1, 2, 3, 3}。首先，构造出这个 RDD，代码如下：

```
//构造一个RDD
val rdd = sc.parallelize(List(1,2,3,3))
```

接下来，学习普通 RDD 上的一些常见的 Transformation 转换操作方法。

1. map(func)

算子 map()是使用 func 函数转换每个 RDD 元素并返回一个新的 RDD 的操作，如图 3-10 所示。

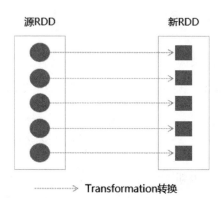

图 3-10　在 RDD 上执行 map 转换操作

应用 map()转换，代码如下：

```
val rdd1 = rdd.map(x => x + 1)       //Transformation
rdd1.collect
```

执行以上代码，输出的结果如下：

```
Array(2,3,4,4)
```

2. flatMap(func)

flatMap(func)转换操作使用 func 函数来转换 RDD 中的每个元素，并将一个或多个元素返到新的 RDD 中，如图 3-11 所示。

应用 flatMap()转换，代码如下：

```
val rdd2 = rdd.flatMap(x => x.to(3))    //Transformation
rdd2.collect
```

执行以上代码，输出的结果如下：

```
Array(1,2,3,2,3,3,3)
```

3. filter(func)

filter(func)转换操作使用 func 函数来过滤 RDD 中的每个元素，当 func 的返回值为 true 时，则被判断的元素添加到新的 RDD 中；当 func 的返回值为 false 时，丢弃被判断的元素，因此经过 filter()转换之后的新 RDD 总是源 RDD 的一个子集，如图 3-12 所示。

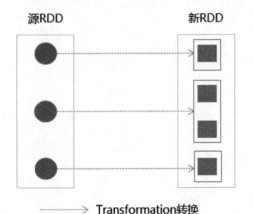

图 3-11 在 RDD 上执行 flatMap 转换操作

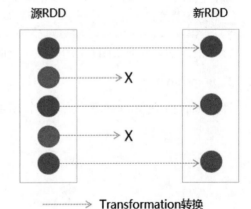

图 3-12 在 RDD 上执行 filter 转换操作

应用 filter()转换，代码如下：

```
val rdd3 = rdd.filter(x => x!=1)       //Transformation
rdd3.collect
```

执行以上代码，输出的结果如下：

```
Array(2,3,3)
```

4. distinct([numPartitions])

这个转换操作用来对 RDD 中的元素去重,源 RDD 中重复的元素在新 RDD 中只保留唯一的一个。在应用这个转换操作时,可以通过参数指定分区数,该转换的原理如图 3-13 所示。

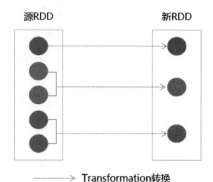

图 3-13 在 RDD 上执行 distinct 转换操作

应用 distinct()转换,代码如下:

```
val rdd4 = rdd.distinct()    //transformation
rdd4.collect
```

执行以上代码,输出的结果如下:

```
Array(1,2,3)
```

5. sample(withReplacement, fraction, seed)

sample(withReplacement, fraction, seed)是一个抽样方法,返回这个 RDD 的一个抽样子集,其中各参数的含义如下。

(1) withReplacement:是否可以对元素进行多次采样(采样后替换)。

(2) fraction:抽样因子。对于 without replacement,表示每个元素被选中的概率,fraction 值必须是[0,1];对于 with replacement,表示每个元素被选择的期望次数,fraction 值必须大于或等于 0。

(3) seed:用于随机数生成器的种子。

例如,抽取 50%的样本子集并保存到新的 RDD 中,代码如下:

```
val rdd5 = rdd.sample(false,0.5)         //transformation
rdd5.collect                             //结果:Array(1,3)
```

执行以上代码,输出的结果如下:

```
Array(2, 3)
```

注意:这个操作并不能保证精确地提供给定 RDD 的计数的比例。

6. keyBy(func):RDD[(K,T)]

当在类型为 T 的 RDD 上调用时,返回一个(K, T)元组对作为元素的新 RDD。通过应用 func()函数创建这个新 RDD 中元素的元组。例如，对于由单词组成的一个 RDD，应用 keyBy()函数转换将 RDD 的元素转换为一个二元组，单词的首字母为 key，单词本身为 value，代码如下：

```
val x = sc.parallelize(Array("Angola", "Brazil", "Aruba", "Benin"))
val y = x.keyBy(w => w.charAt(0))
println(y.collect().mkString(", "))
```

以上代码指定将 RDD 中的元素（单词）转换为键-值对形式的元组，其中 K 来自单词的首字母。执行上面的代码，输出的结果如下：

```
Array((A,Angola), (B,Brazil), (A,Aruba), (B,Benin))
```

7. groupBy(func)，groupBy(func, numPartitions)，groupBy(func, partitioner)

当在类型为 T 的 RDD 数据集上调用 groupBy 操作时，返回一个由(K, Iterable[T])元组元素组成的新 RDD。每个组由一个 key 和一系列映射到该 key 的元素组成。每个组内元素的顺序不可以得到保证，甚至在每次计算结果 RDD 时可能会有所不同。这种方法有可能会引起数据 shuffle。例如，构造一个简单的 RDD，并应用 groupBy()转换，代码如下：

```
val x = sc.parallelize(Array("Angola", "Brazil", "Aruba", "Benin"))
val y = x.groupBy(word => word.charAt(0))      //按单词的首字母分组
//val y = x.groupBy(_.charAt(0))                //另一个短的语法
y.collect
```

在上面的代码中，在 x 这个 RDD 上调用 groupBy()转换操作，它会按所有单词元素的首字母进行分组，把首字母相同的单词分到同一个组中，以首字母为 key，组成新的键-值对 RDD，输出的结果如下：

```
Array((B,CompactBuffer(Benin, Brazil)), (A,CompactBuffer(Angola, Aruba)))
```

8. sortBy(func,[ascending],[numPartitions])

这个操作对源 RDD 中的元素按给定的 func 函数进行排序，并可以指定正序或倒序（默认为正序），返回一个新的 RDD，新的 RDD 中包含排好序的元素。例如，构造一个简单的 RDD，并使用 sortBy()转换进行排序和倒序操作，代码如下：

```
//构造一个 RDD
val data = List(3,1,90,3,5,12)
val rdd = sc.parallelize(data)
rdd.collect

//默认升序
rdd.sortBy(x => x).collect           //Array(1, 3, 3, 5, 12, 90)

//降序
rdd.sortBy(x => x, false).collect    //Array(90, 12, 5, 3, 3, 1)
```

也可以在排序的同时改变分区数量，代码如下：

```
val result1 = rdd.sortBy(x => x, false)
println(result1.partitions.size)     //默认分区数，2
```

```
//将分区数改为1
val result2 = rdd.sortBy(x => x, false, 1)
println(result2.partitions.size)         //现在的分区数,1
```

在上面的示例中,对 rdd 中的元素进行排序,并对排序后的 RDD 分区的个数进行了修改,其中 result1 就是排序后的 RDD,默认的分区个数是 2,result2 是修改分区以后的 RDD,分区数变为了 1。

9. randomSplit(weights,seed)

这个转换操作使用提供的权重随机拆分源 RDD,以数组形式返回拆分后的子 RDD(拆分后的 RDD 组成的数组并返回),其中各参数的含义如下。

(1)weights:分割的权重,如果它们的和不等于 1,则将被标准化。

(2)seed:随机种子。

例如,构造一个简单的 RDD,并按 80:20 的比例将其分割为两个子 RDD,代码如下:

```
//构造一个RDD
val rdd1 = spark.sparkContext.parallelize(Array(1,2,3,4,5,6,7,8,9,10))

//按80:20分割数据集
val splitedRDD = rdd1.randomSplit(Array(0.8,0.2))

//查看
splitedRDD(0).collect              //Array(1, 2, 3, 4, 6, 7, 8, 9)
splitedRDD(1).collect              //Array(5, 10)
```

Spark Core 模块也支持对两个 RDD 执行集合运算。为了便于演示,首先创建两个 RDD,分别包含元素{1, 2, 3, 3}和{3, 4, 5},代码如下:

```
//构造两个RDD
val rdd1 = sc.parallelize(List(1,2,3,3))
rdd1.collect                       //Array(1, 2, 3, 3)

val rdd2 = sc.parallelize(List(3,4,5))
rdd2.collect                       //Array(3, 4, 5)
```

对两个 RDD 可执行如下集合操作。

10. union(otherDataset)

返回此 RDD 与另一个 RDD 的并集。返回的新 RDD 包含两个 RDD 中的全部元素(相当于 SQL 语言中的 union all)。例如,将 rdd1 和 rdd2 合并在一起,代码如下:

```
val rdd3 = rdd1.union(rdd2)
rdd3.collect                       //Array(1, 2, 3, 3, 3, 4, 5)
```

11. intersection(otherDataset)

返回此 RDD 与另一个 RDD 的交集。输出将不包含任何重复元素,即使输入 RDD 中有重复元素也一样。例如,要统计 rdd1 和 rdd2 的交集,代码如下:

```
val rdd4 = rdd1.intersection(rdd2)
rdd4.collect                              //Array(3)
```

12. subtract(otherDataset)

返回一个 RDD，其中包含其他 RDD 中不存在的元素。相当于执行两个 RDD 的差集。例如，找出 rdd1 中存在而 rdd2 中不存在的元素，代码如下：

```
val rdd5 = rdd1.subtract(rdd2)
rdd5.collect                              //Array(2, 1)
```

13. cartesian(otherDataset)

返回这个 RDD 和另一个 RDD 的笛卡儿积，即返回的新 RDD 包含所有元素对(a, b)，其中 a 来自当前这个 RDD，b 来自另一个 RDD。例如，对 rdd1 和 rdd2 执行笛卡儿连接，代码如下：

```
val rdd6 = rdd1.cartesian(rdd2)
rdd6.collect.foreach(println)
```

执行上面的代码，输出的结果如下：

```
(1,3)
(2,3)
(1,4)
(1,5)
(2,4)
(2,5)
(3,3)
(3,3)
(3,4)
(3,5)
(3,4)
(3,5)
```

14. zip(otherDataset)

将这个 RDD 与另一个 RDD 执行"拉链"操作，返回的新 RDD 中包含的元素为键-值对，每个键-值对中第 1 个元素来自第 1 个 RDD，第 2 个元素来自另一个 RDD。两个 RDD 必须有相同数量的元素。例如，将 rdd1 和 rdd2 组合在一起，创建一个新的 rdd3[(key,value)]，其中 key 来自 rdd1，value 来自 rdd2，代码如下：

```
val rdd1 = spark.sparkContext.parallelize(Array("aa","bb","cc"))
val rdd2 = spark.sparkContext.parallelize(Array(1,2,3))

val rdd3 = rdd1.zip(rdd2)
rdd3.collect                              //Array((aa,1), (bb,2), (cc,3))
```

3.4.3 RDD Action 操作

Action 操作指的是一个将结果返给驱动程序或将结果写入存储的操作，并开始执行一个计算，如 count()和 first()。

一旦创建了 RDD，就只有在执行 Action 时才会执行各种 Transformation 转换。可以将一个 Action 的执行结果写回存储系统，或者返回驱动程序，以便在本地进行进一步的计算。常用的 Action 操作函数见表 3-1。

表 3-1 常用的 Action 操作函数

Action 操作函数	描述
reduce(func)	使用函数 func 对 RDD 中的元素进行聚合计算
collect()	将 RDD 操作的所有结果返给驱动程序。这通常对产生足够小的 RDD 的操作很有用
count()	返回 RDD 中的元素数量
first()	返回 RDD 的第 1 个元素。它的工作原理类似于 take(1)函数
take(n)	返回 RDD 的前 n 个元素。它首先扫描一个分区，然后使用该分区的结果来估计满足该限制所需的其他分区的数量。这种方法应该只在预期得到的数组很小的情况下使用，因为所有的数据都会被加载到驱动程序的内存中
top(n)	按照指定的隐式排序从这个 RDD 中取出最大的 n 个元素，并维护排序。这与 takeOrdered 相反。这种方法应该只在预期得到的数组很小的情况下使用，因为所有的数据都会被加载到驱动程序的内存中
takeSample	返回一个数组，其中包含来自 RDD 的元素的抽样子集
takeOrdered(n)	返回 RDD 的前 n 个（最小的）元素，并维护排序。这和 top 是相反的。这种方法应该只在预期得到的数组很小的情况下使用，因为所有的数据都会被加载到驱动程序的内存中
saveAsTextFile(path)	将 RDD 的元素作为文本文件（或文本文件集）写入本地文件系统、HDFS 或任何其他 Hadoop 支持的文件系统的给定目录中。Spark 将对每个元素调用 toString，将其转换为文件中的一行文本
foreach(func)	在 RDD 的每个元素上运行函数 func
aggregate	类似于 reduce，执行聚合运算，但它可以返回具有与输入元素数据类型不同的结果。这个函数用于聚合每个分区的元素，然后使用给定的 combine 组合函数和一个中性的"零值"，对所有分区的结果进行聚合

下面通过几个示例来掌握几个常用的 Action 类型函数的用法。

【示例 3-1】假设有一个包含元素{1, 2, 3, 3}的 RDD，计算该 RDD 中所有元素的和。
实现代码如下：

```scala
//第 3 章/action_reduce.scala

//构造 RDD
val rdd = sc.parallelize(List(1,2,3,3))

//使用 reduce(func)进行聚合计算
val result = rdd.reduce((x,y) => x + y)
//val result = rdd.reduce(_+_)           //等价，简洁写法
```

```
//输出结果
println(result)                                    //9
```
执行上面的代码，输出的结果如下：
```
9
```

【示例3-2】假设有一个包含整数元素的RDD，使用Spark RDD的aggregate()函数计算该RDD中所有元素的和。

分析：这个aggregate()函数的签名如下。
```
def aggregate[U](zeroValue: U)(seqOp: Function2[U, T, U], combOp:
Function2[U, U, U]): U
```
其中各参数的含义分别如下。

（1）zeroValue：当执行seqOp运算时每个分区用来累积结果的初始值，当执行combOp运算时为不同分区合并结果的初始值。它通常是中性元素（例如求和计算时初始值为0或求积计算时初始值为1）。

（2）seqOp：用于在分区内累积结果的运算函数。

（3）combOp：用于合并来自不同分区结果的合并运算函数。

实现代码如下：
```
//第3章/action_aggregate.scala

//构造RDD[Int]
val listRdd = spark.sparkContext.parallelize(List(1,2,3,4,5,3,2))

//定义分区计算函数
def param0= (accu:Int, v:Int) => accu + v

//定义分区结果合并计算函数
def param1= (accu1:Int,accu2:Int) => accu1 + accu2

//执行aggregate计算
val result = listRdd.aggregate(0)(param0,param1)

println("输出: " + result)
```
执行上面的代码，输出的结果如下：
```
20
```

【示例3-3】假设有一个包含整数元素的RDD，使用Spark RDD的aggregate()函数计算该RDD中所有元素的平均值。

分析：这个aggregate()函数类似于reduce()和fold()函数，但reduce()和fold()函数有一个问题，那就是它们的返回值必须与RDD中元素的数据类型相同，而aggregate()函数就打破了这个限制，它可以返回不同数据类型的结果。这个示例要求计算RDD中所有元素的平均值，需要先计算出两个值，一个是RDD的各元素的累加值，另一个是元素计数。对于加法计算，zeroValue要初始化为(0, 0)。

实现代码如下：

```scala
//第3章/action_aggregate2.scala

//构造一个RDD，指定有两个分区
val rdd = sc.parallelize(List(1,2,3,3), 2)

//查看分区数
println(s"分区数：${rdd.partitions.size}")

//定义分区进行累加(总和，数量)操作的函数
def seqOp = (acc:(Int,Int), input:Int) => (acc._1+input, acc._2+1)

//定义合并各个分区结果的函数
def combOp = (acc1:(Int,Int), acc2:(Int,Int)) => (acc1._1+acc2._1, acc1._2+acc2._2)

//执行聚合计算
val result = rdd.aggregate((0,0))(seqOp, combOp)

//计算平均值
val avg = result._1/result._2.toDouble

//输出
println(s"RDD中所有元素的平均值是：${avg}")
```

执行上面的代码，输出的结果如下：

```
分区数：2
RDD中所有元素的平均值是：2.25
```

3.4.4　RDD 上的描述性统计操作

Spark 在包含数值元素的 RDD 上提供了许多描述性统计操作方法，这非常有利于数据分析师执行数据探索任务。描述性统计都是在数据的单次传递中计算的，代码如下：

```scala
//第3章/action_describe.scala

//构造一个RDD
val rdd1 = sc.parallelize(1 to 20 by 2)

println(rdd1.sum)           //统计RDD中所有元素的和
println(rdd1.max)           //统计RDD中元素的最大值
println(rdd1.min)           //统计RDD中元素的最小值
println(rdd1.count)         //统计RDD中元素的数量
println(rdd1.mean)          //统计RDD中元素的平均值
println(rdd1.variance())    //统计RDD中所有元素的方差
println(rdd1.stdev())       //统计RDD中所有元素的标准差
```

执行上面的代码，输出的结果如下：

```
100.0
19
```

```
1
10
10.0
33.0
5.744562646538029
```

也可以用直方图统计数据分布，代码如下：

```
rdd1.histogram(Array(1.0, 8.0, 20.9))        //Array(4, 6)
```

上面代码的意思是，统计 rdd1 中[1.0, 8.0)和[8.0, 20.9) 范围内元素的数量，结果前者有 4 个，后者有 6 个。也可以指定平均划分的区域数量，代码如下：

```
rdd1.histogram(3)       //结果：(Array(1.0, 7.0, 13.0, 19.0),Array(3, 3, 4))
```

结果的意思是，[1.0, 7.0)范围内有 3 个元素，[7.0, 13.0)范围内有 3 个元素，[13.0, 19.0)范围内有 4 个元素。

如果需要多次调用描述性统计方法，则可以使用 StatCounter 对象。可以通过调用 stats()方法返回一个 StatCounter 对象实现，代码如下：

```
//第 3 章/action_describe.scala

//通过调用 stats()方法返回一个 StatCounter 对象实现
val status = rdd1.stats()

//调用各种描述性统计方法
println(status.count)
println(status.mean)
println(status.stdev)
println(status.max)
println(status.min)
println(status.sum)
println(status.variance)
```

执行以上代码，输出的结果如下：

```
10
10.0
5.744562646538029
19.0
1.0
100.0
33.0
```

3.5　Pair RDD

Spark 中有一类特殊的 RDD，其元素是以<key,value>对的形式出现，被称为"Pair RDD"，或者键-值对 RDD。针对键-值对 RDD，Spark 专门提供了一些操作，这些操作只在键-值对的 RDD 上可用。

3.5.1　创建 Pair RDD

Spark 在包含键-值对的 Pair RDD 上提供了专门的 Transformation API，包括 reduceByKey()、groupByKey()、sortByKey() 和 join() 等。Pair RDD 能够在 key 上并行操作，或者跨网络重新组织数据。Pair RDD 常被用于执行聚合操作，以及常被用来完成初始的 ETL（Extract-Transform-Load，抽取-转换-加载）以获取 key-value 格式数据。

注意：除了 count() 操作外，大多数操作通常涉及 shuffle，因为与 key 相关的数据可能驻留在不同的分区上。

创建 Pair RDD 的方式有多种。假设在 HDFS 的 /data/spark/ 目录下存在一个 word.txt 文件，其内容如下：

```
good good study
day day up
```

可以从 word.txt 文件中加载数据以创建 Pair RDD，代码如下：

```scala
//第 3 章/pair_rdd01.scala

//将文件内容加载到 RDD
val lines = sc.textFile("/data/spark/word.txt")   //默认为 HDFS 上的路径
//val lines = sc.textFile("file://data/spark/word.txt")   //本地文件系统

//通过转换，生成键-值对 RDD
val pairRDD = lines.
    flatMap(line => line.split(" ")).
    map(word => (word,1))

//输出查看
pairRDD.collect
```

执行以上代码，输出内容如下：

```
Array((good,1), (good,1), (study,1), (day,1), (day,1), (up,1))
```

可以看到，转换之后的 RDD 中的元素是二元组，其中第 1 个元组元素称为 key，第 2 个元组元素称为 value。

当然，也可以通过并行化内存集合创建 Pair RDD，或通过对已有 RDD 的转换来创建一个新的 Pair RDD，代码如下：

```scala
//第 3 章/pair_rdd02.scala

//并行化一个内存集合
val rdd = sc.parallelize(Seq("Hadoop","Spark","Hive","Spark"))

//通过 map 转换构造一个键-值对 RDD
val pairRDD = rdd.map(word => (word,1))

//输出结果
pairRDD.collect
```

执行上面的代码,输出的结果如下:
```
Array((Hadoop,1), (Spark,1), (Hive,1), (Spark,1))
```
也可以使用 keyBy()转换来构建一个 Pair RDD,代码如下:
```
//构造一个RDD
val word = sc.parallelize(List("black", "blue", "white", "green", "grey"))

//将每个元素的长度值作为 key
val kvWord = word.keyBy(_.length)

//查看
kvWord.collect
```
执行以上代码,输出的结果如下:
```
Array((5,black), (4,blue), (5,white), (5,green), (4,grey))
```

3.5.2 操作 Pair RDD

在 Pair RDD 上除了可执行普通 RDD 上一样的 Transformation 转换操作外,Spark 还专门为这一类 RDD 提供了专用的 Transformation 转换操作和 Action 操作(这些专用转换操作只能作用于键-值对 RDD,不能在普通 RDD 上调用)。

为了演示这些专用于键-值对 RDD 上的转换操作,首先构造一个 Pair RDD,代码如下:
```
//构造键-值对RDD
val pairRDD = sc.parallelize(Seq((1,2),(3,4),(3,6)))
pairRDD.collect
```
执行以上代码,输出的结果如下:
```
Array((1,2), (3,4), (3,6))
```
基于上面构造的键-值对 RDD,下面介绍一些专用的 Transformation 操作。

1. keys

这个转换操作会返回一个包含所有 key 的新 RDD,代码如下:
```
val rdd1 = pairRDD.keys
rdd1.collect
```
执行以上代码,输出的结果如下:
```
Array(1, 3, 3)
```

2. values

这个转换操作会返回一个包含所有 value 的新 RDD,代码如下:
```
val rdd2 = pairRDD.values
rdd2.collect
```
执行以上代码,输出的结果如下:
```
Array(2, 4, 6)
```

3. mapValues(func)

将 func 函数应用到 Pair RDD 中的每个元素上,只对每个键-值对中的 value 执行 map 转换,不对 key 做任何改变,代码如下:

```
val rdd3 = pairRDD.mapValues(x => x*x)    //value => value * value
rdd3.collect
```

执行以上代码,输出的结果如下:

```
Array((1,4), (3,16), (3,36))
```

4. flatMapValues(func)

将 func 函数应用到 Pair RDD 中的每个元素上,只对每个键-值对中的 value 执行 flatMap 转换,不对 key 做任何改变。这也保留了原始的 RDD 分区,代码如下:

```
val rdd4 = pairRDD.flatMapValues(x => (x to 5))
rdd4.collect
```

执行以上代码,输出的结果如下:

```
Array((1,2), (1,3), (1,4), (1,5), (3,4), (3,5))
```

5. sortByKey([ascending], [numPartitions])

对键-值对 RDD 中的元素按照 key 进行排序,默认为升序排序。可通过布尔类型的 ascending 参数指定排序顺序,代码如下:

```
//val rdd5 = pairRDD.sortByKey()          //升序
val rdd5 = pairRDD.sortByKey(false)       //降序

rdd5.collect
```

执行以上代码,输出的结果如下:

```
Array((3,4), (3,6), (1,2))
```

6. groupByKey([numPartitions])

将 RDD 中的元素按 key 进行分组,具有相同 key 的元素值被分到一个序列中。例如,对(K,V)类型的 RDD 调用此转换方法时,返回一个(K, Iterable<V>)类型的新 RDD,代码如下:

```
val rdd6 = pairRDD.groupByKey()

rdd6.collect
```

执行以上代码,输出的结果如下:

```
Array((1,CompactBuffer(2)), (3,CompactBuffer(4, 6)))
```

注意:groupByKey()是一种宽依赖的操作,会导致对 RDD 执行哈希分区,从多个分区 shuffle 数据,并且不使用分区本地的 combiner 来减少数据传输,因此开销很大。当需要对分组数据进行进一步聚合时,不建议使用。

7. subtractByKey(other)

subtractByKey(other)转换操作对两个 RDD 按 key 求差集，返回的新 RDD 中只包含那些 key 仅在当前 RDD 中有而在另一个 RDD 中没有的那些键-值对，代码如下：

```
//构造另一个RDD
val other = sc.parallelize(Seq((3,5),(4,6)))

//求两个RDD的key差集
val rdd7 = rdd5.subtractByKey(other)

//查看
rdd7.collect
```

执行以上代码，输出的结果如下：

```
Array((1,2))
```

8. join(other)

对两个 RDD 按 key 进行连接。例如，当对类型(K,V)和(K,W)的 RDD 调用此操作时，会返回(K,(V,W))类型的 RDD，其中包含每个 key 的所有元素对，代码如下：

```
//构造两个键-值对RDD
val pairRDD1 = sc.parallelize(Seq((1,2),(3,4),(3,6),(5,8)))
val pairRDD2 = sc.parallelize(Seq((1,3),(3,7),(4,9)))

//执行join连接
val joinedRDD = pairRDD1.join(pairRDD2)

//查看
joinedRDD.collect
```

执行以上代码，输出的结果如下：

```
Array((1,(2,3)), (3,(4,7)), (3,(6,7)))
```

9. leftOuterJoin(other)

对两个 RDD 按 key 执行左外连接，返回的新 RDD 中包含左侧 RDD 中的所有 key 和右侧 RDD 中满足连接条件的 key，代码如下：

```
//构造两个键-值对RDD
val pairRDD1 = sc.parallelize(Seq((1,2),(3,4),(3,6),(5,8)))
val pairRDD2 = sc.parallelize(Seq((1,3),(3,7),(4,9)))

//执行左外连接
val joinedRDD = pairRDD1.leftOuterJoin(pairRDD2)

//查看
joinedRDD.collect
```

执行以上代码，输出的结果如下：

```
Array((1,(2,Some(3))), (3,(6,Some(7))), (3,(4,Some(7))), (5,(8,None)))
```

10. rightOuterJoin(other)

对两个 RDD 按 key 执行右外连接，返回的新 RDD 中包含右侧 RDD 中的所有 key 和左侧 RDD 中满足连接条件的 key，代码如下：

```
//构造两个键-值对 RDD
val pairRDD1 = sc.parallelize(Seq((1,2),(3,4),(3,6),(5,8)))
val pairRDD2 = sc.parallelize(Seq((1,3),(3,7),(4,9)))

//执行右外连接
val joinedRDD = pairRDD1.rightOuterJoin(pairRDD2)

//查看
joinedRDD.collect
```

执行以上代码，输出的结果如下：

```
Array((4,(None,9)), (1,(Some(2),3)), (3,(Some(4),7)), (3,(Some(6),7)))
```

11. fullOuterJoin(other)

对两个 RDD 按 key 执行全外连接，返回的新 RDD 中包含两侧 RDD 中的所有 key，代码如下：

```
//构造两个键-值对 RDD
val pairRDD1 = sc.parallelize(Seq((1,2),(3,4),(3,6),(5,8)))
val pairRDD2 = sc.parallelize(Seq((1,3),(3,7),(4,9)))

//执行全外连接
val joinedRDD = pairRDD1.fullOuterJoin(pairRDD2)

//查看
joinedRDD.collect.foreach(println)
```

执行以上代码，输出的结果如下：

```
(4,(None,Some(9)))
(1,(Some(2),Some(3)))
(3,(Some(6),Some(7)))
(3,(Some(4),Some(7)))
(5,(Some(8),None))
```

3.5.3　关于 reduceByKey()操作

在对 Pair RDD 执行 groupByKey()转换操作时，会导致数据 shuffling，并且不会进行分区内优化，因此当需要对数据进行聚合操作时，不建议使用该操作。对于聚合操作，Spark 提供了另外几个带有优化措施的方法，它们是 reduceByKey()、aggregateByKey()和 combinerByKey()，分别适用于不同的场景。首先来了解 reduceByKey()转换操作。

Pair RDD 的 reduceByKey()操作使用 reduce()函数合并每个 key 的值，这也是一个宽依赖操作，因此也有可能发生跨分区的数据 shuffling，但是与 groupByKey()转换操作不同的是，当reduceByKey()操作重复地应用于具有多个分区的同一组RDD数据时，它首先使用reduce()函数在各个分区本地执行合并，然后跨分区发送记录以准备最终结果。也就是说，在跨分区发送数据之前，它还使用相同的reduce()函数在本地合并数据，以减少数据shuffling，以便优化传输性能。

例如，有一组数据，代码如下：

```
val data = Array(("a", 1), ("b", 1), ("a", 1), ("a", 1), ("b", 1),
                 ("b", 1), ("a", 1), ("b", 1), ("a", 1), ("b", 1))
```

将其构造为具有3个分区的RDD，代码如下：

```
val rdd = sc.parallelize(data, 3)
```

如果现在需要对该RDD中的数据求和，则可以应用reduceByKey()操作，其工作过程如图3-14所示。

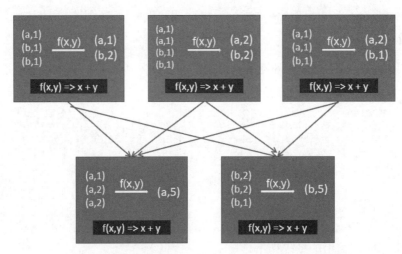

图3-14 在键-值对RDD上执行reduceByKey()转换操作

可以看到，这个RDD有3个分区。数据首先在各个分区执行reduceByKey()操作，然后各个分区计算结果再通过shuffle，相同的key会分配到同一子分区中，再在子分区上执行reduceByKey()操作。由于在shuffling之前各个分区先进行了本地合并，所以会极大地减少shuffling的数据量。

以上过程的完整实现代码如下：

```
//第3章/reducebykey.scala

//有一组数据
val data = Array(
        ("a", 1), ("b", 1), ("a", 1), ("a", 1), ("b", 1),
        ("b", 1), ("a", 1), ("b", 1), ("a", 1), ("b", 1))
//将其构造为具有3个分区的一个RDD
val rdd = sc.parallelize(data, 3)
```

```
//在 RDD 上应用 reduceByKey 操作
val rdd2 = rdd.reduceByKey((a, b) => (a + b))
//val rdd2 = rdd.reduceByKey(_ + _)   //简捷写法

//输出计算结果
rdd2.collect.foreach(println)
```

执行上面的代码,输出的结果如下:

```
(a,5)
(b,5)
```

当计算逻辑比较复杂时,也可以单独定义 reduce 函数,代码如下:

```
//单独定义一个计算函数
def sumFunc(a:Int, b:Int) = a + b

//应用 sumFunc
val rdd3 = rdd.reduceByKey(sumFunc)

//输出
rdd3.collect.foreach(println)
```

执行上面的代码,输出的结果如下:

```
(a,5)
(b,5)
```

当应用 reduceByKey()操作执行聚合计算时,要求计算结果数据类型必须与输入元素的数据类型保持一致,也就是说不能改变 RDD 中元素的数据类型。例如上例中,父 RDD 的元素数据类型为元组(String, Int),那么子 RDD 中的计算结果的数据类型也必须是元组(String, Int)。如果希望在计算过程中改变 RDD 元素的数据类型(也就是子 RDD 的元素类型和父 RDD 的元素类型不同),就不能使用 reduceByKey()转换操作了,这时可以使用 Spark 提供的另一个聚合转换操作 aggregateByKey()。

3.5.4 关于 aggregateByKey()操作

Spark 提供的 aggregateByKey()转换操作会聚合每个 key 的值,使用给定的聚合函数和一个中性的"零值",并为该 key 返回不同类型的值。

这个 aggregateByKey()函数共可接受 3 个参数。

(1) zeroValue:它是累加值或累加器的初值。如果聚合类型是对所有的值求和,则它可以是 0。如果聚合目标是找出最小值,则这个值可以是 Double.MaxValue。如果聚合目标是找出最大值,则这个值可以使用 Double.MinValue。或者,如果只是想要一个各自的集合作为每个 key 的输出,则可以使用一个空的 List 或 Map 对象。

(2) seqOp:是聚合单个分区的所有值的操作。它将一种类型[V]的数据转换/合并为另一种类型[U]的序列操作函数。

(3) combOp:类似于 seqOp,进一步聚合来自不同分区的所有聚合值。它将多个转换后的类型[U]合并为一个单一类型[U]的组合操作函数。

例如，有一组代表学生课目和成绩的数据：

```
val data = Array(
    ("Joseph", "Maths", 83), ("Joseph", "Physics", 74),
    ("Joseph", "Chemistry", 91), ("Joseph", "Biology", 82),
    ("Jimmy", "Maths", 69), ("Jimmy", "Physics", 62),
    ("Jimmy", "Chemistry", 97), ("Jimmy", "Biology", 80),
    ("Tina", "Maths", 78), ("Tina", "Physics", 73),
    ("Tina", "Chemistry", 68), ("Tina", "Biology", 87),
    ("Thomas", "Maths", 87), ("Thomas", "Physics", 93),
    ("Thomas", "Chemistry", 91),("Thomas", "Biology", 74),
    ("Cory", "Maths", 56), ("Cory", "Physics", 65),
    ("Cory", "Chemistry", 71), ("Cory", "Biology", 68),
    ("Jackeline", "Maths", 86), ("Jackeline", "Physics", 62),
    ("Jackeline", "Chemistry", 75), ("Jackeline", "Biology", 83),
    ("Juan", "Maths", 63), ("Juan", "Physics", 69),
    ("Juan", "Chemistry", 64), ("Juan", "Biology", 60)
)
```

其中元素三元组分别代表一个学生的姓名、课目、成绩。每个学生都有四门课的成绩，分别是 Maths、Physics 和 Chemistry 和 Biology。现在要求找出每个学生的最好成绩，以（姓名,成绩）的形式输出，该怎样实现呢？

从需求可以得知，最终输出 RDD 的元素类型为 Tuple2(String, Double)，而输入 RDD 的元素类型为 Tuple3(String, String, Double)，因此不适用 reduceByKey()操作，这时可以使用 aggregateByKey()方法。实现过程如图 3-15 所示（RDD 元素被随机分布到各个分区）。

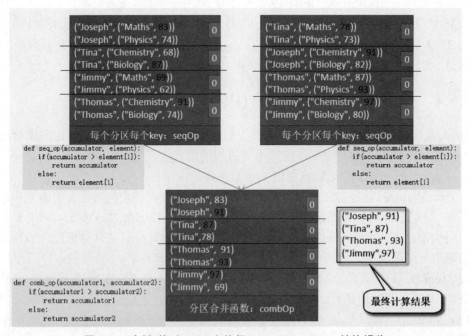

图 3-15　在键-值对 RDD 上执行 aggregateByKey()转换操作

以上过程的实现代码如下:

```scala
//第3章/aggregatebykey.scala

//将学生课目和成绩数据构造为键-值对RDD: PairRDD[(String,(String,Double))]
val data = Array(
    ("Joseph", "Maths", 83), ("Joseph", "Physics", 74),
    ("Joseph", "Chemistry", 91), ("Joseph", "Biology", 82),
    ("Jimmy", "Maths", 69), ("Jimmy", "Physics", 62),
    ("Jimmy", "Chemistry", 97), ("Jimmy", "Biology", 80),
    ("Tina", "Maths", 78), ("Tina", "Physics", 73),
    ("Tina", "Chemistry", 68), ("Tina", "Biology", 87),
    ("Thomas", "Maths", 87), ("Thomas", "Physics", 93),
    ("Thomas", "Chemistry", 91),("Thomas", "Biology", 74),
    ("Cory", "Maths", 56), ("Cory", "Physics", 65),
    ("Cory", "Chemistry", 71), ("Cory", "Biology", 68),
    ("Jackeline", "Maths", 86), ("Jackeline", "Physics", 62),
    ("Jackeline", "Chemistry", 75), ("Jackeline", "Biology", 83),
    ("Juan", "Maths", 63), ("Juan", "Physics", 69),
    ("Juan", "Chemistry", 64), ("Juan", "Biology", 60)
)
val studentRDD = sc.parallelize(data, 2)

//定义Sequence Operation和Combiner Operations
//Sequence Operation : 从单个分区查找最高成绩
def seqOp = (accumulator: Int, element: (String, Int)) =>
    if(accumulator > element._2) accumulator else element._2

//Combiner Operation : 从所有分区累加器中找出最高成绩
def combOp = (accumulator1: Int, accumulator2: Int) =>
    if(accumulator1 > accumulator2) accumulator1 else accumulator2

//Zero Value: 零值设为0
val zeroVal = 0

//执行转换
val aggrRDD = studentRDD.
    map(t => (t._1, (t._2, t._3))).      //构造为[(String,(String,Double))]
    aggregateByKey(zeroVal)(seqOp, combOp)   //执行聚合操作

//查看输出结果
aggrRDD.collect.foreach(println)
```

执行以上代码,输出的结果如下:

```
(Juan,69)
(Tina,87)
(Thomas,93)
(Jimmy,97)
(Jackeline,86)
(Joseph,91)
(Cory,71)
```

在上例的基础上，修改代码，要求计算所有学生的平均成绩。要计算平均成绩，意味着要计算出每个学生的总成绩及课目数，两者相除，其结果就是该学生的平均成绩，代码如下：

```scala
//第3章/aggregatebykey2.scala

//定义各分区执行的操作
def seqOp = (accumulator: (Int, Int), element: (String, Int)) =>
            (accumulator._1 + element._2, accumulator._2 + 1)

//定义分区间结果合并的操作
def combOp = (accumulator1: (Int, Int), accumulator2: (Int, Int)) =>
             (accumulator1._1 + accumulator2._1, accumulator1._2 + accumulator2._2)

//Zero Value: 在此示例中，零值将是包含空白主题名和0的元组
val zeroVal = (0, 0)

//在这里，aggregateByKey()将返回总成绩和科目数
//因此，需要使用单独的map()函数将其转换为百分比值
val aggrRDD = studentRDD.map(t => (t._1, (t._2, t._3)))
                        .aggregateByKey(zeroVal)(seqOp, combOp)
                        .map(t => (t._1, t._2._1/t._2._2*1.0))

//查看输出
aggrRDD.collect.foreach(println)
```

执行以上代码，输出的结果如下：

```
(Juan,64.0)
(Tina,76.0)
(Thomas,86.0)
(Jimmy,77.0)
(Jackeline,76.0)
(Joseph,82.0)
(Cory,65.0)
```

注意：aggregateByKey()是一种宽依赖操作，会导致数据跨分区的shuffling，但它是一个对性能进行优化了的Transformation操作，在shuffling之前会先在各个分区上执行聚合，再shuffling聚合之后的数据。与reduceByKey()相比，当聚合需求要求输入和输出RDD类型不同时，应该使用aggregateByKey()；当聚合需求要求输入和输出RDD类型相同时，应该使用reduceByKey()。

3.5.5 关于combineByKey()操作

除了reduceByKey()和aggregateByKey()这两个转换操作外，Spark还为Pair RDD提供了一个更加通用的combineByKey()操作。它使用一组自定义的聚合函数组合每个key的元素。在其内部combineByKey()操作会按分区合并元素。

Pair RDD 的 combineByKey()转换与 Hadoop MapReduce 编程中的 combiner 非常相似。它也是一个宽依赖操作，在最后阶段需要 shuffle 数据。

Pair RDD 的 combineByKey()操作使用以下 3 个函数作为参数。

（1）createCombiner：在第 1 次遇到一个 Key 时创建组合器函数，将 RDD 数据集中的 V 类型 value 值转换为 C 类型值（V => C）。

（2）mergeValue：合并值函数，当再次遇到相同的 Key 时，将 createCombiner 的 C 类型值与这次传入的 V 类型值合并成一个 C 类型值（C,V）=> C。

（3）mergeCombiner：合并组合器函数，将 C 类型值两两合并成一个 C 类型值。

这个转换操作是 Spark RDD 中最复杂最难理解的一个操作，下面通过一个应用示例来理解它的用法。

首先，假设有一组销售数据，数据采用键-值对的形式（公司，收入），代码如下：

```
val data = Array(
    ("company-1",92),("company-1",85),("company-1",82),
    ("company-1",93),("company-1",86),("company-1",83),
    ("company-2",78),("company-2",96),("company-2",85),
    ("company-3",88),("company-3",94),("company-3",80)
)
```

现在要求使用 combineByKey()来统计出每个公司的总收入和平均收入。实现原理和过程如图 3-16 所示（考虑到 RDD 中的元素是随机分布在各个分区上的）。

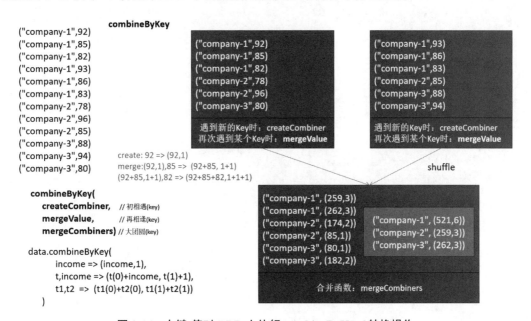

图 3-16　在键-值对 RDD 上执行 combineByKey()转换操作

基于图 3-16 中的转换步骤，进一步理解 combineByKey()操作各个函数参数的含义。

1. createCombiner 函数

createCombiner 函数函数是 combineByKey()操作的第 1 个参数,它是每个 key 的第 1 个聚合步骤。当在一个分区中发现任何新的 key 时,该函数将被执行。这个函数的执行在每个单独的值上对一个节点的分区都是局部的,它类似于 aggregateByKey()函数的第 1 个参数(zeroVal 零值)。

2. mergeValue 函数

mergeValue 函数是 combineByKey()操作的第 2 个参数,它在将同一 key 的下一个值赋给 combiner 时执行。它还在节点的每个分区上本地执行,并合并同一个 key 的所有值。这个函数的参数是一个累加器和一个新的输入值,它在现有累加器中合并输入的新值。这个函数类似于 aggregateByKey()转换操作的第 2 个参数(seqOp)。

3. mergeCombiners 函数

mergeCombiners 函数是 combineByKey()操作的第 3 个参数,用于组合如何跨分区合并单个 key 的两个累加器(combiner)以生成最终的预期结果。它的参数是两个累加器(combiner),用于合并来自不同分区的单个 key 的结果。这个函数类似于 aggregateByKey()函数的第 3 个参数(combOp)。

基于上面的理解,使用 combineByKey()实现的代码如下:

```scala
//第 3 章/combinebykey.scala

//各个公司的销售数据
val data = Array(
    ("company-1",92),("company-1",85),("company-1",82),
    ("company-1",93),("company-1",86),("company-1",83),
    ("company-2",78),("company-2",96),("company-2",85),
    ("company-3",88),("company-3",94),("company-3",80)
)

//构造一个 RDD,并指定两个分区
val rdd = sc.parallelize(data, 2)

//执行聚合转换
val cbk = rdd.combineByKey(
        (income) => (income,1),
        (acc:(Int,Int), income) => (acc._1+income, acc._2+1),
        (acc1:(Int,Int), acc2:(Int,Int) ) => (
                    acc1._1+acc2._1, acc1._2+acc2._2 )
        )

//聚合结果
cbk.collect.foreach(println)
```

执行上面的代码,输出的结果如下:

```
(company-2,(259,3))
(company-3,(262,3))
(company-1,(521,6))
```

上面输出的数据形式是"(公司,(总销售额,月份数量))",而要求是统计出每个公司的总收入和平均收入,因此需要以上面的计算结果进一步转换,实现代码如下:

```
//提取每个元素的值,并计算出平均收入
val res = cbk.map{
    case (key, value) => (key, value._1, value._1/value._2.toFloat)
}

//输出结果
res.collect.foreach(println)
```

执行上面的代码,最终计算结果如下:

```
(company-2,259,86.333336)
(company-3,262,87.333336)
(company-1,521,86.833336)
```

注意:combineByKey()是一种宽依赖操作,它在聚合的最后阶段会shuffle数据并创建另一个RDD。它是一个通用的转换,groupByKey()、reduceByKey()和aggregateByKey()转换的内部实现都使用了combineByKey()。combineByKey()转换操作可灵活执行map端或reduce端combine,因此combineByKey()转换的使用更加复杂。要实现combineByKey(),总是需要实现3个函数:createCombiner、mergeValue、mergeCombiner。

3.6 持久化 RDD

Spark 中最重要的功能之一是跨操作在内存中持久化(或缓存)数据集。当持久化一个 RDD 时,每个节点在内存中存储它计算的任何分区,并在该数据集(或从该数据集派生的数据集)上的其他操作中重用它们。这使后续的操作要快得多(通常超过 10 倍)。缓存是迭代算法和快速交互使用的关键工具。

3.6.1 缓存 RDD

在 Spark 中,RDD 采用惰性求值的机制,每次遇到 Action 操作,Spark 都会从头重新计算 RDD 及其所有的依赖。这对于迭代计算而言,代价是很大的,因为迭代计算经常需要多次重复使用同一组数据。例如,多次计算同一个 RDD,代码如下:

```
//构造一个RDD
val list = List("Hadoop", "Spark", "Hive")
val input = sc.parallelize(list)

//对RDD执行map转换操作
val result = input.map(x => x.toUpperCase)

//Action操作,触发一次真正从头到尾的计算
```

```
println(result.count())

//Action 操作，触发一次真正从头到尾的计算
println(result.collect().mkString(","))
```

在上面的代码中，有两次 Action 操作：count()和 collect()。每遇到一次 Action 操作，都重新构造 RDD、执行 map 转换。可以通过 Spark 中的持久化（缓存）机制避免这种重复计算的开销。

缓存机制是 Spark 提供的一种将数据缓存到内存（或磁盘）的机制，主要用途是使中间计算结果可以被重用。要缓存 RDD，常用到两个函数：cache()和 persist()。

可以使用 persist()方法将一个 RDD 标记为持久化。之所以说"标记为持久化"，是因为出现 persist()语句的地方，并不会马上计算生成 RDD 并把它持久化，而是要等到遇到第 1 个 Action 操作触发真正计算以后，才会把计算结果进行持久化。持久化后的 RDD 分区将会被保留在计算节点的内存中，被后面的 Action 操作重复使用。

如果一个 RDD 数据集被要求参与几个 Action 操作，则持久化该 RDD 数据集会节省大量的时间、CPU 周期、磁盘输入/输出和网络带宽。容错机制也适用于缓存分区。当由于节点故障而丢失任何分区时，它将使用血统图重新计算。

PySpark 还会在随机操作（如 reduceByKey()）中自动保存一些中间数据，甚至不需要用户调用 persist()。这样做是为了避免节点在 shuffling 期间失败时重新计算整个输入。

重写前面的示例代码，加入对 RDD 进行缓存的代码，代码如下：

```
//构造一个 RDD
val list = List("Hadoop","Spark","Hive")
val input = sc.parallelize(list)

//对 RDD 执行 map 转换
val result = input.map(x => x.toUpperCase)

//会调用 persist(MEMORY_ONLY)
//但是，语句执行到这里，并不会缓存 RDD，这时 RDD 还没有被计算生成
result.persist(StorageLevel.MEMORY_ONLY)    //= result.cache()

//第 1 次 Action 操作，触发一次真正从头到尾的计算
//这时才会执行上面的 rdd.cache()，把这个 RDD 放到缓存中
println(result.count())

//第 2 次 Action 操作，不需要触发从头到尾的计算，只需重复使用上面缓存中的 RDD
println(result.collect().mkString(","))

//把持久化的 RDD 从缓存中移除
result.unpersist()
```

如果可用内存不足，Spark 则会将持久的分区溢写到磁盘上。开发人员可以使用 unpersist()删除不需要的 RDD。Spark 会自动监控缓存，并使用 LRU（Least Recently Used，最近最少使用）算法删除旧分区。

Spark 的缓存不仅能将数据缓存到内存，也能使用磁盘，甚至同时使用内存和磁盘，这种缓存的不同存储方式，称作"StorageLevel（存储级别）"。使用 persist()方法指定存储级别，代码如下：

`rdd.persist(StorageLevel.MEMORY_ONLY)。`

缓存方法 cache()本质上就是 persist(StorageLevel.MEMORY_ONLY)，也就是说 persist()方法可以指定存储级别，而 cache()方法不行。Spark 所支持的各种持久化级别见表 3-2。

表 3-2　Spark 所支持的各种持久化级别

持久化级别	内存使用情况	CPU 时间	位于内存中	位于磁盘中	说　　明
MEMORY_ONLY	高	低	是	否	将 RDD 作为反序列化的 Java 对象存储在 JVM 中。如果 RDD 不适合内存，则一些分区将不会被缓存，并在每次需要它们时动态地重新计算。这是默认级别
MEMORY_ONLY_SER （Java 和 Scala）	低	高	是	否	将 RDD 存储为序列化的 Java 对象(每个分区 1 字节数组)。这比反序列化对象更节省空间，特别是在使用快速序列化器时，但读取时需要更多 CPU
MEMORY_AND_DISK	高	中等	部分	部分	将 RDD 作为反序列化的 Java 对象存储在 JVM 中。如果 RDD 不适合内存，则将不适合的分区存储在磁盘上，并在需要时从那里读取它们
MEMORY_AND_DISK_SER （Java 和 Scala）	低	高	部分	部分	类似于 MEMORY_ONLY_SER，但是将不适合内存的分区溢出到磁盘，而不是在每次需要它们时动态地重新计算它们
DISK_ONLY	低	高	否	是	仅在磁盘上存储 RDD 分区
MEMORY_ONLY_2, MEMORY_AND_DISK_2 等					分别与上面的 MEMORY_ONLY 和 MEMORY_AND_DISK 的级别相同，但是在两个集群节点上复制每个分区
OFF_HEAP(实验)					与 MEMORY_ONLY_SER 类似，但将数据存储在堆外内存中。这需要启用堆外内存

注意：在 Python 中，存储的对象将始终使用 Pickle 库进行序列化，所以是否选择序列化级别并不重要。Python 中可用的存储级别包括 MEMORY_ONLY、MEMORY_ONLY_2、MEMORY_AND_DISK、MEMORY_AND_DISK_2、DISK_ONLY 和 DISK_ONLY_2。

3.6.2 RDD 缓存策略

RDD 块可以在多个存储（内存、磁盘、堆外）中以序列化或非序列化格式缓存。

（1）MEMORY_ONLY：数据仅以非序列化格式缓存在内存中。

（2）MEMORY_AND_DISK：数据缓存在内存中。如果没有足够的内存可用，则将从内存中清除的块序列化到磁盘。当重新计算很昂贵且内存资源稀缺时，建议使用这种操作模式。

（3）DISK_ONLY：数据仅以序列化格式缓存在磁盘上。

（4）OFF_HEAP：与 MEMORY_ONLY_SER 类似，但将数据存储在堆外内存中，例如，在 Alluxio 上。这需要启用堆外内存。

上面的缓存策略还可以使用序列化来以序列化格式存储数据。序列化增加了处理成本，但减少了大型数据集的内存占用。将"_SER"后缀附加到上述策略名上表示以使用序列化，例如，MEMORY_ONLY_SER、MEMORY_AND_DISK_SER、DISK_ONLY 和 OFF_HEAP 始终以序列化格式写入数据。

数据也可以通过在 StorageLevel 上添加"_2"后缀来表示将缓存复制到另一个节点，例如 MEMORY_ONLY_2、MEMORY_AND_DISK_SER_2。当集群的一个节点（或执行器）出现故障时，复制有助于加速恢复。

那么如何选择合适的缓存策略呢？

Spark 的不同存储级别意味着在内存使用和 CPU 效率之间提供不同的权衡。建议通过以下步骤来选择一个：

（1）如果 RDD 适合默认存储级别（MEMORY_ONLY），则保留它们。这是 CPU 效率最高的选项，允许 RDD 上的操作尽可能快地运行。

（2）如果没有，则可以尝试使用 MEMORY_ONLY_SER 并选择一个快速序列化库，以使对象更节省空间，但访问速度仍然相当快（适用于 Java 和 Scala）。

（3）不要溢出到磁盘，除非计算数据集的函数开销很大，或者它们过滤了大量数据。否则重新计算分区的速度可能与从磁盘读取分区的速度一样快。

（4）如果希望快速恢复故障（例如，如果使用 Spark 为来自 Web 应用程序的请求提供服务），则应使用复制的存储级别。通过重新计算丢失的数据，所有存储级别都提供了完全的容错能力，但是复制的存储级别允许在 RDD 上继续运行任务，而无须等待重新计算丢失的分区。

Spark 自动监视每个节点上的缓存使用情况，并以最近最少使用（LRU）的方式删除旧的数据分区。如果希望手动删除一个 RDD，而不是等待它被从缓存中删除，则可以使用 RDD.unpersist() 方法。

3.6.3 检查点 RDD

通过连接任意数量的 Transformation 转换操作，RDD lineage（血统）可以任意增长。Spark 提供了一种方法，可以将整个 RDD 持久化到稳定的存储器中，存储的数据包括 RDD 计算后的数据和分区器，然后，在发生节点故障时，Spark 不需要从头开始计算丢失的 RDD 碎片，而是从存储的快照那里开始计算 lineage 中其余的部分。这个特性称为检查点（Check Point）。

简单来讲，检查点是一种截断 RDD 依赖链并把 RDD 数据持久化到存储系统（通常是 HDFS 或本地）的过程。它的主要作用是截断 RDD 依赖关系，防止任意增长的 lineage 导致的堆栈溢出。在检查点之后，RDD 的依赖项及它的父 RDD 的信息会被擦除，因为重新计算不再需要它们了。

必须先调用 SparkContext.setCheckpointDir()方法设置保存数据的目录，然后通过调用 checkpoint()操作来对 RDD 设置检查点，这时该 RDD 将被保存到检查点目录中的一个文件中，所有对其父 RDD 的引用将被删除。

必须在 RDD 上执行任何作业之前调用 checkpoint()。当在 RDD 上调用 checkpoint()方法时，仅是将此 RDD 标记为检查点。必须在 RDD 上调用了 Action 操作才能完成检查点。

例如，设置 RDD 检查点，代码如下：

```
val sc = spark.sparkContext                                 //获取 SparkContext 实例

sc.checkpoint("hdfs://xueai8:8020/ck/rdd")   //设置 RDD 的检查点目录

val data = sc.textFile("hdfs://master:8020/input")          //加载数据源

val rdd = data.map(...).reduceByKey(...)          //执行一系列 Transformation

rdd.checkpoint        //标记对 RDD 做 checkpoint,不会真正执行,直到遇到第 1 个 action
rdd.count             //第 1 个 Action 算子，触发之前的代码执行
```

在使用检查点时，需要注意它与缓存的区别。缓存采用临时保存，Executor 挂掉会导致数据丢失，但是数据可以重新计算，而检查点则用于截断依赖链，可靠方式下 Executor 挂掉不会丢失数据，数据一旦丢失不可恢复。

尽管 Spark 会自动管理（包括创建和回收）由 cache()和 persist()方法持久化的数据，但是检查点持久化的数据需由用户自己管理。检查点会清除 RDD 的血统信息，避免血统过长导致序列化开销增大，而 cache()和 persist()不会清除 RDD 的血统。

3.7 RDD 数据分区

数据分区（partition）是 Spark 中的重要概念，是 Spark 在集群中的多个节点之间划分数据的机制。分区是 RDD 的最小单元，RDD 是由分布在各个节点上的分区组成的。Spark

使用分区来管理数据，分区的数量决定了任务（Task）的数量，每个任务对应着一个数据分区。这些分区有助于并行化分布式数据处理。

3.7.1 获取和指定 RDD 分区数

默认情况下，为每个 HDFS block 块创建一个分区，该分区默认为 128MB（Spark 2.x）。例如，当从本地文件系统将一个文本文件加载到 Spark 时，文件的内容被分成几个分区，这些分区均匀地分布在集群中的节点上。在同一个节点上可能会出现不止一个分区。所有这些分区的总和形成了 RDD。这就是"弹性分布数据集"中"分布"一词的来源。

例如，加载一个数据集，并指定该数据集被分割为 3 个分区，分别存储在集群的 3 个节点上，如图 3-14 所示。

每个 RDD 维护一个分区的列表和一个可选的首选位置列表，用于计算分区。分区的首选位置是一个分区的数据所驻留的主机名或 executors 的列表，这样计算就可以更靠近数据了（如果 Spark 获得了首选位置的列表，Spark 调度程序会试着在数据实际存在的执行器上运行任务，这样就不需要进行数据传输了。这对性能有很大的影响）。可以从 RDD 的 partitions 字段获得 RDD 分区的列表。它是一个数组 Array，所以可以通过读取 RDD 的 partitions.size 字段来获得该 RDD 分区的数量。

例如，查看 RDD 的分区数量，代码如下：

```
//构造一个 Pair RDD
val pairs = sc.parallelize(List((1, 1), (2, 2), (3, 3)))

//查看分区数量
pairs.partitions.size            //2
```

也可以在创建 RDD 时指定分区的数量。例如，在调用 textFile()和 parallelize()方法创建 RDD 时，可以手动指定分区个数。这两种方法的签名如下：

```
def textFile(path: String, minPartitions: Int = defaultMinPartitions): RDD[String]

def parallelize[T](seq: Seq[T], numSlices: Int = defaultParallelism)(implicit arg0: ClassTag[T]): RDD[T]
```

如果在创建 RDD 时未指定 RDD 的分区数量，则 Spark 将使用默认的分区数，默认值为 spark.default.parallelism 配置的参数值。

3.7.2 调整 RDD 分区数

从数据源生成 RDD 时，数据通常被随机分配到不同的分区或者保持数据源的分区。RDD 分区数的多少，会对 Spark 程序的执行产生一定的影响。因为除了影响整个集群中的数据分布之外，它还直接决定了将要运行 RDD 转换的任务的数量。

如果分区数量太少，则直接的影响是集群计算资源不能被充分利用。例如分配 8 个核，但分区数量为 4，则将有一半的核没有被利用。此外，因为数据集可能会变得很大，当无法装入 executor 的内存中时可能会导致内存问题。

如果分区数量太多，虽然计算资源能够充分利用，但会导致 task 数量过多，而 task 数量过多会影响执行效率，主要是 task 在序列化和网络传输过程会带来较大的时间开销。

根据 Spark RDD Programming Guide 上的建议，集群节点的每个核分配 2~4 个分区比较合理，也就是说，建议将分区数设置为集群中 CPU 核数的 3~4 倍，如图 3-17 所示。

> One important parameter for parallel collections is the number of *partitions* to cut the dataset into. Spark will run one task for each partition of the cluster. Typically you want 2-4 partitions for each CPU in your cluster. Normally, Spark tries to set the number of partitions automatically based on your cluster. However, you can also set it manually by passing it as a second parameter to `parallelize` (e.g. `sc.parallelize(data, 10)`). Note: some places in the code use the term slices (a synonym for partitions) to maintain backward compatibility.

图 3-17　Spark RDD Programming Guide 建议的分区数

在某些情况下，为了更有效地分配工作负载或避免内存问题，需要显式地重新划分 RDD 的分区。例如，从 HDFS 上加载压缩文件时，因为压缩文件不能被分片，所以只能有一个 RDD 分区，即使在 sc.textFile("xxx.gz", 100) 中指定了分区数也不会真的分区。在这种情况下，就需要对 RDD 调用 repartition(N) 方法进行重分区。

最主要的两种调整数据分区的方法是 coalesce() 和 repartition() 函数。函数 coalesce() 用于减少或增加分区的数量。完整的方法签名如下：

```
coalesce (numPartitions: Int, shuffle: Boolean = false)
```

第 2 个（可选）布尔参数 shuffle 用于指定是否应该执行 shuffle（默认值为 false）。也就是说，coalesce() 是否触发 RDD 的 shuffling，这取决于 shuffle 标志（默认禁用，即 false）。在减少分区时，coalesce() 并没有对所有数据进行移动，仅仅是在原来分区的基础上进行了合并而已，这样的操作可以减少数据的移动，所以效率较高。

函数 repartition() 也可以增加或减少分区数量。完整的方法签名如下：

```
repartition(numPartitions: Int)
```

如果查看 repartition() 函数的源码，则可以看到 repartition() 是直接调用 coalesce(numPartitions, shuffle=true) 的，repartition(N) 方法相当于 coalesce(N, true) 方法。不同的是，调用 repartition() 函数时，还会产生 shuffle 操作（该操作与 HiveQL 的 DISTRIBUTE BY 操作类似），而 coalesce() 函数可以控制是否 shuffling，但当 shuffle 为 false 时，只能减小分区数，而无法增大分区数，因此，如果是要减少 RDD 的分区数，建议使用 coalesce() 而不是 repartition() 函数，这样可以避免数据跨分区间的 shuffling。

例如，对一个 RDD 重新进行分区，并查看分区数，代码如下：

```
//在创建时，将分区数指定为 4
val rdd1 = sc.parallelize(Seq(1,2,3,4,5,6,7,8), 4)

println(rdd1.partitions.size)        //rdd1 的分区数量，目前为 4
```

```
//对于通过转换得到的新RDD
val rdd2 = rdd1.map((x) => x*x)        //转换得到rdd2

//直接调用 repartition 方法重新分区
val rdd3 = rdd2.repartition(8)         //重新分区，得到rdd3
println(rdd3.partitions.size)          //rdd3 的分区数量，这时为 8
```

执行上面的代码，输出内容如下：

```
4
8
```

3.7.3 内置数据分区器

对于 Pair RDD，Spark 提供了一个 partitionBy()方法进行重分区。当在 Pair RDD 上调用 partitionBy()方法进行重分区时，需要向它传递一个分区程序（也被称为分区器），这是一个 org.apache.spark.Partitioner 对象，该分区器对象将一个分区索引赋给每个 RDD 元素（在每个 key 和分区 ID 间建立起映射，分区 ID 的值从 0 到 numPartitions −1）。如果分区器与之前使用的分区器相同，则保留分区，RDD 保持不变。否则就会安排一次 shuffle，并创建一个新的 RDD。

Spark 内置提供了两个 Partitioner 分区器实现，分别如下：

（1）org.apache.spark.HashPartitioner。

（2）org.apache.spark.RangePartitioner。

其中 HashPartitioner 是 Spark 的默认分区器，它基于一个元素的 Java 散列码（或者键-值对 RDD 中 key 的散列码）计算的分区索引。计算公式如下：

```
partitionIndex = key hashCode % numberOfPartitions
```

分区索引是准随机的，因此，分区的大小很可能不会完全相同，然而，在具有相对较少分区的大型数据集中，该算法可能会在其中均匀地分布数据。

当使用 HashPartitioner 时，RDD 的默认分区数量是由配置参数 spark.default.parallelism 决定的。如果该参数没有被用户指定，则它将被设置为集群中的核的数量。

RangePartitioner 分区程序将已排序的 RDD 的数据划分为大致相等的范围。它对传递给它的 RDD 的内容进行了抽样，并根据抽样数据确定了范围边界。不过一般不太可能直接使用 RangePartitioner 分区器。

例如，调用 partitionBy()方法对指定的 Pair RDD 进行重分区，并指定分区数，代码如下：

```
//构造一个 Pair RDD
val pairs = sc.parallelize(List((1, 1), (2, 2), (3, 3)))
println("初始分区数: " + pairs.partitions.size)         //初始分区数量

import org.apache.spark.HashPartitioner
```

```
//重分区为两个分区，使用指定的分区器
val partitionedRDD = pairs.partitionBy(new HashPartitioner(4))

//持久化，以便后续操作重复使用
partitionedRDD.persist()

//查看所使用的分区器
println("分区器: " + partitionedRDD.partitioner)

//重分区后的分区数量
println("重分区数: " + partitionedRDD.partitions.size)
```

执行上面的代码，输出的结果如下：

```
初始分区数: 2
分区器: Some(org.apache.spark.HashPartitioner@4)
重分区数: 4
```

在上面的代码中，传给 partitionBy() 的参数值为 4，代表重分区的数量，它将控制将来在该 RDD 上执行操作时有多少并行任务数。一般来讲，这个值至少与集群中核的数量一样多。

需要提醒的是，只有当一个 RDD 数据集在面向 key 的操作中被重用多次的情况下（例如 join() 操作），通过 partitionBy() 控制分区才有意义。例如，不带 partitionBy() 的 Pair RDD join 过程如图 3-18 所示。

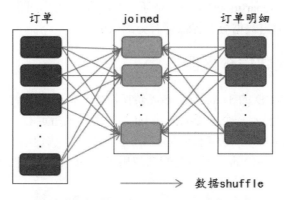

图 3-18 不带 **partitionBy()** 的 **Pair RDD join** 过程

从图 3-18 可以看出，因为 Pair RDD 中的 key 默认为随机分布的，所以同一个 key 可分散分布在一个 Pair RDD 中各个分区中，如果不事先用 partitionBy() 进行重分区，在执行 join 连接时，就会产生大量的数据跨分区 shuffle，严重影响性能。

在这种情况下，可先在 Pair RDD 上调用 partitionBy() 控制重分区，让相同的 key 事先分布到同一个分区上，然后执行 join 连接，过程如图 3-19 所示。

从图 3-19 可以看出，由于订单数据集先按 key 进行了分区，相同 key 的数据位于同一个分区上，因此在执行 join 连接时，这些数据在本地被引用，不会产生数据 shuffle。

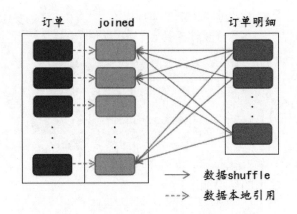

图 3-19　带 partitionBy() 的 Pair RDD join 过程

另外，partitionBy() 也经常用于存储 RDD 时控制输出结果的文件数量。因为默认情况下 RDD 的每个分区对应一个输出文件，当分区数量很大时，写到文件中会产生大量小文件，所以可以在写到文件之前通过 partitionBy() 来调整分区数，从而控制输出结果文件的数量。

从前面的内容可知，repartition() 方法和 partitionBy() 方法都可以调整 RDD 的分区数，但这两种方法是有区别的。repartition() 不能指定分区函数，它主要用于根据内核数量和数据量指定分区的数量。repartition() 实际上是通过在现有值上添加随机键来在内部使用 Pair RDD，因此它不能对输出数据分布提供强有力的保证。

3.7.4　自定义数据分区器

在前面讲到 Pair RDD 的 Transformation 转换操作时，会发现大多数 Pair RDD 转换操作有两个额外的重载方法：一个接受额外的 Int 参数（代表指定的分区数量），另一个则接受一个（定制）Partitioner 类型的附加参数，其中第 1 种方法使用默认的 HashPartitioner。例如，下面两行代码是等价的，因为它们都应用了 100 个分区的 HashPartitioner，代码如下：

```
rdd.foldByKey(afunction, 100)          //使用默认的 HashPartitioner
rdd.foldByKey(afunction, new HashPartitioner(100))
```

如果 Pair RDD 转换操作没有指定一个分区器，则所使用的分区数量将是父 RDD 的最大分区数。如果所有父 RDD 都没有定义分区器，则将使用 HashPartitioner 分区器，分区数由 spark.default.parallelism 参数指定。

当需要精确地控制 Pair RDD 的分区规则和分区逻辑时，可以自定义分区器的实现，通常使用默认的 HashPartitioner 并根据某种算法更改 key 的散列码。

例如，编写自定义分区器对 Pair RDD 进行重分区，将 Pair RDD 中的所有偶数写到一个输出文件，然后将所有奇数写到另一个输出文件。

首先，自定义一个分区程序，继承自 org.apache.spark.Partitioner 类，代码如下：

```
//自定义分区器
class NumberPartitioner(numParts:Int)
                  extends org.apache.spark.Partitioner{

    //覆盖分区数
    override def numPartitions: Int = numParts

    //重写分区号获取函数
    override def getPartition(key: Any): Int = {
        key.toString.toInt % numParts         //取 key 的最后一位数字作为分区 id
    }
}
```

在 Pair RDD 上调用相应的转换操作时，指定使用上面自定义的分区器。例如，构造了一个 Pair RDD，然后应用自定义的 NumberPartitioner 分区器，并将结果保存到存储文件中，代码如下：

```
//构造一个 RDD
val rdd1 = spark.sparkContext.parallelize(1 to 100000)

//使用自定义的分区器进行重分区，并指定分区数量
val rdd2 = rdd1.map((_,null)).partitionBy(new NumberPartitioner(2))
println("自定义分区数: " + rdd2.getNumPartitions)        //2

//将结果输到文件中存储
rdd2.map(_._1).saveAsTextFile("/data/spark/output")
```

因为在自定义分区器中指定了两个分区，所以存储文件相应地也有两个，如图 3-20 所示。

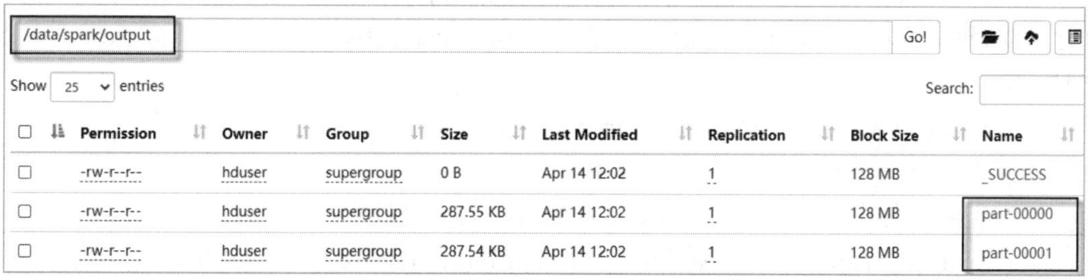

图 3-20　一个分区对应一个存储文件

加载上面存储的结果文件 part-00000，会发现都是偶数，代码如下：

```
sc.textFile("/data/spark/output/part-00000")
  .map(_.toInt)
  .takeOrdered(10)
```

执行上面的代码，输出内容如下：

```
Array(2, 4, 6, 8, 10, 12, 14, 16, 18, 20)
```

加载上面存储的结果文件 part-00001,会发现都是奇数,代码如下:

```
sc.textFile("/data/spark/output/part-00001")
  .map(_.toInt)
  .takeOrdered(10)
```

执行上面的代码,输出内容如下:

```
Array(1, 3, 5, 7, 9, 11, 13, 15, 17, 19)
```

3.7.5 避免不必要的 shuffling

Spark RDD 的 shuffling 是一种重新分配或重新对数据进行分区的机制,以便数据在不同的分区上分组。根据数据大小,可能需要使用 spark.sql.shuffle.partitions 配置或通过代码来减少或增加 RDD/DataFrame 的分区数量。

当需要将来自多个分区的数据组合起来以构建新的分区时,就会发生数据 shuffle。例如,当按 key 对元素进行分组时,Spark 需要检查 RDD 的所有分区,找到具有相同 key 的元素,然后对它们进行物理分组,从而形成新的分区。

Spark RDD 中的某些操作会触发 shuffle 事件,例如 groupByKey()、reduceByKey()、join()、union()、groupBy()、aggregateByKey()等转换操作。在 shuffle 之前和之后的任务分别被称为 map 和 reduce 任务。map 任务的结果被写到中间文件(通常只针对操作系统的文件系统缓存),并通过 reduce 任务读取。Spark shuffling 是一项昂贵的操作,因为它在执行器之间甚至在集群的工作节点之间移动数据,涉及磁盘 I/O、数据序列化和反序列化及网络 I/O,所以在 Spark 作业中尽量减少 shuffling 的次数是很重要的。当在 Spark 作业上遇到性能问题时,应该查看涉及转换的 Spark 转换。

例如,在创建 RDD 时,Spark 并不需要将所有 key 的数据存储在一个分区中,因为在创建 RDD 时,无法为 RDD 设置 key,因此,当运行 reduceByKey()操作来聚合 key 上的数据时,Spark 会执行以下操作:

(1) Spark 首先在所有分区上运行 map 任务,按照每个 key 将所有值分组。

(2) 将 map 任务的结果保存在内存中。

(3) 当结果与内存不匹配时,Spark 将数据存储到磁盘中。

(4) Spark 将 map 后的数据跨分区进行 shuffling,有时还会将 shuffling 后的数据存储到磁盘中,以便在需要重新计算时重用。

(5) 运行垃圾回收。

(6) 最后基于 key 在每个分区上运行 reduce 任务。

虽然 reduceByKey()会触发数据 shuffle,但不会改变分区数,因为子 RDD 从父 RDD 继承了分区大小。

尽管大多数 RDD 转换不需要进行 shuffle,但在特定条件下其中一些转换会引起数据的跨分区 shuffling,因此,为了最小化 shuffle 出现的次数,需要了解引起数据 shuffling 的这些条件。

（1）在明确改变分区时会进行 shuffling。

当使用一个自定义分区器时，总会引发 shuffling。当在转换过程中使用一个与之前的 HashPartitioner 不同的 HashPartitioner 时（分区数量不同），也会引起 shuffling。例如，下面的代码总会引起 shuffling：

```
rdd.aggregateByKey(zeroValue, 100)(seqFunc, comboFunc).collect()

rdd.aggregateByKey(zeroValue, new CustomPartitioner())(seqFunc, comboFunc).collect()
```

（2）由删除分区引起的 shuffling。

有时，转换会导致 shuffling，尽管使用的是默认的分区器。例如，在下面的代码中，第 2 行不会引起 shuffling，但是第 3 行则会引起：

```
val rdd:RDD[Int] = sc.parallelize(1 to 10000)

//不会引起 shuffling
rdd.map(x => (x, x*x)).map(_.swap).count()

//会引起 shuffling
rdd.map(x => (x, x*x)).reduceByKey((v1, v2)=>v1+v2).count()
```

在 Spark 中有两个 shuffle 实现，分别是基于 sort 排序的实现和基于 hash 的实现。从 Spark 1.2 开始默认使用的是基于 sort 排序的 shuffle，因为它的内存更高效，文件更少。可以通过设置 spark.shuffle.manager 参数的值为 hash 或 sort 来定义要使用哪个 shuffle 实现。

可通过 spark.shuffle.consolidateFiles 参数指定在一个 shuffle 期间是否要合并中间文件。出于性能上的考虑，如果使用 EXT4 或 XFS 文件系统，建议将这个参数设为 true（默认值为 false）。

一般来讲，shuffling 可能需要大量的内存，用于聚合和 join 分组。设置 spark.shuffle.spill 参数可指定用于这些任务的内存数量是否应该被限制（默认值为 true）。在这种情况下，任何多余的数据都会溢写到磁盘上。对于内存的限制由参数 spark.shuffle.memoryFraction 指定（默认为 0.2）。此外，spark.shuffle.spill.compress 参数告诉 Spark 是否为溢写的数据启用压缩（默认情况下也是 true）。

溢写阈值不应该设得太高，否则会导致内存溢出异常，但是如果溢写阈值设得太低了，溢写就会频繁发生，所以找到一个好的平衡点很重要。在大多数情况下，保持默认值应该就很合适。

此外，以下这些参数也很有用。

（1）spark.shuffle.compress：指定是否压缩中间文件（默认为 true）。

（2）spark.shuffle.spill.batchSize：指定当溢写到磁盘时将被序列化或反序列化的对象的数量。默认值为 10 000。

（3）spark.shuffle.service.port：如果启用了外部 shuffle 服务，则指定服务器将侦听的端口。

3.7.6 基于数据分区的操作

截至目前，学习到的 RDD Transformation 操作都被应用于 RDD 中的每个元素上，如 map()、flatMap()、filter()等。Spark 另外还提供了一些操作，可以将这些操作应用于每个分区上，而不是每个元素上。这样的操作有 mapPartitions()、mapPartitionsWithIndex()及 glom()、foreachPartition()等。这些基于分区的操作函数见表 3-3。

表 3-3 基于分区的操作函数

函 数 名	传入参数	返回内容	函数签名
mapPartitions	该分区内元素的迭代器	返回元素的迭代器	f: (Iterator[T]) -> Iterator[U]
mapPartitionsWithIndex	集成分区号和该分区内元素的迭代器	返回元素的迭代器	f: (Int, Iterator[T]) -> Iterator[U]
foreachPartition	元素的迭代器	无	f: (Iterator[T]) -> Unit
glom	无	返回通过将每个分区中的所有元素合并为一个数组而创建的 RDD	RDD[Arry[T]]

接下来详细学习这些基于分区的操作函数。

1. mapPartitions()

Spark 的 mapPartitions()是一个转换操作函数，应用于 RDD 中的每个分区。Spark 中的 RDD 存储在分区中，而 mapPartitions()用于在 Spark 架构中的 RDD 分区上应用一个函数。它可以作为 map()和 foreach()的替代方法。

简单来讲，map()转换作用于 RDD 上的每个元素，即对 RDD 中的元素进行一对一转换，而 mapPartitions()作用于 RDD 上的每个分区，即对每个分区进行一对一转换。查阅文档，该方法的签名如下：

```
mapPartitions[U](f: FlatMapFunction[Iterator[(K, V)], U]): JavaRDD[U]
```

可以看到，mapPartitions()转换将函数 f 应用在每个分区上，每个分区中的数据行以迭代器的形式传入函数 f，返回一个新的 Iterator。mapPartitions()函数将结果保存在内存中，直到所有的行都在分区中处理完毕。另外，mapPartitions()函数还接受一个额外的可选参数 preservePartitioning，默认为 false。如果它被设置为 true，新的 RDD 将保留父 RDD 的分区。如果它被设置为 false，则分区器将被移除。

使用 mapPartitions()可以帮助我们更有效地解决一些问题：需要一次性调用每个分区的数据模型的大量初始化，可以通过 mapPartitions()来完成。例如，在连接到数据库时，因为打开数据库连接是开销很大的操作，所以需要在每个分区上应用一次 mapPartitions()，这样数据库连接便是基于数据分区的，而不是基于每个元素。

应用 mapPartitions()转换函数的模板代码如下：
```
//构造一个RDD
val rdd1 = spark.sparkContext.parallelize(Array(1,2,3,4,5,6,7,8,9,10), 2)

//mapPartitions 转换
val rdd2 = rdd1.mapPartitions(iter => Iterator(iter.toArray))

//输出结果
rdd2.collect.foreach(item => println(item.toList))
```
执行以上代码，输出内容如下：
```
List(1, 2, 3, 4, 5)
List(6, 7, 8, 9, 10)
```
当处理逻辑比较复杂时，也可以将其放到单独的方法中，代码如下：
```
//自定义函数，传入每个分区的迭代器，返回每个分区的元素和
def f(i:Iterator[Int]) = {
    //每个分区的每个元素翻倍
    Tuple1(i.sum).productIterator         //返回值需要是Iterator类型
}

//在mapPartitions 转换中应用自定义函数 f
val rdd3 = rdd1.mapPartitions(f)

//输出结果
println(rdd3.collect())
```
执行以上代码，输出的结果如下：
```
Array(15, 40)
```

2. mapPartitionsWithIndex()

与 mapPartitions()类似，但又有所不同，mapPartitionsWithIndex()的 map 函数另外还接受传入的分区索引：（Int，Iterator T）=Iterator U，然后，分区的索引就可以在 map 函数中使用了，用于跟踪原始分区。

这个转换都接受一个额外的可选参数 preservePartitioning，默认为 false。如果它被设置为 true，新的 RDD 将保留父 RDD 的分区。如果它被设置为 false，则分区器将被移除。

应用 mapPartitionsWithIndex()转换函数的示例代码如下：
```
//构造一个RDD，将分区数指定为2
val x = spark.sparkContext.parallelize(Array(1,2,3,4,5,6,7,8,9,10), 2)

//定义函数 f，它有两个传入参数：分区索引和分区元素迭代器
def f(partitionIndex:Int, i:Iterator[Int]) = {
    Tuple1(partitionIndex, i.sum).productIterator
}

//执行转换
val y = x.mapPartitionsWithIndex(f)

//查看
y.collect.foreach(println)
```

执行以上代码,输出的结果如下:

```
(0,15)
(1,40)
```

可以应用 mapPartitionsWithIndex() 来删除 RDD 数据集中的标题行(第 0 号分区 0 行),代码如下:

```
//可用来删除标题行
val z = x.mapPartitionsWithIndex { (idx, iter) =>
    if (idx == 0) iter.drop(1) else iter
}

//查看结果
z.collect
```

执行以上代码,输出的结果如下:

```
Array(2, 3, 4, 5, 6, 7, 8, 9, 10)
```

3. glom()

glom() 方法用于合并每个分区内的所有元素,并返回由这些合并后的元素创建的新 RDD。新 RDD 中元素的数量等于其分区的数量。在这个过程中,分区器会被删除。

简单使用 glom() 方法示例,代码如下:

```
//构造 RDD
val x = spark.sparkContext.parallelize(Array(1,2,3,4,5,6,7,8,9,10), 2)

x.glom().collect    //Array(Array(1, 2, 3, 4, 5), Array(6, 7, 8, 9, 10))
```

在下面的示例代码中,创建一个具有 30 个分区的 RDD 并对其执行 glom() 转换。新 RDD 中的数组对象的计数,包含来自每个分区的数据,也是 30 个,代码如下:

```
//用随机生成 500 个 100 以内的整数构造一个 list
val list = List.fill(500)(scala.util.Random.nextInt(100))

//构造一个 RDD,并用 glom()方法收集分区数据
val rdd = sc.parallelize(list, 30).glom()

//返回 RDD 数据
//rdd.collect()

//统计 RDD 中元素的数量
println(rdd.count())
```

输出的结果如下:

```
30
```

可以看出,glom() 方法将每个分区中的元素聚集到一个数组中,并以这些数组作为新 RDD 的元素。因为有 30 个分区,所以新 RDD 中的元素个数是 30 个。glom() 可以作为一种快速的方法将所有 RDD 的元素放到一个数组中。

4. foreachPartition()

foreachPartition() 是一个 Action 操作,用于执行分区迭代,其方法签名如下:

```
foreachPartition(f: VoidFunction[Iterator[T]]): Unit
```

它将函数 f 作用于 RDD 的每个分区上，因此传入的是一个包含分区元素的迭代器。它是一个 Action 操作，因此不返回值，而是在每个分区上执行输入函数。

这种方法类似于 foreach()方法，但是对性能进行了优化。在 Spark 中，foreachPartition() 用于初始化繁重的操作（如数据库连接），并希望在每个分区初始化一次，而 foreach()用于在 RDD、DataFrame 和 DataSet 分区的每个元素上应用一个函数。

使用 foreachPartition 的模板代码如下：

```
//构造一个RDD
val rdd = spark.sparkContext.parallelize(Seq(1,2,3,4,5,6,7,8,9))

//调用foreachPartition方法
rdd.foreachPartition(partition => {
    //初始化任何数据库连接
    partition.foreach(fun=>{
        //应用函数，例如数据落地
    })
})
```

3.8 深入理解 RDD 执行过程

Spark 程序的执行是基于 DAG 图（Directed Acyclic Graphs，有向无环图）的，其中 RDD 是顶点，依赖是边。每次在一个 RDD 上执行一个转换时，就会创建一个新的顶点（一个新的 RDD）和一条新的边（一个依赖）。新的 RDD 依赖于旧的 RDD，因此边的方向是从子 RDD 指向父 RDD 的。这个依赖图也称为 RDD lineage（RDD 血统）。

3.8.1 Spark RDD 调度过程

Spark 对 RDD 执行调度的过程如图 3-21 所示。

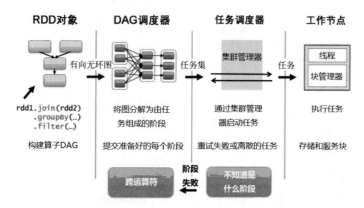

图 3-21 Spark RDD 执行调度过程

当程序执行时，Spark Driver 程序会创建 RDD 并生成 DAG 计算图。当用户运行一个 Action 操作时（如 collect()、count()等），这个 DAG 图会被提交给 DAGScheduler，然后由 DAGScheduler 将 DAG 分解为包含多个任务的阶段（Stages，即 TaskSet），再将 TaskSet 发送至 TaskScheduler，由 TaskScheduler 来调度每个任务（Task），并分配到工作节点上执行，最后得到计算结果。

下面以单词计数为例，来理解 RDD 的执行过程，代码如下：

```
//构造一个RDD
val words = Array("good good study","day day up")
val rdd = spark.sparkContext.parallelize(words)

//对 RDD 执行 Transformation 转换
val wordCounts = rdd
  .flatMap(_.split(" "))
  .map((_,1))
  .reduceByKey(_ + _)

//执行 Action 操作
wordCounts.collect()
```

3.8.2　Spark 执行模型

上述代码段的执行分为两个阶段进行。

1. 逻辑计划阶段

在此阶段，Spark 通过构建一个 DAG 图来跟踪驱动程序中 RDD 的转换，并根据 RDD 之间的依赖维护一个称为 Lineage Graph 的依赖关系网，通常称为"血统图"，可以将它看作 RDD 的逻辑执行计划。可以使用 toDebugString 来查看 Lineage 图，代码如下：

```
wordCounts.toDebugString
```

可以看到输出的这个 RDD 的递归依赖关系，如图 3-22 所示。

```
(2) ShuffledRDD[104] at reduceByKey at <console>:36 []
 +-(2) MapPartitionsRDD[103] at map at <console>:35 []
    |  MapPartitionsRDD[102] at flatMap at <console>:34 []
    |  ParallelCollectionRDD[101] at parallelize at <console>:31 []
```

图 3-22　Spark RDD 调试信息

存在两类基本的依赖：窄依赖和宽依赖。窄依赖可以进一步被分为 one-to-one 依赖和 range 依赖。range 依赖只用于 union 转换，它们在单个的依赖中合并多个父 RDD。one-to-one 依赖被用在所有其他不要求 shuffle 的情况下。

（1）窄依赖：每个父 RDD 的一个分区最多被子 RDD 的一个分区所使用，即 RDD 之间是一对一的关系。在窄依赖的情况下，如果下一个 RDD 被执行时，某个分区执行失败

（数据丢失），则只需重新执行父 RDD 的对应分区便可以进行数恢复。例如 map()、filter()、union()等算子都会产生窄依赖。

（2）宽依赖：是指一个父 RDD 的分区会被子 RDD 的多个分区所使用，即 RDD 之间是一对多的关系。当遇到宽依赖操作时，数据会产生 shuffle。在宽依赖情况下，如果下一个 RDD 被执行时，某个分区执行失败（数据丢失），则需要将父 RDD 的所有分区全部重新执行才能进行数据恢复。例如 groupByKey()、reduceByKey()、sortByKey()等操作都会产生宽依赖。

RDD 依赖关系如图 3-23 所示。

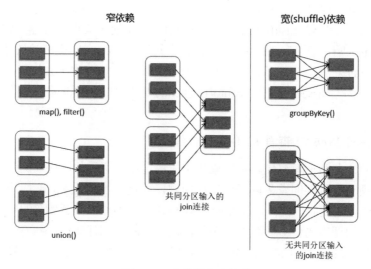

图 3-23　窄依赖和宽依赖

根据这两种依赖关系，相应地 RDD 中的转换操作可以进一步分为两种类型：

（1）窄转换。这些转换操作可以作为一批任务提交执行，并且不需要在分区之间移动数据（shuffle），例如 map()、filter()等。

（2）宽转换。这些转换操作都需要对数据进行 shuffle，因此，这些操作需要分阶段执行，后一阶段的执行依赖于前一阶段的执行结果，例如 reduceByKey()、sortByKey()等。

2．物理计划阶段

在这个阶段，一旦在 RDD 上执行一个 Action 操作，SparkContext 就会相应地产生一个 Job。例如，在单词计数程序中执行 collect()方法时，代码如下：

```
wordCounts.collect()
```

Spark 会将 DAG 图提交给 DAGScheduler，DAGScheduler 将 DAG 图划分为两个阶段（map 阶段和 reduce 阶段），称为 Stages。每个阶段由基于输入数据分区的任务（称为 Task）组成。例如，Spark 在对 Job 中的所有操作划分 Stages 时，一般会按照倒序进行，依据 RDD

之间的依赖关系（宽依赖或窄依赖）进行划分，即从 Action 开始，当遇到窄依赖类型的操作时，划分到同一个执行阶段；当遇到宽依赖操作时，划分一个新的执行阶段，并且新的阶段为之前阶段的父阶段，之前的阶段称作子阶段，然后以此类推递归执行。子阶段需要等待所有的父阶段执行完之后才可以执行，这时各个 Stage 之间根据依赖关系构成了一个大粒度的 DAG。DAGScheduler 将操作符连接在一起形成管道以优化图执行，这种优化是提高 Spark 性能的关键。

在进入下一个阶段（宽转换）之前，DAGScheduler 将检查是否有任何要 shuffle 的分区数据，以及它所依赖的父操作结果是否有任何缺失。如果有缺失，则它通过 DAG 来重新执行这部分操作，这使它具有容错能力。

这些 Stages 被传递给 TaskScheduler。TaskScheduler 通过集群管理器（YARN、Mesos、Spark Standalone）启动任务。TaskScheduler 并不知道 Stages 之间的依赖关系。

TaskScheduler 将任务分配给工作节点上的 Executors（执行器）去执行。每个任务都被分配给 Executor 的 CoarseGrainedExecutorBackend。它从 Namenode 获取块信息，执行计算并返回结果。

3. 通过 Spark Web UI 查看 RDD 执行过程

Spark Web UI 有助于理解代码执行流和完成特定任务所需的时间。通过执行过程可视化，有助于发现在执行期间发生的任何潜在问题，并进一步优化 Spark 应用程序。

当 Spark 作业完成后，就可以在 Spark Web UI 中看到作业的详细信息，例如阶段的数量、作业执行期间调度的任务的数量。例如，在前面的单词计数程序执行完毕后，打开浏览器，访问地址"http://主机:4040/jobs/"，可以看到作业的执行信息，如图 3-24 所示。

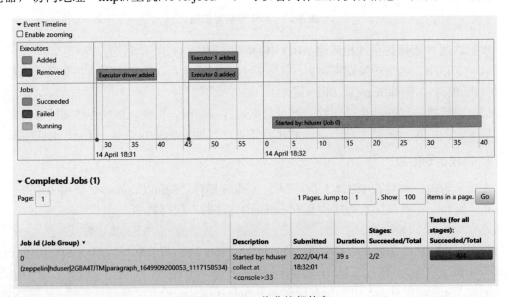

图 3-24　Spark 作业执行信息

还可以看到每个阶段的执行时间，如图 3-25 所示。

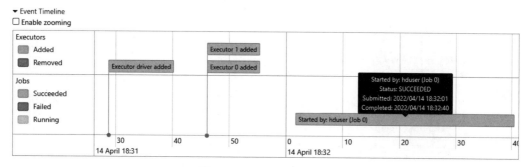

图 3-25　DAG 可视化（1）

单击已完成的作业（在 Completed Jobs 选项下），可以查看 DAG 可视化，例如，作为它的一部分的不同的宽变换和窄变换，如图 3-26 所示。

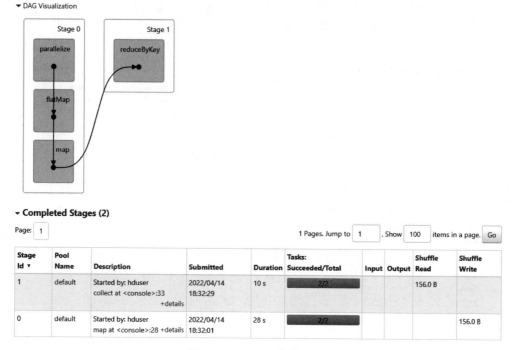

图 3-26　DAG 可视化（2）

在单击作业的某个特定阶段时，它将显示数据块位于何处、数据大小、使用的执行程序、使用的内存和完成特定任务所需的时间、发生的 shuffle 次数等详细信息，如图 3-27 所示。

此外，可以单击 Executors 选项卡查看使用的 Executor，如图 3-28 所示。

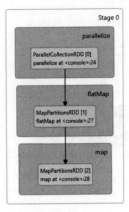

▶ Show Additional Metrics
▼ Event Timeline
☐ Enable zooming

图 3-27 某个特定阶段的信息

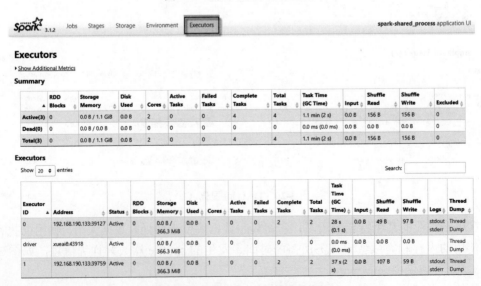

图 3-28 查看 Executor 执行器信息

3.9 Spark 资源管理

Spark 应用程序的资源被作为 Executors（JVM 进程）和 CPU（任务槽，Task Slots）进程调度，然后将内存分配给它们。当前正在运行的集群的集群管理器和 Spark 调度程序会授予用于执行 Spark 作业的资源。

集群管理器启动 Driver 所请求的 Executor 进程，并且当运行在集群部署模式下时启动 Driver 进程本身。集群管理器还可以重新启动和停止它已经启动的进程，并且可以设置 Executor 进程可以使用的最大 CPU 数量。

一旦应用程序的 Driver 和 Executors 开始运行，Spark 调度程序就会直接与它们通信，并决定哪些 Executors 将运行哪些 Tasks，这称为"作业调度"，它会影响集群中的 CPU 资源使用情况。间接地，它也会影响内存使用，因为在单个 JVM 中运行的任务越多，使用的堆内存就越多，然而，内存并不是像 CPU 那样在任务级别上直接管理的。Spark 管理由集群管理器分配的 JVM 堆内存，并根据需要可将其分成几部分。

在集群中运行的每个 Spark 应用程序都会被分配一组专用的 Executors。如果在单个集群中运行多个 Spark 应用程序（或其他类型的应用程序），它们就会争夺集群的资源。

因此，存在以下两个层次的 Spark 资源调度。

（1）集群资源调度：为不同的 Spark 应用程序的 Executors 分配资源。

（2）Spark 资源调度：用于在单个应用程序内调度 CPU 和内存资源。

3.9.1 CPU 资源分配策略

Spark 分配 CPU 资源的方式有两种：FIFO（先进先出）调度和 Fair Scheduling（公平调度）。可通过 Spark 参数 spark.scheduler.mode 设置调度器模式，它有两个可能的值：FAIR 和 FIFO。

1. FIFO 调度策略

这属于"先到先服务"调度策略。请求资源的第 1 个作业占用了所有必需的（和可用的）Executor Task Slots（假设每个 Job 作业只包含一个 Stage 阶段），如图 3-29 所示。

FIFO 调度是默认的调度器模式，如果应用程序是一个只运行一个作业的单用户应用程序，则它的效果最好。

2. FAIR 调度策略

这属于"公平"调度策略，它以循

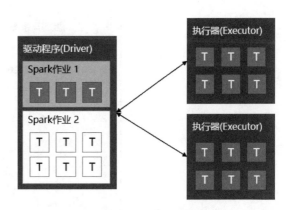

图 3-29　FIFO 调度策略

环的方式在竞争的 Spark 作业中均匀地分配可用资源（Executor 线程）。对于同时运行多个作业的多用户应用程序来讲，这是一个更好的选择，如图 3-30 所示。

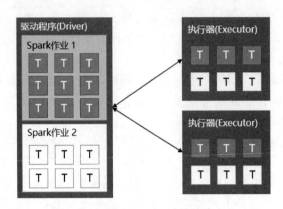

图 3-30　FAIR 调度策略

3.9.2　Spark 内存管理

使用 spark.executor.memory 参数可设置要分配给 Executors 的内存数量，大小可以使用 g 和 m 作为后缀。默认的 Executor 内存大小是 1GB。

Spark 保留了部分内存，用于存储缓存的数据和临时 shuffle 数据。Spark 为这些缓存数据和 shuffle 数据设置堆（Heap），使用的参数包括 spark.storage.memoryFraction（默认为 0.6）和 spark.shuffle.memoryFraction（默认为 0.2）。因为堆的这些部分可以在 Spark 能够测量和限制它们之前增长，所以必须设置两个额外的安全参数：spark.storage.safetyFraction（默认为 0.9）和 spark.shuffle.safetyFraction（默认为 0.8）。安全系数将按指定的数量降低内存比例，如图 3-31 所示。

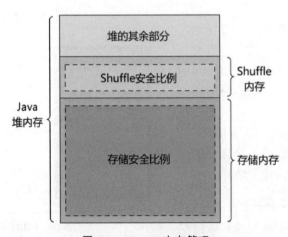

图 3-31　Spark 内存管理

默认存储的堆的实际部分是 0.6×0.9（安全系数乘以存储内存系数），等于 54%。类似地，用于 shuffle 数据的堆的一部分是 0.2×0.8（安全系数乘以 shuffle 内存系数），它等于 16%，然后，剩下 30%的堆预留给其他 Java 对象和运行任务所需的资源（然而，应该只期望 20%）。

可使用 spark.driver.memory 设置 driver 内存。当使用 spark-shell 和 spark-submit 脚本（在集群和客户端部署模式中）启动应用程序时，这个参数就应用了。

如果从另一个应用程序（客户端模式）以编程方式启动，则可使用-Xmx Java 选项设置包含进程的 Java 堆的大小上限。

1. Spark on YARN 模式下 Executors 的内存组成

当以 Spark on YARN 模式在 YARN 上运行 Spark 程序时，Spark 的 Executors 的内存布局如图 3-32 所示。

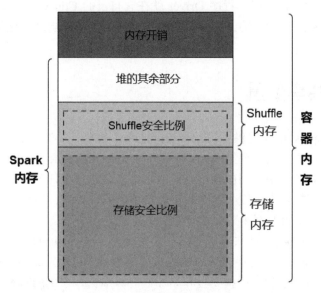

图 3-32　Spark on YARN 模式下 Executors 的内存布局

如果设置 spark.executor.memoryOverhead 的值不够高，则会导致难以诊断的问题。确保指定至少 1024MB。

当应用程序在 YARN 集群模式下运行时，spark.yarn.driver.memoryOverhead 决定了驱动程序 Driver 的 Container 的内存开销。spark.yarn.am.memoryOverhead 决定了在客户端模式下 Application Master 的内存开销。此外，以下这些 YARN 参数（在 yarn-site.xml 文件中指定）会影响内存分配。

（1）yarn.scheduler.maximum-allocation-mb：确定 YARN Containers 的内存上限限制。资源管理器不允许分配更大的内存。默认值为 8192MB。

（2）yarn.scheduler.minimum-allocation-mb：确定资源管理器可以分配的最小内存数量。资源管理器只以这个参数的倍数来分配内存。默认值为 1024 MB。

（3）yarn.nodemanager.resource.memory-mb：确定 YARN 在一个节点上可以使用的内存的最大数量。默认值为 8192 MB。

yarn.nodemanager.resource.memory-mb 应该设置为节点上可用的内存数量减去操作系统所需的内存。yarn.scheduler.maximum-allocation-mb 应该设置为相同的值。因为 YARN 将把所有的分配请求都四舍五入到 yarn.scheduler.minimum-allocation-mb 的倍数，该参数应该设置为一个小到不要浪费不必要的内存的值（例如，256MB）。

2. 以编程方式配置 Executor 资源

当在用 spark.executor.cores、spark.executor.instances 和 spark.executor.memory 以编程方式创建 SparkContext 上下文对象时，也可以指定 Executor 资源，但是，如果应用程序在 YARN 集群模式下运行，则驱动程序 Driver 就在它自己的 Container 中运行，与 Executor 的 Containers 并行启动运行，因此这些参数将不会产生任何影响。在这种情况下，最好在 spark-defaults.conf 文件中或在命令行上设置这些参数。

3.10 使用共享变量

除了 RDD，Spark 中还提供了另一种数据抽象"共享变量"。共享变量可以在并行操作中使用。默认情况下，当传递给 Spark 操作（如 map()或 reduce()）的函数在远程集群节点上执行时，它会将函数中使用的每个变量的副本发送给每个任务，这些变量被复制到每台机器上，而对远程机器上的变量的更新不会传播回驱动程序，但是有时需要在任务之间共享同一个变量，或者在任务和驱动程序之间共享同一个变量。

Spark 支持两种类型的共享变量：广播变量（Broadcast Variable）和累加器（Accumulator）。广播变量可在所有节点的内存中缓存一个值，累加器是只"增加"到其中的变量，例如计数器和求和。广播变量和累加器能够维护一个全局状态，或者在 Spark 程序中的任务和分区之间共享数据。

3.10.1 广播变量

默认情况下，在驱动程序中创建的变量，被序列化并由 Executor 所需的任务一起发送，但是一个驱动程序可以在几个作业中重用相同的变量，并且一些任务可能会被发送到同一个 Executor，作为同一作业的一部分，因此，一个大变量可能会被串行化并在网络上传输超过必要的次数。在这些情况下，最好使用广播变量。

当运行一个定义和使用了广播变量的 Spark RDD 作业时，Spark 按以下过程执行：

（1）Spark 将作业分解为具有分布式 shuffling 的 Stages，并在 Stage 中执行操作。

（2）后续的 Stage 也被分解为多个任务（Tasks）。

（3）Spark 自动广播每个阶段中任务所需的公共数据（可重用）。

（4）被广播的数据会以序列化的格式进行缓存，在执行每个任务之前反序列化。

这意味着，只有当跨多个阶段的任务需要相同的数据，或者以反序列化的形式缓存数据非常重要时，显式地创建广播变量才有用。

使用广播变量，Spark 在每台机器上保持一个缓存的只读变量，而不是将其副本与任务一起发送。Spark 会尝试使用高效的广播算法来分发广播变量，而且只传输一次。

广播变量是用 SparkContext.broadcast(value)方法创建的，它返回一个 Broadcast 类型的对象。值可以是任何可序列化的对象，然后，Executors 可以使用 Broadcast.value()方法进行读取。广播变量可以从整个集群中共享和访问，但它们不能被 Executors 修改。

例如，创建了一个广播变量并在 map()转换操作中读取该广播变量的值用于计算，代码如下：

```scala
//第 3 章/broadcastDemo.scala

val broads = sc.broadcast(3)              //创建广播变量，变量可以是任意类型

val lists = List(1,2,3,4,5)               //创建一个测试的 List
val listRDD = sc.parallelize(lists)       //构造一个 RDD

val results = listRDD.map(x => x * broads.value)  //map 转换，读取广播变量

println("结果是：")
results.collect.foreach(println)          //遍历结果
```

执行上面的代码，输出内容如下：

```
结果是：
3
6
9
12
15
```

广播变量的值可以是任意可序列化的数据。例如，将一个数组作为广播变量，代码如下：

```scala
//第 3 章/broadcastDemo2.scala

val broadcastVar = sc.broadcast(Array(1, 2, 3))   //创建广播变量

val lists = List(1,2,3,4,5)                       //创建一个测试的 List
val listRDD = sc.parallelize(lists)               //构造一个 RDD

//flatMap 转换，读取广播变量，参与计算
val results = listRDD.flatMap(x => broadcastVar.value.map(_*x))

println("结果是：")
results.collect.foreach(println)                  //遍历结果
```

执行以上代码，输出内容如下：

```
结果是:
1
2
3
2
4
6
3
6
9
4
8
12
5
10
15
```

需要注意，广播变量不会通过 sc.broadcast(var)调用后发送给执行程序，而是在它们第1次被使用时被发送给执行程序。

在一个 Spark 程序中，可以有多个广播变量，代码如下：

```scala
//第3章/broadcastDemo3.scala

val states = Map(("豫","河南省"),("粤","广东省"),("鲁","山东省"))
val countries = Map(("CHINA","中国"),("USA","美国"))

//创建广播变量
val broadcastStates = spark.sparkContext.broadcast(states)
val broadcastCountries = spark.sparkContext.broadcast(countries)

//构造RDD
val data = Seq(("张三","CHINA","豫"),
    ("李四","CHINA","粤"),
    ("王老五","CHINA","豫"),
    ("赵小六","CHINA","鲁")
)

val rdd = spark.sparkContext.parallelize(data)

//使用广播变量
val rdd2 = rdd.map(f=>{
    val country = f._2
    val state = f._3
    val fullCountry = broadcastCountries.value.get(country).get
    val fullState = broadcastStates.value.get(state).get
    (f._1,fullCountry,fullState)
})

//输出结果
println(rdd2.collect().mkString("\n"))
```

执行以上代码，输出的结果如下：

(张三,中国,河南省)
(李四,中国,广东省)
(王老五,中国,河南省)
(赵小六,中国,山东省)

当不再需要广播变量时，可以调用 destroy()方法销毁它。所有关于它的信息都将被删除（从 Executors 和驱动程序），并且该变量将不可用。如果试图在调用 destroy()之后访问它，则将抛出一个异常。

另一种方法是调用 unpersist()方法，它只从 Executors 的缓存中删除变量值。如果尝试在 unpersist()之后使用它，则它将再次被发送到 Executors。

广播哈希连接（Broadcast Hash Join）指的是，当有两个 RDD 执行 join 连接时，先将其中一个较小的 RDD 广播到每个工作节点，然后，它与较大的 RDD 的每个分区进行 map 端连接。例如，将一个较大的 rddA 的每个分区和一个较小的 rddB 中相关的值进行 join 连接，如图 3-33 所示。

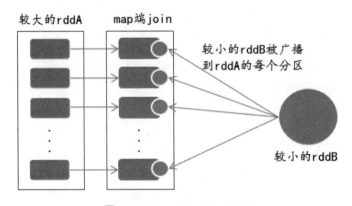

图 3-33　Broadcast Hash Join

如果一个 RDD 可以装入内存，或者经过处理后可以使其装入内存，则进行 Broadcast Hash Join 总是有益的，因为它不需要 shuffle。

Spark Core 模块中没有 Broadcast Hash Join 的实现，不过可以手动实现 Broadcast Hash Join 的一个版本，方法是将较小的 RDD 收集到驱动程序中（通过调用 collect()方法）作为一个 map，然后广播结果，并使用 mapPartitions()方法来组合元素。

在下面的示例中手工实现了 Broadcast Hash Join，用于连接较大和较小的 RDD。在这个示例中不考虑那些 key 不在两个 RDD 中出现的元素，代码如下：

```
//第3章/broadcastHashJoinDemo.scala

//构造一个较大的RDD
val bigData = Array(("a",1),("b",2),("c",3),("a",4),("b",5),("c",6))
```

```
val bigRDD = spark.sparkContext.parallelize(bigData, 2)

//构造一个较小的RDD
val smallData = Array(("a",11),("b",12))
val smallRDD = spark.sparkContext.parallelize(smallData, 2)

//将较小的RDD回收到driver端
val smallRDDLocal = smallRDD.collectAsMap()

//创建广播变量
bigRDD.sparkContext.broadcast(smallRDDLocal)

//执行mapPartitions转换
val joinedRDD = bigRDD.mapPartitions(iter => {
    iter.flatMap{
       case (k, v1) =>
          smallRDDLocal.get(k) match {
             case None => Seq.empty[(String, (Int, Int))]
             case Some(v2) => Seq((k, (v1, v2)))
          }
    }
}, preservesPartitioning = true)

//输出结果
joinedRDD.collect().foreach(println)
```

执行上面的代码,输出的结果如下:

```
(a,(1,11))
(b,(2,12))
(a,(4,11))
(b,(5,12))
```

在上例的基础上,实现一个Broadcast Hash Join通用函数,可用于连接较大的RDD和较小的RDD。它的行为反映了Spark中默认的join操作,代码如下:

```
//第3章/broadcastHashJoinDemo2.scala

import org.apache.spark.rdd._
import scala.reflect.ClassTag

//定义一个通用的Broadcast Hash Join函数
def manualBroadCastHashJoin[K : Ordering : ClassTag,
                            V1 : ClassTag,
                            V2 : ClassTag](
                               bigRDD : RDD[(K, V1)],
                               smallRDD : RDD[(K, V2)]): RDD[(K,(V1, V2))] = {
   val smallRDDLocal: scala.collection.Map[K, V2] = smallRDD.collectAsMap()
   bigRDD.sparkContext.broadcast(smallRDDLocal)
   val joinedRDD = bigRDD.mapPartitions(iter => {
      iter.flatMap{
         case (k,v1 ) =>
```

```
                smallRDDLocal.get(k) match {
                    case None => Seq.empty[(K, (V1, V2))]
                    case Some(v2) => Seq((k, (v1, v2)))
                }
            }
        }, preservesPartitioning = true)
        joinedRDD
}
//end:coreBroadCast[]
```

接下来构造一大一小两个 RDD，应用上面定义的函数实现 Broadcast Hash Join 连接，代码如下：

```
//第 3 章/broadcastHashJoinDemo.scala

//构造一个较大的 RDD
val bigData = Array(("a",1),("b",2),("c",3),("a",4),("b",5),("c",6))
val bigRDD = spark.sparkContext.parallelize(bigData, 2)

//构造一个较小的 RDD
val smallData = Array(("a",11),("b",12))
val smallRDD = spark.sparkContext.parallelize(smallData, 2)

//执行 Broadcast Hash Join
val joinedRDD = manualBroadCastHashJoin[String, Int, Int](bigRDD, smallRDD)

//输出结果
joinedRDD.collect().foreach(println)
```

执行上面的代码，输出的结果如下：

```
(a,(1,11))
(b,(2,12))
(a,(4,11))
(b,(5,12))
```

有时并不是所有较小的 RDD 都能装入内存，但是有些 key 在大 RDD 中有较大比例，所以希望只广播最常见的 key。在这种情况下，可以在大型 RDD 上使用 countByKeyApprox() 来大致了解哪些 key 适宜广播，然后，只针对这些 key 过滤出较小的 RDD，并在 HashMap 中本地收集结果。使用 sc.broadcast() 方法可以广播 HashMap，以便每个工作节点只有一个副本，并对 HashMap 手动执行连接，然后，使用相同的 HashMap 可以筛选大型 RDD，使其不包含大量重复的键，并执行标准连接，将其与手工连接的结果联合起来。这种方法非常复杂，但可以处理用其他方法无法处理的高度倾斜的数据。

3.10.2 累加器

累加器（Accumulators）用来把 Executor 端的变量信息聚合到 Driver 端。在 Driver 程序中定义的变量，在 Executor 端的每个任务（Task）都会得到这个变量的一份新的副本，每个任务更新这些副本的值后，都会传回 Driver 端进行合并。

累加器是跨 Executors 之间共享的分布式只写共享变量，每个 Executor 之间是不可见的，只有 Driver 端可以对其进行读操作。可以使用它们实现 Spark 作业中的全局求和与计数。

可以创建命名或未命名的累加器。一个指定的累加器将显示在 Web UI 中，用于修改该累加器的阶段。Spark 在 Tasks 表中显示由任务修改的每个累加器的值。在 UI 中跟踪累加器对于理解运行阶段的进度非常有用（注意：Python 中还不支持这一点），如图 3-34 所示。

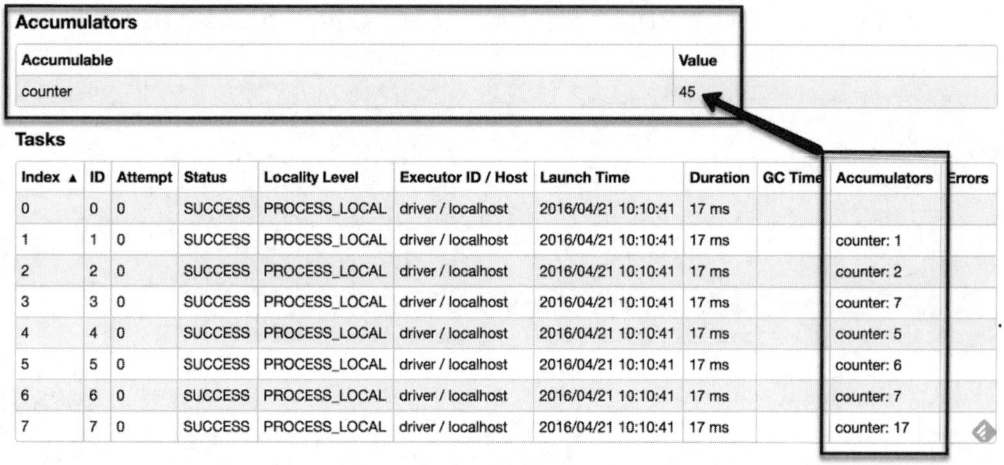

Accumulators

Accumulable	Value
counter	45

Tasks

Index ▲	ID	Attempt	Status	Locality Level	Executor ID / Host	Launch Time	Duration	GC Time	Accumulators	Errors
0	0	0	SUCCESS	PROCESS_LOCAL	driver / localhost	2016/04/21 10:10:41	17 ms			
1	1	0	SUCCESS	PROCESS_LOCAL	driver / localhost	2016/04/21 10:10:41	17 ms		counter: 1	
2	2	0	SUCCESS	PROCESS_LOCAL	driver / localhost	2016/04/21 10:10:41	17 ms		counter: 2	
3	3	0	SUCCESS	PROCESS_LOCAL	driver / localhost	2016/04/21 10:10:41	17 ms		counter: 7	
4	4	0	SUCCESS	PROCESS_LOCAL	driver / localhost	2016/04/21 10:10:41	17 ms		counter: 5	
5	5	0	SUCCESS	PROCESS_LOCAL	driver / localhost	2016/04/21 10:10:41	17 ms		counter: 6	
6	6	0	SUCCESS	PROCESS_LOCAL	driver / localhost	2016/04/21 10:10:41	17 ms		counter: 7	
7	7	0	SUCCESS	PROCESS_LOCAL	driver / localhost	2016/04/21 10:10:41	17 ms		counter: 17	

图 3-34　在 Web UI 中显示的累加器

可以通过调用 SparkContext.longAccumulator()或 SparkContext.doubleAccumulator()来创建数值累加器，以分别累积 Long 或 Double 类型的值，然后，可以在任务中调用累加器的 add()方法来修改值，但是，无法在任务中读取它的值。只有 Driver 端驱动程序可以通过调用累加器的 value()方法读取累加器的值。

例如，在下面的示例中，定义了一个累加器 acc，初始值为 0，然后在 executors 上执行时通过调用 acc.add()方法累加值，代码如下：

```
//第 3 章/accumulatorDemo1.scala

//创建一个累加器
val acc = sc.longAccumulator("My Accumulator")
println("累加器的初始值：" + acc.value)      //0

//构造一个 RDD
val list = sc.parallelize(Array(1, 2, 3, 4), 2)

//在 executors 上执行
list.foreach(x => acc.add(x))

//在 driver 端读累加器的值
println("现在累加器的值：" + acc.value)      //10
```

执行以上代码，输出的结果如下：

```
累加器的初始值：0
现在累加器的值：10
```

在 Spark 的执行模型中，只有当计算被触发（例如，执行一个 Action 操作）时，Spark 才会添加累加器。

在下面这个示例中，要求删除指定数组中既能被 2 整除也能被 3 整除的数，并统计删除了多少个数。这里使用累加器来统计删除了多少个数，代码如下：

```scala
//第3章/accumulatorDemo2.scala

//定义计数器
val acc = spark.sparkContext.longAccumulator("my_counter")

//构造RDD
val rdd1 = spark.sparkContext.parallelize(1 to 12)

//执行转换
val rdd2 = rdd1.filter(number => {
  if(number%2 ==0 && number%3==0){
    acc.add(1)
    false
  }else{
    true
  }
})

rdd2.collect().foreach(println)

//在driver端才能读取累加器的值
println(s"删除了符合条件的元素有${acc.value}个")
```

执行以上代码，输出的结果如下：

```
1
2
3
4
5
7
8
9
10
11
删除了符合条件的元素有两个
```

虽然前面的代码使用了内置的对 Long 类型累加器进行支持，但是用户也可以通过子类化 AccumulatorV2[Input Type, ValueType]来创建自己的累加器类型。AccumulatorV2[Input Type, ValueType]抽象类有以下几个必须重写的方法。

（1）reset()：用于将累加器重置为 0。

(2) add(): 用于向累加器添加另一个值。

(3) merge(): 用于将另一个相同类型的累加器合并到这个累加器中。必须重写的其他方法包含在 API 文档中。

例如，在下面的示例中，自定义了一个累加器，它继承自 AccumulatorV2，可以累加类型 MyPoint，代码如下：

```scala
//第 3 章/customerAccumulator.scala

//自定义类
class MyPoint(var x: Int, var y: Int) extends Serializable{
  def reset(): Unit = {
    x = 0
    y = 0
  }

  def add(p:MyPoint): MyPoint = {
    x = x + p.x
    y = y + p.y
    return this
  }
}

//自定义累加器
import org.apache.spark.util.AccumulatorV2

class PointAccumulatorV2 extends AccumulatorV2[MyPoint, MyPoint]{
    private val point:MyPoint = new MyPoint(0,0)

    def reset(): Unit = {
       point.reset()
    }

    def add(v: MyPoint): Unit = {
       point.add(v)
    }
    def value():MyPoint = {
      return point
    }
    def isZero(): Boolean = {
       return (point.x == 0 && point.y == 0)
    }
    def copy():AccumulatorV2[MyPoint, MyPoint] = {
       return new PointAccumulatorV2()
    }
    def merge(other:AccumulatorV2[MyPoint, MyPoint]) = {
       point.add(other.value)
    }
}
```

在使用累加器之前，需要先对它进行注册，代码如下：

```
//创建一个累加器实例
val accum = new PointAccumulatorV2

//注册累加器
sc.register(accum, "myAccum")

//构造一个RDD
var rdd = sc.parallelize(1 to 10)

//在executor中执行时修改累加器的值
var res = rdd.map(x => accum.add(new MyPoint(x,x)))

//执行Action操作才会真正生成累加器
println(res.count)                  //10

//在driver端查看累加器中的值
println("x: " + accum.value.x)      //165
println("y: " + accum.value.y)      //165
```

执行以上代码，输出的结果如下：

```
10
x: 165
y: 165
```

注意：上面的代码在 spark-shell 中执行一切正常，但是如果在 Zeppelin Notebook 中执行，则会抛出 java.io.IOException: java.lang.ClassNotFoundException: PointAccumulatorV2 异常信息。

3.11 Spark RDD 编程案例

本节通过几个不同类型的应用场景的编程案例，进一步深入理解 Spark RDD。

3.11.1 合并小文件

在使用 Hadoop 时，经常会遇到小文件问题。当系统中有大量小文件时，读写这些小文件会遇到严重的性能问题。

在使用 Spark 时，使用 SparkContext 的 wholeTextFiles()方法和 colleasc()方法，可以实现对小文件的合并，其中 wholeTextFiles()的方法签名如下：

```
def wholeTextFiles(path: String, minPartitions: Int = defaultMinPartitions):
RDD[(String, String)]
```

该方法的第1个参数 path 代表输入文件目录，它可以是以逗号分隔的输入文件列表；第2个参数是建议的最小分区数。返回的 RDD 包含以文件路径和对应的文件内容组成的元组元素。

该方法从 HDFS、本地文件系统（所有的节点都可访问）或任何 Hadoop 支持的文件系统 URI 读取一个文本文件目录。每个读取为单行记录，以键-值对的形式返回，其中 key 是每个文件的路径，value 是每个文件的内容。这些文本文件必须以 UTF-8 编码。

例如，在 HDFS 上有以下这些文件：

```
hdfs://a-hdfs-path/part-00000
hdfs://a-hdfs-path/part-00001
...
hdfs://a-hdfs-path/part-nnnnn
```

可将这些文件内容加载到 RDD 中，代码如下：

```
val rdd = sparkContext.wholeTextFiles("hdfs://a-hdfs-path")
```

返回的 RDD 将包含以下内容：

```
(a-hdfs-path/part-00000, its content)
(a-hdfs-path/part-00001, its content)
...
(a-hdfs-path/part-nnnnn, its content)
```

下面通过一个示例来掌握此用法。首先在本地文件系统准备 3 个文本文件：file1.txt、file2.txt 和 file3.txt。

file1.txt 文件的内容如下：

```
good good study
```

file2.txt 文件的内容如下：

```
day day up
```

file3.txt 文件的内容如下：

```
to be or not to be,this is a question.
```

将上述 3 个文本文件上传到 HDFS 的 /data/spark/inputs/ 目录下（如果 HDFS 上没有该目录，先自行创建），命令如下：

```
$ hdfs dfs -put file1.txt /data/spark/inputs/
$ hdfs dfs -put file2.txt /data/spark/inputs/
$ hdfs dfs -put file3.txt /data/spark/inputs/
```

最后编写代码，合并这 3 个小文件，代码如下：

```
//第 3 章/mergeFiles.scala

//加载数据源，构造 RDD
val textFiles = spark.sparkContext.wholeTextFiles("/data/spark/inputs/*")

//查看内容
textFiles.collect.foreach(println)
```

执行以上代码，输出内容如下：

```
(hdfs://xueai8:8020/data/spark/inputs/file1.txt,good good study)
(hdfs://xueai8:8020/data/spark/inputs/file2.txt,day day up)
(hdfs://xueai8:8020/data/spark/inputs/file3.txt,to be or not to be,this is a question.)
```

取元素二元组中第 2 个字段的值(文件内容)，合并到一个文件中，然后写到 HDFS 上，代码如下：

```
textFiles
    .map(_._2)
    .coalesce(1)
    .saveAsTextFile("/data/spark/outputs")
```

这时查看 HDFS 的/data/spark/outputs 目录，可以看到只有一个文件了，如图 3-35 所示。

图 3-35　将多个小文件合并到一个文件中

最后，使用下面的 HDFS Shell 命令，将原目录删除即可，命令如下：

```
$ hdfs dfs -rm -r /data/spark/inputs
```

3.11.2　二次排序实现

所谓二次排序，指的是对于 Tuple2(key,value)类型的数据，不但按 key 排序，而且每个 key 对应的 value 也是有序的。

假设有一个文本文件 data.txt，内容如下：

```
2018,5,22
2019,1,24
2018,2,128
2019,3,56
2019,1,3
2019,2,-43
2019,4,5
2019,3,46
2018,2,64
2019,1,4
2019,1,21
2019,2,35
2019,2,0
```

其中每行文本是用逗号分隔的年、月和总数量。现在想要对这些数据排序，期望的输出结果如下：

```
2018-2  64,128
2018-5  22
2019-1  3,4,21,24
```

```
2019-2    -43,0,35
2019-3    46,56
2019-4    5
```

Spark 二次排序的解决方案如下：需要将年和月组合起来构成一个key，将第3列作为value，并使用 groupByKey 函数将同一个 key 的所有 value 全部分组到一起，然后对同一个 key 的所有 value 值列表进行排序即可。

首先，将本地的 data.txt 文件上传到 HDFS 的指定位置，命令如下：

```
$ hdfs dfs -put data.txt /data/spark/input/
```

然后，编写 Spark 程序代码，进行二次排序，代码如下：

```scala
//第3章/secondarySort.scala

//加载数据集
val inputPath = "/data/spark/input/data.txt"
val inputRDD = sc.textFile(inputPath)

//实现二次排序
val sortedRDD = inputRDD
    .map(line => {
            val arr = line.split(",")              //按逗号分隔
            val key = arr(0) + "-" + arr(1)        //组合年和月
            val value = arr(2)                     //值
            (key,value)                            //返回键-值对
    })
    .groupByKey()                    //按 key 分组
    .map(t => (t._1,
        t._2.toList.sortWith(_.toInt < _.toInt).mkString(",")))
    .sortByKey(true)                 //true:升序；false:降序

//结果输出
sortedRDD.collect.foreach(t => println(t._1 + "\t" + t._2))
```

执行以上代码，输出的结果如下：

```
2018-2    64,128
2018-5    22
2019-1    3,4,21,24
2019-2    -43,0,35
2019-3    46,56
2019-4    5
```

3.11.3 Top N 实现

推荐领域有一个著名的开放测试数据集 movielens，其中包含电影评分数据集 ratings.csv 及电影数据集 movies.csv。

其中电影数据集 movies.csv 包含3个字段：movieId、title 和 genres，分别表示电影 ID、电影名称和影片类型。查看 movies.csv 中的前 10 行数据记录，内容如下：

```
movieId,title,genres
1,Toy Story (1995),Adventure|Animation|Children|Comedy|Fantasy
2,Jumanji (1995),Adventure|Children|Fantasy
3,Grumpier Old Men (1995),Comedy|Romance
4,Waiting to Exhale (1995),Comedy|Drama|Romance
5,Father of the Bride Part II (1995),Comedy
6,Heat (1995),Action|Crime|Thriller
7,Sabrina (1995),Comedy|Romance
8,Tom and Huck (1995),Adventure|Children
9,Sudden Death (1995),Action
```

电影评分数据集 ratings.csv 包含 4 个字段：userId、movieId、rating 和 timestamp，分别表示用户 ID、电影 ID、用户对该电影的评分和时间戳。查看 ratings.csv 中的前 10 行数据记录，内容如下：

```
userId,movieId,rating,timestamp
1,1,4.0,964982703
1,3,4.0,964981247
1,6,4.0,964982224
1,47,5.0,964983815
1,50,5.0,964982931
1,70,3.0,964982400
1,101,5.0,964980868
1,110,4.0,964982176
1,151,5.0,964984041
```

假设现在有一个需求，要求统计评分最高的 10 部电影，以[电影名, 评分]的形式输出。那么使用 Spark RDD 该怎么实现？

这是一个典型的 Top N 问题，对 ratings.csv 抽取 movieId 和 rating 两个字段，按 rating 评分值字段倒序取前 10 条（也就是评分最高的 10 部电影）。另外，因为输出时需要输出电影名称而不是电影的 ID（普通观众并不知道电影 ID 代表哪部电影），因此需要对两个数据集执行 join 连接操作，以获取评分最高的 10 部电影的名称。

实现过程和代码如下。

（1）加载电影数据集文件数据，构造 RDD，代码如下：

```scala
//第 3 章/topN.scala

//加载电影数据集，构造 RDD
val movies = "file:///home/hduser/data/spark/movielens/movies.csv"
val moviesRDD = sc.textFile(movies)

//电影数据集中数据总记录数量
println("电影数据集的总数据量: " + moviesRDD.count)

//查看前两条数据
moviesRDD.take(2).foreach(println)
```

执行以上代码，输出内容如下：

```
电影数据集的总数据量: 9743
```

```
movieId,title,genres
1,Toy Story (1995),Adventure|Animation|Children|Comedy|Fantasy
```

可以看出,该文件共包含 9743 行记录,其中第 1 行是标题行,实际包含电影记录数量为 9742 条。

(2) 加载评分数据集文件数据,构造 RDD,代码如下:

```
//加载评分数据集,构造 RDD
val ratings = "file://home/hduser/data/spark/movielens/ratings.csv"
val ratingsRDD = sc.textFile(ratings)

//评分数据集中数据总记录数量
println("评分数据集的总数据量: " + ratingsRDD.count)

//查看前两条数据
ratingsRDD.take(2).foreach(println)
```

执行以上代码,输出内容如下:

```
评分数据集的总数据量: 100837
userId,movieId,rating,timestamp
1,1,4.0,964982703
```

可以看出,该文件共包含 100 837 行记录,其中第 1 行是标题行,实际包含评分记录数量为 100 836 条。

(3) 对数据进行初步处理,过滤掉标题行,将 movieId 和 rating 两个字段抽取出来,组成新 RDD 的元素,代码如下:

```
//数据转换操作
import org.apache.spark.rdd._

val rdd1 = ratingsRDD.
    filter(line => !line.startsWith("userId")).    //去掉标题行
    map(line => {
        val fileds = line.split(",")                //拆分一行记录
        (fileds(1).trim.toInt, fileds(2).trim.toDouble) //(movieId, score)
    })

println("去掉标题行后的数据量: " + rdd1.count)
rdd1.take(5).foreach(println)
```

执行以上代码,输出的结果如下:

```
去掉标题行后的数据量: 100836
(1,4.0)
(3,4.0)
(6,4.0)
(47,5.0)
(50,5.0)
```

(4) 因为需要求每部电影的平均评分,所以要以 movieId 为 key 进行分组聚合计算,同时统计每部电影的总评分和打分的人数,这里选择 aggregateByKey 进行转换操作(算子),代码如下:

```
//零值设置
val zeroVal = (0.0, 0)   //(评分和, 评分量)

//在单个分区上聚合数据的函数
def seqOp = (acc: (Double,Int), e: Double) =>
  (acc._1+e, acc._2+1)     //(totalScore,totalCnt)

//合并来自各个分区的聚合结果的函数
def combOp = (acc1: (Double,Int), acc2: (Double,Int)) =>
  (acc1._1 + acc2._1, acc1._2 + acc2._2)

//计算
val rdd2 = rdd1
  .aggregateByKey(zeroVal)(seqOp, combOp)   //返回结果类似于(3586,(4.0,1))
  //计算每部电影的平均评分, 注意, 以平均分为 key
  .map(t => (t._2._1/t._2._2, t._1))
  .sortByKey(false)            //按平均评分倒序
  .map(t => (t._2, t._1))   //顺序反转, movieId 在前, 平均评分在后

//取 Top 10 记录
val Top10 = rdd2.take(10)

//输出 Top 10
Top10.foreach(println)
```

执行上面的代码, 输出的结果如下（因为只输出 Top 10, 平均评分都是 5.0, 所以可以尝试取 Top 500, 这样就会看到不同的平均评分了）:

```
(26169,5.0)
(69211,5.0)
(172589,5.0)
(136355,5.0)
(143511,5.0)
(7071,5.0)
(70451,5.0)
(25887,5.0)
(184245,5.0)
(86237,5.0)
```

可以看出, 经过这一步处理后, 返回的是一个元组, 元组元素为"(movieId,平均得分)", 但是这个结果对用户来讲并不友好, 因为它只显示了电影的 ID。下一步需显示对应的电影名称及其平均得分。

（5）第（4）步在计算结果中显示的是电影的 ID, 这对用户是非常不友好的。为此, 需要结合 moviesRDD 提取电影名称（title）。先对 moviesRDD 进行预处理, 提取"(movieId, title)"的唯一值, 代码如下:

```
//先对电影数据集进行处理
val uniqueMovies = moviesRDD
  .filter(line => !line.startsWith("movieId"))     //去掉标题行
  .map(line=>{
```

```
            val fileds = line.split(",")      //拆分一行记录
            (fileds(0).toInt,fileds(1))       //返回(movieId, title)
        })
        .reduceByKey((m1, m2) => m1)          //去重
uniqueMovies.count                            //9742
```

执行以上代码，输出的结果如下：

```
(1,Toy Story (1995))
(2,Jumanji (1995))
(3,Grumpier Old Men (1995))
(4,Waiting to Exhale (1995))
(5,Father of the Bride Part II (1995))
```

（6）接下来对 Top 10 与 uniqueMovies 这两个 RDD 执行 join 连接操作，获取平均评分前 10 名的电影名称和平均评分，代码如下：

```
val Top 10Movie = spark.sparkContext.parallelize(Top 10)
    //连接后：(movieId, (avgScore, title))
    .join(uniqueMovies)                       //例如(143511,(5.0,Human (2015)))
    //转换为(title, avgScore)
    .map(t => (t._2._2, t._2._1))             //例如 Human (2015),5.0

//输出最终结果
Top 10Movie
    .collect()
    .foreach(t => println(t.productIterator.mkString(",")))
```

执行以上代码，输出平均评分最高的 10 部电影如下：

```
Branded to Kill (Koroshi no rakuin) (1967),5.0
Boy Eats Girl (2005),5.0
Winter in Prostokvashino (1984),5.0
Big Top Scooby-Doo! (2012),5.0
Human (2015),5.0
"Woman Under the Influence,5.0
Max Manus (2008),5.0
Tales of Manhattan (1942),5.0
De platte jungle (1978),5.0
Connections (1978),5.0
```

3.11.4 酒店数据预处理

现有某平台旅游酒店用户评分和评论数据集，内容如下：

```
SEQ,酒店,国家,省份,城市,商圈,星级,业务部门,房间数,图片数,评分,评论数
aba_2066,马尔康嘉绒大酒店,中国,四川,阿坝,,四星级/高档,OTA,85,,4.143799782,108
aba_2069,阿坝马尔康县澜峰大酒店,中国,四川,阿坝,,低星,115,,3.977069009,129
aba_2094,阿坝鑫鸿大酒店,中国,四川,阿坝,四姑娘山,二星及其他,低星,,,,,
aba_2096,九寨沟管理局荷叶迎宾馆,中国,四川,阿坝,九寨沟沟口,二星及其他,低星,
    49,,3.972340107,394
aba_2097,九寨沟风景名胜区管理局贵宾楼饭店,中国,四川,阿坝,九寨沟沟口,三星级/舒适,低星,
    50,,4.12789011,585
```

```
aba_2098,九寨沟九鑫山庄,中国,四川,阿坝,九寨沟沟口,四星级/高档,OTA,
    60,,4.04046011,161
aba_2102,九寨沟冈拉美朵酒店,中国,四川,阿坝,九寨沟沟口,四星级/高档,OTA,
    198,,3.471659899,12
aba_2109,若尔盖大藏酒店古格王朝店,中国,四川,阿坝,西部旅游牧场,五星级/豪华,
    OTA,94,,3.263220072,62
aba_2111,若尔盖大藏酒店圣地店,中国,四川,阿坝,西部旅游牧场,四星级/高档,OTA,
    188,,3.921580076,119
aba_2117,九寨沟保利新九寨宾馆,中国,四川,阿坝,漳扎镇,五星级/豪华,OTA,329,,
    4.353809834,269
aba_2134,九寨沟名人酒店,中国,四川,阿坝,漳扎镇,三星级/舒适,低星,292,,
    4.539999962,57
aba_2150,九寨沟仁智度假酒店,中国,四川,阿坝,九寨沟沟口,三星级/舒适,低星,137,,
    3.782749891,173
aba_2152,九寨沟药泉山庄,中国,四川,阿坝,黄龙机场、川主寺,四星级/高档,OTA,128,,
    3.821099997,310
aba_2156,松潘黄龙寺华龙山庄,中国,四川,阿坝,黄龙风景区,四星级/高档,OTA,154,,
    3.315500021,107
aba_2213,阿坝山之旅背包客栈,中国,四川,阿坝,四姑娘山,二星及其他,客栈,20,,,13
aba_2217,阿坝若尔盖大酒店,中国,四川,阿坝,,二星及其他,低星,71,,4.267769814,2
aba_2233,九寨沟云天海大酒店,中国,四川,阿坝,九寨沟沟口,三星级/舒适,低星,100,,
    2.783930063,
aba_2243,阿坝九旅假日酒店,中国,四川,阿坝,九寨沟沟口,四星级/高档,OTA,140,,
    3.00515008,49
aba_2248,九寨沟川主寺岷江源大酒店,中国,四川,阿坝,黄龙机场、川主寺,,低星,228,,
    3.211859941,109
```

现需要对该数据集进行预处理,要求将数据集中任意关键字段为空的条目剔除,并通过打印语句输出删除条目数,然后需要将关键字段定义为{星级}。

1. 分析

这是某年全国职业院校大数据技术竞赛的题目。这里包含了两层含义,其一是过滤掉"星级"字段缺失的记录;其二是使用累加器统计被过滤掉的记录数量(因为 RDD 是分区计算的,所以需要全局的累加器来统计)。

2. 实现代码

这个任务可以使用 IntelliJ IDEA 编写 Spark 应用程序来处理,实现代码如下:

```scala
//第3章/accDemo.scala
import org.apache.spark.sql.SparkSession

object AccDemo {
  def main(args: Array[String]): Unit = {
    val spark = SparkSession.builder()
            .master("local[*]")
            .appName("accumulator")
            .getOrCreate()
```

```
        //定义计数器
        val starCounter = spark.sparkContext.longAccumulator("star_counter")

        //加载外部数据源，构造 RDD
        val rdd1 = spark.sparkContext.textFile("/input/hotel/sample2.csv")

        //过滤掉标题行
        val rdd2 = rdd1.filter(line => !line.startsWith("SEQ"))

        //将关键字段有缺失值的记录删除，即将字段{星级、评论数、评分}中任意字段为空的数据
        //删除，并打印输出删除条目数，使用计数器统计
        val rdd3 = rdd2.map(_.split(",", -1))
          .filter(arr => {
            //如果"星级"字段为空
            if(arr(6)==null || arr(6).trim.isEmpty){
              starCounter.add(1)      //全局计数器 + 1
              false
            }else{
              true
            }
        })

        //显示
        println(s"过滤前记录数为${rdd2.count}，过滤后记录数为${rdd3.count}")
        println(s"删除的星级字段缺失的记录数是：${starCounter.value}")
    }
}
```

执行以上代码，输出的结果如下：

```
过滤前记录数为 19，过滤后记录数为 17
删除的星级字段缺失的记录数是：2
```

如果把关键字段定义为{星级、评分、评论数}这 3 个字段，并分别统计每个字段的缺失值，则代码如下：

```
//第 3 章/accDemo2.scala

import org.apache.spark.sql.SparkSession

object AccDemo2 {
    def main(args: Array[String]): Unit = {
        val spark = SparkSession.builder()
            .master("local[*]")
            .appName("accumulator")
            .getOrCreate()

        //定义 3 个计数器
        val starCounter = spark.sparkContext.longAccumulator("star_counter")
        val scoreCounter = spark.sparkContext.longAccumulator("score_counter")
        val commCounter = spark.sparkContext.longAccumulator("comm_counter")
```

```
//加载外部数据源,构造RDD
val rdd1 = spark.sparkContext.textFile("input/hotel/sample2.csv")

//过滤掉标题行
val rdd2 = rdd1.filter(line => !line.startsWith("SEQ"))

//将关键字段有缺失值的记录删除
//即将字段{星级、评论数、评分}中任意字段为空的数据删除
//并打印输出删除条目数,使用计数器统计
val rdd3 = rdd2.map(_.split(",",-1))
  .filter(arr => {
    var flag = true           //定义标志变量
    //如果"星级"字段为空
    if(arr(6)==null || arr(6).trim.isEmpty){
      starCounter.add(1)      //全局计数器 + 1
      flag = false
    }

    if(arr(10)==null || arr(10).trim.isEmpty){
      scoreCounter.add(1)     //全局计数器 + 1
      flag = false
    }

    if(arr(11)==null || arr(11).trim.isEmpty){
      commCounter.add(1)      //全局计数器 + 1
      flag = false
    }

    flag
  })

  //显示
  println(s"过滤前记录数为${rdd2.count},过滤后记录数为${rdd3.count}")
  println(s"删除的星级字段缺失的记录数是: ${starCounter.value}")
  println(s"删除的评分字段缺失的记录数是: ${scoreCounter.value}")
  println(s"删除的星级字段缺失的记录数是: ${commCounter.value}")
 }
}
```

执行以上代码,输出的结果如下:

过滤前记录数为19,过滤后记录数为14
删除的星级字段缺失的记录数是:2
删除的评分字段缺失的记录数是:2
删除的星级字段缺失的记录数是:2

第 4 章 Spark SQL

CHAPTER 4

Spark SQL 是 Spark 用于处理结构化和半结构化数据的接口，允许使用关系操作符表示分布式内存计算。结构化数据指的是任何有模式的数据，如 JSON、Hive 表、Parquet。模式意味着每条记录拥有一组已知的字段。半结构化数据是指模式和数据之间没有分离。

通过 Spark SQL，用户可以使用熟悉的 SQL 或 DataFrame API 查询 Spark 程序中的结构化数据。DataFrame 和 SQL 提供了一种通用的方式访问各种数据源，包括 Hive、Avro、Parquet、ORC、JSON 和 JDBC。甚至可以跨这些数据源连接数据。

与 Spark RDD API 不同，Spark SQL 提供的接口为 Spark 提供了有关数据结构和正在执行的计算的更多信息。在内部，Spark SQL 使用这些额外的信息来执行额外的优化。Spark SQL 包括一个基于成本的优化器、柱状存储和代码生成，以提高查询的速度。

Spark 无疑是 Apache 软件基金会能够构想出的最成功的项目之一，并且引入了 Spark SQL，将关系处理与 Spark 的函数式编程 API 集成在一起，因此，Spark SQL 被设计用来集成关系型处理和函数式编程的功能，这样复杂的逻辑就可以在分布式计算设置中实现、优化和扩展。Spark SQL 模块的构成及在整个 Spark 体系结构中的位置如图 4-1 所示。

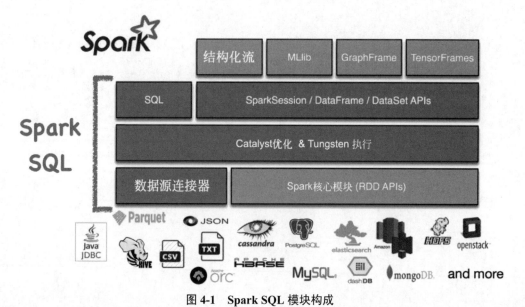

图 4-1 Spark SQL 模块构成

4.1 Spark SQL 数据抽象

尽管 Spark 提供了一个函数式编程 API 来操作分布式的数据集合，但使用 RDD 进行编码有些复杂和混乱，而且有时执行很慢，而 Spark SQL 提供了使用结构化和半结构化数据的 3 个主要功能：

（1）它提供了由 Python、Java 和 Scala 语言支持的 DataFrame/DataSet 抽象，以简化使用结构化数据集的工作。DataFrame/DataSet 类似于关系数据库中的表。

（2）它可以读写各种结构化格式的数据（如 JSON、Hive 表、Parquet）。

（3）它允许在 Spark 程序的内部使用 SQL 查询数据，以及通过标准数据库连接器（JDBC/ODBC）连接到 Spark SQL 的外部工具（如 Tableau 等商业智能工具）中使用 SQL 查询数据。

Spark SQL 提供了一个统一的接口，使用专门的 DataFrameReader 和 DataFrameWriter 对象在分布式存储系统（如 Cassandra 或 HDFS（Hive、Parquet、JSON））中访问数据。Spark SQL 允许对 Hadoop HDFS 或与 Hadoop 兼容的文件系统（如 S3）中的大量数据执行类似 SQL 的查询。它可以访问来自不同数据源（文件或表）的数据。

Spark SQL 的主要数据抽象是 DataSet，它表示结构化数据（具有已知模式的记录）。这种结构化数据表示 DataSet 支持使用存储在 JVM 堆外的托管对象中的压缩柱状格式的紧凑二进制表示。它可以通过减少内存使用和 GC 来加快计算速度。

DataSet 是一个带有 Transformation 和 Action 的结构化查询执行管道的编程接口（就像在旧的 Spark Core 中的 RDD API 一样）。在内部，结构化查询是（逻辑和物理）关系运算符和表达式的 Catalyst 树。

DataSet 的一个特定实现 DataSet[Row]，被称为 DataFrame。Spark DataFrame 是一个分布式的记录集合，被组织成命名的列，如图 4-2 所示。

DataFrame 可以从各种各样的数据源构建，例如结构化数据文件、Hive 中的表、外部数据库或现有的 RDD。

DataFrame API 支持 Scala、Java、Python 和 R 等多种语言。在 Scala 和 Java 中，DataFrame 由 Row 构成的 DataSet 表示，通常写为 DataSet[Row]。在 Scala API 中，DataFrame 只是 DataSet[Row]的一种类型别名，而在 Java API 中，用户需要使用 DataSet[Row]来表示 DataFrame。

Spark SQL 的核心是 Catalyst 优化器，它利用 Scala 的高级特性（例如模式匹配）来提供可扩展的查询优化器。开发人员编写的基于 DataFrame 的高级代码被转换为 Catalyst 表达式，然后通过这个执行和优化管道转换为低级 Java 字节码。

DataFrame 在 RDD 的基础上提高了性能，并且提供了如下两个强大的功能：

（1）自定义内存管理（又名 Project Tungsten）。

数据以二进制格式存储在堆外内存中（Off-Heap Memory）。这节省了大量的内存空间。此外,也不涉及垃圾收集开销。通过提前了解数据的模式并有效地以二进制格式存储,还可以避免昂贵的 Java 序列化。

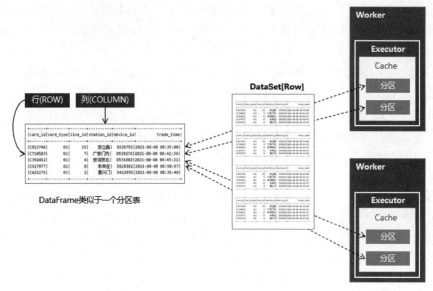

图 4-2　Spark DataFrame 类似一个分区表

（2）优化的执行计划（又名 Catalyst 优化器）。

使用 Spark Catalyst 优化器创建用于执行的查询计划。在经过一些步骤准备好优化的执行计划之后,最终的执行仅在 RDD 内部进行,但这对用户是完全隐藏的。

注意：Spark 2.0 版本对 API 进行了统一,并扩展了 SQL 功能,包括对子查询的支持。在 Spark 2.0 中,DataFrame API 已与 DataSet API 合并,从而统一了跨 Spark 库的数据处理功能。DataFrame、DataSet 和 SQL 查询共享相同的执行和优化管道；因此,使用这些结构中的任何一个(或使用任何受支持的编程 API)都不会影响性能。

4.2　Spark SQL 架构组成

Spark SQL 模块为 Spark 机器学习应用程序、流应用程序、图应用程序和许多其他类型的应用程序体系结构提供了基础,其组件如图 4-3 所示。

Spark SQL 包括以下几个组件。

（1）Spark SQL DataFrame：RDD 有一些缺点。首先,没有处理结构化数据的相关准备,也

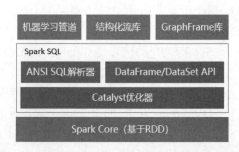

图 4-3　基于 Spark SQL 的组件

没有用于处理结构化数据的优化引擎。其次，基于属性，开发人员必须优化每个 RDD。Spark DataFrame 是一个分布式的数据集合，按顺序排列成指定的列。Spark DataFrame 与关系型数据库中的表非常相似。

（2）Spark SQL DataSet：DataSet 是在 Spark 1.6 版本中引入的接口。这个接口综合了 RDD 的优点及 Spark SQL 优化执行引擎的优点。为了实现 JVM 对象和表格表示之间的转换，使用了编码器的概念。使用 JVM 对象，可以接收数据集，并且必须使用 map()、filter() 等函数转换来修改它们。DataSet API 在 Scala 语言和 Java 语言中都可用，但在 Python 语言中不支持。

（3）Spark Catalyst 优化器：Catalyst 优化器是 Spark SQL 中使用的优化器，所有由 Spark SQL 和 DataFrame DSL 编写的查询都是通过这个工具进行优化的。这个优化器比 RDD 更好，因此系统的性能可以得到提高。

4.3　Spark SQL 编程模型

所有的 Spark SQL 应用程序都以特定的步骤来工作，这些工作步骤如图 4-4 所示。

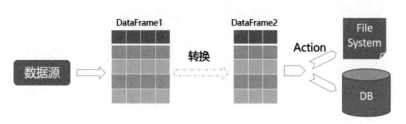

图 4-4　Spark SQL 程序执行步骤

也就是说，每个 Spark SQL 应用程序都由相同的基本部分组成：
（1）从数据源加载数据，构造 DataFrame/DataSet。
（2）对 DataFrame/DataSet 执行转换（Transformation）操作。
（3）将最终的 DataFrame/DataSet 存储到指定位置。

根据这个步骤，下面实现一个完整的 Spark SQL 应用程序。在这个示例中，从一个 JSON 格式的文件中加载人员信息，并进行统计。建议按以下步骤执行。
（1）首先，启动 IntelliJ IDEA，创建一个新的 Spark Maven 项目。
（2）准备源数据文件。

Spark 安装目录中自带了一个 people.json 文件，位于 examples/src/main/resources/ 子目录下。其内容如下：

```
{"name":"Michael"}
{"name":"Andy", "age":30}
{"name":"Justin", "age":19}
```

将 people.json 文件复制到项目的 resources 目录下。

(3) 在项目的 src 目录上右击，根据提示选择相应的项，创建一个 object 类型的 Scala 源文件，并编辑代码如下：

```scala
//第 4 章/src/Demo01.scala

import org.apache.spark.sql.SparkSession

/**
 * Spark SQL 编程模型
 * 实现：加载 JSON 文件中的人员信息，并进行计算。
 */
object Demo01 {
  def main(args: Array[String]): Unit = {

    //1. 创建 SparkSession
    val spark = SparkSession
      .builder()
      .master("local[*]")
      .appName("Spark SQL basic example")
      .getOrCreate()

    //用于隐式转换，如将 RDD 转换为 DataFrame
    import spark.implicits._

    //2. 加载数据源，构造 DataFrame
    val input = "./src/main/resources/people.json"
    val df = spark.read.json(input)

    //3. 执行转换操作
    // 找出年龄超过 21 岁的人
    val resultDF = df.where($"age" > 21)
    resultDF.show()    //显示 DataFrame 数据

    //4. 将结果保存到 CSV 文件中
    val output = "tmp/people-output"
    resultDF.write.format("csv").save(output)
  }
}
```

(4) 执行以上代码，在控制台中可以看到程序执行的输出结果，如图 4-5 所示。

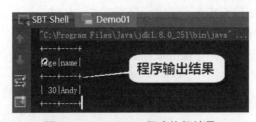

图 4-5 Spark SQL 程序执行结果

（5）打开项目中新生成的 tmp/people-output 目录，查看存储结果的 CSV 文件，如图 4-6 所示。

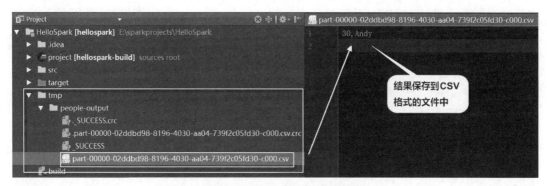

图 4-6 Spark SQL 程序执行结果存储文件

在本例中，使用了开发工具所在的本地文件系统。读者可自行修改代码，切换到使用 HDFS 文件系统，执行过程和结果是一样的。

1. 读数据格式

在 Spark SQL 中，所有读取数据的 API 都遵循以下调用格式：

```
DataFrameReader
  .format(...)
  .option("key", "value")
  .schema(...)
  .load()
```

例如，将一个 CSV 文件内容读到 DataFrame 中，代码如下：

```
spark.read
    .format("csv")
    .option("mode", "FAILFAST")              //读取模式
    .option("inferSchema", "true")           //是否自动推断 Schema
    .option("path", "path/to/file(s)")       //文件路径
    .schema(someSchema)                      //使用预定义的 Schema
    .load()
```

其中读取模式 mode 有 3 种可选项，见表 4-1。

表 4-1 3 种读取模式

读 取 模 式	描　　述
permissive	当遇损坏的记录时，将其所有字段设置为 null，并将所有损坏的记录放在名为 _corruption t_record 的字符串列中
dropMalformed	删除模式不正确的行
failFast	遇到格式不正确的数据时立即失败

2. 写数据格式

在 Spark SQL 中，所有写数据的 API 都遵循以下调用格式：

```
DataFrameWriter
.format(...)
.option(...)
.partitionBy(...)
.bucketBy(...)
.sortBy(...)
.save()
```

例如

```
dataframe.write
    .format("csv")                                  //输出格式
    .option("mode", "OVERWRITE")                    //写模式
    .option("dateFormat", "yyyy-MM-dd")             //日期格式
    .option("path", "path/to/file(s)")              //写入路径
    .save()
```

其中写数据模式 mode 有 4 种可选项，如表 4-2 所示。

表 4-2 4 种写数据模式

写 模 式	描 述
SaveMode.ErrorIfExists	如果给定的路径已经存在文件，则抛出异常。这是默认的模式
SaveMode.Append	以追加的方式写入数据
SaveMode.Overwrite	以覆盖的方式写入数据
SaveMode.ignore	如果给定的路径已经存在文件，则不进行任何操作

4.4 程序入口 SparkSession

在 4.3 节的示例程序中，在代码的开始，首先创建了一个 SparkSession 的实例 spark。为什么要首先创建 spark 实例呢？

在 Spark 2.0 中，SparkSession 表示在 Spark 中操作数据的统一入口点。要创建一个基本的 SparkSession，需要使用 SparkSession.builder()，代码如下：

```
import org.apache.spark.sql.SparkSession

val spark = SparkSession
  .builder()
  .appName("Spark SQL basic example")
  .getOrCreate()

//用于隐式转换，如将 RDDs 转换为 DataFrames
import spark.implicits._
```

在 Spark 程序中，使用构建器设计模式实例化 SparkSession 对象，然而，在 REPL 环境中(在 spark-shell 会话中），SparkSession 会被自动创建，并通过名为 spark 的实例对象提供给用户使用。也就是说，如果使用 spark-shell 交互式地执行 Spark 程序语句，则可以直接调用 spark 实例对象，如图 4-7 所示。

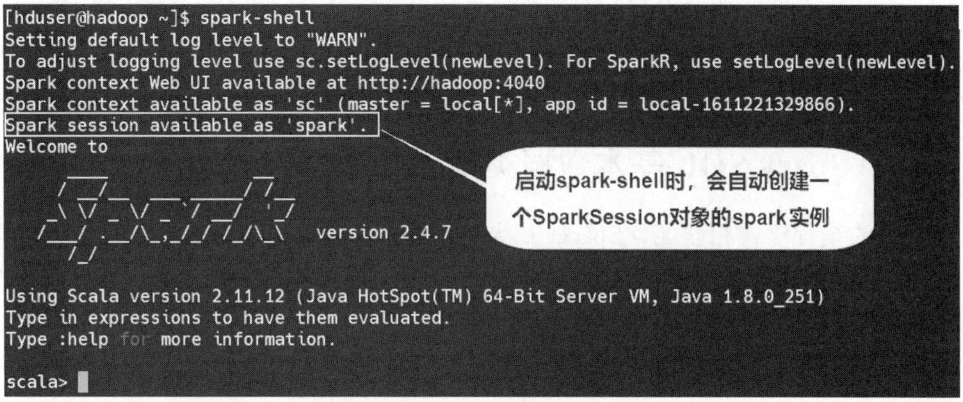

图 4-7　spark-shell 启动时会自动创建 SparkSession 的实例 spark

SparkSession 对象可以用来配置 Spark 的运行时配置属性。例如，Spark 和 YARN 管理的两个主要资源是 CPU 和内存。如果想为 Spark executor 设置内核数量和堆大小，则可以通过分别设置 spark.executor.cores 和 spark.executor.memory 属性实现这一点。例如，分别将 PySpark executor 运行时属性的核设置为 2 个，将内存设置为 4GB，代码如下：

```
spark.conf.set("spark.executor.cores", "2")
spark.conf.set("spark.executor.memory", "4g")
```

在 Spark SQL 中创建 SparkSession 对象的常用模板代码如下：

```
import org.apache.spark.SparkSession
import org.apache.spark.SparkContext

val conf = SparkConf()
conf.set("spark.executor.cores", "num_cores")
conf.set("spark.executor.instances", "num_executors")
conf.set("spark.locality.wait", "0")
conf.set("spark.serializer","org.apache.spark.serializer.KryoSerializer")

val spark = SparkSession
    .builder()
    .appName(application_name)
    .master("local[*]")
    .config(conf=conf)
    .getOrCreate()
```

可以使用 SparkSession 对象从各种源读取数据，例如 CSV、JSON、JDBC、Stream 等。此外，它还可以用来执行 SQL 语句、注册用户定义函数（UDF）和使用 DataSet 和 DataFrame。

4.5 Spark SQL 支持的数据类型

Spark DataFrame/DataSet 就像分布在内存中的表,具有指定的列和模式,其中每列都有特定的数据类型:整数、字符串、数组、map、real、日期、时间戳等。从用户角度来看,Spark DataFrame 就像一张具有行和列的二维表。

在使用 Spark 定义模式之前,首先来了解一下有哪些可用的泛型和结构化数据类型,然后学习如何使用模式创建 DataFrame。

4.5.1 Spark SQL 基本数据类型

与支持的编程语言相匹配,Spark 支持基本的内部数据类型。这些数据类型可以在 Spark 应用程序中声明,也可以在模式中定义。例如,在 Scala 中,用户可以定义或声明一个特定的列,类型为 String、Byte、Long 或 Map 等。

Spark 支持的基本 Scala 数据类型,除了 DecimalType,其余都是类 DataTypes 的子类型,见表 4-3。

表 4-3 Spark 支持的基本数据类型

数据类型	在 Scala 中赋的值	用来实例化的 API
ByteType	Byte	DataTypes.ByteType
ShortType	Short	DataTypes.ShortType
IntegerType	Int	DataTypes.IntegerType
LongType	Long	DataTypes.LongType
FloatType	Float	DataTypes.FloatType
DoubleType	Double	DataTypes.DoubleType
StringType	String	DataTypes.StringType
BooleanType	Boolean	DataTypes.BooleanType
DecimalType	java.math.BigDecimal	DecimalType

4.5.2 Spark SQL 复杂数据类型

对于复杂的数据分析,一般不太可能只处理简单或基本的数据类型。有时遇到的数据是复杂的,通常是结构化的或嵌套的,需要 Spark 处理这些复杂的数据类型。Spark 支持的 Scala 结构化数据类型见表 4-4。

表 4-4 Spark 支持的复杂数据类型

数据类型	在 Scala 中赋的值	用来实例化的 API
BinaryType	Array[Byte]	DataTypes.BinaryType
TimestampType	java.sql.Timestamp	DataTypes.TimestampType

续表

数据类型	在 Scala 中赋的值	用来实例化的 API
DateType	java.sql.Date	DataTypes.DateType
ArrayType	scala.collection.Seq	DataTypes.createArrayType(ElementType)
MapType	scala.collection.Map	DataTypes.createMapType(keyType,valueType)
StructType	org.apache.spark.sql.Row	StructType(ArrayType[fieldTypes])
StructField	对应这个字段类型的值类型	StructField(name, dataType,[nullable])

4.5.3 模式

Spark 中的模式（Schema）为一个 DataFrame/DataSet 定义了列名和关联数据类型。最常见的情况是，当从外部数据源读取结构化数据时，需要用到模式。预先定义模式，而不是采用"读时模式"方法，有以下 3 个好处。

（1）减轻了 Spark 推断数据类型的责任。

（2）可以防止 Spark 仅为了读取文件的大部分内容以确定模式而创建单独的作业，这对于大型数据文件来讲是昂贵和耗时的。

（3）如果数据与模式不匹配，则可以尽早发现错误。

因此，当用户想从数据源读取大文件时，最好预先定义模式。Spark 支持以下两种方式定义模式：

（1）第 1 种是通过编程方式定义 Schema。

（2）第 2 种是使用数据定义语言（Data Definition Language, DDL）字符串。

其中第 2 种定义方式要简单得多，也更容易阅读。

要以编程方式为 DataFrame 定义一个模式，假设这个 DataFrame 有 3 个命名列，如 author、title 和 pages，则可以使用 Spark DataFrame API，代码如下：

```
import org.apache.spark.sql.types._

val schema = StructType(Array(
    StructField("author", StringType, false),
    StructField("title", StringType, false),
    StructField("pages", IntegerType, false)
))
```

如果用 DDL 方式来定义同样的 Schema 则要简单得多。下面是用 DDL 定义同样的 Schema 的示例代码：

```
val schema = "author STRING, title STRING, pages INT"
```

4.5.4 列对象和行对象

在 Spark SQL 中，列是具有 public 方法的对象，由 Column 类型表示。用户可以按名称列出所有列，并且可以使用关系表达式或计算表达式对列的值执行操作。还可以在列上使用逻辑或数学表达式，代码如下：

```
import org.apache.spark.sql.functions._

blogsDF.select(col("Hits") * 2).show(2)
```

DataFrame 中的 Column 对象不能单独存在；每列都是记录（record）中一行的一部分，所有的行一起构成一个 DataFrame。正如将在本章后面看到的那样，DataFrame 实际上是 Scala 中的 DataSet[Row]。

Spark 中的行是一个通用的 Row 对象，包含一个或多个列。每列可以是相同的数据类型（例如，整数或字符串），也可以是不同的类型（整数、字符串、映射、数组等）。因为 Row 是 Spark 中的一个对象，是字段的一个有序集合，因此可以在 Spark 中实例化 Row，并通过从 0 开始的索引访问它的字段，代码如下：

```
import org.apache.spark.sql.Row

//创建一个 Row
val blogRow = Row(6,"读者学苑","http://www.xueai8.com", 255568,"2022-04-18",
Array("大数据", "数据分析"))

//使用索引访问单个项
println("行中索引为 1 的元素值: " + blogRow(1))
```

执行以上代码，输出内容如下：

```
行中索引为 1 的元素值：读者学苑
```

当需要快速交互和探索数据时，可以使用 Row 对象来创建 DataFrame，代码如下：

```
val rows = Seq(("张三", "北京"), ("李四", "上海"))
val authorsDF = rows.toDF("Author", "State")
authorsDF.show()
```

但在实践中，通常希望从文件中读取 DataFrame。在大多数情况下，因为数据文件非常大，所以定义并使用一个模式是创建 DataFrame 的一种更快、更有效的方法。

4.6 创建 DataFrame

作为 Spark 2.0 及以后的应用程序统一入口，开发 Spark 应用程序的第 1 步就是要创建一个 SparkSession 实例。为了访问 DataFrame API，需要将 SparkSession 作为入口点，如图 4-8 所示。

Spark DataFrame 的创建类似于 RDD 的创建。在本节中，将演示如何从各种数据源创建 DataFrame。有以下几种方式可用来创建 DataFrame：

(1) 简单创建单列和多列 DataFrame。
(2) 将已经存在的 RDD 转换为一个 DataFrame。

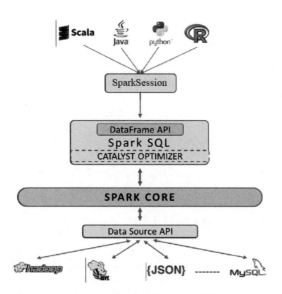

图 4-8　SparkSession 作为 Spark 应用程序的入口

(3) 运行 SQL 查询返回一个 DataFrame。
(4) 将外部数据源的数据加载到一个 DataFrame。

4.6.1　简单创建单列和多列 DataFrame

SparkSession 有一个函数叫 range()，可以很容易地创建带有列名 id 和类型 LongType 的单列 DataFrame，代码如下：

```
//第4章/createSimpleRdd01.scala

//创建单列DataFrame，默认列名是id，类型是LongType
val df0 = spark.range(5).toDF()
df0.printSchema
df0.show
```

执行以上代码，输出的结果如下：

```
root
 |-- id: long (nullable = false)

+---+
| id|
+---+
|  0|
|  1|
|  2|
|  3|
|  4|
+---+
```

还可以在创建 DataFrame 时指定列名，代码如下：

```
val df1 = spark.range(5).toDF("num")
df1.printSchema
df1.show
```

执行以上代码，输出的结果如下：

```
root
 |-- num: long (nullable = false)

+---+
|num|
+---+
|  0|
|  1|
|  2|
|  3|
|  4|
+---+
```

另外，还可以指定范围的起始（含）和结束值（不含），代码如下：

```
val df2 = spark.range(5,10).toDF("num")
df2.show
```

执行以上代码，输出的结果如下：

```
+---+
|num|
+---+
|  5|
|  6|
|  7|
|  8|
|  9|
+---+
```

另外，还可以指定步长，代码如下：

```
val df3 = spark.range(5,15,2).toDF("num")
df3.show
```

执行以上代码，输出的结果如下：

```
+---+
|num|
+---+
|  5|
|  7|
|  9|
| 11|
| 13|
+---+
```

通过将一个元组集合转换为一个 DataFram，可创建多列 DataFrame。这需要使用 SparkSession 对象的 toDF() 方法。toDF() 方法将列标签列表作为可选的参数，以指定转换后的 DataFrame 的标题行，代码如下：

```
//第4章/createSimpleRdd02.scala

//用于隐式转换，如将 RDD 转换为 DataFrame
import spark.implicits._

//元组序列
val movies = Seq(("马特·达蒙", "谍影重重:极限伯恩", 2007L),
                 ("马特·达蒙", "心灵捕手", 1997L))

//将元组转换为 DataFrame
val moviesDF = movies.toDF("演员", "电影", "年份")

//输出模式
moviesDF.printSchema

//显示
moviesDF.show
```

执行以上代码，输出的结果如下：

```
root
 |-- 演员: string (nullable = true)
 |-- 电影: string (nullable = true)
 |-- 年份: long (nullable = false)

+---------+-----------------+----+
|     演员|             电影|年份|
+---------+-----------------+----+
|马特·达蒙|谍影重重:极限伯恩|2007|
|马特·达蒙|         心灵捕手|1997|
+---------+-----------------+----+
```

通过元组来创建单列或多列 DataFrame，每个元组类似于一行。可以选择标题列；否则 Spark 会创建一些模糊的名称，例如_1、_2。列的类型推断是隐式的。例如，与上面的代码类似，但是不指定标题列，代码如下：

```
//第4章/createSimpleRdd03.scala

//用于隐式转换，如将 RDD 转换为 DataFrame
import spark.implicits._

//元组序列
val movies = Seq(("马特·达蒙", "谍影重重:极限伯恩", 2007L),
                 ("马特·达蒙", "心灵捕手", 1997L))

//将元组转换为 DataFrame
val moviesDF = movies.toDF()    //注意这里

//输出模式
moviesDF.printSchema
```

```
//显示
moviesDF.show
```

执行以上代码，输出的结果如下：

```
root
 |-- _1: string (nullable = true)
 |-- _2: string (nullable = true)
 |-- _3: long (nullable = false)

+---------+------------------+----+
|       _1|                _2|  _3|
+---------+------------------+----+
| 马特·达蒙|谍影重重：极限伯恩|2007|
| 马特·达蒙|          心灵捕手|1997|
+---------+------------------+----+
```

4.6.2 从 RDD 创建 DataFrame

从一个已经存在的 RDD 创建 DataFrame 有以下 3 种方式：
（1）使用包含 Row 数据（以元组的形式）的 RDD。
（2）使用 case class 以反射的形式。
（3）明确指定一个模式（Schema）。

从已经存在的 RDD 创建 DataFrame 有多种方式，但是这些方法都必须提供一个 Schema。要么显式地提供，要么隐式地提供。

在下面的示例代码中，显式地调用 RDD 的 toDF()函数，将 RDD 转换为 DataFrame，使用指定的列名。列的类型是从 RDD 中的数据推断出来的，代码如下：

```scala
//第 4 章/createDataFrameFromRdd01.scala

//用于隐式转换，如将 RDD 转换为 DataFrame
import spark.implicits._

//定义一个集合
val persons = List(("张三",23),("李四",18),("王老五",35))

//创建 RDD，类型为 RDD[(String, Int)]
val personRDD = spark.sparkContext.parallelize(persons)

//从 RDD 转换为 DataFrame
val personsDF = personRDD.toDF("name", "age")

//查看 Schema 和数据
personsDF.printSchema()
personsDF.show()
```

执行以上代码，输出内容如下：

```
root
 |-- name: string (nullable = true)
 |-- age: integer (nullable = false)

+------+---+
| name|age|
+------+---+
| 张三 | 23|
| 李四 | 18|
|王老五| 35|
+------+---+
```

在上面的代码中创建了一个 RDD，它包含"(String,Int)"类型的元组元素，然后调用它的 toDF()方法，将该 RDD 转换为一个 DataFrame。需要注意，toDF()方法采用的是元组列表，而不是标量元素。每个元组类似于一行。可以指定列名，否则 Spark 会自行创建一些模糊的名称，例如_1、_2。列的类型推断是隐式的。

如果不想让 Spark 对列的类型进行隐式推断（例如，隐式推断的类型不符合用户预期），则 Spark SQL 支持两种不同的方法，可将现有的 RDD 转换为具有特定 Schema 模式的 DataFrame：

（1）第 1 种方法使用反射技术来推断 RDD 的模式，该模式包含特定类型（case class）的对象。

（2）第 2 种方法是通过一个编程接口，先构造出一个模式，然后将该模式应用到现有的 RDD 中。

虽然第 2 种方法比较冗长，但是在事先不知道列类型的情况下，这种方法允许用户自行构造 DataFrame。

首先看第 1 种方法。使用反射模式，可以实现从 RDD 到 DataFrame 的隐式转换。先定义一个 case class：

```
case class Person(name:String,age:Long)
```

然后构造一个 RDD[Person]类型的 RDD，并显式地调用 toDF()方法将其转换为 DataFrame，这时会利用反射获取模式，代码如下：

```scala
//第 4 章/createDataFrameFromRdd02.scala

import spark.implicits._

//定义 case class 列表数据
val people = List(
  Person("张三",29),
  Person("李四",30),
  Person("王老五",19)
)

//构造 RDD[Person]，并通过反射转换为指定模式的 DataFrame
val peopleDF = spark.sparkContext.parallelize(people).toDF()
```

```
//查看模式和内容
peopleDF.printSchema()
peopleDF.show()
```

执行以上代码,输出内容如下:

```
root
 |-- name: string (nullable = true)
 |-- age: long (nullable = false)

+------+---+
| name|age|
+------+---+
| 张三 | 29|
| 李四 | 30|
|王老五 | 19|
+------+---+
```

如果在上面的示例代码中的集合元素不是 Person 对象,而是字符串,则需要用户自己通过转换来构造出 Person 集合来,代码如下:

```
//第4章/createDataFrameFromRdd03.scala

//创建一个 List 集合
val people = List("张三,29","李四,30","王老五,19")

//构造一个 RDD,经过转换为 RDD[Person]类型后,再转换为 DataFrame
val peopleRDD = spark.sparkContext.parallelize(people)
val peopleDF = peopleRDD
    .map(_.split(","))
    .map(x => Person(x(0),x(1).trim.toLong))
    .toDF()

//查看模式和内容
peopleDF.printSchema()
peopleDF.show()
```

执行上面的代码,输出内容如下:

```
root
 |-- name: string (nullable = true)
 |-- age: long (nullable = false)

+------+---+
| name|age|
+------+---+
| 张三 | 29|
| 李四 | 30|
|王老五 | 19|
+------+---+
```

上面这种通过反射获取模式的方式,虽然使用起来比较简单,但反射会影响程序的性能，特别是读取大型数据文件时，因此，当用户想从数据源读取大文件时，最好采用预先定义模式。在下面的示例中，先定义好了一个 Schema（模式），然后使用该模式创建一个 DataFrame，这时需要使用 SparkSession 的 createDataFrame()方法来创建，代码如下：

```scala
//第4章/createDataFrameFromRdd04.scala

import org.apache.spark.sql.Row
import org.apache.spark.sql.types._

import spark.implicits._

//构造一个RDD
val peopleRDD = spark.sparkContext.parallelize(
   Seq(
       Row("张三",30),
       Row("李四",25),
       Row("王老五",35)
   )
)

//定义字段类型
val fields = Seq(
   StructField("name", StringType, nullable = true),
   StructField("age", IntegerType, nullable = true)
)
//构造一个Schema(模式)
val schema = StructType(fields)

//从给定的RDD应用给定的Schema创建一个DataFrame
val peopleDF = spark.createDataFrame(peopleRDD, schema)

//查看DataFrame Schema
peopleDF.printSchema

//输出
peopleDF.show
```

执行上面的代码，输出内容如下：

```
root
 |-- name: string (nullable = true)
 |-- age: integer (nullable = true)

+------+---+
| name|age|
+------+---+
| 张三 | 30|
| 李四 | 25|
|王老五 | 35|
+------+---+
```

4.6.3 读取外部数据源创建 DataFrame

Spark 提供了一个接口 DataFrameReader，用来以各种格式（如 JSON、CSV、Parquet、Text、Avro、ORC 等）从众多的数据源将数据读到 DataFrame。同样，要将 DataFrame 以特定格式写回数据源，写数据时 Spark 使用 DataFrameWriter 接口。

将存储系统中的文件加载到 DataFrame，可通过 SparkSession 的 read 字段，它是 DataFrameReader 的一个实例。它有个 load()方法，可直接从配置的数据源加载数据。另外它还有 5 个快捷方法：text()、csv()、json()、orc()和 parquet()，相当于先调用 format()方法再调用 load()方法。

为了演示如何将外部数据源加载到 DataFrame，本节使用 Spark 自带的各种格式的数据源文件。这些数据源文件位于$SPARK_HOME/examples/src/main/resources/目录下。首先将它们上传到 HDFS 的/data/spark/resources/目录下。在命令行终端执行 HDFS Shell 命令，将文件上传到 HDFS，命令如下：

```
$ hdfs dfs -put ~/bigdata/spark-3.1.2/examples/src/main/resources /data/
spark/resources
```

1. 读取文本文件创建 DataFrame

文本文件是最常见的数据存储文件。Spark DataFrame API 允许开发者将文本文件的内容转换成 DataFrame。

例如，将 Spark 自带的文本文件 people.txt 读到一个 DataFrame 中，代码如下：

```
//第 4 章/loadFromText.scala

val file = "/data/spark/resources/people.txt"
val txtDF = spark.read.format("text").load(file)      //加载文本文件
//val txtDF = spark.read.text(file)                   //等价于上一句，快捷方法

txtDF.printSchema                                     //打印 Schema
txtDF.show                                            //输出
```

执行以上代码，输出内容如下：

```
root
 |-- value: string (nullable = true)

+-----------+
|      value|
+-----------+
|Michael, 29|
|   Andy, 30|
| Justin, 19|
+-----------+
```

Spark 会自动推断出模式，并相应地创建一个单列（列名为 value）的 DataFrame，因此，没有必要为文本数据定义模式。不过，当加载大数据文件时，显式地定义一个 Schema 要比让 Spark 进行推断效率更高。

2. 读取 CSV 文件创建 DataFrame

在 Spark 3.x 中，加载 CSV 文件是非常简单的。例如，要将 Spark 自带的 people.csv 文件加载到一个 DataFrame 中，代码如下：

```scala
//第 4 章/loadFromCsv.scala

//数据源文件
val file = "/data/spark/resources/people.csv"

val peopleDF = spark.read.format("csv")
    .option("sep", ";")                     //字段使用";"分隔符
    .option("inferSchema", "true")          //指定自动推断模式
    .option("samplingRatio", 0.001)         //根据抽样进行模式推断
    .option("header", "true")               //说明有标题行
    .load(file)

//或者使用快捷方法
/*
val peopleDF = spark.read
    .option("sep", ";")                     //字段使用";"分隔符
    .option("inferSchema", "true")          //指定模式自动推断
    .option("samplingRatio", 0.001)         //根据抽样进行模式推断
    .option("header", "true")               //说明有标题行
    .csv(file)
*/

peopleDF.printSchema                        //打印 Schema
peopleDF.show()                             //显示
```

执行以上代码，输出内容如下：

```
root
 |-- name: string (nullable = true)
 |-- age: string (nullable = true)
 |-- job: string (nullable = true)

+-----+---+---------+
| name|age|      job|
+-----+---+---------+
|Jorge| 30|Developer|
|  Bob| 32|Developer|
+-----+---+---------+
```

在上面的代码中，使用了模式自动推断。对于大型的数据源，指定一个 Schema 要比让 Spark 进行模式推断效率更高。修改上面的代码，明确提供一个 Schema，代码如下：

```scala
//第 4 章/loadFromCsv2.scala

import org.apache.spark.sql.types._

//数据源文件
```

```
val file = "/data/spark/resources/people.csv"

//构造 Schema
val fields = Seq(
   StructField("p_name", StringType, nullable = true),
   StructField("p_age", LongType, nullable = true),
   StructField("p_job", StringType, nullable = true)
)
val schema = StructType(fields)

//读取数据源,创建 DataFrame
val peopleDF = spark.read
    .option("sep", ";")                       //字段使用";"分隔符
    .option("header", "true")                 //说明有标题行
    .schema(schema)                           //指定使用的 Schema
    .csv(file)

peopleDF.printSchema()                        //打印 Schema
peopleDF.show()                               //显示
```

执行以上代码,输出内容如下:

```
root
 |-- p_name: string (nullable = true)
 |-- p_age: long (nullable = true)
 |-- p_job: string (nullable = true)

+------+-----+---------+
|p_name|p_age|    p_job|
+------+-----+---------+
| Jorge|   30|Developer|
|   Bob|   32|Developer|
+------+-----+---------+
```

可以看出,它返回了由行和命名列组成的 DataFrame,该 DataFrame 具有模式中指定的类型。

注意:如果一个文件是由 DataFrame 以 Parquet 格式写出存储的,则模式被保留为 Parquet 元数据的一部分。在这种情况下,随后读入 DataFrame 不需要手动提供模式。Parquet 是一种流行的柱状格式,是 Spark 的默认格式;它使用 snappy 来压缩数据。

也可以使用 CSV 格式读取 TSV 文件。所谓 TSV 文件,指的是以制表符(Tab)作为字段分隔符的文件。例如,将 Spark 自带的 people.tsv 文件加载到一个 DataFrame 中,代码如下:

```
//第 4 章/loadFromCsv3.scala

val file = "/data/spark/resources/people.tsv"

//读取数据源,创建 DataFrame
val peopleDF = spark.read
    .option("sep", "\t")                      //字段使用 Tab 分隔符
    .option("inferSchema", "true")            //指定模式自动推断
    .option("header", "true")                 //说明有标题行
```

```
      .csv(file)

peopleDF.printSchema()                    //打印 Schema
peopleDF.show()                           //显示
```

执行以上代码,输出内容如下:

```
root
 |-- name: string (nullable = true)
 |-- age: integer (nullable = true)
 |-- job: string (nullable = true)

+-----+---+---------+
| name|age|      job|
+-----+---+---------+
|Jorge| 30|Developer|
|  Bob| 32|Developer|
+-----+---+---------+
```

在使用 SparkSession 对象的 read 字段读取 CSV 文件时,可指定的 option 选项参数和值的列表很长,见表 4-5。

表 4-5 加载 CSV 文件时可指定的 option 选项

option 选项	默 认 值	说　　明	引入版本
sep	,	设置单个字符作为每个字段和值的分隔符	v2.0.0
encoding	UTF-8	根据给定的编码类型解码 CSV 文件	v2.0.0
quote	"	设置用于转义引用值的单个字符,其中分隔符可以是值的一部分。如果要关闭引号,则需要设置的不是 null,而是一个空字符串	v2.0.0
escape	\	设置用于转义已引用值中的引号的单个字符	v2.0.0
comment	空字符串	设置用于跳过以该字符开头的行的单个字符。默认情况下,它是禁用的	v2.0.0
header	false	是否使用第 1 行作为列的名称。不支持两行标题	v2.0.0
inferSchema	false	告诉 Spark 是否尝试基于列值推断列类型(从数据自动推断输入模式)。它需要额外传递一次数据	v2.0.0
ignoreLeadingWhiteSpace	false	指示是否跳过正在读取的值中的前导空格	v2.0.0
ignoreTrailingWhiteSpace	false	指示是否跳过正在读取的值中的尾部空格	v2.0.0
nullValue	空字符串	设置 null 值的字符串表示形式。从 2.0.1 版本开始,这适用于所有受支持的类型,包括字符串类型	v2.0.0

option 选项	默认值	说明	引入版本
nanValue	NaN	设置非数字"值"的字符串表示形式	v2.0.0
positiveInf	Inf	设置正无穷值的字符串表示形式	v2.0.0
negativeInf	–Inf	设置负无穷值的字符串表示形式	v2.0.0
dateFormat	yyyy-MM-dd	设置指示日期格式的字符串。自定义日期格式遵循 java.text.SimpleDateFormat 的格式。这适用于日期(date)类型	v2.0.0
timestampFormat	yyyy-MM-dd'T'HH:mm:ss.SSSXXX	设置指示时间戳格式的字符串。自定义日期格式遵循 java.text.SimpleDateFormat 的格式。适用于时间戳(timestamp)类型	v2.1.0
maxColumns	20480	定义一个记录可以有多少列的硬限制	
maxCharsPerColumn	–1	定义允许读取任何给定值的最大字符数。默认值为 1 000 000	
mode	PERMISSIVE	允许在解析期间处理损坏记录的模式。它支持以下不区分大小写的模式。 - PERMISSIVE：当遇到损坏的记录时，将其他字段设置为 null，并将格式错误的字符串放入由 columnNameOfCorruptRecord 配置的字段中。要保存损坏的记录，用户可以在用户定义的模式中设置一个名为 columnNameOfCorruptRecord 的字符串类型字段。如果一个模式没有该字段，则它将在解析期间删除损坏的记录。当解析后的 CSV tokens 长度小于模式的预期长度时，它会为额外字段设置 null。 - DROPMALFORMED：忽略整个损坏的记录。 - FAILFAST：当遇到损坏的记录时抛出异常	v2.0.0
columnNameOfCorruptRecord		允许重命名新字段，该字段具有由 PERMISSIVE 模式创建的格式错误字符串。这将覆盖 spark.sql.columnNameOfCorruptRecord。 默认值为在 spark.sql.columnNameOfCorruptRecord 中指定的值	v2.2.0
multiLine		解析一条记录，它可能跨越多行	v2.2.0
wholeFile	false	解析一条记录，它可能跨越多行	

3. 读取 JSON 文件创建 DataFrame

Spark SQL 可以自动推断 JSON DataSet 的模式，并将其加载为 DataSet[Row]。这种转换可以在 DataSet[String]或 JSON 文件上使用 SparkSession.read.json()完成。

注意：作为 JSON 文件提供的文件实际上并不是标准意义上的 JSON 文件，而是要求每行必须包含一个单独的、自包含的有效 JSON 对象。对于常规的多行 JSON 文件，将 multiLine 选项设置为 true。

读取 JSON 数据源文件时，Spark 会从 key 中自动推断模式，并相应地创建一个 DataFrame，因此，没有必要为 JSON 数据定义模式。此外，Spark 极大地简化了访问复杂 JSON 数据结构中的字段所需的查询语法。例如，将 Spark 自带的 people.json 文件内容读到一个 DataFrame 中，代码如下：

```
//第 4 章/loadFromJson.scala

//数据源文件
//JSON 数据集路径可以是单个文件，也可以是存储文件的目录
val file = "/data/spark/resources/people.json"

//读取数据源，创建 DataFrame
val df = spark.read.json(file)          //JSON 解析；列名和数据类型隐式地推断

//Schema
df.printSchema()

//显示
df.show()
```

执行以上代码，输出内容如下：

```
root
 |-- age: long (nullable = true)
 |-- name: string (nullable = true)

+----+-------+
| age|   name|
+----+-------+
|null|Michael|
|  30|   Andy|
|  19| Justin|
+----+-------+
```

当然，也可以明确指定一个 Schema，覆盖 Spark 的推断 Schema，代码如下：

```
//第 4 章/loadFromJson2.scala

import org.apache.spark.sql.types._

//数据源文件
```

```
//JSON 数据集路径可以是单个文件,也可以是存储文件的目录
val file = "/data/spark/resources/people.json"

//创建 Schema。字段名称要与 JSON 对象的 key 名称保持一致
val fields = Seq(
    StructField("name",StringType,nullable = true),
    StructField("age",IntegerType,nullable = true)
)
val schema = StructType(fields)

//读取数据源,创建 DataFrame,使用自定义的 Schema
val df = spark.read.schema(schema).json(file)

//Schema
df.printSchema()

//显示
df.show()
```

执行以上代码,输出内容如下:

```
root
 |-- name: string (nullable = true)
 |-- age: integer (nullable = true)

+-------+----+
|   name| age|
+-------+----+
|Michael|null|
|   Andy|  30|
| Justin|  19|
+-------+----+
```

如果在加载 JSON 文件时 JSON 格式解析错误,则将默认创建的 DataFrame 的相应各行各列值都设为 null。在下面的示例代码中,在 Schema 中将姓名这一列(name)对应的数据类型错误地设为 Boolean 类型,因此 SparkSession 在解析时出现解析错误,因而整个 DataFrame 中的各行各列值均为 null,代码如下:

```
//第 4 章/loadFromJson3.scala

import org.apache.spark.sql.types._

//数据源文件
//JSON 数据集路径可以是单个文件,也可以是存储文件的目录
val file = "/data/spark/resources/people.json"

//创建 Schema。字段名称要与 JSON 对象的 key 名称保持一致
val fields = Seq(
    StructField("name",BooleanType,nullable = true), //错误的类型 BooleanType
    StructField("age",IntegerType,nullable = true)
)
```

```
val schema = StructType(fields)

//读取数据源,创建 DataFrame,使用自定义的 Schema
val df = spark.read.schema(schema).json(file)

//Schema
df.printSchema()

//显示
df.show()
```

执行以上代码,并不会出现任何异常,输出的结果如下:

```
root
 |-- name: boolean (nullable = true)
 |-- age: integer (nullable = true)

+----+----+
|name| age|
+----+----+
|null|null|
|null|  30|
|null|  19|
+----+----+
```

可以看到,name 列的值全部被解析为 null,但是这样处理 JSON 格式解析错误并不是一种好的方式,因为它经常会掩盖错误事实,让用户迷惑,所以最好的方式是,如果 JSON 格式解析错误,就直接抛出异常(快速失败),而不是全都设为 null 值。例如,在将 Spark 自带的 people.json 文件读到一个 DataFrame 时,指定快速失败的处理方式,代码如下:

```
//第 4 章/loadFromJson4.scala

import org.apache.spark.sql.types._

//数据源文件
//JSON 数据集路径可以是单个文件,也可以是存储文件的目录
val file = "/data/spark/resources/people.json"

//创建 Schema。字段名称要与 JSON 对象的 key 名称保持一致
val fields = Seq(
  StructField("name",BooleanType,nullable = true),   //错误的类型
  StructField("age",IntegerType,nullable = true)
)

//读取数据源,创建 DataFrame,使用自定义的 Schema
val df = spark.read
  .option("mode","failFast")                         //指定 failFast 模式
  .schema(StructType(fields))
  .json(file)

//Schema
```

```
df.printSchema()

//显示
df.show()      //当执行一个Action操作时,Spark将抛出一个RuntimeException
```

在上面的代码中,将 mode 指定为 failFast 模式,这是用来告诉 Spark,当面对解析错误时,快速失败,而不是生成 null 值。执行以上代码,会抛出异常信息,异常信息如下:

```
root
 |-- name: boolean (nullable = true)
 |-- age: integer (nullable = true)

org.apache.spark.SparkException: Job aborted due to stage failure: Task 0
in stage 7.0 failed 4 times, most recent failure: Lost task 0.3 in stage
7.0 (TID 10) (192.168.190.133 executor 1): org.apache.spark.SparkException:
Malformed records are detected in record parsing. Parse Mode: FAILFAST. To
process malformed records as null result, try setting the option 'mode' as
'PERMISSIVE'.
    at ...
Caused by: org.apache.spark.sql.catalyst.util.BadRecordException: java.lang.
RuntimeException: Failed to parse a value for data type boolean (current
token: VALUE_STRING).
    at ...
    ... 25 more
Caused by: java.lang.RuntimeException: Failed to parse a value for data type
boolean (current token: VALUE_STRING).
    at ...
    ... 27 more
```

可以看到,在 JSON 格式解析错误时会立即抛出异常信息,从而不会对用户产生误导。

4. 读取 Parquet 文件创建 DataFrame

Apache Parquet 是一种高效的、压缩的、面向列的开源数据存储格式。它提供了多种存储优化,允许读取单独的列而非整个文件,这不仅节省了存储空间而且提升了读取效率。它是 Spark 默认的文件格式,支持非常有效的压缩和编码方案,也可用于 Hadoop 生态系统中的任何项目,可以大大提高这类应用程序的性能。

Apache Spark 提供了对读取和写入 Parquet 文件的支持,这些文件自动保存原始数据的模式。Parquet 是一种非常流行的格式,在 Spark 中有一些额外的选项可以用于读写 Parquet 文件。在编写 Parquet 文件时,出于兼容性考虑,所有列都会自动转换为 nullable。

例如,将 Spark 自带的 Parquet 文件 users.parquet 的内容读到一个 DataFrame 中,然后打印其 Schema 并输出数据,代码如下:

```
//第4章/loadFromParquet.scala

//读取Parquet文件
val parquetFile = "/data/spark/resources/users.parquet"

//Parquet是默认的格式,因此当读取时不需要指定格式
```

```
val usersDF = spark.read.load(parquetFile)

//如果想要更加明确,则可以指定parquet函数
//val usersDF = spark.read.parquet(parquetFile)

//输出模式和内容
usersDF.printSchema()
usersDF.show()
```

执行以上代码,输出的结果如下:

```
root
 |-- name: string (nullable = true)
 |-- favorite_color: string (nullable = true)
 |-- favorite_numbers: array (nullable = true)
 |    |-- element: integer (containsNull = true)

+------+--------------+----------------+
|  name|favorite_color|favorite_numbers|
+------+--------------+----------------+
|Alyssa|          null|  [3, 9, 15, 20]|
|   Ben|           red|              []|
+------+--------------+----------------+
```

5. 读取 ORC 文件创建 DataFrame

ORC(Optimized Row Columnar)是一种流行的大数据文件存储格式。ORC 文件是一种自描述的、类型感知的列存储格式数据文件,它针对大型数据的读写进行了优化,具有高性能、可压缩性强等特点,被许多顶级 Apache 产品支持,例如 Hive、Crunch、Cascading、Spark 等,是大数据中常用的文件格式。Apache Spark 提供了对读取和写入 ORC 文件的支持。

例如,将 Spark 自带的 ORC 文件 users.orc 读到一个 DataFrame 中,代码如下:

```
//第4章/loadFromOrc.scala

//数据源文件
val orcFile = "/data/spark/resources/users.orc"

//读取ORC文件,构造DataFrame
val orcDF = spark.read.format("orc").load(orcFile)
//val orcDF = spark.read.orc(orcFile)      //简洁写法

//输出模式和内容
orcDF.printSchema()
orcDF.show()
```

执行以上代码,输出的结果如下:

```
root
 |-- name: string (nullable = true)
 |-- favorite_color: string (nullable = true)
 |-- favorite_numbers: array (nullable = true)
 |    |-- element: integer (containsNull = true)
```

```
+------+---------------+----------------+
|  name|favorite_color|favorite_numbers|
+------+---------------+----------------+
|Alyssa|           null|  [3, 9, 15, 20]|
|   Ben|            red|              []|
+------+---------------+----------------+
```

从 Spark 2.3 开始，Spark 支持一个带有新的 ORC 文件格式的向量化 ORC 读取器，配置见表 4-6。

表 4-6 读取 ORC 文件时的配置

属 性 名	默认值	含 义
spark.sql.orc.impl	native	The name of ORC 实现的名称。可以是 native 或 hive。native 意味着本地 ORC 支持，它构建在 Apache ORC 1.4 之上。hive 意味着 Hive 1.2.1 中的 ORC 库
spark.sql.orc.enableVectorizedReader	true	在 native 实现中启用向量化 ORC 解码。如果值为 false，则在 native 实现中使用新的非向量化 ORC reader

当 spark.sql.orc.impl 设置为 native 及 spark.sql.orc.enableVectorizedReader 设置为 true 时，会将向量化的读取器用于本地 ORC 表，例如，使用 USING ORC 子句创建的表。对于 Hive ORC serde 表，当 spark.sql.hive.convertMetastoreOrc 也被设置为 true 时，也会使用向量化的读取器。

6. 使用 JDBC 从数据库创建 DataFrame

Spark SQL 还包括一个可以使用 JDBC 从其他关系型数据库读取数据的数据源。开发人员可以使用 JDBC 创建来自其他数据库的 DataFrame，只要确保预定数据库的 JDBC 驱动程序是可访问的（需要在 Spark 类路径中包含特定数据库的 JDBC 驱动程序）。

注意：Spark 安装程序默认为没有提供数据库驱动，所以在使用前需要将相应的数据库驱动上传到$SPARK_HOME 目录下的 jars 目录中。本书示例使用的是 MySQL 数据库，所以使用前需要将相应的 mysql-connector-java-x.x.x.jar 包上传到$SPARK_HOME/jars 目录下。

如果是在 Maven 项目中开发，则需要在 pom.xml 文件中添加 MySQL 驱动程序依赖，内容如下：

```
<dependency>
    <groupId>mysql</groupId>
    <artifactId>mysql-connector-java</artifactId>
    <version>8.0.18</version>
</dependency>
```

如果是在 SBT 项目中开发，则需要在 build.sbt 文件中添加 MySQL 驱动程序依赖，内容如下：

```
libraryDependencies += "mysql" % "mysql-connector-java" % "8.0.18"
```

注意：笔者本地安装的 MySQL 数据库的版本为 8.0.18，读者在应用时修改为自己的本地数据库版本。

在下面的示例中，通过 JDBC 读取 MySQL 数据库中的一个 people 数据表，并创建 DataFrame。为此，先在 MySQL 中执行如下脚本，创建一个名为 xueai8 的数据和一个名为 people 的数据表，并向表中插入一些样本记录，命令如下：

```
mysql> create database xueai8;
mysql> use xueai8;
mysql> create table people(id int not null primary key, name varchar(20), age int);
mysql> insert into people values(1,"张三",23),(2,"李四",18),(3,"王老五",35);
mysql> select * from people;
```

然后编写代码，从 MySQL 中将 people 表中的数据读到 DataFrame 中，代码如下：

```scala
//第4章/loadFromJdbc1.scala

//定义MySQL 8的数据库连接URL，数据库名为xueai8
val DB_URL= "jdbc:mysql://localhost:3306/xueai8?useSSL=false" +
    "&serverTimezone=Asia/Shanghai&allowPublicKeyRetrieval=true"

//读取JDBC数据源
val peopleDF = spark.read.format("jdbc")
    .option("driver", "com.mysql.cj.jdbc.Driver")   //MySQL 8数据库驱动程序类名
    .option("url", DB_URL)                          //连接url
    .option("dbtable", "people")                    //要读取的表
    .option("user", "root")                         //连接账户，需修改为自己的
    .option("password","admin")                     //连接密码，需修改为自己的
    .option("fetchsize","50")                       //每轮读取多少行
    .load()

//输出模式和内容
peopleDF.printSchema()
peopleDF.show()
```

注意：上面的连接配置是 MySQL 8 数据库的配置。如果要连接的是 MySQL 5 数据库，则连接 URL 不需要指定时区信息，并且驱动程序名称为 com.mysql.jdbc.Driver。另外，数据库连接账号和密码应修改为自己本机数据的账号和密码。

执行以上代码，输出的结果如下：

```
root
 |-- id: integer (nullable = true)
 |-- name: string (nullable = true)
 |-- age: integer (nullable = true)

+---+------+---+
| id|  name|age|
+---+------+---+
|  1|  张三| 23|
|  2|  李四| 18|
|  3|王老五| 35|
+---+------+---+
```

也可以将 JDBC 参数放到一个 Map 集合中，作为 options 的参数传入，代码如下：

```scala
//第 4 章/loadFromJdbc2.scala

//定义 MySQL 8 的数据库连接 URL，数据库名为 xueai8
val DB_URL= "jdbc:mysql://localhost:3306/xueai8?useSSL=false" +
    "&serverTimezone=Asia/Shanghai&allowPublicKeyRetrieval=true"

val jdbcMap = Map(
    "url" -> DB_URL,                             //JDBC url
    "driver" -> "com.mysql.cj.jdbc.Driver",      //驱动程序
    "dbtable" -> "people",                       //要读取的数据表
    "user" -> "root",                            //MySQL 账号
    "password" -> "admin",                       //MySQL 密码
    "fetchsize" -> "50"                          //每轮读取多少行
)

//读取 JDBC 数据源，创建 DataFrame
val peopleDF = spark.read.format("jdbc").options(jdbcMap).load()

//输出模式和内容
peopleDF.printSchema()
peopleDF.show()
```

也可以直接使用 jdbc() 快捷方法从关系型数据库中加载数据。例如，改写上面的示例，代码如下：

```scala
//第 4 章/loadFromJdbc3.scala

//定义 MySQL 8 的数据库连接 URL，数据库名为 xueai8
val DB_URL= "jdbc:mysql://localhost:3306/xueai8?useSSL=false" +
    "&serverTimezone=Asia/Shanghai&allowPublicKeyRetrieval=true"

//创建一个 Properties()对象来保存参数
import java.util.Properties
val props = new Properties()

props.put("user", "root")                        //账号
props.put("password", "admin")                   //密码

val driverClass = "com.mysql.cj.jdbc.Driver"     //MySQL 8 驱动
//val driverClass = "com.mysql.jdbc.Driver"      //MySQL 5 驱动
props.put("Driver", driverClass)
//props.setProperty("Driver", driverClass)       //等价于上一句

//使用快捷方式
val peopleDF = spark.read.jdbc(DB_URL, "people", props)

//输出模式和内容
peopleDF.printSchema()
peopleDF.show()
```

注意：根据 MySQL 数据库的版本不同（主要是 MySQL 8 有较大的变化），相应的驱动程序名和 JDBC 的连接 URL 也发生变化。

如果读写的是 MySQL 5.x：

（1）驱动程序为 com.mysql.jdbc.Driver。

（2）连接 URL 为 jdbc:mariadb://localhost:3306/db?useSSL=false。

如果读写的是 MySQL 8.x：

（1）驱动程序为 com.mysql.cj.jdbc.Driver。

（2）连接 URL 为 jdbc:mysql://localhost:3306/db?useSSL=false&serverTimezone=Asia/Shanghai&allowPublicKeyRetrieval=true。

还可以使用 query 选项指定用于将数据读入 Spark 的查询语句。指定的查询将被圆括号括起来，并在 FROM 子句中用作子查询。Spark 还将为子查询子句分配一个别名。例如，Spark 将向 JDBC 源发出如下形式的查询：

```
SELECT <columns> FROM (<user_specified_query>) spark_gen_alias
```

使用此选项时有以下限制：

（1）不允许同时指定 dbtable 和 query 选项。

（2）不允许同时指定 query 和 partitionColumn 选项。当需要指定 partitionColumn 选项时，可以使用 dbtable 选项指定子查询，分区列可以使用作为 dbtable 的一部分提供的子查询别名进行限定。

例如，通过 PySpark SQL 读取 MySQL 中的数据，使用 query 选项，代码如下：

```scala
//第4章/loadFromJdbc4.scala

//定义 MySQL 8 的数据库连接 URL，数据库名为 xueai8
val DB_URL= "jdbc:mysql://localhost:3306/xueai8?useSSL=false" +
    "&serverTimezone=Asia/Shanghai&allowPublicKeyRetrieval=true"

//定义查询语句
val query = "select name,age from people where age>20"

val jdbcMap = Map(
    "url" -> DB_URL,                              //JDBC url
    "driver" -> "com.mysql.cj.jdbc.Driver",       //驱动程序
    "query" -> query,                             //查询语句
    "user" -> "root",                             //MySQL 账号
    "password" -> "admin"                         //MySQL 密码
)

//读取 JDBC 数据源，创建 DataFrame
val peopleDF = spark.read.format("jdbc").options(jdbcMap).load()

//输出 Schema 和内容
peopleDF.printSchema()
peopleDF.show()
```

执行以上代码,输出的结果如下:

```
root
 |-- name: string (nullable = true)
 |-- age: integer (nullable = true)


+------+---+
|  name|age|
+------+---+
|    张三| 23|
|  王老五| 35|
+------+---+
```

7. 读取图像文件创建 DataFrame

随着用于图像分类和对象检测的深度学习框架的发展,Apache Spark 中对标准图像处理的需求也越来越大。图像处理和预处理有其特定的挑战,例如,图像有不同的格式(如 jpeg、png 等)、大小和颜色方案,而且没有简单的方法来测试正确性(静默失败)。图像数据源通过提供标准表示来解决这些问题,用户可以针对特定图像表示的细节进行编码和抽象。

Apache Spark 2.3 提供了 ImageSchema.readImages API,它最初是在 MMLSpark 库中开发的,在 Apache Spark 2.4 中成为一个内置的数据源,因此更容易使用。ImageDataSource 实现了一个 Spark SQL 数据源 API,用于将图像数据作为 DataFrame 加载。

要想使用图像数据源,需要添加相关的依赖。如果使用的是 SBT 项目构建和管理工具,则在 build.sbt 文件中添加的依赖如下:

```
libraryDependencies += "org.apache.spark" %% "spark-mllib" % "3.1.2"
```

如果使用的是 Maven 项目构建和管理工具,则在 pom.xml 文件中添加的依赖如下:

```xml
<dependency>
    <groupId>org.apache.spark</groupId>
    <artifactId>spark-mllib_2.12</artifactId>
    <version>3.1.2</version>
    <scope>provided</scope>
</dependency>
```

例如,将 images 目录下的所有图像读到 DataFrame 中,代码如下:

```scala
//第4章/loadFromImages.scala

//加载图像数据源
val file = "/data/spark/resources/images/*"
val imageDF = spark
    .read
    .format("image")
    .option("dropInvalid", "true")
    .load(file)

//输出
```

```
imageDF.printSchema()
imageDF
   .select("image.origin", "image.width", "image.height",
           "image.nChannels","image.mode")
   .show()
```

执行以上代码，输出的结果如下：

```
root
 |-- image: struct (nullable = true)
 |    |-- origin: string (nullable = true)
 |    |-- height: integer (nullable = true)
 |    |-- width: integer (nullable = true)
 |    |-- nChannels: integer (nullable = true)
 |    |-- mode: integer (nullable = true)
 |    |-- data: binary (nullable = true)

+--------------------+-----+------+---------+----+
|              origin|width|height|nChannels|mode|
+--------------------+-----+------+---------+----+
|hdfs://xueai8:802...| 2436|  1366|        3|  16|
|hdfs://xueai8:802...| 1280|   720|        3|  16|
|hdfs://xueai8:802...|  320|   213|        4|  24|
|hdfs://xueai8:802...| 1280|   720|        3|  16|
+--------------------+-----+------+---------+----+
```

从上面的示例可以看出，使用图像数据源可以从目录加载图像文件，加载的 DataFrame 有一个 StructType 列（image），其中包含作为 image Schema 存储的图像数据。该 image 列的 Schema 构成如下。

（1）origin：StringType（代表该头像的文件路径）。

（2）height：IntegerType（图像的高度）。

（3）width：IntegerType（图像的宽度）。

（4）nChannels：IntegerType（图像通道的数量）。

（5）mode：IntegerType（OpenCV 兼容的类型）。

（6）data：BinaryType（以 OpenCV 兼容顺序的图像字节：在大多数情况下按行排列的 BGR）。

关于 nChannels，指的是颜色通道的数量。典型值：1 为灰度图像，3 为彩色图像（例如，RGB），4 为带有 alpha 通道的彩色图像。

关于 mode，这是一个整数标志，提供如何解释数据字段的信息。它用于指定数据存储的数据类型和通道顺序。该字段的值被期望（但不是强制的）映射到 OpenCV 类型之一，OpenCV 类型定义为 1、2、3 或 4 个通道和像素值的一些数据类型，见表 4-7。

关于 data：以二进制格式存储的图像数据。图像数据表示为一个三维数组，其维度形状（高度、宽度、nChannels）和类型为 t 的数组值由 mode 字段指定。数组按行主顺序存储。

表 4-7　OpenCV 中类型到数字的映射（数据类型 × 通道数量）

类型	C1	C2	C3	C4
CV_8U	0	8	16	24
CV_8S	1	9	17	25
CV_16U	2	10	18	26
CV_16S	3	11	19	27
CV_32U	4	12	20	28
CV_32S	5	13	21	29
CV_64F	6	14	22	30

对于可加载的图像，Spark 规范明确针对当前行业中最常见的用例：二进制、int32、int64、浮点或双数据的多通道矩阵，可以轻松地放入 JVM 堆中。以下是一些限制：

（1）图像的总大小应该限制在 2GB 以下（大致）。

（2）颜色通道的含义是特定于应用程序的，不受标准的约束（与 OpenCV 标准一致）。

（3）不支持气象学、医学等领域使用的专门格式。

（4）这种格式专门用于图像，并不试图解决 Spark 中表示 n 维张量的更一般的问题。

8. 读取 Avro 创建 DataFrame

从 Apache Spark 2.4 开始，Spark SQL 为读取和写入 Apache Avro 数据提供了内置支持，新增一个基于 Databricks 的 spark-avro 模块的原生 Avro 数据源。

因为 spark-avro 模块是外部的，所以 DataFrameReader 或 DataFrameWriter 中没有 .avro API。要想使用 Avro 数据源，需要添加相关的依赖。如果使用的是 SBT 项目构建和管理工具，则在 build.sbt 文件中添加的依赖如下：

```
libraryDependencies += "org.apache.spark" %% "spark-avro" % "3.1.2"
```

如果使用的是 Maven 项目构建和管理工具，则在 pom.xml 文件中添加的依赖如下：

```xml
<dependency>
    <groupId>org.apache.spark</groupId>
    <artifactId>spark-avro_2.12</artifactId>
    <version>3.1.2</version>
</dependency>
```

如果使用 spark-submit 命令启动应用程序，则使用 --packages 选项添加 spark-avro_2.12 及其依赖，命令如下：

```
$ ./bin/spark-submit --packages org.apache.spark:spark-avro_2.12:3.1.2 ...
```

同样，如果使用 spark-shell 命令交互式执行程序代码，则可以使用 --packages 选项添加 spark-avro_2.12 及其依赖，命令如下：

```
$ ./bin/spark-shell --packages org.apache.spark:spark-avro_2.12:3.1.2 ...
```

在下面的示例中，将 Spark 自带的 user.avro 数据源文件读到 DataFrame 中。要以 Avro 格式加载/保存数据，需要将数据源的 format 选项指定为 avro，代码如下：

```
//第4章/loadFromAvro.scala

//数据源文件
val file = "/data/spark/resources/users.avro"

//读取Avro文件,创建DataFrame
val usersDF = spark.read.format("avro").load(file)

//输出
usersDF.printSchema()
usersDF.show()
```

执行以上代码,输出的结果如下:

```
root
 |-- name: string (nullable = true)
 |-- favorite_color: string (nullable = true)
 |-- favorite_numbers: array (nullable = true)
 |    |-- element: integer (containsNull = true)

+------+--------------+----------------+
|  name|favorite_color|favorite_numbers|
+------+--------------+----------------+
|Alyssa|          null|  [3, 9, 15, 20]|
|   Ben|           red|              []|
+------+--------------+----------------+
```

Avro 的数据源选项可以使用 DataFrameReader 或 DataFrameWriter 上的.option 方法设置,可使用的配置项见表 4-8。

表 4-8 Avro 数据源配置选项

属 性 名	默 认 值	含 义	应用范围
avroSchema	None	用户以 JSON 格式提供的可选 Avro 模式。记录字段的数据类型和命名在从 Avro 读取时应与 Avro 数据类型匹配,在写入 Avro 文件时应与 Spark 的内部数据类型(例如 StringType、IntegerType)匹配;否则读/写操作将失败	read write
recordName	topLevelRecord	写入结果中的顶级记录名,这在 Avro 规范中是必需的	write
recordNamespace	""	在写入结果中的记录命名空间	write
ignoreExtension	true	该选项控制在 read 中忽略没有.avro 扩展名的文件。如果启用了该选项,则加载所有文件(有和没有.avro 扩展名)	read
compression	snappy	compression 选项允许指定写入中使用的压缩编解码器。目前支持的编解码器有 uncompressed、snappy、deflate、bzip2 和 xz。如果未设置此选项,则会考虑配置 spark.sql.avro.compression.codec 配置项	write

Avro 的配置可以在 SparkSession 上使用 setConf()方法进行设置，或者使用 SQL 运行 SET key=value 命令进行设置，可设置的配置见表 4-9。

表 4-9 Avro 配置

属 性 名	默认值	含 义
spark.sql.legacy.replaceDatabricksSparkAvro.enabled	true	如果将其设置为 true，则数据源提供程序 com.databricks.spark.avro 被映射到内置但属于外部的 Avro 数据源模块，以便向后兼容
spark.sql.avro.compression.codec	snappy	用于写 Avro 文件的压缩编解码器。支持的编解码器：uncompressed、deflate、snappy、bzip2 和 xz。默认的编解码器是 snappy
spark.sql.avro.deflate.level	−1	写 Avro 文件中使用的 deflate codec 的压缩水平（级别）。有效值必须在 1～9（含 9）或−1。默认值为−1，对应于当前实现中的 6 级

Avro 包提供了两个函数 from_avro()和 to_avro()，用于在 DataFrame 中读取和写入 Avro 数据，而不仅是文件，其中，to_avro()函数以 Avro 格式将列编码为二进制，from_avro() 函数将 Avro 二进制数据解码为列。这两个函数都可将一列转换为另一列，输入/输出 SQL 数据类型可以是复杂类型或基本类型。

通常在使用 Spark SQL 以 Avro 格式读写 Kafka 中的数据时，使用这两个函数进行转换。当从 Kafka 这样的流源读取或写入数据时，使用 Avro 记录作为列非常有用。每个 Kafka 键值记录都将使用一些元数据进行扩充，例如将摄入时间戳添加到 Kafka 中，以及将偏移量添加到 Kafka 中等。在以 Avro 格式读写 Kafka 中的数据时，需要注意以下几点：

（1）如果包含数据的 value 字段在 Avro 中，则可以使用 from_avro()来提取数据、充实数据、清理数据，然后将数据下推到 Kafka，或者将数据写到文件中。

（2）to_avro()可以用来将结构转换成 Avro 记录。在将数据写入 Kafka 时，如果需要将多个列重新编码为一个列，则此方法特别有用。

这两个函数目前只能在 Scala 和 Java 中使用。

9. 读取二进制文件创建 DataFrame

从 Spark 3.0 开始，Spark 支持二进制文件数据源，它读取二进制文件，并将每个文件转换成一条记录，其中包含文件的原始内容和元数据。它生成一个 DataFrame，包含以下列和可能的分区列。

（1）path：StringType。

（2）modificationTime：TimestampType。

（3）length：LongType。

（4）content：BinaryType。

要读取整个二进制文件，需要将数据源 format 指定为 binaryFile。如果要加载具有匹配给定全局模式的路径的文件，同时保持分区发现的行为，则可以使用通用数据源选项 pathGlobFilter。例如，从输入目录中读取所有 JPG 文件，代码如下：

```scala
//第4章/loadFromBinary.scala

//二进制文件路径
val filePath = "/data/spark/resources/images/*"

//读取JPG文件
val binDf = spark
  .read
  .format("binaryFile")
  .option("pathGlobFilter", "*.jpg")
  .load(filePath)

//输出
binDf.printSchema()
binDf.show()
```

执行以上代码，输出的结果如下：

```
root
 |-- path: string (nullable = true)
 |-- modificationTime: timestamp (nullable = true)
 |-- length: long (nullable = true)
 |-- content: binary (nullable = true)

+--------------------+--------------------+------+--------------------+
|                path|    modificationTime|length|             content|
+--------------------+--------------------+------+--------------------+
|hdfs://xueai8:802...|2022-02-11 10:14:...|358076|[FF D8 FF E0 00 1...|
|hdfs://xueai8:802...|2022-02-11 10:14:...|232093|[FF D8 FF E0 00 1...|
|hdfs://xueai8:802...|2022-02-11 10:14:...|128326|[FF D8 FF E0 00 1...|
+--------------------+--------------------+------+--------------------+
```

4.7 操作 DataFrame

在 Spark 2.0 中，Spark 将 DataFrame 看作 DataSet 的一个特例，即 DataSet[Row]。DataSet/DataFrame 为结构化数据操作提供了一种特定于领域的语言（DSL），其中 DataFrame 所支持的这些操作也称为"无类型转换"，同时提供对 Scala、Java、Python 和 R 语言的支持，而 DataSet 所支持的这些操作也称为"类型化转换"，只提供强类型的 Scala 和 Java 语言支持的 API。

本节将重点讨论可以在 DataFrame 上执行的各种操作。这些操作被分为两类，Transformation 和 Action。开发人员使用多个操作来选择、过滤、转换、聚合和排序在 DataFrame 中的数据。底层的 Catalyst 优化器确保了这些操作的高效执行。

4.7.1 列的多种引用方式

在学习操作 DataFrame 之前，需要掌握 Spark 所提供的引用 DataFrame 列的多种方式。首先创建一个 DataFrame，代码如下：

```
import spark.implicits._

//使用元组序列创建一个 DataFrame
val kvDF = Seq((1,2),(3,4)).toDF("key","value")

//输出
kvDF.printSchema()
kvDF.show()
```

执行以上代码，输出内容如下：

```
root
 |-- key: integer (nullable = false)
 |-- value: integer (nullable = false)

+---+-----+
|key|value|
+---+-----+
|  1|    2|
|  3|    4|
+---+-----+
```

要显示一个 DataFrame 的所有列名，可以调用其 columns 属性，代码如下：

```
kvDF.columns                    //输出：Array(key, value)
```

以不同的方式选择特定的列，代码如下：

```
//导入函数所在的包
import org.apache.spark.sql.functions._

//以不同的方式选择特定的列
kvDF.select("key").show              //列为字符串类型
kvDF.select(col("key")).show         //col 是内置函数，它返回 Column 类型
kvDF.select(column("key")).show      //column 是内置函数，它返回 Column 类型
kvDF.select(expr("key")).show        //expr 与 col 的调用方法相同
kvDF.select($"key").show             //Scala 中构造 Column 类型的语法糖
kvDF.select('key).show               //同上

//也可以使用 DataFrame 的 col 函数
kvDF.select(kvDF.col("key")).show
```

以上所有的输出内容均相同，输出内容如下：

```
+---+
|key|
+---+
|  1|
|  3|
+---+
```

对于列的各种引用方式，区别如下：

（1）如果只是为了获取特定列的值，则直接以字符串类型引用列名即可。

（2）如果引用列是为了任何形式的计算，包括排序、类型转换、别名、比较、计算列等，则需要应用上述任一函数将列转换为 Column 对象。

（3）语法糖形式的引用（$和'）仅支持 Scala 语言。

例如，下面的代码选择 key 列，并增加一个新的列，新列的值由 key 列计算得来，类型为 boolean 值，表示 key 列的值是否大于 1，代码如下：

```
kvDF.select('key, 'key > 1).show
```

执行上面这行代码，输出内容如下：

```
+---+---------+
|key|(key > 1)|
+---+---------+
|  1|    false|
|  3|     true|
+---+---------+
```

也可以给列取一个别名，代码如下：

```
//给列取一个别名
kvDF.select('key, 'key > 1 as "v").show

//或（等价）
kvDF.select('key, ('key > 1).alias("v")).show
```

执行上面的代码，输出内容相同，内容如下：

```
+---+-----+
|key|    v|
+---+-----+
|  1|false|
|  3| true|
+---+-----+
```

4.7.2 对 DataFrame 执行 Transformation 转换操作

DataFrame API 提供了许多用来执行关系运算的函数，这些函数模拟了 SQL 关系操作。

（1）选择数据：select()。

（2）删除某列：drop()。

（3）过滤数据：where()和 filter()，两者是同义的。

（4）限制返回的数量：limit()。

（5）重命名列：withColumnRenamed()。

（6）增加一个新的列：withColumn()。

（7）数据分组：groupBy()。

（8）数据排序：orderBy()和sort()，两者是等价的。

在进一步演示 DataFrame 的各种操作方法之前，先准备好要用到的数据。这里将使用一部电影数据集 movies.csv，应将它上传到 HDFS 的/data/spark/目录下，其中部分数据格式如下：

```
actor,title,year
"McClure, Marc (I)",Freaky Friday,2003
"McClure, Marc (I)",Coach Carter,2005
"McClure, Marc (I)",Superman II,1980
"McClure, Marc (I)",Apollo 13,1995
"McClure, Marc (I)",Superman,1978
"McClure, Marc (I)",Back to the Future,1985
"McClure, Marc (I)",Back to the Future Part III,1990
"Cooper, Chris (I)","Me, Myself & Irene",2000
"Cooper, Chris (I)",October Sky,1999
"Cooper, Chris (I)",Capote,2005
"Cooper, Chris (I)",The Bourne Supremacy,2004
```

可以看到，数据集有 3 个字段，即 actor、title 和 year。数据集的第 1 行是标题行。

首先，将数据集加载到 DataFrame 中，代码如下：

```scala
//第 4 章/transformationDemo1.scala

//数据源文件
val file = "/data/spark/movies/movies.csv"

//将文件读到 DataFrame 中
val movies = spark.read
    .option("header","true")
    .option("inferSchema","true")
    .csv(file)

//输出
movies.printSchema()
movies.show(5)                      //只查看前 5 行
```

执行以上代码，输出内容如下：

```
root
 |-- actor: string (nullable = true)
 |-- title: string (nullable = true)
 |-- year: integer (nullable = true)

+-----------------+-------------+----+
|            actor|        title|year|
+-----------------+-------------+----+
|McClure, Marc (I)|Freaky Friday|2003|
```

```
|McClure, Marc (I)|  Coach Carter|2005|
|McClure, Marc (I)|  Superman II|1980|
|McClure, Marc (I)|    Apollo 13|1995|
|McClure, Marc (I)|      Superman|1978|
+-----------------+-------------+----+
only showing top 5 rows
```

接下来,学习 DataFrame 的各种转换操作函数。

1. select(cols: Column*)

类似于 SQL 中的 select 子句,用来选择指定的列。例如,选取 movies 中的 title 和 year 两列,并返回一个新的 DataFrame,代码如下:

```
movies.select("title","year").show(5)
```

执行以上代码,输出内容如下:

```
+-------------+----+
|        title|year|
+-------------+----+
|Freaky Friday|2003|
| Coach Carter|2005|
| Superman II|1980|
|   Apollo 13|1995|
|     Superman|1978|
+-------------+----+
only showing top 5 rows
```

使用列计算将电影上映的年份转换到年代,并赋予一个别名,代码如下:

```
import spark.implicits._
movies.select($"title",($"year"-$"year" % 10).as("decade")).show(5)
```

执行以上代码,输出内容如下:

```
+-------------+------+
|        title|decade|
+-------------+------+
|Freaky Friday|  2000|
| Coach Carter|  2000|
|  Superman II|  1980|
|   Apollo 13|  1990|
|     Superman|  1970|
+-------------+------+
only showing top 5 rows
```

2. selectExpr(exprs: String*)

用来选择一组 SQL 表达式,即使用 SQL 表达式来选择列。例如,在下面的代码中,用通配符星号(*)来表示选择所有的列,并增加一个新的列 decade,新列的值是通过对 year 列值计算得到的电影上映的年代,代码如下:

```
movies.selectExpr("*","(year - year % 10) as decade").show(5)
```

执行以上代码,输出内容如下:

```
+----------------+-------------+----+------+
|           actor|        title|year|decade|
+----------------+-------------+----+------+
|McClure, Marc (I)|Freaky Friday|2003|  2000|
|McClure, Marc (I)| Coach Carter|2005|  2000|
|McClure, Marc (I)|  Superman II|1980|  1980|
|McClure, Marc (I)|    Apollo 13|1995|  1990|
|McClure, Marc (I)|     Superman|1978|  1970|
+----------------+-------------+----+------+
only showing top 5 rows
```

在 selectExpr()方法中，不但支持使用 SQL 表达式，还支持直接使用 SQL 内置函数。例如，使用 SQL 表达式和内置函数，来查询电影数量和演员数量这两个值，代码如下：

```
movies
    .selectExpr("count(distinct(title)) as movies",
                "count(distinct(actor)) as actors")
    .show()
```

执行以上代码，输出内容如下：

```
+------+------+
|movies|actors|
+------+------+
|  1409|  6527|
+------+------+
```

可以看出，数据集中的电影共有 1409 部，演员共有 6527 名。

3. filter(condition: Column)，where(condition: Column)

使用给定的条件过滤 DataFrame 中的行。这两个函数是等价的，相当于 SQL 中的 where 子句。例如，要找出 2000 年以前上映的电影，需要在 filter()函数或 where()函数中指定过滤条件，代码如下：

```
movies.filter($"year" < 2000).show(5)
movies.where($"year" < 2000).show(5)              //等价
```

执行以上代码，输出内容如下：

```
+----------------+--------------------+----+
|           actor|               title|year|
+----------------+--------------------+----+
|McClure, Marc (I)|         Superman II|1980|
|McClure, Marc (I)|           Apollo 13|1995|
|McClure, Marc (I)|            Superman|1978|
|McClure, Marc (I)|  Back to the Future|1985|
|McClure, Marc (I)|Back to the Futur...|1990|
+----------------+--------------------+----+
only showing top 5 rows
```

如果想找出 2000 年及以后上映的电影，则代码如下：

```
movies.filter($"year" >= 2000).show(5)
movies.where($"year" >= 2000).show(5)
```

执行以上代码,输出内容如下:

```
+----------------+--------------------+----+
|           actor|               title|year|
+----------------+--------------------+----+
|McClure, Marc (I)|       Freaky Friday|2003|
|McClure, Marc (I)|        Coach Carter|2005|
|Cooper, Chris (I)|  Me, Myself & Irene|2000|
|Cooper, Chris (I)|              Capote|2005|
|Cooper, Chris (I)|The Bourne Supremacy|2004|
+----------------+--------------------+----+
only showing top 5 rows
```

如果要找出 2000 年上映的电影,则代码如下:

```
movies.filter($"year" === 2000).show(5)
movies.where($"year" === 2000).show(5)         //等价
```

注意,在 Scala 中,相等比较要求 3 个等号。

执行以上代码,输出内容如下:

```
+----------------+--------------------+----+
|           actor|               title|year|
+----------------+--------------------+----+
|Cooper, Chris (I)|  Me, Myself & Irene|2000|
|Cooper, Chris (I)|         The Patriot|2000|
|  Jolie, Angelina|Gone in Sixty Sec...|2000|
|    Yip, Françoise|    Romeo Must Die|2000|
|   Danner, Blythe|   Meet the Parents|2000|
+----------------+--------------------+----+
only showing top 5 rows
```

要找出非 2000 年上映的电影,代码如下:

```
movies.select("title","year").filter('year =!= 2000).show(5)
movies.select("title","year").where('year =!= 2000).show(5)
```

注意,在 Scala 中,不相等比较使用的操作符是 =!=。

执行以上代码,输出内容如下:

```
+-------------+----+
|        title|year|
+-------------+----+
|Freaky Friday|2003|
| Coach Carter|2005|
|  Superman II|1980|
|    Apollo 13|1995|
|     Superman|1978|
+-------------+----+
only showing top 5 rows
```

找出 2001—2002 年间上映的电影,代码如下:

```
movies.filter('year.isin(2001,2002)).show(5,false)
movies.where('year.isin(2001,2002)).show(5,false)         //等价
```

执行以上代码,输出内容如下:

```
+------------------+------------------------------------+----+
|actor             |title                               |year|
+------------------+------------------------------------+----+
|Cooper, Chris (I) |The Bourne Identity                 |2002|
|Cassavetes, Frank |John Q                              |2002|
|Knight, Shirley (I)|Divine Secrets of the Ya-Ya Sisterhood|2002|
|Jolie, Angelina   |Lara Croft: Tomb Raider             |2001|
|Cueto, Esteban    |Collateral Damage                   |2002|
+------------------+------------------------------------+----+
only showing top 5 rows
```

可使用 OR 和 AND 运算符组合一个或多个比较表达式。例如,要找出 2000 年及以后上映且名称长度少于 5 个字符的电影,代码如下:

```
movies.filter('year >= 2000 && length('title) < 5).show(5)
movies.where('year >= 2000 && length('title) < 5).show(5)        //等价
movies.where('year >= 2000 and length('title) < 5).show(5)       //等价
```

执行以上代码,输出内容如下:

```
+---------------+-----+----+
|          actor|title|year|
+---------------+-----+----+
|Jolie, Angelina| Salt|2010|
| Cueto, Esteban|  xXx|2002|
|  Butters, Mike|  Saw|2004|
| Franko, Victor|   21|2008|
| Ogbonna, Chuk| Salt|2010|
+---------------+-----+----+
only showing top 5 rows
```

另一种实现相同结果的方法是调用 where() 函数两次,代码如下:

```
movies.where('year >= 2000)
      .where(length('title) < 5)
      .show(5)
```

4. distinct(),dropDuplicates()

返回一个新数据集,其中仅包含此数据集中的唯一行。distinct() 是 dropDuplicates() 的别名。例如,想知道数据集中共有多少条唯一行(去重),代码如下:

```
movies.distinct().count()
movies.dropDuplicates().count()        //与上句等价
```

执行以上代码,输出内容如下:

```
31394
```

但是,如果想要知道数据集中共包含多少部电影,则需要基于 title 字段进行唯一值统计,代码如下:

```
movies.select("title").distinct().count()
```

```
//其实也可以使用 SQL 的 distinct 函数
//movies.selectExpr("count(distinct(title)) as movies").show()
```
执行以上代码，输出内容如下：
```
1409
```

5. dropDuplicates(colNames: Array[String])

仅考虑列的子集，返回删除（按列的子集）重复行的新数据集。例如，同样需要统计数据集中共包含多少部电影，代码如下：
```
movies.dropDuplicates(Array("title")).count()
```
执行以上代码，输出内容如下：
```
1409
```

6. sort(sortExprs: Column*)，orderBy(sortExprs: Column*)

相当于 SQL 中的 order by 子句，它返回按指定的列排序后的新数据集。例如，要按电影名称长度顺序及上映年份倒序显示，代码如下：
```
movies.dropDuplicates(Seq("title", "year"))           //去重
    .selectExpr("title", "length(title) as title_length", "year")
    .orderBy('title_length.asc, 'year.desc)           //asc:升序;desc:倒序
    .show(10,false)
```
执行以上代码，输出内容如下：
```
+-----+------------+----+
|title|title_length|year|
+-----+------------+----+
|Up   |2           |2009|
|21   |2           |2008|
|12   |2           |2007|
|RV   |2           |2006|
|X2   |2           |2003|
|Rio  |3           |2011|
|Hop  |3           |2011|
|300  |3           |2006|
|Saw  |3           |2004|
|Elf  |3           |2003|
+-----+------------+----+
only showing Top 10 rows
```

7. groupBy(cols: Column*)，groupBy(col1: String, cols: String*)

相当于 SQL 中的 group by 子句，按指定的列对数据进行分组，以便执行聚合统计操作。例如，统计每年上映的电影数量并按数量倒序显示，代码如下：
```
movies.groupBy("year")
    .count()
    .orderBy($"count".desc)
    .show(10)
```

执行以上代码，输出内容如下：

```
+----+-----+
|year|count|
+----+-----+
|2006| 2078|
|2004| 2005|
|2007| 1986|
|2005| 1960|
|2011| 1926|
|2008| 1892|
|2009| 1890|
|2010| 1843|
|2002| 1834|
|2001| 1687|
+----+-----+
only showing Top 10 rows
```

如果要统计上映电影数量超过2000部的年份，则代码如下：

```
movies.groupBy("year")
    .count()
    .where($"count" > 2000)
    .show()
```

执行以上代码，输出内容如下：

```
+----+-----+
|year|count|
+----+-----+
|2006| 2078|
|2004| 2005|
```

8. limit(n: Int)

通过获取前 *n* 行返回一个新的 DataSet，相当于 SQL 中的 limit 子句。通常将其与 orderBy 配合实现 Top N 算法。例如，统计上映电影数量最多的 5 个年份，代码如下：

```
movies.groupBy("year")
    .count()
    .orderBy($"count".desc)
    .limit(5)
    .show()
```

执行以上代码，输出内容如下：

```
+----+-----+
|year|count|
+----+-----+
|2006| 2078|
|2004| 2005|
|2007| 1986|
|2005| 1960|
|2011| 1926|
+----+-----+
```

查询电影名称最长的 5 部电影,代码如下:

```
movies.dropDuplicates("title")                          //去重
    .selectExpr("title", "length(title) as title_length")
    .orderBy(col("title_length").desc)                  //倒序
    .limit(5)
    .show(false)
```

执行以上代码,输出内容如下:

```
+-----------------------------------------------------------+------------+
|title                                                      |title_length|
+-----------------------------------------------------------+------------+
|Borat: Cultural Learnings of America for Make Benefit
 Glorious Nation of Kazakhstan                             |83          |
|The Chronicles of Narnia: The Lion, the Witch and the Wardrobe|62       |
|Hannah Montana & Miley Cyrus: Best of Both Worlds Concert  |57          |
|The Chronicles of Narnia: The Voyage of the Dawn Treader   |56          |
|Istoriya pro Richarda, milorda i prekrasnuyu Zhar-ptitsu   |56          |
+-----------------------------------------------------------+------------+
```

9. union(other: DataSet[T]),unionAll(other: DataSet[T])

按列位置合并两个数据集的行,相当于 SQL 中的 union all。假设现在想在名称为 "12" 的电影中添加一个缺失的演员信息,可以采用如下的方法:另外创建一个 DataFrame,其中包含所缺失的演员信息,然后将原数据集与这个含有缺失演员信息的新数据集执行一个 union 操作,将两个数据集合并在一起。这样就将缺失的演员信息添加到了原数据集中。

首先获得电影 "12" 的数据集,代码如下:

```
val shortNameMovieDF = movies.where($"title" === "12")
shortNameMovieDF.show()
```

执行以上代码,输出内容如下:

```
+--------------------+-----+----+
|               actor|title|year|
+--------------------+-----+----+
|    Efremov, Mikhail|   12|2007|
|     Stoyanov, Yuriy|   12|2007|
|     Gazarov, Sergey|   12|2007|
|Verzhbitskiy, Viktor|   12|2007|
+--------------------+-----+----+
```

可以看到,该部电影有 4 名演员,但是缺少了演员 Brychta,Edita 的信息。接下来,另外创建一个 DataFrame,包含所缺失演员 Brychta,Edita 的信息,然后将这个 DataFrame 和上面的 DataFrame 进行合并,这样就将演员 Brychta,Edita 的信息合并到上面的数据集中了,代码如下:

```
//另创建一个 DataFrame
val forgottenActor = Seq(("Brychta, Edita", "12", 2007L))
val forgottenActorDF = forgottenActor.toDF("actor","title","year")

//通过合并两个 DataFrame,实现添加缺失的演员姓名
```

```
val completeShortNameMovieDF = shortNameMovieDF.union(forgottenActorDF)
completeShortNameMovieDF.show()
```

执行以上代码，输出内容如下：

```
+--------------------+-----+----+
|               actor|title|year|
+--------------------+-----+----+
|    Efremov, Mikhail|   12|2007|
|     Stoyanov, Yuriy|   12|2007|
|     Gazarov, Sergey|   12|2007|
|Verzhbitskiy, Viktor|   12|2007|
|       Brychta, Edita|  12|2007|
+--------------------+-----+----+
```

如果要实现类 SQL 中 UNION 的功能，则可在 union()操作之后使用 distinct()操作，这样就可以去掉重复的数据了。

10. withColumn(colName: String, col: Column)

通过添加列或替换具有相同名称的现有列，返回一个新的数据集。例如，向 movies 数据集增加一个新列 decade，该列的值由基于 year 列的表达式计算得到，计算的结果是该电影上映的年代，代码如下：

```
movies.withColumn("decade", $"year" - $"year" % 10).show(5)
```

执行以上代码，输出内容如下：

```
+-----------------+-------------+----+------+
|            actor|        title|year|decade|
+-----------------+-------------+----+------+
|McClure, Marc (I)| Freaky Friday|2003|  2000|
|McClure, Marc (I)|  Coach Carter|2005|  2000|
|McClure, Marc (I)|   Superman II|1980|  1980|
|McClure, Marc (I)|     Apollo 13|1995|  1990|
|McClure, Marc (I)|       Superman|1978|  1970|
+-----------------+-------------+----+------+
only showing top 5 rows
```

如果传给 withColumn 函数的列名与现有的列名相同，则意味着用新值替换旧值。例如，将 year 列的值替换为年代（原值为年份），代码如下：

```
movies.withColumn("year", $"year" - $"year" % 10).show(5)
```

执行以上代码，输出内容如下：

```
+-----------------+-------------+----+
|            actor|        title|year|
+-----------------+-------------+----+
|McClure, Marc (I)| Freaky Friday|2000|
|McClure, Marc (I)|  Coach Carter|2000|
|McClure, Marc (I)|   Superman II|1980|
|McClure, Marc (I)|     Apollo 13|1990|
|McClure, Marc (I)|       Superman|1970|
+-----------------+-------------+----+
only showing top 5 rows
```

11. withColumnRenamed(existingName: String, newName: String)

返回一个重命名列的新数据集。如果模式不包含 existingName，则不进行任何操作。例如，将 movies 数据集中的列名改为新的名称，代码如下：

```
movies
    .withColumnRenamed("actor", "actor_name")
    .withColumnRenamed("title", "movie_title")
    .withColumnRenamed("year", "produced_year")
    .show(5)
```

执行以上代码，输出内容如下：

```
+----------------+-------------+-------------+
|      actor_name| movie_title|produced_year|
+----------------+-------------+-------------+
|McClure, Marc (I)|Freaky Friday|         2003|
|McClure, Marc (I)| Coach Carter|         2005|
|McClure, Marc (I)|   Superman II|         1980|
|McClure, Marc (I)|    Apollo 13|         1995|
|McClure, Marc (I)|     Superman|         1978|
+----------------+-------------+-------------+
only showing top 5 rows
```

12. drop(col: Column)，drop(colNames: String*)

返回一个删除了指定列的新 DataSet。如果要被删除的列不存在，则不进行任何操作。例如，删除 movies 数据集中指定的列，代码如下：

```
//删除指定的列，包括一个并不存在的列 me
val dropedMovies = movies.drop("actor", "me")

dropedMovies.printSchema()
dropedMovies.show(5)
```

执行以上代码，输出内容如下：

```
root
 |-- title: string (nullable = true)
 |-- year: integer (nullable = true)

+-------------+----+
|        title|year|
+-------------+----+
|Freaky Friday|2003|
| Coach Carter|2005|
|  Superman II|1980|
|   Apollo 13|1995|
|     Superman|1978|
+-------------+----+
only showing top 5 rows
```

从输出结果可以看出，actor 列被删除了，而 me 这一列在原 DataFrame 中并不存在，所以删除不存在的列没有任何影响。

13. sample(fraction: Double)

数据抽样。通过随机种子对部分行进行抽样（不进行替换），返回一个新的 DataSet。抽样比例由 fraction 参数决定，它用于指定要生成的行的比例，范围为[0.0,1.0]。这种方法有多个不同的重载版本。

（1）sample(fraction: Double, seed: Long)：指定种子。
（2）sample(withReplacement: Boolean, fraction: Double)：指定是否替换，使用随机种子。
（3）sample(withReplacement: Boolean, fraction: Double, seed: Long)：同上，但指定种子。

例如，指定无放回抽样和抽样比例，代码如下：

```
movies.sample(false, 0.0003).show(false)
```

执行以上代码，输出内容如下（随机的）：

```
+--------------------+------------------------------------------+----+
|actor               |title                                     |year|
+--------------------+------------------------------------------+----+
|Baskin, Elya        |Austin Powers: International Man of Mystery|1997|
|Rapace, Noomi       |Prometheus                                |2012|
|Hatcher, Teri       |Coraline                                  |2009|
|Damon, Matt         |The Bourne Supremacy                      |2004|
|Novak, B.J.         |Knocked Up                                |2007|
|Masamune, Tohoru    |Jumper                                    |2008|
|Cruise, Tom         |Collateral                                |2004|
|Harvey, Steve (I)   |Racing Stripes                            |2005|
|Giuliani, Rudolph W.|Anger Management                          |2003|
|Sloan, Amy          |The Aviator                               |2004|
|Krumholtz, David (I)|The Mexican                               |2001|
+--------------------+------------------------------------------+----+
```

执行有放回抽样并指定比例因子、种子的抽样，代码如下：

```
movies.sample(true, 0.0003, 123456).show()
```

执行以上代码，输出内容如下（保持不变的）：

```
+-----------------+--------------------+----+
|            actor|               title|year|
+-----------------+--------------------+----+
|    Piddock, Jim |        The Prestige|2006|
|    Reed, Tanoai |  The Stepford Wives|2004|
|    Moyo, Masasa |    Angels & Demons |2009|
| Zemeckis, Leslie|             Beowulf|2007|
|    Huston, Danny|X-Men Origins: Wo...|2009|
|    Pompeo, Ellen|          Old School|2003|
|Utt, Kenneth (I) |        Philadelphia|1993|
|Cannon, Kevin (I)|             Cop Out|2010|
+-----------------+--------------------+----+
```

14. randomSplit(weights: Array[Double])，randomSplit(weights: Array[Double], seed: Long)

使用提供的权重随机分割数据集。如果分割的权重和不等于1，则将被标准化。该操作返回一个切分后的 DataSet 集合数组。例如，将 movies 分割为 3 个数据集，所占数据比例分别为 0.6、0.3 和 0.1，代码如下：

```
//按比例分割数据集
val sMovies = movies.randomSplit(Array(0.6, 0.3, 0.1))

//看一看各部分计数之和是否等于1
//数据集总数量
println("总数据量: " + movies.count())

//分割后的第1个数据集的数量
println("第1个子数据集的数量: " + sMovies(0).count())

//3个数据集之和
val total = sMovies(0).count() + sMovies(1).count() + sMovies(2).count()
println("3个子数据集之和: " + total.toInt)
```

执行以上代码，输出内容如下：

```
总数据量: 31394
第1个子数据集的数量: 18906
3个子数据集之和: 31394
```

4.7.3 对 DataFrame 执行 Action 操作

与 RDD 类似，DataSet/DataFrame 上的 Transformation 都是延迟执行的。只有当在 DataFrame 上执行 Action 操作时，才会触发真正的计算。这些 Action 操作相对都比较简单，关于这些 Action 操作方法的使用，代码如下：

```
//查看前5条数据。当列内容较长时，第2个参数用于指定是否截断显示，false为不截断
movies.show(5,false)

//返回数据集中的数量
movies.count

//返回数据集中第1条数据
movies.first()

//等价于first方法
movies.head()

//以Array的形式返回数据集中的前3条数据
movies.head(3)

//以Array的形式返回数据集中的前3条数据
movies.take(3)

//以List的形式返回数据集中的前3条数据
movies.takeAsList(3)
```

```
//返回一个包含数据集中所有行的数组
movies.collect

//以 List 的形式返回一个包含数据集中所有行的数组
movies.collectAsList

//以 Array 的形式返回数据集的数据类型
df.types

//以 Array 的形式返回数据集的列名
df.columns
```

4.7.4 对 DataFrame 执行描述性统计操作

Spark 还为 DataSet/DataFrame 提供了一个 describe()函数,用来计算数字列和字符串列的基本统计信息,包括 count、mean、stddev、min 和 max。如果没有给定列,则此函数将计算所有数值列或字符串列的统计信息。该函数返回的也是一个 DataFrame,其方法签名如下:

```
def describe(cols: String*): DataFrame
```

这种方法是一个 Action 类型操作,经常用于对数据集执行探索性数据分析,代码如下:

```
val descDF = movies.describe()

descDF.printSchema()
descDF.show()
```

执行以上代码,输出内容如下:

```
root
 |-- summary: string (nullable = true)
 |-- actor: string (nullable = true)
 |-- title: string (nullable = true)
 |-- year: string (nullable = true)

+-------+------------------+--------------------+------------------+
|summary|             actor|               title|              year|
+-------+------------------+--------------------+------------------+
|  count|             31394|               31394|             31393|
|   mean|              null|   312.61538461538464| 2002.7964514382188|
| stddev|              null|   485.70434143901151| 6.3771353799933117|
|    min|    Aaron, Caroline|'Crocodile' Dunde...|              1961|
|    max|  von Sydow, Max (I)|                 xXx|              2012|
+-------+------------------+--------------------+------------------+
```

要计算 year 字段的基本统计信息,则代码如下:

```
val descDF = movies.describe("year")

descDF.printSchema()
descDF.show()
```

执行以上代码，输出内容如下：

```
root
 |-- summary: string (nullable = true)
 |-- year: string (nullable = true)

+-------+------------------+
|summary|              year|
+-------+------------------+
|  count|             31393|
|   mean|2002.7964514382188|
| stddev| 6.377135379933117|
|    min|              1961|
|    max|              2012|
+-------+------------------+
```

如果要计算 year 和 actor 这两列的基本统计信息，则代码如下：

```
val descDF = movies.describe("year","actor")

descDF.printSchema()
descDF.show()
```

执行以上代码，输出内容如下：

```
root
 |-- summary: string (nullable = true)
 |-- year: string (nullable = true)
 |-- actor: string (nullable = true)

+-------+------------------+------------------+
|summary|              year|             actor|
+-------+------------------+------------------+
|  count|             31393|             31394|
|   mean|2002.7964514382188|              null|
| stddev| 6.377135379933117|              null|
|    min|              1961|    Aaron, Caroline|
|    max|              2012|von Sydow, Max (I)|
+-------+------------------+------------------+
```

注意：这个函数用于探索性数据分析，它不能保证结果数据集模式的向后兼容性。如果希望以编程方式计算汇总统计信息，则可使用 agg() 函数。

Spark 还提供了一个与 describe()函数类似的 summary()函数，用于提供数据集的摘要信息。如果没有给出统计信息，则这个函数将计算 count、mean、stddev、min、近似四分位数（25%、50%和75%的百分位数）和 max 值。例如，对 movies 调用 summary()函数，代码如下：

```
val summaryDF = movies.summary()

summaryDF.printSchema()
summaryDF.show()
```

执行以上代码，输出内容如下：

```
root
 |-- summary: string (nullable = true)
 |-- actor: string (nullable = true)
 |-- title: string (nullable = true)
 |-- year: string (nullable = true)

+-------+--------------------+--------------------+------------------+
|summary|               actor|               title|              year|
+-------+--------------------+--------------------+------------------+
|  count|               31394|               31394|             31393|
|   mean|                null|   312.6153846153846|2002.7964514382188|
| stddev|                null|   485.7043414390151| 6.377135379933117|
|    min|      Aaron, Caroline|'Crocodile' Dunde...|              1961|
|    25%|                null|                21.0|              1999|
|    50%|                null|                21.0|              2004|
|    75%|                null|               300.0|              2008|
|    max|    von Sydow, Max (I)|                 xXx|              2012|
+-------+--------------------+--------------------+------------------+
```

也可以指定想要的统计信息，则代码如下：

```
val summaryDF = movies.summary("count", "min", "25%", "75%", "max")

summaryDF.printSchema()
summaryDF.show()
```

执行以上代码，输出内容如下：

```
root
 |-- summary: string (nullable = true)
 |-- actor: string (nullable = true)
 |-- title: string (nullable = true)
 |-- year: string (nullable = true)

+-------+--------------------+--------------------+-----+
|summary|               actor|               title| year|
+-------+--------------------+--------------------+-----+
|  count|               31394|               31394|31393|
|    min|      Aaron, Caroline|'Crocodile' Dunde...| 1961|
|    25%|                null|                21.0| 1999|
|    75%|                null|               300.0| 2008|
|    max|    von Sydow, Max (I)|                 xXx| 2012|
+-------+--------------------+--------------------+-----+
```

要对指定的列做一个摘要，首先选择这些列，然后执行 summary()方法，代码如下：

```
val summaryDF = movies.select("year").summary()

summaryDF.printSchema()
summaryDF.show()
```

执行以上代码，输出内容如下：

```
root
 |-- summary: string (nullable = true)
 |-- year: string (nullable = true)
```

```
+-------+------------------+
|summary|              year|
+-------+------------------+
|  count|             31393|
|   mean|2002.7964514382188|
| stddev| 6.377135379933117|
|    min|              1961|
|    25%|              1999|
|    50%|              2004|
|    75%|              2008|
|    max|              2012|
+-------+------------------+
```

4.7.5 取 DataFrame Row 中特定字段

有时，用户需要从 DataFrame 中获取特定行的特定字段的值，这时有两种方法可以调用：getAs()方法和模式匹配。

1. getAs()方法

该方法的一般使用模板代码如下：

```
df.foreach{ line =>
   val col1 = line.getAs[String]("col1")
   val col2 = line.getAs[String]("col2")
}
```

它有两个重载的方法，分别如下。

1）getAs[T](fieldName: String): T

这种方法返回给定字段名称的值。对于原始数据类型，如果值为 null，则它返回对应原始数据类型的零值，例如对于 Int 类型返回 0。可以使用 isNullAt()来确保该值不为 null。调用该方法时，如果数据类型不匹配，则抛出 ClassCastException 异常；如果指定的字段名不存在，则抛出 IllegalArgumentException 异常；当 Schema 模式没有定义时，则抛出 UnsupportedOperationException 异常。

2）getAs[T](i: Int): T

返回指定位置 i 处的值。对于原始数据类型，如果值为 null，则它返回对应原始数据类型的零值，例如对于 Int 类型返回 0。可以使用 isNullAt()来确保该值不为 null。调用该方法时，如果数据类型不匹配，则抛出 ClassCastException 异常。

在下面的示例中，要求加载 Spark 自带的数据文件 people.json，并输出每个字段的值。首先将 people.json 文件读到一个 DataFrame 中，代码如下：

```
//加载数据源，构造 DataFrame
val input = "/data/spark/resources/people.json"
val df = spark.read.json(input)

df.printSchema()
df.show()
```

执行以上代码,输出内容如下:

```
root
 |-- age: long (nullable = true)
 |-- name: string (nullable = true)

+----+-------+
| age|   name|
+----+-------+
|null|Michael|
|  30|   Andy|
|  19| Justin|
+----+-------+
```

然后调用 getAs()方法,根据指定的字段名称返回每行该字段的值,代码如下:

```
//取每个字段的值
df.foreach{ row =>
    val col1 = row.getAs[String]("name")
    val col2 = row.getAs[Long]("age")
    println(col1 + "\t" + col2)
}
```

执行以上代码,输出内容如下:

```
Michael 0
Andy    30
Justin  19
```

或者,也可以按字段的位置取每个字段的值,代码如下:

```
//取每个字段的值
df.foreach{ row =>
    val col1 = row.getAs[String](1)
    val col2 = row.getAs[Long](0)
    println(col1 + "\t" + col2)
}
```

2. 模式匹配

也可以使用模式匹配来取每行每个字段的值,一般模板代码如下:

```
df.foreach{
    case Row(col1:String, col2:Long) =>
        println(col1 + "\t" + col2)
    case _ =>
}
```

例如,重构前面的代码,加载 Spark 自带的数据文件 people.json,并输出每个字段的值,但这一次使用模式匹配的方式,代码如下:

```
import org.apache.spark.sql._

//加载数据源,构造 DataFrame
val input = "/data/spark/resources/people.json"
val df = spark.read.json(input)
```

```
//取每个字段的值
df.collect().foreach{
    case Row(col1:Long,col2:String) =>
        println(col1 + "\t" + col2)
    case _ =>
}
```

执行以上代码，输出内容如下：

```
30    Andy
19    Justin
```

可以看出，与 getAs() 方法相比，模式匹配无法匹配和读取 null 值。

4.7.6 操作 DataFrame 示例

下面通过一个示例来进一步理解和掌握 DataFrame 的 Transformation 和 Action 操作。假设现在给出一个员工信息名单 employees.csv，包含员工薪资信息，内容如下：

```
张三,paramedic i/c,fire,f,salary,,91080.00,
李四,lieutenant,fire,f,salary,,114846.00,
王老五,sergeant,police,f,salary,,104628.00,
赵六,police officer,police,f,salary,,96060.00,
钱七,clerk iii,police,f,salary,,53076.00,
周扒皮,firefighter,fire,f,salary,,87006.00,
吴用,law clerk,law,f,hourly,35,,14.51
```

其中各个字段的含义依次如下："姓名,职业,部门,全职或兼职,固定薪资或计时薪资,工作时长,年薪,每小时工资"。现在要求找出其中收入最高的前 3 名员工（Top N 问题）。

需要注意，在数据集中，"吴用"没有年薪，是计时薪资（hourly，时薪），所以有最后一个"每小时工资"的字段，而其他人都是固定薪资（salary），所以有"年薪"字段，但是没有"每小时工资"字段。

首先，使用 HDFS Shell 命令，将该数据文件 employees.csv 上传到 HDFS 上，命令如下：

```
$ hdfs dfs -put employees.csv /data/spark/
```

然后编写 Spark 程序，代码如下：

```
//第 4 章/employeesDemo.scala

//定义文件路径
val file = "/data/spark/employees.csv"

//指定列名
val columns = Array("uname", "designation", "department", "jobtype", "NA",
"NA2", "salary", "NA3")

//加载数据文件，并构造 DataFrame
val df = spark.read
    .option("inferSchema", "true")
    .option("header","false")
```

```
        .csv(file)
        .toDF(columns:_*)

//按salary列降序排列,并显示
import spark.implicits._
df.orderBy($"salary".desc).limit(3).show()
```

执行以上代码,输出内容如下:

```
+------+--------------+----------+-------+------+----+--------+----+
| uname|   designation|department|jobtype|    NA| NA2|  salary| NA3|
+------+--------------+----------+-------+------+----+--------+----+
|  李四|    lieutenant|      fire|      f|salary|null|114846.0|null|
|王老五|      sergeant|    police|      f|salary|null|104628.0|null|
|  赵六|police officer|    police|      f|salary|null| 96060.0|null|
+------+--------------+----------+-------+------+----+--------+----+
```

4.8 存储 DataFrame

有时,需要将 DataFrame 中的数据写到外部存储系统中,例如,在一个典型的 ETL 数据处理作业中,处理结果通常需要被写到一些存储系统中,如本地文件系统、HDFS、Hive 或 Amazon S3。下面学习如何存储这些 DataFrame。

4.8.1 写出 DataFrame

在 Spark SQL 中,org.apache.spark.sql.DataFrameWriter 类负责将 DataFrame 中的数据写入外部存储系统。在 DataFrame 中有一个变量 write,它实际上就是 DataFrameWriter 类的一个实例。与 DataFrameWriter 交互的模式和与 DataFrameReader 的交互模式有点类似。

与 DataFrameWriter 交互的常见模式,代码如下:

```
movies.write                    //DataFrameWriter 实例对象
    .format(...)                //指定存储格式
    .mode(...)                  //指定写出模式: append 或 overwrite
    .option(...)                //指定选项
    .partitionBy(...)           //指定分区列
    .bucketBy(...)              //指定分桶
    .sortBy(...)                //排序
    .save(path)                 //保存到指定路径(文件夹)
```

与 DataFrameReader 相似,可以使用 JSON、ORC 和 Parquet 文件存储格式,默认格式是 Parquet。需要注意的是,save()函数的输入参数是目录名,而不是文件名,它可将数据直接保存到文件系统,如 HDFS、Amazon S3 或者一个本地路径 URL。

这些方法大多有相应的快捷方式,如 df.write.csv()、df.write.json()、df.write.orc()、df.write.parquet()、df.write.jdbc()等。这些方法相当于先调用 format()方法,再调用 save()方法。

在下面这个示例中，先读取 Spark 自带的数据文件 people.json，然后进行简单计算，并把结果 DataFrame 保存到 CSV 格式的存储文件中，最后将这个文件加载到 RDD 中，代码如下：

```scala
#第4章/dataframeWriteDemo1.scala

//加载数据源，构造 DataFrame
val input = "/data/spark/resources/people.json"
val df = spark.read.json(input)

//找出 age 不是 null 的信息，保存到 CSV 文件中
import spark.implicits._

//写出到存储文件中
val output = "/data/spark/people-csv-output"
df.where($"age".isNotNull).write.format("csv").save(output)
//df.where($"age".isNotNull).write.csv(output)    //与上一句等价

//将保存的 CSV 数据再次加载到 RDD 中
val textFile = spark.sparkContext.textFile(output)
textFile.collect.foreach(println)
```

执行以上代码，输出内容如下：

```
30,Andy
19,Justin
```

打开浏览器，访问 HDFS 的 Web UI（网址为 http://localhost:9870），然后浏览 HDFS 文件系统中的结果文件存储目录（在上面的代码中指定的/data/spark/people-csv-output），可以看到的存储内容如图 4-9 所示。

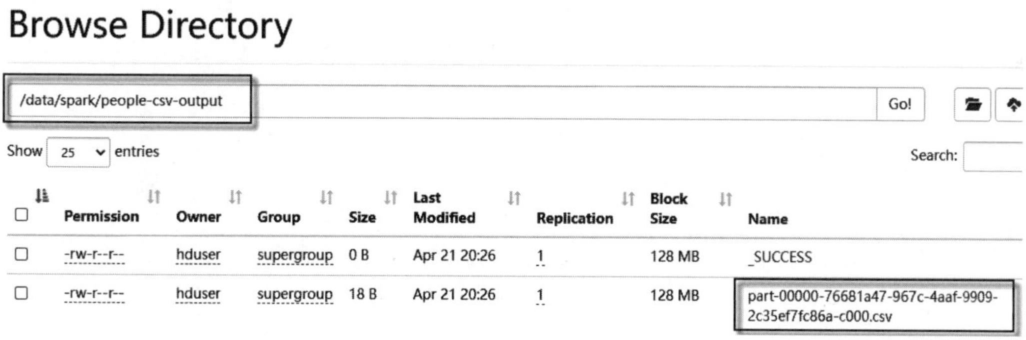

图 4-9 DataFrame 存储的结果文件

注意：write.format()支持输出 JSON、Parquet、JDBC、ORC、Libsvm、CSV、Text 等格式文件，如果要输出 Text 文本文件，则可以采用 write.format("text")，但是，需要注意，只有 select() 中只存在一个列时，才允许保存成文本文件，如果存在两个列，例如 select("name", "age")，就不能保存成文本文件了。

Spark 支持通过 JDBC 方式连接到其他数据库,并将 DataFrame 存储到数据库中。在下面的这个示例中,将分析结果 DataFrame 写到 MySQL 数据库中。

首先,在 MySQL 中新建一个测试 Spark 程序的数据库,数据库的名称是 spark,数据表的名称是 student。登录 MySQL 数据库,执行以下 SQL 语句:

```
mysql> create database spark;
mysql> use spark;
mysql> create table student (id int(4), name char(20), gender char(4), age int(4));
mysql> insert into student values(1,'张三','F',23);
mysql> insert into student values(2,'李四','M',18);
mysql> select * from student;
```

然后,编写 Spark 应用程序,构造一个 DataFrame,包含两行学生信息,代码如下:

```scala
//第4章/dataframeWriteDemo2.scala

//首先导入依赖的包
import java.util.Properties
import org.apache.spark.sql.types._
import org.apache.spark.sql.Row
import org.apache.spark.rdd.RDD

//下面设置两条数据表示两个学生信息
val studentRDD: RDD[Row] = spark.sparkContext
    .parallelize(Array("3,王老五,男,46","4,赵小花,女,27"))
    .map(_.split(","))
    .map(stu => Row(stu(0).toInt, stu(1).trim, stu(2).trim, stu(3).toInt))

//下面要设置模式信息
val schema = StructType(
    Seq(
      StructField("id", IntegerType, nullable = true),
      StructField("name", StringType, nullable = true),
      StructField("gender", StringType, nullable = true),
      StructField("age", IntegerType, nullable = true)
    )
)

//建立起 Row 对象和模式之间的对应关系,也就是把数据和模式对应起来
val studentDF = spark.createDataFrame(studentRDD, schema)
studentDF.printSchema()
studentDF.show()
```

执行以上代码,输出内容如下:

```
root
 |-- id: integer (nullable = true)
 |-- name: string (nullable = true)
 |-- gender: string (nullable = true)
 |-- age: integer (nullable = true)
```

```
+---+------+------+---+
| id| name|gender|age|
+---+------+------+---+
|  3| 王老五|     F| 44|
|  4| 赵小虎|     M| 27|
+---+------+------+---+
```

接下来，编写 Spark 代码，连接 MySQL 数据库并且将上面的 studentDF 写入 MySQL 保存。在本例中向 spark.student 表中插入两条记录，代码如下：

```
//下面创建一个 prop 变量，用来保存 JDBC 连接参数
val prop = new Properties()
prop.put("user", "root")                              //表示用户名是 root
prop.put("password", "admin")                         //表示密码是 hadoop
prop.put("driver","com.mysql.jdbc.Driver")            //表示驱动程序

//连接数据库，采用 append 模式，表示将记录追加到数据库 spark 的 student 表中
val DB_URL= "jdbc:mysql://localhost:3306/spark?useSSL=false"
studentDF.write
    .mode("append")
    .jdbc(DB_URL, "student", prop)
```

执行以上代码，然后到 MySQL 中查询 spark.student 表，SQL 语句如下：

```
mysql> select * from spark.student;
```

可以看到以下的查询结果，证明数据已经被正确地写入了数据库中：

```
+------+------------+--------+------+
|  id  |    name    | gender | age  |
+------+------------+--------+------+
|   1  |    张三     |   F    |  23  |
|   2  |    李四     |   M    |  18  |
|   3  |    王老五    |   男   |  46  |
|   4  |    赵小花    |   女   |  27  |
+------+------------+--------+------+
```

4.8.2　存储模式

DataFrameWriter 类中的一个重要选项是 save mode，它表示存储模式。在将 DataFrame 中的数据写到存储系统上时，默认行为是创建一个新表。如果指定的输出目录或同名表已经存在，则抛出错误消息。可以使用 Spark SQL 的 SaveMode 特性来更改此行为。

Spark 所支持的各种存储模式见表 4-10。

表 4-10　Spark SQL 写出 DataFrame 时支持的存储格式

Scala/Java	任何语言	说　　明
SaveMode.Append	"append"	当将一个 DataFrame 保存到数据源时，如果数据/表已经存在，则该 DataFrame 的内容将被追加到已经存储的数据/表中

续表

Scala/Java	任何语言	说明
SaveMode.Overwrite	"overwrite"	当将一个 DataFrame 保存到数据源时，如果数据/表已经存在，已经存在的数据/表将被该 DataFrame 的内容所覆盖
SaveMode.ErrorIfExists(默认)	"error"或"errorIfExists"(默认)	当将一个 DataFrame 保存到数据源时，如果数据已经存在，将抛出一个异常
SaveMode.Ignore	"ignore"	当将一个 DataFrame 保存到数据源时，如果数据/表已经存在，save 操作不会保存该 DataFrame 的内容，并且不会改变已经存在的数据/表。这类似于 SQL 中的 create table if not exists

这些保存模式不使用任何锁定，也不是原子性的。此外，在执行 overwrite 时，将在写入新数据之前删除原数据。

例如，将数据以 CSV 格式写出，但使用 "#" 作为分隔符，代码如下：

```
movies.write
    .format("csv")
    .option("sep", "#")
    .save("/tmp/output/csv")
```

如果使用 overwrite 模式写出数据，并同时写出标题行，则代码如下：

```
movies.write
    .format("csv")
    .mode("overwrite")
    .option("sep", "#")
    .option("header", "true")
    .save("/tmp/output/csv")
```

4.8.3 控制 DataFrame 的输出文件数量

Spark SQL DataFrame API 在转换操作执行 shuffling 时会增加分区。诸如 join()、union() 这些 DataFrame 操作及所有聚合函数操作会触发数据 shuffle。

例如，要查看一个 DataFrame 拥有的分区数量，代码如下：

```
import spark.implicits._

//构造 DataFrame
val simpleData = Seq(
    ("张三","销售部","北京",90000,34,10000),
    ("李四","销售部","北京",86000,56,20000),
    ("王老五","销售部","上海",81000,30,23000),
    ("赵老六","财务部","上海",90000,24,23000),
    ("钱小七","财务部","上海",99000,40,24000),
    ("周扒皮","财务部","北京",83000,36,19000),
    ("孙悟空","财务部","北京",79000,53,15000),
```

```
    ("朱八戒","市场部","上海",80000,25,18000),
    ("沙悟净","市场部","北京",91000,50,21000)
)
val df = simpleData.
    toDF("employee_name","department","city","salary","age","bonus")

println(s"shuffle 前的分区数:${df.rdd.getNumPartitions}")

//groupBy 操作会触发数据 shuffle
val df2 = df.groupBy("city").count()

println(s"shuffle 后的分区数:${df2.rdd.getNumPartitions}")
```

执行以上代码,输出内容如下:

```
shuffle 前的分区数:2
shuffle 后的分区数:200
```

从输出结果可以看出,经过 groupBy()聚合操作及数据跨分区 shuffling 后,DataFrame 的默认分区数变为 200。当 Spark 操作执行数据 shuffling(join()、union()、aggregation 函数)时,DataFrame 会自动将分区数增加到 200。这个默认的 shuffle 分区数来自 Spark SQL 配置 spark.sql.shuffle.partitions,默认设置为 200。

可以使用 SparkSession 对象的 conf()方法或使用 Spark Submit 命令配置来修改这个默认的 shuffle 分区数,代码如下:

```
spark.conf.set("spark.sql.shuffle.partitions",100)
println(df.groupBy("city").count().rdd.partitions.length)   //100
```

保存到指定输出目录的文件数量与 DataFrame 拥有的分区数量是相对应的。例如 DataFrame 有 100 个分区,那么将其存储到 HDFS 中时就会生成 100 个存储文件,每个分区对应一个存储文件。在下面的示例中,将一部电影数据集加载到 DataFrame,并将其重分区为 4 个分区,然后保存到指定的位置,代码如下:

```
//将电影数据集文件加载到 DataFrame 中
val file = "/data/spark/movies/movies.csv"
val movies = spark.read
    .option("header","true")
    .option("inferSchema","true")
    .csv(file)

val movies2 = movies.repartition(4)       //将 DataFrame 重分区为 4 个分区
println("movies2 的分区数是:" + movies2.rdd.getNumPartitions)

//保存 movies2
movies2.write.csv("/data/spark/movies-out-partitions")
```

执行以上代码,输出内容如下:

```
movies2 的分区数是:4
```

打开浏览器,访问地址 http://localhost:9870,查看保存数据的 HDFS 文件目录,如图 4-10 所示。

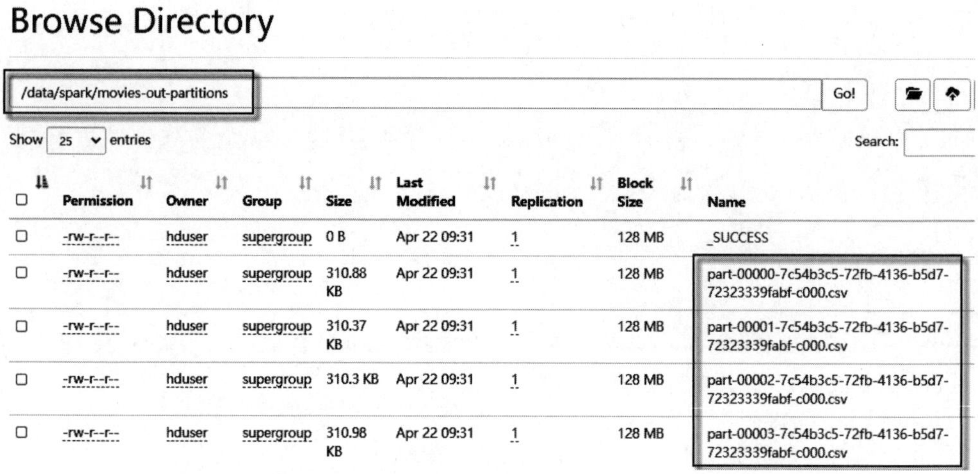

Browse Directory

`/data/spark/movies-out-partitions`

Permission	Owner	Group	Size	Last Modified	Replication	Block Size	Name
-rw-r--r--	hduser	supergroup	0 B	Apr 22 09:31	1	128 MB	_SUCCESS
-rw-r--r--	hduser	supergroup	310.88 KB	Apr 22 09:31	1	128 MB	part-00000-7c54b3c5-72fb-4136-b5d7-72323339fabf-c000.csv
-rw-r--r--	hduser	supergroup	310.37 KB	Apr 22 09:31	1	128 MB	part-00001-7c54b3c5-72fb-4136-b5d7-72323339fabf-c000.csv
-rw-r--r--	hduser	supergroup	310.3 KB	Apr 22 09:31	1	128 MB	part-00002-7c54b3c5-72fb-4136-b5d7-72323339fabf-c000.csv
-rw-r--r--	hduser	supergroup	310.98 KB	Apr 22 09:31	1	128 MB	part-00003-7c54b3c5-72fb-4136-b5d7-72323339fabf-c000.csv

图 4-10 DataFrame 存储的结果文件数量与分区数量相对应

在某些情况下，DataFrame 的内容并不大，并且需要写入单个文件中。这时可以使用 coalesce()转换函数，将 DataFrame 的分区数量减少到 1，然后把它写出。coalesce()函数的定义如下：

```
coalesce(numPartitions: Int, shuffle: Boolean = false)
```

这种方法会返回一个新的 RDD，该 RDD 被缩减为 numPartitions 所指定的分区数，其中第 1 个参数 numPartitions 用于指定要被缩减到的目标分区数，每两个参数指定允不允许数据 shuffle。这导致了一个窄依赖，例如，如果从 1000 个分区缩减到 100 个分区，不会产生数据 shuffle，而是每 100 个新分区将占用 10 个当前分区。如果请求比当前分区数更多的分区，则将保持当前的分区数。

使用 shuffle = true，实际上可以合并到更多的分区。如果有少量的分区(例如 100 个)，并且可能有几个分区异常大，则这是有用的。调用 coalesce(1000, shuffle = true)将导致 1000 个分区，数据使用散列分区器分布。传入的可选分区合并器必须是可序列化的。

假设现在想将前面代码中的 movies2 写入一个文件中存储，那么可以对上面的示例代码进行修改，最后一行代码如下：

```
//将 movies2 保存到一个存储文件中
movies2.coalesce(1).write.csv("/data/spark/movies-out-partitions")
```

1. 分区多少对性能的影响

根据数据集大小、CPU 核数量和内存大小，Spark shuffling 可能对作业有利也可能有害。当处理较少的数据量时，通常应该减少 shuffle 分区，否则将最终得到许多分区文件，每个分区中的记录数量较少。这将导致运行许多需要处理的数据较少的任务。另一方面，当有太多的数据和较少的分区数时同样会导致长时间运行少量任务，这时也可能会得到内存错误，所以分区过少或过多，都可能是不利的。

（1）分区过少：将无法充分利用群集中的所有可用的 CPU 核。

（2）分区过多：产生非常多的小任务，从而会产生过多的开销。

在这两者之间，第 1 个对性能的影响相对比较大。对于小于 1000 个分区数的情况而言，调度太多的小任务所产生的影响相对较小，但是，如果有成千上万个分区，则 Spark 会变得非常慢。

获得正确的 shuffle 分区大小总是很棘手的，需要多次尝试不同的值来获得最优的分区数。当在 Spark 作业上遇到性能问题时，这是需要查找的关键属性之一。

2. 如何设置分区数量

假设要对一个大数据集进行操作，该数据集的分区数也比较大，那么当进行一些操作之后，例如 filter()过滤操作、sample()抽样操作，这些操作可能会使结果数据集的数据量大幅减少，但是 Spark 却不会对分区进行调整，由此会造成大量的分区没有数据，并且会向 HDFS 读取和写入大量的空文件，效率会很低，这种情况就需要我们重新调整分区数量，以此来提升效率。

通常情况下，当结果集的数据量减少时，其对应的分区数也应当相应地减少。那么该如何确定具体的分区数呢？

Spark 中的 shuffle 分区数是静态的，它不会随着不同的数据大小而变化。在上文提到：默认情况下，控制 shuffle 分区数的参数 spark.sql.shuffle.partitions 的值为 200，这将导致以下问题：

（1）对于较小的数据，200 分区过多了，由于调度开销，通常会导致处理速度变慢。

（2）对于大数据，200 分区则太少了，无法有效地使用群集中的所有资源。

一般情况下，可以通过将集群中的 CPU 数量乘以 2、3 或 4 来确定分区的数量。如果要将数据写到文件系统中，则可以选择一个分区大小，以创建合理大小的文件。

3. 如何控制输出文件的大小

如果写入时产生的小文件数量过多，则会产生大量的元数据开销。Spark 和 HDFS 一样，都不能很好地处理这个问题，这被称为"小文件问题"。同时数据文件也不能过大，否则在查询时会有不必要的性能开销，因此要把文件大小控制在一个合理的范围内。

在 Spark 中，有两种方法可以控制输出文件的大小：

（1）可以通过分区数量来控制生成文件的数量，从而间接地控制文件的大小。

（2）Spark 2.2 引入了一个新的 maxRecordsPerFile 参数，以更自动化的方式控制文件大小。通过它可以控制写入文件的记录数来控制文件的大小。

例如，可以在写出 DataFrame 结果时，指定如下的选项，Spark 将确保每个输出文件最多包含 5000 条记录，代码如下：

```
df.write.option("maxRecordsPerFile", 5000)
```

4.8.4 控制 DataFrame 实现分区存储

在 Spark SQL 中写出数据时也可以使用分区技术。DataFrameWriter 有一个 partitionBy() 方法，它用于指定数据在写入磁盘前如何先进行分区。

例如，在 movies 这个 DataFrame 中，year 列是最适合用来进行分区的候选列。假设要写出 movies 这个 DataFrame，用 year 列来分区，DataFrameWriter 将把所有具有相同年份的电影都写在同一个目录中。输出文件夹中的目录数量将对应于 movies 这个 DataFrame 中的年份的数量，代码如下：

```
//将电影数据集文件加载到 DataFrame 中
val file = "/data/spark/movies/movies.csv"
val movies = spark.read
    .option("header","true")
    .option("inferSchema","true")
    .csv(file)

//输出 movies DataFrame，使用 Parquet 格式并按 year 列进行分区
val outpath = "/data/spark/movies-out-partitions"
movies.repartition($"year")    //先重分区，更有效率
    .write
    .partitionBy("year")    //按 year 字段动态分区
    .save(outpath)
```

注意，在上面的代码中，先用 repartition()方法在内存中按 year 列进行分区，然后调用 repartitionBy()方法按相同的 year 列进行物理分区。这是通常的做法，有利于优化物理分区速度。

执行以上代码，然后在 HDFS 的 Web UI（http://localhost:9870）查看存储目录，可以看到分区存储结果，如图 4-11 所示。

图 4-11 使用 partitionBy 指定分区存储

可以看到，/data/spark/movies-out-partitions 目录下包含多个子目录，从 year=1961 到 year=2012。图 4-11 中只截取了部分内容。

4.9 使用类型化的 DataSet

在 Spark 中，DataSet 是强类型的、分布式的、表状的对象集合，具有良好定义的行和列。DataSet 具有一个 Schema，用来定义列的名称和数据类型。DataSet 提供编译时类型安全，这意味着 Spark 在编译时会检查 DataSet 元素的类型。

DataSet 是类型化的分布式数据集合，它统一了 DataFrame 和 RDD API。DataSet 需要结构化/半结构化数据，模式（Schema）和编码器（Encoder）是数据集的一部分。

从 Spark2.0 开始，Spark 整合了 DataSet 和 DataFrame，前者是有明确类型的数据集，后者是无明确类型的数据集。DataFrame 实际上是 DataSet[Row]类型，DataFrame 每行的类型是 Row（Row 是一个通用的非类型化 JVM 对象，不解析时无法得知每行的字段名和对应的字段类型）。相反，DataSet 是强类型 JVM 对象的集合。两者之间的关系如图 4-12 所示。

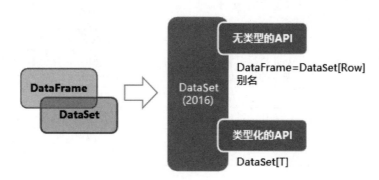

图 4-12　Spark 2.0 统一了 DataSet 和 DataFrame API

4.9.1 了解 DataSet

DataSet 是在 Spark 1.6 中添加的一个新接口，它是 DataFrame API 的扩展，是新的抽象。它同时提供了 RDD（强类型，能够使用强大的 lambda 函数）的优点和 Spark SQL 的优化执行引擎的优点。可以从 JVM 对象构造 DataSet，然后使用函数转换（map()、flatMap()、filter() 等）进行操作。DataSet API 可在 Scala 和 Java 中使用。Python 不支持 DataSet API。

DataSet 由类型化对象组成，这意味着转换语法错误（例如方法名中的拼写错误）和分析错误（例如不正确的输入变量类型）可以在编译时捕获。DataFrame 由无类型的 Row 对象组成，这意味着在编译时只能捕获语法错误。Spark SQL 由一个字符串组成，这意味着语法错误和分析错误只在运行时被捕获。DataSet 可以更快地捕获错误，从而节省开发人员的时间，即使是在使用 IDE 时的输入错误也能被捕获。SQL、DataFrame 和 DataSet 这三类 API 的错误捕获级别见表 4-11。

表 4-11 错误捕获级别

捕获级别	SQL	DataFrame	DataSet
语法错误	运行时捕获	编译时捕获	编译时捕获
分析错误	运行时捕获	运行时捕获	编译时捕获

在 DataSet 中使用 case class 来定义数据模式的结构。使用 case class 很容易处理 DataSet。case class 中不同属性的名称直接映射到 DataSet 中的字段名称。它给人一种使用 RDD 的感觉，但实际上它与 DataFrame 的工作原理是一样的。

关于 RDD、DataFrame 和 DataSet 的比较，见表 4-12。

表 4-12 RDD、DataFrame 和 DataSet 的比较

RDD	DataFrame	DataSet
函数式转换 API	关系型转换 API	函数式 & 关系型转换 API
非类型安全的，编译器不会进行类型检查	非类型安全的，编译器不会进行类型检查	类型安全的，编译时进行类型检查
无内置优化	Catalyst 查询优化	Catalyst 查询优化
内存 JVM 对象导致 GC 和 Java 序列化开销	Tungsten 执行引擎，使用二进制内存数据表示的堆外内存管理	Tungsten 执行引擎，使用二进制内存数据表示的堆外内存管理
	JIT 代码生成	JIT 代码生成
	无反序列化的排序/shuffling	无反序列化的排序/shuffling
	无编码器	有编码器，生成字节码并提供对属性的按需访问，无须反序列化

DataFrame 实际上被视为通用 Row 对象的数据集，即 DataFrame=DataSet[Row]，因此，可以通过调用 DataFrame 上的 as() 方法，将 DataFrame 转换为 DataSet，例如，df.as[MyClass]。

4.9.2 创建 DataSet

创建 DataSet 的方法有多种：
（1）第 1 种方法是使用 DataFrame 类的 as() 函数将 DataFrame 转换为 DataSet。
（2）第 2 种方法是使用 SparkSession.createDataSet() 函数从本地集合对象中创建 DataSet。
（3）第 3 种方法是使用 toDS() 隐式转换程序。

在创建 DataSet 的这 3 种方法中，第 1 种方法是最流行的方法。

不管使用哪一种方法，通常需要做的第一件事都是定义一个领域特定的对象来表示每行数据（需要先定义一个代表数据模式的 case class）。

1. 方法一

下面的代码使用第 1 种方法，将一个 DataFrame 转换为 DataSet。首先定义一个代表每行数据的 case class，代码如下：

```
//定义Movie case class
case class Movie(actor_name:String, movie_title:String, produced_year:Long)
```

然后将 movies.parquet 文件中的数据读到一个 DataFrame 中，代码如下：

```
//定义文件路径
val parquetFile = "/data/spark/movies/movies.parquet"

//读到 DataFrame 中
val movies = spark.read.parquet(parquetFile)

//输出 Schema 和内容
movies.printSchema()
movies.show(5)
```

执行以上代码，输出结果及执行信息如图 4-13 所示（以交互式方式执行时）。

```
root
 |-- actor_name: string (nullable = true)
 |-- movie_title: string (nullable = true)
 |-- produced_year: long (nullable = true)

+----------------+----------------+-------------+
|      actor_name|     movie_title|produced_year|
+----------------+----------------+-------------+
|McClure, Marc (I)|    Coach Carter|         2005|
|McClure, Marc (I)|     Superman II|         1980|
|McClure, Marc (I)|       Apollo 13|         1995|
|McClure, Marc (I)|        Superman|         1978|
|McClure, Marc (I)|Back to the Future|       1985|
+----------------+----------------+-------------+
only showing top 5 rows

parquetFile: String = /data/spark/movies/movies.parquet
movies: org.apache.spark.sql.DataFrame = [actor_name: string, movie_title: string ... 1 more field]
```

图 4-13　输出结果及执行信息

从图 4-13 的输出结果及执行信息中可以看出，调用 spark.read.parquet()方法将 Parquet 文件加载到一个 org.apache.spark.sql.DataFrame 类型的对象中。接下来，可以在该 DataFrame 上调用 as()方法，将其转换到强类型的 DataSet[Movie]，代码如下：

```
import spark.implicits._

//将 DataFrame 转换到强类型的 DataSet
val moviesDS = movies.as[Movie]
```

执行上面的代码，可以看到执行信息（以交互式方式执行时）如图 4-14 所示。

```
import spark.implicits._
moviesDS: org.apache.spark.sql.Dataset[Movie] = [actor_name: string, movie_title: string ... 1 more field]
```

图 4-14　将 DataFrame 转换为强类型的 DataSet[Movie]

从图 4-14 中可以看到，moviesDS 是一个 org.apache.spark.sql.DataSet[Movie] 类型的实例对象，然后就可以在该强类型对象上调用类似 RDD 上的 Transformation 转换操作。例如，要找出 2010 年上映的电影信息，代码如下：

```
//过滤 2010 生产的电影
moviesDS.filter(movie => movie.produced_year == 2010).show(5)
```

执行以上代码，输出内容如下：

```
+--------------------+--------------------+-------------+
|          actor_name|         movie_title|produced_year|
+--------------------+--------------------+-------------+
|    Cooper, Chris (I)|            The Town|         2010|
|      Jolie, Angelina|                Salt|         2010|
|      Jolie, Angelina|         The Tourist|         2010|
|      Danner, Blythe|      Little Fockers|         2010|
|   Byrne, Michael (I)|Harry Potter and ...|         2010|
+--------------------+--------------------+-------------+
only showing top 5 rows
```

如果要显示 moviesDS 中第一部电影的名称，代码如下：

```
//显示 moviesDS 中第一部电影的 title
println(moviesDS.first.movie_title)
```

执行以上代码，输出内容如下：

```
Coach Carter
```

如果在上面的代码中，拼写错了 movie_title，例如错误地写成 movie_tile，会怎样？拼写错误的 movie_tile 代码如下：

```
//试一下：拼写错 movie_title，得到编译时错误
moviesDS.first.movie_tile
```

这时会得到编译时错误信息，内容如下：

```
<console>:30: error: value movie_tile is not a member of Movie
       moviesDS.first.movie_tile
```

可以使用 map() 转换执行投影操作，只保留想要的字段，代码如下：

```
//使用 map transformation 执行投影(projection)
//map: 返回一个新的 DataSet,其中包含对每个元素应用'func'的结果
val titleYearDS = moviesDS.map(m => ( m.movie_title, m.produced_year))

titleYearDS.printSchema()
titleYearDS.show(5)
```

执行以上代码，输出内容如下：

```
root
 |-- _1: string (nullable = true)
 |-- _2: long (nullable = false)

+-----------------+----+
|               _1|  _2|
+-----------------+----+
|     Coach Carter|2005|
|      Superman II|1980|
|        Apollo 13|1995|
|         Superman|1978|
|Back to the Future|1985|
+-----------------+----+
only showing top 5 rows
```

相比 DataFrame，DataSet 支持类型安全的 Transformation 转换。例如，对于 movies（这是一个 DataFrame），如果编写了错误的 Transformation 转换，则在编译时是无法发现的，代码如下：

```
//两个字符串列是无法执行减法操作的，但编译时不会发现
val df1 = movies.select('movie_title - 'movie_title)
```

只有到运行时才会发现错误，代码如下：

```
//对于 DataFrame 来讲，直到运行时才会检查出问题
df1.show(5)
```

执行 Action 操作，运行并输出结果如下：

```
+---------------------------+
|(movie_title - movie_title)|
+---------------------------+
|                       null|
|                       null|
|                       null|
|                       null|
|                       null|
+---------------------------+
only showing top 5 rows
```

但是，如果在 moviesDS（这是一个 DataSet）上执行同样的减法操作，代码如下：

```
val df2 = moviesDS.map(m => m.movie_title - m.movie_title)
```

则在编译时就会检查出问题，给出的错误信息如下：

```
<console>:28: error: value - is not a member of String
       val df2 = moviesDS.map(m => m.movie_title - m.movie_title)
```

对于 DataSet 来讲，Transformation 转换是类型安全的，它会在编译时失败，因为在字符串类型的列上执行了减法操作。

在 DataSet 上执行 Action 操作时，会将 case class 对象作为行返给驱动程序。例如，在 moviesDS 上调用 take() 操作，会将 Movie 对象行返给驱动程序，代码如下：

```
moviesDS.take(10).foreach(println)
```

执行以上代码,输出内容如下:

```
Movie(McClure, Marc (I),Coach Carter,2005)
Movie(McClure, Marc (I),Superman II,1980)
Movie(McClure, Marc (I),Apollo 13,1995)
Movie(McClure, Marc (I),Superman,1978)
Movie(McClure, Marc (I),Back to the Future,1985)
Movie(McClure, Marc (I),Back to the Future Part III,1990)
Movie(Cooper, Chris (I),Me, Myself & Irene,2000)
Movie(Cooper, Chris (I),October Sky,1999)
Movie(Cooper, Chris (I),Capote,2005)
Movie(Cooper, Chris (I),The Bourne Supremacy,2004)
```

DataSet 也可以从 JSON 文件中创建,类似于 DataFrame。需要注意,JSON 文件可能包含几条记录,但是每条记录必须在一行上。如果源 JSON 有换行符,则必须以编程的方式删除它们。JSON 记录可能有数组,并且可以嵌套。它们不需要有统一的模式。

在下面的示例中,从 JSON 文件创建一个 DataSet,JSON 文件中的每条记录有一个附加的标签和一个数据数组。

首先,同样需要定义一个 case class,代码如下:

```
//定义 Movie case class
case class Movie(actor_name:String, movie_title:String, produced_year:Long)
```

然后将 movies.json 文件读到 DataFrame 中,并使用 as()方法将其转换为 DataSet[Movie],代码如下:

```
//设置文件路径
val file = "/data/spark/movies/movies.json"

import spark.implicits._

//使用 case class 从 JSON 创建 DataSet
val movies = spark.read.json(file).as[Movie]

//查看数据
movies.printSchema()
movies.show(10)
```

执行以上代码,输出内容如下:

```
root
 |-- actor_name: string (nullable = true)
 |-- movie_title: string (nullable = true)
 |-- produced_year: long (nullable = true)

+-----------------+--------------------+-------------+
|       actor_name|         movie_title|produced_year|
+-----------------+--------------------+-------------+
|McClure, Marc (I)|        Coach Carter|         2005|
|McClure, Marc (I)|         Superman II|         1980|
|McClure, Marc (I)|          Apollo 13|         1995|
```

```
|McClure, Marc (I)|            Superman|1978|
|McClure, Marc (I)|  Back to the Future|1985|
|McClure, Marc (I)|Back to the Futur...|1990|
|Cooper, Chris (I)|  Me, Myself & Irene|2000|
|Cooper, Chris (I)|         October Sky|1999|
|Cooper, Chris (I)|              Capote|2005|
|Cooper, Chris (I)|The Bourne Supremacy|2004|
+-----------------+--------------------+----+
only showing Top 10 rows
```

2. 方法二

创建 DataSet 的第 2 种方法是使用 SparkSession.createDataset()函数从本地集合对象中创建 DataSet。下面的代码演示了这种方法。

首先，定义一个 case class，代码如下：

```
//定义 Movie case class
case class Movie(actor:String, title:String, year:Long)
```

然后，创建一个本地集合，使用 SparkSession.createDataset()从本地集合创建，代码如下：

```
//定义一个本地集合
val localMovies = Seq(
    Movie("郭涛", "疯狂的石头", 2018L),
    Movie("黄渤", "疯狂的石头", 2018L)
)

//常见类型的编码器是通过导入 spark.implicit._ 自动提供的
import spark.implicits._

//使用 SparkSession.createDataset()函数从本地集合对象中创建 DataSet
val localMoviesDS = spark.createDataset(localMovies)

//输出
localMoviesDS.printSchema()
localMoviesDS.show()
```

在上面的代码中，case class 会添加结构信息。Spark 使用这种结构来创建最优的数据布局和编码。执行以上代码，输出内容如下：

```
root
 |-- actor: string (nullable = true)
 |-- title: string (nullable = true)
 |-- year: long (nullable = false)

+-----+----------+----+
|actor|     title|year|
+-----+----------+----+
| 郭涛|疯狂的石头|2018|
| 黄渤|疯狂的石头|2018|
+-----+----------+----+
```

3. 方法三

创建 DataSet 的第 3 种方法是使用 toDS()隐式转换程序。下面的示例演示了这种方法。首先，还是先创建一个 case class，代码如下：

```
//定义 Movie case class
case class Movie(actor:String, title:String, year:Long)
```

然后，创建一个本地集合，使用 toDS()从本地集合创建，代码如下：

```
//定义一个本地集合
val localMovies = Seq(
    Movie("郭涛", "疯狂的石头", 2018L),
    Movie("黄渤", "疯狂的石头", 2018L)
)

//常见类型的编码器是通过导入 spark.`implicit`._自动提供的
import spark.implicits._

//toDS 隐式方法创建一个 DataSet
val localMoviesDS2 = localMovies.toDS()

//输出
localMoviesDS2.printSchema()
localMoviesDS2.show()
```

执行以上代码，输出内容如下：

```
root
 |-- actor: string (nullable = true)
 |-- title: string (nullable = true)
 |-- year: long (nullable = false)


+-----+----------+----+
|actor|     title|year|
+-----+----------+----+
| 郭涛|疯狂的石头|2018|
| 黄渤|疯狂的石头|2018|
+-----+----------+----+
```

4. 其他方法

除了上面这几种方法，还可以通过 toDS()方法将一个 Range 类型转换为 DataSet，代码如下：

```
import spark.implicits._

//从一个简单的集合来创建一个 DataSet
val ds1 = List.range(1,5).toDS()

ds1.printSchema()
ds1.show()
```

执行以上代码，输出内容如下：

```
root
 |-- value: integer (nullable = false)


+-----+
|value|
+-----+
|    1|
|    2|
|    3|
|    4|
+-----+
```

或者将一个 RDD 转换为 DataSet,代码如下:

```
import spark.implicits._

//从 RDD 到 DataSet 间的转换,使用类型推断
val phones = List(
    ("小米","中国",3999.00),
    ("华为","中国",4999.00),
    ("苹果","美国",5999.00),
    ("三星","韩国",1999.00),
    ("诺基亚","荷兰",999.00)
)
val phones_ds = spark.sparkContext.parallelize(phones).toDS()

phones_ds.printSchema()
phones_ds.show()
```

执行以上代码,输出内容如下:

```
root
 |-- _1: string (nullable = true)
 |-- _2: string (nullable = true)
 |-- _3: double (nullable = false)

+------+----+------+
|    _1|  _2|    _3|
+------+----+------+
|  小米|中国|3999.0|
|  华为|中国|4999.0|
|  苹果|美国|5999.0|
|  三星|韩国|1999.0|
|诺基亚|荷兰| 999.0|
+------+----+------+
```

4.9.3 操作 DataSet

对 DataSet 的操作同样分为 Transformation 转换操作和 Action 行为操作,而且非常类似 DataFrame 和 RDD。常用的 DataSet Transformation 操作见表 4-13。

表 4-13 常用的 DataSet Transformation 操作

Transformation 操作	描 述
map	返回将输入函数应用到每个元素之后的新 DataSet
filter	返回一个新 DataSet，它包含输入函数为 true 的元素
groupByKey	返回一个 KeyValueGroupedDataset，其数据按给定的 key 函数分组

常用的 DataSet Action 操作见表 4-14。

表 4-14 常用的 DataSet Action 操作

Action 操作	描 述
show(n)	显示 DataSet 中前 n 行数据
take(n)	以数组形式返回 DataSet 中前 n 个对象
count	返回 DataSet 中的行数

DataSet 既支持类型化的操作，也支持非类型化的操作。

1. 类型化操作

DataSet 的类型化操作包含转换类型操作、过滤类型操作、集合类型操作、分段类型操作、排序类型操作、分区类型操作、去重类型操作、set 集合类型操作等。

1) 转换类型操作

转换类型操作主要包括 flatMap()、map()、mMapPartitions()、transform()和 as()操作。

（1）flatMap()：可以通过 flatMap()将数据块转换为数组，然后展开数组并将其放入 DataSet 中，代码如下：

```
import org.apache.spark.sql.Dataset

val ds: Dataset[String] = Seq("hello spark","hello hadoop").toDS()
ds.flatMap(item => item.split(" ")).show
```

执行上面的代码，输出内容如下：

```
+------+
| value|
+------+
| hello|
| spark|
| hello|
|hadoop|
+------+
```

（2）map()：可以将数据集中的每个数据转换为另一种形式（通过传入计算函数实现），代码如下：

```
//case class
case class Person(name:String, age:Int)
```

```
import org.apache.spark.sql.Dataset

val ds2 = Seq(Person("张三",15),Person("李四",20)).toDS()
```

执行上面的代码,输出内容如下:

```
+----+---+
|name|age|
+----+---+
| 张三| 30|
| 李四| 40|
+----+---+
```

(3) mapPartitions(): mapPartitions()和 map()是一样的,但是 map()的处理单元是每个数据,而 mapPartitions()的处理单元是每个分区,代码如下:

```
//case class
case class Person(name:String, age:Int)

import org.apache.spark.sql.Dataset

val ds3 = Seq(Person("张三",15),Person("李四",20)).toDS()
ds3.mapPartitions(
    //iter 对于每个 Executor 的内存来讲不能太大,否则它将会 OOM
    //每个元素都被转换为一个新的集合
    iter => {
      val result = iter.map(item => Person(item.name,item.age*2))
      result
    }
  ).show()
```

执行上面的代码,输出内容如下:

```
+----+---+
|name|age|
+----+---+
| 张三| 30|
| 李四| 40|
+----+---+
```

(4) transform(): map()、mapPartitions()和 transform()都用于转换,其中 map()和 mapPartitions()函数用于数据转换,而 transform()函数用于整个 DataSet 数据集转换。最大的区别是 transform()可以直接获取 DataSet 进行操作,代码如下:

```
val ds = spark.range(10)
ds.transform(item => item.withColumn("double",'id * 2))
  .show()
```

执行上面的代码,输出内容如下:

```
+---+------+
| id|double|
+---+------+
|  0|     0|
|  1|     2|
```

```
| 2|    4|
| 3|    6|
| 4|    8|
| 5|   10|
| 6|   12|
| 7|   14|
| 8|   16|
| 9|   18|
+---+-----+
```

(5) as(): as[Type]操作符的主要功能是将弱类型 DataSet 转换为强类型 DataSet。它有许多适用场景,但最常见的场景是读取数据时,因为 DataFrameReader 系统主要将读取的数据转换为 DataFrame 的形式。如果以后需要使用 DataSet 的强类型 API,则需要转换 DataFrame。可以使用 as[Type]操作符来完成此转换操作,代码如下:

```
//定义Movie case class
case class Movie(actor_name:String, movie_title:String, produced_year:Long)

//定义文件路径
val parquetFile = "/data/spark/movies/movies.parquet"

//读到DataFrame中
val movies = spark.read.parquet(parquetFile)

import spark.implicits._

//将DataFrame转换到强类型的DataSet
val moviesDS = movies.as[Movie]

//查看数据
moviesDS.printSchema()
moviesDS.show(10)
```

执行上面的代码,输出内容如下:

```
root
 |-- actor_name: string (nullable = true)
 |-- movie_title: string (nullable = true)
 |-- produced_year: long (nullable = true)

+------------------+--------------------+-------------+
|        actor_name|         movie_title|produced_year|
+------------------+--------------------+-------------+
|McClure, Marc (I)|        Coach Carter|         2005|
|McClure, Marc (I)|         Superman II|         1980|
|McClure, Marc (I)|           Apollo 13|         1995|
|McClure, Marc (I)|            Superman|         1978|
|McClure, Marc (I)|  Back to the Future|         1985|
|McClure, Marc (I)|Back to the Futur...|         1990|
|Cooper, Chris (I)|  Me, Myself & Irene|         2000|
|Cooper, Chris (I)|         October Sky|         1999|
```

```
|Cooper, Chris (I)|              Capote|        2005|
|Cooper, Chris (I)|The Bourne Supremacy|        2004|
+-----------------+--------------------+------------+
only showing Top 10 rows
```

2）过滤类型操作

过滤型的操作主要包括 filter()，filter()根据条件对数据集进行过滤。例如，创建一个包含个人信息的 DataSet，然后应用 filter()方法过滤出其中年龄大于 15 岁的个人信息，代码如下：

```
case class Person(name:String, age:Int)

import spark.implicits._

val ds = Seq(Person("张三",15),Person("李四",20)).toDS()

ds.filter(item => item.age>15).show()
```

执行上面的代码，输出内容如下：

```
+----+---+
|name|age|
+----+---+
|李四| 20|
+----+---+
```

3）集合类型操作

Collection 集合类型的操作主要包括 groupByKey()。

groupByKey()：groupByKey()操作的返回结果是 KeyValueGroupedDataSet，而不是 DataSet，因此，必须首先通过 KeyValueGroupedDataSet 中的方法进行聚合，然后在使用 Action 获取结果之前返回 DataSet。

该方法的签名如下：

```
groupByKey[K: Encoder](func: T => K): KeyValueGroupedDataset[K, T]
```

事实上，这也证实了分组必须是聚合的这一事实，代码如下：

```
import spark.implicits._
import org.apache.spark.sql.KeyValueGroupedDataset

case class Person(name:String, age:Int)

val ds = Seq(Person("张三",15),Person("李四",20)).toDS()

val grouped: KeyValueGroupedDataset[String, Person] = ds.groupByKey(item 
=> item.name)

val result: Dataset[(String, Long)] = grouped.count()

result.show()
```

执行上面的代码，输出内容如下：

```
+----+--------+
| key|count(1)|
+----+--------+
|李四|       1|
|张三|       1|
+----+--------+
```

函数 groupByKey()通过输入函数 func 对记录（类型为 T）进行分组，最后返回一个 KeyValueGroupedDataset，以便对其应用聚合，代码如下：

```
case class Product(title: String, quantity: Int, price: Double)

val products = Seq(
    Product("服装", 2, 2000.00),
    Product("服装", 1, 2500.00),
    Product("玩具", 3, 500.00),
    Product("玩具", 2, 500.00),
    Product("玩具", 4, 1000.00)
)
val productsDS = products.toDS().cache()

productsDS.printSchema()

//简单分组统计
productsDS
    .groupByKey(_.title)
    .count()
    .show()

//同时统计多个列
import org.apache.spark.sql.expressions.scalalang.typed
productsDS
    .groupByKey(_.title)
    .agg(
      typed.sum[Product](_.quantity),       //总数量
      typed.avg[Product](_.price),          //均价
      typed.count(_.title)                  //类别数
    )
    .toDF("商品","总价","均价","总数")
    .orderBy($"value".desc)
    .show()
```

在 DataSet 中，每行是什么类型是不一定的，在自定义了 case class 之后可以很自由地获得每行的信息（可以定义字段名和类型）。可以看出，DataSet 在需要访问列中的某个字段时是非常方便的。

执行以上代码，输出的结果如下：

```
root
 |-- title: string (nullable = true)
 |-- quantity: integer (nullable = false)
 |-- price: double (nullable = false)
```

```
+-----+--------+
|value|count(1)|
+-----+--------+
| 服装|       2|
| 玩具|       3|
+-----+--------+

+----+----+------------------+----+
|商品|总价|              均价|总数|
+----+----+------------------+----+
|玩具| 9.0|666.6666666666666 |   3|
|服装| 3.0|            2250.0|   2|
+----+----+------------------+----+
```

4）分段类型操作

分段类型的操作主要包括 randomSplit()和 sample()。

（1）randomSplit()：randomSplit()将根据传入的权重将一个 DataSet 随机地划分为多个数据集。在 randomSplit()中传递的数组中有多少个权重，最终就将生成多少个数据集。这些权重的和应该是 1，否则将被标准化。例如，创建一个 DataSet，然后使用 randomSplit()方法将其按比例分为 3 个子数据集，代码如下：

```
val ds = spark.range(15)
val datasets: Array[Dataset[java.lang.Long]] = ds.randomSplit(Array(5.0,2.0,3.0))
datasets.foreach(_.show())
```

执行上面的代码，输出内容如下：

```
+---+
| id|
+---+
|  1|
| 10|
| 11|
| 12|
| 13|
+---+

+---+
| id|
+---+
|  2|
|  3|
|  4|
+---+
```

```
+---+
| id|
+---+
|  0|
|  5|
|  6|
|  7|
|  8|
|  9|
| 14|
+---+
```

（2）sample()：sample()将在 DataSet 中随机抽样，代码如下：

```
val ds = spark.range(15)

ds.sample(withReplacement = false,fraction = 0.4).show()
```

执行上面的代码，输出内容如下：

```
+---+
| id|
+---+
|  0|
|  5|
|  6|
|  7|
|  9|
| 13|
+---+
```

5）排序类型操作

排序类型的操作主要包括 orderBy()和 sort()。

（1）orderBy()：通过 Column API, orderBy()可以实现正序和负序排列，代码如下：

```
case class Person(name:String, age:Int)

val ds = Seq(
      Person("zhangsan",15),
      Person("lisi",20),
      Person("ss",5)
  ).toDS()

ds.orderBy('age.desc_nulls_first).show()
```

执行上面的代码，输出内容如下：

```
+--------+---+
|    name|age|
+--------+---+
|    lisi| 20|
|zhangsan| 15|
|      ss|  5|
+--------+---+
```

（2）sort()：实际上，orderBy()是 sort()的别名，因此它们实现的功能是相同的，代码如下：

```
case class Person(name:String, age:Int)

val ds = Seq(
     Person("zhangsan",15),
     Person("lisi",20),
     Person("ss",5)
   ).toDS()

ds.sort('age.asc).show()
```

执行上面的代码，输出内容如下：

```
+--------+---+
|    name|age|
+--------+---+
|      ss|  5|
|zhangsan| 15|
|    lisi| 20|
+--------+---+
```

6）分区类型操作

分区类型的操作主要包括 coalesce()和 repartition()。

（1）coalesce()：用于减少分区。这个操作符不同于 RDD 中的 coalesce()。DataSet 中的 coalesce()只能减少分区的数量。DataSet 的 coalesce()方法将直接创建一个逻辑操作，并将 shuffle 设置为 false，代码如下：

```
val ds = spark.range(15)
ds.coalesce(1).explain(true)
```

执行上面的代码，输出内容如下：

```
== Parsed Logical Plan ==
Repartition 1, false
+- Range (0, 15, step=1, splits=Some(2))

== Analyzed Logical Plan ==
id: bigint
Repartition 1, false
+- Range (0, 15, step=1, splits=Some(2))

== Optimized Logical Plan ==
Repartition 1, false
+- Range (0, 15, step=1, splits=Some(2))

== Physical Plan ==
Coalesce 1
+- *(1) Range (0, 15, step=1, splits=2)
```

（2）repartition()：repartition()有两种功能，一种是对特定数量的分区进行重分区；另一种是对特定列进行分区，类似于 SQL 中的 DISTRIBUTE BY。例如，将一个 DataSet 重分区为 4 个分区，代码如下：

```
val ds = Seq(Person("zhangsan", 12),
           Person("zhangsan", 8),
           Person("lisi", 15)).toDS()
ds.repartition(4)
ds.repartition('name)
```

执行上面的代码，输出内容如下：

```
ds: org.apache.spark.sql.Dataset[Person] = [name: string, age: int]
res28: org.apache.spark.sql.Dataset[Person] = [name: string, age: int]
```

7）去重类型操作

去重类型的操作主要包括 dropDuplicates()和 distinct()。

（1）dropDuplicates()：使用 dropDuplicates()删除某些列中的重复行，代码如下：

```
case class Person(name:String, age:Int)

val ds = Seq(
     Person("Zhang San",15),
     Person("Zhang San",15),
     Person("Li Si",20)
   ).toDS()

ds.dropDuplicates("age").show()
```

执行上面的代码，输出内容如下：

```
+---------+---+
|     name|age|
+---------+---+
|    Li Si| 20|
|Zhang San| 15|
+---------+---+
```

（2）distinct()：当 dropDuplicates()中没有指定列名时，意味着去重所有的列。dropDuplicates()方法也有一个别名，称为 distinct()。

因此，使用 distinct()也可以去重复，并且只能根据所有列去重，代码如下：

```
case class Person(name:String, age:Int)

val ds = Seq(
     Person("Zhang San",15),
     Person("Zhang San",15),
     Person("Li Si",20)
   ).toDS()

ds.distinct().show()
```

执行上面的代码，输出内容如下：

```
+---------+---+
|     name|age|
+---------+---+
|Zhang San| 15|
|    Li Si| 20|
+---------+---+
```

8）Set 集合类型操作

Set 集合类型的操作主要包括 except()、intersect()、union()和 limit()。

（1）except()：except 表示获取 ds1 中存在但在 ds2 中不存在的数据，实际上是差集，代码如下：

```
val ds1 = spark.range(10)
val ds2 = spark.range(5,15)

ds1.except(ds2).show()
```

执行上面的代码，输出内容如下：

```
+---+
| id|
+---+
|  0|
|  1|
|  2|
|  3|
|  4|
+---+
```

（2）intersect()：求两个集合的交集，代码如下：

```
val ds1 = spark.range(10)
val ds2 = spark.range(5,15)

ds1.intersect(ds2).show()
```

执行上面的代码，输出内容如下：

```
+---+
| id|
+---+
|  5|
|  6|
|  7|
|  8|
|  9|
+---+
```

（3）union()：求两个集合的并集，代码如下：

```
val ds1 = spark.range(10)
val ds2 = spark.range(5,15)

ds1.union(ds2).show()
```

执行上面的代码,输出内容如下:

```
+---+
| id|
+---+
|  0|
|  1|
|  2|
|  3|
|  4|
|  5|
|  6|
|  7|
|  8|
|  9|
|  5|
|  6|
|  7|
|  8|
|  9|
| 10|
| 11|
| 12|
| 13|
| 14|
+---+
```

(4) limit():限制结果集的数量,代码如下:

```
val ds1 = spark.range(10)
val ds2 = spark.range(5,15)

ds2.limit(5).show()
```

执行上面的代码,输出内容如下:

```
+---+
| id|
+---+
| 10|
| 11|
| 12|
| 13|
| 14|
+---+
```

2. 无类型操作

DataSet 也支持无类型操作,这些操作包括选择类型的操作、裁剪类型的操作、聚合类型的操作。

1) 选择类型操作

选择类型的操作主要有 select()、selectExpr()、withColumn()和 withColumnRenamed()。

（1）select()：select()用于选择出现在结果集中的列，代码如下：

```
case class Person(name:String, age:Int)

val ds = Seq(
     Person("zhangsan",15),
     Person("lisi",20),
     Person("cq",18)
   ).toDS()

ds.select("name").show()
```

执行上面的代码，输出内容如下：

```
+--------+
|    name|
+--------+
|zhangsan|
|    lisi|
|      cq|
+--------+
```

（2）selectExpr()：在 SQL 语句中，可以经常在 select 子句中使用 count()、range()和其他函数，也可以在 select expr 中使用这些 SQL 表达式。同时，可以通过带有 expr()函数的 select 实现类似的效果，代码如下：

```
case class Person(name:String, age:Int)

val ds = Seq(
     Person("zhangsan",15),
     Person("lisi",20),
     Person("cq",18)
   ).toDS()

ds.selectExpr("sum(age)").show()

import org.apache.spark.sql.functions._
ds.select(expr("sum(age)")).show()
```

执行上面的代码，输出内容如下：

```
+--------+
|sum(age)|
+--------+
|      53|
+--------+

+--------+
|sum(age)|
+--------+
|      53|
+--------+
```

（3）withColumn()：通过 Column 对象在 DataSet 中创建新列或修改原始列，代码如下：

```
case class Person(name:String, age:Int)

val ds = Seq(
    Person("zhangsan",15),
    Person("lisi",20),
    Person("cq",18)
  ).toDS()

import org.apache.spark.sql.functions._

ds.withColumn("random",expr("rand(1000)")).show()
ds.withColumn("random",rand(1000)).show()
ds.withColumn("name_new",'name).show()
ds.withColumn("name_joke",'name==="").show()
```

执行上面的代码,输出内容如下:

```
+--------+---+--------------------+
|    name|age|              random|
+--------+---+--------------------+
|zhangsan| 15|0.032708201585009866|
|    lisi| 20|  0.3092807517624554|
|      cq| 18| 0.12566721461312746|
+--------+---+--------------------+

+--------+---+--------------------+
|    name|age|              random|
+--------+---+--------------------+
|zhangsan| 15|0.032708201585009866|
|    lisi| 20|  0.3092807517624554|
|      cq| 18| 0.12566721461312746|
+--------+---+--------------------+

+--------+---+--------+
|    name|age|name_new|
+--------+---+--------+
|zhangsan| 15|zhangsan|
|    lisi| 20|    lisi|
|      cq| 18|      cq|
+--------+---+--------+

+--------+---+---------+
|    name|age|name_joke|
+--------+---+---------+
|zhangsan| 15|    false|
|    lisi| 20|    false|
|      cq| 18|    false|
+--------+---+---------+
```

(4) withColumnRenamed():修改列名,代码如下:

```
case class Person(name:String, age:Int)
```

```
val ds = Seq(
    Person("zhangsan",15),
    Person("lisi",20),
    Person("cq",18)
).toDS()

import org.apache.spark.sql.functions._

ds.withColumnRenamed("name","new_name").show()
```

执行上面的代码，输出内容如下：

```
+--------+---+
|new_name|age|
+--------+---+
|zhangsan| 15|
|    lisi| 20|
|      cq| 18|
+--------+---+
```

2）裁剪类型操作

裁剪类型的操作主要包括 drop()。

drop()：减少一个列，代码如下：

```
case class Person(name:String, age:Int)

val ds = Seq(
    Person("zhangsan",15),
    Person("lisi",20),
    Person("cq",18)
).toDS()

import org.apache.spark.sql.functions._

ds.drop("name").show()
```

执行上面的代码，输出内容如下：

```
+---+
|age|
+---+
| 15|
| 20|
| 18|
+---+
```

3）聚合类型操作

聚合类型的操作主要包括 groupBy()。

groupBy()：根据给定的行分组，代码如下：

```
case class Person(name:String, age:Int)

val ds = Seq(
```

```
        Person("zhangsan",15),
        Person("lisi",20),
        Person("cq",18)
).toDS()

import org.apache.spark.sql.functions._

ds.groupBy('name).agg(mean("age")).show()
```

执行上面的代码，输出内容如下：

```
+--------+--------+
|    name|avg(age)|
+--------+--------+
|      cq|    18.0|
|zhangsan|    15.0|
|    lisi|    20.0|
+--------+--------+
```

3. DataSet API 的限制

尽管 DataSet API 是使用 RDD 和 DataFrame 的最佳方式，但它在当前的开发阶段仍然有一些限制：

（1）在查询数据集时，所选字段应该像 case class 一样给定特定的数据类型，否则输出将变成 DataFrame。

（2）Python 和 R 本质上是动态语言，因此不支持类型化的 DataSet。

4.9.4 类型安全检查

当 DataSet 应用在以下几个场景时，有助于更早地发现异常和错误：

（1）当在 filter 或 map 函数中应用 lambda 表达式时。

（2）当查询不存在的列时。

（3）如果转换回 RDD，DataFrame 和 DataSet 是否保留模式。

在下面的示例中，对 DataSet 在类型安全方面与 DataFrame 进行比较。首先创建一个 case class，作为域对象的类型，代码如下：

```
case class Employ(name: String, age: Int, id: Int, department: String)
```

接下来，创建一个 RDD，并构造一个 DataFrame，然后转换为 DataSet，代码如下：

```
//样本数据
val empData = Seq(
    Employ("张三", 24, 132, "人力"),
    Employ("李四", 26, 131, "工程师"),
    Employ("王老五", 25, 135, "数据科学家")
)

//需要引入隐式转换类
import spark.implicits._
```

```
//创建RDD
val empRDD = spark.sparkContext.makeRDD(empData)

//从RDD创建DataFrame
val empDF = empRDD.toDF()

//从RDD创建DataSet
val empDS = empRDD.toDS()

//输出
empDS.printSchema()
empDS.show()
```

执行以上代码，输出内容如下：

```
root
 |-- name: string (nullable = true)
 |-- age: integer (nullable = false)
 |-- id: integer (nullable = false)
 |-- department: string (nullable = true)

+------+---+---+----------+
|  name|age| id|department|
+------+---+---+----------+
|  张三| 24|132|      人力|
|  李四| 26|131|    工程师|
|王老五| 25|135|数据科学家|
+------+---+---+----------+
```

开发人员在使用这两个 API 时面临不同场景。例如，想要找出年龄大于 24 岁的员工信息，对于 empDS，可以使用 filter()方法进行过滤操作，代码如下：

```
val empDSResult = empDS.filter(employ => employ.age>24)
empDSResult.show()
```

执行以上代码，可以得到的输出内容如下：

```
+------+---+---+----------+
|  name|age| id|department|
+------+---+---+----------+
|  李四| 26|131|    工程师|
|王老五| 25|135|数据科学家|
+------+---+---+----------+
```

对于 empDF，使用 filter()转换应用 lambda 表达式则要麻烦得多，代码如下：

```
val empDFResult = empDF.filter(employ => employ.getAs[Int]("age")>24)
```

而使用关系型 API 则更加简单方便，代码如下：

```
val empDFResult = empDF.filter(col("age")>24)
```

另一个应用场景是，在不存在的列上进行查询时表现也不同。例如，在 empDF 上查询不存在的 salary 列时，会产生运行时错误，代码如下：

```
val empDFResult1 = empDF.select("salary")
empDFResult1.show()
```

执行以上代码，会抛出的错误信息如下：

```
org.apache.spark.sql.AnalysisException: cannot resolve '`salary`' given
input columns: [age, department, id, name];
'Project ['salary]
......
```

而当在 DataSet 上查询不存在的 salary 列时，在编译时就会发现错误，代码如下：

```
val empDSResult1 = empDS.map(employ => employ.salary)
```

抛出的编译时错误信息如下：

```
<console>:34: error: value salary is not a member of Employ
       val empDSResult1 = empDS.map(employ => employ.salary)
```

第 3 个应用场景是，当转换回 RDD 时，DataFrame 不会保留 Schema，而 DataSet 会保留 Schema。例如，从 empDF 转换回 RDD，代码如下：

```
val rddFromDataFrame = empDF.rdd              //返回的是 RDD[Row]
rddFromDataFrame.map(employ => employ.name).foreach(println)
```

执行以上代码时，会产生的错误信息如下：

```
<console>:37: error: value name is not a member of org.apache.spark.sql.Row
       rddFromDataFrame.map(employ => employ.name).foreach(println)
                                             ^
```

因为从 DataFrame 转换为 RDD 时，没有保留 Schema，所以调用 employ.name 时会产生错误。如果从 empDS 转换回 RDD，则代码如下：

```
val rddFromDataset = empDS.rdd                //返回的是 RDD[Employ]
rddFromDataset
    .map(employ => employ.name)
    .collect
    .foreach(println)
```

如果执行以上代码，则可正常执行，输出内容如下：

```
张三
李四
王老五
```

4.9.5 编码器

DataSet API 的核心是编码器（Encoder）。编码器负责 JVM 对象和表格表示之间的转换，这种表示以 Tungsten 二进制格式存储，绕过了 JVM 的内存管理和垃圾收集，提高了内存利用率。

Java 虚拟机（JVM）将内存分为栈（stack）和堆（heap）。新对象总是在堆空间中创建，对这些对象的引用存储在栈内存中。Java 垃圾收集（GC）是 Java 程序执行自动内存管理的过程。当堆上的某些对象不再需要时，垃圾收集器会找到它们并删除它们以释放堆内存空间。Java 垃圾收集是一个自动过程。

由于存储在堆上的所有数据都受到垃圾收集的影响，如果存储在堆上的数据变得更大，则垃圾收集器消耗的时间将更长。垃圾收集器有一个垃圾收集暂停，也称为 stop-the-world 事件，这意味着在某个时刻，所有应用程序线程将被挂起，直到垃圾收集器处理完堆中的所有对象。在暂停期间，所有操作都被挂起，因此在 GC 暂停期间，集群上的节点可以表现宕机到其他节点。

使用堆大小大于 1GB 可以获得显著的暂停，然而，当今的服务器应用程序可能需要远远超过 4GB 的堆内存。为了克服这个问题，Spark 没有在堆上创建基于 JVM 的对象（它们是垃圾收集的对象）来存储 DataSet 或 DataFrame。相反，它将堆外 Java 内存分配给它们。堆外内存位于 JVM 之外，不受垃圾收集的影响。为了将任意对象保存到这个非托管的堆外内存中，必须使用序列化（serialization）。序列化是一种将对象转换为字节流的机制。反序列化是反向过程，在此过程中使用字节流重新创建实际的 JVM 对象，因此，应用程序将对象序列化到堆外内存中，然后，可以使用反序列化读取对象。

此外，Spark 基于分布式进行计算，因此数据经常在集群中的计算机节点之间通过网络传输，需要将数据序列化为二进制格式，以便在网络上共享。当需要将一个 JVM 对象从一个节点发送到另一个节点时，发送方首先将其序列化为一字节数组，然后将其发送到接收节点。当接收节点接收到二进制格式的数据时，它将把数据反序列化为一个 JVM 对象。

Spark 有自己的 C 风格的内存访问，专门用于解决它所支持的工作流。DataSet API 中的 Encoders 可以有效地序列化和反序列化 JVM 对象，以生成紧凑的字节码。由此产生紧凑的内存表示占用更少的内存，并具有高效的内存管理，并且导致在 shuffle 操作期间网络负载减小。较小的字节码确保更快的执行速度。

编码器生成紧凑的字节码，直接在序列化的对象上操作，如过滤、排序和散列，并提供对单个属性的按需访问，而无须将字节反序列化回对象，从而提高性能。在缓存 DataSet 时，尽早了解模式会导致内存中更优的布局。对比缓存时 DataSet 和 RDD 对内存的使用情况，如图 4-15 所示（来自 Databricks）。

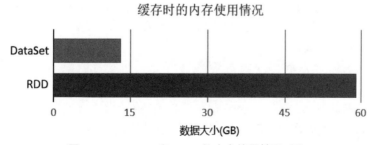

图 4-15　DataSet 和 RDD 的内存使用情况对比

编码器是 Spark SQL 2.0 中序列化和反序列化（SerDe）框架的基本概念。Spark SQL 使用了 SerDe 框架实现 I/O，使其在时间和空间上都非常高效。

在 Spark SQL 2.0 中，编码器被建模为 Encoder[T] trait，代码如下：

```
trait Encoder[T] extends Serializable {
  def schema: StructType
  def clsTag: ClassTag[T]
}
```

类型 T 代表 Encoder[T]可以处理的记录类型。一种类型为 T 的编码器，即 Encoder [T]，用于转换（编码和解码）任何 JVM 对象或类型为 T 的原语（可以是域对象）到 Spark SQL 的内部二进制行格式表示的 InternalRow（使用 Catalyst 表达式和代码生成）。

编码器是任何具有 Encoder[T]的 DataSet[T]的组成（和内部）部分，Encoder[T]用于序列化和反序列化该数据集的记录。DataSet[T]类型是一个 Scala 类型构造函数，带有类型参数 T。Encoder[T]也是，它处理 T 与内部表示的序列化和反序列化。

编码器知道记录的模式，因此与默认的 Java 或 Kryo 序列化器相比，它们提供更快的序列化和反序列化。

例如，定义一个 case class，代表大数据集中记录（record）的域对象，代码如下：

```
case class Person(id: Long, name: String)
```

创建一个 Person 的编码器，代码如下：

```
import org.apache.spark.sql.Encoders

val personEncoder = Encoders.product[Person]        //Encoder[Person]
personEncoder
```

上面的 personEncoder 的数据类型是 org.apache.spark.sql.Encoder[Person]。执行上面的代码，输出内容如下：

```
personEncoder: org.apache.spark.sql.Encoder[Person] = class[id[0]: bigint, name[0]: string]
```

查看编码器的模式，代码如下：

```
personEncoder.schema
```

执行上面的代码，输出内容如下：

```
res5: org.apache.spark.sql.types.StructType = StructType(StructField(id,LongType,false), StructField(name,StringType,true))
```

查看编码器的类标签，代码如下：

```
personEncoder.clsTag
```

执行以上代码，输出内容如下：

```
res6: scala.reflect.ClassTag[Person] = Person
```

编码器 Encoder 也被称为"DataSet 中 serde 表达式的容器"。在 Spark SQL 2/3 中，Encoder trait 的唯一实现是 ExpressionEncoder。ExpressionEncoder 是 Spark 2.0 中唯一一个 Encoder trait 的实现，具有附加属性，即 flat，以及一个或多个 serializers 和一个 deserializer 表达式，其定义如下：

```
case class ExpressionEncoder[T](
```

```
    schema: StructType,
    flat: Boolean,
    serializer: Seq[Expression],
    deserializer: Expression,
    clsTag: ClassTag[T])
extends Encoder[T]
```

所以可以把 personEncoder 转换为 ExpressionEncoder 类型，代码如下：

```
import org.apache.spark.sql.catalyst.encoders.ExpressionEncoder

val personExprEncoder = personEncoder.asInstanceOf[ExpressionEncoder[Person]]
```

查看编码器的 Serializer、Deserializer 和 namedExpressions 部分，代码如下：

```
//编码器的 Serializer 部分
personExprEncoder.serializer

//编码器的 Deserializer 部分
personExprEncoder.deserializer

//编码器的 namedExpressions 部分
personExprEncoder.namedExpressions
```

编码器按名称将 DataSet 的列映射到 JVM 对象的字段上。通过 Encoders，可以将 JVM 对象连接到数据源（CSV、JDBC、Parquet、Avro、JSON、Cassandra、Elasticsearch、MemSQL），反之亦然。

注意：在 Spark SQL 2.0 中，DataFrame 类型仅仅是 DataSet[Row] 的类型别名，RowEncoder 是编码器。

4.10 临时视图与 SQL 查询

Spark SQL 支持直接应用标准 SQL 语句进行查询。当在 Spark SQL 中编写 SQL 命令时，它们会被翻译为 DataFrame 上的关系操作。在 SQL 语句内，可以访问所有 SQL 表达式和内置函数。

4.10.1 在 Spark 程序中执行 SQL 语句

Spark 提供了以下几种在 Spark 中运行 SQL 语句的方法：
（1）Spark SQL CLI（./bin/spark-sql）。
（2）JDBC/ODBC 服务器。
（3）在 Spark 中应用程序以编程方式运行 SQL。
前两种方法提供了与 Apache Hive 的集成，以利用 Hive 的元数据。Spark SQL 支持使用基本 SQL 语法或 HiveQL 编写的 SQL 查询的执行。
在 Spark Shell 或 Zeppelin Notebook 中，会自动导入 spark.sql，所以可以直接使用该函

数用来编写 SQL 命令，代码如下：

```
spark.sql("select current_date() as today , 1 + 100 as value").show()
```

SparkSession 的 sql()函数可执行给定的 SQL 查询，该查询会返回一个 DataFrame。

本节只讨论最后一种方法，即在 Spark 应用程序中以编程方式运行 SQL。下面执行一个不带注册视图的 SQL 语句的简单示例，代码如下：

```
val infoDF = spark.sql("select current_date() as today , 1 + 100 as value")
infoDF.show()
```

执行以上代码，输出内容如下：

```
+----------+-----+
|     today|value|
+----------+-----+
|2021-07-28|  101|
+----------+-----+
```

除了可以使用 Spark read API 将文件加载到 DataFrame 并对其进行查询外，Spark 也可以使用 SQL 语句直接查询该数据文件，代码如下：

```
val hdfsFile = "/data/spark/resources/users.parquet"
val sqlDF = spark.sql(s"SELECT * FROM parquet.`${hdfsFile}`")
sqlDF.show()
```

执行以上代码，输出内容如下：

```
+------+--------------+----------------+
|  name|favorite_color|favorite_numbers|
+------+--------------+----------------+
|Alyssa|          null|  [3, 9, 15, 20]|
|   Ben|           red|              []|
+------+--------------+----------------+
```

4.10.2 注册临时视图并执行 SQL 查询

DataFrame 和 DataSet 本质上就像数据库中的表一样，可以通过 SQL 语句来查询它们。不过，在可以发出 SQL 查询来操作它们之前，需要将它们注册为一个临时视图，然后就可以使用 SQL 查询从临时表中查询数据了。每个临时视图都有一个名字，通过视图的名字来引用该 DataFrame，该名字在 select 子句中用作表名。

例如，要查询电影数据集 movies.parquet 文件，代码如下：

```
//定义文件路径
val parquetFile = "/data/spark/movies/movies.parquet"

//读到 DataFrame 中
val movies = spark.read.parquet(parquetFile)

//现在将 movies DataFrame 注册为一个临时视图
movies.createOrReplaceTempView("movies")
```

```
//从视图 movies 中查询 2009 年后上映并且演员名称包含 Jolie 的电影
val sql = """
   select *
   from movies
   where actor_name like '%Jolie%' and produced_year > 2009
"""
spark.sql(sql).show()
```

执行以上代码,输出内容如下:

```
+---------------+---------------+-------------+
|     actor_name|    movie_title|produced_year|
+---------------+---------------+-------------+
|Jolie, Angelina|           Salt|         2010|
|Jolie, Angelina|Kung Fu Panda 2|         2011|
|Jolie, Angelina|    The Tourist|         2010|
```

也可以在 sql() 函数中混合使用 SQL 语句和 DataFrame 转换 API。例如,查询电影数据集,找出参演影片超过 30 部的演员,代码如下:

```
//定义文件路径
val parquetFile = "/data/spark/movies/movies.parquet"

//读到 DataFrame 中
val movies = spark.read.parquet(parquetFile)

//现在将 movies DataFrame 注册为一个临时视图
movies.createOrReplaceTempView("movies")

//查询参演影片超过 30 部的演员
import spark.implicits._
spark.sql("select actor_name, count(*) as count from movies group by actor_name")
    .where('count > 30)
    .orderBy('count.desc)
    .show()
```

执行以上代码,输出内容如下:

```
+------------------+-----+
|        actor_name|count|
+------------------+-----+
|   Tatasciore, Fred|   38|
|      Welker, Frank|   38|
|Jackson, Samuel L.|   32|
|     Harnell, Jess|   31|
+------------------+-----+
```

当 SQL 语句较长时,可以利用"""(3 个双引号)来格式化多行 SQL 语句。例如,查询电影数据集,使用子查询来计算每年拍摄的电影数量,代码如下:

```
//定义文件路径
val parquetFile = "/data/spark/movies/movies.parquet"

//读到 DataFrame 中
```

```
val movies = spark.read.parquet(parquetFile)

//现在将movies DataFrame注册为一个临时视图
movies.createOrReplaceTempView("movies")

//使用子查询来计算每年拍摄的电影数量（利用"""来格式化多行SQL语句）
import spark.implicits._
spark.sql("""select produced_year, count(*) as count
            from (select distinct movie_title, produced_year from movies)
            group by produced_year
         """)
    .orderBy('count.desc)
    .show(5)
```

执行以上代码，输出内容如下：

```
+-------------+-----+
|produced_year|count|
+-------------+-----+
|         2011|   86|
|         2004|   86|
|         2006|   86|
|         2005|   85|
|         2008|   82|
+-------------+-----+
only showing top 5 rows
```

注意：Spark实现了ANSI SQL:2003修订版（最流行的RDBMS服务器支持）的一个子集。此外，Spark 2.0通过包含一个新的ANSI SQL解析器扩展了Spark SQL功能，支持子查询和SQL:2003标准。更具体地说，子查询支持现在包括相关/不相关的子查询，以及WHERE / HAVING子句中的IN / NOT IN和EXISTS / NOT EXISTS谓词。

4.10.3 使用全局临时视图

Spark为临时视图提供了两个级别的范围。一个是Spark会话级别，当DataFrame在这个级别上注册时，只有在同一个会话中发出的查询才能引用该DataFrame。当Spark会话关闭时，会话范围的级别将消失。另一个作用域级别是全局级别，这意味着可以在所有Spark会话中将这些视图用于SQL语句。

Spark SQL中的临时视图是会话范围的，如果创建它的会话终止，它就会消失。如果希望拥有一个在所有会话之间共享的临时视图，并一直保持活动状态，直到Spark应用程序终止，则可以创建一个全局临时视图。如果要将全局临时视图绑定到系统保留的数据库global_temp，则必须使用限定名来引用它，例如"SELECT * FROM global_temp.view1"。

如果要注册全局临时视图，则可使用createOrReplaceGlobalTempView方法。例如，将movies注册为名为movies_g的全局临时视图，代码如下：

```
movies.createOrReplaceGlobalTempView("movies_g")
```

从全局视图中查询，需要使用关键字global_temp来作为视图名称的前缀，代码如下：

```
spark.sql("select count(*) as total from global_temp.movies_g").show
```

在下面这个示例中,演示了如何跨多个会话使用全局临时视图进行查询。首先,将 Spark 自带的 people.json 文件读到一个 DataFrame 中,代码如下:

```
//将 JSON 数据读到 DataFrame
val df = spark.read.json("/data/spark/resources/people.json")

//简单探索
df.printSchema()
df.show()
```

执行以上代码,输出内容如下:

```
root
 |-- age: long (nullable = true)
 |-- name: string (nullable = true)

+----+-------+
| age|   name|
+----+-------+
|null|Michael|
|  30|   Andy|
|  19| Justin|
+----+-------+
```

然后,将 df 注册为全局视图,名称为 people,代码如下:

```
//将 DataFrame 注册为全局临时视图
df.createGlobalTempView("people")
```

因为已将全局临时视图绑定到一个系统保留的数据库 global_temp 上,因此查询时需要在数据表前加上此数据库前缀,代码如下:

```
spark.sql("SELECT * FROM global_temp.people").show()
```

执行以上代码,输出的结果如下:

```
+----+-------+
| age|   name|
+----+-------+
|null|Michael|
|  30|   Andy|
|  19| Justin|
+----+-------+
```

全局临时视图是跨会话的,所以可以在另一个会话中使用此全局临时视图,代码如下:

```
spark.newSession().sql("SELECT * FROM global_temp.people").show()
```

执行以上代码,输出的结果如下:

```
+----+-------+
| age|   name|
+----+-------+
|null|Michael|
|  30|   Andy|
|  19| Justin|
+----+-------+
```

4.10.4 直接使用数据源注册临时视图

在前面的示例中，都是先将数据加载到一个 DataFrame 中，然后将该 DataFrame 注册为临时视图或全局视图。除此之外，也可以使用 SparkSession 的 sql() 方法从注册的数据源直接加载数据注册临时视图。

例如，注册一个 Parquet 文件并加载它的内容，代码如下：

```
//从 Parquet 数据源创建临时视图
spark.sql("""
    create temporary view usersParquet
    using org.apache.spark.sql.parquet
    options(path '/data/spark/resources/users.parquet')
""")

//查询临时视图
spark.sql("select * from usersParquet").show()
```

执行以上代码，输出的结果如下：

```
+------+--------------+----------------+
|  name|favorite_color|favorite_numbers|
+------+--------------+----------------+
|Alyssa|          null|  [3, 9, 15, 20]|
|   Ben|           red|              []|
+------+--------------+----------------+
```

下面是另一个使用内置数据源的例子：从 JDBC 注册一个临时视图，然后使用 SQL 语句查询该临时视图（这里连接的是 MySQL 5 数据库，读者如果使用的是其他版本的 MySQL，则应自行修改为相应版本的连接参数），代码如下：

```
//数据库连接 URL
val DB_URL= "jdbc:mysql://localhost:3306/xueai8"

//从 JDBC 数据源创建临时视图
spark.sql(s"""
    create temporary view moviesjdbc
    using org.apache.spark.sql.jdbc
    options(
        url '${DB_URL}',
        dbtable 'people',
        user 'root',
        password 'admin'
    )
""")

//在临时视图上执行查询操作
spark.sql("select * from moviesjdbc").show()
```

执行上面的代码，输出的结果如下：

```
+---+------+---+
| id|  name|age|
+---+------+---+
|  1|  张三| 23|
|  2|  李四| 18|
|  3|王老五| 35|
+---+------+---+
```

4.10.5 查看和管理表目录

当将 DataFrame/DataSet 注册为临时视图时，Spark 会将该视图的定义存储在"表目录 (Table Catalog)"中。所有已注册的视图都保存在这个元数据目录中，Spark 提供了一个工具，用于管理这个表目录。这个管理工具是作为 Catalog 类实现的，通过 SparkSession 的 catalog 字段访问。

可以使用该 Catalog 对象来查看当前注册的表有哪些（显示 catalog 目录中的表），如果没有，则返回一个空的 list 列表。例如，查看当前注册的临时视图有哪些，代码如下：

```
//查看当前数据库中注册的表有哪些
spark.catalog.listTables().show()
```

执行上面的代码，输出内容如下：

```
+------------+--------+-----------+---------+-----------+
|        name|database|description|tableType|isTemporary|
+------------+--------+-----------+---------+-----------+
|   moviesjdbc|    null|       null|TEMPORARY|       true|
| usersparquet|    null|       null|TEMPORARY|       true|
+------------+--------+-----------+---------+-----------+
```

另外，还可以使用该 Catalog 对象来检查一个指定的视图所具有的列。例如，要查看 usersParquet 视图的列信息，代码如下：

```
//返回给定表/视图或临时视图的所有列
spark.catalog.listColumns("usersParquet").show()
```

执行上面的代码，输出内容如下：

```
+----------------+-----------+----------+--------+-----------+--------+
|            name|description|  dataType|nullable|isPartition|isBucket|
+----------------+-----------+----------+--------+-----------+--------+
|            name|       null|    string|    true|      false|   false|
|  favorite_color|       null|    string|    true|      false|   false|
|favorite_numbers|       null|array<int>|    true|      false|   false|
+----------------+-----------+----------+--------+-----------+--------+
```

如果要获得所有可用的 SQL 函数列表，则可使用 Catalog 对象的 listFunctions()方法，代码如下：

```
//返回在当前数据库中注册的函数列表
spark.catalog.listFunctions().show(false)
```

执行上面的代码，输出内容如下：

```
+----------+--------+-----------+--------------------+-----------+
|      name|database|description|           className|isTemporary|
+----------+--------+-----------+--------------------+-----------+
|         !|    null|       null|org.apache.spark....|       true|
|         %|    null|       null|org.apache.spark....|       true|
|         &|    null|       null|org.apache.spark....|       true|
|         *|    null|       null|org.apache.spark....|       true|
|         +|    null|       null|org.apache.spark....|       true|
|         -|    null|       null|org.apache.spark....|       true|
|         /|    null|       null|org.apache.spark....|       true|
|         <|    null|       null|org.apache.spark....|       true|
|        <=|    null|       null|org.apache.spark....|       true|
|       <=>|    null|       null|org.apache.spark....|       true|
|         =|    null|       null|org.apache.spark....|       true|
|        ==|    null|       null|org.apache.spark....|       true|
|         >|    null|       null|org.apache.spark....|       true|
|        >=|    null|       null|org.apache.spark....|       true|
|         ^|    null|       null|org.apache.spark....|       true|
|       abs|    null|       null|org.apache.spark....|       true|
|      acos|    null|       null|org.apache.spark....|       true|
|     acosh|    null|       null|org.apache.spark....|       true|
|add_months|    null|       null|org.apache.spark....|       true|
| aggregate|    null|       null|org.apache.spark....|       true|
+----------+--------+-----------+--------------------+-----------+
only showing top 20 rows
```

还可以使用 cacheTable()、uncacheTable()、isCached()和 clearCache()方法来管理这些元数据,包括缓存临时视图、删除临时视图和清除缓存等。

4.11 缓存 DataFrame/DataSet

可以在内存中对 DataFrame 进行持久化/缓存,就像 RDD 一样。在 DataFrame 类中也可以使用同样熟悉的持久性 API(persist()和 unpersist()),然而,DataFrame 的缓存与 RDD 有很大的不同。Spark SQL 知道 DataFrame 中数据的模式(Schema),因此它以一种列格式组织数据,并应用任何适用的压缩来最小化空间的使用。最终的结果是,当两个都由同一个数据文件支持时,在内存中存储 DataFrame 所需的空间比存储 RDD 所需的空间要少得多。

Spark 缓存和持久化是用于迭代和交互 Spark 应用程序的 DataFrame/DataSet 优化技术,以提高作业的性能。

4.11.1 缓存方法

Spark SQL 提供了 cache()和 persist()方法,用来缓存 DataFrame/DataSet。使用 cache()和 persist()方法,Spark 提供了一种优化机制来存储 Spark DataFrame 的中间计算结果,以

便在后续操作中重用。

当持久化一个数据集时,每个节点将它的分区数据存储在自己的内存中,并在该数据集的其他操作中重用它们。Spark 在节点上的持久化数据是容错的,这意味着如果数据集的任何分区丢失,则它将自动使用创建它的原始转换重新计算。

例如,缓存一个 DataFrame,代码如下:

```
//读取 CSV 文件源,创建 DataFrame
val file = "/data/spark/movies/movies.csv"
val df = spark.read
  .options(Map("inferSchema"->"true","delimiter"->",","header"->"true"))
  .csv(file)

//缓存
import spark.implicits._
val df2 = df.where($"year" === "2012").cache()
df2.show(5, truncate = false)

println(s"2012 年上映的电影数量有${df2.count()}部")

val df3 = df2.where($"title" === "The Hunger Games")
println("电影饥饿游戏的主演是: " + df3.collect()(0)(0))        //df3.first.get(0)
```

执行以上代码,输出内容如下:

```
+-----------------+-----------------+----+
|actor            |title            |year|
+-----------------+-----------------+----+
|Cassavetes, Frank|Battleship       |2012|
|Danner, Blythe   |The Lucky One    |2012|
|Manji, Rizwan    |The Dictator     |2012|
|Harrelson, Woody |The Hunger Games |2012|
|Hardy, Tom (I)   |This Means War   |2012|
+-----------------+-----------------+----+
only showing top 5 rows

2012 年上映的电影数量有 601 部
电影饥饿游戏的主演是: Harrelson, Woody
```

DataFrame 或 DataSet 的 cache()方法在内部调用了 persist()方法,默认情况下采用的存储级别是 MEMORY_AND_DISK,因为重新计算底层表的内存列表示是非常昂贵的。注意,这与 RDD.cache()的默认缓存级别 MEMORY_ONLY 不同。

该 persist()方法调用 sparkSession.sharedState.cacheManager.cacheQuery()缓存 DataFrame 或 DataSet 的结果集。不管是 cache()还是 persist()方法,都是延迟计算的(只有在执行 Action 类型的操作时才进行计算)。实际上,cache()是 persist(StorageLevel.MEMORY_AND_DISK) 方法的别名。

4.11.2 缓存策略

存储级别指定如何及在何处持久化或缓存 Spark DataFrame 和 DataSet。Spark SQL 支持的所有不同存储级别都可以在 org.apache.spark.storage.StorageLevel 类上获得：

（1）MEMORY_ONLY。这是 RDD cache()方法的默认行为，并将 DataFrame 作为反序列化对象存储到 JVM 内存中。当没有足够的可用内存时，它将不保存某些分区的 DataFrame，并在需要时重新计算这些 DataFrame。这需要更多的内存，但与 RDD 不同的是，速度将比 MEMORY_AND_DISK 级别慢，因为它重新计算未保存的分区，并且重新计算底层表的内存列表示非常昂贵。

（2）MEMORY_ONLY_SER。这与 MEMORY_ONLY 相同，但不同之处在于它将 RDD/DataFrame 作为序列化对象存储到 JVM 内存中。它比 MEMORY_ONLY 占用更少的内存(节省空间)，因为它将对象以序列化的方式进行保存，并且为了反序列化多占用了几个 CPU 周期。

（3）MEMORY_ONLY_2。与 MEMORY_ONLY 存储级别相同，但将每个分区复制到两个集群节点。

（4）MEMORY_ONLY_SER_2。与 MEMORY_ONLY_SER 存储级别相同，但将每个分区复制到两个集群节点。

（5）MEMORY_AND_DISK。这是 DataFrame 或 DataSet 的默认行为。在这个存储级别中，数据帧将作为反序列化对象存储在 JVM 内存中。当所需的存储大于可用内存时，它将一些多余的分区存储到磁盘中，并在需要时从磁盘读取数据。当涉及 I/O 时，它会较慢。

（6）MEMORY_AND_DISK_SER。这与 MEMORY_AND_DISK 存储级别相同，不同之处在于，当空间不可用时，它会在内存和磁盘上序列化 DataFrame 对象。

（7）MEMORY_AND_DISK_2。与 MEMORY_AND_DISK 存储级别相同，但会将每个分区复制到两个集群节点。

（8）MEMORY_AND_DISK_SER_2。与 MEMORY_AND_DISK_SER 存储级别相同，但将每个分区复制到两个集群节点。

（9）DISK_ONLY。在这个存储级别中，DataFrame 仅存储在磁盘上，CPU 计算时间长，因为涉及 I/O。

（10）DISK_ONLY_2。与 DISK_ONLY 存储级别相同，但会将每个分区复制到两个集群节点。

Spark 会自动监控每个 persist()和 cache()的调用，并检查每个节点上的使用情况，如果没有使用或使用最近最少使用（LRU）算法，则删除持久化数据。也可以使用 unpersist()方法手动删除。unpersist()将 DataSet/DataFrame 标记为非持久的，并从内存和磁盘中删除它的所有块，代码如下：

```
val dfPersist = dfPersist.unpersist()
```

4.11.3 缓存表

也可以使用 Spark 的 catalog 来缓存 DataFrame，以易读的名字持久化表。例如，将一个 DataFrame 持久化到表中，代码如下：

```
//创建一个 DataFrame
val numDF = spark.range(1000).toDF("id")

//注册为一个视图
numDF.createOrReplaceTempView("num_df")

//使用 Spark catalog 来缓存该 numDF，使用名字"num_df"
spark.catalog.cacheTable("num_df")

//通过 count action 操作来强制持久化
numDF.count
```

需要注意，cache()和 persist()持久化是延迟执行的，而 cacheTable()方法是立即执行的。

4.12 Spark SQL 编程案例

本节通过对几个案例的学习，掌握使用 Spark SQL DSL API 和 SQL 进行大数据分析的方法。

4.12.1 实现单词计数

到目前为止，已经了解到，在 Spark 中，对数据的处理，有 3 种方案：
（1）使用 RDD。
（2）使用 DataFrame/DataSet 关系型 API。
（3）使用 Spark SQL 语句。

强烈建议不要直接使用 RDD，而是使用 DataFrame/DataSet 的关系型 API 或 SQL 来对数据进行分析计算。只有这样，才能充分利用 Spark SQL 的 Catalyst 优化器来对分析过程自动进行优化。

在下面的这个示例中，使用 DataFrame 和 SQL 两种方式实现单词计数功能。要求统计某个英文文本中单词出现的次数，找出出现次数最高的 3 个单词。

实现过程和代码如下。

（1）准备数据文件。自行创建一个纯文本文件 word.txt，并编辑内容如下：

```
good good study
day day up
to be or not to be
this is a question
```

将该文件上传到 HDFS 的/data/spark/目录下。

（2）第 1 种方法，使用关系型 API 实现单词计数，代码如下：

```
//定义文件路径
val filePath = "/data/spark/word.txt"

//加载到 DataSet
```

```
val wordDS = spark.read.textFile(filePath)

import spark.implicits._

//对 DataSet 进行一系列处理,产生一个包含最终结果的 DataSet
val wordDF = wordDS
    .flatMap(_.split("\\s+"))
    .filter(_.size>0)
    .groupByKey(_.toLowerCase)
    .count
    .toDF("word","count")

//获得前 3 个出现次数最多的词
val top3 = wordDF.orderBy($"count".desc).limit(3)

//输出结果
top3.show()
```

隐式对象是在 SparkSession 内部定义的,继承了 SQLImplicits 抽象类,提供了隐式转换,用于将 Scala 对象(包括 RDD)转换为 DataSet、DataFrame、Columns。

执行以上代码,输出的结果如下:

```
+----+-----+
|word|count|
+----+-----+
| day|    2|
|good|    2|
|  up|    1|
+----+-----+
```

(3)第 2 种方法,使用 SQL 语句,代码如下:

```
//定义文件路径
val filePath = "/data/spark/word.txt"
val wordDS = spark.read.textFile(filePath)

import spark.implicits._

//对 DataSet 进行一系列处理,产生一个包含最终结果的 DataSet
val wordDF = wordDS
    .flatMap(_.split("\\s+"))
    .filter(_.size>0)
    .toDF("word")

//注册为临时 view
wordDF.createOrReplaceTempView("wc_tb")

//执行 SQL 查询,分析产生结果
val sql = """
    select word,count(1) as count
      from wc_tb
      group by word
```

```
    order by count desc
    limit 3
"""

val resultDF = spark.sql(sql)
resultDF.show()
```

执行以上代码,输出的结果如下:

```
+----+-----+
|word|count|
+----+-----+
| day|    2|
|good|    2|
|  up|    1|
+----+-----+
```

4.12.2 用户数据集分析

Spark SQL 支持两种不同的方法将现有的 RDD 转换为 DataFrame/DataSet。

(1)第 1 种方法使用反射推断模式。这种方法使代码更简洁,在编写 Spark 应用程序时如果已经了解模式,则这种方法可以很好地工作。

(2)第 2 种方法是通过编程接口自己构造模式,然后将其应用于现有的 RDD。虽然此方法更冗长,但它允许用户在列及其类型直到运行时才知道时构造数据集。

1. 使用反射推断模式创建 RDD

Spark SQL 的 Scala 接口支持将包含 case class 的 RDD 自动转换为 DataFrame。case class 定义了表的模式,它的参数名使用反射读取,并成为列的名称。case class 还可以嵌套或包含复杂类型,如 Seq 或 Array。可以隐式地将此 RDD 转换为 DataFrame,然后将其注册为表。表可以在后续的 SQL 语句中使用。

下面这个示例演示了如何使用反射推断模式创建 RDD,以此来分析 Spark 安装包自带的 people.txt 文件内容。首先定义一个名为 Person 的 case class,代码如下:

```
case class Person(name:String, age:Int)
```

然后将文件加载到一个 RDD,再转换为 DataFrame,代码如下:

```
//用于从 RDD 到 DataFrame 的隐式转换
import spark.implicits._    //导入隐式类, rdd->DataFrame 需要

//使用反射推断模式创建 RDD
//从一个文本文件创建 RDD[Person],并将它转换为一个 DataFrame
val file = "/data/spark/resources/people.txt"   //定义文件路径
val peopleDF = spark.sparkContext
   .textFile(file)             //RDD[String]
   .map(_.split(","))          //RDD[Array[String]]
   .map(atts => Person(atts(0), atts(1).trim.toInt))  //RDD[Person]
   .toDF()                     //DataFrame
```

```
peopleDF.printSchema()
peopleDF.show()
```

执行以上代码,输出内容如下:

```
root
 |-- name: string (nullable = true)
 |-- age: integer (nullable = false)

+-------+---+
|   name|age|
+-------+---+
|Michael| 29|
|   Andy| 30|
| Justin| 19|
+-------+---+
```

将该 DataFrame 注册为一个名为 people 的临时视图,代码如下:

```
peopleDF.createOrReplaceTempView("people")
```

使用 SparkSession 的 sql()方法来运行 SQL 语句,结果集中每行的列可以通过字段索引访问。例如,从上面的临时表中查询出年龄位于 13~19 岁的用户,代码如下:

```
//执行 SQL 查询
val sqlStr = "SELECT name, age FROM people WHERE age BETWEEN 13 AND 19"
val teenagersDF = spark.sql(sqlStr)

//结果集中每行的列可以通过字段索引访问
teenagersDF.map(teenager => "Name: " + teenager(0)).show()
```

执行以上代码,输出的结果如下:

```
+------------+
|       value|
+------------+
|Name: Justin|
+------------+
```

或者也可以通过字段名访问,代码如下:

```
teenagersDF.map(teenager =>
    "Name: " + teenager.getAs[String]("name")).show()
```

执行以上代码,输出的结果如下:

```
+------------+
|       value|
+------------+
|Name: Justin|
+------------+
```

如果想要以 Map[K,V]形式返回查询到的内容,则需要自定义编码器,因为 Dataset[Map[K,V]]没有预定义的编码器,代码如下:

```
//明确定义编码器
implicit val mapEncoder = org.apache.spark.sql.Encoders.kryo[Map[String, Any]]
```

```
//row.getValuesMap[T]一次检索多个列并保存到一个 Map[String, T]
val result = teenagersDF
    .map(teenager => teenager.getValuesMap[Any](List("name", "age")))
    .collect()

result.foreach(m => println(m))
result.foreach(m => println(m.mkString("\n")))
```

执行以上代码,输出的结果如下:

```
Map(name -> Justin, age -> 19)
name -> Justin
age -> 19
```

2. 以编程方式指定模式

当不能提前定义 case class 时(例如,记录的结构编码在一个字符串中,或者文本数据集将被解析,字段将针对不同的用户以不同的方式投影),可以通过以下 3 个步骤以编程方式创建 DataFrame。

(1)从原始 RDD 创建一个 Row RDD。
(2)创建由 StructType 表示的模式,该模式与第(1)步创建的 RDD 中的 Row 结构相匹配。
(3)通过 SparkSession 提供的 createDataFrame()方法将模式应用到 Row RDD。

下面这个示例同样用于分析 Spark 安装包自带的 people.txt 文件内容,但是以编程方式指定模式。首先将 people.txt 加载到一个 RDD 中,代码如下:

```
//定义文件路径
val file = "/data/spark/resources/people.txt"

//创建一个 RDD
val peopleRDD = spark.sparkContext.textFile(file)
```

然后定义由 StructType 表示的模式,包含字段名称和类型,代码如下:

```
import org.apache.spark.sql._
import org.apache.spark.sql.types._

//定义字段和 Schema
val fields = Array(
    StructField("name",StringType,nullable = true),
    StructField("age",IntegerType,nullable = true)
)
val schema = StructType(fields)
```

然后将该模式应用到 RDD,创建一个 DataFrame,代码如下:

```
//将 RDD[String]转换为 RDD[Row]
val rowRDD = peopleRDD
    .map(_.split(","))
    .map(atts => Row(atts(0), atts(1).trim.toInt))

//将这个 Schema 应用到该 RDD
val peopleDF = spark.createDataFrame(rowRDD, schema)
```

最后，将 peopleDF 注册到一个临时视图，并执行 SQL 语句进行相应的查询。查询的结果也是一个 DataFrame，支持所有正常的 RDD 操作，代码如下：

```
//使用该 DataFrame 创建一个临时视图
peopleDF.createOrReplaceTempView("people")

//SQL 可以在使用 DataFrames 创建的临时视图上运行
val sqlStr = "SELECT name, age FROM people WHERE age BETWEEN 13 AND 19"
val results = spark.sql(sqlStr)

import spark.implicits._    //注意，需要导入隐式类

//SQL 查询的结果是 DataFrame，支持所有正常的 RDD 操作
//可以通过字段索引或字段名访问结果中一行的列
results.map(atts => "Name: " + atts(0)).show()
results.map(atts => "Name: " + atts.getAs[String]("name")).show()
```

执行以上代码，输出的结果如下：

```
+------------+
|       value|
+------------+
|Name: Justin|
+------------+


+------------+
|       value|
+------------+
|Name: Justin|
+------------+
```

4.12.3 电商用户评论数据集分析

在 Kaggle 上获取了 JSON 格式的亚马逊用户评论数据集，该数据集包含电子产品类别的大约 169 万条亚马逊评论。在本案例中直接读取 JSON 数据集来创建 Spark SQLDataFrame。

首先从 JSON 文件中将数据文件读到 DataFrame 中，代码如下：

```
//数据集路径
val filePath = "/data/spark/amazon/Electronics_5.json"

//加载数据文件，创建 DataFrame
val reviewsDF = spark.read.json(filePath).cache()

//简单查看 Schema
reviewsDF.printSchema()
```

执行以上代码，输出内容如下：

```
root
 |-- asin: string (nullable = true)
```

```
|-- helpful: array (nullable = true)
|    |-- element: long (containsNull = true)
|-- overall: double (nullable = true)
|-- reviewText: string (nullable = true)
|-- reviewTime: string (nullable = true)
|-- reviewerID: string (nullable = true)
|-- reviewerName: string (nullable = true)
|-- summary: string (nullable = true)
|-- unixReviewTime: long (nullable = true)
```

为了便于使用 SQL 进行分析，将 reviewsDF 注册到一个临时视图中，代码如下：

```
reviewsDF.createOrReplaceTempView("reviewsTable")
```

然后执行 SQL 查询，找出综合评分（overall 字段）大于 3 分的评论信息，代码如下：

```
val sqlStr =
    """
      SELECT asin, overall, reviewTime, reviewerID, reviewerName
      FROM reviewsTable
      WHERE overall >= 3
    """
val selectedDF = spark.sql(sqlStr)
selectedDF.show(5)
```

执行以上代码，输出内容如下：

```
+----------+-------+-----------+--------------+--------------------+
|      asin|overall| reviewTime|    reviewerID|        reviewerName|
+----------+-------+-----------+--------------+--------------------+
|0528881469|    5.0| 06 2, 2013|AO94DHGC771SJ|             amazdnu|
|0528881469|    3.0| 09 9, 2010|A3N7T0DY83Y4IG|       C. A. Freeman|
|0594451647|    5.0| 01 3, 2014|A2JXAZZI9PHK9Z|Billy G. Noland "...|
|0594451647|    5.0| 05 4, 2014|AAZ084UMH8VZ2|D. L. Brown "A Kn...|
|0594451647|    4.0|07 11, 2014|AEZ3CR6BKIROJ|        Mark Dietter|
+----------+-------+-----------+--------------+--------------------+
only showing top 5 rows
```

从数据集的 Schema 可以得知，helpful 列是数组类型的元素。要获取这样的列值，可以使用 getItem() 方法，代码如下：

```
import spark.implicits._

val selectedJSONArrayElementDF = reviewsDF
    .select($"asin", $"overall", $"helpful")
    .where($"helpful".getItem(0) < 3)

selectedJSONArrayElementDF.show(5)
```

执行以上代码，输出内容如下：

```
+----------+-------+-------+
|      asin|overall|helpful|
+----------+-------+-------+
|0528881469|    5.0| [0, 0]|
|0528881469|    1.0| [0, 0]|
```

```
|0594451647|    2.0| [0, 0]|
|0594451647|    5.0| [0, 0]|
|0594451647|    4.0| [0, 0]|
+----------+-------+-------+
only showing top 5 rows
```

4.12.4 航空公司航班数据集分析

本案例将使用美国交通部的一些航班信息，探索导致航班延误的航班属性。航班延误的标准是航班延误超过40min。使用 Spark DataSet，将探索这些航班数据来回答以下问题：

（1）哪家航空公司的航班延误次数最多？
（2）每周哪几天的航班延误次数最多？
（3）哪些始发机场的航班延误次数最多？
（4）每天什么时候的航班延误次数最多？

航班数据保存在 JSON 文件 flights20170102.json 中，每个航班记录包含的信息见表4-15。

表4-15 航班记录各字段含义说明

属 性	含 义
id	ID，由承运人、日期、出发地、目的地、航班号组成
dofW	星期几（1 = Monday 星期一，7 = Sunday 星期日）
carrier	承运人代码
origin	起始机场代码
dest	目的地机场代码
crsdephour	规定起飞时间 hour（scheduled departure hour）
crsdeptime	规定起飞时间 time（scheduled departure time）
depdelay	起飞延误分钟数（departure delay in minutes）
crsarrtime	预定到达时间（scheduled arrival time）
arrdelay	到达延误分钟数（arrival delay minutes）
crselapsedtime	飞行时间
dist	距离（distance）

每条航班信息的格式如下：

```
{
  "_id": "AA_2017-01-01_ATL_LGA_1678",
  "dofW": 7,
  "carrier": "AA",
  "origin": "ATL",
  "dest": "LGA",
  "crsdephour": 17,
  "crsdeptime": 1700,
```

```
    "depdelay": 0.0,
    "crsarrtime": 1912,
    "arrdelay": 0.0,
    "crselapsedtime": 132.0,
    "dist": 762.0
}
```

将 flights20170102.json 文件上传到 HDFS 的/data/spark/flightdelay/目录下，然后按以下步骤实现。

（1）首先导入本案例中所需要的依赖包和类，代码如下：

```
import org.apache.spark.ml.feature.Bucketizer
import org.apache.spark.sql._
import org.apache.spark.sql.functions._
import org.apache.spark.sql.types._
```

（2）定义一个 case class，表示航班信息的数据类型（可将其理解为域对象），代码如下：

```
case class Flight(_id: String, dofW: Integer, carrier: String,
                  origin: String, dest: String, crsdephour: Integer,
                  crsdeptime: Double, depdelay: Double, crsarrtime:
                  Double, arrdelay: Double, crselapsedtime: Double,
                  dist: Double)
    extends Serializable
```

（3）定义 Schema，用于创建 DataFrame，代码如下：

```
val schema = StructType(Array(
    StructField("_id", StringType, true),
    StructField("dofW", IntegerType, true),
    StructField("carrier", StringType, true),
    StructField("origin", StringType, true),
    StructField("dest", StringType, true),
    StructField("crsdephour", IntegerType, true),
    StructField("crsdeptime", DoubleType, true),
    StructField("depdelay", DoubleType, true),
    StructField("crsarrtime", DoubleType, true),
    StructField("arrdelay", DoubleType, true),
    StructField("crselapsedtime", DoubleType, true),
    StructField("dist", DoubleType, true)
))
```

（4）将数据集加载到 DataSet 中，应用上一步定义的 Schema，并进行简单探索，代码如下：

```
//定义数据集路径
var file: String = "/data/spark/flightdelay/flights20170102.json"

//将数据加载到 DataSet
import spark.implicits._
val df: Dataset[Flight] = spark.read
    .format("json")
    .option("inferSchema", "false")
    .schema(schema)
```

```
            .load(file)
            .as[Flight]

println(s"数据集中包含记录数量${df.count()}条")

//查看前 5 条记录
df.show(5)
```

执行以上代码,输出内容如下:

```
数据集中包含记录数量 41348 条
+--------------------+----+-------+------+----+-----------+----------+--------+---------+--------+-------------+-----+
|                 _id|dofW|carrier|origin|dest|crsdephour|crsdeptime|depdelay|crsarrtime|arrdelay|crselapsedtime| dist|
+--------------------+----+-------+------+----+-----------+----------+--------+---------+--------+-------------+-----+
|ATL_BOS_2017-01-0...|   7|     DL|   ATL| BOS|         11|    1141.0|     0.0|   1409.0|     0.0|        148.0|946.0|
|ATL_BOS_2017-01-0...|   7|     WN|   ATL| BOS|         13|    1335.0|     0.0|   1600.0|     0.0|        145.0|946.0|
|ATL_BOS_2017-01-0...|   7|     DL|   ATL| BOS|         14|    1416.0|     0.0|   1644.0|     0.0|        148.0|946.0|
|ATL_BOS_2017-01-0...|   7|     DL|   ATL| BOS|         16|    1616.0|    15.0|   1849.0|     0.0|        153.0|946.0|
|ATL_BOS_2017-01-0...|   7|     WN|   ATL| BOS|         18|    1845.0|     0.0|   2110.0|     0.0|        145.0|946.0|
+--------------------+----+-------+------+----+-----------+----------+--------+---------+--------+-------------+-----+
only showing top 5 rows
```

(5) 查看上午 10 点起飞的航班有哪些。这里只取前三条,代码如下:

```
println("过滤上午 10 点起飞的航班。take 3")
df.filter(flight => flight.crsdephour == 10)
  .take(3)
  .foreach(println)
```

执行以上代码,输出内容如下:

```
过滤上午 10 点起飞的航班。take 3
Flight(ATL_BOS_2017-01-02_10_DL_1202,1,DL,ATL,BOS,10,1004.0,10.0,1236.0,3.0,152.0,946.0)
Flight(ATL_BOS_2017-01-03_10_DL_1202,2,DL,ATL,BOS,10,1004.0,0.0,1236.0,0.0,152.0,946.0)
Flight(ATL_BOS_2017-01-03_10_WN_235,2,WN,ATL,BOS,10,1000.0,12.0,1230.0,6.0,150.0,946.0)
```

(6) 按承运人(carrier)分组统计数量,代码如下:

```
println("按承运人(carrier)分组统计:")
df.groupBy("carrier").count().show()
```

执行以上代码,输出内容如下:

```
按承运人(carrier)分组统计:
+-------+-----+
|carrier|count|
+-------+-----+
|     UA|18873|
|     AA|10031|
|     DL|10055|
|     WN| 2389|
+-------+-----+
```

（7）按目的地统计起飞延误超过 40min 的航班信息，并按照延误时间倒序排序，延误时间最长的前三趟航班，代码如下：

```
println("按目的地统计超过 40min 的起飞延误，并按延误时间倒序排序：")
df.filter($"depdelay" > 40)
  .groupBy("dest")
  .count()
  .orderBy(desc("count"))
  .show(3)
```

执行以上代码，输出内容如下：

```
按目的地统计超过 40min 的起飞延误，并按延误时间倒序排序：
+----+-----+
|dest|count|
+----+-----+
| SFO|  711|
| EWR|  620|
| ORD|  593|
+----+-----+
only showing top 3 rows
```

（8）接下来执行数据探索。首先以列格式在内存中缓存 DataFrame，并创建临时视图或缓存临时视图，代码如下：

```
df.cache                                    //缓存

//创建临时表视图
df.createOrReplaceTempView("flights")

//以列格式在内存中缓存表
spark.catalog.cacheTable("flights")         //缓存表
```

（9）回答问题：显示前 5 个延误时间最长的航班信息。

使用 DataFrame API，代码如下：

```
//使用 DataFrame Transformation
df.select($"carrier", $"origin", $"dest", $"depdelay", $"crsdephour")
  .filter($"depdelay" > 40)
  .orderBy(desc("depdelay"))
  .show(5)
```

也可以使用 SQL 语句，代码如下：

```
//使用 SQL
spark.sql(
    """
      |select carrier,origin, dest, depdelay,crsdephour
      |from flights
      |where depdelay > 40
      |order by depdelay desc
      |limit 5
    """.stripMargin).show
```

两种方式的执行结果相同，输出内容如下：

```
+-------+------+----+--------+----------+
|carrier|origin|dest|depdelay|crsdephour|
+-------+------+----+--------+----------+
|     AA|   SFO| ORD|  1440.0|         8|
|     DL|   BOS| ATL|  1185.0|        17|
|     UA|   DEN| EWR|  1138.0|        12|
|     DL|   ORD| ATL|  1087.0|        19|
|     UA|   MIA| EWR|  1072.0|        20|
+-------+------+----+--------+----------+
```

（10）回答问题：显示承运人的平均起飞延误时间，代码如下：

```
println("承运人的平均起飞延误")
df.groupBy("carrier")
  .agg(avg("depdelay"))
  .show
```

执行以上代码，输出内容如下：

```
承运人的平均起飞延误
+-------+------------------+
|carrier|      avg(depdelay)|
+-------+------------------+
|     UA|17.477878450696764|
|     AA| 10.45768118831622|
|     DL|15.316061660865241|
|     WN|13.491000418585182|
+-------+------------------+
```

（11）回答问题：统计一周内每天的平均起飞延误时间，代码如下：

```
println("一周内每天的平均起飞延误时间")
spark.sql(
    """
      |SELECT dofW, avg(depdelay) as avgdelay
      |FROM flights
      |GROUP BY dofW
      |ORDER BY avgdelay desc
    """.stripMargin).show
```

执行以上代码，输出内容如下：

```
一周内每天的平均延误起飞时间
+----+------------------+
|dofW|          avgdelay|
+----+------------------+
|   7| 18.754202401372211|
|   1|17.404697121580984|
|   6|16.829883897021706|
|   4|14.609846568875401|
|   2| 13.50552611657835|
|   5| 13.10193236714976|
|   3| 11.3646816543676|
+----+------------------+
```

（12）回答问题：按承运人统计起飞延误（延误≥40min）次数。

使用 DataFrame API，代码如下：

```
println("按承运人统计起飞延误")

//使用 Transformation API
df.filter($"depdelay" > 40)
  .groupBy("carrier")
  .count
  .orderBy(desc("count"))
  .show()
```

也可以使用 SQL 语句，代码如下：

```
//使用 SQL 语句
spark.sql(
  """
    |select carrier, count(depdelay) count
    |from flights
    |where depdelay > 40
    |group by carrier
  """.stripMargin).show()
```

两种方式的执行结果相同，输出内容如下：

```
按承运人统计起飞延误
+-------+-----+
|carrier|count|
+-------+-----+
|     UA| 2420|
|     DL| 1043|
|     AA|  757|
|     WN|  244|
+-------+-----+
```

（13）回答问题：按出发机场统计延误次数，代码如下：

```
println("如果按出发地机场延误分钟>40，则起飞延误的次数是多少")
spark.sql(
  """
    |select origin, count(depdelay) count
    |from flights
    |where depdelay > 40
    |group by origin
    |ORDER BY count(depdelay) desc
  """.stripMargin).show
```

执行以上代码，输出内容如下：

```
如果按出发地机场延误分钟>40，则起飞延误的次数是多少
+------+-----+
|origin|count|
+------+-----+
|   ORD|  679|
|   ATL|  637|
```

```
|  SFO|  542|
|  EWR|  518|
|  DEN|  484|
|  IAH|  447|
|  LGA|  432|
|  MIA|  429|
|  BOS|  296|
+-----+-----+
```

（14）回答问题：按每周每天统计起飞延误次数。

使用 DataFrame API，代码如下：

```
println("按每周天数统计起飞延误次数,延误以>40 计")

//使用 Transformation API
df.filter($"depdelay" > 40)
  .groupBy("dofW")
  .count
  .orderBy("dofW")
  .show()
```

也可以使用 SQL 语句，代码如下：

```
//使用 SQL
spark.sql(
  """
    |select dofW, count(depdelay) count
    |from flights
    |where depdelay > 40
    |group by dofW
    |order by dofW
  """.stripMargin).show()
```

执行以上代码，输出结果相同，输出内容如下：

按每周天数统计起飞延误次数,延误以>40 计
```
+----+-----+
|dofW|count|
+----+-----+
|   1|  940|
|   2|  712|
|   3|  482|
|   4|  626|
|   5|  579|
|   6|  424|
|   7|  701|
+----+-----+
```

（15）回答问题：按一天中的不同时段统计起飞延误次数，代码如下：

```
println("按小时统计起飞延误次数")
spark.sql(
  """
    |select crsdephour, count(depdelay) count
    |from flights
```

```
    |where depdelay > 40
    |group by crsdephour
    |order by crsdephour
  """.stripMargin).show(24)
```

执行以上代码，输出内容如下：

```
按小时统计起飞延误次数
+---------+-----+
|crsdephour|count|
+---------+-----+
|        0|    9|
|        1|    1|
|        5|   15|
|        6|   68|
|        7|  112|
|        8|  190|
|        9|  175|
|       10|  284|
|       11|  280|
|       12|  227|
|       13|  336|
|       14|  353|
|       15|  331|
|       16|  351|
|       17|  474|
|       18|  396|
|       19|  371|
|       20|  230|
|       21|  160|
|       22|   65|
|       23|   27|
|       24|    9|
+---------+-----+
```

（16）回答问题：按航线统计延误次数，代码如下：

```
println("按航线统计延误次数")
spark.sql(
  """
    |select origin,dest,count(depdelay) count
    |from flights
    |where depdelay > 40
    |group by origin,dest
    |ORDER BY count(depdelay) desc
  """.stripMargin).show
```

执行以上代码，输出内容如下：

```
按航线统计延误次数
+------+----+-----+
|origin|dest|count|
+------+----+-----+
|   DEN| SFO|  172|
```

```
|  ORD| SFO|  168|
|  ATL| LGA|  155|
|  ATL| EWR|  141|
|  SFO| DEN|  134|
|  LGA| ATL|  130|
|  ORD| EWR|  122|
|  SFO| ORD|  118|
|  EWR| ORD|  115|
|  ORD| LGA|  100|
|  IAH| SFO|   98|
|  IAH| EWR|   94|
|  MIA| LGA|   92|
|  ORD| ATL|   88|
|  LGA| ORD|   88|
|  ATL| SFO|   87|
|  LGA| MIA|   81|
|  SFO| EWR|   79|
|  EWR| SFO|   77|
|  EWR| ATL|   76|
+-----+----+-----+
only showing top 20 rows
```

4.12.5 数据增量抽取和全量抽取

所谓全量抽取，指的是将数据源（例如，数据库）中整个表的数据都读取出来。所谓增量抽取，指的是每次只读取数据源中新增加的数据。

在下面的例子中，实现了 MySQL 数据的全量抽取、增量抽取。建议按以下步骤操作。

（1）在 MySQL 中创建一个数据表 mytb，并插入一些初始数据，执行的 SQL 语句如下：

```
#创建表
create table olist.mytb(
    mid int,
    mdate date
);

#插入初始数据
insert into olist.mytb
values
(1,"2022-04-05"),
(2,"2022-04-06");
```

（2）将 mytb 表中初始的数据全部抽取（全量抽取）到 DataFrame 中，代码如下：

```
//抽取
//从 MariaDB/MySQL 服务器的一张表中读取数据
val mysqlURL= "jdbc:mysql://localhost:3306/olist"
val df = spark.read
    .format("jdbc")
    .option("url", mysqlURL)
    .option("driver", "com.mysql.jdbc.Driver")
```

```
        .option("dbtable", "mytb")         //注意:全量抽取,用 dbtable 选项
        .option("user", "root")            //MySQL 的账号
        .option("password","admin")        //MySQL 的密码
        .load()

df.show()
```

执行以上代码,输出的结果如下:

```
+---+----------+
|mid|     mdate|
+---+----------+
|  1|2022-04-05|
|  2|2022-04-06|
+---+----------+
```

(3)假设 MySQL 的 mytb 表中又新增加了数据,使用 SQL 的 insert 语句模拟,SQL 语句如下:

```
mysql>insert into olist.mytb values(3,"2022-04-07");
```

(4)接下来的任务是将 MySQL 的 mytb 表中新增数据抽取出来(增量抽取),代码如下:

```
//增量抽取(2022-04-06 之后的数据)
//从 MariaDB/MySQL 服务器的一张表中读取数据
val mysqlURL= "jdbc:mysql://localhost:3306/olist"
val lastDate = "2022-04-06"
val sql = s"select * from mytb where mdate>'${lastDate}'"
val df = spark.read
    .format("jdbc")
    .option("url", mysqlURL)
    .option("driver", "com.mysql.jdbc.Driver")
    .option("query", sql)          //注意:增量抽取,用 query 选项和 sql 条件查询语句
    .option("user", "root")
    .option("password","admin")
    .load()

df.show()
```

执行以上代码,输出的结果如下:

```
+---+----------+
|mid|     mdate|
+---+----------+
|  3|2022-04-07|
+---+----------+
```

可以看出,当执行全量查询时,使用 dbtable 选项;当执行增量查询时,使用 query 选项和条件查询语句。

第 5 章 Spark SQL（高级）
CHAPTER 5

为了帮助执行复杂的分析，Spark SQL 提供了一组强大而灵活的聚合函数、连接多个数据集的函数、一组内置的高性能函数和一组高级分析函数。本章将详细介绍这些主题。另外本章还将介绍 Spark SQL 模块的一些高级功能，并解释 Catalyst 优化器和 Tungsten 引擎所提供的优化和执行效率。

5.1　Spark SQL 函数

Spark DataFrame/DataSet API 设计目的是在数据集中操作或转换单行数据，如过滤或分组。如果想要转换一个数据集中的每行的列的值，例如，将字符串从大写字母转换为驼峰命名形式，就需要使用一个函数实现。Spark SQL 内置了一组常用的函数，同时也提供了用户自定义新函数的简单方法。

为了有效地使用 Spark SQL 执行分布式数据操作，必须熟练使用 Spark SQL 函数。Spark SQL 提供了超过 200 个内置函数，它们被分组到不同的类别中。

按类别来分，SQL 函数可分为以下四类。

（1）标量函数：每行返回单个的值。
（2）聚合函数：每一组行返回单个的值。
（3）窗口函数：每一组行返回多个值。
（4）用户自定义函数（UDF）：包括自定义的标量函数和聚合函数。

标量函数和聚合函数位于 org.apache.spark.sql.functions 包内。在使用前，需要先导入它，代码如下：

```
import org.apache.spark.sql.functions._
```

如果使用 spark-shell 或 Zeppelin Notebook 进行交互式分析，则会自动导入该包。

5.2　内置标量函数

Spark SQL 提供了大量的标量函数，主要包括以下几类。
（1）数学计算：例如 abs()、hypot()、log()、cbrt()等。

（2）字符串操作：例如 length()、trim()、concat()等。
（3）日期操作：例如 year()、date_add()等。

下面详细来介绍这些函数及其用法。

5.2.1 日期时间函数

Spark 内置的日期时间函数大致可分为以下 3 个类别：
（1）执行日期时间格式转换的函数。
（2）执行日期时间计算的函数。
（3）从日期时间戳中提取特定值（如年、月、日等）的函数。

日期和时间转换函数有助于将字符串转换为日期、时间戳或 UNIX 时间戳，反之亦然。在内部，它使用 Java 日期格式模式语法。这些函数使用的默认的日期格式是"yyyy-mm-dd HH:mm:ss"，因此，如果日期或时间戳列的日期格式不同，则需要向这些转换函数传入指定的模式。

1. 将字符串转换为日期或时间戳

例如，将字符串类型的日期和时间戳转换为 Spark SQL 的 date 和 timestamp 类型，代码如下：

```scala
//第5章/functions_date.scala

import spark.implicits._

//1. 日期和时间转换函数：这些函数使用的默认的日期格式是 yyyy-mm-dd HH:mm:ss
//构造一个简单的 DataFrame，注意最后两列不遵循默认日期格式
val testDate = Seq((1, "2019-01-01", "2019-01-01 15:04:58", "01-01-2019", "12-05-2018 45:50"))
val testDateTSDF = testDate.toDF("id", "date", "timestamp", "date_str", "ts_str")

//将这些字符串转换为 date、timestamp 和 UNIX timestamp
//并指定一个自定义的 date 和 timestamp 格式
val testDateResultDF = testDateTSDF.select(
    to_date('date).as("date1"),
    to_timestamp('timestamp).as("ts1"),
    to_date('date_str,"MM-dd-yyyy").as("date2"),
    to_timestamp('ts_str,"MM-dd-yyyy mm:ss").as("ts2"),
    unix_timestamp('timestamp).as("unix_ts")
)

testDateResultDF.printSchema       //date1 和 ts1 分别为日期和时间戳类型
testDateResultDF.show(false)
```

执行以上代码，输出的结果如下：

```
root
 |-- date1: date (nullable = true)
```

```
|-- ts1: timestamp (nullable = true)
|-- date2: date (nullable = true)
|-- ts2: timestamp (nullable = true)
|-- unix_ts: long (nullable = true)

+----------+-------------------+----------+-------------------+----------+
|date1     |ts1                |date2     |ts2                |unix_ts   |
+----------+-------------------+----------+-------------------+----------+
|2019-01-01|2019-01-01 15:04:58|2019-01-01|2018-12-05 00:45:50|1546326298|
+----------+-------------------+----------+-------------------+----------+
```

2. 将日期或时间戳转换为字符串

将日期或时间戳转换为时间字符串是很容易的，可以使用 date_format()函数和定制日期格式，或者使用 from_unixtime()函数将 UNIX 时间戳（以秒为单位）转换成字符串。参看将日期和时间戳转换为格式字符串的转换例子，代码如下：

```scala
//第5章/functions_date.scala

testDateResultDF.select(
    date_format('date1, "dd-MM-yyyy").as("date_str"),
    date_format('ts1, "dd-MM-yyyy HH:mm:ss").as("ts_str"),
    from_unixtime('unix_ts,"dd-MM-yyyy HH:mm:ss").as("unix_ts_str")
).show()
```

执行以上代码，输出的结果如下：

```
+----------+-------------------+-------------------+
| date_str |             ts_str|        unix_ts_str|
+----------+-------------------+-------------------+
|01-01-2019|01-01-2019 15:04:58|01-01-2019 15:04:58|
+----------+-------------------+-------------------+
```

3. 日期计算函数

日期-时间计算函数有助于计算两个日期或时间戳的相隔时间，以及执行日期或时间算术运算。关于日期-时间计算的示例，代码如下：

```scala
//第5章/functions_date.scala

import spark.implicits._

//2. 日期-时间(date-time)计算函数
val employeeData = Seq(
    ("黄渤", "2016-01-01", "2017-10-15"),
    ("王宝强", "2017-02-06", "2017-12-25")
  ).toDF("name", "join_date", "leave_date")

employeeData.show()

//执行 date 和 month 计算
employeeData.select(
```

```
    'name,
    datediff('leave_date, 'join_date).as("days"),
    months_between('leave_date, 'join_date).as("months"),
    last_day('leave_date).as("last_day_of_mon")
).show()

//执行日期加、减计算
val oneDate = Seq("2019-01-01").toDF("new_year")
oneDate.select(
    date_add('new_year, 14).as("mid_month"),
    date_sub('new_year, 1).as("new_year_eve"),
    next_day('new_year, "Mon").as("next_mon")
).show()
```

执行上面的代码，输出的结果如下：

```
+------+----------+----------+
|  name|join_date |leave_date|
+------+----------+----------+
|  黄渤|2016-01-01|2017-10-15|
|王宝强|2017-02-06|2017-12-25|
+------+----------+----------+


+------+----+-----------+---------------+
|  name|days|     months|last_day_of_mon|
+------+----+-----------+---------------+
|  黄渤| 653| 21.4516129|     2017-10-31|
|王宝强| 322|10.61290323|     2017-12-31|
+------+----+-----------+---------------+


+----------+------------+----------+
| mid_month|new_year_eve|  next_mon|
+----------+------------+----------+
|2019-01-15|  2018-12-31|2019-01-07|
+----------+------------+----------+
```

4. 转换不规范的日期

有时，采集到的数据是不受控制的，得到的日期可能是不规范的，这就需要将这些不规范的日期转换为规范的表示形式。对不规范日期进行转换，代码如下：

```
//第5章/functions_date.scala

import spark.implicits._

//转换不规范的日期
val df = Seq(
    "Nov 05, 2018 02:46:47 AM",
    "Nov 5, 2018 02:46:47 PM"
  ).toDF("times")

df.withColumn("times2",
```

```
        from_unixtime(
            unix_timestamp($"times", "MMM d, yyyy hh:mm:ss a"),
            "yyyy-MM-dd HH:mm:ss.SSSSSS"
        )
).show(false)
```

执行以上代码,输出的结果如下:

```
+------------------------+--------------------------+
|times                   |times2                    |
+------------------------+--------------------------+
|Nov 05, 2018 02:46:47 AM|2018-11-05 02:46:47.000000|
|Nov 5, 2018 02:46:47 PM |2018-11-05 14:46:47.000000|
+------------------------+--------------------------+
```

5. 处理时间序列数据

在处理时间序列数据（time-series data）时，经常需要提取日期或时间戳值的特定字段（如年、月、小时、分钟和秒）。例如，当需要按季度、月或周对所有股票交易进行分组时，可以从交易日期提取该信息，并按这些值分组。从日期或时间戳中提取字段，代码如下:

```
//第5章/functions_date.scala

import spark.implicits._

//3. 提取日期或时间戳值的特定字段（如年、月、小时、分钟和秒）
//从一个日期值中提取指定的字段
val valentimeDateDF = Seq("2019-02-14 13:14:52").toDF("date")
valentimeDateDF.select(
    year('date).as("year"),             //年
    quarter('date).as("quarter"),       //季
    month('date).as("month"),           //月
    weekofyear('date).as("woy"),        //周
    dayofmonth('date).as("dom"),        //日
    dayofyear('date).as("doy"),         //天
    hour('date).as("hour"),             //小时
    minute('date).as("minute"),         //分
    second('date).as("second")          //秒
  ).show()
```

执行以上代码,输出的结果如下:

```
+----+-------+-----+---+---+---+----+------+------+
|year|quarter|month|woy|dom|doy|hour|minute|second|
+----+-------+-----+---+---+---+----+------+------+
|2019|      1|    2|  7| 14| 45|  13|    14|    52|
+----+-------+-----+---+---+---+----+------+------+
```

5.2.2 字符串函数

毫无疑问，数据集的大多数列是字符串类型的。Spark SQL 内置的字符串函数提供了

操作字符串类型列的通用和强大的方法。一般来讲，这些函数分为以下两类。

（1）执行字符串转换的函数。

（2）执行字符串提取（或替换）的函数，使用正则表达式。

1. 字符串去空格

最常见的字符串转换包括去空格、填充、大写转换、小写转换和字符串连接等。演示使用各种内置字符串函数转换字符串的各种方法，代码如下：

```scala
//第5章/functions_string.scala

import spark.implicits._

//使用各种内置字符串函数转换字符串的各种方法
val sparkDF = Seq(" Spark ").toDF("name")

//去空格
sparkDF.select(
    trim($"name").as("trim"),       //去掉"name"列两侧的空格
    ltrim($"name").as("ltrim"),     //去掉"name"列左侧的空格
    rtrim($"name").as("rtrim")      //去掉"name"列右侧的空格
   ).show()
```

执行上面的代码，输出的结果如下：

```
+-----+------+------+
| trim| ltrim| rtrim|
+-----+------+------+
|Spark|Spark |  Spark|
+-----+------+------+
```

2. 字符串填充

用给定的填充字符串将字符串填充到指定长度，代码如下：

```scala
//第5章/functions_string.scala

import spark.implicits._

//使用各种内置字符串函数转换字符串的各种方法
val sparkDF = Seq(" Spark ").toDF("name")

//首先去掉"Spark"前后的空格，然后填充到8个字符长
sparkDF
    .select(trim($"name").as("trim"))       //去掉两侧的空格
    .select(
      lpad($"trim", 8, "-").as("lpad"),     //宽度为8，不够，左侧填充"-"
      rpad($"trim", 8, "=").as("rpad")      //宽度为8，不够，右侧填充"="
    )
    .show()
```

执行以上代码，输出的结果如下：

```
+--------+--------+
|    lpad|    rpad|
+--------+--------+
|---Spark|Spark===|
+--------+--------+
```

3. 字符串转换

字符串之间可以使用转换函数进行多种转换，代码如下：

```scala
//第5章/functions_string.scala

import spark.implicits._

//使用concat-WS、upper、lower和reverse转换一个字符串
val sentenceDF = Seq(("Spark", "is", "excellent"))
    .toDF("subject", "verb", "adj")

sentenceDF
    //用空格连接多列值
    .select(concat_ws(" ",$"subject", $"verb", $"adj").as("sentence"))
    .select(
      lower($"sentence").as("lower"),            //转小写
      upper($"sentence").as("upper"),            //转大写
      initcap($"sentence").as("initcap"),        //转首字母大写
      reverse($"sentence").as("reverse")         //翻转
    )
    .show()

//从一个字符转换到另一个字符
sentenceDF
    .select($"subject", translate($"subject","pr","oc").as("translate"))
    .show()
```

执行以上代码，输出的结果如下：

```
+-----------------+-----------------+-----------------+-----------------+
|            lower|            upper|          initcap|          reverse|
+-----------------+-----------------+-----------------+-----------------+
|spark is excellent|SPARK IS EXCELLENT|Spark Is Excellent|tnellecxe si krapS|
+-----------------+-----------------+-----------------+-----------------+

+-------+---------+
|subject|translate|
+-------+---------+
|  Spark|    Soack|
+-------+---------+
```

4. 字符串提取

如果想从字符串列值中替换或提取子字符串，可以使用函数 regexp_extract()和 regexp_replace()。这两个函数通过传入的正则表达式实现替换或提取。Spark 利用 Java 正

则表达式库实现这两个字符串函数的底层实现。

如果要提取字符串的某一部分子字符串，则可使用 regexp_extract()函数。regexp_extract()函数的输入参数是字符串列、匹配的模式和组索引。在字符串中可能会有多个匹配模式，因此，需要组索引（从 0 开始）来确定是哪一个。如果没有指定模式的匹配，则该函数返回空字符串。使用 regexp_extract 函数()，代码如下：

```
//第5章/functions_string.scala

import spark.implicits._

//使用regexp_extract字符串函数来提取"fox"，使用一个模式
val str = "A fox saw a crow sitting on a tree singing \"Caw! Caw! Caw!\""
val strDF = Seq(str).toDF("comment")

//使用一个模式进行提取
strDF
    .select(regexp_extract($"comment", "[a-z]*o[xw]",0).as("substring"))
    .show()
```

执行以上代码，输出的结果如下：

```
+---------+
|substring|
+---------+
|      fox|
+---------+
```

5. 字符串替换

如果要替换字符串的某一部分（子字符串），则可使用 regexp_replace()函数。regexp_replace()字符串函数的输入参数是字符串列、匹配的模式及替换的值。使用 regexp_replace()函数，代码如下：

```
//第5章/functions_string.scala

import spark.implicits._

//用regexp_replace字符串函数将fox和Caw替换为animal
val str = "A fox saw a crow sitting on a tree singing \"Caw! Caw! Caw!\""
val strDF = Seq(str).toDF("comment")

//下面两行可生成相同的输出
strDF
    .select(regexp_replace($"comment","fox|crow","animal").as("new_comm"))
    .show(false)

strDF
    .select(regexp_replace($"comment","[a-z]*o[xw]","animal").as("new_comm"))
    .show(false)
```

执行以上代码，输出的结果如下：

```
+------------------------------------------------------------+
|new_comm                                                    |
+------------------------------------------------------------+
|A animal saw a animal sitting on a tree singing "Caw! Caw! Caw!"|
+------------------------------------------------------------+

+------------------------------------------------------------+
|new_comm                                                    |
+------------------------------------------------------------+
|A animal saw a animal sitting on a tree singing "Caw! Caw! Caw!"|
+------------------------------------------------------------+
```

例如，使用regexp_replace()函数从混乱的数据中抽取出手机号，代码如下：

```
import spark.implicits._

val telDF = Seq("135a-123b4-c5678").toDF("tel")
telDF.withColumn("phone",regexp_replace('tel,"-|\\D","")).show()
```

执行以上代码，输出的结果如下：

```
+----------------+----------+
|             tel|     phone|
+----------------+----------+
|135a-123b4-c5678|13512345678|
+----------------+----------+
```

5.2.3 数学计算函数

Spark SQL 还提供了许多对数值类型列进行计算的函数，其中最常使用的是 round() 函数，它对传入的列值执行一个四舍五入法。这个函数有以下两种签名方法。

（1）def round(e: Column, scale: Int): Column。

（2）def round(e: Column): Column。

使用 round() 函数对不同格式的数值执行四舍五入法，代码如下：

```
//第 5 章/functions_math.scala

import spark.implicits._

val numberDF = Seq((3.14159, -3.14159)).toDF("pie", "-pie")
numberDF
  .select(
      $"pie",
      round($"pie").as("pie0"),          //整数四舍五入
      round($"pie", 2).as("pie1"),       //四舍五入，保留小数点后 2 位
      round($"pie", 4).as("pie2"),       //四舍五入，保留小数点后 4 位
      $"-pie",
      round($"-pie").as("-pie0"),        //整数四舍五入
      round($"-pie", 2).as("-pie1"),     //四舍五入，保留小数点后 2 位
      round($"-pie", 4).as("-pie2")      //四舍五入，保留小数点后 4 位
  )
  .show()
```

执行以上代码，输出的结果如下：

```
+-------+----+----+------+--------+-----+-----+-------+
|    pie|pie0|pie1|  pie2|    -pie|-pie0|-pie1|  -pie2|
+-------+----+----+------+--------+-----+-----+-------+
|3.14159| 3.0|3.14|3.1416|-3.14159| -3.0|-3.14|-3.1416|
+-------+----+----+------+--------+-----+-----+-------+
```

还有两个数学函数也经常用到，分别是 ceil() 和 floor()，它们的方法签名如下。

（1）def ceil(e: Column): Column。

（2）def floor(e: Column): Column。

关于这两个函数的使用，代码如下：

```
//第5章/functions_math.scala

import spark.implicits._

val numberDF = Seq((3.14159, -3.14159)).toDF("v1", "v2")
numberDF
  .select(
    $"v1",
    ceil($"v1"),              //向上取整
    floor($"v1"),             //向下取整
    $"v2",
    ceil($"v2"),              //向上取整
    floor($"v2")              //向下取整
  )
  .show()
```

执行以上代码，输出的结果如下：

```
+-------+--------+---------+--------+--------+---------+
|     v1|CEIL(v1)|FLOOR(v1)|      v2|CEIL(v2)|FLOOR(v2)|
+-------+--------+---------+--------+--------+---------+
|3.14159|       4|        3|-3.14159|      -3|       -4|
+-------+--------+---------+--------+--------+---------+
```

5.2.4 集合元素处理函数

集合被用来处理复杂的数据类型，如 arrays、maps 和 structs。本节将介绍两种特定类型的集合函数。第1种方法是使用 array 数据类型，第2种方法是处理为 JSON 数据格式。

6min

1. 数组处理函数

Spark DataFrame 支持复杂数据类型，也就是列值可以是一个集合。可以使用与数组相关的集合函数来轻松地获取数组的大小、检查值的存在或者对数组进行排序。下面的代码包含了处理各种数组相关函数的示例，代码如下：

```
//第5章/functions_array.scala

import spark.implicits._
```

```
//创建一个任务集 DataFrame
val tasksDF = Seq(("星期天", Array("抽烟", "喝酒", "去烫头"))).toDF("day", "tasks")

//tasksDF 的 Schema
tasksDF.printSchema()

//获得该数组的大小,对其进行排序,并检查在该数组中是否存在一个指定的值
tasksDF
    .select(
      $"day",
      size($"tasks").as("size"),                              //数组大小
      sort_array($"tasks").as("sorted_tasks"),                //对数组排序
      array_contains($"tasks", "去烫头").as("是否去烫头")      //是否包含
    )
    .show(false)

//使用 explode 表函数将为数组中的每个元素创建一个新行
tasksDF.select($"day", explode($"tasks") as "task").show()
```

执行以上代码,输出的结果如下:

```
root
 |-- day: string (nullable = true)
 |-- tasks: array (nullable = true)
 |    |-- element: string (containsNull = true)

+------+----+--------------------+----------+
|day   |size|sorted_tasks        |是否去烫头|
+------+----+--------------------+----------+
|星期天|3   |[去烫头, 喝酒, 抽烟]|true      |
+------+----+--------------------+----------+


+------+------+
|   day|  task|
+------+------+
|星期天|  抽烟|
|星期天|  喝酒|
|星期天|去烫头|
+------+------+
```

2. json 处理函数

许多非结构化数据集是以 JSON 的形式存在的。对于 JSON 数据类型的列,使用相关的集合函数将 JSON 字符串转换成 struct(结构体)数据类型。主要的函数是 from_json()、get_json_object()和 to_json()。一旦 JSON 字符串被转换为 PySpark struct 数据类型,就可以轻松地提取这些值了。下面的代码演示了 from_json()和 to_json()函数的使用方法。

首先构造一个带有 JSON 字符串内容的 DataFrame,代码如下:

```
//第 5 章/functions_json.scala

import spark.implicits._
```

```
//创建一个字符串，它包含JSON格式的字符串内容
val todos = """{"day": "星期天","tasks": ["抽烟", "喝酒", "去烫头"]}"""
val todoStrDF = Seq(todos).toDF("todos_str")

//查看Schema
todoStrDF.printSchema()
```

执行以上代码，输出的结果如下：

```
root
 |-- todos_str: string (nullable = true)
```

为了将一个 JSON 字符串转换为一个 Spark 结构体数据类型，需要将其结构描述给 Spark，为此需要定义一个 Schema 模式，并在 from_json()函数中应用，代码如下：

```
import org.apache.spark.sql.types._

val todoSchema = new StructType()
    .add("day", StringType)
    .add("tasks", ArrayType(StringType))

//使用 from_json 来转换 JSON string
val todosDF = todoStrDF
    .select(from_json($"todos_str", todoSchema).as("todos"))

//todos 是一个 struct 数据类型，包含两个字段：day 和 tasks
todosDF.printSchema()
todosDF.show()
```

执行以上代码，输出的结果如下：

```
root
 |-- todos: struct (nullable = true)
 |    |-- day: string (nullable = true)
 |    |-- tasks: array (nullable = true)
 |    |    |-- element: string (containsNull = true)

+------------------------------+
|                         todos|
+------------------------------+
|    {星期天, [抽烟, 喝酒, 去烫头]}|
+------------------------------+
```

可以使用 Column 类的 getItem()函数检索出结构体数据类型的值，代码如下：

```
//使用 Column 类的 getItem 函数检索出 struct 数据类型的值
todosDF
    .select(
      $"todos".getItem("day"),
      $"todos".getItem("tasks"),
      $"todos".getItem("tasks")(0).as("first_task")
    )
    .show(false)
```

执行以上代码，输出的结果如下：

```
+---------+--------------------+----------+
|todos.day|todos.tasks         |first_task|
+---------+--------------------+----------+
|星期天    |[抽烟, 喝酒, 去烫头]  |抽烟       |
+---------+--------------------+----------+
```

也可以使用 to_json() 函数将一个 Spark 结构体数据类型转换为 JSON 格式字符串,代码如下:

```
//使用 to_json 函数将一个 Spark struct 数据类型转换为 JSON 字符串
todosDF.select(to_json($"todos")).show(false)
```

执行以上代码,输出的结果如下:

```
+-------------------------------------------------+
|structstojson(todos)                             |
+-------------------------------------------------+
|{"day":"星期天","tasks":["抽烟","喝酒","去烫头"]}     |
+-------------------------------------------------+
```

5.2.5 其他函数

除了前面几节介绍的函数外,还有一些函数在特定的场景下非常有用。本节将介绍 monotonically_increasing_id()、when()、coalesce()和 lit()函数。

1. monotonically_increasing_id()函数

有时需要为数据集中的每行生成单调递增的唯一但不一定是连续的 id。例如,如果一个 DataSet 有 2 亿行,并且是分区存储的,则应如何确保这些 id 值是唯一的并且同时增加呢? Spark SQL 提供了一个 monotonically_increasing_id()函数,它生成 64 位整数作为 id 值。使用 monotonically_increasing_id()函数,代码如下:

```scala
//第 5 章/functions_increasing_id.scala

import spark.implicits._

//首先生成一个 DataFrame,并将它的值分散到 5 个分区中
val numDF = spark.range(1,11,1,5)

//验证的确被分散到 5 个分区
println("分区数为" + numDF.rdd.getNumPartitions)

//现在生成单调递增的值,并查看所在的分区
import org.apache.spark.sql.functions._
numDF.select(
    $"id",
    monotonically_increasing_id().as("m_ii"),
    spark_partition_id().as("partition")
).show()
```

执行上面的代码,输出的结果如下:

```
分区数为5
+---+-----------+---------+
| id|       m_ii|partition|
+---+-----------+---------+
|  1|          0|        0|
|  2|          1|        0|
|  3| 8589934592|        1|
|  4| 8589934593|        1|
|  5|17179869184|        2|
|  6|17179869185|        2|
|  7|25769803776|        3|
|  8|25769803777|        3|
|  9|34359738368|        4|
| 10|34359738369|        4|
+---+-----------+---------+
```

2. when()函数

在 DataFrame 中,如果需要根据条件列表来评估一个值并返回一个值,则可以使用 when()函数。例如,使用 when()函数将数字值转换为字符串,代码如下:

```scala
//第5章/functions_when.scala

import spark.implicits._

//创建一个具有从1到7的值的DataFrame来表示一周中的每天
val dayOfWeekDF = spark.range(1,8)

//将每个数值转换成字符串
import org.apache.spark.sql.functions._
dayOfWeekDF.select(
    $"id",
    when($"id" === 1, "星期一")
      .when($"id" === 2, "星期二")
      .when($"id" === 3, "星期三")
      .when($"id" === 4, "星期四")
      .when($"id" === 5, "星期五")
      .when($"id" === 6, "星期六")
      .when($"id" === 7, "星期日")
      .as("星期")
  ).show()
```

执行以上代码,输出的结果如下:

```
+---+------+
| id|  星期|
+---+------+
|  1|星期一|
|  2|星期二|
|  3|星期三|
```

```
|   4|  星期四|
|   5|  星期五|
|   6|  星期六|
|   7|  星期日|
+---+------+
```

处理默认情况时，可以使用 Column 类的 otherwise()函数，代码如下：

```scala
//第5章/functions_when.scala

import spark.implicits._

//创建一个具有从1到7的值的 DataFrame 来表示一周中的每天
val dayOfWeekDF = spark.range(1,8)

//将每个数值转换为字符串
import org.apache.spark.sql.functions._
dayOfWeekDF.select(
    $"id",
    when($"id" === 6, "周末")
      .when($"id" === 7, "周末")
      .otherwise("工作日")
      .as("day_type")
  ).show()
```

执行以上代码，输出的结果如下：

```
+---+--------+
| id|day_type|
+---+--------+
|  1|   工作日|
|  2|   工作日|
|  3|   工作日|
|  4|   工作日|
|  5|   工作日|
|  6|    周末|
|  7|    周末|
+---+--------+
```

3. coalesce()函数和 lit()函数

在处理数据时，正确处理 null 值是很重要的。Spark SQL 提供了一个名为 coalesce 的函数，该函数可接收一个或多个列值，并返回第 1 个非空值，而 coalesce 函数中的每个参数都必须是 Column 类型，所以如果想传入字面量值，则需要使用 lit()函数，将字面量值包装为 Column 类的实例。在下面的代码中演示了 coalesce()和 lit()函数的使用。

例如，创建一个 case class，代表一部电影域对象，代码如下：

```scala
//第5章/functions_coalesce_lit.scala

case class Movie(actor_name:String, movie_title:String, produced_year:Long)
```

然后构造一个带有 null 值的 DataFrame，并使用 coalesce()函数来处理列中的 null 值。

注意其中 lit() 函数的用法，其作用是将一个普通字符串转换为 Column 类型，代码如下：

```scala
//第5章/functions_coalesce_lit.scala

import spark.implicits._
import org.apache.spark.sql.functions._

//构造一个 DataFrame，带有 null 值
val badMoviesDF = Seq(
    Movie(null, null, 2018L),
    Movie("黄渤", "一出好戏", 2018L)
  ).toDF()

badMoviesDF.show()

//使用 coalesce 来处理 title 列中的 null 值
badMoviesDF
    .select(
      coalesce($"actor_name", lit("路人甲")).as("演员"),
      coalesce($"movie_title", lit("烂片")).as("电影"),
      coalesce($"produced_year", lit("烂片")).as("年份")
    )
    .show()
```

执行以上代码，输出的结果如下：

```
+----------+-----------+-------------+
|actor_name|movie_title|produced_year|
+----------+-----------+-------------+
|      null|       null|         2018|
|      黄渤|     一出好戏|         2018|
+----------+-----------+-------------+

+------+--------+----+
|  演员|    电影|年份|
+------+--------+----+
|路人甲|    烂片|2018|
|  黄渤|一出好戏|2018|
+------+--------+----+
```

5.2.6　函数应用示例

前面几节讲解了 Spark SQL 常用的一些内置函数。本节通过一个示例演示如何使用其中一些函数实现 Spark DataFrame 二次排序。

假设在 HDFS 的 /data/spark/ 路径下有一个输入文件 data.txt，内容如下：

```
2018,5,22
2019,1,24
2018,2,128
2019,3,56
```

```
2019,1,3
2019,2,-43
2019,4,5
2019,3,46
2018,2,64
2019,1,4
2019,1,21
2019,2,35
2019,2,0
```

其中每行由逗号分隔的分别是年、月和总数。现在想要对这些数据排序，期望的输出结果是先按年月进行排序，在相同年月的情况下，数值列表按大小排序，代码如下：

```
2018-2    64,128
2018-5    22
2019-1    3,4,21,24
2019-2    -43,0,35
2019-3    46,56
2019-4    5
```

建议按以下步骤实现。

（1）将数据集加载到一个 DataFrame 中，代码如下：

```scala
//第5章/secondary_sort.scala

val inputPath = "/data/spark/data.txt"
val inputDF = spark.read
              .option("inferSchema","true")
              .option("header","false")
              .csv(inputPath)
              .toDF("year","month","cnt")

inputDF.show()
```

执行以上代码，输出的结果如下：

```
+----+-----+---+
|year|month|cnt|
+----+-----+---+
|2018|    5| 22|
|2019|    1| 24|
|2018|    2|128|
|2019|    3| 56|
|2019|    1|  3|
|2019|    2|-43|
|2019|    4|  5|
|2019|    3| 46|
|2018|    2| 64|
|2019|    1|  4|
|2019|    1| 21|
|2019|    2| 35|
|2019|    2|  0|
+----+-----+---+
```

（2）将 year 和 month 组合为一列，并取名为 ym，代码如下：

```
import org.apache.spark.sql.functions._

import spark.implicits
val df2 = inputDF.select(concat_ws("-",$"year",$"month").as("ym"), $"cnt")
df2.printSchema()
df2.show
```

执行以上代码，输出的结果如下：

```
root
 |-- ym: string (nullable = false)
 |-- cnt: integer (nullable = true)

+------+---+
|    ym|cnt|
+------+---+
|2018-5| 22|
|2019-1| 24|
|2018-2|128|
|2019-3| 56|
|2019-1|  3|
|2019-2|-43|
|2019-4|  5|
|2019-3| 46|
|2018-2| 64|
|2019-1|  4|
|2019-1| 21|
|2019-2| 35|
|2019-2|  0|
+------+---+
```

（3）先按 ym 进行分组聚合，然后对每一组的 cnt 列进行排序，并输出，代码如下：

```
df2.groupBy("ym")
   .agg(sort_array(collect_list("cnt")).as("cnt"))
   .orderBy("ym")
   .show
```

执行以上代码，输出的结果如下：

```
+------+--------------+
|    ym|           cnt|
+------+--------------+
|2018-2|     [64, 128]|
|2018-5|          [22]|
|2019-1|[3, 4, 21, 24]|
|2019-2|  [-43, 0, 35]|
|2019-3|      [46, 56]|
|2019-4|           [5]|
+------+--------------+
```

（4）最后，可以把上面的代码写到一个 ETL 处理逻辑中，代码如下：

```
import org.apache.spark.sql.functions._

val inputPath = "/data/spark/data.txt"

import spark.implicits._
spark.read
    .option("inferSchema","true")
    .option("header","false")
    .csv(inputPath)
    .toDF("year","month","cnt")
    .select(concat_ws("-",$"year",$"month").as("ym"),$"cnt")
    .groupBy("ym")
    .agg(sort_array(collect_list("cnt")).as("cnt"))
    .orderBy("ym")
    .show
```

5.2.7 Spark 3 数组函数

Spark 3 增加了一些新的数组函数,其中 transform()和 aggregate()数组函数是功能特别强大的通用函数。它们提供的功能相当于 Scala 中的 map()和 fold(),使 ArrayType 列更容易处理。

1. exists()

如果函数对数组中的任何值都有返回值 true,则 exists()函数就返回值 true。下面创建一个 DataFrame,然后运行 org.apache.spark.sql.functions.exists()函数,用于将一个 even_exists 列附加到 DataFrame 上。

首先,构造一个 DataFrame,代码如下:

```
//第5章/functions_exists.scala

//数据集
val data = List(
    ("a", Array(3, 4, 5)),
    ("b", Array(8, 12)),
    ("c", Array(7, 13)),
    ("d", null),
)

//列名
val columns = List("person_id","best_numbers")

//创建 DataFrame
val df = data.toDF(columns:_*)

//查看
df.printSchema
df.show
```

执行以上代码，输出的结果如下：
```
root
 |-- person_id: string (nullable = true)
 |-- best_numbers: array (nullable = true)
 |    |-- element: integer (containsNull = false)

+---------+------------+
|person_id|best_numbers|
+---------+------------+
|        a|   [3, 4, 5]|
|        b|     [8, 12]|
|        c|     [7, 13]|
|        d|        null|
+---------+------------+
```

然后，判断 best_numbers 中包含偶数元素的记录，代码如下：
```
import org.apache.spark.sql.functions._
import org.apache.spark.sql.functions.exists

//增加一个新列，判断'best_numbers'列是否包含偶数值
val resDF = df
    .withColumn("even_exists", exists(col("best_numbers"), _ % 2 === 0))

resDF.show
```

执行以上代码，输出的结果如下：
```
+---------+------------+-----------+
|person_id|best_numbers|even_exists|
+---------+------------+-----------+
|        a|   [3, 4, 5]|       true|
|        b|     [8, 12]|       true|
|        c|     [7, 13]|      false|
|        d|        null|       null|
+---------+------------+-----------+
```

2. forall()

用来遍历数组中的每个元素。下面的代码演示了该函数的用法。首先构造一个 DataFrame，代码如下：
```
//第5章/functions_forall.scala

//数据集
val data = List(
    (Array("ants", "are", "animals")),
    (Array("italy", "is", "interesting")),
    (Array("brazilians", "love", "soccer")),
    (null),
  )
```

```
//创建 DataFrame
val df = data.toDF("words")

df.printSchema
df.show(false)
```

执行以上代码,输出的结果如下:

```
root
 |-- words: array (nullable = true)
 |    |-- element: string (containsNull = true)

+-------------------------+
|words                    |
+-------------------------+
|[ants, are, animals]     |
|[italy, is, interesting] |
|[brazilians, love, soccer]|
|null                     |
+-------------------------+
```

然后用 forall()函数来标识所有单词以字母 a 开头的数组,代码如下:

```
import org.apache.spark.sql.functions._
import org.apache.spark.sql.functions.exists

val resDF = df
    .withColumn("start_with_a", forall(col("words"), _.startsWith("a")))

resDF.show(false)
```

执行以上代码,输出的结果如下:

```
+-------------------------+------------+
|words                    |start_with_a|
+-------------------------+------------+
|[ants, are, animals]     |true        |
|[italy, is, interesting] |false       |
|[brazilians, love, soccer]|false       |
|null                     |null        |
+-------------------------+------------+
```

3. filter()

过滤数组中的每个元素。下面的代码演示了该函数的用法。首先构造一个 DataFrame,代码如下:

```
//第5章/functions_filter.scala

val data = List(
   (Array("bad", "bunny", "is", "funny")),
   (Array("food", "is", "bad", "tasty")),
   (null),
   )
```

```
//创建 DataFrame
val df = data.toDF("words")

df.printSchema
df.show(false)
```

执行以上代码,输出的结果如下:

```
root
 |-- words: array (nullable = true)
 |    |-- element: string (containsNull = true)

+----------------------+
|words                 |
+----------------------+
|[bad, bunny, is, funny]|
|[food, is, bad, tasty] |
|null                  |
+----------------------+
```

然后过滤掉数组中 bad 这个单词,代码如下:

```
import org.apache.spark.sql.functions._
import org.apache.spark.sql.functions.exists

val resDF = df
   .withColumn("filtered_words", filter(col("words"), _ =!= "bad"))
resDF.show(false)
```

执行以上代码,输出的结果如下:

```
+----------------------+------------------+
|words                 |filtered_words    |
+----------------------+------------------+
|[bad, bunny, is, funny]|[bunny, is, funny]|
|[food, is, bad, tasty] |[food, is, tasty] |
|null                  |null              |
+----------------------+------------------+
```

4. transform()

相当于 map()操作,用来转换数组中的每个元素。下面的代码演示了该函数的用法。首先构造一个 DataFrame,代码如下:

```
//第5章/functions_transform.scala

val data = List(
   (Array("New York", "Seattle")),
   (Array("Barcelona", "Bangalore")),
   (null),
  )

//创建 DataFrame
```

```
val df = data.toDF("places")

df.printSchema
df.show(false)
```

执行以上代码，输出的结果如下：

```
root
 |-- places: array (nullable = true)
 |    |-- element: string (containsNull = true)

+----------------------+
|places                |
+----------------------+
|[New York, Seattle]   |
|[Barcelona, Bangalore]|
|null                  |
+----------------------+
```

然后调用 transform() 转换函数，将 places 列中的每个数组元素与另一个字符串连接起来，代码如下：

```
import org.apache.spark.sql.functions._
import org.apache.spark.sql.functions.exists

val resDF = df
    .withColumn("fun_places", transform(col("places"), concat(_, lit(" is fun!")) ))
resDF.show(false)
```

执行以上代码，输出的结果如下：

```
+----------------------+------------------------------------------+
|places                |fun_places                                |
+----------------------+------------------------------------------+
|[New York, Seattle]   |[New York is fun!, Seattle is fun!]       |
|[Barcelona, Bangalore]|[Barcelona is fun!, Bangalore is fun!]    |
|null                  |null                                      |
+----------------------+------------------------------------------+
```

5. aggregate()

相当于 Scala 中的 fold() 操作，用来执行数组元素的聚合。下面的代码演示了该函数的用法。首先构造一个 DataFrame，代码如下：

```
//第5章/functions_aggregate.scala

val data = List(
    (Array(1, 2, 3, 4)),
    (Array(5, 6, 7)),
    (null),
 )

//创建 DataFrame
```

```
val df = data.toDF("numbers")

df.printSchema
df.show(false)
```

执行以上代码，输出的结果如下：

```
root
 |-- numbers: array (nullable = true)
 |    |-- element: integer (containsNull = false)

+------------+
|numbers     |
+------------+
|[1, 2, 3, 4]|
|[5, 6, 7]   |
|null        |
+------------+
```

然后调用 aggregate() 函数对数组元素执行聚合运算，代码如下：

```
import org.apache.spark.sql.functions._
import org.apache.spark.sql.functions.exists

val resDF = df
    .withColumn("numbers_sum", aggregate(col("numbers"), lit(0), _ + _ ))
resDF.show
```

执行以上代码，输出的结果如下：

```
+------------+-----------+
|     numbers|numbers_sum|
+------------+-----------+
|[1, 2, 3, 4]|         10|
|   [5, 6, 7]|         18|
|        null|       null|
+------------+-----------+
```

6. zip_with()

拉链操作，用来组合两个数组中的相同索引位置处的元素。下面的代码演示了该函数的用法。首先构造一个 DataFrame，代码如下：

```
//第5章/functions_zip_with.scala

val data = List(
    (Array("a", "b"), Array("c", "d")),
    (Array("x", "y"), Array("p", "o")),
    (null, Array("e", "r"))
 )

//列名
val columns = List("letters1","letters2")
```

```
//创建 DataFrame
val df = data.toDF(columns:_*)

df.printSchema
df.show(false)
```

执行上面的代码,输出内容如下:

```
root
 |-- letters1: array (nullable = true)
 |    |-- element: string (containsNull = true)
 |-- letters2: array (nullable = true)
 |    |-- element: string (containsNull = true)

+--------+--------+
|letters1|letters2|
+--------+--------+
|[a, b]  |[c, d]  |
|[x, y]  |[p, o]  |
|null    |[e, r]  |
+--------+--------+
```

然后执行拉链操作,使用 zip_with()函数将 letters1 列和 letters2 列的数组元素一一对应地组合在一起,代码如下:

```
import org.apache.spark.sql.functions._
import org.apache.spark.sql.functions.exists

val resDF = df.withColumn(
  "zipped_letters",
  zip_with(
    col("letters1"),
    col("letters2"),
    concat_ws("***", _, _)
  )
)
resDF.show(false)
```

上面的语句也可以使 lambda 表达式实现,两者等价,代码如下:

```
import org.apache.spark.sql.functions._
import org.apache.spark.sql.functions.exists
import org.apache.spark.sql._

val resDF = df.withColumn(
  "zipped_letters",
  zip_with(
    col("letters1"),
    col("letters2"),
    (left: Column, right: Column) => concat_ws("***", left, right)
  )
)

resDF.show(false)
```

执行以上代码,输出的结果如下:

```
+--------+--------+--------------+
|letters1|letters2|zipped_letters|
+--------+--------+--------------+
|  [a, b]|  [c, d]|[a***c, b***d]|
|  [x, y]|  [p, o]|[x***p, y***o]|
|    null|  [e, r]|          null|
+--------+--------+--------------+
```

5.3 聚合与透视函数

对大数据进行分析通常需要对数据进行聚合操作。聚合通常需要某种形式的分组,要么在整个数据集上,要么在一个或多个列上,然后对它们应用聚合函数,例如对每个组进行求和、计数或求平均值等。Spark SQL 提供了许多常用的聚合函数。

5.3.1 聚合函数

在 Spark 中,所有的聚合都是通过函数来完成的。聚合函数被用来在一组行上执行聚合,不管那组行是由 DataFrame 中的所有行还是一组子行组成的。Spark 中常见的聚合函数见表 5-1。

表 5-1 常见聚合函数

聚 合 函 数	描 述
count(col)	返回每组中成员数量
countDistinct(col)	返回每组中成员唯一数量
approx_count_distinct(col)	返回每组中成员唯一近似数量
min(col)	返回每组中给定列的最小值
max(col)	返回每组中给定列的最大值
sum(col)	返回每组中给定列的值的和
sumDistinct(col)	返回每组中给定列的唯一值的和
avg(col)	返回每组中给定列的值的平均值
skewness(col)	返回每组中给定列的值的分布的偏度
kurtosis(col)	返回每组中给定列的值的分布的峰度
variance(col)	返回每组中给定列的值的无偏方差
stddev(col)	返回每组中给定列的值的标准差
collect_list(col)	返回每组中给定列的值的集合。返回的集合可能包含重复的值
collect_set(col)	返回每组中给定列的唯一值的集合

为了演示这些函数的用法,在下面的示例中将使用"2018 年 11 月 14 日深圳市价格定期监测信息"数据集。这个数据集包含一些主要副食品的监测信息,以 CSV 格式存储在

文件中。将该数据集上传到 HDFS 分布式文件系统的/data/spark 目录下。

首先读取价格监测信息数据集，并创建 DataFrame，代码如下：

```scala
//第5章/functions_agg.scala

//读取数据源文件，创建 DataFrame
val filePath = "/data/spark/2018年11月14日深圳市价格定期监测信息.csv"
val priceDF = spark
  .read
  .option("header","true")
  .option("inferSchema","true")
  .csv(filePath)

priceDF.printSchema()
priceDF.show(5)
```

执行以上代码，输出的结果如下：

```
root
 |-- RECORDID: string (nullable = true)
 |-- JCLB: string (nullable = true)
 |-- JCMC: string (nullable = true)
 |-- BQ: double (nullable = true)
 |-- SQ: double (nullable = true)
 |-- TB: double (nullable = true)
 |-- HB: double (nullable = true)

+--------------------+----+------+------+------+------+------+
|            RECORDID|JCLB|  JCMC|    BQ|    SQ|    TB|    HB|
+--------------------+----+------+------+------+------+------+
|537B9A6E0C836F36E...|null|  椰菜|  2.27| 2.261|-0.067| 0.004|
|537B9A6E0C846F36E...|null|东北米| 2.863| 2.843|-0.067| 0.007|
|537B9A6E0C856F36E...|null|早籼米|  3.08| 3.044| 0.219| 0.012|
|537B9A6E0C866F36E...|null|晚籼米| 3.217|  3.22| 0.081|-0.001|
|537B9A6E0C876F36E...|null|泰国香米| 9.48|  9.48| 0.047|   0.0|
+--------------------+----+------+------+------+------+------+
only showing top 5 rows
```

数据集中每行代表一条商品的价格监测信息，其中各个字段的含义如下。

（1）JCLB：监测类别。

（2）JCMC：监测名称。

（3）BQ：本期价格。

（4）SQ：上期价格。

（5）TB：同比价格变化。

（6）HB：环比价格变化。

对数据集进行简单探索。首先，找出这个数据集总共有多少行，代码如下：

```scala
println("监测的商品数量有" + priceDF.count())
```

执行以上代码，输出的结果如下：

监测的商品数量有 331

下面使用一些常用的聚合函数进行统计。

1. count(col)

统计指定列的数量。例如，统计数据集中商品（JCMC）的数量和监测类别（JCLB）的数量，代码如下：

```
import org.apache.spark.sql.functions._

//商品数量
priceDF.select(count("JCMC").as("监测商品")).show()

//当统计一列中的项目数量时，count（col）函数不包括计数中的 null 值
priceDF.select(count("JCLB").as("监测类别")).show()

//判断"JCLB"列为 null 值的有多少
val nullJclb = priceDF.where(col("JCLB").isNull).count()
println("\"JCLB\"列为 null 值的有" + nullJclb)
```

执行以上代码，输出的结果如下：

```
+--------+
|监测商品|
+--------+
|     331|
+--------+


+--------+
|监测类别|
+--------+
|     298|
+--------+

"JCLB"列为 null 值的有 33
```

2. countDistinct(col)

countDistinct(col)只计算每组的唯一项。例如，统计总共有多少个商品类别，多少种商品，代码如下：

```
//去重：统计检测的商品类别有多少
//priceDF.select("JCLB").distinct.show()
//priceDF.select("JCLB").distinct.count()

//去重：统计检测的商品有多少种
//priceDF.select("JCMC").distinct.show()
//priceDF.select("JCMC").distinct.count()

//countDistinct(col)：它只计算每组的唯一项，不包括 null
priceDF.select(
   countDistinct("JCLB"),
   countDistinct("JCMC"),
```

```
    count("*")
).show()

priceDF.select(
    countDistinct("JCLB").as("监测类别"),
    countDistinct("JCMC").as("监测商品"),
    count("*").as("总数量")
).show()
```

执行以上代码，输出的结果如下：

```
+---------------------+---------------------+--------+
|count(DISTINCT JCLB)|count(DISTINCT JCMC)|count(1)|
+---------------------+---------------------+--------+
|                   6|                  35|     331|
+---------------------+---------------------+--------+

+--------+--------+------+
|监测类别|监测商品|总数量|
+--------+--------+------+
|       6|      35|   331|
+--------+--------+------+
```

3. approx_count_distinct (col, max_estimated_error=0.05)

近似唯一计数。在一个大数据集里计算每组中唯一项的确切数量是一个成本很高且很耗时的操作。在某些用例中，有一个近似唯一的计数就足够了。例如，在线广告业务中，每小时有数亿个广告曝光并且需要生成一份报告来显示每个特定类型的成员段的独立访问者的数量。Spark 实现了 approx_count_distinct()函数，用来统计近似唯一计数。因为唯一计数是一个近似值，所以会有一定的误差。这个函数允许指定一个可接受估算误差的值。

使用 approx_count_distinct()函数，代码如下：

```
//统计 price DataFrame 的"JCMC"列。默认估算误差是 0.05 (5%)
priceDF.select(
    count("JCMC"),
    countDistinct("JCMC"),
    approx_count_distinct("JCMC", 0.05)
).show()
```

执行以上代码，输出的结果如下：

```
+------------+---------------------+---------------------------+
|count(JCMC)|count(DISTINCT JCMC)|approx_count_distinct(JCMC)|
+------------+---------------------+---------------------------+
|         331|                  35|                         33|
+------------+---------------------+---------------------------+
```

4. min(col), max(col)

获取 col 列的最小值和最大值。例如，统计本期价格的最大值和最小值，代码如下：

```
//统计本期价格(BQ)的最大值和最小值
priceDF.select(
```

```
       min("BQ").as("最便宜的"),
       max("BQ").as("最贵的")
    ).show()
```

执行以上代码，输出的结果如下：

```
+--------+------+
|最便宜的|最贵的|
+--------+------+
|     1.7|43.515|
+--------+------+
```

5. sum(col)

sum(col)函数用于计算一个数字列中的值的总和。例如，计算本期价格之和，代码如下：

```
priceDF.select(sum("BQ")).show()
```

执行以上代码，输出的结果如下：

```
+-----------------+
|          sum(BQ)|
+-----------------+
|2904.4770000000003|
+-----------------+
```

6. sumDistinct(col)

sumDistinct(col)函数只汇总了一个数字列的不同值。例如，计算本期价格（唯一值）之和，代码如下：

```
priceDF.select(sumDistinct("BQ")).show
```

执行以上代码，输出的结果如下：

```
+-----------------+
| sum(DISTINCT BQ)|
+-----------------+
|2544.9589999999994|
+-----------------+
```

7. avg(col)

avg(col)函数用于计算一个数字列的平均值。这个方便的函数简单地取总并除以项目的数量。例如，计算本期的平均价格，代码如下：

```
priceDF.select(avg("BQ"), sum("BQ") / count("BQ")).show
```

执行以上代码，输出的结果如下：

```
+-----------------+---------------------+
|          avg(BQ)|(sum(BQ) / count(BQ))|
+-----------------+---------------------+
|8.774854984894262|    8.774854984894262|
+-----------------+---------------------+
```

8. skewness(col), kurtosis(col)

在统计领域中，峰度（Kurtosis）与偏度（Skewness）是测量数据正态分布特性的两个指标。了解偏度和峰度这两个统计量的含义很重要，在对数据进行正态转换时，需要将其作为参考，选择合适的转换方法。

偏态是一种度量数据集的值分布对称性的度量：

（1）当偏度≈0时，可认为分布是对称的，服从正态分布。

（2）当偏度>0时，分布为右偏，即拖尾在右边，峰尖在左边，也称为正偏态。

（3）当偏度<0时，分布为左偏，即拖尾在左边，峰尖在右边，也称为负偏态。

数据分布的左偏或右偏，指的是数值拖尾的方向，而不是峰的位置，如图5-1所示。

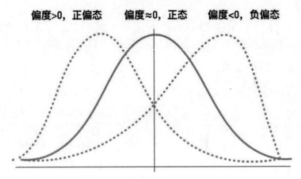

图 5-1　数据分布的偏态特征

峰度是对分布曲线性状的度量，用来衡量数据分布的平坦度（曲线有可能是正常的、平坦的或尖的）。正的Kurtosis表示曲线是细而尖的，负的Kurtosis表示曲线是宽而平的。正态分布的峰度值为3。不同大小的峰度值的含义如下：

（1）当峰度≈0时，可认为分布的峰态合适，服从正态分布（不胖不瘦）。

（2）当峰度>0时，分布的峰态陡峭（高尖）。

（3）当峰度<0时，分布的峰态平缓（矮胖）。

这几种峰态特征分布如图5-2所示。

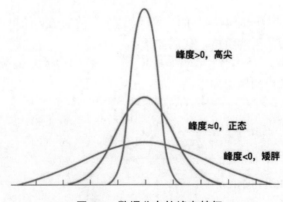

图 5-2　数据分布的峰态特征

例如，要计算 BQ 列（表示本期价格列）的偏度和峰度，代码如下：

```
priceDF.select(skewness("BQ"), kurtosis("BQ")).show
```

执行以上代码，输出的结果如下：

```
+------------------+------------------+
|      skewness(BQ)|      kurtosis(BQ)|
+------------------+------------------+
|2.2954037816302346|4.8142280508421511|
+------------------+------------------+
```

结果表明，BQ 列的分布是不对称的，呈正偏态分布（右边的尾巴比左边的尾巴长或宽）。峰度值表明分布曲线是高而尖的。

9. variance(col), stddev(col)

在统计学中，方差（variance）和标准偏差（stddev）用于测量数据的分散性或分布。换句话说，它们被用来说明 values 到平均值的平均距离。当方差值较低时，意味着该值接近均值。方差和标准差是相关的，后者是前者的平方根。

函数 variance() 和 stddev() 分别用于计算方差和标准差。Spark 提供了这些函数的两种不同实现：一种是利用抽样来加速度计算，另一种使用全样数据。例如，要计算 priceDF 这个 DataFrame 中的 BQ 列的方差和标准偏差，代码如下：

```
//使用方差和标准差的两种变化
priceDF.select(variance("BQ"), var_pop("BQ"), stddev("BQ"), stddev_pop("BQ")).show
```

执行以上代码，输出的结果如下：

```
+------------------+------------------+------------------+------------------+
|      var_samp(BQ)|       var_pop(BQ)|   stddev_samp(BQ)|    stddev_pop(BQ)|
+------------------+------------------+------------------+------------------+
| 85.96212834860381| 85.70242403335124| 9.271576368051111| 9.257560371574751|
+------------------+------------------+------------------+------------------+
```

从输出结果来看，priceDF 这个 DataFrame 中的 BQ（本期价格）值很分散。

5.3.2 分组聚合

分组聚合不会在 DataFrame 中对全局组执行聚合，而是在 DataFrame 中的每个子组中执行聚合。通常分组执行聚合的过程分为两步。第 1 步是通过 groupBy（col1、col2、……）转换来执行分组，也就是指定要按哪些列分组。与其他返回 DataFrame 的转换不同，这个 groupBy()转换会返回一个 RelationalGroupedDataset 类的实例。类 RelationalGroupedDataset 提供了一组标准的聚合函数，可以将它们应用到每个子组中。这些聚合函数有 avg(cols)、count()、mean(cols)、min(cols)、max(cols)和 sum(cols)。除了 count()函数外，其余所有的函数都在数字列上执行。

例如，要按 JCLB（检测类别）列分组并执行一个 count()聚合（groupBy()列将自动包含在输出中），代码如下：

```
//按检测的商品大类分组统计
priceDF.groupBy("JCLB").count().show(false)
```

执行以上代码,输出的结果如下:

```
+------+-----+
|JCLB  |count|
+------+-----+
|粮食   |55   |
|食用油 |37   |
|水产品 |27   |
|null  |33   |
|肉奶蛋 |7    |
|蔬菜   |117  |
|肉蛋奶 |55   |
+------+-----+
```

按 JCLB 和 JCMC 分组之后,执行 count() 聚合,并按统计数量降序排序,代码如下:

```
//按商品大类和商品小类分组统计
priceDF
    .groupBy($"JCLB", $"JCMC")
    .count
    .orderBy($"count".desc)
    .show(false)

priceDF
    .groupBy($"JCLB", $"JCMC")
    .count
    .where($"JCMC" === "花生油")
    .orderBy($"count".desc)
    .show(false)
```

执行以上代码,输出的结果如下:

```
+------+----------+-----+
|JCLB  |JCMC      |count|
+------+----------+-----+
|粮食   |东北米    |10   |
|食用油 |花生油    |10   |
|蔬菜   |蔬菜均价  |9    |
|蔬菜   |大白菜    |9    |
|粮食   |散装面粉  |9    |
|食用油 |菜籽油    |9    |
|蔬菜   |萝卜      |9    |
|粮食   |袋装面粉  |9    |
|蔬菜   |菠菜      |9    |
|蔬菜   |芹菜      |9    |
|蔬菜   |其中,青椒 |9    |
|蔬菜   |西红柿    |9    |
|蔬菜   |黄瓜      |9    |
|蔬菜   |茄子      |9    |
|水产品 |大头鱼    |9    |
|食用油 |调和油    |9    |
```

```
|粮食    |泰国香米    |9    |
|粮食    |早籼米      |9    |
|水产品  |草鱼        |9    |
|粮食    |晚籼米      |9    |
+------+----------+-----+
only showing top 20 rows

+------+------+-----+
|JCLB  |JCMC  |count|
+------+------+-----+
|食用油|花生油|10   |
|null  |花生油|1    |
```

有时需要在同一时间对每个组执行多个聚合。例如，除了计数外，还想知道最小值和最大值。RelationalGroupedDataset 类提供一个名为 agg() 的功能强大的函数，它接受一个或多个列表达式，这意味着可以使用任何聚合函数。这些聚合函数返回 Column 类的一个实例，这样就可以使用所提供的函数来应用任何列表达式了。一个常见的需求是在聚合完成后重命名列，使之更短、更可读、更易于引用。

例如，按 JCLB 分组之后，执行多个聚合，代码如下：

```
//在同一时间对每个组执行多个聚合
import org.apache.spark.sql.functions._

priceDF
    .na.drop
    .groupBy("JCLB")
    .agg(
      count("BQ").as("本期数量"),
      min("BQ").as("本期最低价格"),
      max("BQ").as("本期最高价格"),
      avg("BQ").as("本期平均价格")
    )
    .show()
```

执行以上代码，输出的结果如下：

```
+------+--------+------------+------------+------------------+
| JCLB |本期数量|本期最低价格|本期最高价格|   本期平均价格   |
+------+--------+------------+------------+------------------+
| 粮食 |    55  |     2.844  |      9.48  |  4.258527272727273|
|食用油|    37  |     4.5    |     11.43  |  7.179567567567568|
|水产品|    27  |     8.33   |     37.26  | 17.51562962962963 |
|肉奶蛋|     7  |     2.99   |     42.93  | 19.195714285714285|
| 蔬菜 |   117  |     1.7    |     8.765  |  4.372042735042735|
|肉蛋奶|    55  |     2.8    |    43.515  | 18.250181818181815|
+------+--------+------------+------------+------------------+
```

这个 agg() 函数提供了一种通过基于字符串的 key-value 映射来表达列表达式的附加方法。key 是列名，而 value 是一个聚合函数，它可以是 avg、max、min、sum 或 count，代码如下：

```
priceDF.na.drop
    .groupBy("JCLB")
    .agg(
      "BQ" -> "count",
      "BQ" -> "min",
      "BQ" -> "max",
      "BQ" -> "avg"
    )
    .show()
```

执行以上代码，输出的结果如下：

```
+------+--------+-------+-------+------------------+
|  JCLB|count(BQ)|min(BQ)|max(BQ)|           avg(BQ)|
+------+--------+-------+-------+------------------+
|  粮食|      55|  2.844|   9.48|  4.258527272727273|
|食用油|      37|    4.5|  11.43|  7.179567567567568|
|水产品|      27|   8.33|  37.26| 17.515629629629632|
|肉奶蛋|       7|   2.99|  42.93|19.195714285714285|
|  蔬菜|     117|    1.7|  8.765|  4.372042735042735|
|肉蛋奶|      55|    2.8| 43.515|18.250181818181815|
+------+--------+-------+-------+------------------+
```

函数 collect_list(col) 和 collect_set(col) 用于在应用分组后收集特定组的所有值。一旦每个组的值被收集到一个集合中，那么就可以自由地以任何选择的方式对其进行操作。这两个函数的返回集合之间有一个小的区别，那就是唯一性。collect_list() 函数返回一个可能包含重复值的集合，collect_set() 函数返回一个只包含唯一值的集合。

例如，使用 collection_list() 函数来收集每种商品大类下的商品名称，代码如下：

```
import org.apache.spark.sql.functions._
import spark.implicits._

priceDF.na.drop
    .groupBy('JCLB.as("监测类别"))
    .agg(collect_set("JCMC").as("监测的商品"))
    .withColumn("监测商品数量",size('监测的商品))
    .show()
```

执行以上代码，输出的结果如下：

```
+--------+--------------------------+------------+
|监测类别|                监测的商品|监测商品数量|
+--------+--------------------------+------------+
|    粮食|  [早籼米, 散装面粉, 泰国香米,...]|           6|
|  食用油|  [花生油, 菜籽油, 豆油, 调和油]|           4|
|  水产品|       [带鱼, 草鱼, 大头鱼]|           3|
|  肉奶蛋|  [其中:精瘦肉, 鸡蛋, 牛肉, ...]|           7|
|    蔬菜|  [蔬菜均价, 西红柿, 大白菜, ...]|          14|
|  肉蛋奶|  [其中:精瘦肉, 鸡蛋, 牛肉, ...]|           8|
+--------+--------------------------+------------+
```

5.3.3 数据透视

数据透视是一种通过聚合和旋转把数据行转换成数据列的技术，它是一种将行转换成列同时应用一个或多个聚合的方法。这样一来，分类值就会从行转到单独的列中。这种技术通常用于数据分析或报告。

数据透视过程从一个或多个列的分组开始，然后在一个列上旋转，最后在一个或多个列上应用一个或多个聚合，因此，当透视数据时，需要确定 3 个要素：要在行（分组元素）中看到的元素，要在列（扩展元素）上看到的元素，要在数据部分看到的元素（聚合元素）。

在下面的例子中，有一个包含学生信息的数据集，每行包含学生姓名、性别、体重、毕业年份。现在想要知道每个毕业年份不同性别的平均体重。

首先，定义一个 case class，代表学生的域对象，代码如下：

```
case class Student(
    name:String,
    gender:String,
    weight:Int,
    graduation_year:Int)
```

然后创建一个 DataFrame，并调用 pivot()函数在 gender 列上旋转，统计不同性别的平均体重，代码如下：

```
import spark.implicits._

//转换为DataFrame
val studentsDF = Seq(
    Student("刘宏明", "男", 180, 2015),
    Student("赵薇", "女", 110, 2015),
    Student("黄海波", "男", 200, 2015),
    Student("杨幂", "女", 109, 2015),
    Student("楼一萱", "女", 105, 2015),
    Student("龙梅子", "女", 115, 2016),
    Student("陈知远", "男", 195, 2016)
).toDF()

//studentsDF.show()

//计算每年不同性别的平均体重
studentsDF.groupBy("graduation_year").pivot("gender").avg("weight").show()
```

执行以上代码，输出的结果如下：

```
+---------------+-----+-----+
|graduation_year|   女|   男|
+---------------+-----+-----+
|           2015|108.0|190.0|
|           2016|115.0|195.0|
+---------------+-----+-----+
```

可以利用 agg()函数来执行多个聚合，这会在结果表中创建更多的列，代码如下：

```
studentsDF
    .groupBy("graduation_year")
    .pivot("gender")
    .agg(
      min("weight").as("min"),
      max("weight").as("max"),
      avg("weight").as("avg")
    ).show()
```

执行以上代码，输出的结果如下：

```
+---------------+------+------+------+------+------+------+
|graduation_year|女_min|女_max|女_avg|男_min|男_max|男_avg|
+---------------+------+------+------+------+------+------+
|           2015|   105|   110| 108.0|   180|   200| 190.0|
|           2016|   115|   115| 115.0|   195|   195| 195.0|
+---------------+------+------+------+------+------+------+
```

如果 pivot()的列有许多不同的值，则可以选择性地选择生成聚合的值，代码如下：

```
studentsDF
    .groupBy("graduation_year")
    .pivot("gender", Seq("男"))
    .agg(
      min("weight").as("min"),
      max("weight").as("max"),
      avg("weight").as("avg")
    ).show()
```

执行以上代码，输出的结果如下：

```
+---------------+------+------+------+
|graduation_year|男_min|男_max|男_avg|
+---------------+------+------+------+
|           2015|   180|   200| 190.0|
|           2016|   195|   195| 195.0|
+---------------+------+------+------+
```

为 pivot()列指定一个 distinct 值的列表实际上会加速旋转过程。

5.3.4 谓词子查询

Spark SQL 支持一个非常有用的特性：谓词子查询。谓词子查询是指操作数为子查询的谓词。Spark 2.0 支持 EXISTS 和 IN 两种基本形式。Spark 2.0 目前只支持 WHERE 子句中的谓词子查询。

这意味着可以使用来自另一个查询的内容方便地过滤主查询中的行。当只需从一张表中选择行，而另一张表中存在匹配内容时，这将非常有用。

例如，使用谓词子查询根据在另一张表 interesting_users 中找到的用户筛选 clickstream 表中的用户，代码如下：

```
select count(1)
from clickstream
where user in (select distinct user from interesting_users)
```

5.4 高级分析函数

Spark SQL 提供了许多的高级分析函数，如多维聚合函数、时间窗口聚合函数和窗口分析函数等。本节就介绍这些高级分析函数的使用。

5.4.1 使用多维聚合函数

在高级分析函数中，第 1 个是关于多维聚合的，它对于涉及分层数据分析的用例非常有用，在这种情况下，通常需要在一组分组列中计算子总数和总数。常用的多维聚合函数包括 rollup()和 cube()，它们基本上是在多列上进行分组的高级版本，通常用于在这些列的组合和排列中生成子总数和总数。

1. rollup()

当使用分层数据时，例如不同部门和分部的销售收入数据等，rollup()可以很容易地计算出它们的子总数和总数。rollup()按给定的列集的层次结构，并且总是在层次结构中的第 1 列启动 rolling up 过程。使用 rollup()函数的代码如下：

```
//第 5 章/functions_orders.csv

//读取超市订单汇总数据
val filePath = "/data/spark/超市订单.csv"
val ordersDF = spark
  .read
  .option("header", "true")
  .option("inferSchema","true")
  .csv(filePath )

println("订单数量: " + ordersDF.count())
ordersDF.printSchema()
```

执行以上代码，输出的结果如下：

```
订单数量: 10000
root
 |-- 行 ID: integer (nullable = true)
 |-- 订单 ID: string (nullable = true)
 |-- 订单日期: string (nullable = true)
 |-- 发货日期: string (nullable = true)
 |-- 邮寄方式: string (nullable = true)
 |-- 客户 ID: string (nullable = true)
 |-- 客户名称: string (nullable = true)
 |-- 细分: string (nullable = true)
```

```
 |-- 城市: string (nullable = true)
 |-- 省/自治区: string (nullable = true)
 |-- 国家: string (nullable = true)
 |-- 地区: string (nullable = true)
 |-- 产品 ID: string (nullable = true)
 |-- 类别: string (nullable = true)
 |-- 子类别: string (nullable = true)
 |-- 产品名称: string (nullable = true)
 |-- 销售额: double (nullable = true)
 |-- 数量: integer (nullable = true)
 |-- 折扣: double (nullable = true)
 |-- 利润: double (nullable = true)
```

查看前 10 条数据，代码如下：

```
ordersDF.show(10)
```

执行以上代码，输出结果如图 5-3 所示。

行 ID	订单 ID	订单日期	发货日期	邮寄方式	客户 ID	客户名称	细分	城市	省/自治区	国家	地区	产品 ID	类别	子类别	产品名称	销售额	数量	折扣	利润
1	US-2017-1357144	2017/4/27	2017/4/29	二级	曾惠-14485	曾惠	公司	杭州	浙江	中国	华东	办公用-用品-10002717	办公用品	用品	Fiskars 剪刀，蓝色	129.696	2	0.4	-60.704
2	CN-2017-1973789	2017/6/15	2017/6/19	标准级	许安-10165	许安	消费者	内江	四川	中国	西南	办公用-信封-10004832	办公用品	信封	GlobeWeis 搭扣信封，红色	125.44	2	0.0	42.56
3	CN-2017-1973789	2017/6/15	2017/6/19	标准级	许安-10165	许安	消费者	内江	四川	中国	西南	办公用-装订-10001505	办公用品	装订机	Cardinal 孔加固材料，回收	31.92	2	0.4	4.2
4	US-2017-3017568	2017/12/9	2017/12/13	标准级	宋良-17170	宋良	公司	镇江	江苏	中国	华东	办公用-用品-10003746	办公用品	用品	Kleencut 开信刀，工业	321.216	4	0.4	-27.104
5	CN-2016-2975416	2016/5/31	2016/6/2	二级	万兰-15730	万兰	消费者	汕头	广东	中国	中南	办公用-器具-10003452	办公用品	器具	KitchenAid 搅拌机，黑色	1375.92	3	0.0	550.2
6	CN-2015-4497736	2015/10/27	2015/10/31	标准级	俞明-18325	俞明	消费者	景德镇	江西	中国	华东	技术-设备-10001640	技术	设备	柯尼卡 打印机，红色	11129.58	9	0.0	3783.78
7	CN-2015-4497736	2015/10/27	2015/10/31	标准级	俞明-18325	俞明	消费者	景德镇	江西	中国	华东	办公用-装订-10001029	办公用品	装订机	Ibico 订书机，实惠	479.92	2	0.0	172.76
8	CN-2015-4497736	2015/10/27	2015/10/31	标准级	俞明-18325	俞明	消费者	景德镇	江西	中国	华东	家具-椅子-10000578	家具	椅子	SAFCO 扶手椅，可调	8659.84	4	0.0	2684.08
9	CN-2015-4497736	2015/10/27	2015/10/31	标准级	俞明-18325	俞明	消费者	景德镇	江西	中国	华东	办公用-纸张-10001629	办公用品	纸张	Green Bar 计划信息表，多色	588.0	5	0.0	46.9
10	CN-2015-4497736	2015/10/27	2015/10/31	标准级	俞明-18325	俞明	消费者	景德镇	江西	中国	华东	办公用-系固-10004801	办公用品	系固件	Stockwell 橡皮筋，整包	154.28	2	0.0	33.88

only showing top 10 rows

图 5-3　订单数据

对数据进行过滤，让数据集的体积更小，以便更容易地看到 rollup()的结果，代码如下：

```
val twoSummary = ordersDF
    .select($"地区", $"省/自治区/直辖市", $"订单 ID")
    .where($"地区" === "华东" || $"地区" === "华北")

//看一看数据是什么样子的
println("数据量: " + twoSummary.count)
twoSummary.show()
```

执行以上代码，输出的结果如下：

```
数据量: 4327
+----+---------+---------------+
| 地区| 省/自治区/直辖市|    订单 ID|
+----+---------+---------------+
| 华东|      浙江|US-2017-1357144|
| 华东|      江苏|US-2017-3017568|
| 华东|      江西|CN-2015-4497736|
| 华东|      江西|CN-2015-4497736|
| 华东|      江西|CN-2015-4497736|
| 华东|      江西|CN-2015-4497736|
| 华东|      江西|CN-2015-4497736|
| 华东|      山东|CN-2015-2752724|
| 华东|      山东|CN-2015-2752724|
```

```
| 华东|        山东|CN-2015-2752724|
| 华东|        江苏|US-2016-2511714|
| 华东|        江苏|US-2016-2511714|
| 华东|        上海|CN-2017-5631342|
| 华东|        上海|CN-2017-5631342|
| 华东|        上海|CN-2017-5631342|
| 华东|        上海|CN-2017-5631342|
| 华东|        上海|CN-2017-5631342|
| 华东|        上海|CN-2017-5631342|
| 华东|        浙江|US-2016-4150614|
+----+----------+---------------+
only showing top 20 rows
```

接下来按地区、省/自治区/直辖市执行 rollup() 操作，然后计算计数的总和，最后按 null 排序，代码如下：

```
twoSummary
    .rollup($"地区", $"省/自治区/直辖市")
    .agg(count("订单 ID") as "total")
    .orderBy($"地区".asc_nulls_last, $"省/自治区/直辖市".asc_nulls_last)
    .show()
```

执行以上代码，输出的结果如下：

```
+----+----------+-----+
| 地区|省/自治区/直辖市|total|
+----+----------+-----+
| 华东|        上海|  292|
| 华东|        安徽|  347|
| 华东|        山东|  914|
| 华东|        江苏|  583|
| 华东|        江西|  139|
| 华东|        浙江|  424|
| 华东|        福建|  259|
| 华东|      null| 2958|
| 华北|       内蒙古|  224|
| 华北|        北京|  252|
| 华北|        天津|  304|
| 华北|        山西|  201|
| 华北|        河北|  388|
| 华北|      null| 1369|
|null|      null| 4327|
+----+----------+-----+
```

这个输出显示了华东区和华北区的每个城市的子总数，而总计显示在最后一行，并带有在"地区"和"省/自治区/直辖市"的列上的 null 值。注意带有 asc_nulls_last 选项进行排序，因此 Spark SQL 会将 null 值排序到最后位置。

2. cube()

cube() 函数可以看作 rollup() 函数的更高级版本。它在分组列的所有组合中执行聚合操作,因此,结果包括 rollup()提供的及其他组合所提供的聚合。在上面的"地区"和"省/自治区/直辖市"的例子中,结果将包括每个"省/自治区/直辖市"的聚合。使用 cube()函数的方法类似于使用 rollup()函数,代码如下:

```
//执行 cube
twoSummary
    .cube($"地区", $"省/自治区/直辖市")
    .agg(count("订单 ID") as "total")
    .orderBy($"地区".asc_nulls_last, $"省/自治区".asc_nulls_last)
    .show(30)
```

执行以上代码,输出的结果如下:

```
+----+---------------+-----+
|地区|省/自治区/直辖市|total|
+----+---------------+-----+
|华东|           上海|  292|
|华东|           安徽|  347|
|华东|           山东|  914|
|华东|           江苏|  583|
|华东|           江西|  139|
|华东|           浙江|  424|
|华东|           福建|  259|
|华东|           null| 2958|
|华北|           内蒙古| 224|
|华北|           北京|  252|
|华北|           天津|  304|
|华北|           山西|  201|
|华北|           河北|  388|
|华北|           null| 1369|
|null|           上海|  292|
|null|           内蒙古| 224|
|null|           北京|  252|
|null|           天津|  304|
|null|           安徽|  347|
|null|           山东|  914|
|null|           山西|  201|
|null|           江苏|  583|
|null|           江西|  139|
|null|           河北|  388|
|null|           浙江|  424|
|null|           福建|  259|
|null|           null| 4327|
+----+---------------+-----+
```

在结果表格中,在"地区"列中有 null 值的行表示一个地区中所有城市的聚合,因此,一个 cube()的结果总会比 rollup()的结果有更多的行。

5.4.2 使用时间窗口聚合

在高级分析函数中,第 2 个功能是基于时间窗口执行聚合操作,这在处理来自物联网设备的事务或传感器值等时间序列数据时非常有用。

这些时序数据由一系列的时间顺序数据点组成。这种数据集在金融或电信等行业很常见。例如，股票市场交易数据集有交易日期、开盘价、收盘价、交易量和每个股票代码的其他信息，例如京东股票的历史数据，如图 5-4 所示。

JD历史数据

时间范围：每日　　　　　　　　　　　　　　　　　下载数据　2022/04/05 - 2022/05/05

日期	收盘	开盘	高	低	交易量	涨跌幅
2022年5月4日	63.18	60.23	63.33	59.11	9.32M	1.62%
2022年5月3日	62.17	62.80	63.96	61.65	9.10M	-1.89%
2022年5月2日	63.37	61.64	63.55	60.97	10.17M	2.77%
2022年4月29日	61.66	65.08	65.29	61.56	19.76M	6.66%
2022年4月28日	57.81	57.82	58.80	56.15	8.07M	0.68%
2022年4月27日	57.42	55.49	58.89	55.35	14.69M	7.91%
2022年4月26日	53.21	54.13	55.49	52.92	10.16M	-0.95%
2022年4月25日	53.72	50.83	54.02	50.59	9.09M	3.23%
2022年4月22日	52.04	52.48	54.27	51.33	12.52M	2.64%
2022年4月21日	50.70	52.89	53.68	50.25	10.45M	-5.67%
2022年4月20日	53.75	56.60	56.60	53.03	9.30M	-5.52%
2022年4月19日	56.89	56.35	56.92	55.07	8.44M	-1.06%
2022年4月18日	57.50	55.44	58.13	55.34	9.22M	1.66%
2022年4月14日	56.56	57.34	57.97	56.27	7.35M	-2.95%
2022年4月13日	58.28	57.74	59.70	56.99	11.72M	3.19%
2022年4月12日	56.48	57.48	58.31	56.44	7.96M	-0.62%
2022年4月11日	56.83	55.78	58.23	54.85	8.77M	0.51%
2022年4月8日	56.54	56.75	57.85	55.82	10.47M	-0.98%
2022年4月7日	57.10	57.85	57.85	56.21	10.82M	-3.34%
2022年4月6日	59.07	59.50	59.75	57.59	8.98M	-3.04%
2022年4月5日	60.92	62.32	62.40	60.37	6.88M	-3.78%

最高：65.29　　　　最低：50.25　　　　差价：15.04　　　　平均：57.39　　　　涨跌幅：-0.21

图 5-4　股票历史价格时序数据

在 Spark 2.0 中引入了时间窗口的聚合，使其能够轻松地处理时间序列数据。可以使用时间窗口聚合分析时序数据，例如京东股票的周平均收盘价，或者京东股票跨周的月平均收盘价。

时间窗口函数有几个版本，但是它们都需要一个时间戳类型列和一个窗口长度，该窗口长度可以指定为几秒、几分钟、几小时、几天或几周。窗口长度代表一个时间窗口，它有一个开始时间和结束时间，它被用来确定一个特定的时间序列数据应该属于哪个"桶"。

有两种类型的时间窗口：滚动窗口和滑动窗口。与滚动窗口（也叫固定窗口）相比，滑动窗口需要提供额外的输入参数，用来说明在计算下一个"桶"时，一个时间窗口应该滑动多少。

下面的例子将使用京东股票历史交易数据。例如，要计算京东股票的周平均价格，代码如下：

```scala
//第 5 章/functions_jd_stock.scala

//加载京东股票历史交易数据
val csvPath = "/data/spark/jd/jd-formated.csv"
val jdDF = spark
    .read
    .option("header", "true")
    .option("inferSchema","true")
    .csv(csvPath)

//显示该 Schema，第 1 列是交易日期
jdDF.printSchema()
jdDF.show(10)
```

执行以上代码，输出的结果如下：

```
root
 |-- Date: string (nullable = true)
 |-- Close: double (nullable = true)
 |-- Volume: integer (nullable = true)
 |-- Open: double (nullable = true)
 |-- High: double (nullable = true)
 |-- Low: double (nullable = true)

+----------+-----+-------+------+-----+-----+
|      Date|Close| Volume|  Open| High|  Low|
+----------+-----+-------+------+-----+-----+
|2022-02-15|76.13|6766205| 75.35|76.35| 74.8|
|2022-02-14|74.45|5244967| 73.94|74.62|73.01|
|2022-02-11|73.98|6673354| 75.97|76.55|73.55|
|2022-02-10| 76.4|6432184|75.955|78.39|75.24|
|2022-02-09|78.29|7061571| 76.83|78.67|76.61|
|2022-02-08|75.36|7903249| 73.12|76.07|72.05|
|2022-02-07|73.15|6135832| 74.09|74.99|72.81|
|2022-02-04|73.77|6082889| 71.94|74.95|71.86|
|2022-02-03|71.85|7493688| 72.08| 73.3|71.33|
|2022-02-02|73.21|5887066| 75.58|75.71|72.41|
+----------+-----+-------+------+-----+-----+
only showing Top 10 rows
```

注意，其中 Date 字段被自动推断为 String 类型，因此需要对交易日期进行整理（将 Date 字段的字符串类型转换为 Date 类型），代码如下：

```scala
import org.apache.spark.sql.functions._

val jdStock = jdDF.withColumn("Date", to_date(col("Date")))
```

```
jdStock.printSchema()
jdStock.show(10)
```

执行以上代码,输出的结果如下:

```
root
 |-- Date: date (nullable = true)
 |-- Close: double (nullable = true)
 |-- Volume: integer (nullable = true)
 |-- Open: double (nullable = true)
 |-- High: double (nullable = true)
 |-- Low: double (nullable = true)

+----------+-----+-------+------+-----+-----+
|      Date|Close| Volume|  Open| High|  Low|
+----------+-----+-------+------+-----+-----+
|2022-02-15|76.13|6766205| 75.35|76.35| 74.8|
|2022-02-14|74.45|5244967| 73.94|74.62|73.01|
|2022-02-11|73.98|6673354| 75.97|76.55|73.55|
|2022-02-10| 76.4|6432184|75.955|78.39|75.24|
|2022-02-09|78.29|7061571| 76.83|78.67|76.61|
|2022-02-08|75.36|7903249| 73.12|76.07|72.05|
|2022-02-07|73.15|6135832| 74.09|74.99|72.81|
|2022-02-04|73.77|6082889| 71.94|74.95|71.86|
|2022-02-03|71.85|7493688| 72.08| 73.3|71.33|
|2022-02-02|73.21|5887066| 75.58|75.71|72.41|
+----------+-----+-------+------+-----+-----+
only showing Top 10 rows
```

可以看到,Date 字段的数据类型已经被转换为了 date 日期类型。接下来,使用时间窗口函数来计算京东股票的周平均收盘价,代码如下:

```
//使用窗口函数计算 groupBy 变换内的周平均价格
//这是一个滚动窗口的例子,也就是固定窗口
val jdWeeklyAvg = jdStock
    .groupBy(window(col("Date"), "1 week"))
    .agg(avg("Close").as("weekly_avg"))

//结果模式有窗口启动和结束时间
jdWeeklyAvg.printSchema()

//按开始时间顺序显示结果,并采用四舍五入法精确到小数点后 2 位
jdWeeklyAvg
    .orderBy("window.start")
    .selectExpr("window.start", "window.end", "round(weekly_avg, 2) as weekly_avg")
    .show(10)
```

执行以上代码,输出的结果如下:

```
root
 |-- window: struct (nullable = false)
 |    |-- start: timestamp (nullable = true)
 |    |-- end: timestamp (nullable = true)
 |-- weekly_avg: double (nullable = true)
```

```
+-------------------+-------------------+----------+
|              start|                end|weekly_avg|
+-------------------+-------------------+----------+
|2017-02-09 08:00:00|2017-02-16 08:00:00|     30.23|
|2017-02-16 08:00:00|2017-02-23 08:00:00|     30.29|
|2017-02-23 08:00:00|2017-03-02 08:00:00|     30.65|
|2017-03-02 08:00:00|2017-03-09 08:00:00|     30.93|
|2017-03-09 08:00:00|2017-03-16 08:00:00|     31.41|
|2017-03-16 08:00:00|2017-03-23 08:00:00|     31.09|
|2017-03-23 08:00:00|2017-03-30 08:00:00|     31.45|
|2017-03-30 08:00:00|2017-04-06 08:00:00|     31.65|
|2017-04-06 08:00:00|2017-04-13 08:00:00|     32.43|
|2017-04-13 08:00:00|2017-04-20 08:00:00|     33.22|
+-------------------+-------------------+----------+
only showing Top 10 rows
```

上面的例子使用了一个星期的滚动窗口，其中交易数据没有重叠，因此，每个交易只使用一次来计算移动平均值，而下面的例子使用了滑动窗口来计算京东股票的月平均收盘价，每周计算一次。这意味着在计算平均每月的移动平均值时，一些交易数据将被多次使用。在这个滑动窗口中，窗口的大小是四个星期，每个窗口一次滑动一个星期，代码如下：

```
//使用时间窗口函数来计算京东股票的月平均收盘价
//4 周窗口长度，每次滑动 1 周
val jdMonthlyAvg = jdStock
    .groupBy(window(col("Date"), "4 week", "1 week"))
    .agg(avg("Close").as("monthly_avg"))

//按开始时间显示结果
jdMonthlyAvg
    .orderBy("window.start")
    .selectExpr("window.start", "window.end", "round(monthly_avg, 2) as monthly_avg")
    .show(10)
```

执行以上代码，输出的结果如下：

```
+-------------------+-------------------+-----------+
|              start|                end|monthly_avg|
+-------------------+-------------------+-----------+
|2017-01-19 08:00:00|2017-02-16 08:00:00|      30.23|
|2017-01-26 08:00:00|2017-02-23 08:00:00|      30.28|
|2017-02-02 08:00:00|2017-03-02 08:00:00|      30.46|
|2017-02-09 08:00:00|2017-03-09 08:00:00|      30.62|
|2017-02-16 08:00:00|2017-03-16 08:00:00|      30.85|
|2017-02-23 08:00:00|2017-03-23 08:00:00|      31.02|
|2017-03-02 08:00:00|2017-03-30 08:00:00|      31.22|
|2017-03-09 08:00:00|2017-04-06 08:00:00|       31.4|
|2017-03-16 08:00:00|2017-04-13 08:00:00|      31.66|
|2017-03-23 08:00:00|2017-04-20 08:00:00|      32.13|
+-------------------+-------------------+-----------+
only showing Top 10 rows
```

由于滑动窗口的间隔是一个星期，所以这个结果表显示的两个连续行的起始时间间隔是一个星期的间隔。在连续两行之间，有大约三周的重叠交易，这意味着一个交易被多次使用，以此来计算移动平均值。

5.4.3 使用窗口分析函数

第三类高级分析函数是在逻辑分组中执行聚合的函数，这个逻辑分组被称为窗口，这些函数被称为窗口函数。有时需要对一组数据进行操作，并为每组输入行返回一个值，而窗口函数提供了这种独特的功能，使其易于执行计算，如移动平均、累计和或每行的 rank。使用窗口函数，能够轻松地执行例如移动平均、累计和/或每行的排名这样的计算。它们显著地提高了 Spark 的 SQL 和 DataFrame API 的表达能力。

注意：窗口函数是 SQL-2003 标准中定义的一项新特性，并在 SQL-2011、SQL-2016 中又加以完善，添加了若干拓展。窗口函数不同于用户熟悉的常规函数及聚合函数，它为每行数据进行一次计算，特点是输入多行（一个窗口）、返回一个值。

使用窗口函数有以下两个主要步骤：

（1）第 1 步是定义一个窗口规范，该规范定义了称为 frame 的行逻辑分组，这是每行被计算的上下文。

（2）第 2 步是应用一个合适的窗口函数。

窗口规范定义了窗口函数将使用的 3 个重要组件：

（1）第 1 个组件被称为 partition by，指定用来对行进行分组的列（一个或多个列）。

（2）第 2 个组件称为 order by，它定义了如何根据一个或多个列来排序各行，以及顺序是升序或降序的。

（3）最后一个组件称为 frame，它定义了窗口相对于当前行的边界。换句话说，frame 限制了在计算当前行的值时包括哪些行。可以通过行索引或 order by 表达式的实际值来指定在 window frame 中包含的一系列行。

最后一个组件 frame 是可选的，有的窗口函数需要，有的窗口函数或场景不需要。窗口规范是使用在 org.apache.spark.sql.expressions.Window 类中定义的函数构建的。rowsBetween 和 rangeBetween 函数分别用来定义行索引和实际值的范围。

窗口函数可分为 3 种类型：排序函数、分析函数和聚合函数。关于排序函数的描述，见表 5-2。

表 5-2 排序函数

函数名称	描述
rank	返回一个 frame 内行的排名和排序，基于一些排序规则
dense_rank	类似于 rank，但是在不同的排名之间没有间隔，紧密衔接显示
ntile(n)	在一个有序的窗口分区中返回 ntile 分组 ID。例如，如果 n 是 4，则前 25%行得到的 ID 值为 1，第 2 个 25%行得到的 ID 值为 2，以此类推
row_number	返回一个序列号，每个 frame 从 1 开始

关于分析函数的描述，见表 5-3。

表 5-3 分析函数

函数名称	描述
cume_dist	返回一个 frame 的值的累积分布。换句话说，低于当前行的行的比例
lag(col,offset)	返回当前行之前 offset 行的列值
lead(col,offset)	返回当前行之后 offset 行的列值

下面通过一个小的样本数据集 MonthlySales.csv 来演示窗口函数的功能。这个小的样本数据集文件包含两个产品(P1 和 P2)的月销售数据，共 24 个观察数据，内容如下：

```
Product,Month,Sales
P1,1,66
P1,2,24
P1,3,54
P1,4,0
P1,5,56
P1,6,34
P1,7,48
P1,8,46
P1,9,76
P1,10,12
P1,11,8
P1,12,24
P2,1,98
P2,2,16
P2,3,78
P2,4,66
P2,5,14
P2,6,76
P2,7,62
P2,8,92
P2,9,60
P2,10,68
P2,11,10
P2,12,82
```

将该数据文件上传到 HDFS 的/data/spark/目录下。

现要求计算过去 3 个月中每个月的平均销售额。首先创建一个 DataFrame，包含两个产品的月销售数据，代码如下：

```scala
//第 5 章/functions_sales.scala

val file = "/data/spark/MonthlySales.csv"
val monthlySales = spark
    .read
    .option("header","true")
```

```
        .option("inferSchema","true")
        .csv(file)

monthlySales.printSchema()
monthlySales.show()
```

执行以上代码，输出的结果如下：

```
root
 |-- Product: string (nullable = true)
 |-- Month: integer (nullable = true)
 |-- Sales: integer (nullable = true)

+-------+-----+-----+
|Product|Month|Sales|
+-------+-----+-----+
|     P1|    1|   66|
|     P1|    2|   24|
|     P1|    3|   54|
|     P1|    4|    0|
|     P1|    5|   56|
|     P1|    6|   34|
|     P1|    7|   48|
|     P1|    8|   46|
|     P1|    9|   76|
|     P1|   10|   12|
|     P1|   11|    8|
|     P1|   12|   24|
|     P2|    1|   98|
|     P2|    2|   16|
|     P2|    3|   78|
|     P2|    4|   66|
|     P2|    5|   14|
|     P2|    6|   76|
|     P2|    7|   62|
|     P2|    8|   92|
+-------+-----+-----+
only showing top 20 rows
```

然后定义一个窗口规范，为每个产品创建一个包含 3 个月的滑动窗口，并在该滑动窗口上定义一个求移动的平均值的计算，代码如下：

```
import org.apache.spark.sql.expressions.Window

//准备 Windowspec，为每个产品创建一个包含 3 个月的滑动窗口
//负数下标表示在当前行之上（前）的行
val w = Window.partitionBy("Product").orderBy("Month").rangeBetween(-2,0)

//在该滑动窗口上定义计算，在本例中是一个移动的平均值
val f = avg("Sales").over(w)

//应用该滑动窗口和计算，检查结果
```

```
monthlySales
    .select($"Product",$"Sales",$"Month",bround(f,2).alias("MovingAvg"))
    .orderBy($"Product",$"Month")
    .show()
```

以上代码也可以进行精简合并,代码如下:

```
import org.apache.spark.sql.expressions.Window

//定义窗口规范
val w = Window.partitionBy("Product").orderBy("Month").rangeBetween(-2,0)

monthlySales
    .select($"Product",
            $"Sales",
            $"Month",
            bround(avg("Sales").over(w),2).alias("MovingAvg"))
    .orderBy($"Product",$"Month")
    .show()
```

执行以上代码,输出的结果如下:

```
+-------+-----+-----+---------+
|Product|Sales|Month|MovingAvg|
+-------+-----+-----+---------+
|     P1|   66|    1|     66.0|
|     P1|   24|    2|     45.0|
|     P1|   54|    3|     48.0|
|     P1|    0|    4|     26.0|
|     P1|   56|    5|    36.67|
|     P1|   34|    6|     30.0|
|     P1|   48|    7|     46.0|
|     P1|   46|    8|    42.67|
|     P1|   76|    9|    56.67|
|     P1|   12|   10|    44.67|
|     P1|    8|   11|     32.0|
|     P1|   24|   12|    14.67|
|     P2|   98|    1|     98.0|
|     P2|   16|    2|     57.0|
|     P2|   78|    3|     64.0|
|     P2|   66|    4|    53.33|
|     P2|   14|    5|    52.67|
|     P2|   76|    6|     52.0|
|     P2|   62|    7|    50.67|
|     P2|   92|    8|    76.67|
+-------+-----+-----+---------+
only showing top 20 rows
```

下面这个示例进一步演示了窗口函数的用法。现在假设有两个用户 user01 和 user02,这两个用户的购物交易数据如下:

用户ID	交易日期	交易金额
user01	2018-07-02	13.35

```
user01         2018-07-06         27.33
user01         2018-07-04         21.72
user02         2018-07-07         69.74
user02         2018-07-01         59.44
user02         2018-07-05         80.14
```

有了这个购物交易数据，可尝试使用窗口函数来回答以下问题：

(1) 对于每个用户，最高的交易金额是多少？
(2) 每个用户的交易金额和最高交易金额之间的差是多少？
(3) 每个用户的交易金额相对上一次交易的变化是多少？
(4) 每个用户的移动平均交易金额是多少？
(5) 每个用户的累计交易金额是多少？

首先，构造一个 DataFrame，包含这个小型购物交易数据，代码如下：

```
//第5章/functions_transactions.scala

val txDataDF= Seq(
      ("user01", "2018-07-02", 13.35),
      ("user01", "2018-07-06", 27.33),
      ("user01", "2018-07-04", 21.72),
      ("user02", "2018-07-07", 69.74),
      ("user02", "2018-07-01", 59.44),
      ("user02", "2018-07-05", 80.14)
  ).toDF("uid", "tx_date", "amount")

txDataDF.printSchema()
txDataDF.show()
```

执行以上代码，输出内容如下：

```
root
 |-- uid: string (nullable = true)
 |-- tx_date: string (nullable = true)
 |-- amount: double (nullable = false)

+------+----------+------+
|   uid|   tx_date|amount|
+------+----------+------+
|user01|2018-07-02| 13.35|
|user01|2018-07-06| 27.33|
|user01|2018-07-04| 21.72|
|user02|2018-07-07| 69.74|
|user02|2018-07-01| 59.44|
|user02|2018-07-05| 80.14|
+------+----------+------+
```

下面应用窗口函数来回答这些问题。

(1) 为了回答第 1 个问题，可以将 rank()窗口函数应用于一个窗口规范，该规范按用户 ID 对数据进行分区，并按交易金额对其进行降序排序。rank()窗口函数根据每个 frame 中每行的排序顺序给每行分配一个排名，代码如下：

```
//导入Window类
import org.apache.spark.sql.expressions.Window

//定义窗口规范，按用户id分区，按数量降序排序
val forRankingWindow = Window.partitionBy("uid").orderBy(desc("amount"))

//增加一个新列，以包含每行的等级，应用rank函数对每行分级(rank)
val txDataWithRankDF = txDataDF
    .withColumn("rank", rank().over(forRankingWindow))
//txDataWithRankDF.show

//根据等级过滤行，以找到第一名并显示结果
txDataWithRankDF.where($"rank" === 1).show
```

执行以上代码，输出的结果如下：

```
+------+----------+------+----+
|   uid|   tx_date|amount|rank|
+------+----------+------+----+
|user02|2018-07-05| 80.14|   1|
|user01|2018-07-06| 27.33|   1|
+------+----------+------+----+
```

可以看出，用户user01的最高交易金额是27.33，用户user02的最高交易金额是80.14。

（2）解决第2个问题的方法是在每个分区的所有行的amount列上应用max()函数。除了按用户ID分区外，它还需要定义一个包含每个分区中所有行的frame边界。要定义这个frame，可以使用Window.rangeBetween()函数，以Window.unboundedPreceding作为开始值，以Window.unboundedFollowing作为结束值，代码如下：

```
//使用rangeBetween来定义frame边界，它包含每个frame中的所有行
val w = Window
    .partitionBy("uid")
    .orderBy(desc("amount"))
    .rangeBetween(Window.unboundedPreceding,Window.unboundedFollowing)

//将max函数应用于amount列，然后计算差异
val amountDifference = max(txDataDF("amount")).over(w) - txDataDF("amount")

//增加amount_diff列，使用上面定义的逻辑
val txDF = txDataDF.withColumn("amount_diff", round(amountDifference, 3))

//显示结果
txDF.show
```

执行以上代码，输出的结果如下：

```
+------+----------+------+-----------+
|   uid|   tx_date|amount|amount_diff|
+------+----------+------+-----------+
|user02|2018-07-05| 80.14|        0.0|
|user02|2018-07-07| 69.74|       10.4|
|user02|2018-07-01| 59.44|       20.7|
```

```
|user01|2018-07-06| 27.33|        0.0|
|user01|2018-07-04| 21.72|       5.61|
|user01|2018-07-02| 13.35|      13.98|
+------+----------+------+-----------+
```

（3）解决第 3 个问题的方法是使用每个分区的当前行的 amount 列减去上一行的 amount 列。获取上一行的指定字段用 lag()函数。除了按用户 ID 分区外，它还需要定义一个包含每个分区中所有行的 frame 边界。默认 frame 包括所有前面的行和当前行，代码如下：

```
//定义 window specification
val w = Window.partitionBy("uid").orderBy("tx_date")

//计算交易量的变动值
val amountVariant = txDataDF("amount") - lag(txDataDF("amount"),1).over(w)

//增加 amount_diff 列，使用上面定义的逻辑
val lagDF = txDataDF.withColumn("amount_var", round(amountVariant, 3))

//显示结果
lagDF.show
```

执行以上代码，输出的结果如下：

```
+------+----------+------+----------+
|   uid|   tx_date|amount|amount_var|
+------+----------+------+----------+
|user02|2018-07-01| 59.44|      null|
|user02|2018-07-05| 80.14|      20.7|
|user02|2018-07-07| 69.74|     -10.4|
|user01|2018-07-02| 13.35|      null|
|user01|2018-07-04| 21.72|      8.37|
|user01|2018-07-06| 27.33|      5.61|
+------+----------+------+----------+
```

（4）为了计算每个用户按交易日期顺序移动的平均移动数量，将利用 avg()函数来根据 frame 中的一组行计算每行的平均数量。这里希望每个 frame 都包含 3 行：当前行、前面的一行和后面的一行。与前面的例子类似，窗口规范将按用户 ID 对数据进行分区，但是每个 frame 中的行将按交易日期排序，代码如下：

```
//定义窗口规范
val w = Window
   .partitionBy("uid")
   .orderBy("tx_date")
   .rowsBetween(Window.currentRow-1,Window.currentRow+1)

//在窗口上将 avg()函数应用到 amount 列，并将移动平均量按四舍五入法精确到两位小数
val avgDF = txDataDF.withColumn("mov_avg",round(avg("amount").over(w), 2))

//显示结果
avgDF.show
```

执行以上代码，输出的结果如下：

```
+-----+----------+------+----------+
| uid | tx_date  |amount|moving_avg|
+-----+----------+------+----------+
|user02|2018-07-01| 59.44|     69.79|
|user02|2018-07-05| 80.14|     69.77|
|user02|2018-07-07| 69.74|     74.94|
|user01|2018-07-02| 13.35|     17.54|
|user01|2018-07-04| 21.72|      20.8|
|user01|2018-07-06| 27.33|     24.53|
+-----+----------+------+----------+
```

（5）为了计算每个用户的交易金额的累积总和，将把sum()函数应用于一个frame，该frame由所有行到当前行组成。其partitionBy()和orderBy()方法与移动平均示例相同，代码如下：

```
//定义每个frame的窗口规范包括所有以前的行和当前行
val w = Window.partitionBy("uid")
    .orderBy("tx_date")
    .rowsBetween(Window.unboundedPreceding,Window.currentRow)

//将sum()函数应用于窗口规范
val sumDF = txDataDF.withColumn("culm_sum", round(sum("amount").over(w),2))

//显示结果
sumDF.show
```

执行以上代码，输出的结果如下：

```
+-----+----------+------+--------+
| uid | tx_date  |amount|culm_sum|
+-----+----------+------+--------+
|user02|2018-07-01| 59.44|   59.44|
|user02|2018-07-05| 80.14|  139.58|
|user02|2018-07-07| 69.74|  209.32|
|user01|2018-07-02| 13.35|   13.35|
|user01|2018-07-04| 21.72|   35.07|
|user01|2018-07-06| 27.33|    62.4|
+-----+----------+------+--------+
```

窗口规范的默认frame包括所有前面的行和当前行。对于前面的例子，没有必要指定frame，所以应该得到相同的结果。

前面的窗口函数示例是使用DataFrame API编写的。也可以通过SQL实现相同的目标，使用关键字：PARTITION BY、ORDER BY、ROWS BETWEEN和RANGE BETWEEN。frame边界可以使用以下关键字来指定：UNBOUNDED PRECEDING、UNBOUNDED FOLLOWING、CURRENT ROW、<value> PRECEDING和<value> FOLLOWING。

使用SQL的窗口函数的示例，代码如下：

```
//第5章/functions_transactions_sql.scala

//将txDataDF注册为一个临时视图，叫作tx_data
```

```
txDataDF.createOrReplaceTempView("tx_data")

//使用 RANK 窗口函数来找出前两个最高的交易量
spark.sql("""
    select uid, tx_date, amount, rank
    from(
     select uid, tx_date, amount,
            RANK() OVER (PARTITION BY uid ORDER BY amount DESC) as rank
     from tx_data
    ) where rank=1
""").show

//与最大交易金额的差额
spark.sql("""
    select uid, tx_date, amount, round((max_amount - amount),2) as amount_diff
    from(
     select uid, tx_date, amount,
            MAX(amount) OVER (PARTITION BY uid ORDER BY amount DESC
            RANGE BETWEEN UNBOUNDED PRECEDING AND UNBOUNDED FOLLOWING) as max_amount
     from tx_data)
""").show

//与上一次交易金额的差额
spark.sql("""
    select uid, tx_date, amount, round(amount - lag_amount, 2) as amount_var
    from(
     select uid, tx_date, amount,
            lag(amount,1) OVER (PARTITION BY uid ORDER BY tx_date) as lag_amount
     from tx_data)
""").show

//移动平均
spark.sql("""
    select uid, tx_date, amount, round(moving_avg,2) as moving_avg
    from(
     select uid, tx_date, amount,
           AVG(amount) OVER (PARTITION BY uid ORDER BY tx_date
           ROWS BETWEEN 1 PRECEDING AND 1 FOLLOWING ) as moving_avg
     from tx_data)
""").show

//累加和
spark.sql("""
    select uid, tx_date, amount, round(culm_sum,2) as moving_avg
    from(
     select uid, tx_date, amount,
           SUM(amount) OVER (PARTITION BY uid ORDER BY tx_date
           ROWS BETWEEN UNBOUNDED PRECEDING AND CURRENT ROW) as culm_sum
     from tx_data)
""").show
```

执行上面的代码,输出结果应当与使用 DataFrame API 进行窗口操作的结果一样。当使用 SQL 中的窗口函数时,必须在单个语句中指定 partition by、order by 和 frame 窗口。

5.5 用户自定义函数(UDF)

尽管 Spark SQL 为大多数常见用例提供了大量内置函数,但总会在一些情况下,这些函数无法提供用户的用例所需要的功能。Spark SQL 提供了一个相当简单的工具来编写用户定义的函数(UDF),并在 Spark 数据处理逻辑或应用程序中使用它们,就像使用内置函数一样。

UDF 实际上是使用户可以扩展 Spark 的功能以满足特定需求的一种方式。Spark 的 UDF 可以用 Python、Java 或 Scala 来写,它们可以利用和集成任何必要的库。

注意:从概念上讲,UDF 只是一些常规的函数,它们接受一些输入并提供输出。尽管 UDF 可以用 Scala、Java 或 Python 编写,但是必须注意当 UDF 用 Python 编写时的性能差异。UDF 必须在使用 Spark 之前注册,因此 Spark 知道将它们发送到 Executor,以便使用和执行。鉴于 Executor 是用 Scala 编写的 JVM 进程,它们可以在同一个进程中在本地执行 Scala 或 Java UDF。如果一个 UDF 是用 Python 编写的,则 Executor 就不能在本地执行它,因此它必须生成一个单独的 Python 进程来执行 Python UDF。除了生成 Python 过程的成本外,在数据集的每行中都要对数据进行序列化,这也是一个很大的开销。

使用 UDF 涉及以下 3 个步骤:

(1)第 1 步是编写一个函数并进行测试。
(2)第 2 步是通过将函数名及其签名传递给 Spark 的 udf 函数来注册该函数。
(3)最后一步是在 DataFrame 代码或发出 SQL 查询时使用 UDF。在 SQL 查询中使用 UDF 时,注册过程略有不同。

下面的代码用一个简单的 UDF 将数字等级转换为考查等级,它演示了前面提到的 3 个步骤。

首先定义一个 case class,代码如下:

```
case class Student(name:String, score:Int)
```

然后创建一个包含学生成绩的 DataFrame,代码如下:

```
//第 5 章/functions_udf.scala

import spark.implicits._

//创建学生成绩 DataFrame
val studentDF = Seq(
    Student("张三", 85),
    Student("李四", 90),
    Student("王老五", 55)
).toDF()
```

```
studentDF.printSchema()
studentDF.show()
```

执行以上代码,输出内容如下:

```
root
 |-- name: string (nullable = true)
 |-- score: integer (nullable = false)

+------+-----+
|  name|score|
+------+-----+
|  张三|   85|
|  李四|   90|
| 王老五|   55|
+------+-----+
```

将 studentDF 注册到名为 students 的临时视图,代码如下:

```
studentDF.createOrReplaceTempView("students")
```

接下来创建一个普通的 Scala 函数,用来将成绩转换到考查等级,代码如下:

```
//导入依赖包
import org.apache.spark.sql.functions._

//一个普通的Scala函数,用来进行成绩转换
def convertGrade(score:Int) : String = {
    score match {
      case `score` if score > 100 => "作弊"
      case `score` if score >= 90 => "优秀"
      case `score` if score >= 80 => "良好"
      case `score` if score >= 70 => "中等"
      case _ => "不及格"
    }
}

//注册为一个UDF(在DataFrame/Dataset API 中使用时的注册方法)
val convertGradeUDF = udf(convertGrade(_:Int):String)
```

最后,可以像使用普通 Spark 内置函数一样使用该 UDF,将成绩转换为字母等级,代码如下:

```
studentDF
    .select($"name",$"score", convertGradeUDF($"score").as("grade"))
    .show()
```

执行以上代码,输出的结果如下:

```
+------+-----+------+
|  name|score| grade|
+------+-----+------+
|  张三|   85|  良好|
|  李四|   90|  优秀|
| 王老五|   55|不及格|
+------+-----+------+
```

当在 SQL 查询中使用 UDF 时，注册过程与上面略有不同，代码如下：

```
//注册为 UDF，在 SQL 中使用
spark.udf.register("convertGrade", convertGrade(_: Int): String)

//在 SQL 查询中调用 UDF
spark.sql("""
    select name, score, convertGrade(score) as grade
    from students"""
).show()
```

执行以上代码，输出的结果如下：

```
+------+-----+------+
| name|score| grade|
+------+-----+------+
|  张三|   85|  良好|
|  李四|   90|  优秀|
|王老五|   55|不及格|
+------+-----+------+
```

5.6 数据集的 join 连接

Spark SQL 支持对两个或多个 DataFrame/DataSet 执行各种类型的 join 连接操作。本节将介绍这些连接类型和使用方法，并介绍在 Spark SQL 内部执行 join 连接的一些细节。

5.6.1 join 表达式和 join 类型

执行两个数据集的连接需要指定以下两部分内容：

（1）第 1 部分内容是连接表达式，它指定来自每个数据集的哪些列应该用于确定来自两个数据集的哪些行将被包含在连接后的数据集中（确定连接列/等值列）。

（2）第 2 部分内容是连接类型，它决定了连接后的数据集中应该包含哪些内容。

在 Spark SQL 中所支持的 join 类型见表 5-4。

表 5-4 Spark SQL 所支持的 join 类型

类型	描述
内连接（又叫等值连接）	当连接表达式的计算结果为 true 时，返回来自两个数据集的行
左外连接	甚至当连接表达式的计算结果为 false 时，也返回来自左侧数据集的行
右外连接	甚至当连接表达式的计算结果为 false 时，也返回来自右侧数据集的行
外连接	甚至当连接表达式的计算结果为 false 时，也返回来自两侧数据集的行
左反连接	当连接表达式的计算结果为 false 时，只返回来自左侧数据集的行
左半连接	当连接表达式的计算结果为 true 时，只返回来自左侧数据集的行
交叉连接（又名笛卡儿连接）	返回左数据集中每行和右数据集中每行合并后的行。行的数量将是两个数据集的乘积

左半连接和左反连接是唯一的只有值来自左表的连接类型。左半连接与过滤左表中只有在右表中存在键的行相同。左反连接也只返回来自左表的数据，但只返回右表中不存在的记录。

DataFrame 支持自连接，但是连接结果集中会得到重复的列名，因此当执行自连接时，需要为连接列取一个别名。一旦为每个 DataFrame 设置了别名，在结果中就可以使用 dfName.colName 访问每个 DataFrame 的各个列了，代码如下：

```
val joined = df.as("a").join(df.as("b")).where($"a.name" === $"b.name")
```

在 Spark SQL 中，通过调用 queryExecution.executedPlan()方法可以看到正在执行的连接类型。如果其中一张表比另一张表小得多，则可能需要广播散列连接（Broadcast Hash Join）。可以向 Spark SQL 提供一个提示：某个 DataFrame 应该在 join 连接之前通过在 DataFrame 上调用 broadcast 对其进行广播以便通过 join 连接，代码如下：

```
df1.join(broadcast(df2), "key")
```

Spark 还自动使用 spark.sql.conf. autoBroadcastJoinThreshold 来确定是否应该广播表。

5.6.2 执行 join 连接

为了演示如何在 Spark SQL 中使用 join 连接，需要先准备两个小型的 DataFrame。第 1 个 DataFrame 代表一个员工列表，每行包含员工姓名和所属部门。第 2 个 DataFrame 包含一个部门列表，每行包含一个部门 ID 和部门名称。

为此，首先创建两个 case class，分别代表员工类和部门类，代码如下：

```scala
//员工类
case class Employee(emp_name:String, dept_no:Long)

//部门类
case class Dept(id:Long, name:String)
```

然后创建两个测试 DataFrame，代码如下：

```scala
//第5章/functions_join.scala

val employeeDF = Seq(
   Employee("刘宏明", 31),
   Employee("赵薇", 33),
   Employee("黄海波", 33),
   Employee("杨幂", 34),
   Employee("楼一萱", 34),
   Employee("龙梅子", null.asInstanceOf[Int])
).toDF

val deptDF = Seq(
   Dept(31, "销售部"),
   Dept(33, "工程部"),
   Dept(34, "财务部"),
   Dept(35, "市场营销部")
).toDF
```

将这两个 DataFrame 注册为临时视图，代码如下：

```
employeeDF.createOrReplaceTempView("employees")
deptDF.createOrReplaceTempView("departments")
```

然后就可以使用 SQL 来执行 join 连接测试了。

1. 内连接

内连接是最常用的连接类型，它使用相等比较的连接表达式，包含来自两个数据集与连接条件相匹配的列。连接的数据集只有当连接表达式结果为真时才包含行。没有匹配列值的行将被排除在连接数据集之外。在 Spark SQL 中，内连接是默认连接类型。

对 employeeDF 和 deptDF 这两个数据集按 id 执行内连接，代码如下：

```
//定义相等比较的join表达式
val joinExpression = employeeDF.col("dept_no") === deptDF.col("id")

//执行该join
employeeDF.join(deptDF, joinExpression, "inner").show

//也可以不需要指定该join类型，因为"inner"是默认的
//employeeDF.join(deptDF, joinExpression).show

//使用SQL执行join连接
spark.sql("select * from employees JOIN departments on dept_no == id").show
```

执行上面的代码，输出的结果如下：

```
+--------+-------+---+------+
|emp_name|dept_no| id|  name|
+--------+-------+---+------+
|  刘宏明|     31| 31|销售部|
|    赵薇|     33| 33|工程部|
|  黄海波|     33| 33|工程部|
|    杨幂|     34| 34|财务部|
|  楼一萱|     34| 34|财务部|
+--------+-------+---+------+

+--------+-------+---+------+
|emp_name|dept_no| id|  name|
+--------+-------+---+------+
|  刘宏明|     31| 31|销售部|
|    赵薇|     33| 33|工程部|
|  黄海波|     33| 33|工程部|
|    杨幂|     34| 34|财务部|
|  楼一萱|     34| 34|财务部|
+--------+-------+---+------+
```

连接表达式可以在 join()转换中指定，也可以使用 where()变换。如果列名是唯一的，则可以使用简写引用 join 表达式中的列。如果没有，则需要通过 col()函数指定特定列来自哪个 DataFrame，代码如下：

```
//join 表达式的简写版本
employeeDF.join(deptDF, col("dept_no") === col("id")).show

//在 join transformation 内指定 join 表达式
employeeDF.join(deptDF, employeeDF.col("dept_no") === deptDF.col("id")).show

//使用 where transformation 指定 join 表达式
employeeDF.join(deptDF).where($"dept_no" === $"id").show
```

执行上面的代码,输出的结果如下:

```
+--------+-------+---+------+
|emp_name|dept_no| id| name|
+--------+-------+---+------+
|  刘宏明|     31| 31|销售部|
|    赵薇|     33| 33|工程部|
|  黄海波|     33| 33|工程部|
|    杨幂|     34| 34|财务部|
|  楼一萱|     34| 34|财务部|
+--------+-------+---+------+

+--------+-------+---+------+
|emp_name|dept_no| id| name|
+--------+-------+---+------+
|  刘宏明|     31| 31|销售部|
|    赵薇|     33| 33|工程部|
|  黄海波|     33| 33|工程部|
|    杨幂|     34| 34|财务部|
|  楼一萱|     34| 34|财务部|
+--------+-------+---+------+

+--------+-------+---+------+
|emp_name|dept_no| id| name|
+--------+-------+---+------+
|  刘宏明|     31| 31|销售部|
|    赵薇|     33| 33|工程部|
|  黄海波|     33| 33|工程部|
|    杨幂|     34| 34|财务部|
|  楼一萱|     34| 34|财务部|
+--------+-------+---+------+
```

2. 左外连接

这种 join 类型的连接后的数据集包括来自内连接的所有行加上来自左边数据集的连接表达式的计算结果为 false 的所有行。对于那些不匹配的行,它将为右边的数据集的列填充 null 值。例如,对 employeeDF 和 deptDF 执行左外连接,代码如下:

```
//连接类型既可以是"left_outer",也可以是"leftouter"
employeeDF.join(deptDF, $"dept_no" === $"id", "left_outer").show

//使用 SQL
```

```
spark.sql("""
    select *
    from employees
        LEFT OUTER JOIN departments
        on dept_no == id
""").show
```

执行上面的代码，输出的结果如下：

```
+--------+-------+----+------+
|emp_name|dept_no|  id|  name|
+--------+-------+----+------+
|  刘宏明|     31|  31|销售部|
|    赵薇|     33|  33|工程部|
|  黄海波|     33|  33|工程部|
|    杨幂|     34|  34|财务部|
|  楼一萱|     34|  34|财务部|
|  龙梅子|      0|null|  null|
+--------+-------+----+------+


+--------+-------+----+------+
|emp_name|dept_no|  id|  name|
+--------+-------+----+------+
|  刘宏明|     31|  31|销售部|
|    赵薇|     33|  33|工程部|
|  黄海波|     33|  33|工程部|
|    杨幂|     34|  34|财务部|
|  楼一萱|     34|  34|财务部|
|  龙梅子|      0|null|  null|
+--------+-------+----+------+
```

3. 右外连接

这种 join 类型的行为类似于左外连接类型的行为，除了将相同的处理应用于右边的数据集之外。换句话说，连接后的数据集包括来自内连接的所有行加上来自右边数据集的连接表达式的计算结果为 false 的所有行。对于那些不匹配的行，它将为左边数据集的列填充 null 值。例如，对 employeeDF 和 deptDF 执行右外连接，代码如下：

```
//连接类型既可以是"right_outer"，也可以是"rightouter"
employeeDF.join(deptDF, $"dept_no" === $"id", "right_outer").show

//使用SQL
spark.sql("""
    select *
    from employees
        RIGHT OUTER JOIN departments
        on dept_no == id
""").show
```

执行上面的代码，输出的结果如下：

```
+--------+-------+----+----------+
|emp_name|dept_no| id|      name|
+--------+-------+----+----------+
|   刘宏明|     31|  31|     销售部|
|   黄海波|     33|  33|     工程部|
|     赵薇|     33|  33|     工程部|
|   楼一萱|     34|  34|     财务部|
|     杨幂|     34|  34|     财务部|
|    null|   null|  35|   市场营销部|
+--------+-------+----+----------+

+--------+-------+----+----------+
|emp_name|dept_no| id|      name|
+--------+-------+----+----------+
|   刘宏明|     31|  31|     销售部|
|   黄海波|     33|  33|     工程部|
|     赵薇|     33|  33|     工程部|
|   楼一萱|     34|  34|     财务部|
|     杨幂|     34|  34|     财务部|
|    null|   null|  35|   市场营销部|
+--------+-------+----+----------+
```

4. 全外连接

这种join类型的行为实际上与将左外连接和右外连接的结果结合起来是一样的。例如，对 employeeDF 和 deptDF 执行全外连接，代码如下：

```
//使用join转换
employeeDF.join(deptDF, $"dept_no" === $"id", "outer").show

//使用SQL
spark.sql("""
    select *
    from employees
        FULL OUTER JOIN departments
        on dept_no == id
""").show
```

执行上面的代码，输出的结果如下：

```
+--------+-------+----+----------+
|emp_name|dept_no| id|      name|
+--------+-------+----+----------+
|   龙梅子|      0|null|      null|
|     杨幂|     34|  34|     财务部|
|   楼一萱|     34|  34|     财务部|
|   刘宏明|     31|  31|     销售部|
|     赵薇|     33|  33|     工程部|
|   黄海波|     33|  33|     工程部|
|    null|   null|  35|   市场营销部|
+--------+-------+----+----------+
```

```
+--------+-------+----+----------+
|emp_name|dept_no|  id|      name|
+--------+-------+----+----------+
|   龙梅子|      0|null|      null|
|     杨幂|     34|  34|    财务部|
|   楼一萱|     34|  34|    财务部|
|   刘宏明|     31|  31|    销售部|
|     赵薇|     33|  33|    工程部|
|   黄海波|     33|  33|    工程部|
|    null|   null|  35|市场营销部|
+--------+-------+----+----------+
```

5. 左反连接

这种 join 类型能够发现来自左边数据集的哪些行在右边的数据集上没有任何匹配的行，而连接后的数据集只包含来自左边数据集的列。

例如，对 employeeDF 和 deptDF 执行左反连接，代码如下：

```
//使用join转换
employeeDF.join(deptDF, $"dept_no" === $"id", "left_anti").show

//使用SQL
spark.sql("""
    select *
    from employees
        LEFT ANTI JOIN departments
        on dept_no == id
""").show
```

注意：没有右反连接（right anti-join）类型。

执行上面的代码，输出的结果如下：

```
+--------+-------+
|emp_name|dept_no|
+--------+-------+
|   龙梅子|      0|
+--------+-------+


+--------+-------+
|emp_name|dept_no|
+--------+-------+
|   龙梅子|      0|
+--------+-------+
```

6. 左半连接

这种 join 类型的行为类似于内连接类型，除了连接后的数据集不包括来自右边数据集的列。可以将这种 join 类型看作与左反连接相反的类型，在这里，连接后的数据集只包含匹配的行。

例如，对 employeeDF 和 deptDF 执行左半连接，代码如下：

```
//使用join转换
employeeDF.join(deptDF, $"dept_no" === $"id", "left_semi").show

//使用SQL
spark.sql("""
    select *
    from employees
        LEFT SEMI JOIN departments
        on dept_no == id
""").show
```

执行上面的代码，输出的结果如下：

```
+--------+-------+
|emp_name|dept_no|
+--------+-------+
|   刘宏明|     31|
|    赵薇|     33|
|  黄海波|     33|
|    杨幂|     34|
|  楼一萱|     34|
+--------+-------+

+--------+-------+
|emp_name|dept_no|
+--------+-------+
|   刘宏明|     31|
|    赵薇|     33|
|  黄海波|     33|
|    杨幂|     34|
|  楼一萱|     34|
+--------+-------+
```

7. 交叉连接

交叉连接又称笛卡儿连接。例如，对 employeeDF 和 deptDF 执行交叉连接，代码如下：

```
//使用crossJoin transformation 并显示该count
println(employeeDF.crossJoin(deptDF).count)   //Long = 24

//使用SQL，并显示前30行以观察连接后的数据集中所有的行
spark.sql("select * from employees CROSS JOIN departments").show(30)
```

执行上面的代码，输出的结果如下：

```
24
+--------+-------+---+----------+
|emp_name|dept_no| id|      name|
+--------+-------+---+----------+
|   刘宏明|     31| 31|     销售部|
|   刘宏明|     31| 33|     工程部|
|   刘宏明|     31| 34|     财务部|
|   刘宏明|     31| 35|  市场营销部|
```

```
|    赵薇|  33|  31|       销售部|
|    赵薇|  33|  33|       工程部|
|    赵薇|  33|  34|       财务部|
|    赵薇|  33|  35|    市场营销部|
|  黄海波|  33|  31|       销售部|
|  黄海波|  33|  33|       工程部|
|  黄海波|  33|  34|       财务部|
|  黄海波|  33|  35|    市场营销部|
|    杨幂|  34|  31|       销售部|
|    杨幂|  34|  33|       工程部|
|    杨幂|  34|  34|       财务部|
|    杨幂|  34|  35|    市场营销部|
|  楼一萱|  34|  31|       销售部|
|  楼一萱|  34|  33|       工程部|
|  楼一萱|  34|  34|       财务部|
|  楼一萱|  34|  35|    市场营销部|
|  龙梅子|   0|  31|       销售部|
|  龙梅子|   0|  33|       工程部|
|  龙梅子|   0|  34|       财务部|
|  龙梅子|   0|  35|    市场营销部|
+--------+-----+----+-----------+
```

5.6.3 处理重复列名

有时，在连接两个具有同名列的 DataFrame 之后，会出现一个意想不到的问题。当这种情况发生时，连接后的 DataFrame 会有多个同名的列。在这种情况下，在对连接后的 DataFrame 进行某种转换时，就不太方便引用其中一列。

例如，向 deptDF 增加一个新的列，列名为 dept_no，值来自 id 列，代码如下：

```
//向 deptDF 增加一个新的列，列名为 dept_no
val deptDF2 = deptDF.withColumn("dept_no", 'id)

deptDF2.printSchema
```

执行上面的代码，输出的结果如下：

```
root
 |-- id: long (nullable = false)
 |-- name: string (nullable = true)
 |-- dept_no: long (nullable = false)
```

现在，使用 employeeDF 连接 deptDF2，基于 dept_no 列进行连接，代码如下：

```
val dupNameDF = employeeDF
    .join(deptDF2, employeeDF.col("dept_no") === deptDF2.col("dept_no"))

dupNameDF.printSchema
```

执行上面的代码，输出的结果如下：

```
root
 |-- emp_name: string (nullable = true)
 |-- dept_no: long (nullable = false)
 |-- id: long (nullable = false)
 |-- name: string (nullable = true)
 |-- dept_no: long (nullable = false)
```

注意，dupNameDF 现在有两个名称相同的列，都叫 dept_no。当试图在 dupNameDF 中投影 dept_no 列时，Spark 会抛出一个错误。例如，选择 dept_no 列，代码如下：

```
dupNameDF.select("dept_no")
```

执行上面的代码，会抛出异常信息，内容如下：

```
org.apache.spark.sql.AnalysisException: Reference 'dept_no' is ambiguous,
could be: dept_no, dept_no.
 at org.apache.spark.sql.catalyst.expressions.package$AttributeSeq.
 resolve(package.scala:363)
 ...
```

这个异常信息的意思是，因为 dept_no 列是模糊的（不知道应该引用哪个 DataFrame 中的 dept_no 列），所以无法正确地执行，抛出异常。

要解决这个问题，可以有以下几种方法。

1. 使用原始的 DataFrame

连接后的 DataFrame 会记得在连接过程中哪些列来自哪个原始的 DataFrame。为了消除某个特定列来自哪个 DataFrame 的歧义，可以告诉 Spark 以其原始的 DataFrame 名称作为前缀，代码如下：

```
//解决方法一：明确来自哪个 DataFrame
//dupNameDF.select(employeeDF.col("dept_no")).show
dupNameDF.select(deptDF2.col("dept_no")).show
```

执行上面的代码，会发现可以正常执行而没有抛出异常。

2. 在 join 之前重命名列

为了避免列名称的模糊性问题，另一种方法是使用 withColumnRenamed()转换来重命名其中一个 DataFrames 中的列，代码如下：

```
//解决方法二：join 之前重命名列
val deptDF3 = deptDF2.withColumnRenamed("dept_no","dept_id")
deptDF3.printSchema

val dupNameDF2 = employeeDF.join(deptDF3, 'dept_no === 'dept_id)
dupNameDF2.printSchema

dupNameDF2.select("dept_no").show
```

执行上面的代码，会发现可以正常执行而没有抛出异常，因为在 join 时已经不存在重名的列了。

3. 使用一个连接后的列名

在两个 DataFrames 中，当连接的列名是相同的时，在 join()函数中指定一个连接列名即可，这会自动从连接后的 DataFrame 中删除重复列名，但是，如果这是一个自连接，也就是说连接一个 DataFrame 本身，就没有办法引用其他重复的列名了。在这种情况下，需要使用第 1 种技术来重命名一个 DataFrame 的列，代码如下：

```
val noDupNameDF = employeeDF.join(deptDF2, "dept_no")

noDupNameDF.printSchema
noDupNameDF.show
```

执行上面的代码，输出的结果如下：

```
root
 |-- dept_no: long (nullable = false)
 |-- emp_name: string (nullable = true)
 |-- id: long (nullable = false)
 |-- name: string (nullable = true)

+-------+--------+---+------+
|dept_no|emp_name| id|  name|
+-------+--------+---+------+
|     31|  刘宏明| 31|销售部|
|     33|    赵薇| 33|工程部|
|     33|  黄海波| 33|工程部|
|     34|    杨幂| 34|财务部|
|     34|  楼一萱| 34|财务部|
+-------+--------+---+------+
```

5.6.4 join 连接策略

可以说，join 连接是 Spark 中最昂贵的操作之一。不过，有两种不同的策略可以用来连接两个数据集，它们是 Shuffle Hash Join 和 Broadcast Hash Join。选择特定策略的主要标准是基于两个数据集的大小。当两个数据集的大小都很大时，就会使用 Shuffle Hash Join 策略。当其中一个数据集的大小足够小且可以容纳进 Executors 的内存时，就会使用 Broadcast Hash Join 策略。下面的部分给出了每个连接策略的工作细节。

1. Shuffle Hash Join

Shuffle Hash Join 的实现由两个步骤组成。首先计算在每个数据集的每行的连接表达式中的列的哈希值，然后将这些具有相同哈希值的行 shuffle 到同一分区。为了确定某一行将被移动到哪个分区，Spark 执行一个简单的算术操作，它通过分区的数量来计算哈希值的模。一旦第 1 步完成，第 2 步就将那些具有相同列哈希值的行的列组合起来。在较高的层次上，这两个步骤与 MapReduce 编程模型的步骤很相似。Shuffle Hash Join 中 shuffling 的过程如图 5-5 所示。

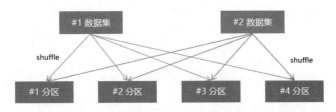

图 5-5　Shuffle Hash Join 中 shuffling 的过程

正如前面提到的，这是一个昂贵的操作，因为它需要通过网络在多台机器上移动大量数据。当在网络上移动数据时，数据通常会经过数据序列化和反序列化过程。想象一下，在两个大型数据集上执行一个 join 连接，其中每个数据集的大小为 100 GB。在这个场景中，它需要移动大约 200 GB 的数据。在 join 连接两个大型数据集时，不可能完全避免 Shuffle Hash Join，但重要的是要注意在可能的情况下减少连接它们的频率。

2. Broadcast Hash Join

只有当其中一个数据集足够小到可以装入内存时，这种 join 策略才适用。已经知道 Shuffle Hash Join 是一项昂贵的操作，Broadcast Hash Join 避免了对两个数据集都进行 shuffling，而是只对较小的数据进行 shuffling。与 Shuffle Hash Join 策略类似，这个策略也包含两个步骤。第 1 步是将整个小数据集的副本广播到较大数据集的每个分区上。第 2 步是遍历较大数据集中的每行，并在较小的数据集中按匹配列值查找对应的行。

Broadcast Hash Join 连接策略中广播较小数据集的过程如图 5-6 所示。

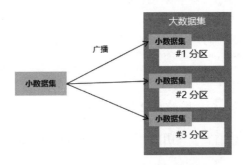

图 5-6　Broadcast 中广播小数据集的过程

很容易理解，在可能的情况下，首选 Broadcast Hash Join。Spark SQL 在大多数情况下可以根据在读取数据时对数据集的一些统计数据自动判断是否使用 Broadcast Hash Join 或 Shuffle Hash Join，然而，也可以在使用 join 转换时，明确提供一个提示给 Spark SQL，以便使用 Broadcast Hash Join，代码如下：

```
//提供一个提示，使用一个 Broadcast Hash Join 来广播 deptDF
import org.apache.spark.sql.functions.broadcast
//输出执行计划以验证 Broadcast Hash Join 策略被使用了
```

```
val joined = employeeDF.join(broadcast(deptDF), col("dept_no") === col("id"))
joined.show()
```

或者使用 SQL 语句执行连接，代码如下：

```
//使用SQL
val joined = spark.sql("""
    select /*+ MAPJOIN(departments) */ *
    from employees
        JOIN departments on dept_no == id
""")
joined.show()
```

执行上面的代码，输出的结果如下：

```
+--------+-------+---+------+
|emp_name|dept_no| id|  name|
+--------+-------+---+------+
|  刘宏明 |    31 | 31| 销售部|
|    赵薇 |    33 | 33| 工程部|
|  黄海波 |    33 | 33| 工程部|
|    杨幂 |    34 | 34| 财务部|
|  楼一萱 |    34 | 34| 财务部|
+--------+-------+---+------+
```

可以在 DataFrame 上调用 explain()方法输出物理执行计划，代码如下：

```
joined.explain()
```

执行上面的代码，输出的物理执行计划如图 5-7 所示。

```
== Physical Plan ==
*(1) BroadcastHashJoin [dept_no#536L], [id#544L], Inner, BuildRight, false
:- *(1) LocalTableScan [emp_name#535, dept_no#536L]
+- BroadcastExchange HashedRelationBroadcastMode(List(input[0, bigint, false]),false), [id=#1205]
    +- LocalTableScan [id#544L, name#545]
```

图 5-7 查看物理执行计划

从输出的物理计划可以看到，利用到了 Broadcast Hash Join。

5.7 读写 Hive 表

Spark SQL 还支持读取和写入存储在 Apache Hive 中的数据。Spark 支持两种 SQL 方言：Spark 的 SQL 方言和 Hive 查询语言(HQL)。Spark SQL 支持 HiveQL 语法，同时支持 Hive SerDes 和 UDF，可以访问现有的 Hive 仓库。

用户可以通过 Spark SQL 访问已经存在的 Hive 表并使用已经存在的、社区构建的 Hive UDF。没有部署 Hive 的用户仍然可以启用 Hive 支持。

注意：Spark 中与每个表相关联的是它的相关元数据，它是关于表及其数据的信息：

模式、描述、表名、数据库名、列名、分区、实际数据所在的物理位置等。所有这些都存储在一个中央元存储中。Spark 默认情况下没有为 Spark 表提供单独的元存储，而是使用 Apache Hive Metastore，位于/user/hive/warehouse，用于持久化关于表的所有元数据，但是，可以通过将 Spark 配置变量 spark.sql.warehouse.dir 设置为另一个位置来更改默认位置，该位置可以设置为本地或外部分布式存储。

5.7.1 Spark SQL 的 Hive 配置

要通过 Spark SQL 访问 Hive 数据表，需要先进行以下配置。

1. 配置 Hive 支持

要配置 Hive 支持，需要将文件 hive-site.xml、core-site.xml（用于安全配置）和 hdfs-site.xml（用于 HDFS 配置）复制到 Spark 安装的 conf 目录下。

注意：如果没有配置 hive-site.xml 文件，则上下文会自动在当前目录中创建 metastore_db，并创建一个由 spark.sql.warehouse.dir 配置的目录，它默认为启动 Spark 应用程序的当前目录下的 spark-warehouse 目录。注意，从 Spark 2.0.0 开始，hive-site.xml 文件中的 hive.metastore.warehouse.dir 属性已被弃用，取而代之的是 spark.sql.warehouse.dir，以指定 warehouse 中数据库的默认位置。需要向启动 Spark 应用程序的用户授予写权限。

在 hive-site.xml 文件中，需要包含以下几项。

（1）hive.metastore.warehouse.dir：数据仓库目录，用来存放托管表对应的数据文件。

（2）spark.sql.hive.metastore.version：指定 Hive Metastore 的版本。在 Spark 3.x 中，Hive Metastore 的默认版本为 2.3.7。

（3）spark.sql.hive.metastore.jar：设置为 Maven(Spark 检索 JAR)或 Hive JAR 存在的系统路径。

（4）hive.metastore.uris：Hive Metastore Server 的 thrift URL。

（5）MySQL 连接的相关配置。

2. 复制 JDBC 驱动

因为 Spark SQL 需要访问 Hive Metastore，所以需要将 JDBC 驱动配置在 Driver 及所有 Executors 的 classpath 中。最简单的方式是在提交应用程序或启动 Spark Shell 时，使用 --jars 选项指定要提供的 JAR 文件。或者，将 MySQL 的 JDBC 驱动 JAR 包复制到$SPARK_HOME/jars/目录下。

3. 启动 Hive Metastore 服务

Hive Metastore 服务是 Spark SQL 应用程序将要连接到的服务，用于获取 Hive 表的元数据。启动 Hive Metastore 服务的命令如下：

```
$ cd $Hive_home/bin
$ hive --service metastore
```

5.7.2 Spark Maven 项目的 Hive 配置

如果创建 Spark Maven 项目来读写 Hive 表，则需要在项目的 pom.xml 文件中添加相关依赖项，依赖项的配置内容如下：

```xml
<!-- spark-hive -->
<dependency>
    <groupId>org.apache.spark</groupId>
    <artifactId>spark-hive_2.11</artifactId>
    <version>3.1.2</version>
    <scope>compile</scope>
</dependency>

<!-- hive -->
<dependency>
    <groupId>org.apache.hive</groupId>
    <artifactId>hive-JDBC</artifactId>
    <version>3.1.2</version>
</dependency>
<dependency>
    <groupId>org.apache.hive</groupId>
    <artifactId>hive-exec</artifactId>
    <version>3.1.2</version>
</dependency>

<!-- MySQL JDBC -->
<dependency>
    <groupId>mysql</groupId>
    <artifactId>mysql-connector-java</artifactId>
    <version>5.1.49</version>
</dependency>
```

然后将 Hive 的配置文件 hive-site.xml 复制到项目的 src/main/resources 目录下。

最后在代码中创建 SparkSession 实例，并启用 Hive 支持，代码如下：

```scala
object BigDataPlay {

  def main(args: Array[String]): Unit = {

    //创建 SparkSession
    val spark = SparkSession.builder()
        .master("local[*]")
        .appName("bigdata play")
        .enableHiveSupport()              /* 启用 Hive 支持 */
        .getOrCreate()

    //查看 Hive 中的数据库
    spark.sql("show databases").show()
  }
}
```

5.7.3 Spark SQL 读写 Hive 表

可以使用 SparkSession 或 DataFrameReader 的 table()方法从一个 Hive Metastore 的注册表中加载一个 DataFrame。例如，从 Hive 中读取表 test 中的数据，代码如下：

```
val df = spark.read.table("test")
df.show
```

或者，也可以在 SparkSession 上直接调用 table()方法，代码如下：

```
val df = spark.table("test")
df.show
```

将 DataFrame/DataSet 保存到 Hive 表中，使用 DataFrameWriter 的 saveAsTable()方法，它会将 DataFrame 数据保存到 Hive 表中，并在 Hive Metastore 中注册。例如，将一个 DataFrame 保存到表 test_tb 中，代码如下：

```
df.write.saveAsTable("test")
```

默认情况下，如果不指定表路径，Spark 将把数据写到 warehouse 目录下的默认表路径。当删除表时，默认的表路径也将被删除。

当创建一个 Hive 表时，需要定义这个表如何从/向文件系统读取/写入数据，即"输入格式"和"输出格式"。还需要定义此表应该如何将数据反序列化为 Row，或将 Row 序列化为数据，即 serde。例如，创建一个 Hive 表，以 Parquet 格式存储，代码如下：

```
CREATE TABLE src(id int) USING hive OPTIONS(fileFormat 'parquet')
```

在使用 create 语句创建 Hive 表时，OPTIONS 可以用来指定存储格式（"serde"、"input format"、"output format"）。可以指定的 OPTIONS 选项见表 5-5。

表 5-5 Spark SQL 所支持的 OPTIONS 选项

属 性 名	含 义
fileFormat	文件格式是一种存储格式规范的包，包括"serde"、"input format"和"output format"。目前支持 6 种文件格式:"sequencefile"、"rcfile"、"orc"、"parquet"、"textfile"和"avro"
inputFormat, outputFormat	这两个选项用于指定相应的'InputFormat '和' OutputFormat '类的名称。 例如，"org.apache.hadoop.hive.ql.io.orc.OrcInputFormat"。 这两个选项必须成对出现，如果已经指定了 fileFormat 选项，则不能指定它们
serde	此选项用于指定 serde 类的名称。 当指定"fileFormat"选项时，如果给定的"fileFormat"已经包含 serde 的信息，则不需要指定此选项。 目前"sequencefile"、"textfile"和"rcfile"不包含 serde 信息，可以对这 3 种文件格式使用此选项
fieldDelim, escapeDelim, collectionDelim, mapkeyDelim, lineDelim	这些选项只能用于"textfile"文件格式，它们定义如何将带分隔符的文件读入行

默认情况下，Spark 将以纯文本的形式读取表文件。需要注意，在创建表时还不支持 Hive 存储处理程序，可以在 Hive 端使用存储处理程序创建一张表，并使用 Spark SQL 读取它。

下面通过一个示例程序，演示如何使用 Spark SQL 读取 Hive 元数据及读写 Hive 表数据。建议按以下步骤操作。

（1）确保已经启动了 Hive Metastore Server。

如果没有启动，则可在终端窗口中使用如下命令启动：

```
$ hive --service metastore
```

（2）查看 spark.sql.catalogImplementation 内部属性，该属性应该是 hive，代码如下：

```
spark.conf.get("spark.sql.catalogImplementation")
```

（3）查看当前有哪些数据库和数据表。

列出目前所有已知的数据库，代码如下：

```
spark.catalog.listDatabases.show
```

列出所有已知的表（包括 Hive 表和临时表），代码如下：

```
spark.catalog.listTables.show
```

列出 default 数据库中的表，代码如下：

```
spark.sharedState.externalCatalog.listTables("default")
```

（4）通过 Spark 创建 Hive 表 src，代码如下：

```
spark.sql("CREATE TABLE IF NOT EXISTS src (key INT, value STRING) USING hive")
```

（5）将 Spark 自带的数据文件 kv1.txt 加载到上面所创建的 Hive 表中，代码如下：

```
val file = "/data/spark/resources/kv1.txt"
spark.sql(s"LOAD DATA INPATH '${file}' INTO TABLE src")
```

（6）用 HiveQL 进行查询，代码如下：

```
spark.sql("SELECT * FROM src").show(10)
```

执行以上查询代码，输出内容如下：

```
+---+-------+
|key|  value|
+---+-------+
|238|val_238|
| 86| val_86|
|311|val_311|
| 27| val_27|
|165|val_165|
|409|val_409|
|255|val_255|
|278|val_278|
| 98| val_98|
|484|val_484|
+---+-------+
only showing Top 10 rows
```

HiveQL 还支持聚合查询，代码如下：

```
spark.sql("SELECT COUNT(*) as cnt FROM src").show()
```

执行以上查询代码，输出内容如下：

```
+----+
| cnt|
+----+
|1500|
+----+
```

SQL 查询的结果是 DataFrame 类型，并且支持所有正常的函数，代码如下：

```
//查询结果是 DataFrame[Row]类型
val sqlDF = spark.sql("SELECT key, value FROM src WHERE key < 10 ORDER BY key")

import org.apache.spark.sql.Row

//DataFrame 中的项是 Row 类型的，可以按顺序访问每列
val stringsDS = sqlDF.map {
  case Row(key: Int, value: String) => s"Key: $key, Value: $value"
}
stringsDS.show()
```

执行以上代码，输出内容如下：

```
+--------------------+
|               value|
+--------------------+
|Key: 0, Value: val_0|
|Key: 0, Value: val_0|
|Key: 0, Value: val_0|
|Key: 0, Value: val_0|
|Key: 0, Value: val_0|
|Key: 0, Value: val_0|
|Key: 0, Value: val_0|
|Key: 0, Value: val_0|
|Key: 0, Value: val_0|
|Key: 2, Value: val_2|
|Key: 2, Value: val_2|
|Key: 2, Value: val_2|
|Key: 4, Value: val_4|
|Key: 4, Value: val_4|
|Key: 4, Value: val_4|
|Key: 5, Value: val_5|
|Key: 5, Value: val_5|
|Key: 5, Value: val_5|
|Key: 5, Value: val_5|
|Key: 5, Value: val_5|
+--------------------+
only showing top 20 rows
```

（7）定义另一个域对象（case class），代码如下：

```
case class Record(key: Int, value: String)
```

另外创建一个 DataFrame，并将其与已有的 src 表执行 join 连接，代码如下：

```
//创建另一个 DataFrame
val recordsDF = spark
    .createDataFrame((1 to 100).map(i => Record(i, s"val_$i")))

//也可这样创建
//val recordsDF = (1 to 100).map(i => Record(i, s"val_$i")).toDF()

//使用 DataFrame 在 SparkSession 中创建临时视图
recordsDF.createOrReplaceTempView("records")

//然后，将新的 DataFrame 与存储在 Hive 中的数据 join 连接起来
spark.sql("SELECT * FROM records r JOIN src s ON r.key = s.key")
    .show()
```

执行以上代码，输出的结果如下：

```
+---+------+---+------+
|key| value|key| value|
+---+------+---+------+
| 86|val_86| 86|val_86|
| 27|val_27| 27|val_27|
| 98|val_98| 98|val_98|
| 66|val_66| 66|val_66|
| 37|val_37| 37|val_37|
| 15|val_15| 15|val_15|
| 82|val_82| 82|val_82|
| 17|val_17| 17|val_17|
| 57|val_57| 57|val_57|
| 20|val_20| 20|val_20|
| 92|val_92| 92|val_92|
| 47|val_47| 47|val_47|
| 72|val_72| 72|val_72|
|  4| val_4|  4| val_4|
| 35|val_35| 35|val_35|
| 54|val_54| 54|val_54|
| 51|val_51| 51|val_51|
| 65|val_65| 65|val_65|
| 83|val_83| 83|val_83|
| 12|val_12| 12|val_12|
+---+------+---+------+
only showing top 20 rows
```

（8）将 DataFrame 写入 Hive 表中，代码如下：

```
//创建一个 Hive 托管的 Parquet 表，使用 HQL 语法
spark.sql("CREATE TABLE hive_records(key int, value string) STORED AS PARQUET")

//将 src 表数据加载到 DataFrame
val df = spark.table("src")

import org.apache.spark.sql.SaveMode
```

```
//saveAsTable()方法：会将 DataFrame 数据保存到 Hive 表中
df.write.mode(SaveMode.Overwrite).saveAsTable("hive_records")    //覆盖写入

//插入之后，Hive 托管的表现在有数据了
spark.sql("SELECT * FROM hive_records").show()
//spark.table("hive_records").show()        //也可以实现该功能
```

执行以上代码，输出的结果如下：

```
+---+-------+
|key|  value|
+---+-------+
|238|val_238|
| 86| val_86|
|311|val_311|
| 27| val_27|
|165|val_165|
|409|val_409|
|255|val_255|
|278|val_278|
| 98| val_98|
|484|val_484|
|265|val_265|
|193|val_193|
|401|val_401|
|150|val_150|
|273|val_273|
|224|val_224|
|369|val_369|
| 66| val_66|
|128|val_128|
|213|val_213|
+---+-------+
only showing top 20 rows
```

（9）也可以通过 Spark SQL 读写 Hive 外部表，代码如下：

```
//准备一个 HDFS 目录，DataSet 数据将以 Parquet 格式写入该目录
val dataDir = "/data/parquet_data"         //用作外部表目录
spark.range(10).write.parquet(dataDir)

//可以使用 HDFS 命令查看外部表目录：hdfs dfs -ls /data/parquet_data

//创建一个 Hive external Parquet 表
spark.sql(s"CREATE EXTERNAL TABLE hive_bigints(id bigint) STORED AS PARQUET
LOCATION '$dataDir'")

 //Hive 外部表应该已经有数据了
spark.sql("SELECT * FROM hive_bigints").show()
//spark.table("hive_bigints").show()
```

（10）Spark SQL 也支持 Hive 动态分区表。从 Spark 2.1 开始，持久数据源表将每个分区的元数据存储在 Hive Metastore 中。例如，将 DataFrame 按 key 列动态分区存储到 Hive 分区表中，代码如下：

```
//打开 Hive 动态分区的标志
spark.sqlContext.setConf("hive.exec.dynamic.partition", "true")
spark.sqlContext.setConf("hive.exec.dynamic.partition.mode", "nonstrict")

//使用 DataFrame API 创建一个 Hive 分区表
df.write.partitionBy("key").format("hive").saveAsTable("hive_part_tbl")

//可以使用 HDFS 命令查看物理分区
//$ hdfs dfs -ls /user/hive/warehouse/hive_part_tbl

//分区列'key'将被移动到 Schema 的末尾
sql("SELECT * FROM hive_part_tbl").show()

//按 key 查找
sql("SELECT * FROM hive_part_tbl where key=33").show()
```

5.7.4 分桶、分区和排序

使用 Spark SQL 读写 Hive 表数据时，对于基于文件的输出，还可以进行分桶、分区和排序。DataFrameWriter 提供了 bucketBy()和 sortBy()方法，用于控制输出文件的目录结构。如果要对输出结果分桶存储，则可以使用 "write.bucketBy(<分桶数>,<分桶字段>)" 方法。

接下来使用 MovieLens 数据集，将其写入 Hive 中，应用分桶、分区和排序方法。这里主要涉及其中的 movies、ratings 和 tags 数据集。

首先，确保已经启动了 Hive Metastore Server。如果没有启动，则可在终端窗口中使用如下命令启动：

```
$ hive --service metastore
```

因为要实现加载数据集并写入 Hive，所以需要先创建一个 Hive 数据库 movies_db。当然可以在 Hive 中创建，不过这里使用 Spark SQL 来完成，代码如下：

```
spark.sql("create database if not exists movies_db")
```

同样，如果想要查看现有的数据库信息或数据表信息，则代码如下：

```
//查看当前有哪些数据库
spark.sql("show databases").show()

//查看指定数据库中有哪些数据表
spark.sql("show tables in movies_db").show()
```

如果要查看这个数据库的详细信息，则代码如下：

```
spark.sql("DESCRIBE DATABASE EXTENDED movies_db").show(false)
```

执行上面的代码，输出内容如下：

```
+--------------+--------------------------------------------------+
|info_name     |info_value                                        |
+--------------+--------------------------------------------------+
|Database Name |movies_db                                         |
|Comment       |                                                  |
|Location      |hdfs://xueai8:8020/user/hive/warehouse/movies_db.db|
|Owner         |hduser                                            |
|Properties    |                                                  |
+--------------+--------------------------------------------------+
```

现在可以切换到该数据 movies_db，这样后续的 SQL 操作都将作用于此数据库，代码如下：

```
spark.sql("use movies_db")
```

在 movies_db 数据库中创建数据表 movies，并查看该表的详细信息，代码如下：

```
//执行 Hive QL 语法，创建 Hive 表 movies
val sql = """
create table if not exists movies (
   movieId int,
   title string,
   genres string
) row format delimited fields terminated by ','
 stored as textfile
 tblproperties("skip.header.line.count"="1")
"""
spark.sql(sql)

//查看表的详细信息
spark.sql("describe formatted movies").show(truncate = false)
```

执行以上代码，输出内容如图 5-8 所示。

```
+----------------------------+------------------------------------------------------------+-------+
|col_name                    |data_type                                                   |comment|
+----------------------------+------------------------------------------------------------+-------+
|movieId                     |int                                                         |null   |
|title                       |string                                                      |null   |
|genres                      |string                                                      |null   |
|                            |                                                            |       |
|# Detailed Table Information|                                                            |       |
|Database                    |movies_db                                                   |       |
|Table                       |movies                                                      |       |
|Owner                       |hduser                                                      |       |
|Created Time                |Fri May 06 14:26:11 CST 2022                                |       |
|Last Access                 |UNKNOWN                                                     |       |
|Created By                  |Spark 3.1.2                                                 |       |
|Type                        |MANAGED                                                     |       |
|Provider                    |hive                                                        |       |
|Table Properties            |[skip.header.line.count=1, transient_lastDdlTime=1651818371]|       |
|Location                    |hdfs://xueai8:8020/user/hive/warehouse/movies_db.db/movies  |       |
|Serde Library               |org.apache.hadoop.hive.serde2.lazy.LazySimpleSerDe          |       |
|InputFormat                 |org.apache.hadoop.mapred.TextInputFormat                    |       |
|OutputFormat                |org.apache.hadoop.hive.ql.io.HiveIgnoreKeyTextOutputFormat  |       |
|Storage Properties          |[serialization.format=, field.delim=,]                      |       |
|Partition Provider          |Catalog                                                     |       |
+----------------------------+------------------------------------------------------------+-------+
```

图 5-8 查看 Hive 表的详细信息

将movies.csv数据文件加载到该表中,并测试是否加载成功,代码如下:

```
//将数据加载到Hive表中
val file = "/data/spark/ml-latest-small/movies.csv"
spark.sql(s"load data local inpath '${file}' into table movies")

//用HiveQL进行查询
spark.sql("SELECT * FROM movies").show()
```

执行以上代码,输出内容如下:

```
+-------+--------------------+--------------------+
|movieId|               title|              genres|
+-------+--------------------+--------------------+
|   null|               title|              genres|
|      1|    Toy Story (1995)|Adventure|Animati...|
|      2|      Jumanji (1995)|Adventure|Childre...|
|      3|Grumpier Old Men ...|      Comedy|Romance|
|      4|Waiting to Exhale...|Comedy|Drama|Romance|
|      5|Father of the Bri...|              Comedy|
|      6|         Heat (1995)|Action|Crime|Thri...|
|      7|      Sabrina (1995)|      Comedy|Romance|
|      8| Tom and Huck (1995)|   Adventure|Children|
|      9| Sudden Death (1995)|              Action|
|     10|    GoldenEye (1995)|Action|Adventure|...|
|     11|"American President|         The (1995)"|
|     12|Dracula: Dead and...|       Comedy|Horror|
|     13|        Balto (1995)|Adventure|Animati...|
|     14|        Nixon (1995)|               Drama|
|     15|Cutthroat Island ...|Action|Adventure|...|
|     16|       Casino (1995)|         Crime|Drama|
|     17|Sense and Sensibi...|       Drama|Romance|
|     18|   Four Rooms (1995)|              Comedy|
|     19|Ace Ventura: When...|              Comedy|
+-------+--------------------+--------------------+
only showing top 20 rows
```

前面都是通过执行Hive QL语句先创建Hive表,然后将数据加载到表中。实际上,Spark SQL也支持ETL过程,即在Spark SQL中抽取数据文件后创建并写入Hive表。例如,在Spark SQL中先将ratings.csv抽取到一个DataFrame中,然后调用DataFrameWriter的saveAsTable()方法创建并写入指定的Hive表中。这样的ETL实现代码如下:

```
import org.apache.spark.sql.types._
import org.apache.spark.sql.SaveMode

//指定Schema
val ratingsFields = Array(
    StructField("userId", IntegerType, nullable=true),
    StructField("movieId", IntegerType, nullable=true),
    StructField("rating", DoubleType, nullable=true),
    StructField("timestamp", StringType, nullable=true)
)
```

```
val ratingsSchema = StructType(ratingsFields)

//ETL 过程
val ratingsFile = "file://home/hduser/data/spark/ml-latest-small/ratings.csv"
spark
    .read
    .option("header","true")
    .schema(ratingsSchema)
    .csv(ratingsFile)
    .write
    .mode(SaveMode.Overwrite)
    .saveAsTable("ratings")             //覆盖

//测试写入是否成功
spark.table("ratings").show
```

执行上面的代码，输出内容如下：

```
+------+-------+------+---------+
|userId|movieId|rating|timestamp|
+------+-------+------+---------+
|     1|      1|   4.0|964982703|
|     1|      3|   4.0|964981247|
|     1|      6|   4.0|964982224|
|     1|     47|   5.0|964983815|
|     1|     50|   5.0|964982931|
|     1|     70|   3.0|964982400|
|     1|    101|   5.0|964980868|
|     1|    110|   4.0|964982176|
|     1|    151|   5.0|964984041|
|     1|    157|   5.0|964984100|
|     1|    163|   5.0|964983650|
|     1|    216|   5.0|964981208|
|     1|    223|   3.0|964980985|
|     1|    231|   5.0|964981179|
|     1|    235|   4.0|964980908|
|     1|    260|   5.0|964981680|
|     1|    296|   3.0|964982967|
|     1|    316|   3.0|964982310|
|     1|    333|   5.0|964981179|
|     1|    349|   4.0|964982563|
+------+-------+------+---------+
only showing top 20 rows
```

使用 Spark SQL 和 Hive，可以很容易地实现类似 CTAS 的功能。例如，在 movies 表上执行查询操作，找出同一类型影片（genres 字段）超过 500 部的记录，并将查询结果写入一个结果 Hive 表 genres_by_count 中，代码如下：

```
val countByGenresSql = """
    select genres, count(*) as count
    from movies
    group by genres
```

```
            having count(*) > 500
            order by count desc
"""

//CTAS 过程
spark
    .sql(countByGenresSql)
    .write
    .mode("overwrite")
    .saveAsTable("genres_by_count")

//查看一下，是否数据已经被正确地保存到 Hive 表中
spark.table("genres_by_count").show
```

执行以上代码，输出的结果如下：

```
+--------------+-----+
|        genres|count|
+--------------+-----+
|Comedy|Romance|  600|
| Drama|Romance|  528|
|  Comedy|Drama|  676|
|   Documentary|  544|
|        Comedy| 1588|
|         Drama| 1552|
+--------------+-----+
```

Spark SQL 支持临时视图和 Hive 表进行混合使用。例如，从 tags 数据集创建一个临时视图，然后将它与 Hive 中的 movies 表和 rating 表连接（join 连接）起来，代码如下：

```
//指定字段和 Schema
val tagsFields = Array(
    StructField("userId", IntegerType, nullable=true),
    StructField("movieId", IntegerType, nullable=true),
    StructField("tag", StringType, nullable=true),
    StructField("timestamp", StringType, nullable=true),
)
val tagsSchema = StructType(tagsFields)

//创建 DataFrame
val tagsFile = "/data/spark/ml-latest-small/tags.csv"
val tagsDF = spark
    .read
    .option("header","true")
    .schema(tagsSchema)
    .csv(tagsFile)

//接下来，将该 DataFrame 注册为临时表
tagsDF.createOrReplaceTempView("tags_tb")

//三表执行内连接
val joinSql = """
```

```
        select m.title, m.genres, r.movieId, r.userId, r.rating,
            r.timestamp as ratingTimestamp, t.tag, t.timestamp as tagTimestamp
        from ratings as r
            inner join tags_tb t on r.movieId = t.movieId and r.userId = t.userId
            inner join movies as m on r.movieId = m.movieId
"""
val joinedDF = spark.sql(joinSql)

//查看前 5 条记录
joinedDF
    .select("title","genres","rating","tag")
    .show(5, truncate = false)
```

执行以上代码,输出内容如下:

```
+--------------------+------+------+---------------+
|title               |genres|rating|tag            |
+--------------------+------+------+---------------+
|Step Brothers (2008)|Comedy|5.0   |will ferrell   |
|Step Brothers (2008)|Comedy|5.0   |will ferrell   |
|Step Brothers (2008)|Comedy|5.0   |Highly quotable|
|Step Brothers (2008)|Comedy|5.0   |Highly quotable|
|Step Brothers (2008)|Comedy|5.0   |funny          |
+--------------------+------+------+---------------+
only showing top 5 rows
```

修改上面的示例代码,让 movies 结果集以 Parquet 格式按电影名称(title 列)分桶存储,代码如下:

```
//将电影数据集文件加载到 DataFrame 中
val file = "/data/spark/ml-latest-small/movies.csv"
val movies = spark.read
    .option("header","true")
    .option("inferSchema","true")
    .csv(file)

//输出 movies DataFrame,使用 Parquet 格式并按 title 列存储在 5 个桶中
//不能直接将 bucket 数据保存到任何其他文件中,只能持久化到 Hive 的表中
movies
    .write
    .format("parquet")
    .mode("overwrite")
    .bucketBy(5,"title")
    .saveAsTable("movies_tb")
```

执行以上代码,将 movies 数据集以 Parquet 格式存储到 movies_tb 表中,分为 5 个桶文件存储,然后打开浏览器,查看该表对应的桶文件,如图 5-9 所示。

也可以通过程序查看存储的元数据信息,代码如下:

```
import org.apache.spark.sql.catalyst.TableIdentifier

//获取 movies_tb 表的元数据
```

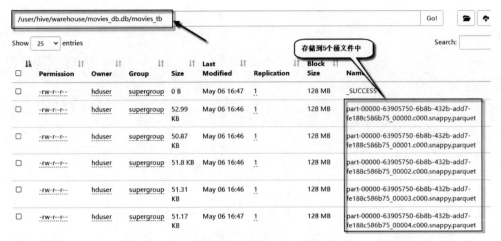

图 5-9 数据集在 Hive 中分桶存储

```
val metadata = spark
  .sessionState
  .catalog
  .getTableMetadata(TableIdentifier("movies_tb"))

//输出分桶规范信息
metadata.bucketSpec.foreach(println)

//输出分桶数据
println(metadata.bucketSpec.get.numBuckets)    //5

//输出分桶字段名称
println(metadata.bucketSpec.get.bucketColumnNames(0))    //"title"
```

执行以上代码，输出内容如下：

```
5 buckets, bucket columns: [title]
5
title
```

有可能对单个表同时使用分区和分桶及排序。例如，存储 movies 电影数据集，按电影风格（genres 字段）分区存储，在同一分区下按电影名称（title 字段）分 5 个桶存储，并且按 movieId 排序。对以上代码进行修改，修改后的代码如下：

```
//将电影数据集文件加载到 DataFrame 中
val file = "/data/spark/ml-latest-small/movies.csv"
val movies = spark.read
  .option("header","true")
  .option("inferSchema","true")
  .csv(file)

//输出 movies DataFrame
//按 genres 字段分区，同一分区下按 title 字段分 5 个桶存储
import spark.implicits._
```

```
movies
  .write
  .format("parquet")
  .mode("overwrite")
  .partitionBy("genres")
  .bucketBy(5,"title")
  .sortBy("movieId")
  .saveAsTable("movies_tb2")
```

执行以上代码，然后打开浏览器，查看对应的存储目录结构，结果如图 5-10 所示。

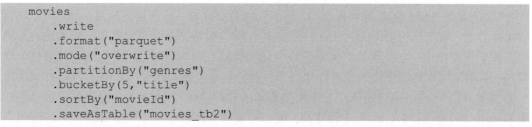

图 5-10　数据集先分区后分桶进行存储

注意：

（1）目前 bucketBy 需要和 saveAsTable() 结合使用，而不能和 save() 一起使用。

（2）目前只支持 sortBy() 和 bucketBy() 结合使用，并且排序列不可以是分区列的一部分。

5.8　查询优化器 Catalyst

Spark SQL 的核心是 Catalyst 优化器，它以一种新颖的方式利用高级编程语言的特性（例如 Scala 的模式匹配和准引号）来构建可扩展的查询优化器。

Catalyst 支持基于规则和基于成本的优化，其工作过程如图 5-11 所示。

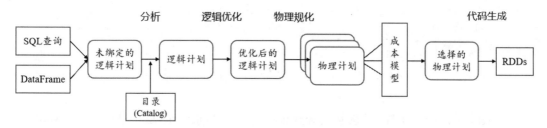

图 5-11　Catalyst 优化器工作过程

Catalyst 是 Spark SQL 查询优化器，用于获取查询计划，并将其转换为 Spark 可以运行的执行计划。它在确保用 DataFrame API 或 SQL 编写的数据处理逻辑高效、快速地运行方面发挥了重要作用。它的设计的目的是最小化端到端查询响应时间，并具有可扩展性，这样 Spark 用户可以将用户代码注入优化器，以执行自定义优化。

当用户在 DataFrame/DataSet 上应用关系和函数转换时，Spark SQL 建立了一个语法树来表示用户的查询计划，称为逻辑计划。Spark 能够在逻辑计划上应用许多优化，还可以使用基于成本的模型在相同逻辑计划的多个物理计划之间进行选择。

5.8.1 窄转换和宽转换

本节将介绍 Spark 如何将 DataSet 的 Transformation 和 Action 转换为执行模型。为了理解应用程序如何在 Spark 集群上运行，需要了解 Spark 如何将 DataSet 的 Transformation 和 Action 转换为执行模型。

Spark 的 DataSet Transformation 分为两种类型：窄转换和宽转换。在解释执行模型之前，首先需要了解这两种转换类型，如图 5-12 所示。

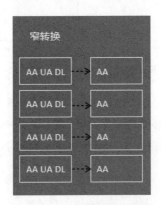

图 5-12 DataSet 的窄转换和宽转换

回顾一下，Transformation 将从现有 DataSet 创建一个新的 DataSet。在从现有 DataSet 创建新的 DataSet 时，窄转换不必在分区之间移动数据。例如，在执行 filter() 和 select() 转换时，就属于窄转换，代码如下：

```
//select 和 filter 是窄转换(Narrow Transformations)
df.select($"carrier",$"origin", $"dest", $"depdelay", $"crsdephour")
  .filter($"carrier" === "AA")
  .show(2)
```

在一个称为流水线的过程中，可以对内存中的 DataSet 执行多个窄转换，从而使窄转换非常有效。

在创建新的 DataSet 时，宽转换会导致数据在分区之间移动，这个过程被称为 shuffle。通过宽转换 shuffle，数据通过网络发送到其他节点并写到磁盘，从而导致网络和磁盘进行输入和输出操作,并使 shuffle 成为一项昂贵的操作。例如,groupBy()、agg()、sortBy()和 orderBy()就属于宽转换,代码如下:

```
df.groupBy("carrier").count.show
```

5.8.2 Spark 执行模型

在深入理解 Spark 执行模型之前,先回顾一下 Spark 的架构体系,如图 5-13 所示。

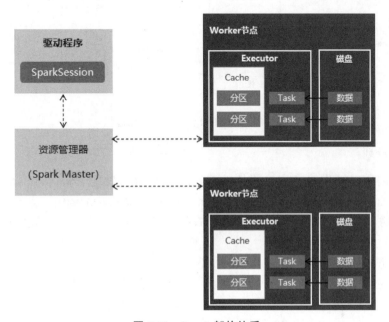

图 5-13 Spark 架构体系

Spark 执行模型可以定义为 3 个阶段:首先需要创建逻辑计划,然后将其转换为物理计划,最后在集群上执行任务。Spark Catalyst 将用户写入的数据处理逻辑转换为逻辑计划,然后使用启发式方法对其进行优化,再将逻辑计划转换为物理计划,最后根据物理计划生成代码。这个过程可以用图 5-14 进行描述。

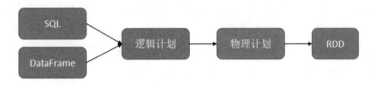

图 5-14 Spark 执行模型

在 Catalyst 中,针对逻辑执行计划和物理执行计划,分别提供了逻辑优化和物理优化。
(1)逻辑优化:这包括优化器将筛选谓词下推到数据源的能力,以及执行跳过不相关数据的能力。例如，对于 Parquet 文件，可以跳过整个块，并且可以通过字典编码将字符串上的比较转换为开销更小的整数比较。

（2）物理优化：这包括智能地在广播连接和 shuffle 连接之间选择连接以减少网络流量的能力，执行低级别的优化，例如消除昂贵的对象分配和减少虚拟函数调用。

Catalyst 优化器的优化过程如图 5-15 所示。

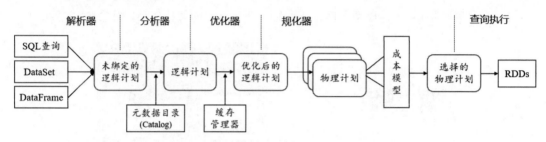

图 5-15　Catalyst 优化器的优化过程

1. 逻辑计划

逻辑计划是以操作符和表达式树的形式对用户数据处理逻辑进行内部表示。通过 DataFrame/DataSet（或 SQL 查询）上的 Transformation 转换构造逻辑计划，一开始逻辑计划是未解析的。Spark 优化器是多阶段的，在执行任何优化之前，它需要解析表达式的引用和类型。

Catalyst 解析和优化逻辑计划的过程如下：

（1）初始查询计划本质上是一个未绑定元数据的逻辑计划，也就是说，在这个阶段用户不知道数据集或包含在数据集中的列，用户也不知道列的类型。

（2）然后，Catalyst 解析抽象语法树，这会生成一个解析后（但未绑定元数据目录）的逻辑计划。

（3）接下来，Catalyst 结合元数据目录来分析未绑定的逻辑计划并以此解析引用，包括检查表名、列名和限定名（称为关系）的名称，以确保它们是有效的，这会创建一个绑定的逻辑计划。

（4）将一组基于规则和成本的优化应用于逻辑计划，基于规则的优化包括常量折叠、项目修剪、谓词下推等。在这一步中，Catalyst 试图通过重新排列和组合较低级别的操作来优化计划，并最终产生一个优化的逻辑计划。例如，它可能决定在 join 操作之前移动一个 filter 操作，以减少 join 中涉及的数据量。基于规则的优化列表定义在 org.apache.spark.sql.catalyst.optimizer.Optimizer 类中。

（5）在 Spark 2.2 中引入了基于成本的优化，使 Catalyst 能够更智能地根据正在处理的数据统计并选择正确的 join 连接类型。基于成本的优化依赖于参与过滤器或连接条件的列的详细统计信息，这就是引入统计信息收集框架的原因。统计信息的示例包括基数、不同值的数量、最大/最小、平均/最大长度等。

（6）这两种类型（基于规则和基于成本）的优化都遵循尽早删除不必要数据和最小化每个操作符成本的原则。

2. 物理计划

逻辑计划优化后，Spark 将生成物理计划。物理计划会标识将执行计划的资源，例如内存分区和计算任务。物理计划阶段有基于规则和基于成本的优化，包括将投影和过滤合并成一个单一的操作及将投影或过滤谓词下推到支持这个特性的数据源，例如 Parquet。此阶段最重要的优化之一是将谓词下推到数据源级别。

Action 操作将触发逻辑 DAG 转换为物理执行计划，Catalyst 优化器可能会从优化的逻辑计划中计算出多个物理计划，并根据成本模型选择最佳方案。基于成本的优化器（Cost-Based Optimizer，CBO）的第 1 个版本是在 Spark 2.2 中发布的。

Catalyst 执行的最后一步是生成成本最低的物理计划的 Java 字节码。

5.8.3 Catalyst 实践

本节将演示如何使用 DataFrame 类的 explain()函数来显示逻辑和物理计划。通过调用 explain(true)方法，可以看到一个 DataSet/DataFrame 的逻辑计划和物理计划；如果调用的是 explain(false)方法，则只显示物理计划。

在下面的示例中，重现 4.12.4 节中的航班数据集分析过程，并通过查看其逻辑和物理执行计划，来深入理解 Catalyst 优化过程。首先导入必要的依赖包，并定义一个代表航班域对象的 case class，代码如下：

```scala
//第 5 章/catalyst_demo.scala

import org.apache.spark.sql.types._
import org.apache.spark.sql._
import org.apache.spark.sql.functions._

case class Flight(
    _id: String,
    dofW: Long,
    carrier: String,
    origin: String,
    dest: String,
    crsdephour: Long,
    crsdeptime:Double,
    depdelay: Double,
    crsarrtime: Double,
    arrdelay: Double,
    crselapsedtime: Double,
    dist: Double
) extends Serializable
```

然后将数据源文件加载到一个 DataSet 中，并执行过滤操作，代码如下：

```
var file = "/data/flightdelay/flights20170102.json"
val ds = spark
    .read
    .format("json")
    .option("inferSchema", "true")
    .load(file)
    .as[Flight]

val ds2 = ds.filter($"depdelay" > 40)
ds2.take(1)
```

在 ds2 上调用 explain()方法,查看逻辑和物理执行计划,代码如下:

```
ds2.explain(true)
```

执行上面的代码,输出内容如下:

```
== Parsed Logical Plan ==
'Filter ('depdelay > 40)
+- Relation[_id#7,arrdelay#8,...] json

== Analyzed Logical Plan ==
_id: string, arrdelay: double, ...
Filter (depdelay#14 > cast(40 as double))
+- Relation[_id#7,arrdelay#8,...] json

== Optimized Logical Plan ==
Filter (isnotnull(depdelay#14) AND (depdelay#14 > 40.0))
+- Relation[_id#7,arrdelay#8,...] json

== Physical Plan ==
*(1) Filter (isnotnull(depdelay#14) AND (depdelay#14 > 40.0))
+- FileScan json [_id#7,arrdelay#8,...] Batched: false, DataFilters:
[isnotnull(depdelay#14), (depdelay#14 > 40.0)], Format: JSON, Location:
InMemoryFileIndex[/data/flightdelay/flights20170102.json], PartitionFilters: [],
PushedFilters: [IsNotNull(depdelay), GreaterThan(depdelay,40.0)], ReadSchema:
struct<_id:string,arrdelay:double,carrier:string,crsarrtime:double,
crsdephour:bigint,crsdeptime:d...
```

从上面的输出内容可以看到,ds2 的 DAG 由一个 FileScan 和一个在 depdelay 列上的 Filter 过滤器组成,如图 5-16 所示。

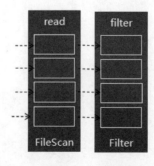

图 5-16　ds2 的 DAG 物理执行计划

接下来执行一个聚合统计操作，并查看 ds3 的执行计划，代码如下：

```
//分组聚合统计
val ds3 = ds2.groupBy("carrier").count
ds3.collect

//查看物理计划
ds3.explain
```

执行上面的代码，输出内容如下：

```
== Physical Plan ==
*(2) HashAggregate(keys=[carrier#710], functions=[count(1)])
+- Exchange hashpartitioning(carrier#710, 200), ENSURE_REQUIREMENTS, [id=#216]
   +- *(1) HashAggregate(keys=[carrier#710], functions=[partial_count(1)])
      +- *(1) Project [carrier#710]
         +- *(1) Filter (isnotnull(depdelay#715) AND (depdelay#715 > 40.0))
            +- FileScan json [carrier#710,depdelay#715] Batched: false,
DataFilters: [isnotnull(depdelay#715), (depdelay#715 > 40.0)], Format: JSON,
Location: ...
```

从上面的输出内容可以看到，在 explain 执行之后，ds3 的物理计划由一个 FileScan、Filter、Project、HashAggregate、Exchange 和 HashAggregate 组成，其中 Exchange 是 groupBy 转换引起的 shuffle。在对 Exchange 中的数据进行 shuffling 之前，Spark 对每个分区执行散列聚合（Hash Aggregation）。在 Exchange 之后，将对以前的子聚合进行散列聚合。这个 ds3 生成的 DAG 如图 5-17 所示。

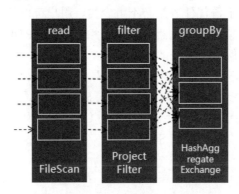

图 5-17　ds2 的 DAG 物理执行计划

需要注意，如果 ds2 被缓存，则将在 DAG 中进行内存扫描，而不是在文件中扫描。例如，修改上一步的代码，增加了对 ds2 的缓存，代码如下：

```
val ds2 = ds.filter($"depdelay" > 40)

//缓存
ds2.cache

//执行聚合统计
```

```
val ds3 = ds2.groupBy("carrier").count
ds3.collect

//查看执行计划
ds3.explain
```

执行上面的代码，输出的结果如下：

```
== Physical Plan ==
*(2) HashAggregate(keys=[carrier#710], functions=[count(1)])
+- Exchange hashpartitioning(carrier#710, 200), ENSURE_REQUIREMENTS, [id=#245]
   +- *(1) HashAggregate(keys=[carrier#710], functions=[partial_count(1)])
      +- InMemoryTableScan [carrier#710]
         +- InMemoryRelation [_id#708, ...], StorageLevel(disk, memory,
deserialized, 1 replicas)
            +-*(1) Filter (isnotnull(depdelay#715) AND (depdelay#715 > 40.0))
               +- FileScan json [_id#708,arrdelay#709,...] Batched:
false, DataFilters: [isnotnull(depdelay#715), (depdelay#715 > 40.0)], Format:
JSON, Location:...
```

从上面的输出内容可以看出，在这个 DAG 中，如果 ds2 被缓存，则 Catalyst 将使用内存扫描而不是文件扫描。

5.8.4 可视化 Spark 程序执行

Spark Web UI 接口对于调优任务是必不可少的。用户可以使用该接口来监视 Spark 应用程序的执行，显示的信息包括调度器阶段和任务的列表、RDD 大小和内存使用的摘要、环境信息及关于正在运行的执行程序的信息。

可以使用以下 URL 在 Web 浏览器中实时查看关于 Spark 作业的有用信息，Web UI 的网址为 http://<driver-node>:4040。对于已经完成的 Spark 应用程序，可以使用 Spark 历史服务器在 Web 浏览器中查看这些信息，这时使用的 URL 网址为 http://<server-url>:18080。

下面通过一个示例来直观地研究 Spark SQL 应用程序的执行。首先创建两组数据集，代码如下：

```
//两组数据集

//第 1 组：较小的数据集
val t1 = spark.range(7)
val t2 = spark.range(13)
val t3 = spark.range(19)

//第 2 组：较大的数据集
val t4 = spark.range(1e8.toLong)
val t5 = spark.range(1e8.toLong)
val t6 = spark.range(1e3.toLong)
```

查看对这两组数据集执行 JOIN 查询的物理执行计划，代码如下：

```
val query = t1
    .join(t2).where(t1("id") === t2("id"))
    .join(t3).where(t3("id") === t1("id"))
    .explain()
```

执行以上代码，输出内容如图 5-18 所示。

```
== Physical Plan ==
*(3) BroadcastHashJoin [id#99L], [id#103L], Inner, BuildRight, false
:- *(3) BroadcastHashJoin [id#99L], [id#101L], Inner, BuildLeft, false
:  :- BroadcastExchange HashedRelationBroadcastMode(List(input[0, bigint, false]),false), [id=#126]
:  :  +- *(1) Range (0, 7, step=1, splits=2)
:  +- *(3) Range (0, 13, step=1, splits=2)
+- BroadcastExchange HashedRelationBroadcastMode(List(input[0, bigint, false]),false), [id=#130]
   +- *(2) Range (0, 19, step=1, splits=2)
```

图 5-18　ds2 的 DAG 物理执行计划

可视化 Spark Web UI 仪表板上的 Spark 作业信息，为此需要调用 Action 操作触发程序的执行，代码如下：

```
val query = t1
    .join(t2).where(t1("id") === t2("id"))
    .join(t3).where(t3("id") === t1("id"))
    .count()
```

执行上面的代码，然后在 Spark Web UI 中查看事件的时间轴信息，如图 5-19 所示。

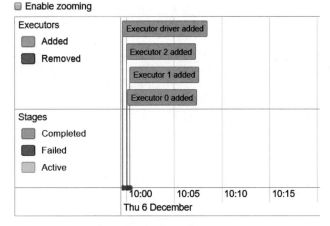

图 5-19　事件的时间轴信息

生成的带有阶段和 shuffle 的 DAG 可视化如图 5-20 所示。

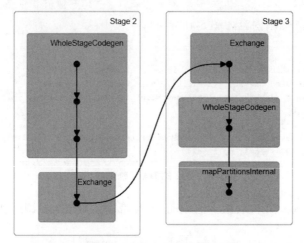

图 5-20　DAG 可视化图

有关作业摘要，包括执行持续时间、成功任务、任务总数等信息，如图 5-21 所示。

图 5-21　作业摘要信息

单击"SQL 标签"查看详细的执行流程，如图 5-22 所示。

接下来，将对较大的数据集运行相同的查询，代码如下：

```
val query = t4
    .join(t5).where(t4("id") === t5("id"))
    .join(t6).where(t4("id") === t6("id"))
    .explain()
```

执行上面的代码，输出内容如图 5-23 所示。

从输出内容可以看到，上一个例子中的 BroadcastHashJoin 现在更改为 SortMergeJoin，原因是输入数据集的大小增加了。

执行 Action 操作，触发作业执行，代码如下：

```
val query = t4
    .join(t5).where(t4("id") === t5("id"))
    .join(t6).where(t4("id") === t6("id"))
    .count
```

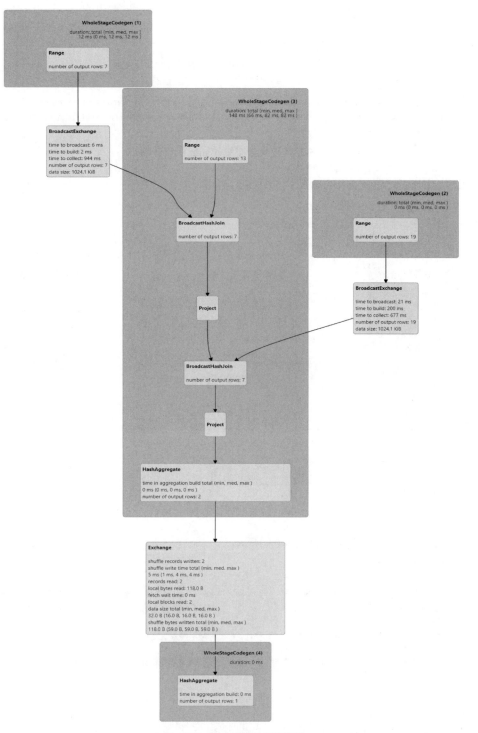

图 5-22 详细执行过程查看

```
== Physical Plan ==
*(6) BroadcastHashJoin [id#105L], [id#109L], Inner, BuildRight, false
:- *(6) SortMergeJoin [id#105L], [id#107L], Inner
:  :- *(2) Sort [id#105L ASC NULLS FIRST], false, 0
:  :  +- Exchange hashpartitioning(id#105L, 200), ENSURE_REQUIREMENTS, [id=#236]
:  :     +- *(1) Range (0, 100000000, step=1, splits=2)
:  +- *(4) Sort [id#107L ASC NULLS FIRST], false, 0
:     +- ReusedExchange [id#107L], Exchange hashpartitioning(id#105L, 200), ENSURE_REQUIREMENTS, [id=#236]
+- BroadcastExchange HashedRelationBroadcastMode(List(input[0, bigint, false]),false), [id=#249]
   +- *(5) Range (0, 1000, step=1, splits=2)
```

图 5-23　较大数据集的 join 执行计划

执行上面的代码，生成的 DAG 如图 5-24 所示。

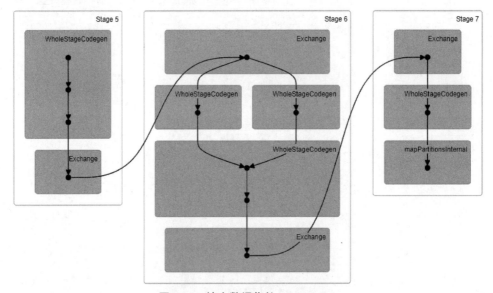

图 5-24　较大数据集的 join DAG

查看作业执行汇总信息，如图 5-25 所示。

Stage Id	Description	Submitted	Duration	Tasks: Succeeded/Total	Input	Output	Shuffle Read	Shuffle Write
7	Started by: hduser count at <console>:29 +details	2018/12/06 11:48:46	0.3 s	1/1			11.5 KB	
6	Started by: hduser count at <console>:29 +details	2018/12/06 11:48:22	23 s	200/200			959.7 MB	11.5 KB
5	Started by: hduser count at <console>:29 +details	2018/12/06 11:47:23	59 s	6/6				479.8 MB

图 5-25　较大数据集的作业执行汇总信息

最后，查看此时的 SQL 执行细节，如图 5-26 所示。

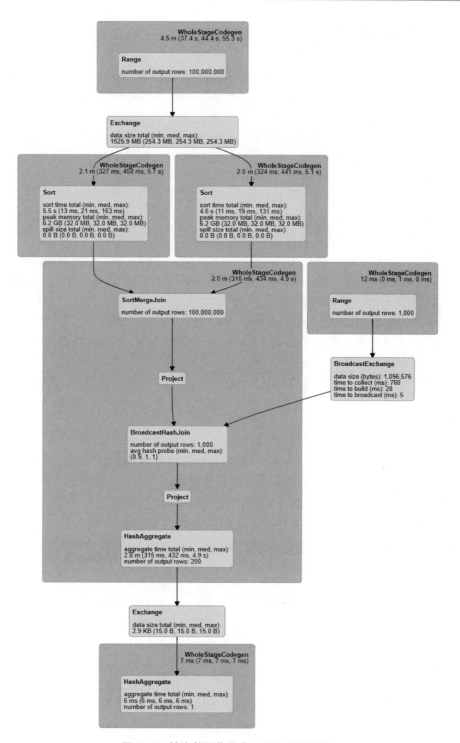

图 5-26　较大数据集的作业执行详细信息

5.9 项目 Tungsten

从 2015 年开始，Spark 设计者注意到，因为硬件方面的进步（例如 10 Gb/s 的网络连接和高速 SSD），Spark 工作负载越来越多地受到 CPU 和内存瓶颈的影响，而不是 I/O 和网络通信。在 Spark 2.0 之前，大部分 CPU 周期花在了无用的工作上，例如进行虚函数调用或将中间数据读写到 CPU 缓存或内存。项目 Tungsten 的创建就是为了提高 Spark 应用程序中内存和 CPU 的使用效率。它基于现代编译器和大规模并行处理（MPP）技术的思想，并将性能提升到接近现代硬件的极限。

在 Tungsten 项目中采用了以下几个措施：

（1）通过堆外管理技术来显式地管理内存，以紧凑的二进制格式存储 Java 对象以减少 GC（垃圾收集）开销。密集的内存数据格式以减少溢出效应（例如，Parquet 格式）和对于理解数据类型（在 DataFrame、DataSet 和 SQL）的操作符直接工作在内存二进制格式上而不是序列化/反序列化等。内存数据的柱状布局避免了不必要的 I/O，并提高了现代 CPU 和 GPU 上的分析处理性能。

（2）使用智能缓存感知（cache-aware）算法和数据结构来利用内存层次结构。为了提高数据处理的速度，通过更有效地使用 L1/L2/L3 CPU 缓存，Spark 算法和数据结构利用内存层次与缓存感知计算。

（3）通过将多个操作符组合到单个 Java 字节码函数中（函数折叠），从而优化了 CPU 使用。该技术消除了虚拟函数的调用，并使用 CPU 寄存器存储中间数据，这进而大大提高了 CPU 效率和运行时性能。

（4）向量化技术。这利用了现代 CPU 设计，允许 CPU 对向量进行操作，向量是来自多个记录的列值的数组。在向量化中，引擎以列格式将多行进行批处理，每个操作符在批处理中迭代数据，但是，它仍然需要将中间数据放在内存中，而不是将它们保存在 CPU 寄存器中，因此，只在不可能完成全阶段代码生成时才使用向量化。

下面的例子显示在 DataFrame 中过滤和汇总整数的物理计划。通过检查物理计划来了解全阶段代码生成计划。在 explain()输出中，当一个操作符被标记为星号时，就意味着该操作的全阶段代码生成已经启用，代码如下：

```
spark.range(1000)
    .filter("id > 100")
    .selectExpr("sum(id)")
    .explain()
```

执行上面的代码，输出的物理计划如图 5-27 所示。

```
== Physical Plan ==
*(2) HashAggregate(keys=[], functions=[sum(id#264L)])
+- Exchange SinglePartition, ENSURE_REQUIREMENTS, [id=#442]
   +- *(1) HashAggregate(keys=[], functions=[partial_sum(id#264L)])
      +- *(1) Filter (id#264L > 100)
         +- *(1) Range (0, 1000, step=1, splits=2)
```

图 5-27　查看物理计划中的全阶段代码生成

全阶段代码生成将过滤和汇总整数的逻辑组合成一个单一的 Java 函数。

5.10　Spark 性能调优

Spark 提供了许多配置来改进和调优 Spark SQL 工作负载的性能，这些配置可以通过编程方式完成，也可以使用 Spark submit 在全局级别应用。对于某些工作负载，可以通过在内存中缓存数据或启用一些实验性选项来提高性能。

1. 在内存中缓存数据

Spark SQL 可以通过调用 spark.catalog.cachetable("tableName")或 dataframe.cache()方法来使用内存中的列格式来缓存表，然后，Spark SQL 将只扫描所需的列，并自动调整压缩，以最小化内存使用和 GC 压力。可以调用 spark.catalog.uncacheTable("tableName")从内存中删除该表。

可以使用 SparkSession 上的 setConf()方法或使用 SQL 运行 SET key=value 命令来配置内存缓存，这些配置项见表 5-6。

表 5-6　Spark SQL 内存缓存配置项

属　性　名	默认	含　义
spark.sql.inMemoryColumnarStorage.compressed	true	当设置为 true 时，Spark SQL 将根据数据的统计信息自动为每列选择一个压缩编解码器
spark.sql.inMemoryColumnarStorage.batchSize	10 000	控制用于列缓存的批的大小。更大的批处理大小可以提高内存利用率和压缩，但是在缓存数据时存在 OOM 风险

当从 DataFrame/SQL 缓存数据时，建议尽可能地使用内存中的柱状格式。当对列执行 DataFrame/SQL 操作时，Spark 只检索所需的列，这将减少数据检索和内存使用。

通过将 spark.sql.inMemoryColumnarStorage.compressed 配置设置为 true，可以使 Spark 使用内存中的柱状存储，代码如下：

```
spark.conf.set("spark.sql.inMemoryColumnarStorage.compressed", true)
```

2. 使用 Spark CBO

Spark CBO 指的是基于成本的优化器。当使用多个 join 连接时，应使用基于成本的优化器，因为它可以基于表和列统计信息，以便改进查询计划。默认情况下是启用的，如果已被禁用，则可以通过将 spark.sql.cbo.enabled 设置为 true 来启用它，代码如下：

```
spark.conf.set("spark.sql.cbo.enabled", true)
```

在执行 JOIN 查询之前，需要运行 ANALYZE TABLE 命令，指出要连接的所有列。此命令为基于成本的优化器收集表和列的统计信息，以找出最佳查询计划，代码如下：

```
ANALYZE TABLE table_name COMPUTE STATISTICS FOR COLUMNS col1,col2
```

3. 为 Shuffle 分区使用最优值

当执行触发数据 shuffle 的操作（如聚合和 join 连接）时，Spark 默认创建 200 个分区。这是因为配置属性 spark.sql.shuffle.partitions 默认被设置为 200。之所以设置这个默认值，是因为 Spark 不知道 shuffle 操作后要使用的最佳分区大小。大多数情况下，这个值会导致性能问题，因此，需要根据数据大小更改它。如果用户有非常庞大的数据，就需要有更大的分区数字，如果有较小的数据集，就需要较小的分区数字，代码如下：

```
spark.conf.set("spark.sql.shuffle.partitions",30)         //默认值为 200
```

用户需要调优此值和其他值，直到达到性能基准。

4. 尽可能使用 Broadcast Hash Join

当 join 数据可以装入内存时，使用 Broadcast Hash Join。

在 Spark 中可用的所有不同的 join 策略中，广播散列连接（Broadcast Hash Join）提供了更好的性能。只有当一个连接表足够小且能够在广播阈值范围内装入内存时，才可以使用此策略。

当数据很大时，可以使用下面的配置来增加该阈值大小（默认为 100 MB），代码如下：

```
spark.conf.set("spark.sql.autoBroadcastJoinThreshold",10485760)
```

5. 在 SQL 上使用 coalesce() & repartition()

在使用 Spark SQL 查询时，可以通过在查询中使用 COALESCE、REPARTITION 和 REPARTITION_BY_RANGE 来根据数据大小增加或减少分区，代码如下：

```
SELECT /*+ COALESCE(3) */ * FROM EMP_TABLE
SELECT /*+ REPARTITION(3) */ * FROM EMP_TABLE
SELECT /*+ REPARTITION(c) */ * FROM EMP_TABLE
SELECT /*+ REPARTITION(3, dept_col) */ * FROM EMP_TABLE
SELECT /*+ REPARTITION_BY_RANGE(dept_col) */ * FROM EMP_TABLE
SELECT /*+ REPARTITION_BY_RANGE(3, dept_col) */ * FROM EMP_TABLE
```

6. 启用"自适应查询执行"

自适应查询执行是 Spark 3.0 的一个特性，它通过在运行时使用每个阶段完成后收集的统计数据重新优化查询计划，从而提高查询性能。可以通过将 spark.sql.adaptive.enabled 配置属性设置为 true 来启用它，代码如下：

```
spark.conf.set("spark.sql.adaptive.enabled",true)
```

7. 启用 Shuffle 后分区合并

在 Spark 3.0 中，在作业的每个阶段之后，Spark 通过查看完成阶段的指标来动态地确定分区的最佳数量。为了使用它，需要启用以下配置，代码如下：

```
spark.conf.set("spark.sql.adaptive.coalescePartitions.enabled",true)
```

8. 优化倾斜连接

有时，可能会遇到分区中分布不均匀的数据，这称为数据倾斜。诸如 join 之类的操作在这些分区上执行时会变得非常慢。通过启用 AQE（自适应查询执行），Spark 检查 stage 统计信息，确定是否存在倾斜连接，并通过将较大的分区划分为较小的分区（与其他表/DataFrame 上的分区大小匹配）来优化它，代码如下：

```
spark.conf.set("spark.sql.adaptive.skewJoin.enabled",true)
```

9. 合理的配置选项

还可以使用多个选项来调优查询执行的性能，这些选项见表 5-7。

表 5-7　Spark SQL 调优查询执行性能的配置选项

属 性 名	默认	含 义
spark.sql.files.maxPartitionBytes	134 217 728 (128 MB)	在读取文件时，要塞进单个分区中的最大字节数
spark.sql.files.openCostInBytes	4 194 304 (4 MB)	打开一个文件的估计成本，通过同时扫描的字节数来衡量。这是在将多个文件放入一个分区时使用的。最好是过度估计，那么带有小文件的分区将比带有大文件的分区(这是优先计划的)更快
spark.sql.broadcastTimeout	300	广播连接中广播等待时间的超时(s)
spark.sql.autoBroadcastJoinThreshold	10 485 760 (10 MB)	配置将在执行 join 时广播到所有工作节点的表的最大大小(以字节为单位)。通过将此值设置为 -1，可以禁用广播
spark.sql.shuffle.partitions	200	在为连接或聚合重组(shuffling)数据时，配置要使用的分区数量

在将来的版本中，随着对自动执行进行更多优化，这些选项可能会被弃用。

上面介绍了通过不同的配置，以提高 Spark SQL 查询和应用程序的性能。在具体应用中，需要调优这些配置的值及执行器 CPU 内核和执行器内存，直到满足需求为止。

5.11　Spark SQL 编程案例

本节通过对几个案例的学习，掌握使用 Spark SQL 进行大数据分析的复杂用法。

5.11.1　电影数据集分析

本节使用 Spark SQL 实现对电影数据集进行分析。在这里使用推荐领域的一个著名的开放测试数据集 MovieLens。MovieLens 数据集包括电影元数据信息和用户属性信息。本例将使用其中的 users.dat 和 ratings.dat 两个数据集。

要求使用 Spark DataSet API 统计看过"Lord of the Rings,The(1978)"的用户的年龄和性别分布。为了简单起见，直接使用该影片的 id（2116）。建议按以下步骤开发执行。

(1) 定义两种类型 case class 类，分别定义用户和评分的 Schema，代码如下：

```scala
//第5章/examples_movies.scala

case class User(
    userID:Long,
    gender:String,
    age:Integer,
    occupation:String,
    zipcode:String
)

case class Rating(
    userID:Long,
    movieID:Long,
    rating:Integer,
    timestamp:Long
)
```

(2) 读取用户数据集 users.dat，并注册为临时表 users，代码如下：

```scala
import org.apache.spark.sql.functions._

//定义文件路径
val usersFile = "/data/spark/ml-1m/users.dat"

//获得RDD
val rawUserRDD = sc.textFile(usersFile)

//支持 RDD 到 DataFrame 的隐式转换
import spark.implicits._

//对 RDD 进行转换操作，最后转换为 DataFrame
val userDF = rawUserRDD
    .map(_.split("::"))
    .map(x=>User(x(0).toLong, x(1), x(2).toInt, x(3), x(4)))
    .toDF()

//查看用户数据
userDF.printSchema()
userDF.show(5)
```

执行以上代码，输出的结果如下：

```
root
 |-- userID: long (nullable = false)
 |-- gender: string (nullable = true)
 |-- age: integer (nullable = true)
 |-- occupation: string (nullable = true)
 |-- zipcode: string (nullable = true)
```

```
+------+------+---+----------+-------+
|userID|gender|age|occupation|zipcode|
+------+------+---+----------+-------+
|     1|     F|  1|        10|  48067|
|     2|     M| 56|        16|  70072|
|     3|     M| 25|        15|  55117|
|     4|     M| 45|         7|  02460|
|     5|     M| 25|        20|  55455|
+------+------+---+----------+-------+
only showing top 5 rows
```

(3) 读取评分数据集 ratings.dat，并注册成临时表 ratings，代码如下：

```
//定义文件路径
val ratingsFile = "/data/spark/ml-1m/ratings.dat"

//生成 RDD
val rawRatingRDD = sc.textFile(ratingsFile)

//对 RDD 进行转换，最后转换为 DataFrame
val ratingDF = rawRatingRDD
    .map(_.split("::"))
    .map(x=>Rating(x(0).toLong, x(1).toLong,x(2).toInt,x(3).toLong))
    .toDF

//查看
ratingDF.printSchema()
ratingDF.show(5)
```

执行以上代码，输出的结果如下：

```
root
 |-- userID: long (nullable = false)
 |-- movieID: long (nullable = false)
 |-- rating: integer (nullable = true)
 |-- timestamp: long (nullable = false)


+------+-------+------+---------+
|userID|movieID|rating|timestamp|
+------+-------+------+---------+
|     1|   1193|     5|978300760|
|     1|    661|     3|978302109|
|     1|    914|     3|978301968|
|     1|   3408|     4|978300275|
|     1|   2355|     5|978824291|
+------+-------+------+---------+
only showing top 5 rows
```

(4) 将两个 DataFrame 注册为临时表，对应的表名分别为 users 和 ratings，代码如下：

```
userDF.createOrReplaceTempView("users")
ratingDF.createOrReplaceTempView("ratings")
```

（5）通过 SQL 处理临时表 users 和 ratings 中的数据，并输出最终结果。为了简单起见，避免三表连接操作，这里直接使用了 movieID，代码如下：

```
val MOVIE_ID = "2116"
val sqlStr = s"""
    select age,gender,count(*) as total_people
    from users as u join ratings as r on u.userid=r.userid
    where movieid=${MOVIE_ID} group by gender,age
"""
val resultDF = spark.sql(sqlStr)

//显示 resultDF 的内容
resultDF.show()
```

执行以上代码，输出的结果如下：

```
+---+------+------------+
|age|gender|total_people|
+---+------+------------+
| 18|     M|          72|
| 18|     F|           9|
| 56|     M|           8|
| 45|     M|          26|
| 45|     F|           3|
| 25|     M|         169|
| 56|     F|           2|
|  1|     M|          13|
|  1|     F|           4|
| 50|     F|           3|
| 50|     M|          22|
| 25|     F|          28|
| 35|     F|          13|
| 35|     M|          66|
+---+------+------------+
```

（6）以交叉表的形式统计不同年龄不同性别的用户数，代码如下：

```
import org.apache.spark.sql.functions._

resultDF
  .groupBy("age")
  .pivot("gender")
  .agg(sum("total_people").as("cnt"))
  .show()
```

执行以上代码，输出的结果如下：

```
+---+---+---+
|age|  F|  M|
+---+---+---+
|  1|  4| 13|
| 35| 13| 66|
| 50|  3| 22|
| 45|  3| 26|
```

```
| 25| 28|169|
| 56|  2|  8|
| 18|  9| 72|
+---+---+---+
```

（7）在 Zeppelin 中，支持查询结果的可视化显示。在 Zeppelin Notebook 的单元格中，执行以下语句，可视化显示数据（注意，第 1 行必须输入%sql），代码如下：

```
%sql
select age,gender,count(*) as total_people
from users as u join ratings as r
on u.userid=r.userid
where movieid=${MOVIE_ID=2116}
group by gender,age
```

执行以上代码，输出结果如图 5-28 所示。

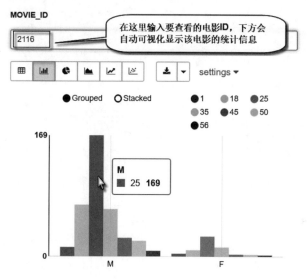

图 5-28　在 Zeppelin Notebook 中可视化查询结果

用户可以在输入框输入要查看的影片 ID，下方会自动可视化显示该影片的统计信息。

5.11.2　电商数据集分析

本示例使用著名的电商数据集 Northwind。Northwind 是一个最初由 Microsoft 创建的示例数据库，包含一个名为 Northwind Traders 的虚拟公司的销售数据，该公司从世界各地进口和出口特色食品。

现要求通过分析该电商数据集，回答以下问题：

（1）每个客户下了多少订单？
（2）每个国家的订单有多少？

(3) 每月/年有多少订单?
(4) 每个客户的年销售总额是多少?
(5) 客户每年的平均订单是多少?

要回答以上问题,需要用到其中的订单表和订单明细表。建议按以下步骤操作。

(1) 将订单数据集加载到 DataFrame 中,代码如下:

```
//第 5 章/examples_nw.scala

//1. 读取源数据文件
val filePath = "file://home/hduser/data/spark/nw/"
//将订单数据集加载到 DataFrame 中
val orders = spark.read
    .option("header","true")
    .option("inferSchema","true")
    .csv(filePath + "NW-Orders-01.csv")

println("订单有" + orders.count() + "行")
orders.printSchema()
orders.show(3)
```

执行以上代码,输出内容如下:

```
订单有830 行
root
 |-- OrderID: integer (nullable = true)
 |-- CustomerID: string (nullable = true)
 |-- EmployeeID: integer (nullable = true)
 |-- OrderDate: string (nullable = true)
 |-- ShipCountry: string (nullable = true)

+-------+----------+----------+---------+-----------+
|OrderID|CustomerID|EmployeeID|OrderDate|ShipCountry|
+-------+----------+----------+---------+-----------+
|  10248|     VINET|         5| 1996-7-2|     France|
|  10249|     TOMSP|         6| 1996-7-3|    Germany|
|  10250|     HANAR|         4| 1996-7-6|     Brazil|
+-------+----------+----------+---------+-----------+
only showing top 3 rows
```

(2) 将订单明细数据集加载到 DataFrame 中,代码如下:

```
val orderDetails = spark.read
    .option("header","true")
    .option("inferSchema","true")
    .csv(filePath + "NW-Order-Details.csv")

println("订单明细有" + orderDetails.count() + "行")
orderDetails.printSchema()
orderDetails.show(3)
```

执行以上代码,输出内容如下:

```
订单明细有 2155 行
root
 |-- OrderID: integer (nullable = true)
 |-- ProductId: integer (nullable = true)
 |-- UnitPrice: double (nullable = true)
 |-- Qty: integer (nullable = true)
 |-- Discount: double (nullable = true)

+-------+---------+---------+---+--------+
|OrderID|ProductId|UnitPrice|Qty|Discount|
+-------+---------+---------+---+--------+
|  10248|       11|     14.0| 12|     0.0|
|  10248|       42|      9.8| 10|     0.0|
|  10248|       72|     34.8|  5|     0.0|
+-------+---------+---------+---+--------+
only showing top 3 rows
```

（3）回答第 1 个问题：每个客户下了多少订单？代码如下：

```
val orderByCustomer = orders.groupBy("CustomerID").count()
orderByCustomer.sort(col("count").desc).show(10)
```

执行以上代码，输出内容如下：

```
+----------+-----+
|CustomerID|count|
+----------+-----+
|     SAVEA|   31|
|     ERNSH|   30|
|     QUICK|   28|
|     HUNGO|   19|
|     FOLKO|   19|
|     HILAA|   18|
|     RATTC|   18|
|     BERGS|   18|
|     BONAP|   17|
|     WARTH|   15|
+----------+-----+
only showing Top 10 rows
```

（4）回答第 2 个问题：每个国家的订单有多少？代码如下：

```
val orderByCountry = orders.groupBy("ShipCountry").count()
orderByCountry.sort(col("count").desc).show(10)
```

执行以上代码，输出内容如下：

```
+-----------+-----+
|ShipCountry|count|
+-----------+-----+
|    Germany|  122|
|        USA|  122|
|     Brazil|   82|
|     France|   77|
|         UK|   56|
```

```
|      Venezuela|  46|
|        Austria|  40|
|         Sweden|  37|
|         Canada|  30|
|         Mexico|  28|
+---------------+----+
only showing Top 10 rows
```

对于后面 3 个问题,需要对数据进行以下转换:

① 向 Orders DataFrame 增加一个 OrderTotal 列,为此,需要计算每个明细的实际金额,然后根据 order id 统计每张订单的总金额,并对 order details & orders 进行等值内连接,增加订单总金额。另外还要检查是否有任何 null 列。

② 增加一个 date 列。

③ 增加 month 和 year 列,以便按月进行统计。

(5) 向 order details 中增加每行的小计(每个订单明细的实际金额),代码如下:

```
import spark.implicits._

val orderDetails1 = orderDetails
    .select($"OrderID",
            (($"UnitPrice" * $"Qty") - ($"UnitPrice" * $"Qty") * $"Discount").as("OrderPrice"))

orderDetails1.show(10)
```

执行以上代码,输出内容如下:

```
+-------+----------+
|OrderID|OrderPrice|
+-------+----------+
|  10248|     168.0|
|  10248|      98.0|
|  10248|     174.0|
|  10249|     167.4|
|  10249|    1696.0|
|  10250|      77.0|
|  10250|    1261.4|
|  10250|     214.2|
|  10251|     95.76|
|  10251|     222.3|
+-------+----------+
only showing Top 10 rows
```

(6) 根据 order id 统计每张订单的总金额,代码如下:

```
//按订单编号执行聚合统计
val orderTot = orderDetails1
    .groupBy("OrderID")
    .agg(sum("OrderPrice").as("OrderTotal"))

//查看
orderTot
```

```
    .select($"OrderID",bround($"OrderTotal",2))
    .sort("OrderID")
    .show(10)
```

执行以上代码,输出内容如下:

```
+-------+--------------------+
|OrderID|bround(OrderTotal, 2)|
+-------+--------------------+
|  10248|               440.0|
|  10249|              1863.4|
|  10250|              1552.6|
|  10251|              654.06|
|  10252|              3597.9|
|  10253|              1444.8|
|  10254|              556.62|
|  10255|              2490.5|
|  10256|               517.8|
|  10257|              1119.9|
+-------+--------------------+
only showing Top 10 rows
```

(7) 对 order details & orders 进行等值内连接,增加订单总金额,代码如下:

```
val orders1 = orders
    .join(orderTot, orders("OrderID").equalTo(orderTot("OrderID")), "inner")
    .select(
        orders("OrderID"),
        orders("CustomerID"),
        orders("OrderDate"),
        orders("ShipCountry").alias("ShipCountry"),
        orderTot("OrderTotal").alias("Total")
    )

orders1.sort("CustomerID").show()
```

执行以上代码,输出内容如下:

```
+-------+----------+----------+-----------+------------------+
|OrderID|CustomerID| OrderDate|ShipCountry|             Total|
+-------+----------+----------+-----------+------------------+
|  11011|     ALFKI|  1998-4-7|    Germany|             933.5|
|  10692|     ALFKI| 1997-10-1|    Germany|             878.0|
|  10702|     ALFKI|1997-10-11|    Germany|             330.0|
|  10835|     ALFKI| 1998-1-13|    Germany|             845.8|
|  10643|     ALFKI| 1997-8-23|    Germany|             814.5|
|  10952|     ALFKI| 1998-3-14|    Germany|             471.2|
|  10308|     ANATR| 1996-9-16|     Mexico|              88.8|
|  10926|     ANATR|  1998-3-2|     Mexico|             514.4|
|  10759|     ANATR|1997-11-26|     Mexico|             320.0|
|  10625|     ANATR|  1997-8-6|     Mexico|            479.75|
|  10507|     ANTON| 1997-4-13|     Mexico|           749.0625|
|  10365|     ANTON|1996-11-25|     Mexico|403.20000000000005|
|  10535|     ANTON| 1997-5-11|     Mexico|           1940.85|
```

```
| 10573|     ANTON| 1997-6-17|          Mexico|             2082.0|
| 10682|     ANTON| 1997-9-23|          Mexico|              375.5|
| 10677|     ANTON| 1997-9-20|          Mexico|            813.365|
| 10856|     ANTON| 1998-1-26|          Mexico|              660.0|
| 10383|     AROUT|1996-12-14|              UK|              899.0|
| 10355|     AROUT|1996-11-13|              UK|              480.0|
| 10453|     AROUT| 1997-2-19|              UK|              407.7|
+------+----------+----------+----------------+-------------------+
only showing top 20 rows
```

（8）检查 Total 列是否有任何 null 值，代码如下：

```
orders1.filter(orders1("Total").isNull).show()
```

执行以上代码，输出内容如下：

```
+-------+----------+---------+-----------+-----+
|OrderID|CustomerID|OrderDate|ShipCountry|Total|
+-------+----------+---------+-----------+-----+
+-------+----------+---------+-----------+-----+
```

可以看出，计算的订单总金额没有空值。

（9）增加一个 date 列，代码如下：

```
val orders2 = orders1.withColumn("Date",to_date(orders1("OrderDate")))

orders2.printSchema()
orders2.show(5)
```

执行以上代码，输出内容如下：

```
root
 |-- OrderID: integer (nullable = true)
 |-- CustomerID: string (nullable = true)
 |-- OrderDate: string (nullable = true)
 |-- ShipCountry: string (nullable = true)
 |-- Total: double (nullable = true)
 |-- Date: date (nullable = true)


+-------+----------+---------+-----------+------------------+----------+
|OrderID|CustomerID|OrderDate|ShipCountry|             Total|      Date|
+-------+----------+---------+-----------+------------------+----------+
|  10248|     VINET| 1996-7-2|     France|             440.0|1996-07-02|
|  10249|     TOMSP| 1996-7-3|    Germany|            1863.4|1996-07-03|
|  10250|     HANAR| 1996-7-6|     Brazil|1552.6000000000001|1996-07-06|
|  10251|     VICTE| 1996-7-6|     France|            654.06|1996-07-06|
|  10252|     SUPRD| 1996-7-7|    Belgium|            3597.9|1996-07-07|
+-------+----------+---------+-----------+------------------+----------+
only showing top 5 rows
```

（10）增加 month 和 year 列，从 Orderdate 列抽取，代码如下：

```
val orders3 = orders2
    .withColumn("Month",month($"OrderDate"))
    .withColumn("Year",year($"OrderDate"))

orders3.show(10)
```

执行以上代码，输出内容如下：

```
+-------+----------+---------+-----------+------------------+----------+-----+----+
|OrderID|CustomerID|OrderDate|ShipCountry|             Total|      Date|Month|Year|
+-------+----------+---------+-----------+------------------+----------+-----+----+
|  10248|     VINET| 1996-7-2|     France|             440.0|1996-07-02|    7|1996|
|  10249|     TOMSP| 1996-7-3|    Germany|            1863.4|1996-07-03|    7|1996|
|  10250|     HANAR| 1996-7-6|     Brazil|1552.6000000000001|1996-07-06|    7|1996|
|  10251|     VICTE| 1996-7-6|     France|            654.06|1996-07-06|    7|1996|
|  10252|     SUPRD| 1996-7-7|    Belgium|            3597.9|1996-07-07|    7|1996|
|  10253|     HANAR| 1996-7-8|     Brazil|1444.8000000000002|1996-07-08|    7|1996|
|  10254|     CHOPS| 1996-7-9|Switzerland|  556.6199999999999|1996-07-09|    7|1996|
|  10255|     RICSU|1996-7-10|Switzerland|            2490.5|1996-07-10|    7|1996|
|  10256|     WELLI|1996-7-13|     Brazil|             517.8|1996-07-13|    7|1996|
|  10257|     HILAA|1996-7-14|  Venezuela|            1119.9|1996-07-14|    7|1996|
+-------+----------+---------+-----------+------------------+----------+-----+----+
only showing Top 10 rows
```

（11）回答第3个问题：每月/年有多少订单金额？代码如下：

```
val ordersByYM = orders3
    .groupBy("Year","Month")
    .agg(sum("Total").as("Total"))

ordersByYM
    .select($"Year",$"Month",bround($"Total",2) as "Total")
    .sort($"Year",$"Month")
    .show()
```

执行以上代码，输出内容如下：

```
+----+-----+--------+
|Year|Month|   Total|
+----+-----+--------+
|1996|    7| 30741.9|
|1996|    8|22726.88|
|1996|    9| 27691.4|
|1996|   10|38380.12|
|1996|   11|45694.44|
|1996|   12|52494.33|
|1997|    1|51612.97|
|1997|    2|38483.64|
|1997|    3|40918.82|
|1997|    4|57116.71|
|1997|    5|50270.33|
|1997|    6|34392.08|
|1997|    7|52744.68|
|1997|    8|46991.78|
|1997|    9|57723.23|
|1997|   10|62253.63|
|1997|   11|51294.81|
```

```
|1997|   12| 67920.23|
|1998|    1|107049.96|
|1998|    2| 85240.83|
+----+----+---------+
only showing top 20 rows
```

（12）回答第 4 个问题：每个客户的年销售总额是多少？代码如下：

```
var ordersByCY = orders3
    .groupBy("CustomerID","Year")
    .agg(sum("Total").as("Total"))

ordersByCY.sort($"CustomerID",$"Year").show()
```

执行以上代码，输出内容如下：

```
+----------+----+------------------+
|CustomerID|Year|             Total|
+----------+----+------------------+
|     ALFKI|1997|            2022.5|
|     ALFKI|1998|            2250.5|
|     ANATR|1996|              88.8|
|     ANATR|1997|            799.75|
|     ANATR|1998|             514.4|
|     ANTON|1996| 403.20000000000005|
|     ANTON|1997|          5960.7775|
|     ANTON|1998|             660.0|
|     AROUT|1996|            1379.0|
|     AROUT|1997|  6406.900000000001|
|     AROUT|1998|           5604.75|
|     BERGS|1996|            4324.4|
|     BERGS|1997|13849.015000000001|
|     BERGS|1998|           6754.1625|
|     BLAUS|1997|            1079.8|
|     BLAUS|1998|            2160.0|
|     BLONP|1996|            9986.2|
|     BLONP|1997|           7817.88|
|     BLONP|1998|             730.0|
|     BOLID|1996|             982.0|
+----------+----+------------------+
only showing top 20 rows
```

（13）回答第 5 个问题：客户每年的平均订单金额是多少？代码如下：

```
ordersByCY = orders3
    .groupBy("CustomerID","Year")
    .agg(avg("Total").as("Avg"))

ordersByCY
    .select($"CustomerID",$"Year",bround($"Avg",2))
    .sort($"CustomerID",$"Year")
    .show()
```

执行以上代码，输出内容如下：

```
+----------+----+--------------+
|CustomerID|Year|bround(Avg, 2)|
+----------+----+--------------+
|     ALFKI|1997|        674.17|
|     ALFKI|1998|        750.17|
|     ANATR|1996|          88.8|
|     ANATR|1997|        399.88|
|     ANATR|1998|         514.4|
|     ANTON|1996|         403.2|
|     ANTON|1997|       1192.16|
|     ANTON|1998|         660.0|
|     AROUT|1996|         689.5|
|     AROUT|1997|        915.27|
|     AROUT|1998|       1401.19|
|     BERGS|1996|       1441.47|
|     BERGS|1997|        1384.9|
|     BERGS|1998|       1350.83|
|     BLAUS|1997|        269.95|
|     BLAUS|1998|         720.0|
|     BLONP|1996|       3328.73|
|     BLONP|1997|       1116.84|
|     BLONP|1998|         730.0|
|     BOLID|1996|         982.0|
+----------+----+--------------+
only showing top 20 rows
```

5.12 Spark SQL 分析案例

随着互联网和电商的发展，人们习惯于网上购物。在国内，电商平台深受欢迎，每年的"双11"、"双12"等活动，大量的用户在淘宝平台浏览商品，或收藏或加入购物车或直接购买，淘宝平台也因此积累了海量的用户网上购物行为数据。本案例通过对淘宝用户行为数据的分析，探索用户购买的规律，了解商品的受欢迎程度，结合店铺的营销策略，以实现更加精细和精准的运营、让业务更好地增长。

5.12.1 用户行为数据集说明

本案例所使用的淘宝用户行为数据集来自阿里云天池公开数据集（由阿里巴巴公司提供）。本数据集包含了 2017 年 11 月 25 日至 2017 年 12 月 3 日，有行为的约一百万名随机用户的所有行为（行为包括单击、购买、加购、喜欢）。数据集的组织形式和 MovieLens-20M 类似，即数据集的每行表示一条用户行为，由用户 ID、商品 ID、商品类目 ID、行为类型和时间戳组成，并以逗号分隔。关于数据集中每列的详细描述见表 5-8。

注意到，其中用户行为类型共 4 种，每种行为类型所代表的含义见表 5-9。

原始的淘宝用户行为数据集解压缩后大小约为 3.41GB，包含数据记录数约为一亿条。

为了便于读者测试使用,笔者抽取其中 1%(约一百万条记录)组成新的测试数据集,并命名为 UserBehavior_100w.csv。读者可从本书配套的源码包中找到这个数据集文件,并将该数据集上传到 HDFS 的指定路径上,例如,/data/spark/UserBehavior_100w.csv。

表 5-8 淘宝用户行为数据集说明

列 名 称	数据类型	含 义
用户 ID	整数类型	序列化后的用户 ID
商品 ID	整数类型	序列化后的商品 ID
商品类目 ID	整数类型	序列化后的商品所属类目 ID
行为类型	字符串型	枚举类型,包括('pv', 'buy', 'cart', 'fav')
时间戳	长整数类型	行为发生的时间戳,单位秒

表 5-9 用户行为枚举值含义说明

行为类型	说 明
pv	商品详情页 pv,等价于单击页面行为
buy	商品购买行为
cart	将商品加入购物车的行为
fav	收藏商品的行为

5.12.2 分析需求说明

首先明确分析目的,即要对电商用户行为进行分析。那么应该主要从用户维度、产品维度、用户行为的维度进行分析,如图 5-29 所示。

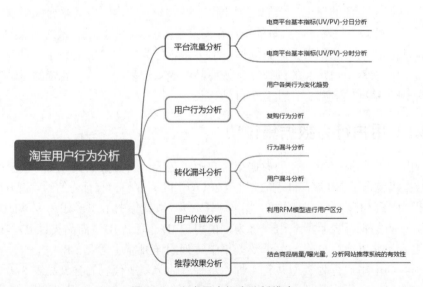

图 5-29 电商用户行为分析维度

1. 平台流量分析

主要分析电商平台的基本指标，包括 UV 分析和 PV 分析。

其中 UV 指的是独立访客（Unique Visitor），是指通过互联网访问、浏览这个网页的自然人。如果一天内同一名访客多次访问，则仅计算一个 UV。例如，用户张三在上午浏览了一次淘宝上的商品网页，然后退出登录，在下午和晚上又多次进入淘宝网站浏览商品。虽然张三一天内多次访问了淘宝网页，但淘宝后台在统计 UV 数时，只算为一个 UV。UV 指标主要用来衡量用户的覆盖度。

PV 指的是页面浏览量（Page View）或单击量，用户每次对网站中的每个网页访问均被记录 1 个 PV。用户对同一页面的多次访问，会累计访问量。PV 指标主要用来衡量网站网页的受关注程度。

2. 用户行为分析

分析用户最活跃的日期及活跃时段；从成交量、人均购买次数、复购率等指标探索用户对商品的购买偏好，了解商品的销售规律。

3. 转化漏斗分析

漏斗模型是一套流程式数据分析模型，它能够科学地反映用户行为状态及从起点到终点各阶段用户转化率情况，是一种重要的分析模型。漏斗分析模型已经广泛应用于网站和 App 用户行为分析的流量监控、电商行业、零售的购买转化率、产品营销和销售等日常数据运营与数据分析的工作中。

例如漏斗模型在电商网站中的应用，用户从首页进入最终完成支付的行为，大多需要经过几个环节，从商品/浏览分类→查看商品详情→加入购物车→生成订单→开始支付→完成支付→回购商品。这其中的每个环节都有一定的转化率，需要做的工作是监控用户在流程上各个层次的行为路径，寻找每个层级的可优化点，提高用户在每个层级之间的转化率，最终提高 GMV。

本案例从收藏转化率、购物车转化率、成交转化率等方面，对用户行为从浏览到购买进行漏斗分析。

4. 用户价值分析

RFM 模型是衡量客户价值和客户创利能力的重要工具和手段，由于本数据集缺少消费金额信息，因此尝试从最近一次购买时间（近度）和购买次数（频度 F）入手对本时段用户进行一次粗略的价值分析。

5. 推荐效果分析

面对电商平台海量的商品信息，用户的浏览和购买等行为往往会受到电商平台推荐系统的影响。为了判断推荐系统的有效性，需要从不同维度评估不同的指标。

5.12.3 数据探索和预处理

数据分析的第 1 步是对数据进行探索，对要分析的数据有个大概的了解。弄清数集质量、大小、特征和样本数量、数据类型、数据的概率分布等，然后基于对数据的了解，在正式进行分析之前，还需要对数据进行预处理，包括错误数据处理、缺失值处理、属性选择与转换处理等。

因为本案例的主要任务是对淘宝用户行为数据集进行分析，因此选择 Zeppelin Notebook 作为大数据交互分析工具。关于 Zeppelin Notebook 的使用，可参考 2.4 节的内容。

首先将 UserBehavior_100w.csv 文件内容加载到一个 DataFrame 中，代码如下：

```
import org.apache.spark.sql.types._

val filePath = "file://home/hduser/data/spark/UserBehavior_100w.csv"

//构造 Schema
val fields = Seq(
    StructField("userid", IntegerType, nullable = true),
    StructField("itemid", IntegerType, nullable = true),
    StructField("category", IntegerType, nullable = true),
    StructField("behavior", StringType, nullable = true),
    StructField("ts", LongType, nullable = true),
)
val schema = StructType(fields)

//读取数据源，创建 DataFrame
val data = spark.read
    .option("header", "false")           //说明有标题行
    .schema(schema)                       //指定使用的 Schema
    .csv(filePath)

data.printSchema()                        //打印 Schema
data.show()                               //显示
```

执行以上代码，输出内容如下：

```
root
 |-- userid: integer (nullable = true)
 |-- itemid: integer (nullable = true)
 |-- category: integer (nullable = true)
 |-- behavior: string (nullable = true)
 |-- ts: long (nullable = true)


+-------+-------+--------+--------+----------+
| userid| itemid|category|behavior|        ts|
+-------+-------+--------+--------+----------+
|      1|3911125|  982926|      pv|1512044958|
|    100|3658601| 2342116|      pv|1511757958|
|   1000|1186274| 2939262|     fav|1511744330|
|   1000|1925478|  171529|      pv|1511944345|
```

```
|1000011| 106154| 4339722|      pv|1511581267|
|1000011| 453400| 4756105|      pv|1511948184|
| 100002| 324802| 2096639|      pv|1511705000|
| 100002|2968522| 4806751|      pv|1511862934|
|1000040|3381470| 4145813|      pv|1511596534|
|1000040|2044513|  570735|      pv|1511763134|
|1000040|1371392| 1575622|      pv|1511882223|
|1000040|3163184| 1575622|      pv|1511882497|
|1000040|1783293| 1575622|      pv|1511882507|
|1000040|   8592| 4756105|      pv|1512063066|
|1000040|5155007| 4145813|      pv|1512202659|
|1000045|3199716| 3607361|      pv|1511595415|
|1000054|4138494| 4874384|      pv|1512013826|
|1000054|4763785| 4181361|      pv|1512083033|
|1000059|4541139|  801221|      pv|1511622625|
|1000059|2761782| 4756105|      pv|1511792872|
+-------+-------+--------+--------+----------+
only showing top 20 rows
```

从输出结果可以看到，ts 字段值的格式并不是标准的时间格式，所以需要对其进行格式转换处理，代码如下：

```
import org.apache.spark.sql.functions._

//增加一个新的列 ts_datetime
val data2 = data
    .withColumn("ts_datetime",to_timestamp(from_unixtime($"ts")))

data2.printSchema()
data2.show()
```

执行以上代码，输出内容如下：

```
root
 |-- userid: integer (nullable = true)
 |-- itemid: integer (nullable = true)
 |-- category: integer (nullable = true)
 |-- behavior: string (nullable = true)
 |-- ts: long (nullable = true)
 |-- ts_datetime: timestamp (nullable = true)


+-------+-------+--------+--------+----------+-------------------+
| userid| itemid|category|behavior|        ts|        ts_datetime|
+-------+-------+--------+--------+----------+-------------------+
|      1|3911125|  982926|      pv|1512044958|2017-11-30 20:29:18|
|    100|3658601| 2342116|      pv|1511757958|2017-11-27 12:45:58|
|   1000|1186274| 2939262|     fav|1511744330|2017-11-27 08:58:50|
|   1000|1925478|  171529|      pv|1511944345|2017-11-29 16:32:25|
|1000011| 106154| 4339722|      pv|1511581267|2017-11-25 11:41:07|
|1000011| 453400| 4756105|      pv|1511948184|2017-11-29 17:36:24|
| 100002| 324802| 2096639|      pv|1511705000|2017-11-26 22:03:20|
| 100002|2968522| 4806751|      pv|1511862934|2017-11-28 17:55:34|
```

```
|1000040|3381470|  4145813|        pv|1511596534|2017-11-25 15:55:34|
|1000040|2044513|   570735|        pv|1511763134|2017-11-27 14:12:14|
|1000040|1371392|  1575622|        pv|1511882223|2017-11-28 23:17:03|
|1000040|3163184|  1575622|        pv|1511882497|2017-11-28 23:21:37|
|1000040|1783293|  1575622|        pv|1511882507|2017-11-28 23:21:47|
|1000040|   8592|  4756105|        pv|1512063066|2017-12-01 01:31:06|
|1000040|5155007|  4145813|        pv|1512202659|2017-12-02 16:17:39|
|1000045|3199716|  3607361|        pv|1511595415|2017-11-25 15:36:55|
|1000054|4138494|  4874384|        pv|1512013826|2017-11-30 11:50:26|
|1000054|4763785|  4181361|        pv|1512083033|2017-12-01 07:03:53|
|1000059|4541139|   801221|        pv|1511622625|2017-11-25 23:10:25|
|1000059|2761782|  4756105|        pv|1511792872|2017-11-27 22:27:52|
+-------+-------+---------+----------+----------+-------------------+
only showing top 20 rows
```

下面通过在 data2 数据集上调用 describe()方法执行描述性统计分析,代码如下:

```
data2.describe().show()
```

执行以上代码,输出内容如下:

```
+-------+------------------+------------------+------------------+--------+-------------------+
|summary|            userid|            itemid|          category|behavior|                 ts|
+-------+------------------+------------------+------------------+--------+-------------------+
|  count|           1002020|           1002020|           1002020| 1002020|            1002020|
|   mean| 507810.5550448095|2581137.6960210376|2696309.0145216663|    null|1.511948747697995E9|
| stddev| 294243.9926429888|1488145.6014356988|1464295.5803918217|    null|  6339425.696395243|
|    min|                 1|                14|                80|     buy|        -2134949234|
|    max|           1018009|           5163067|           5161669|      pv|         1761303078|
+-------+------------------+------------------+------------------+--------+-------------------+
```

从以上输出内容可以看出,所有字段的 count 统计值均为 1 002 020,这意味着所有字段都没有缺失值,因为描述性统计的 count 统计不会统计 null 值。

接下来,查看数据集中所有行为记录覆盖的日期时间范围,代码如下:

```
data2.select(min("ts_datetime"), max("ts_datetime")).show()
```

执行以上代码,输出内容如下:

```
+-------------------+-------------------+
|   min(ts_datetime)|   max(ts_datetime)|
+-------------------+-------------------+
|1902-05-08 06:32:46|2025-10-24 18:51:18|
+-------------------+-------------------+
```

从输出结果可以看到,数据集用户行为记录时间存在异常值。数据集时间跨度为 2017 年 11 月 25 日至 2017 年 12 月 3 日,不在这一时间段的数据应被视作异常值,直接过滤掉,代码如下:

```
val data3 = data2
    .where($"ts_datetime">="2017-11-25")
    .where($"ts_datetime"<"2017-12-04")

println(data3.count)
```

执行以上代码，输出内容如下：

```
1001460
```

从输出内容可以得知，共过滤掉了异常记录数 560 条（1002020 – 1001460）。再次检查 data3 中数据集涵盖的日期时间范围，代码如下：

```
data3.select(min("ts_datetime"), max("ts_datetime")).show()
```

执行以上代码，输出内容如下：

```
+-------------------+-------------------+
|   min(ts_datetime)|   max(ts_datetime)|
+-------------------+-------------------+
|2017-11-25 00:00:00|2017-12-03 23:59:58|
+-------------------+-------------------+
```

从输出内容可以看出，过滤掉异常记录后的数据集包含共 9 天的用户行为数据。

接下来检查一下数据集中是否有重复的记录，代码如下：

```
//查看是否有重复数据
val countRecord = data3.distinct.count
println("去重后的记录数量: " + countRecord)
```

执行以上代码，输出内容如下：

```
去重后的记录数量: 1001460
```

从输出结果可以得知，数据集中并不包含重复的数据记录。

最后，统计一下用户不同行为类型的数量，代码如下：

```
//查看用户行为分类
data3.groupBy("behavior").count.show()
```

执行以上代码，输出内容如下：

```
+--------+------+
|behavior| count|
+--------+------+
|     buy| 19967|
|     fav| 28965|
|    cart| 55636|
|      pv|896892|
+--------+------+
```

从输出内容中可以看出，用户行为类型共包含 4 种：buy、fav、cart 和 pv，其中 pv 行为（用户浏览页面）占了大部分。

因为后面在进行分析时，需要按天进行统计或按小时进行统计，所以需要对数据集进行预处理操作，增加两个新的属性字段，分别代表日期（年、月、日）和小时。从 ts_datetime 列分别抽取日期和时间，添加到这两个新列 ts_date 和 ts_hour 中，代码如下：

```
val df = data3
    .withColumn("ts_date", to_date($"ts_datetime"))
    .withColumn("ts_hour", hour($"ts_datetime"))

df.printSchema()
df.show()
```

执行以上代码，输出内容如图 5-30 所示。

```
root
 |-- userid: integer (nullable = true)
 |-- itemid: integer (nullable = true)
 |-- category: integer (nullable = true)
 |-- behavior: string (nullable = true)
 |-- ts: long (nullable = true)
 |-- ts_datetime: timestamp (nullable = true)
 |-- ts_date: date (nullable = true)
 |-- ts_hour: integer (nullable = true)

+-------+-------+--------+--------+----------+-------------------+----------+-------+
| userid| itemid|category|behavior|        ts|        ts_datetime|   ts_date|ts_hour|
+-------+-------+--------+--------+----------+-------------------+----------+-------+
|      1|3911125|  982926|      pv|1512044958|2017-11-30 20:29:18|2017-11-30|     20|
|    100|3658601| 2342116|      pv|1511757958|2017-11-27 12:45:58|2017-11-27|     12|
|   1000|1186274| 2939262|     fav|1511744330|2017-11-27 08:58:50|2017-11-27|      8|
|   1000|1925478|  171529|      pv|1511944345|2017-11-29 16:32:25|2017-11-29|     16|
|1000011| 106154| 4339722|      pv|1511581267|2017-11-25 11:41:07|2017-11-25|     11|
|1000011| 453400| 4756105|      pv|1511948184|2017-11-29 17:36:24|2017-11-29|     17|
| 100002| 324802| 2096639|      pv|1511705000|2017-11-26 22:03:20|2017-11-26|     22|
| 100002|2968522| 4806751|      pv|1511862934|2017-11-28 17:55:34|2017-11-28|     17|
|1000040|3381470| 4145813|      pv|1511596534|2017-11-25 15:55:34|2017-11-25|     15|
|1000040|2044513|  570735|      pv|1511763134|2017-11-27 14:12:14|2017-11-27|     14|
|1000040|1371392| 1575622|      pv|1511882223|2017-11-28 23:17:03|2017-11-28|     23|
|1000040|3163184| 1575622|      pv|1511882497|2017-11-28 23:21:37|2017-11-28|     23|
|1000040|1783293| 1575622|      pv|1511882507|2017-11-28 23:21:47|2017-11-28|     23|
|1000040|   8592| 4756105|      pv|1512063066|2017-12-01 01:31:06|2017-12-01|      1|
|1000040|5155007| 4145813|      pv|1512202659|2017-12-02 16:17:39|2017-12-02|     16|
|1000045|3199716| 3607361|      pv|1511595415|2017-11-25 15:36:55|2017-11-25|     15|
|1000054|4138494| 4874384|      pv|1512013826|2017-11-30 11:50:26|2017-11-30|     11|
|1000054|4763785| 4181361|      pv|1512083033|2017-12-01 07:03:53|2017-12-01|      7|
|1000059|4541139|  801221|      pv|1511622625|2017-11-25 23:10:25|2017-11-25|     23|
|1000059|2761782| 4756105|      pv|1511792872|2017-11-27 22:27:52|2017-11-27|     22|
+-------+-------+--------+--------+----------+-------------------+----------+-------+
only showing top 20 rows
```

图 5-30　从 ts_datetime 列提取两个新的属性列

至此，已经完成了数据探索和预处理任务。本阶段最后一步，将 df 注册为临时视图，以便接下来应用 Spark SQL 进行业务分析，代码如下：

```
df.createOrReplaceTempView("user_behaviors")
```

执行以上代码，df 被注册到一个名为 user_behaviors 的临时视图中。

5.12.4　平台流量分析

最常见的分析指标是电商平台的基本指标：UV 分析和 PV 分析。

1. 分日 UV/PV 分析

首先分析平台的每日 UV 数和 PV 数，其中 PV 数根据用户行为字段 behavior='pv'进行统计，代码如下：

```
val sql = """
   select ts_date as `日期`,
          count(distinct userid) `访客数UV`,
              sum(case when behavior='pv' then 1 else 0 end) `单击数PV`
    from user_behaviors
    group by ts_date
    order by `日期`
"""

spark.sql(sql).show()
```

执行以上代码,输出内容如下:

```
+----------+-------+-------+
|      日期|访客数UV|单击数PV|
+----------+-------+-------+
|2017-11-25|  88266|  93762|
|2017-11-26|  90372|  95706|
|2017-11-27|  86232|  90543|
|2017-11-28|  84427|  88168|
|2017-11-29|  88200|  92459|
|2017-11-30|  88999|  93271|
|2017-12-01|  92633|  97539|
|2017-12-02| 116441| 122852|
|2017-12-03| 116120| 122592|
+----------+-------+-------+
```

为了更清楚地观察每日 UV/PV 数的变化趋势,可视化这个查询结果。在 Zeppelin Notebook 中应用%sql 解释器可以直接将 Spark SQL 的查询结果可视化,执行过程和结果如图 5-31 所示。

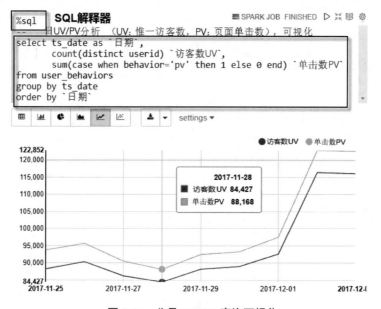

图 5-31　分日 UV/PV 查询可视化

从分析结果来看，可以得出以下结论：

（1）日 PV 数和日 UV 数呈现明显的周期趋势，工作日（11.26-12.1）相较周末（11.24-11.26，12.2-12.3）两指标均出现一定下滑。

（2）12.2-12.3 期间日 UV 数和日 PV 数大幅上涨，推测是电商开启了"双十二"预热活动，带来大量流量涌入。

2. 分时 UV/PV 分析

为了分析平台一天中的流量变化规律，接下来按小时统计 UV/PV 数量，代码如下：

```
val sql = """
    select ts_hour `小时`,
        count(distinct userid) `访客数`,
        sum(case when behavior='pv' then 1 else 0 end) `单击数`
    from user_behaviors
    group by ts_hour
    order by `小时`
"""

spark.sql(sql).show(24)
```

执行以上代码，输出内容如下：

```
+----+------+------+
|小时|访客数|单击数|
+----+------+------+
|   0| 30063| 30449|
|   1| 14129| 14271|
|   2|  7626|  7613|
|   3|  5137|  5079|
|   4|  4510|  4503|
|   5|  5969|  5892|
|   6| 12622| 12289|
|   7| 22782| 22385|
|   8| 31160| 30494|
|   9| 38236| 37116|
|  10| 44632| 43418|
|  11| 43240| 41935|
|  12| 43616| 42504|
|  13| 47262| 46489|
|  14| 47183| 46374|
|  15| 48891| 48257|
|  16| 46944| 46018|
|  17| 43202| 42181|
|  18| 43992| 43330|
|  19| 54198| 54353|
|  20| 65302| 66232|
|  21| 73577| 75676|
|  22| 72183| 74246|
|  23| 54781| 55788|
+----+------+------+
```

为了更清楚地观察分时 UV/PV 数的变化趋势，可视化这个查询结果。执行过程和结果如图 5-32 所示。

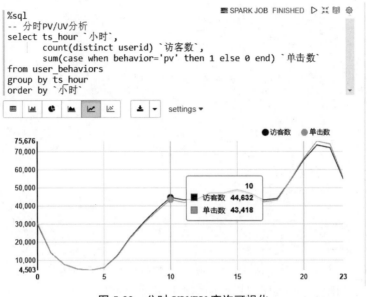

图 5-32　分时 UV/PV 查询可视化

从分析结果来看，可以得出以下结论：

（1）18:00—0:00 为电商平台活跃时段，PV 数和 UV 数都明显增加。

（2）日间网站访问量相对比较平均，夜间 21:00—22:00 会出现访问高峰，PV 数可增长到日间的 150%左右，网站应注意服务器资源的时间分配。

5.12.5　用户行为分析

用户行为分析包括用户跳出率分析、用户购买相关行为分析和复购行为分析。

1. 用户跳出率分析

跳出率是指在只访问了入口页面（如网站首页）就离开的访问量与所产生总访问量的百分比。通俗地讲，跳出率指用户通过搜索关键词来到你的网站，仅浏览了一个页面就离开的访问次数与所有访问次数的百分比。

观察跳出率可以得知用户对网站内容的认可，或者说网站是否对用户有吸引力，而网站的内容是否能够对用户有所帮助（如留住用户）也直接可以在跳出率中看出来，所以跳出率是衡量网站内容质量的重要标准。

跳出率计算公式：跳出率=访问一个页面后离开网站的次数/总访问次数。对跳出率进行分析，代码如下：

```
val sql = """
   select `总用户数`, `只访问一次页面数`, concat(round((`只访问一次页面数` * 100) /
(`总用户数`),2), "%") `跳出率`
     from (select count(pv_cnt) as `只访问一次页面数`
           from (select userid, count(userid) as pv_cnt
                 from user_behaviors
                 where behavior='pv'
                 group by userid
                 having pv_cnt=1)
          ) a CROSS JOIN (select count(distinct userid) as `总用户数` from user_
behaviors) b
"""

spark.sql(sql).show()
```

执行以上代码，输出内容如下：

```
+--------+----------------+------+
|总用户数|只访问一次页面数|跳出率|
+--------+----------------+------+
|  528254|          269538|51.02%|
+--------+----------------+------+
```

从输出内容可以看出，在测试的约 100 万条数据中，跳出率达到了 51.02%，超过了一半，因此营销设计部门有必要研究提升网站的用户黏性问题。

2. 用户购买相关行为分析

统计用户每天的行为量，如每天的购买量、加购物车量和收藏量等指标，可以分析用户的购买行为趋势。对用户日行为量进行统计，代码如下：

```
val sql = """
   select ts_date as `日期`,
          sum(case when behavior='buy' then 1 else 0 end) as `购买`,
          sum(case when behavior='cart' then 1 else 0 end) as `加购物车`,
          sum(case when behavior='fav' then 1 else 0 end) as `收藏`
     from user_behaviors
     group by `日期`
     order by `日期`
"""

spark.sql(sql).show()
```

执行以上代码，输出内容如下：

```
+----------+----+--------+----+
|      日期|购买|加购物车|收藏|
+----------+----+--------+----+
|2017-11-25|2005|    5608|3012|
|2017-11-26|2042|    5831|3027|
|2017-11-27|2188|    5487|2913|
|2017-11-28|2193|    5222|2920|
|2017-11-29|2257|    5550|3053|
```

```
|2017-11-30|2187|        5740|3023|
|2017-12-01|2051|        6279|3146|
|2017-12-02|2538|        8102|3905|
|2017-12-03|2506|        7817|3966|
+----------+----+------------+----+
```

为了更清楚地观察用户日行为量的变化趋势，可视化这个查询结果。执行过程和结果如图 5-33 所示。

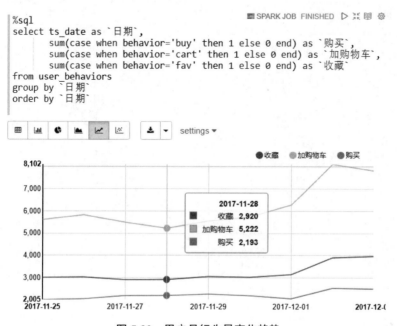

图 5-33　用户日行为量变化趋势

从分析结果来看，可以得出以下结论：

（1）加入购物车这一行为与日 PV 数趋势基本一致，但收藏商品行为和购买行为的变化幅度明显较小。

（2）活动期间（12.2-12.3）加入购物车和收藏行为明显增多，但购买行为变化并不显著。推测该活动只是为"双 12"提前宣传，鼓励大家提前选择商品，等到"双 12"再进行购买。

同样，可以按小时统计用户的行为量，以观察用户在一天中的购买行为方式。对分时行为量进行统计，代码如下：

```
//分时行为量
val sql = """
    select ts_hour `小时`,
         sum(case when behavior='buy' then 1 else 0 end) as `购买`,
         sum(case when behavior='cart' then 1 else 0 end) as `加购物车`,
         sum(case when behavior='fav' then 1 else 0 end) as `收藏`
```

```
    from user_behaviors
    group by `小时`
    order by `小时`
"""

spark.sql(sql).show(24)
```

执行以上代码，输出内容如下：

```
+----+----+--------+----+
|小时|购买|加购物车|收藏|
+----+----+--------+----+
|   0| 536|    1913|1078|
|   1| 224|     802| 521|
|   2| 131|     423| 260|
|   3|  72|     288| 200|
|   4|  53|     285| 142|
|   5|  78|     374| 212|
|   6| 183|     879| 424|
|   7| 319|    1429| 746|
|   8| 628|    1936|1009|
|   9| 991|    2337|1299|
|  10|1267|    2800|1513|
|  11|1241|    2577|1462|
|  12|1193|    2562|1396|
|  13|1255|    2762|1476|
|  14|1211|    2767|1547|
|  15|1176|    2900|1516|
|  16|1165|    2813|1469|
|  17| 951|    2647|1367|
|  18| 946|    2488|1339|
|  19|1153|    3082|1611|
|  20|1403|    3937|1899|
|  21|1401|    4771|2211|
|  22|1355|    4951|2335|
|  23|1035|    3913|1933|
+----+----+--------+----+
```

为了更清楚地观察用户分时行为量的变化趋势，可视化这个查询结果。执行过程和结果如图 5-34 所示。

从分析结果来看，可以得出以下结论：

（1）加入购物车和收藏商品行为在夜间（21:00—22:00）存在高峰，但购买行为与日间没有明显区别。

（2）晚间用户更倾向于浏览商品，日间使用时具有明显的目的性，购买意愿相对强烈。

接下来，分析访问用户每日购买次数，代码如下：

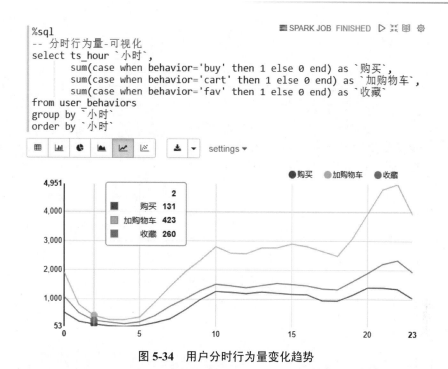

图 5-34　用户分时行为量变化趋势

```
%sql
-- 访问用户日均购买次数
select ts_date as `日期`,
      sum(case when behavior='buy' then 1 else 0 end)/count(distinct userid)
as `访问用户日均购买次数`
from user_behaviors
group by ts_date
order by ts_date
```

执行以上代码，输出结果如图 5-35 所示。

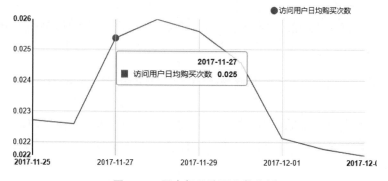

图 5-35　用户每日购买次数分析

再分析付费用户（有过购买行为的用户）的日购买次数，代码如下：

```sql
%sql
-- 付费用户（有过购买行为的用户）日购买次数
select ts_date as `日期`,
       sum(case when behavior='buy' then 1 else 0 end)/count(distinct userid)
as `付费用户日均购买次数`
from user_behaviors
where behavior='buy'
group by ts_date
order by ts_date
```

执行以上代码，输出结果如图 5-36 所示。

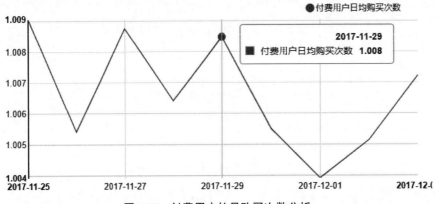

图 5-36 付费用户的日购买次数分析

从分析结果来看，可以得出以下结论：

（1）每天下单购买的客户里人均大约会购买 1 件商品，可以进行"2 件 9 折""满减优惠"等活动刺激客户增加购买量。

（2）日购买次数与日 UV 数和 PV 数出现了负相关，工作日网站流量虽然下降，但人均购买量却有所提升，推测存在"购买冷静期"，即用户倾向于周末选择商品，周中逐渐下定决心购买。

（3）工作日访问客户购买意愿较强，可以提供相应的小幅优惠以鼓励下单购买。

3. 复购行为分析

复购的概念是什么？简单来讲就是用户买了一个商品觉得还不错，再次下单就叫复购。

做电商复购是非常重要的，因为推广需要付出高额的广告费，而复购则省去了广告成本，所以有时复购带来的利润是非常高的，因此复购也是电商用户行为分析中的一个重要的分析指标。

复购率公式：复购率 = 本时段购买次数大于 2 次的用户数/本时段进行过购买的用户数。计算复购率的代码如下：

```
val sql = """
    select concat(round(count(case when buy_times>1 then 1 else null end)*100/
count(*),2),"%") as `复购率`
    from (
        select userid, count(userid) as buy_times
        from user_behaviors
        where behavior='buy'
        group by userid
    )
"""

spark.sql(sql).show()
```

执行以上代码，输出内容如下：

```
+------+
|复购率|
+------+
| 2.14%|
+------+
```

可以看出，本时段（9 天时间）网站复购率仅为 2.14%，说明付费用户的忠诚度相对较低。

注意：因为本案例所使用的数据集为原始数据集的 1%抽样子集，所以分析结果不代表真实结果。这同样适用于其他分析指标。

接下来分析复购频次。复购频次指的是复购一次、复购两次、复购 N 次的数量统计，代码如下：

```
%sql
-- //复购频数
select buy_times,count(*) as `复购频数`
from (
    select userid, count(userid) as buy_times
    from user_behaviors
    where behavior='buy'
    group by userid
)
group by buy_times
order by buy_times
```

执行以上代码，输出结果如图 5-37 所示。

从分析结果来看，可以得出以下结论：

（1）多数用户本时段购买次数在 1 次，复购 2 次以上的客户比例较低，说明平台黏性较小，不足以吸引客户多次购物。

（2）部分高购买次数用户可能是"剁手党"，但也有可能是刷单账号，需要进行进一步识别。

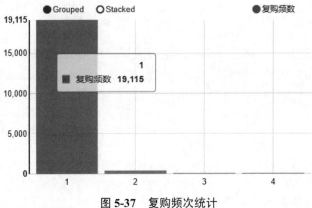

图 5-37 复购频次统计

接下来统计用户的复购间隔时间。复购间隔时间越短,说明用户的购买意愿越强烈,对网站商品越认可,代码如下:

```sql
%sql
-- 复购时间间隔
select buy_interval, count(*) as freq
from (
   select a.userid, datediff(a.ts_date, b.min_date) as buy_interval from user_behaviors as a,
   (select userid, min(ts_date) as min_date, count(userid) as buy_times from user_behaviors where behavior='buy' group by userid having count(*)>1) as b
where a.userid=b.userid and a.behavior='buy') as c
group by buy_interval
order by buy_interval
```

执行以上代码,输出结果如图 5-38 所示。

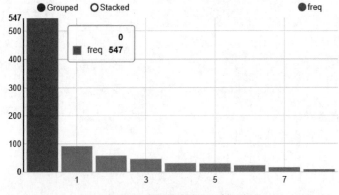

图 5-38 用户的复购间隔时间统计

从统计结果来看,多数用户会在一天内多次购买,相对来讲,复购时间间隔在 1~5 日均有分布,5 日后复购人数已经明显减少,商家应注重一周内对于客户的持续宣传。

5.12.6 转化漏斗分析

漏斗模型是一套流程式数据分析模型，它能够科学地反映用户行为状态及从起点到终点各阶段用户转化率情况，是一种重要的分析模型。漏斗分析模型已经广泛应用于网站和App用户行为分析的流量监控、电商行业、零售的购买转化率、产品营销和销售等日常数据运营与数据分析的工作中。

1. 行为漏斗分析

行为漏斗分析指的是分析用户行为的转化，从浏览→加购物车→收藏→购买的路径分析各环节的转化情况，代码如下：

```
%sql
-- 行为漏斗分析
select behavior, count(behavior) behavior_cnt
from user_behaviors
group by behavior
order by behavior_cnt desc
```

执行以上代码，输出结果如图 5-41 所示。

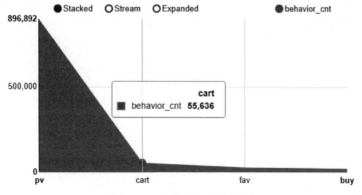

图 5-39　购物行为漏斗分析

因为 Zeppelin Notebook 没有提供漏斗图，所以这里使用面积图来表示用户行业转化。为了计算各环节的转化率，进一步利用窗口函数进行计算，代码如下：

```
val sql = """
    select behavior, behavior_cnt, lag_behavior_cnt, round(behavior_cnt/lag_behavior_cnt, 2) as convert_rate
    from(
        select behavior, behavior_cnt,
            lag(behavior_cnt,1) OVER (PARTITION BY uid ORDER BY behavior_cnt desc) as lag_behavior_cnt
        from (select 1 as uid, behavior, count(behavior) behavior_cnt
            from user_behaviors
            group by behavior)
```

```
        )
    """

spark.sql(sql).show()
```

执行以上代码，输出内容如下：

```
+--------+------------+----------------+------------+
|behavior|behavior_cnt|lag_behavior_cnt|convert_rate|
+--------+------------+----------------+------------+
|      pv|      896892|            null|        null|
|    cart|       55636|          896892|        0.06|
|     fav|       28965|           55636|        0.52|
|     buy|       19967|           28965|        0.69|
+--------+------------+----------------+------------+
```

以上统计的结果是各个环节相对于上一环节的转化率。如果要统计各个行为相对于浏览用户的转化率，则可对代码进行修改，修改后的代码如下：

```
val sql = """
    select behavior, behavior_cnt, max_behavior_cnt, round(behavior_cnt/max_behavior_cnt, 2) as convert_rate
    from(
        select behavior, behavior_cnt,
            max(behavior_cnt) OVER (PARTITION BY uid ORDER BY behavior_cnt desc) as max_behavior_cnt
            from (select 1 as uid, behavior, count(behavior) behavior_cnt
                from user_behaviors
                group by behavior)
    )
"""

spark.sql(sql).show()
```

执行以上代码，输出内容如下：

```
+--------+------------+----------------+------------+
|behavior|behavior_cnt|max_behavior_cnt|convert_rate|
+--------+------------+----------------+------------+
|      pv|      896892|          896892|         1.0|
|    cart|       55636|          896892|        0.06|
|     fav|       28965|          896892|        0.03|
|     buy|       19967|          896892|        0.02|
+--------+------------+----------------+------------+
```

从分析结果来看，可以得出以下结论：

（1）只有0.06%的点击行为最终转变成了加购行为，平均1500次~2000次点击才会促成一次加购物车操作。

（2）只有0.02%的点击行为最终转变成了购买行为，平均5000次点击才会促成一次购买交易。

（3）加入购物车/收藏商品行为向购买行为的转化率相对较高，说明用户对于购物车/收藏单的商品具有较强的购买欲望，可以重点关注这些商品，加大宣传力度。

2. 用户漏斗分析

同样使用漏斗分析方法，分析各个行为转化间的用户数，代码如下：

```sql
%sql
-- 用户漏斗分析
select behavior,count(distinct userid) as user_freq
from user_behaviors
group by behavior
order by user_freq desc
```

执行以上代码，输出结果如图 5-40 所示。

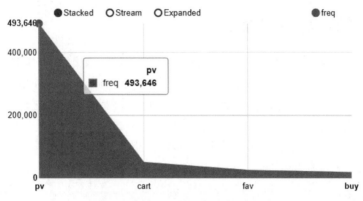

图 5-40　用户购物漏斗分析

计算各个行为间用户的转化率，代码如下：

```
val sql = """
    select behavior, user_freq, lag_user_freq, round(user_freq/lag_user_freq,
2) as user_rate
    from(
        select behavior, user_freq,
            lag(user_freq,1) OVER (PARTITION BY uid ORDER BY user_freq
desc) as lag_user_freq
        from (select 1 as uid, behavior, count(distinct userid) as user_freq
            from user_behaviors
            group by behavior)
    )
"""

spark.sql(sql).show()
```

执行以上代码，输出的结果如下：

```
+--------+---------+-------------+---------+
|behavior|user_freq|lag_user_freq|user_rate|
+--------+---------+-------------+---------+
|      pv|   493646|         null|     null|
|    cart|    51654|       493646|      0.1|
```

```
|      fav|   26204|         51654|    0.51|
|      buy|   19534|         26204|    0.75|
+---------+--------+--------------+--------+
```

从分析结果来看，可以得出以下结论：

（1）平台单击用户中最终有 10%的用户成为购买用户，转化率一般，说明网站的商品还需要进一步优化，以满足绝大多数用户的购买需要，促使用户转换为消费者。

（2）结合行为漏斗，单击行为只有 0.06%变为购买行为，说明多数用户需要浏览大量的商品页面才会完成一次购买行为，网站的推荐系统不够精确，此环节是影响转化率的重点，应进行针对性优化，实现精准推荐，减少用户搜寻心仪商品的时间成本。

5.12.7　用户 RFM 价值分析

RFM 模型是衡量客户价值和客户创利能力的重要工具。该模型通过客户的最近交易行为（Recency）、交易频率（Frequency）及交易金额（Monetary）三项指标来描述该客户的价值状况。一般来讲，会将这 3 项指标分成几个区间进行评分，通过计算评分找到有价值的用户，并对用户进行分类。这 3 项指标的说明如下。

（1）最近一次消费：是指最近一次消费距离上一次消费之间的时间长短。它反映了客户对产品的态度及对品牌价值的信任度，它关乎消费者的存留状况。

（2）消费频率：是指某个特定时间内消费的次数。它直接反映了客户的忠诚度，消费频率越高，忠诚度就越高；忠诚度越高的客户数量越多，公司的竞争优势就越强，市场份额就越大。

（3）消费金额：是指一定时间内消费某产品的金额。它反映的是客户的价值，价值越高，给公司创造的利益就更大。

RFM 模型如图 5-41 所示。

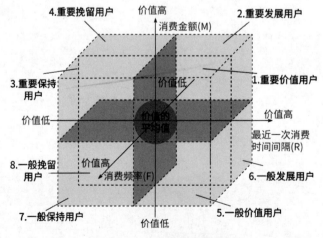

图 5-41　RFM 客户价值模型

因为本案例数据源里没有与金额相关的信息，所以只能通过 R 和 F 来对客户价值进行评分。

（1）R：用户最近一次的购买时间到 12 月 3 日的时间差表示用户最近一次消费间隔。

（2）F：用户在 11 月 25 日到 12 月 3 日购买的次数表示用户消费频率。

（3）M：本次数据集未包含相关字段，故不考虑。

参照 RFM 模型，对用户进行分类，找出有价值的用户。这里只需计算近度 R 和频度 F，代码如下：

```
//计算近度和频度
val sql = """
    select userid, datediff('2017-12-03',max(ts_date)) as Recency, count(userid) as Frequency
    from user_behaviors
    where behavior='buy'
    group by userid
"""

spark.sql(sql).show()
```

执行以上代码，输出内容如下：

```
+------+-------+---------+
|userid|Recency|Frequency|
+------+-------+---------+
|946380|      5|        1|
|987552|      4|        1|
|991295|      2|        1|
|995239|      5|        1|
|112971|      3|        1|
|145095|      3|        1|
|163046|      6|        1|
|166150|      3|        1|
|167532|      2|        1|
|231421|      4|        1|
|418573|      0|        1|
| 42635|      3|        1|
|453185|      8|        1|
|455910|      3|        1|
|460056|      1|        1|
|466710|      2|        1|
|521745|      3|        1|
|555859|      4|        1|
|579502|      7|        1|
|705988|      8|        1|
+------+-------+---------+
only showing top 20 rows
```

基于 R 和 F，划分 R 与 F 等级。以平均数为划分界限，将 Recency 和 Frequency 划分为高低两个等级，代码如下：

```
val sql = """
   with RFdata as
    (select userid, datediff('2017-12-03',max(ts_date)) as Recency, count(userid) as Frequency
     from user_behaviors
     where behavior='buy'
     group by userid)
   select userid,
       (case when Recency < (select avg(Recency) from RFdata) then 0 else 1 end) as R,
       (case when Frequency < (select avg(Frequency) from RFdata) then 0 else 1 end) as F
    from RFdata
"""

spark.sql(sql).show()
```

执行以上代码，输出内容如下：

```
+-------+---+---+
| userid|  R|  F|
+-------+---+---+
|1007334|  1|  0|
| 157793|  1|  0|
|  18498|  0|  0|
| 238193|  1|  0|
| 245534|  0|  0|
| 246115|  1|  0|
| 256761|  0|  0|
| 284449|  1|  0|
| 315781|  1|  0|
| 338794|  1|  0|
| 438386|  0|  0|
| 473385|  0|  0|
| 556576|  1|  0|
| 566063|  1|  0|
| 689665|  1|  0|
| 692396|  1|  0|
| 693448|  1|  0|
| 723845|  0|  0|
| 757731|  0|  0|
| 791874|  0|  0|
+-------+---+---+
only showing top 20 rows
```

建立用户分类打分规则。用户分类规则见表 5-10。

表5-10 用户分类规则表

R	F	客户类型	备注
0	0	挽留客户	可能流失或将要流失的客户
0	1	重要唤回客户	一段时间未活跃的忠诚客户

R	F	客户类型	备注
1	0	重要发展客户	值得发展的新活跃用户
1	1	高价值客户	高活跃高消费次数的理想型客户

根据用户分类规则表,对客户价值进行细分,得出用户价值分类,代码如下:

```sql
%sql
with RFdata as
(select userid, datediff('2017-12-03',max(ts_date)) as Recency, count(userid) as Frequency
    from user_behaviors
    where behavior='buy'
    group by userid)
select
    count(case when R=0 and F=0 then userid end) as `挽回客户`,
    count(case when R=0 and F=1 then userid end) as `重要唤回客户`,
    count(case when R=1 and F=0 then userid end) as `重要发展客户`,
    count(case when R=1 and F=1 then userid end) as `高价值客户`
from (
    select userid,
        (case when Recency < (select avg(Recency) from RFdata) then 0 else 1 end) as R,
        (case when Frequency < (select avg(Frequency) from RFdata) then 0 else 1 end) as F
    from RFdata
)
```

执行以上代码,输出结果如图 5-42 所示。

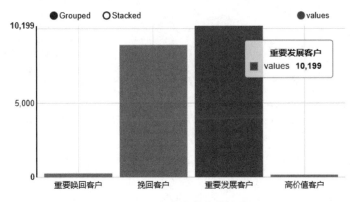

图 5-42　电商用户价值分类

针对不同价值客户,可采用不同营销手段。

(1)挽回客户:调查用户流失原因,优化转化漏斗。

(2)重要唤回客户:采用邮件/短信推送等主动联系方式,争取客户返回平台活跃购物。

(3)重要发展客户:提供价格刺激,增加购买频率。

（4）高价值客户：重点关注并维护，提升服务质量。

注意：由于本数据集统计时间较短，对客户的分类可能不够精确，但数据处理和分类过程仍具有一定参照意义，对于大时间跨度样本依然适用。

5.12.8 推荐效果分析

对于类似淘宝这样的电商平台来讲，用户的浏览、购买等行为严重依赖于网站的推荐系统所做出的推荐和引导，因此，从不同维度分析商品类目和商品的浏览量（PV）、购买量和转化率（CVR）等指标，对于评估网站推荐系统的有效性有着比较重要的意义。

1. 商品类目分析

统计商品类目的浏览量（PV）、购买量和转化率（CVR）。这里重点统计转化率超过1的商品类目，代码如下：

```
val sql = """
   with category_view as
   (select category,
        count(case when behavior='pv' then category end) as pv,
        count(case when behavior='buy' then category end) as buy_times,
        round(count(case when behavior='buy' then category end)/count(case when behavior='pv' then category end),1) as cvr_rate,
        concat(round(count(case when behavior='buy' then category end)*100/count(case when behavior='pv' then category end),1),'%') as CVR
    from user_behaviors
    group by category)
   select category, pv, buy_times, cvr_rate, CVR
   from category_view
   where cvr_rate>1
   order by cvr_rate desc
"""

spark.sql(sql).show()
```

执行以上代码，输出内容如下：

```
+--------+---+---------+--------+-------+
|category| pv|buy_times|cvr_rate|    CVR|
+--------+---+---------+--------+-------+
| 1781126|  1|       11|    11.0|1100.0%|
| 2923285|  1|       10|    10.0|1000.0%|
| 5047668|  1|        5|     5.0| 500.0%|
| 1404020|  6|       19|     3.2| 316.7%|
| 3901193|  5|       12|     2.4| 240.0%|
| 2589085|  1|        2|     2.0| 200.0%|
| 2460565|  4|        5|     1.3| 125.0%|
|  257793| 13|       15|     1.2| 115.4%|
| 1421972| 13|       14|     1.1| 107.7%|
+--------+---+---------+--------+-------+
```

从分析结果来看，可以得出以下结论：

（1）25%的商品类别贡献了80%的销量，基本上符合"二八定律"。

（2）即便是淘宝这样品类齐全的电商平台，多数用户购买商品的类别只会占全部类别的 1/4～1/5，平台应据此确定自身定位，对于这些种类的商品应进行宣传资源的倾斜，提高用户的购买可能。

2. 商品分析

统计所有商品的浏览量(PV)、购买量和转化率(CVR)，代码如下：

```
val sql = """
   with item_view as
   (select itemid,
   count(case when behavior='pv' then itemid end) as pv,
   count(case when behavior='buy' then itemid end) as buy_times,
   round(count(case when behavior='buy' then itemid end)/count(case when behavior='pv' then itemid end),1) as cvr_rate,
   concat(round(count(case when behavior='buy' then itemid end)*100/count(case when behavior='pv' then itemid end),1),'%') as CVR
   from user_behaviors
   group by itemid
   )
   select itemid, pv, buy_times, cvr_rate, CVR
   from item_view
"""

spark.sql(sql).show()
```

执行以上代码，输出内容如下：

```
+-------+---+---------+--------+-----+
| itemid| pv|buy_times|cvr_rate|  CVR|
+-------+---+---------+--------+-----+
|4064224|  0|        0|    null| null|
|1960493|  4|        0|     0.0| 0.0%|
|4079445|  1|        0|     0.0| 0.0%|
|3402643|  2|        0|     0.0| 0.0%|
|3517664|  5|        0|     0.0| 0.0%|
|1604293|  5|        0|     0.0| 0.0%|
|4049453|  1|        0|     0.0| 0.0%|
|2044553|  6|        0|     0.0| 0.0%|
|2479072|  3|        0|     0.0| 0.0%|
|4307719|  3|        0|     0.0| 0.0%|
| 241660|  3|        0|     0.0| 0.0%|
|4110375|  7|        2|     0.3|28.6%|
|3906181| 18|        0|     0.0| 0.0%|
|2959274| 19|        0|     0.0| 0.0%|
|4531890|  2|        0|     0.0| 0.0%|
|3351641|  1|        0|     0.0| 0.0%|
| 145095| 11|        0|     0.0| 0.0%|
|3415447|  5|        0|     0.0| 0.0%|
```

```
|4136579|  1|         0|      0.0| 0.0%|
|2939731|  1|         0|      0.0| 0.0%|
+-------+---+----------+---------+-----+
only showing top 20 rows
```

可以看出大量的商品仅有浏览（PV），而没有人购买。

长尾效应的根本就是强调"个性化""客户力量"和"小利润大市场"，也就是要赚每个人很少的钱，但是要赚很多人的钱。当将市场细分到很细很小时，就会发现这些细小市场的累计会带来明显的长尾效应。

注意：长尾效应，英文名称为 Long Tail Effect。"头"（head）和"尾"（tail）是两个统计学名词。正态曲线中间的突起部分叫"头"；两边相对平缓的部分叫"尾"。从人们需求的角度来看，大多数的需求会集中在头部，而这部分可以称为"流行"，而分布在尾部的需求是个性化的、零散的小量的需求，而这部分差异化的、少量的需求会在需求曲线上形成一条长长的"尾巴"，而所谓长尾效应就在于它的数量上，将所有非流行的市场累加起来就会形成一个比流行市场还大的市场。

接下来分析商品销量的"长尾效应"，代码如下：

```
%sql
with item_view as
(select itemid,
 count(case when behavior='pv' then itemid end) as pv,
 count(case when behavior='buy' then itemid end) as buy_times,
 round(count(case when behavior='buy' then itemid end)/count(case when behavior=
'pv' then itemid end),1) as cvr_rate,
 concat(round(count(case when behavior='buy' then itemid end)*100/count(case
when behavior='pv' then itemid end),1),'%') as CVR
 from user_behaviors
 group by itemid
)
select buy_times,count(buy_times) as item_buy_freq
from item_view
group by buy_times
order by buy_times
```

执行以上代码，输出结果如图 5-43 所示。

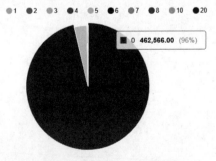

图 5-43 商品销量的"长尾效应"

从分析结果来看，可以得出以下结论：

（1）96%以上的商品无人购买，但是并不代表应该全部下架，其一，数据统计时间过短，某些商品的购买具有季节性，可能在该时间段销量本来就很低；其二，这些商品的存在可以给平台营造一种琳琅满目、应有尽有的购物氛围，并且某些商品本身的作用就是为与主推商品形成对比，促进主推商品销售。

（2）多数商品只被购买过 1~2 次，说明电商平台是依靠大量低销量商品的"长尾效应"发展的，并非是靠少数爆款商品的高销量拉动，因此，平台在关注热销商铺的同时，也应对中小商铺给予照顾与支持，利用"长尾效应"促进市场的均衡发展。

3. 推荐系统有效性分析

面对电商平台海量的商品信息，用户的浏览和购买等行为往往会受到电商平台推荐系统的影响。为了判断推荐系统的有效性，可以分析销量 Top 100 和点击量 Top 100 的商品类目重合度，以及销量 Top 1000 和单击量 Top 1000 商品的重合度，重合度越高，说明电商平台的推荐系统性能越好。

首先分析前销量 Top 100 和点击量 Top 100 的商品类目重合度，代码如下：

```
val sql = """
   with category_view as
   (select category,
          count(case when behavior='pv' then category end) as pv,
          count(case when behavior='buy' then category end) as buy_times,
          round(count(case when behavior='buy' then category end)/count(case when behavior='pv' then category end),1) as cvr_rate,
          concat(round(count(case when behavior='buy' then category end)*100/count(case when behavior='pv' then category end),1),'%') as CVR
    from user_behaviors
    group by category
   )
   select count(a.category)
   from
      (select * from category_view order by pv desc limit 100) as a
      inner join
      (select * from category_view order by buy_times desc limit 100) as b
   on a.category=b.category
"""

spark.sql(sql).show()
```

执行以上代码，输出内容如下：

```
+---------------+
|count(category)|
+---------------+
|             70|
+---------------+
```

从输出结果可以得知，销量 Top 100 有 70 个商品类目同时也在单击量 Top 100 内，这说明平台对于商品类目的推荐相对来讲是比较精确的，网站主推的商品门类大多能满足用户的购买需求。

接下来分析销量 Top 1000 和点击量 Top 1000 商品的重合度，代码如下：

```
val sql = """
   with item_view as
   (select itemid,
    count(case when behavior='pv' then itemid end) as pv,
    count(case when behavior='buy' then itemid end) as buy_times,
    round(count(case when behavior='buy' then itemid end)/count(case when behavior='pv' then itemid end),1) as cvr_rate,
    concat(round(count(case when behavior='buy' then itemid end)*100/count(case when behavior='pv' then itemid end),1),'%') as CVR
    from user_behaviors
    group by itemid
   )
   select count(a.itemid)
   from
     (select * from item_view order by pv desc limit 1000) as a
     inner join
     (select * from item_view order by buy_times desc limit 1000) as b
   on a.itemid=b.itemid
"""

spark.sql(sql).show()
```

执行以上代码，输出内容如下：

```
+--------------+
|count(itemid)|
+--------------+
|          105|
+--------------+
```

从输出结果可以得知，细化到单一商品范围，推荐系统的准确度明显下降，销量和点击量前 1000 位的商品重合度仅有 11% 左右。换言之，大量高曝光度的商品的销量其实很低，网站的推荐系统应该进一步优化，努力提高商品细分领域的推荐精准度。

4. 转化漏斗-推荐系统综合分析

平台转化率损失最大的环节在于浏览商品页(PV)→有购买意向（加购物车/收藏）的转化，这一阶段仅有 0.06% 的单击行为成功完成转化，用户需要浏览大量的商品推荐页才能确定所需的商品，结合推荐系统分析，许多高曝光度的商品销量反而不高，这些商品转移了过多的用户注意力，引起顾客疲倦，降低了购物意愿。

据此，推荐系统是优化转化漏斗的关键，平台应考虑改进推荐算法，优化搜索结果，并且整合页面内容，减少无关信息，突出重点，营造贴心舒适的购物环境。

5.12.9　项目分析总结

本项目从 5 个方面对淘宝平台用户数据进行了较为全面的分析，得出的结论和建议如下：

（1）网站流量高峰集中于周末和夜间，利用这些时段进行直播带货等营销活动吸睛效果最好；"双 12" 预热行为明显带动了网站 PV 数和 UV 数的上升，应注意服务器稳定性和资源的提前分配。

（2）网站用户在工作日白天下单购买意愿更为强烈，购买用户中每人日均购买 1.5 件商品，可以考虑进行多买多减的优惠活动以刺激用户购买；本时段（9 天内）用户复购率达到 66%，说明平台的用户黏性相对较好。

（3）通过 RFM 模型可以对用户进行价值分层，针对不同类型的客户制定专门的营销策略，达到更好的营销效果。

（4）平台商品类目上存在明显的二八定律，80%的用户只会购买 20%左右商品类目下的商品，平台应针对主流商品类目开展重点宣传，确立自身优势商品所在领域。

（5）细分到每件商品而言，平台销量呈现出长尾效应，多数商品只被购买过 1~2 次，说明平台是依靠大量零散销量的小商品蓬勃发展的，应注重对中小商铺的扶持，利用长尾效应扩大市场占有面。

（6）转化漏斗的关键在于商品页 PV 数→形成购买意愿（加购物车/收藏）这一环节，用户大约需要 5000 次单击才能完成一次购买行为，在这一环节大量用户失去耐心放弃购买。结合推荐系统分析可知，平台的推荐系统在大类推荐上效果较好，曝光量与销量匹配度高，但向下延伸至具体商品的推荐上却显得不够精确，大量高曝光度的商品销量较低，转移了用户注意力，进而减弱了购买欲望。平台后续更应该着力与推荐引擎优化，采用更为高效的人工智能算法针对不同用户进行针对性推荐，并预测用户购买概率，以便对商品搜索结果进行排序。

第 6 章　Spark Streaming 流处理

在很多领域，如股市走向分析、气象数据测控、网站用户行为分析等，由于数据产生快，实时性强，数据量大，所以很难统一采集并入库存储后再进行处理，这便导致传统的数据处理架构不能满足需要。流计算的出现，就是为了更好地解决这类数据在处理过程中遇到的问题。与传统架构不同，流计算模型在数据流动的过程中实时地进行捕捉和处理，并根据业务需求对数据进行计算分析，最终把结果保存或者分发给需要的组件。

Apache Spark 的统一数据处理平台受欢迎的一个因素就是能够执行流数据处理和批数据处理。

注意：Spark 流处理引擎有两代：第 1 代流处理引擎基于 RDD，称为 Spark Streaming；第 2 代流处理引擎基于 DataFrame/DataSet，称为 Spark Structured Streaming（Spark 结构化流）。目前 Spark Streaming 处于维护模式，官方的建议是尽可能地使用 Spark Structured Streaming，因此本章不会全面介绍 Spark Streaming，仅介绍其基本概念及简单应用，以期读者对 Spark Streaming 流处理的一些概念有些基本了解即可。

6.1　Spark DStream

第 1 代 Spark 流式处理引擎 Spark Streaming 是在 2012 年引入的，它是 Spark 核心 API 的扩展，支持实时数据流的可扩展、高吞吐量和容错流处理。数据可以从 Kafka、Kinesis 或 TCP Socket 等多种来源获取，也可以通过 map()、reduce()、join() 和 window() 等高级函数表达复杂的算法进行处理。最后，可以将处理过的数据推送到文件系统、数据库和实时仪表板。事实上，用户可以在数据流上应用 Spark 的机器学习和图形处理算法如图 6-1 所示。

图 6-1　Spark Streaming 实时流处理

在内部，Spark Streaming 接收实时的输入数据流，并将数据进行批量处理，再由 Spark 引擎进行处理，最终批量生成结果流。Spark Streaming 的工作原理如图 6-2 所示。

图 6-2　Spark Streaming 工作原理

Spark Streaming 流处理引擎的主要编程抽象称为离散流，即 DStream。它的工作方式是使用微批处理模型将传入的数据流分成批处理，每批都由 RDD 在内部表示，然后由 Spark 批处理引擎处理，如图 6-3 所示。

图 6-3　Spark DStream 离散流

6.2　Spark 流处理示例

对于实时数据处理，Spark 使用"微批处理"。这意味着 Spark 流会将特定时间段内的数据块打包成 RDD，如图 6-4 所示。

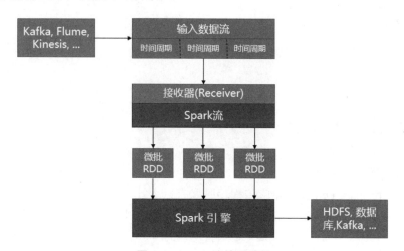

图 6-4　Spark 流处理过程

输入 DStream 表示从流源接收的输入数据流。每个输入的 DStream 都与一个 Receiver 对象相关联，该对象接收来自源的数据并将其存储在 Spark 的内存中进行处理。

一个 DStream 可以从像 Kafka、AWS Kinesis、一个文件或者一个 Socket 套接字这样的输入数据流中创建。在创建 DStream 时，需要的关键信息之一是批间隔，这个批间隔可以为几秒或几毫秒。有了 DStream，就可以在数据输入流上应用一个高级的数据处理函数了，如 map()、filter()、reduce()或 reduceByKey()。此外，还可以执行窗口操作，例如通过提供窗口长度和滑动间隔来计算一个滚动或滑动窗口。

6.2.1 Spark Streaming 编程模型

下面是一个简单的单词记数 Spark Streaming 流处理程序。通过这个程序可以了解一个典型的 Spark Streaming 流应用程序的结构，代码如下：

```scala
//第6章/streaming_wordcount.scala

import org.apache.spark.SparkConf
import org.apache.spark.storage.StorageLevel
import org.apache.spark.streaming.{Seconds, StreamingContext}

//StreamingContext 是流程序的入口点
val ssc = new StreamingContext(sc, Seconds(1))

//定义输入源
val host = "localhost"
val port = 9999
val lines = ssc
  .socketTextStream(host, port, StorageLevel.MEMORY_AND_DISK_SER)

//DStream 处理逻辑
val words = lines.flatMap(_.split(" "))
val wordCounts = words.map(x => (x, 1)).reduceByKey(_ + _)

//输出结果到控制台
wordCounts.print()

//触发流程序执行
ssc.start()

//等待流程序结束
ssc.awaitTermination()
```

在编写一个 Spark DStream 流应用程序时，有以下几个重要的步骤：

（1）DStream 应用程序的入口点是 StreamingContext，其中一个必需的输入参数是批间隔，它定义了一个持续时间段，Spark 用来将这段时间段输入的数据输入 RDD 进行处理。它也代表了一个触发点，在该触发点 Spark 应该执行流应用程序计算逻辑。例如，如果批处理的间隔是 1s，就会触发所有到达 1s 间隔内的数据；在该间隔之后，它将把这批数据转换为 RDD，并根据提供的处理逻辑对其进行处理。

（2）一旦创建了 StreamingContext，下一步就需要通过定义输入源来创建实例 DStream 了。前面的例子将输入源定义为读取文本行的 Socket 套接字。

（3）接下来，将为新创建的 DStream 提供处理逻辑。前面例子中的处理逻辑并不复杂。一旦一个系列的 RDD 在 1s 后可用，那么 Spark 就会执行将每行代码分割成单词的逻辑，将每个单词转换成单词的元组和 1 的计数，最后聚合统计同一个单词的计数。

（4）最后，将计数在控制台上打印出来。

（5）流应用程序是一个长期运行的应用程序，因此，它需要一个信号来启动接收和处理传入的数据流的任务。这个信号是通过调用 StreamingContext 的 start()函数实现的，通常是在文件的末尾完成的，而 awaitTermination()函数用于等待流应用程序的执行停止，以及在流应用程序运行时防止驱动程序退出。在一个典型的流程序中，一旦执行最后一行代码，它就会退出，然而，一个长时间运行的流应用程序需要在启动时继续运行，并且只有当显式地停止它时才会结束。

要测试该流程序的执行，首先需要运行一个 Netcat 服务器，用来模拟 TCP 服务器端。如果用户的 Linux 系统中还没有 Netcat 服务器，则可以使用如下的 Linux 命令安装（这里以 CentOS 为例，需要 root 权限）：

```
$ yum install -y nc
```

要运行 Netcat 服务器端监听 9999 端口，在终端窗口执行的命令如下：

```
$ nc -lk 9999
```

执行上面的命令，会启动 Netcat 服务器，并且运行在 9999 端口，保持其处于运行状态即可，如图 6-5 所示。

图 6-5　运行 Netcat 服务器

接下来就可以执行上面编程的流处理程序代码了。对 Spark 的流程序的测试运行环境，可直接使用 spark-shell 以交互方式执行，因此，打开一个新的终端，执行的命令如下：

```
$ spark-shell
```

执行以上命令，结果如图 6-6 所示。

在 spark-shell 中，支持:paste 输入模式，在此模式下，可以一次输入多行代码。进入:paste 模式，并输入上面示例中的流程序代码，如图 6-7 所示。

按下 Ctrl + D 快捷键，执行这些代码。因为 Spark Streaming 流程序是由 start()方法触发执行的，所以上面的代码此时并没有真正执行，只是进行了编译。继续输入以下代码，调用 start()方法，触发流程序的执行，如图 6-8 所示。

```
[hduser@xueai8 ~]$ spark-shell
SLF4J: Class path contains multiple SLF4J bindings.
SLF4J: Found binding in [jar:file:/home/hduser/bigdata/spark-3.1.2/jars/slf4j-log4j12-1.7.30.
SLF4J: Found binding in [jar:file:/home/hduser/bigdata/hadoop-3.2.2/share/hadoop/common/lib/sl
ass]
SLF4J: See http://www.slf4j.org/codes.html#multiple_bindings for an explanation.
SLF4J: Actual binding is of type [org.slf4j.impl.Log4jLoggerFactory]
22/05/10 08:57:56 WARN NativeCodeLoader: Unable to load native-hadoop library for your platfo
Setting default log level to "WARN".
To adjust logging level use sc.setLogLevel(newLevel). For SparkR, use setLogLevel(newLevel).
Spark context Web UI available at http://xueai8:4040
Spark context available as 'sc' (master = local[*], app id = local-1652144286623).
Spark session available as 'spark'.
Welcome to
      ____              __
     / __/__  ___ _____/ /__
    _\ \/ _ \/ _ `/ __/  '_/
   /___/ .__/\_,_/_/ /_/\_\   version 3.1.2
      /_/

Using Scala version 2.12.10 (Java HotSpot(TM) 64-Bit Server VM, Java 1.8.0_281)
Type in expressions to have them evaluated.
Type :help for more information.

scala>
```

图 6-6 启动 spark-shell 交互执行环境

```
scala> :paste
// Entering paste mode (ctrl-D to finish)

import org.apache.spark.SparkConf
import org.apache.spark.storage.StorageLevel
import org.apache.spark.streaming.{Seconds, StreamingContext}

// StreamingContext是流程序的入口点
val ssc = new StreamingContext(sc, Seconds(1))

// 定义输入源
val host = "localhost"
val port = 9999
val lines = ssc
  .socketTextStream(host, port, StorageLevel.MEMORY_AND_DISK_SER)

// DStream处理逻辑
val words = lines.flatMap(_.split(" "))
val wordCounts = words.map(x => (x, 1)).reduceByKey(_ + _)

// 将结果输出到控制台
wordCounts.print()
```

图 6-7 在:paste 模式下输入多行流程序代码

```
scala> ssc.start()

-------------------------------------------
Time: 1652144931000 ms
-------------------------------------------


-------------------------------------------
Time: 1652144932000 ms
-------------------------------------------
```

图 6-8 调用 start()方法触发流程序运行

此时流程序开始运行,然后转到 Netcat 运行窗口,随意输入一些文字,每行由多个单词组成,单词之间用空格符分隔,如图 6-9 所示。

图 6-9 在 Netcat 窗口输入文本内容

然后转到流程序执行窗口，此时会看到 Spark Streaming 流程序已经接收到了输入的文本内容，并将计算结果输到窗口中，如图 6-10 所示。

图 6-10 流处理程序输出内容

在 spark-shell 交互环境下，不需要执行 ssc.awaitTermination()语句，只有在独立部署的流程序中才需要。要退出流程序的执行，需要连续两次按下 Ctrl + C 快捷键（在生产环境下，流程序是持续运行的）。

6.2.2 实时股票交易分析

假设要为某证券公司开发一个实时股票交易分析程序，数据源是证券公司实时发布的市场交易数据，例如证券、股票等的买卖。通过该实时应用程序，实现回答以下问题：

（1）计算每秒的销售和购买订单数量。

（2）根据购买或出售的总金额来计算前 5 个客户（Top N 问题）。

（3）找出过去一小时内前 5 个交易量最多的股票（Top N 问题）。

证券公司交易数据在 orders.txt 文件中，该数据文件包含 500 000 行数据，每行代表一个买/卖订单。每行包含如下用逗号分隔的元素。

（1）Order 时间戳：格式为 yyyy-mm-dd hh:MM:ss。

（2）Order ID：订单 ID，这是连续递增的整数。

（3）Client ID：客户 ID，这是从 1 到 100 的整数。

（4）Stock 代码：共 80 个股票代码。

（5）股票买卖的数量：从 1 到 1000。

（6）股票买卖的价格：从 1 到 100。

（7）字符 B 或 S：代表一个订单的买（B）/卖（S）事件。

查看文件的前 5 行内容，命令如下：

```
$ head -n 5 ~/spark_demo/streaming/orders.txt
```

可以看到前 5 行交易数据，内容如下：

```
2016-03-22 20:25:28,1,80,EPE,710,51.00,B
2016-03-22 20:25:28,2,70,NFLX,158,8.00,B
2016-03-22 20:25:28,3,53,VALE,284,5.00,B
2016-03-22 20:25:28,4,14,SRPT,183,34.00,B
2016-03-22 20:25:28,5,62,BP,241,36.00,S
```

以上这些格式的数据就是流程序要处理的实时交易数据。为了模拟此实时交易数据源，本示例采用文件数据源，使用如下的思路和操作步骤：

（1）提供了一个名为 splitAndSend.sh 的脚本文件，它会将 orders.txt 这个股票交易订单文件拆分为 50 个小文件，每个小文件 10 000 行。

（2）运行 splitAndSend.sh 脚本文件，它会周期性地（每 3s）将这些小文件移动到一个指定的 HDFS 目录。

（3）编写的 Spark Streaming 流处理程序会监视这些指定的 HDFS 目录。一旦发现有新的数据文件，就实时进行处理，并将结果写到 HDFS 中。

接下来实现这个 Spark Streaming 流处理程序。

1. 计算每 3s 的卖出和买入订单数量

为了简单起见，本示例从一个 HDFS 文件读取数据，并将结果写回 HDFS。确保 Executor 的可用核数不少于 2，因为每个 Spark Streaming Receiver 都必须使用一个核（技术上，它是一个线程）来读取传入的数据，并且至少需要另一个核来执行程序的计算。

操作步骤如下。

（1）启动 spark-shell，并指定核数，命令如下：

```
$ spark-shell --master local[4]
```

（2）创建流上下文环境，代码如下：

```
import org.apache.spark._
import org.apache.spark.streaming._

val ssc = new StreamingContext(sc, Seconds(3))
```

在构造 StreamingContext 时，第 2 个参数为 Duration 对象，除了可以使用 Seconds 外，还可以使用 Milliseconds 和 Minutes 对象来指定一个周期。

上面代码中的 StreamingContext 构造器重用了一个已经存在的 SparkContext 实例 sc，但是如果给定一个 Spark 配置对象，SparkStreaming 还能从一个新的 SparkContext 开始，代码如下：

```
val conf = new SparkConf().setMaster("local[4]").setAppName("App name")
val ssc = new StreamingContext(conf, Seconds(3))
```

注意：以上两行代码可以在单独的应用程序中使用，但是不能在 spark-shell 中使用，因为在同一个 JVM 中不能实例化两个 Spark Context。

（3）创建一个离散流。

StreamingContext 提供了 textFileStream()方法，用于监控一个指定的目录（任何 Hadoop 兼容的目录，如 HDFS、S3、ClusterFS 和本地目录），并读取该目录下新创建的文件。该方法只有一个参数：被监控的目录的名称，但是只对流程序开始后新复制到该目录下的文件进行处理。

创建一个离散流，指定要监控的目录，代码如下：

```
val inputPath = "/data/spark/streaming/orders-input"
val fileStream = ssc.textFileStream(inputPath)
```

（4）使用离散流。构造 DStream，并对其进行转换计算，代码如下：

```
//导入相关的包
import java.sql.Timestamp

//case class 类
case class Order(
    time: java.sql.Timestamp,
    orderId:Long,
    clientId:Long,
    symbol:String,
    amount:Int,
    price:Double,
    buy:Boolean
)

import java.text.SimpleDateFormat

//解析出每行
val orders = fileStream.flatMap( line => {
    //用来解析时间戳
    val dateFormat = new SimpleDateFormat("yyyy-MM-dd hh:mm:ss")

    //每行首先按逗号拆分
    val s = line.split(",")
    try{
       //第 7 个字段应该等于 B 或 S
       assert(s(6)=="B" || s(6)=="S")
       //从解析出来的字段构造一个 Order 对象
       List(Order(new Timestamp(dateFormat.parse(s(0)).getTime()),s(1).toLong,
s(2).toLong, s(3), s(4).toInt,s(5).toDouble, s(6) == "B"))
    }
    catch{
       //如果解析该行时出错
       case e: Throwable => println("Wrong line format(" + e + "): " + line)
       //返回一个空的 list，忽略有问题的行
       List()
    }
})
```

订单 DStream 的每个 RDD 现在都包含 Order 对象。

包含两个元素元组的 DStream 会自动地转换为 PairDStreamFunctions 对象，然后就可以使用处理 Pair RDD 的方法了，如 combineByKey()、reduceByKey()、flatMapValues()、各种 join 连接方法等。如果将 orders 映射到包含订单类型作为 key 并且计数为 value 的元组，就可以使用 reduceByKey()函数。

例如，计算每个订单类型（购买或出售）的次数，代码如下：

```
val numPerType = orders.map(o => (o.buy, 1L)).reduceByKey((c1,c2) => c1 + c2)
numPerType.print()
```

（5）将结果保存到文件。

使用 DStream 的 saveAsTextFiles()方法可以将计算结果保存到一个文件中。每个微批处理中的 RDD 被保存到一个名为<your_prefix>-<time_in_milliseconds>.<your_suffix>或者<your_prefix>-<time_in_milliseconds>的文件夹中。这意味着，每 3s（本例中），就有一个新的目录被创建。每个目录中，对应 RDD 的每个分区，包含一个对应的文件，名为 part-xxxxx，其中 xxxxx 是该分区号。因为在本例中，可以肯定每个 RDD 最多包含两个元素，所以可以将 DStream 重分区为只有一个分区，然后保存到文件中（这样一个目录中就只有一个文件了），代码如下：

```
val resultPath = "/data/spark/streaming/orders-output/result"
numPerType.repartition(1).saveAsTextFiles(resultPath, "txt")
```

（6）启动和停止流计算，代码如下：

```
ssc.start()
```

如果是在独立的应用程序中，则还需要加上等待流程序执行结束的语句，否则驱动程序的主线程会退出，代码如下：

```
ssc.awaitTermination()
```

（7）打开另一个终端窗口，执行 splitAndSend.sh 脚本文件，用来模拟实时订单数据生成，代码如下：

```
$ cd ~/data/spark_demo/streaming/
$ hdfs dfs -mkdir /tmp/streaming
$ ./splitAndSend.sh /data/spark_demo/streaming/orders-input
```

如果流的输入目录是本地，则需要在命令后加上 local 参数，命令如下：

```
$ ./splitAndSend.sh /data/spark/streaming/orders-input local
```

（8）接下来，可以等待所有的文件都被处理完，或者马上从 Shell 停止运算流，代码如下：

```
ssc.stop(false)
```

参数 false 告诉流执行上下文环境不要停止该 sc（SparkContext 实例）。不能重新启动已经停止的流执行上下文环境，但是可以重用已经存在的 sc 来创建一个新的流执行上下文环境。

（9）查看最后结果。在终端窗口执行的 HDFS Shell 命令如下：
```
$ hdfs dfs -cat /data/spark/streaming/orders-output/result-*/*
```
计算结果（部分计算结果）如图 6-11 所示。

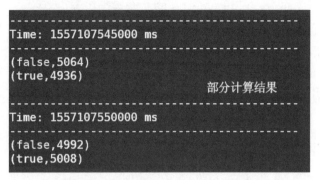

图 6-11　每 3s 的卖出和买入订单数量

2. 根据买入或卖出的总金额来计算前 5 个客户

这是一个 Top N 问题。要根据购买或出售的总金额来计算前 5 名客户，就必须跟踪每个客户购买或出售的总金额，因此，随着时间的推移，流应用程序需要保存之前的计算状态。

换句话说，必须跟踪一个持续时间和不同小批量的状态，这个概念如图 6-12 所示。

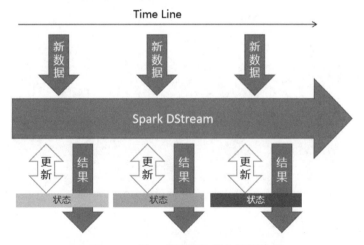

图 6-12　每 3s 的卖出和买入订单数量

新数据定期以小批量的速度出现，每个 DStream 都是一个处理数据并产生结果的程序。通过 Spark Streaming 方法来更新状态，DStream 可以将来自状态的持久数据与当前微批的新数据结合起来，结果是更强大的流程序。

Spark Streaming 提供了两种执行状态计算的方法：updateStateByKey()和 mapWithState()。这两种方法都可以从 PairDStreamFunctions 访问。换句话说，它们只适用于包含 key-value 元组的 DStream，因此，使用这些方法之前，必须创建包含 key-value 元组的 DStream。

查看 API 文档，可以看到 updateStateByKey 函数的定义如下：

```
updateStateByKey[S](updateFunc: (Seq[V], Option[S]) => Option[S])
```

该函数会返回一个新的状态 DStream，其中每个 key 的状态通过对该 key 的前一种状态和新值应用给定的函数来更新。在每一微批处理中，即使没有新值，也会为每种状态调用 updateFunc。使用哈希分区来生成具有 Spark 默认分区数的 RDD。如果 updateFunc 这种状态更新函数返回 null，则相应的状态键-值对将被消除。

在前面已经计算出的 orders 的基础上继续，根据买卖股票的总金额来计算出前 5 个客户，代码如下：

```scala
//创建包含 key-value 元组的 DStream
val amountPerClient = orders.map(o => (o.clientId, o.amount * o.price))

//使用 updateStateByKey()方法
val amountState = amountPerClient.updateStateByKey((vals,totalOpt) => {
  totalOpt match{
      //如果这个 key 的 state 已经存在，则累加上新的值
      case Some(total) => Some(vals.sum + total)
      //否则只返回新值的和
      case None => Some(vals.sum)
  }
})
amountState.print()

//top 5：先按交易额排序，再用拉链方法加上索引，然后取前 5 个
val top5Clients = amountState.transform(
    _.sortBy(_._2,false).zipWithIndex.filter(x=>x._2<5).map(x=>x._1))
top5Clients.print()
```

执行以上代码，部分输出结果如下：

```
-------------------------------------------
Time: 1557112795000 ms
-------------------------------------------
(100,2584474.0)
(84,2433250.0)
(96,2002749.0)
(52,2560673.0)
(56,2505880.0)
(76,2533741.0)
(4,2712570.0)
(16,2754024.0)
(80,2628837.0)
(28,2079145.0)
...
```

```
-------------------------------------------
Time: 1557112795000 ms
-------------------------------------------
(34,3626552.0)
(21,3342417.0)
(92,3306191.0)
(94,3212596.0)
(69,3164793.0)


-------------------------------------------
Time: 1557112800000 ms
-------------------------------------------
(100,4669628.0)
(84,5382475.0)
(96,4124313.0)
(56,4613215.0)
(52,5425443.0)
(76,5310475.0)
(4,5421270.0)
(16,5418655.0)
(28,4822011.0)
(80,5548338.0)
...

-------------------------------------------
Time: 1557112800000 ms
-------------------------------------------
(34,6733866.0)
(69,6610800.0)
(92,6166129.0)
(36,6101743.0)
(64,6002732.0)
```

3. 同时输出卖出/买入多少手及交易额居前 5 个客户 ID

为了在每个微批周期内同时输出两个 DStream 的结果（例如，Top 5 客户和买/卖数量）一次，必须首先将它们合并到单个 DStream 中。两个 DStream 可以使用各种 join()方法、cogroup()方法或 union()方法，按 key 进行合并。

要合并两个 DStreams，它们的元素必须是相同的类型。这里将把 top5Clients 和 numPerType 这两个 DStreams 的元素转换为元组，这个元组的第 1 个元素是 key，代表输出的结果类型，用来描述这条数据显示的是 BUYS(买入)、SELLS(卖出)或 TOP5CLIENTS(Top 5 客户的 ID)，第 2 个元素是字符串列表，形式如下：

```
(卖出,List(4980))
(买入,List(5020))
(TOP5CLIENTS,List(34, 21, 92, 94, 69))
```

实现代码如下：

```
val buySellList = numPerType.map(t =>
    if(t._1) ("买入", List(t._2.toString))
```

```
        else ("卖出", List(t._2.toString))
)

//glom()返回一个新的DStream,其中通过对DStream的每个RDD应用glom()生成每个RDD
//对RDD应用glom()可以将每个分区中的所有元素合并到一个数组中
val top5clList = top5clients
    .repartition(1)
    .map(x => x._1.toString)
    .glom()
    .map(arr => ("TOP5CLIENTS", arr.toList))

//然后,将两个DStream合并在一起
val finalStream = buySellList.union(top5clList)

//保存组合后的DStream
finalStream
    .repartition(1)
    .saveAsTextFiles("/data/spark/streaming/orders-output1", "txt")

//在启动之前,先设置检查点目录
sc.setCheckpointDir("/tmp/streaming/checkpoint")
```

最后,完整的代码如下:

```
//第6章/streaming_orders.scala

import org.apache.spark._
import org.apache.spark.streaming._
import java.sql.Timestamp
import java.text.SimpleDateFormat

//创建流执行环境,读取被监控的流文件,构造DStream
val inputPath = "/data/spark/streaming/orders-input"
val ssc = new StreamingContext(sc, Seconds(5))
val fileStream = ssc.textFileStream(inputPath)

//case class 类
case class Order(
    time: java.sql.Timestamp,
    orderId:Long,
    clientId:Long,
    symbol:String,
    amount:Int,
    price:Double,
    buy:Boolean
)

//解析出每行
val orders = fileStream.flatMap( line => {
    //用来解析时间戳
    val dateFormat = new SimpleDateFormat("yyyy-MM-dd hh:mm:ss")
```

```scala
        //每行首先按逗号拆分
        val s = line.split(",")
        try{
            //第 7 个字段应该等于 B 或 S
            assert(s(6)=="B" || s(6)=="S")
            //从解析出来的字段构造一个 Order 对象
            List(Order( new Timestamp(dateFormat.parse(s(0)).getTime()),
s(1).toLong, s(2).toLong, s(3), s(4).toInt,s(5).toDouble, s(6) == "B"))}
        catch{
            //如果解析该行时出错
            case e: Throwable => println("错误的行格式(" + e + "): " + line)
            //返回一个空的 list，忽略有问题的行
            List()
        }
})
//取返回的 DStream 中每条数据中的"buy"字段，构造(Boolean,1)元组，表示买/卖 1 次
//然后分类统计交易数量
val numPerType = orders.map(o => (o.buy, 1L)).reduceByKey((c1,c2) => c1 + c2)

//取返回的 DStream 中每条数据中的"clientId"字段
//构造(客户 ID,交易金额) 元组，每个客户的交易金额
val amountPerClient = orders.map(o => (o.clientId, o.amount * o.price))

//计算累加交易金额
val amountState = amountPerClient
        .updateStateByKey((vals:Seq[Double],totalOpt:Option[Double]) => {
    totalOpt match{
        //如果这个 key 的 state 已经存在，则累加上新的值
        case Some(total) => Some(vals.sum + total)
        //否则只返回新值的和
        case None => Some(vals.sum)
    }
})
//amountState.print()

//找出交易额最大的前 5 个客户 ID 及其买卖金额
val top5Clients = amountState.transform(
    _.sortBy(_._2,false).zipWithIndex.filter(x=>x._2<5).map(x=>x._1))
//top5Clients.print()

//将买卖表示由 true/false 转换为中文的"买入"和"卖出"，交易数量转换为统一的 List 格式
val buySellList = numPerType.map(t =>
    if(t._1) ("买入", List(t._2.toString))
    else ("卖出", List(t._2.toString))
)

//只保留前 5 个客户的 ID 作为 value，将 key 统一设为"TOP5CLIENTS"
val top5clList = top5Clients
    .repartition(1)
```

```
         .map(x => x._1.toString)
         .glom()
         .map(arr => ("TOP5CLIENTS", arr.toList))

//然后，将两个DStream合并在一起
val finalStream = buySellList.union(top5clList)
finalStream.print()

//保存组合后的DStream
finalStream
    .repartition(1)
    .saveAsTextFiles("/data/spark/streaming/orders-output2", "txt")

//在启动之前，先设置检查点目录
sc.setCheckpointDir("/ck/streaming")

//启动流处理程序
ssc.start()
```

启动了 StreamingContext 后，另打开一个终端窗口，在这个新的终端窗口执行 splitAndSend.sh 脚本，实时生成订单数据。流程序最后输出的结果如下：

```
-------------------------------------------
Time: 1557114450000 ms
-------------------------------------------
(TOP5CLIENTS,List())

-------------------------------------------
Time: 1557114455000 ms
-------------------------------------------
(卖出,List(4980))
(买入,List(5020))
(TOP5CLIENTS,List(34, 21, 92, 94, 69))

-------------------------------------------
Time: 1557114460000 ms
-------------------------------------------
(TOP5CLIENTS,List(34, 21, 92, 94, 69))

-------------------------------------------
Time: 1557114465000 ms
-------------------------------------------
(卖出,List(4926))
(买入,List(5074))
(TOP5CLIENTS,List(34, 69, 92, 36, 64))

-------------------------------------------
Time: 1557114470000 ms
-------------------------------------------
(卖出,List(4965))
(买入,List(5035))
```

```
(TOP5CLIENTS,List(34, 36, 94, 66, 21))
-----------------------------------------
Time: 1557114475000 ms
-----------------------------------------
(TOP5CLIENTS,List(34, 36, 94, 66, 21))
......
```

等待所有的文件都被处理完后,从 Shell 停止运算流:

```
ssc.stop(false)
```

4. 使用 mapWithState()状态处理函数

状态处理函数 mapWithState()是从 Spark 1.6 引入的,与另一种状态处理函数 updateStateByKey()相比要更新一些。它带来了一些性能改进,可以维护的 key 的数量是 updateStateByKey()函数的 10 倍,而且它的速度是 updateStateKey()函数的 6 倍(原因是在没有新 key 到达的情况下避免处理),该方法的签名如下:

```
mapWithState[StateType, MappedType](spec: StateSpec[K, V, StateType,
MappedType])
```

该方法通过对流的每个 key-value 元素应用一个函数来返回一个 MapWithStateDStream, 同时维护每个唯一 key 的一些状态数据。该转换的映射函数和其他规范(例如、分区器、超时、初始状态数据等)可以使用 StateSpec 类指定。

对于流中接收的每个 key-value 元素,都将调用传递给 StateSpec(及随后的 mapWithState)的函数,该函数的输入参数为新收到的 key 和 value,以及每个 key 已经存在的状态。状态数据可以作为映射函数中 state 类型的参数访问。

所提供的函数接收的 State 对象持有 key 的状态,并且有以下几个有用的方法来操纵它。

(1)exists:如果该状态被定义,则返回值为 true。

(2)get:获得该状态值。

(3)remove:为一个 key 删除该状态。

(4)update:为一个 key 更新或设置新状态值。

使用 mapWithState()状态函数的模板代码如下:

```
//维护一个整数状态并返回一个字符串的映射函数
def mappingFunction(key: String,
                    value: Option[Int],
                    state: State[Int]): Option[String] = {
  //使用 state 的 exists()、get()、update()和 remove()来维护状态,并返回必要的字符串
  //...
}
val spec = StateSpec.function(mappingFunction).numPartitions(10)

val mapWithStateDStream =
           keyValueDStream.mapWithState[StateType, MappedType](spec)
```

例如，下面的函数使用 mapWithState()状态函数，它接收的前两个参数是当前流元素的客户 ID（key）和其交易金额（value），第 3 个参数是该客户 ID 之前已经存在的状态，即之前存储的交易金额，代码如下：

```
val updateAmountState =
    (clientId:Long, amount:Option[Double], state:State[Double]) => {

//获取新传入的交易金额。如果是第 1 次遇到这个 key，则为 0
val total = amount.getOrElse(0.toDouble)
//如果有已经的状态（以前的交易金额）
if(state.exists()){
    //累加交易金额
    total += state.get()
}
//用新值更新状态
state.update(total)

//返回带有客户 ID 和新 state 值的 tuple
Some((clientId, total))
}
```

像下面这样使用这个函数：

```
val amountState = amountPerClient
    .mapWithState(StateSpec.function(updateAmountState))
    .stateSnapshots()
```

如果没有最后一个 stateSnapshots()方法，就会得到一个带有客户 ID 和交易总金额的 DStream，但是只针对那些在当前的小批量中订单到达的客户。方法 stateSnapshots()给了一个带有所有状态（所有客户）的 DStream。

5. 使用窗口操作进行时间限制的计算

继续前面的例子，还有最后一个任务要完成：每 5s 报告一次过去一小时内前 5 个交易量最多的股票。这与之前的任务不同，因为它是有时间限制的。在 Spark Streaming 中，这种类型的计算是通过"窗口操作"完成的。一个滑动窗口如图 6-13 所示。

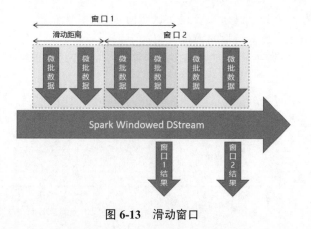

图 6-13 滑动窗口

在这个任务中，窗口的持续时间是一小时，但是滑动周期与微批持续时间相同（都是 5s），因为想要每 5s 报告一次交易量排名前五的股票。

要创建一个窗口的 DStream，可以使用窗口方法 reduceByKeyAndWindow()，它的定义如下：

```
reduceByKeyAndWindow(reduceFunc: (V, V) ⇒ V, windowDuration: Duration): DStream[(K, V)]
```

在这种方法中需要指定 reduce() 函数及窗口持续时间（也可以指定滑动周期，如果它与微批周期不同），所以要计算每个股票每个窗口的交易金额，可以使用如下的代码（放在 finalStream 变量初始化之前）：

```
val stocksPerWindow = orders
    .map(x => (x.symbol, x.amount))
    .window(Minutes(60))
    .reduceByKey((a1:Int, a2:Int) => a1+a2)

/* 或
val stocksPerWindow = orders
    .map(x => (x.symbol, x.amount))
    .reduceByKeyAndWindow((a1:Int, a2:Int) => a1+a2, Minutes(60))
*/

val topStocks = stocksPerWindow
    .transform(_.sortBy(_._2,false)
                .map(_._1)
                .zipWithIndex
                .filter(x=>x._2<5))
    .repartition(1)
    .map(x => x._1.toString)
    .glom()
    .map(arr => ("TOP5STOCKS", arr.toList))

//将这个结果添加到 finalDStream
val finalStream = buySellList.union(top5clList).union(topStocks)

//后面的代码与之前的一样
//保存组合后的 DStream
finalStream
    .repartition(1)
    .saveAsTextFiles("/data/spark/streaming/orders-output3", "txt")

//在启动之前，先设置检查点目录
sc.setCheckpointDir("/ck/streaming")

ssc.start()
```

启动了 StreamingContext 后，另打开一个终端窗口，在这个新的终端窗口执行 splitAndSend.sh 脚本文件，模拟生成交易数据，流程序的输出内容如下：

```
-----------------------------------------
Time: 1557116120000 ms
-----------------------------------------
(卖出,List(4980))
(买入,List(5020))
(TOP5CLIENTS,List(34, 21, 92, 94, 69))
(TOP5STOCKS,List(MNKD, BAC, INTC, FB, CTRE))

-----------------------------------------
Time: 1557116125000 ms
-----------------------------------------
(卖出,List(4926))
(买入,List(5074))
(TOP5CLIENTS,List(34, 69, 92, 36, 64))
(TOP5STOCKS,List(SIRI, BP, CTRE, MNKD, TSU))

-----------------------------------------
Time: 1557116130000 ms
-----------------------------------------
(TOP5CLIENTS,List(34, 69, 92, 36, 64))
(TOP5STOCKS,List(SIRI, BP, CTRE, MNKD, TSU))

-----------------------------------------
Time: 1557116135000 ms
-----------------------------------------
(卖出,List(4965))
(买入,List(5035))
(TOP5CLIENTS,List(34, 36, 94, 66, 21))
(TOP5STOCKS,List(SIRI, INTC, CTRE, IBN, LEU))

-----------------------------------------
Time: 1557116140000 ms
-----------------------------------------
(TOP5CLIENTS,List(34, 36, 94, 66, 21))
(TOP5STOCKS,List(SIRI, INTC, CTRE, IBN, LEU))

-----------------------------------------
Time: 1557116145000 ms
-----------------------------------------
(卖出,List(4964))
(买入,List(5036))
(TOP5CLIENTS,List(36, 34, 15, 64, 9))
(TOP5STOCKS,List(BP, SIRI, LUV, AMD, INTC))
...
```

等待所有的文件都被处理完后,从spark-shell窗口停止运算流,代码如下:

```
ssc.stop(false)
```

完整代码如下:

```
//第6章/streaming_slide_window.scala

import org.apache.spark._
```

```
import org.apache.spark.streaming._
import java.sql.Timestamp
import java.text.SimpleDateFormat

val ssc = new StreamingContext(sc, Seconds(5))
val inputPath = "/data/spark/streaming/orders-input"
val fileStream = ssc.textFileStream(inputPath)

//case class 类
case class Order(
    time: java.sql.Timestamp,
    orderId:Long,
    clientId:Long,
    symbol:String,
    amount:Int,
    price:Double,
    buy:Boolean
)

//解析出每行
val orders = fileStream.flatMap( line => {
    //用来解析时间戳
    val dateFormat = new SimpleDateFormat("yyyy-MM-dd hh:mm:ss")

    //每行首先按逗号拆分
    val s = line.split(",")
    try{
       //第 7 个字段应该等于 B 或 S
       assert(s(6)=="B" || s(6)=="S")
       //从解析出来的字段构造一个 Order 对象
       List(Order(new Timestamp(dateFormat.parse(s(0)).getTime()),
               s(1).toLong,
               s(2).toLong,
               s(3),
               s(4).toInt,
               s(5).toDouble,
               s(6) == "B"))
    }
    catch{
      //如果解析该行时出错
      case e: Throwable => println("Wrong line format(" + e + "): " + line)
      //返回一个空的 list, 忽略有问题的行
      List()
    }
})

val numPerType = orders.map(o => (o.buy, 1L)).reduceByKey((c1,c2) => c1 + c2)

val amountPerClient = orders.map(o => (o.clientId, o.amount * o.price))
```

```scala
val amountState = amountPerClient
        .updateStateByKey((vals:Seq[Double],totalOpt:Option[Double]) => {
    totalOpt match{
        //如果这个 key 的 state 已经存在,则累加上新的值的和
        case Some(total) => Some(vals.sum + total)
        //否则只返回新值的和
        case None => Some(vals.sum)
    }
})
//amountState.print()

//找出前 5 个客户 ID 及买卖金额
val top5Clients = amountState.transform(
        _.sortBy(_._2,false).zipWithIndex.filter(x=>x._2<5).map(x=>x._1))
//top5Clients.print()

val buySellList = numPerType.map(t =>
    if(t._1) ("买入", List(t._2.toString))
    else ("卖出", List(t._2.toString))
)

//只保留前 5 个客户的 ID
val top5clList = top5Clients
    .repartition(1)
    .map(x => x._1.toString)
    .glom()
    .map(arr => ("TOP5CLIENTS", arr.toList))

//窗口的持续时间是一小时,但是滑动周期与 mini-batch 持续时间相同(5s)
val stocksPerWindow = orders
    .map(x => (x.symbol, x.amount))
    .window(Minutes(60))
    .reduceByKey((a1:Int, a2:Int) => a1+a2)

val topStocks = stocksPerWindow.transform(
        _.sortBy(_._2,false).map(_._1).zipWithIndex.filter(x=>x._2<5))
    .repartition(1)
    .map(x => x._1.toString)
    .glom()
    .map(arr => ("TOP5STOCKS", arr.toList))

//将这个结果添加到 finalDStream
val finalStream = buySellList.union(top5clList).union(topStocks)
finalStream.print()

//后面的与之前的一样,保存组合后的 DStream
finalStream
    .repartition(1)
    .saveAsTextFiles("/data/spark/streaming/orders-output3", "txt")
```

```
//在启动之前,先设置检查点目录
sc.setCheckpointDir("/ck/streaming")

//启动流处理程序
ssc.start()
```

在 Spark Streaming 中,window()方法并不是唯一可用的窗口操作。还有很多其他方法,其中有些适用于普通的 DStream,而另一些则仅用于 key-value DStream,这些方法见表 6-1。

表 6-1 可用的窗口操作方法

窗口操作	描述
window(winDur, [slideDurl])	返回一个新的 DStream,其中每个 RDD 包含在该 DStream 上的一个滑动时间窗口中看到的所有元素。slideDurl 默认等于微批处理持续时间
countByWindow(winDur, slideDur)	返回一个新的 DStream,其中每个 RDD 都有一个单独的元素,该元素是通过计算该 DStream 上滑动窗口中的元素数量生成的。哈希分区用于生成具有 Spark 默认分区数的 RDDs
countByValueAndWindow(winDur, slideDur, [numParts])	返回一个新的 DStream,其中每个 RDD 在该 DStream 上的一个滑动窗口中包含 RDDs 中不同元素的计数。哈希分区用于生成带有 numpartition 分区的 RDDs(如果没有指定 numpartition,则使用 Spark 的默认分区数)
reduceByWindow(reduceFunc, winDur, slideDur)	返回一个新的 DStream,其中每个 RDD 都有一个单独的元素,该元素是通过 reduce 该 DStream 上滑动窗口中的所有元素生成的
reduceByWindow(reduceFunc, invReduceFunc, winDur, slideDur)	返回一个新的 DStream,其中每个 RDD 都有一个单独的元素,该元素是通过 reduce 该 DStream 上滑动窗口中的所有元素生成的,然而,reduce 是增量地使用旧窗口的 reduce 后的值: reduce 输入窗口的新值(例如,添加新计数); "反向减少(InverseReduce)"窗口中遗留的旧值(例如,减去旧计数)。这比没有 InverseReduce 函数的 reduceByWindow 更有效,然而,它只适用于"可逆约简函数"(Invertible Reduce Functions)
groupByKeyAndWindow(winDur, [slideDur], [numParts/partitioner]	通过在该 DStream 上的滑动窗口上应用 groupByKey 创建一个新的 DStream。类似于 DStream.groupByKey(),但它应用于一个滑动窗口。还可以指定分区的数量或使用的分区器。 只应用于 Pair DStream
reduceByKeyAndWindow(reduceFunc, winDur, [slideDur], [numParts/partitioner])	通过在滑动窗口上应用 reduceByKey 返回一个新的 DStream。与 DStream.reduceByKey()类似,但应用于一个滑动窗口。还可以指定分区的数量或使用的分区器。 只应用于 Pair DStream

续表

窗口操作	描述
reduceByKeyAndWindow(reduceFunc, invReduceFunc, winDur, [slideDur], [numParts], [filterFunc])	通过在滑动窗口上应用增量 reduceByKey 返回一个新的 DStream。使用旧窗口的约简值计算新窗口的约简值： （1）reduce 输入窗口的新值（例如，添加新计数）。 （2）"反向减少（Inverse Reduce）"窗口中遗留的旧值（例如，减去旧计数）。 这比没有 Inverse Reduce 函数的 reduceByKeyAndWindow 更有效，然而，它只适用于"可逆约简函数"。哈希分区用于生成具有 Spark 默认分区数的 RDDs

在之前的示例中，也可以用这样的方式来替代 reduceByKeyAndWindow()方法：先使用 window()方法，然后使用 reduceByKey()方法。

6.2.3 使用外部数据源 Kafka

Kafka 通常用于构建实时流数据管道，以便可靠地在系统之间移动数据，还用于转换和响应数据流。Kafka 作为集群在一个或多个服务器上运行。Kafka 的一些关键概念描述如下。

（1）Topic：消息发布到的类别或流名称的高级抽象。主题可以有 0、1 或多个消费者，这些消费者订阅发布到该主题的消息。用户为每个新的消息类别定义一个新主题。

（2）Producers：向主题发布消息的客户端。

（3）Consumers：使用来自主题的消息的客户端。

（4）Brokers：复制和持久化消息数据的一个或多个服务器。

此外，生产者和消费者可以同时对多个主题进行读写。每个 Kafka 主题都是分区的，写入每个分区的消息都是顺序的。分区中的消息具有一个偏移量，用来唯一标识分区内的每条消息。

主题的分区是分布式的，每个 Broker 处理对分区共享的请求。每个分区在 Brokers（数量可配置）之间复制。Kafka 集群在一段可配置的时间内保留所有已发布的消息。Apache Kafka 使用 Apache ZooKeeper 作为其分布式进程的协调服务。

注意：Kafka 的数据源可能是在生产型流应用程序中最常用的数据源。为了有效地处理这个数据源，读者需要具有一定的使用 Kafka 的基本知识。

官方提供了 Spark 连接器，用于集成 Kafka。本节将演示如何使用一个 Shell 脚本将股票交易数据发送到 Kafka topic。Spark Streaming 作业将从这个 topic 读取交易数据并将计算出的结果写到另一个 Kafka topic，然后使用 Kafka 的 kafka-console-consumer.sh 脚本来接收并显示结果。处理流程如图 6-14 所示。

图 6-14　**Spark Streaming 集成 Kafka 流程图**

1. 安装和配置 Kafka

要设置 Kafka，首先需要下载它，下载网址为 http://kafka.apache.org/downloads.html。注意，要选择与自己的 Spark 版本相兼容的版本。

然后，将下载的 Kafka 压缩包解压缩到~/bigdata/目录下，命令如下：

```
$ cd ~/bigdata
$ tar -xvfz kafka_2.12-2.4.1.tgz
```

Kafka 依赖于 Apache ZooKeeper，所以在启动 Kafka 之前，要先启动它，命令如下：

```
$ cd ~/bigdata/kafka_2.12-2.4.1
$ ./bin/zookeeper-server-start.sh config/zookeeper.properties &
```

这将在 2181 端口启动 ZooKeeper 进程，并让 ZooKeeper 在后台工作。接下来，启动 Kafka 服务器，命令如下：

```
$ ./bin/kafka-server-start.sh config/server.properties &
```

最后，需要创建用于发送股票买卖数据和计算结果数据的主题，命令如下：

```
$ ./bin/kafka-topics.sh --create --zookeeper localhost:2181 --replication-factor 1 --partitions 1 --topic orders
$ ./bin/kafka-topics.sh --create --zookeeper localhost:2181 --replication-factor 1 --partitions 1 --topic metrics
```

如果想要查看已有的主题，则可执行的命令如下：

```
$ ./bin/kafka-topics.sh --list --zookeeper localhost:2181
```

2. 加载 Kafka 库和连接器

重新启动 spark-shell，在启动时将 Kafka 库和 Spark Kafka 连接器库添加到类路径。需要使用--package 选项参数让 Spark 自动下载这些文件，命令如下：

```
$ spark-shell --master spark://localhost:7077 --packages org.apache.spark:spark-streaming-kafka-0-10_2.12:3.1.2,org.apache.kafka:kafka_2.12:2.4.1
```

如果要创建 Maven 项目，则需要在 pom.xml 文件中添加依赖，内容如下：

```
<dependency>
    <groupId>org.apache.spark</groupId>
    <artifactId>spark-streaming-kafka-0-10_2.12</artifactId>
    <version>3.1.2</version>
</dependency>
<dependency>
    <groupId>org.apache.kafka</groupId>
```

```
        <artifactId>kafka_2.12</artifactId>
        <version>2.4.1</version>
</dependency>
```

如果要创建 SBT 项目,则需要在 build.sbt 文件中添加依赖,内容如下:

```
libraryDependencies += "org.apache.spark" %% "spark-streaming-kafka-0-10" % "3.1.2"
libraryDependencies += "org.apache.kafka" %% "kafka" % "2.4.1"
```

3. 使用 Spark Kafka 连接器

Kafka 的连接器有两个版本。第 1 个是基于 receiver 的连接器,第 2 个是较新的 direct 连接器。在某些情况下,当使用基于 receiver 连接器的连接器时,相同的消息可能会被多次使用,而 direct 连接器使实现传入消息只执行一次处理成为可能。基于 receiver 的连接器效率也较低,因为它需要设置一个 Write-Ahead Log(WAL,预写日志),这会降低计算速度。在本节中将使用 direct 连接器。

下面重构 6.2.3 节中的实时股票交易分析流程序,使用 Kafka 作为数据源和 Data Sink。完整的流处理代码如下:

```scala
//第6章/streaming_kafka.scala

import org.apache.spark.streaming._
import org.apache.kafka.clients.consumer.ConsumerRecord
import org.apache.kafka.clients.producer.{ProducerRecord, KafkaProducer}
import org.apache.kafka.common.serialization._
import org.apache.spark.streaming.kafka010._
import org.apache.spark.streaming.kafka010.LocationStrategies._
import org.apache.spark.streaming.kafka010.ConsumerStrategies.Subscribe

//构造流处理环境
val ssc = new StreamingContext(sc, Seconds(5))

//Bootstrap.servers 是集群中 Kafka broker 的地址,这里改为自己的 IP
val kafkaParams = Map[String, Object](
    "Bootstrap.servers" -> "localhost:9092",
    "key.deserializer" -> classOf[StringDeserializer],
    "value.deserializer" -> classOf[StringDeserializer],
    "auto.offset.reset" -> "latest",
    "group.id" -> "stock-consumer-group"
)
//topic 主题
val topics = Array("orders")
//两个参数泛型:key type 和 value type
val kafkaStream = KafkaUtils.createDirectStream[String, String](
ssc,PreferConsistent,Subscribe[String, String](topics, kafkaParams))

import java.sql.Timestamp

case class Order(
```

```
    time: java.sql.Timestamp,
    orderId:Long,
    clientId:Long,
    symbol:String,
    amount:Int,
    price:Double,
    buy:Boolean
)
import java.text.SimpleDateFormat

//kafkaStream 中的每项都是一个 ConsumerRecord
val orders = kafkaStream.flatMap(record => {
    val dateFormat = new SimpleDateFormat("yyyy-MM-dd hh:mm:ss")
    val s = record.value.split(",")
    try {
        assert(s(6) == "B" || s(6) == "S")
        List(Order(new Timestamp(dateFormat.parse(s(0)).getTime()),
                s(1).toLong,
                s(2).toLong,
                s(3),
                s(4).toInt,
                s(5).toDouble,
                s(6) == "B"))
    }catch {
        case e : Throwable => println("错误的行格式 ("+e+"): " + record.value)
        List()
    }
})

val numPerType = orders.map(o => (o.buy, 1L)).reduceByKey((c1, c2) => c1+c2)

val buySellList = numPerType.map(t =>
    if(t._1) ("买入", List(t._2.toString))
    else ("卖出", List(t._2.toString))
)

val amountPerClient = orders.map(o => (o.clientId, o.amount*o.price))
val amountState = amountPerClient
    .updateStateByKey((vals, totalOpt:Option[Double]) => {
        totalOpt match {
            case Some(total) => Some(vals.sum + total)
            case None => Some(vals.sum)
        }
})

val top5clients = amountState.transform(
        _.sortBy(_._2, false).map(_._1).zipWithIndex.filter(x => x._2 < 5))

val top5clList = top5clients
```

```
        .repartition(1)
        .map(x => x._1.toString)
        .glom()
        .map(arr => ("TOP5CLIENTS", arr.toList))

val stocksPerWindow = orders
    .map(x => (x.symbol, x.amount))
    .reduceByKeyAndWindow((a1:Int, a2:Int) => a1+a2, Minutes(60))

val topStocks = stocksPerWindow.transform(
      _.sortBy(_._2,false).map(_._1).zipWithIndex.filter(x => x._2 < 5))
    .repartition(1)
    .map(x => x._1.toString)
    .glom()
    .map(arr => ("TOP5STOCKS", arr.toList))

val finalStream = buySellList.union(top5clList).union(topStocks)
finalStream.print()

//向Kafka写入,要用到
//org.apache.kafka.clients.producer.KafkaProducer<K,V>类
finalStream.foreachRDD((rdd) => {
   rdd.foreachPartition((iter) => {
      val props = new java.util.Properties
      props.put("Bootstrap.servers", "localhost:9092");
      props.put("key.serializer", "org.apache.kafka.common.serialization.StringSerializer");
      props.put("value.serializer", "org.apache.kafka.common.serialization.StringSerializer");

      val producer = new KafkaProducer[String, String](props);

      iter.foreach({ case (metric, list) =>
         //"metrics"是写回到的Kafka主题
         producer.send(new ProducerRecord[String, String]("metrics", metric, list.toString));
      })
   })
})

sc.setCheckpointDir("/ck/streaming_kafka")

ssc.start()
```

4. 运行程序

另外再打开一个终端窗口,执行脚本 streamOrders.sh。这个脚本会从 orders.txt 文件中流式读取行并发送给 Kafka 的 orders 主题(注意,orders.txt 文件要和 streamOrders.sh 脚本文件放在相同目录下),命令如下:

```
$ chmod +x streamOrders.sh
$ ./streamOrders.sh localhost:9092
```

另外再打开一个终端窗口,在这个终端窗口中,启动 kafka-console-consumer.sh 脚本,它会消费来自 metrics topic 的消息;查看流程序的输出内容,命令如下:

```
$ ./bin/kafka-console-consumer.sh --zookeeper localhost:2181 --topic metrics
```

如果一切正常,则输出的结果如下:

```
TOP5CLIENTS, List(62, 2, 92, 25, 19)
SELLS, List(12)
BUYS, List(20)
TOP5STOCKS, List(CHK, DOW, FB, SRPT, ABX)

TOP5CLIENTS, List(2, 62, 87, 52, 45)
TOP5STOCKS, List(FB, CTRE, AU, PHG, EGO)
SELLS, List(28)
BUYS, List(21)

SELLS, List(37)
BUYS, List(12)
TOP5STOCKS, List(FB, CTRE, SDLP, AU, NEM)
TOP5CLIENTS, List(14, 2, 81, 43, 31)
```

等待所有的文件都被处理完后,从 spark-shell 停止运算流,执行的命令如下:

```
ssc.stop(false)
```

第 7 章 Spark 结构化流

CHAPTER 7

Spark 2.0 引入了更高级别的新的流处理 API，叫作 Structured Streaming，可以称为"结构化流"。结构化流是一种基于 Spark SQL 引擎的快速、可扩展、容错、精确一次的有状态流处理方法。它支持流分析，而无须考虑流的底层机制。

这个可扩展和容错的高级流 API 构建在 Spark SQL 引擎上，与 SQL 查询和 DataFrame/DataSet API 紧密集成。主要优点是使用相同的 Spark DataFrame 和 DataSet API 及 Spark 引擎计算出操作所需的增量和连续执行，简化实时、连续的大数据应用程序的开发。结构化流将批处理和流处理计算统一起来，并可以连接（join）流和批数据。

此外，还可以使用查询管理 API 来管理多个并行运行的流查询。例如，可以列出正在运行的查询、停止和重新启动查询、在失败的情况下检索异常等。

7.1 结构化流简介

Spark 结构化流提供了快速、可扩展、容错、端到端的精确一次性流处理，而用户无须对流进行推理。结构化流操作直接工作在 DataFrame（或 DataSet）上。不再有"流"的概念，只有流式 DataFrame 和普通 DataFrame。流式 DataFrame 是作为 append-only 表实现的。在流数据上的查询会返回新的 DataFrame，使用它们就像在批处理程序中一样。

使用 Spark 结构化流的模型如图 7-1 所示。

在结构化流处理模型中，将实时数据流视为一个不断增长的输入表。当一组新的数据到达时，将这些新到达的数据作为一组新的行添加到输入表。触发器指定检查输入新数据到达的时间间隔。考虑传入数据流的方式，只不过是一个不断追加的表，这样就能够利用现有的用于 DataFrame 和 DataSet 的结构化 API（在 Scala、Java 或 Python 中），把流计算表示为标准的批处理查询，就像在静态表上的查询一样，如映射、筛选和合并，并且随着新的流数据的到来，由结构化流引擎负责将其作为无边界输入表上的增量查询持续地运行。结果表示在每个触发器间隔中根据指定的操作更新的最终表。输出定义了在每个时间间隔内将结果写入数据接收器的部分，如图 7-2 所示。

因此，无论是静态数据还是流式数据，只需像在静态数据表上那样启动类似于批处理的查询，Spark 就会在无界输入表上作为增量查询运行它，因此，开发人员在输入表上定义一个查询，对于静态有界表和动态无界表都是一样的。

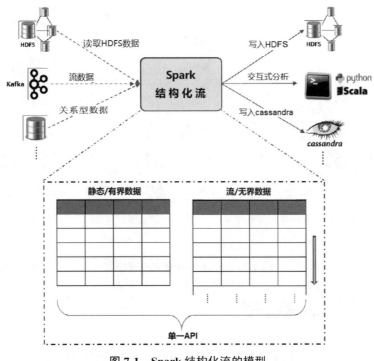

图 7-1　Spark 结构化流的模型

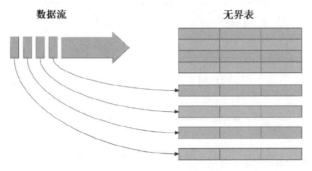

图 7-2　将实时数据流视为一个不断增长的输入表

在结构化流处理模型中，用户使用批处理 API 在输入表上进行查询，Spark SQL Planner（规划器）在流数据上增量执行，整个过程如图 7-3 所示。

传统上，当从流式应用程序将数据发送到外部存储系统时，确保没有重复的数据或数据丢失是开发人员的责任。这是流应用程序开发人员提出的一个痛点。在 Spark 结构流引擎内部，已经提供了一个端到端的、精确一次性的保证，现在同样的保证被扩展到外部存储系统，只要这些系统支持事务。

从根本上说，结构化流由 Spark SQL 的 Catalyst 优化器负责优化，因此，它使开发人员不再担心底层的管道，在处理静态或实时数据流时，使查询更高效。

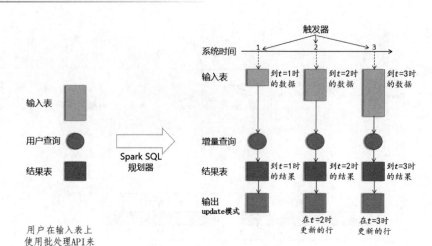

图 7-3 Spark 结构化流处理模型

7.2 结构化流编程模型

假设有一个监听 TCP 套接字的数据服务器，现在想维护该服务器接收的文本数据的运行时单词计数。当查询启动时，Spark 将持续检查来自套接字连接的新数据。如果有新数据，则 Spark 将运行一个增量查询，将以前的运行计数与新数据结合起来，计算更新的计数，如图 7-4 所示。

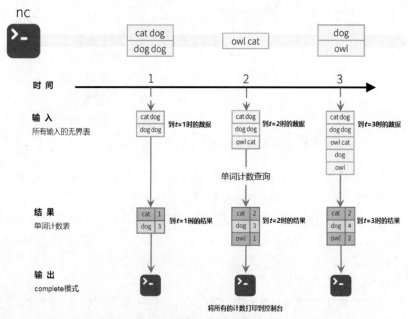

图 7-4 Spark 结构化流实现实时单词计数

可以使用统一入口点 SparkSession 从流源创建流 DataFrame/DataSet，并对它们应用与静态 DataFrame/DataSet 相同的操作。

接下来看一看如何使用 Spark 结构化流来表达这一需求。

首先需要使用 Netcat 作为数据服务器。在 Linux 的终端中，执行命令，启动 Netcat 服务器，使其保持运行，命令如下：

```
$ nc -lk 9999
```

注意：如果没有 Netcat 服务器，则可以使用命令安装：$ sudo yum install -y nc。

然后，编写 Spark 结构化流程序并执行。建议按以下步骤操作。

（1）先导入必要的类并创建本地 SparkSession，代码如下：

```scala
//第7章/sstreaming_wordcount.scala

import org.apache.spark.sql.functions._
import org.apache.spark.sql.SparkSession

val spark = SparkSession
  .builder
  .appName("StructuredNetworkWordCount")
  .getOrCreate()

//用于从 RDD 到 DataFrame 的隐式转换
import spark.implicits._
```

（2）接下来，创建一个流 DataFrame，它表示从所监听的 localhost:9999 服务器接收的文本数据，并对该 DataFrame 进行转换以进行单词计数，代码如下：

```scala
//创建表示从连接到localhost:9999的输入行流的DataFrame
val lines = spark.readStream
  .format("socket")
  .option("host", "localhost")
  .option("port", 9999)
  .load()

//这个line DataFrame表示一个包含流文本数据的无界表
//将行拆分为单词
val words = lines.as[String].flatMap(_.split(" "))

//生成运行时单词计数
//wordCounts 是一个流 DataFrame，它表示流的运行时单词计数
val wordCounts = words.groupBy("value").count()
```

（3）在流数据上设置查询。

在流数据上进行设置，以便每次更新计数时都将完整的计数集打印到控制台（由 outputMode("complete")指定），然后使用 start()启动流计算，代码如下：

```scala
//开始运行将运行时单词计数打印到控制台的查询
val query = wordCounts.writeStream
  .outputMode("complete")
```

```
    .format("console")
    .start()

query.awaitTermination()
```

执行此代码后，流计算将在后台开始，其中 query 对象是该活动流查询的句柄，使用 awaitTermination()等待查询终止，以防止查询处于活动状态时进程退出。对于生产和长期运行的流应用程序，有必要调用 StreamingQuery.awaitTermination()函数，这是一个阻塞调用，它会防止 Driver 驱动程序退出，并允许流查询持续运行和当新数据到达数据源时处理新数据。

（4）切换到 Netcat 窗口，输入几行任意的内容，单词之间用空格分隔。例如，输入以下两行文本内容：

```
good good study
study day day up
```

（5）切换回流程序执行窗口，查看输出结果，输出的结果如下：

```
-------------------------------------------
Batch: 1
-------------------------------------------
+-----+-----+
|value|count|
+-----+-----+
|study|    1|
| good|    2|
+-----+-----+


-------------------------------------------
Batch: 2
-------------------------------------------
+-----+-----+
|value|count|
+-----+-----+
|  day|    2|
|study|    2|
|   up|    1|
| good|    2|
+-----+-----+
```

有时希望停止流查询来改变输出模式、触发器或其他配置。这时可以使用 StreamingQuery.stop()函数来阻止数据源接收新数据，并停止在流查询中逻辑的连续执行。管理流查询的示例代码如下：

```
//这是阻塞调用
mobileSQ.awaitTermination()

//停止流查询
mobileSQ.stop
```

```
//在Spark应用程序中停止所有流查询的另一种方法
for(qs <- spark.streams.active) {
    println(s"停止流查询: ${qs.name} - active: ${qs.isActive}")
    if (qs.isActive) {
       qs.stop
    }
}
```

7.3 结构化流核心概念

Apache Spark 结构化流应用程序包括以下几个主要部分：
（1）指定一个或多个流数据源。
（2）提供了以 DataFrame 转换的形式操纵传入数据流的逻辑。
（3）定义输出模式和触发器（都有默认值，所以是可选的）。
（4）最后指定一个数据接收器（Data Sink）。
以上步骤如图 7-5 所示。

图 7-5 Spark 结构化流应用程序步骤

其中用星号（*）标记的步骤表示非必须执行的步骤。下面的部分将详细描述这些概念。

7.3.1 数据源

对于批处理，数据源是驻留在某些存储系统上的静态数据集，如本地文件系统、HDFS 或 S3，而结构化流的数据源是完全不同的。它们产生的数据是连续的，可能永远不会结束，而且生产速率也会随着时间的变化而变化。

Apache Spark 结构化流提供了以下几个开箱即用的数据源：

（1）Kafka 源：要求 Apache Kafka 的版本是 0.10 或更高版本。这是生产环境中最流行的数据源。连接和读取来自 Kafka 主题的数据需要提供一组特定的设置。

（2）文件源：文件位于本地文件系统、HDFS 或 S3 上。当新的文件被放入一个目录中时，这个数据源将会把它们挑选出来进行处理。支持常用的文件格式，如文本、CSV、JSON、ORC 和 Parquet。在处理这个数据源时，一个好的实践是先完全地写出输入文件，然后将它们移动到这个数据源的路径中（例如，流程序监控的是 HDFS 上的 A 目录，那么先将输入文件写到 HDFS 的 B 目录中，再从 B 目录将它们移动到 A 目录）。

（3）Socket 源：这仅用于测试目的。它从一个监听特定的主机和端口的 Socket 上读取 UTF-8 数据。

(4) Rate 源：这仅用于测试和基准测试。这个源可以被配置为每秒产生许多事件，其中每个事件由时间戳和一个单调递增的值组成。这是学习结构化流时使用的最简单的源。

数据源需要提供的一个重要的属性是一种跟踪流中的读位置的方法，用于结构化的流来传递端到端、精确一次性保证。例如，Kafka 的数据源提供了一个 Kafka 的偏移量来跟踪一个主题分区的读位置。这个属性决定了一个特定的数据源是否具有容错能力。

每个开箱即用数据源的一些选项，见表 7-1。

表 7-1 Spark 结构化流数据源

数 据 源	是否容错	配 置
File	是	path：输入目录的路径 maxFilesPerTrigger：每个触发器读取新行的最大数量 latestFirst：是否处理最新的文件(根据 modification time)
Socket	否	要求有以下参数 host：要连接到的主机 port：要连接到的端口号
Rate	是	rowsPerSecond：每秒生成的行的数量 rampUpTime：在到达 rowsPerSecond 之前的时间，以秒为单位 numPartitions：分区的数量
Kafka	是	kafka.Bootstrap.servers：Kafka brokers 列表，以逗号分隔的 host:port subscribe：主题列表，以逗号分隔

7.3.2 输出模式

输出模式是一种方法，用来告诉 Spark 结构流如何将输出数据写入数据接收器中。这个概念对于 Spark 中的流处理来讲是独一无二的。输出模式有以下 3 个选项。

（1）append 模式：如果没有指定输出模式，这是默认模式。在这种模式下，只有追加到结果表的新行才会被发送到指定的输出接收器。只有自上次触发后在结果表中附加的新行将被写入外部存储器。这仅适用于结果表中的现有行不会更改的查询。

（2）complete 模式：此模式将数据完全从内存写入接收器，即整个结果表将被写到输出接收器。当对流数据执行聚合查询时，就需要这种模式。

（3）update 模式：只有自上次触发后在结果表中更新的行才会被写到输出接收器中。对于那些没有改变的行，它们将不会被写出来。注意，这与 complete 模式不同，因为此模式不输出未更改的行。

7.3.3 触发器类型

触发器是另一个需要理解的重要概念。结构化流引擎使用触发器信息来确定何时在流应用程序中运行所提供的流计算逻辑。不同的触发器类型见表 7-2。

表 7-2 Spark 结构化流触发器类型

类 型	描 述
未指定（默认）	对于默认类型，Spark 将使用微批模型，并且当前一批数据完成处理后，立即处理下一批数据
固定周期	对于这种类型，Spark 将使用微批模型，并基于用户提供的周期处理这批数据。如果因为某些原因导致上一批数据的处理超过了该周期，则前一批数据完成处理后，立即处理下一批数据。换句话说，Spark 将不会等到下一个周期的区间边界
一次性	这个触发器类型意味着用于一次性处理可用的批数据，并且一旦该处理完成，Spark 将立即停止流程序。当数据量特别小时，这个触发器很有用，因此，构建一个集群并每天处理几次数据更划算
持续	这个触发器类型会调用新的持续处理模型，该模型被设计用于非常低延迟需求的特定流应用程序。这是 Spark 2.3 中新的实验性处理模式。这时将支持"最少一次性"保证

7.3.4 数据接收器

数据接收器（Data Sink）是用来存储流应用程序的输出的。不同的数据接收器可以支持不同的输出模式，并且具有不同的容错能力，了解这一点很重要。Apache Spark 结构化流支持以下几种数据接收器。

（1）Kafka Sink：要求 Apache Kafka 的版本是 0.10 或更高版本。有一组特定的设置可以连接到 Kafka 集群。

（2）File Sink：这是文件系统、HDFS 或 S3 的目的地。支持常用的文件格式，如文本、CSV、JSON、ORC、Parquet。

（3）Foreach Sink：这是为了在输出中的行上运行任意计算。

（4）Console Sink：这仅用于测试和调试目的，以及在处理低容量数据时。每个触发器上的输出会被打印到控制台。

（5）Memory Sink：这是在处理低容量数据时进行测试和调试的目的。它使用驱动程序的内存来存储输出。

每个数据接收器的配置选项见表 7-3。

表 7-3 Spark 结构化流 Data Sink 配置选项

名 称	支持的输出模式	是否容错	配 置
File	Append	是	path：这是输入目录的路径。支持所有流行的文件格式。详细信息可查看 DataFrameWriter API
Foreach	Append Update Complete	依情况而定	这是一个非常灵活的接收器，它特定于不同的实现

续表

名　称	支持的输出模式	是否容错	配　　置
Console	Append Update Complete	否	numRows：这是每个触发器输出的行的数量，默认为 20 行 truncate：如果每行太长，是否截断，默认为 true
Memory	Append Complete	否	N/A
Kafka	Append Update Complete	是	kafka.Bootstrap.servers：Kafka brokers 列表，以逗号分隔的 host:port topic：这是写入数据的 Kafka 主题

数据接收器必须支持的一个重要的属性（用于结构化的流交付端到端、精确一次性保证）是处理重做的幂等性。换句话说，它必须能够处理使用相同数据的多个写（在不同的时间发生），结果就像只有一个写一样。多重写用于在故障场景中重新处理数据的结果。

7.3.5　水印

水印（Watermark）是流处理引擎中常用的一种技术，用于处理迟到的数据。流应用程序开发人员可以指定一个阈值，让结构化的流引擎知道数据在事件时间（Event Time）内的预期延迟时间。有了这个信息，超过这个预期延迟时间到达的迟到数据会被丢弃。

更重要的是，结构化流使用指定的阈值来确定何时可以丢弃旧状态。如果没有这些信息，则结构化流将需要无限期地维护所有状态，这将导致流应用程序的内存溢出问题。任何执行某种聚合或连接的生产环境下的结构化流应用程序都需要指定水印。

注意：水印是一个重要的概念，关于这个主题的更多细节将在后面的部分中讨论和说明。

7.4　使用各种流数据源

7.3 节描述了结构化流提供的各个内置源。本节将更详细地介绍这些数据源，并将提供使用它们的示例代码。

7.4.1　使用 Socket 数据源

Socket（套接字）数据源很容易使用，只需提供主机和端口号。不过该数据源仅限于学习和测试使用，不要在生产环境中使用。下面这个示例演示了如何在 Spark 结构化流程序中使用 Socket 数据源。

建议按以下过程和步骤执行。

（1）在启动套接字数据源的流式查询之前，首先使用一个网络命令行实用工具，如 Mac 上的 nc 或 Windows 上的 Netcat，启动一个套接字服务器。打开一个终端窗口，执行下面的命令，启动带有端口号 9999 的套接字服务器：

```
$ nc -lk 9999
```

（2）另外打开第 2 个终端，启动 spark-shell，命令如下：

```
$ spark-shell --master spark://localhost:7077
```

（3）在 spark-shell 中，执行结构化流处理代码，代码如下：

```scala
//第7章/sstreaming_source_socket.scala

//从Socket数据源读取流数据
val socketDF = spark.readStream
    .format("socket")
    .option("host", "localhost")
    .option("port", "9999")
    .load()

//流数据转换
val words = socketDF.as[String].flatMap(_.split(" "))
val wordCounts = words.groupBy("value").count()

//流数据输出
val query = wordCounts.writeStream
    .format("console")
    .outputMode("complete")
    .start()
```

（4）回到第 1 个终端窗口，任意输入一些单词，以空格分隔，并按 Enter 键。多输入几行，然后在第 2 个终端窗口观察流计算输出，如图 7-6 所示。

图 7-6 在 Netcat 运行窗口输入文本内容

在第 2 个终端窗口观察到的输出结果如图 7-7 所示。

图 7-7 流计算结果

（5）当完成测试 Socket 数据源时，可以通过调用 stop()函数来停止流查询。在停止流查询之后，在第 1 个终端中输入任何内容都不会导致在 spark-shell 中显示这些内容，代码如下：

```
query.stop
```

7.4.2 使用 Rate 数据源

与 Socket 数据源类似，Rate 数据源也是为了测试和学习目的而设计的。它支持以下这些选项。

（1）rowsPerSecond：每秒应该生成多少行，例如，指定为 100，默认为 1。如果这个数字很高，就可以提供下一个可选配置 rampUpTime。

（2）rampUpTime：在生成速度变为 rowsPerSecond 之前需要多长时间来提升，例如，5s，默认为 0s。使用比秒更细的粒度将被截断为整数秒。

（3）numPartitions：生成行的分区数。默认为 Spark 默认的并行度。

Rate 源将尽力达到 rowsPerSecond，但是查询可能受到资源限制，可以调整 numPartitions 以帮助达到所需的速度。

Rate 源产生的每段数据只包含两列：时间戳和自动增加的值。下面的示例包含打印 Rate 数据源数据的代码。启动 spark-shell，执行以下代码：

```scala
//第 7 章/sstreaming_source_rate.scala

//配置它以便每秒产生 10 行
val rateSourceDF = spark.readStream
    .format("rate")
    .option("rowsPerSecond","10")
    .load()

//以 update 模式将结果写到控制台，并启动流计算
val query = rateSourceDF.writeStream
    .outputMode("update")
    .format("console")
    .option("truncate", "false")
    .start()
```

执行以上代码，可观察到每秒输出 10 条数据，其中部分批次数据如下：

```
-------------------------------------------
Batch: 1
-------------------------------------------
+-----------------------+-----+
|timestamp              |value|
+-----------------------+-----+
|2021-02-02 17:32:01.264|0    |
|2021-02-02 17:32:01.664|4    |
|2021-02-02 17:32:02.064|8    |
```

```
|2021-02-02 17:32:01.364|1    |
|2021-02-02 17:32:01.764|5    |
|2021-02-02 17:32:02.164|9    |
|2021-02-02 17:32:01.464|2    |
|2021-02-02 17:32:01.864|6    |
|2021-02-02 17:32:01.564|3    |
|2021-02-02 17:32:01.964|7    |
+-----------------------+-----+
```

值得注意的是，value 列中的数字保证在所有分区中都是连续的。例如，查看 3 个分区的输出结果，代码如下：

```scala
//第7章/sstreaming_source_rate2.scala

import org.apache.spark.sql.functions._

//配置它以便每秒产生10行，分3个分区
val rateSourceDF = spark.readStream
    .format("rate")
    .option("rowsPerSecond","10")
    .option("numPartitions",3)
    .load()

//添加分区 id 列来检查
val rateWithPartitionDF = rateSourceDF.withColumn("partition_id", spark_partition_id())
val query = rateWithPartitionDF.writeStream
    .outputMode("update")
    .format("console")
    .option("truncate", "false")
    .start()
```

执行以上代码，观察到的输出结果如下：

```
-------------------------------------------
Batch: 1
-------------------------------------------
+-----------------------+-----+------------+
|timestamp              |value|partition_id|
+-----------------------+-----+------------+
|2021-02-02 17:35:43.461|0    |0           |
|2021-02-02 17:35:43.761|3    |0           |
|2021-02-02 17:35:44.061|6    |0           |
|2021-02-02 17:35:44.361|9    |0           |
|2021-02-02 17:35:43.561|1    |1           |
|2021-02-02 17:35:43.861|4    |1           |
|2021-02-02 17:35:44.161|7    |1           |
|2021-02-02 17:35:43.661|2    |2           |
|2021-02-02 17:35:43.961|5    |2           |
|2021-02-02 17:35:44.261|8    |2           |
+-----------------------+-----+------------+
```

前面的输出显示了这 10 行分布在 3 个分区上，并且这些值是连续的，就好像它们是为单个分区生成的一样。

7.4.3 使用 File 数据源

文件数据源是最容易理解和使用的。Spark Structured Streaming 开箱即用地支持所有常用的文件格式，包括文本、CSV、JSON、ORC 和 Parquet。

File 数据源支持以下选项配置。

（1）path：输入目录的路径，对所有文件格式都通用。

（2）maxFilesPerTrigger：每个触发器中考虑处理的最大新文件数（默认为 no max）。

（3）latestFirst：是否先处理最新的文件，当有大量文件积压时很有用（默认为 false）。

（4）fileNameOnly：是否仅根据文件名而不是根据完整路径检查新文件（默认为 false）。将此值设置为 true 后，具有一样的文件名但是不同前缀协议的文件被认为是相同的文件。例如，下面的这些文件被认为是相同的文件，均为 dataset.txt：

```
"file://dataset.txt"
"s3://a/dataset.txt"
"s3n://a/b/dataset.txt"
"s3a://a/b/c/dataset.txt"
```

使用 File 数据源的流程序模板代码如下：

```
//使用 File 数据源，读取 JSON 文件
val mobileSSDF = spark.readStream
    .schema(mobileDataSchema)
    .json("<directoryname>")

//如果指定 maxFilesPerTrigger
val mobileSSDF = spark.readStream
    .schema(mobileDataSchema)
    .option("maxFilesPerTrigger",1)          //一个文件一个文件地处理
    .json("<directory name>")

//如果想要首先处理新文件
val mobileSSDF = spark.readStream
    .schema(mobileDataSchema)
    .option("latestFirst", "true")           //首先处理新产生的文件
    .json("<directory name>")
```

下面通过一个示例程序来演示如何使用结构化流读取文件数据源。假设移动电话的开关机等事件会保存在 JSON 格式的文件中。现在编写 Spark 结构化流处理程序来读取这些事件并处理。建议按以下步骤操作。

（1）准备数据。

在本示例中，使用文件数据源，该数据源以 JSON 文件的格式记录了一小组移动电话动作事件。每个事件由以下 3 个字段组成。

① id：表示手机的唯一 ID。在样例数据集中，电话 ID 将类似于 phone1、phone2、phone3 这样的字符串。

② action：表示用户所采取的操作，该操作的可能值是 open 或 close。

③ ts：表示用户 action 发生时的时间戳，这是事件时间。

下面准备了 3 个存储移动电话事件数据的 JSON 文件，这 3 个文件的内容如下。

file1.json 文件的内容如下：

```
{"id":"phone1","action":"open","ts":"2018-03-02T10:02:33"}
{"id":"phone2","action":"open","ts":"2018-03-02T10:03:35"}
{"id":"phone3","action":"open","ts":"2018-03-02T10:03:50"}
{"id":"phone1","action":"close","ts":"2018-03-02T10:04:35"}
```

file2.json 文件的内容如下：

```
{"id":"phone3","action":"close","ts":"2018-03-02T10:07:35"}
{"id":"phone4","action":"open","ts":"2018-03-02T10:07:50"}
```

file3.json 文件的内容如下：

```
{"id":"phone2","action":"close","ts":"2018-03-02T10:04:50"}
{"id":"phone5","action":"open","ts":"2018-03-02T10:10:50"}
```

为了模拟数据流的行为，将把这 3 个 JSON 文件复制到指定的目录下。

（2）先导入相关的依赖包，代码如下：

```
import org.apache.spark.sql.SparkSession
import org.apache.spark.sql.types._
import org.apache.spark.sql.functions._
```

（3）为手机事件数据创建模式（Schema）。

默认情况下，结构化流在从基于文件的数据源读取数据时需要一个模式（因为最初目录可能是空的，因此结构化的流无法推断模式），但是，可以将配置参数 spark.sql.streaming.schemaInference 的值设置为 true 来启用模式推断。在这个例子中，将显式地创建一个模式，代码如下：

```
//为手机事件数据创建一个 Schema
val fields = Array(
    StructField("id", StringType, nullable = false),
    StructField("action", StringType, nullable = false),
    StructField("ts", TimestampType, nullable = false)
  )
val mobileDataSchema = StructType(fields)
```

（4）读取文件数据源，创建流 DataFrame，并将"action"列的值转换为大写，代码如下：

```
//监听的文件目录
val dataPath = "/data/spark/mobile"

//读取指定目录下的源数据文件，一次一个
val mobileDF = spark.readStream
    .option("maxFilesPerTrigger", 1)
```

```
        .option("mode","failFast")
        .schema(mobileDataSchema)
        .json(dataPath)

//mobileSSDF.isStreaming
//mobileSSDF.printSchema()

//将所有"action"列值转换为大写
import spark.implicits._
val upperDF = mobileDF.select($"id",upper($"action"),$"ts")
```

（5）将结果 DataFrame 输到控制台显示，代码如下：

```
//将结果输到控制台
val query = upperDF.writeStream
    .format("console")
    .option("truncate","false")
    .outputMode("append")
    .start()
```

（6）执行流处理程序，输出的结果如下：

```
-------------------------------------------
Batch: 0
-------------------------------------------
+------+-------------+-------------------+
|id    |upper(action)|ts                 |
+------+-------------+-------------------+
|phone1|OPEN         |2018-03-02 10:02:33|
|phone2|OPEN         |2018-03-02 10:03:35|
|phone3|OPEN         |2018-03-02 10:03:50|
|phone1|CLOSE        |2018-03-02 10:04:35|
+------+-------------+-------------------+


-------------------------------------------
Batch: 1
-------------------------------------------
+------+-------------+-------------------+
|id    |upper(action)|ts                 |
+------+-------------+-------------------+
|phone3|CLOSE        |2018-03-02 10:07:35|
|phone4|OPEN         |2018-03-02 10:07:50|
+------+-------------+-------------------+


-------------------------------------------
Batch: 2
-------------------------------------------
+------+-------------+-------------------+
|id    |upper(action)|ts                 |
+------+-------------+-------------------+
|phone2|CLOSE        |2018-03-02 10:04:50|
|phone5|OPEN         |2018-03-02 10:10:50|
+------+-------------+-------------------+
```

完整的代码如下：

```scala
//第7章/sstreaming_source_file.scala

//为手机事件数据创建一个Schema
val fields = Array(
    StructField("id", StringType, nullable = false),
    StructField("action", StringType, nullable = false),
    StructField("ts", TimestampType, nullable = false)
  )
val mobileDataSchema = StructType(fields)

//监听的文件目录
val dataPath = "/data/spark/mobile"

//读取指定目录下的源数据文件，一次一个
val mobileDF = spark.readStream
    .option("maxFilesPerTrigger", 1)
    .option("mode","failFast")
    .schema(mobileDataSchema)
    .json(dataPath)

//mobileSSDF.isStreaming
//mobileSSDF.printSchema()

//将所有"action"列值转换为大写
import spark.implicits._
val upperDF = mobileDF.select($"id",upper($"action"),$"ts")

//将结果输到控制台
val query = upperDF.writeStream
    .format("console")
    .option("truncate","false")
    .outputMode("append")
    .start()
```

7.4.4 使用 Kafka 数据源

Kafka 通常用于构建实时流数据管道，以便可靠地在系统之间移动数据，还用于转换和响应数据流。Kafka 作为集群运行在一个或多个服务器上。Kafka 的一些关键概念描述如下。

（1）Topic：主题。消息发布到的类别或流名称的高级抽象。主题可以有 0、1 或多个消费者，这些消费者订阅发布到该主题的消息。用户为每个新的消息类别定义一个新主题。

（2）Producers：生产者。向主题发布消息的客户端。

（3）Consumers：消费者。使用来自主题的消息的客户端。

（4）Brokers：服务器。复制和持久化消息数据的一个或多个服务器。

此外，生产者和消费者可以同时对多个主题进行读写。每个 Kafka 主题都是分区的，写入每个分区的消息都是顺序的。分区中的消息具有一个偏移量，用来唯一标识分区内的每条消息。

Kafka 中主题（Topic）的分区是分布式的，每个 Broker 会处理对分区共享的请求。每个分区在 Brokers（数量可配置的）之间复制。Kafka 集群会在一段时间内（可配置的）保留所有已发布的消息。Kafka 使用 ZooKeeper 作为其分布式进程的协调服务。

注意：Kafka 的数据源可能是在生产型流应用程序中最常用的数据源。为了有效地处理这个数据源，读者需要具有一定的使用 Kafka 的基本知识。

在使用 Kafka 数据源时，Spark 流程序实际上充当了 Kafka 的消费者，因此，流程序所需要的信息与 Kafka 的消费者所需要的信息相似。Kafka 数据源的一些配置选项见表 7-4。

表 7-4　Kafka 数据源配置选项

Option	值	描述
kafka.Bootstrap.servers	host1:port1, host2:port2	Kafka 服务器列表，以逗号分隔
subscribe	topic1, topic2	这个数据源要读取的主题名列表，以逗号分隔
subscribePattern	topic.*	使用正则模式表示要读取数据的主题，比 subscribe 更灵活
assign	{topic1:[1,2], topic2:[3,4]}	指定要读取数据的主题的分区。这个信息必须是 JSON 格式

其中必需的信息是要连接的 Kafka 服务器的列表，以及一个或多个从其读取数据的主题。为了支持选择从哪个主题和主题分区来读取数据的各种方法，它支持 3 种不同的方式来指定这些信息。我们只需选择最适合自身用例的那个。

还有一些可选配置选项，都有自己的默认值，见表 7-5。

表 7-5　Kafka 数据源的可选配置选项

Option	默认值	值	描述

startingOffsets	latest	earliest, latest
		每个主题的开始偏移位置，JSON 格式字符串，例如
		{
		"topic1":{"0":45, "1":-1},
		"topic2":{"0":-2}
		}
			earliest：意味着主题的开始处
			latest：意味着主题中的任何最新数据
			当使用 JSON 字符串格式时，-2 代表在一个特定分区中的 earliest offset，-1 代表在一个特定分区中的 latest offset

续表

Option	默认值	值	描述

endingOffsets	latest	Latest
		JSON 格式字符串，例如
		{
		"topic1":{"0":45, "1":-1},
		"topic2":{"0":-1}
		}
			latest：意味着主题中的最新数据
			当使用 JSON 字符串格式时，–1 代表在一个特定分区中的 latest offset。当然 –2 不适用于此选项

maxOffsetsPerTrigger	none	Long，例如，500	此选项是一种速率限制机制，用于控制每个触发器间隔要处理的记录数量。如果指定了一个值，则它表示所有分区的记录总数，而不是每个分区的记录总数

注意：这里只列出了部分选项。更详细的 option 设置选项可以参考 https://spark.apache.org/docs/latest/structured-streaming-kafka-integration.html。

要设置 Kafka，首先需要下载它，下载网址为 http://kafka.apache.org/downloads.html。注意，要选择与 Spark 版本相兼容的版本。

然后，按以下步骤设置 Kafka。

（1）将下载的 Kafka 压缩包解压缩到~/bigdata/目录下，命令如下：

```
$ cd ~/bigdata
$ tar -xvfz kafka_2.12-2.4.1.tgz
```

（2）Kafka 依赖于 Apache ZooKeeper，所以在启动 Kafka 之前，要先启动它，命令如下：

```
$ cd ~/bigdata/kafka_2.12-2.4.1
$ ./bin/zookeeper-server-start.sh config/zookeeper.properties
```

这将在 2181 端口启动 ZooKeeper 进程，并让 ZooKeeper 在后台工作。

（3）接下来，启动 Kafka 服务器，命令如下：

```
$ ./bin/kafka-server-start.sh config/server.properties
```

（4）创建名为 test 的 Kafka 主题，命令如下：

```
$ ./bin/kafka-topics.sh --create --zookeeper localhost:2181 --replication-factor 1 --partitions 1 --topic test
```

（5）查看已有的主题，命令如下：

```
$ ./bin/kafka-topics.sh --list --zookeeper localhost:2181
```

（6）删除一个主题，命令如下：

```
$ ./bin/kafka-topics.sh --delete --Bootstrap-server localhost:9092 --topic test
```

需要修改启动的配置文件 server.properties，设置 delete.topic.enable=true（默认设置为 false）。

默认情况下，Kafka 的数据源并不是 Spark 的内置数据源，因此如果要开发读取 Kafka 数据的 Spark 结构化流处理程序，则必须将 Kafka 的依赖包添加到 classpath 中。

如果要从 spark-shell 中使用 Kafka 数据源，则需要在启动 spark-shell 时将依赖的 JAR 包添加到 classpath 中。有以下两种方式可以做到：

（1）手动将依赖包添加到 classpath。

首先将 kafka_2.12-2.4.1.jar、spark-streaming-kafka-0-10-assembly_2.12-3.1.2.jar 和 spark-sql-kafka-0-10_2.12-3.1.2.jar 包复制到~/bigdata/spark-3.1.2/jars/目录下，然后启动 spark-shell，命令如下：

```
$ cd ~/bigdata/spark-3.1.2
$ ./spark-shell --master spark://localhost:7077
```

（2）使用 jars 参数让 Spark 自动下载这些文件，命令如下：

```
$ cd ~/bigdata/spark-3.1.2
$ ./spark-shell --master spark://localhost:7077 --jars spark-sql-kafka-0-
10_2.12-3.1.2.jar,kafka-clients-2.4.1.jar,spark-streaming-kafka-0-10-
assembly_2.12-3.1.2.jar
```

或者使用--package 选项参数让 Spark 自动下载这些文件，命令如下：

```
$ cd ~/bigdata/spark-3.1.2
$ ./spark-shell --master spark://localhost:7077 --packages org.apache.
spark:spark-sql-kafka-0-10_2.12:3.1.2,org.apache.kafka:kafka_2.12:2.4.1,
org.apache.spark:spark-streaming-kafka-0-10-assembly_2.12:3.1.2
```

如果使用 IDE 开发并使用 SBT 来管理依赖，则需要在项目的 build.sbt 文件中添加依赖配置以使用 Kafka 数据源，内容如下：

```
libraryDependencies += "org.apache.spark" %% "spark-sql-kafka-0-10" %
"3.1.2"
libraryDependencies += "org.apache.kafka" %% "kafka" % "2.4.1"
```

如果使用 IDE 开发并使用 Maven 来管理依赖，则需要在项目的 pom.xml 文件中添加依赖配置以使用 Kafka 数据源，内容如下：

```
<dependency>
    <groupId>org.apache.spark</groupId>
    <artifactId>spark-sql-kafka-0-10_2.12</artifactId>
    <version>3.1.2</version>
</dependency>
<dependency>
    <groupId>org.apache.kafka</groupId>
    <artifactId>kafka</artifactId>
    <version>2.4.1</version>
</dependency>
```

下面通过一个示例来演示如何编写 Spark 结构化流处理程序来读取 Kafka 中的数据。在这个示例中，编写 Spark 结构化流程序作为 Kafka 的消费者程序，Kafka 作为流数据源。使用 Kafka 自带的生产者脚本向 Kafka 的 test 主题发送内容，而 Spark 结构化流程序会订阅该主题。一旦它收到了订阅的消息，马上输到控制台中。程序的处理流程如图 7-8 所示。

图 7-8　流计算结果

首先，编写 Spark 结构化流程序代码，代码如下：

```scala
//第 7 章/sstreaming_source_kafka.scala

//创建一个流来监听 test topic 的消息
val dataDF = spark.readStream
    .format("kafka")
    .option("kafka.Bootstrap.servers", "localhost:9092")
    .option("subscribe", "test")
    .option("startingOffsets", "earliest")
    .load()

//查看这个 DataFrame 的 Schema
dataDF.printSchema()

//将该流转换为 String 数据类型（key 和 value 都采用字节数组形式）
//kvstream=dataDF.selectExpr("CAST(key as string)","CAST(value as string)")
val kvstream = dataDF.selectExpr("CAST(value as string)","topic","partition","offset")

//将该流写到控制台
val query = kvstream.writeStream
    .outputMode("append")
    .format("console")
    .start()
```

要运行这个流程序，建议按以下步骤进行操作：

（1）启动 ZooKeeper 服务。Kafka 依赖于 Apache ZooKeeper，所以在启动 Kafka 之前，要先启动它。打开一个终端窗口，执行的命令如下：

```
$ cd ~/bigdata/kafka_2.12-2.4.1
$ ./bin/zookeeper-server-start.sh config/zookeeper.properties
```

等待 30s 左右，以便 ZooKeeper 启动。

（2）接下来，启动 Kafka 服务器。另打开一个终端窗口，执行的命令如下：

```
$ cd ~/bigdata/kafka_2.12-2.4.1
$ ./bin/kafka-server-start.sh config/server.properties
```

等待 30s 左右，以便 Kafka 启动。

（3）查看和创建 Kafka 主题（如果已经有了 test 主题，则此步略过）。另外打开第 3 个终端窗口，创建 test 主题，执行的命令如下：

```
$ cd ~/bigdata/kafka_2.12-2.4.1
$ ./bin/kafka-topics.sh --create --zookeeper localhost:2181 --replication-factor 1 --partitions 1 --topic test
```

查看已有的主题，执行的命令如下：

```
$ ./bin/kafka-topics.sh --list --zookeeper localhost:2181
```

（4）启动流程序，开始接收从 Kafka test 主题订阅的消息。

（5）向 Kafka test 主题发送消息。另外打开第 4 个终端窗口，生产消息并发布给 test 主题，执行的命令如下：

```
$ cd ~/bigdata/kafka_2.12-2.4.1
$ ./bin/kafka-console-producer.sh --broker-list localhost:9092 --topic test
```

然后，随意输入一些消息。例如，输入以下内容：

```
> good good study
> day day up
```

（6）回到流程序执行窗口，如果一切正常，则应该可以看到在控制台收到的订阅消息，内容如下：

```
root
 |-- key: binary (nullable = true)
 |-- value: binary (nullable = true)
 |-- topic: string (nullable = true)
 |-- partition: integer (nullable = true)
 |-- offset: long (nullable = true)
 |-- timestamp: timestamp (nullable = true)
 |-- timestampType: integer (nullable = true)

-------------------------------------------
Batch: 0
-------------------------------------------
+-----+-----+---------+------+
|value|topic|partition|offset|
+-----+-----+---------+------+
+-----+-----+---------+------+


-------------------------------------------
Batch: 1
-------------------------------------------
+---------------+-----+---------+------+
|          value|topic|partition|offset|
+---------------+-----+---------+------+
|good good study| test|        0|     0|
+---------------+-----+---------+------+


-------------------------------------------
Batch: 2
-------------------------------------------
+----------+-----+---------+------+
|     value|topic|partition|offset|
+----------+-----+---------+------+
|day day up| test|        0|     1|
+----------+-----+---------+------+
```

从上面的输出内容可以看出，从 Kafka 中读取的数据每列都有固定的格式，见表 7-6。

表 7-6 Kafka 消息格式

列	类型
key	binary
value	binary
topic	string
partition	int
offset	long
timestamp	long
timestampType	int

从 Kafka 读取消息时，有多种不同的方式。指定 Kafka 的主题、分区和从 Kafka 读取消息的偏移量的不同变化方式的模板代码如下：

```
//指定 Kafka topic、partition 和 offset 的各种变化
//从多个主题读取，使用默认的 startingOffsets 和 endingOffsets
val kafkaDF = spark.readStream
    .format("kafka")
    .option("kafka.Bootstrap.servers","server1:9092,server2:9092")
    .option("subscribe", "topic1,topic2")
    .load()

//从多个主题读取，使用 subscribePattern
val kafkaDF = spark.readStream
    .format("kafka")
    .option("kafka.Bootstrap.servers","server1:9092,server2:9092")
    .option("subscribePattern", "topic*")
    .load()

//使用 JSON 格式从一个特定的 offset 读取
//Scala 中的三重引号格式用于转义 JSON 字符串中的双引号
Val kafkaDF = spark.readStream
    .format("kafka")
    .option("kafka.Bootstrap.servers","localhost:9092")
    .option("subscribe", "topic1,topic2")
    .option("startingOffsets", """{"topic1": {"0":51} } """)
    .load()
```

7.5 流 DataFrame 操作

前面的例子表明，一旦配置和定义了数据源，DataStreamReader 将返回一个 DataFrame 的实例。这意味着可以使用大多数熟悉的 DataFrame 关系操作和 Spark SQL 函数来表达应

用程序流计算逻辑,但是要注意,并不是所有的 DataFrame 操作都受流式 DataFrame 支持,例如 limit()、distinct()和 sort()就不能在流式 DataFrame 上使用,这是因为它们在流数据处理的上下文中不适用。

7.5.1 选择、投影和聚合操作

Apache Spark 结构化流的一个优点是具有一组统一的 API,用于 Spark 的批处理和流处理。使用流数据格式的 DataFrame,可以应用任何 select()和 filter()转换,以及任何作用在列上的 Spark SQL 函数。此外,基本聚合和高级分析函数也可用于流 DataFrame。下面通过一个移动电话事件数据流分析示例程序来演示这些用法。

移动电话的开关机等事件会保存在 JSON 格式的文件中。现在编写 Spark 结构化流处理程序来读取这些事件并处理。建议按以下步骤操作。

(1)准备数据。

在本示例中,使用文件数据源,该数据源以 JSON 文件的格式记录了一小组移动电话的动作事件。每个事件由以下 3 个字段组成。

① id:表示手机的唯一 ID。在样例数据集中,电话 ID 类似于 phone1、phone2、phone3 等字符串。

② action:表示用户所采取的操作,该操作的可能值是 open 或 close。

③ ts:表示用户 action 发生时的时间戳,这是事件时间。

已经准备了 3 个存储移动电话事件数据的 JSON 文件,这 3 个文件的内容如下。

file1.json 文件的内容如下:

```
{"id":"phone1","action":"open","ts":"2018-03-02T10:02:33"}
{"id":"phone2","action":"open","ts":"2018-03-02T10:03:35"}
{"id":"phone3","action":"open","ts":"2018-03-02T10:03:50"}
{"id":"phone1","action":"close","ts":"2018-03-02T10:04:35"}
```

file2.json 文件的内容如下:

```
{"id":"phone3","action":"close","ts":"2018-03-02T10:07:35"}
{"id":"phone4","action":"open","ts":"2018-03-02T10:07:50"}
```

file3.json 文件的内容如下:

```
{"id":"phone2","action":"close","ts":"2018-03-02T10:04:50"}
{"id":"phone5","action":"open","ts":"2018-03-02T10:10:50"}
```

为了模拟数据流的行为,应把这 3 个 JSON 文件复制到 HDFS 的指定目录下。

(2)先导入相关的依赖包,代码如下:

```
import org.apache.spark.sql.SparkSession
import org.apache.spark.sql.types._
import org.apache.spark.sql.functions._
```

(3)为手机事件数据创建模式(Schema)。

默认情况下,Spark 结构化流在从基于文件的数据源读取数据时需要一个模式(因为

最初目录可能是空的，因此结构化的流无法推断模式），但是，可以将配置参数 spark.sql.streaming.schemaInference 的值设置为 true 来启用模式推断。在这个例子中，将显式地创建一个模式，代码如下：

```
//为手机事件数据创建一个Schema
val fields = Array(
    StructField("id", StringType, nullable = false),
    StructField("action", StringType, nullable = false),
    StructField("ts", TimestampType, nullable = false)
  )
val mobileDataSchema = StructType(fields)
```

（4）读取流文件数据源，创建 DataFrame，并将 "action" 列值转换为大写，代码如下：

```
//监听的文件目录
val dataPath = "/data/spark/mobile"

//读取指定目录下的源数据文件，一次一个
val mobileDF = spark.readStream
    .option("maxFilesPerTrigger", 1)
    .option("mode","failFast")
    .schema(mobileDataSchema)
    .json(dataPath)
```

（5）执行过滤、投影、聚合等转换操作，代码如下：

```
val mobileDF2 = mobileDF
    .where("action='open' or action='close'")
    .withColumn("action",upper(col("action")))
    .select("id","action","ts")
    .groupBy("action")
    .count
```

（6）将结果 DataFrame 输到控制台，代码如下：

```
//将结果输到控制台
val query = mobileDF2.writeStream
    .format("console")
    .option("truncate","false")
    .outputMode("append")
    .start()
```

（7）执行流处理程序，输出的结果如下：

```
-------------------------------------------
Batch: 0
-------------------------------------------
+------+-----+
|action|count|
+------+-----+
|CLOSE |1    |
|OPEN  |3    |
+------+-----+
```

```
-------------------------------------------
Batch: 1
-------------------------------------------
+------+-----+
|action|count|
+------+-----+
|CLOSE |2    |
|OPEN  |4    |
+------+-----+


-------------------------------------------
Batch: 2
-------------------------------------------
+------+-----+
|action|count|
+------+-----+
|CLOSE |3    |
|OPEN  |5    |
+------+-----+
```

完整的实现代码如下:

```scala
//第7章/sstreaming_operator.scala

import org.apache.spark.sql.SparkSession
import org.apache.spark.sql.functions._
import org.apache.spark.sql.types._

//为手机事件数据创建一个Schema
val fields = Array(
    StructField("id", StringType, nullable = false),
    StructField("action", StringType, nullable = false),
    StructField("ts", TimestampType, nullable = false)
  )
val mobileDataSchema = StructType(fields)

//监听的文件目录
val dataPath = "src/main/data/mobile"

//读取指定目录下的源数据文件，一次一个
val mobileDF = spark.readStream
    .option("maxFilesPerTrigger", 1)
    .option("mode","failFast")
    .schema(mobileDataSchema)
    .json(dataPath)

//选择、投影、聚合等操作
val mobileDF2 = mobileDF
    .where("action='open' or action='close'")
    .withColumn("action",upper(col("action")))
    .select("id","action","ts")
```

```
        .groupBy("action")
        .count

//也可以创建一个视图来应用 SQL 查询
//mobileDF2.createOrReplaceTempView("clean_mobile")
//val sqlDF = spark.sql("select id,action,ts from clean_mobile")
//val sqlDF = spark.sql("select action,count(*) as cnt from clean_mobile
group by action")

//将结果输到控制台（注意，输出模式设置）
val query = mobileDF2.writeStream
    .format("console")
    .option("truncate","false")
    .outputMode("complete")
    .start()
```

在这个示例中，采用的输出模式是 complete。在没有聚合操作的情况下，不能使用 complete 输出模式；在有聚合操作的情况下，不能使用 append 模式。

需要注意，在流 DataFrame 中，不支持以下 DataFrame 转换（因为它们太过复杂，无法维护状态，或者由于流数据的无界性）：

（1）在流 DataFrame 上的多个聚合或聚合链。

（2）limit()和 take() N 行。

（3）distinct()转换。

（4）在没有任何聚合的情况下对流 DataFrame 进行排序。

任何使用不受支持的操作的尝试都会导致一个 AnalysisException 异常及类似 "XYZ 操作不受流 streaming DataFrame/DataSets 支持" 的消息。

7.5.2　执行 join 连接操作

从 Spark 2.3 开始，结构化流支持对两个流 DataFrame 执行 join 连接操作。考虑到流 DataFrame 的无界性，结构化的流必须维护两个流 DataFrame 的历史数据，以匹配任何未来的、尚未收到的数据。

可以用一个流式 DataFrame 来连接另一个静态的 DataFrame 或者流式 DataFrame，然而，join 连接是一个复杂的操作，其中最棘手的问题在于并非所有的数据在连接时都是可用的流 DataFrame，因此，join 连接的结果是在每个触发器点上增量地生成的。

下面通过一个 IoT（物联网）示例来学习两个流 DataFrame 的连接操作。假设在某个数据中心，通过不同的传感器采集不同类型的实时数据，其中第 1 个传感器采集不同机架的实时温度读数；第 2 个传感器采集不同机架的实时负载信息。

这些数据都存储在 JSON 格式的数据文件中，包含数据中心中不同位置机架的温度读数。

file1_temp.json 文件的内容如下：

```
{"temp_location_id":"rack1","temperature":99.5,"temp_taken_time":"2017-06-02T08:01:01"}
{"temp_location_id":"rack2","temperature":100.5,"temp_taken_time":"2017-06-02T08:06:02"}
{"temp_location_id":"rack3","temperature":101.0,"temp_taken_time":"2017-06-02T08:11:03"}
{"temp_location_id":"rack4","temperature":102.0,"temp_taken_time":"2017-06-02T08:16:04"}
```

包含同一数据中心中每台计算机的负载信息，file2_load.json 文件的内容如下：

```
{"load_location_id":"rack1","load":199.5,"load_taken_time":"2017-06-02T08:01:02"}
{"load_location_id":"rack2","load":1105.5,"load_taken_time":"2017-06-02T08:06:04"}
{"load_location_id":"rack3","load":2104.0,"load_taken_time":"2017-06-02T08:11:06"}
{"load_location_id":"rack4","load":1108.0,"load_taken_time":"2017-06-02T08:16:08"}
{"load_location_id":"rack4","load":1108.0,"load_taken_time":"2017-06-02T08:21:10"}
```

现在需要编写一个 Spark 流处理程序，连接这两个流数据集，统计每个机架实时的温度和负载。实现代码如下（注意其中的连接条件和时间约束条件设置）：

```scala
//第 7 章/sstreaming_join.scala

//导入依赖包
import org.apache.spark.sql.functions.expr
import org.apache.spark.sql.types._
import org.apache.spark.sql.functions._

//为 IoT 温度数据创建一个 Schema
val tempDataSchema = new StructType()
    .add("temp_location_id", StringType, nullable = false)
    .add("temperature", DoubleType, nullable = false)
    .add("temp_taken_time", TimestampType, nullable = false)

//为 IoT 负载数据创建一个 schema
val loadDataSchema = new StructType()
    .add("load_location_id", StringType, nullable = false)
    .add("load", DoubleType, nullable = false)
    .add("load_taken_time", TimestampType, nullable = false)

//读取流数据源
val dataPath1 = "/data/spark/iot2/temp-input"
val tempDataDF = spark.readStream
    .option("maxFilesPerTrigger", 1)
    .option("timestampFormat","yyyy-MM-dd hh:mm:ss")   //*
    .option("mode","failFast")
    .schema(tempDataSchema)
    .json(dataPath1)
```

```
val dataPath2 = "/data/spark/iot2/load-input"
val loadDataDF = spark.readStream
    .option("maxFilesPerTrigger", 1)
    .option("timestampFormat","yyyy-MM-dd hh:mm:ss")   //*
    .option("mode","failFast")
    .schema(loadDataSchema)
    .json(dataPath2)

//基于 location id 连接,以及事件时间约束
val sqlExpr =
    """temp_location_id = load_location_id AND
      load_taken_time >= temp_taken_time AND
      load_taken_time <= temp_taken_time + interval 10 minutes
    """
val tempWithLoadDataDF = tempDataDF.join(loadDataDF, expr(sqlExpr))

//将结果输到控制台(注意,因为有聚合操作,所以输出模式必须是"complete")
val query = tempWithLoadDataDF.writeStream
    .format("console")
    .option("truncate","false")
    .outputMode("append")
    .start()
```

注意:在上面的代码中注释标*的代码行不是必需的。如果是非标准时间戳格式,则可使用 option 来指定解析格式。

执行以上代码,可以得到的输出结果如下:

```
+----------------+-----------+-------------------+----------------+------+-------------------+
|temp_location_id|temperature|temp_taken_time    |load_location_id|load  |load_taken_time    |
+----------------+-----------+-------------------+----------------+------+-------------------+
|rack1           |99.5       |2017-06-02 08:01:01|rack1           |199.5 |2017-06-02 08:01:02|
|rack3           |101.0      |2017-06-02 08:11:03|rack3           |2104.0|2017-06-02 08:11:06|
|rack4           |102.0      |2017-06-02 08:16:04|rack4           |1108.0|2017-06-02 08:16:08|
|rack4           |102.0      |2017-06-02 08:16:04|rack4           |1108.0|2017-06-02 08:21:10|
|rack2           |100.5      |2017-06-02 08:06:02|rack2           |1105.5|2017-06-02 08:06:04|
+----------------+-----------+-------------------+----------------+------+-------------------+
```

在这个结果表上,可以进一步执行 Spark SQL 查询操作。

当连接一个静态 DataFrame 和一个流 DataFrame 时,以及当连接两个流 DataFrames 时,外连接会受到更多的限制,相关的一些细节见表 7-7。

表 7-7 两个流 DataFrames 在执行 join 连接时的限制说明

左侧 + 右侧	连接类型	说 明
静态数据 + 流数据	内连接	支持
静态数据 + 流数据	左外连接	不支持
静态数据 + 流数据	右外连接	支持
静态数据 + 流数据	全外连接	不支持

续表

左侧 + 右侧	连接类型	说　明
流数据 + 流数据	内连接	支持
流数据 + 流数据	左外连接	有条件地支持。必须在右侧指定水印及时间约束
流数据 + 流数据	右外连接	有条件地支持。必须在左侧指定水印及时间约
流数据 + 流数据	全连接	不支持

7.6 使用数据接收器

流应用程序的最后一步通常是将计算结果写入一些外部系统或存储系统。Spark 结构化流提供了 5 个内置数据接收器，其中 3 个是用于生产的，两个用于测试目的。下面的部分将详细介绍每个数据接收器。

7.6.1 使用 File Data Sink

File Data Sink（文件数据接收器）是一个非常简单的数据接收器，需要提供的唯一且必需的选项是输出目录。由于 File Data Sink 是容错的，结构化的流将需要一个检查点来写进度信息和其他元数据，以帮助在出现故障时进行恢复。

在下面这个示例中，配置 Rate 数据源，每秒产生 10 行数据，将生成的数据行发送到两个分区，并将数据以 JSON 格式写到指定的目录。

实现代码如下：

```scala
//第 7 章/sstreaming_sink_file.scala

//将数据从 Rate 数据源写到 File Sink
val rateSourceDF = spark.readStream
    .format("rate")
    .option("rowsPerSecond","10")           //每秒产生 10 条数据
    .option("numPartitions","2")            //两个分区
    .load()

val query = rateSourceDF.writeStream
    .outputMode("append")
    .format("json")                         //或"csv"
    .option("path", "tmp/output")           //设置输出目录
    .option("checkpointLocation", "tmp/ck") //设置 checkpoint
    .start()

//如果是在应用程序中执行，则包含这一句
//query.awaitTermination()                  //等待流程序结束
```

由于分区的数量被配置为两个分区，所以每当结构化流在每个触发点上写出数据时，就会将两个文件写到输出目录中，因此，如果检查输出目录，则会看到带有名称的文件，

这些名称以 part-00000 或 part-00001 开头。将 Rate 数据源配置为每秒 10 行，并且有两个分区，因此，每个输出包含 5 行。

part-00000-*.json 文件的内容如下：

```
{"timestamp":"2021-02-03T17:56:46.283+08:00","value":0}
{"timestamp":"2021-02-03T17:56:46.483+08:00","value":2}
{"timestamp":"2021-02-03T17:56:46.683+08:00","value":4}
{"timestamp":"2021-02-03T17:56:46.883+08:00","value":6}
{"timestamp":"2021-02-03T17:56:47.083+08:00","value":8}
{"timestamp":"2021-02-03T17:56:47.283+08:00","value":10}
{"timestamp":"2021-02-03T17:56:47.483+08:00","value":12}
{"timestamp":"2021-02-03T17:56:47.683+08:00","value":14}
{"timestamp":"2021-02-03T17:56:47.883+08:00","value":16}
{"timestamp":"2021-02-03T17:56:48.083+08:00","value":18}
```

part-00001-*.json 文件的内容如下：

```
{"timestamp":"2021-02-03T17:56:46.383+08:00","value":1}
{"timestamp":"2021-02-03T17:56:46.583+08:00","value":3}
{"timestamp":"2021-02-03T17:56:46.783+08:00","value":5}
{"timestamp":"2021-02-03T17:56:46.983+08:00","value":7}
{"timestamp":"2021-02-03T17:56:47.183+08:00","value":9}
{"timestamp":"2021-02-03T17:56:47.383+08:00","value":11}
{"timestamp":"2021-02-03T17:56:47.583+08:00","value":13}
{"timestamp":"2021-02-03T17:56:47.783+08:00","value":15}
{"timestamp":"2021-02-03T17:56:47.983+08:00","value":17}
{"timestamp":"2021-02-03T17:56:48.183+08:00","value":19}
```

7.6.2　使用 Kafka Data Sink

在结构化的流中，将流 DataFrame 的数据写入 Kafka Data Sink（Kafka 数据接收器），要比从 Kafka 的数据源中读取数据简单得多。

Kafka 的 Data Sink 配置选项见表 7-8。

表 7-8　Kafka Data Sink 配置选项

Option	值	描述
kafka.Bootstrap.servers	host1:port1 host2:port2	Kafka 服务器列表，用逗号分隔
topic	字符串，如"topic1"	这是单个的主题(topic)名称
key	一个字符串，或二进制	这个 key 用来决定一个 Kafka 消息应该被发送到哪个分区。所有具有相同 key 的 Kafka 消息将被发送到同一分区。这是一个可选项
value	一个字符串，或二进制	这是消息的内容。对于 Kafka，它只是一字节数组，对 Kafka 没有任何意义

其中有 3 个选项是必需的。重点要理解的是 key 和 value，它们与 Kafka 消息的结构有

关。正如前面提到的，Kafka 的数据单元是一条消息，本质上是一个键-值对。这个 value 的作用就是保存消息的实际内容，而它对 Kafka 没有任何意义。就 Kafka 而言，value 只是一堆字节，然而，key 被 Kafka 认为是一个元数据，它和 value 一起被保存在 Kafka 的信息中。当一条消息被发送到 Kafka 并且一个 key 被提供时，Kafka 将其作为一种路由机制来确定一个特定的 Kafka 消息应该被发送到哪一个分区，按照对该 key 哈希并对 topic 的分区数求余。这意味着所有具有相同 key 的消息都将被路由到同一个分区。如果消息中没有提供 key，则 Kafka 就不能保证消息被发送到哪个分区，而 Kafka 使用了一个循环算法来平衡分区之间的消息。

提供 Topic 主题名称有两种方法。第 1 种方法是在设置 Kafka Data Sink 时在配置中提供主题名称，第 2 种方法是在流 DataFrame 中定义一个名为 topic 的列，该列的值将用作 Topic 主题的名称。

如果名为 key 的列存在于流 DataFrame 中，则该列的值将用作消息的 key。因为该 key 是一个可选的元数据，所以在流 DataFrame 中不是必须有这一列。

另一方面，必须提供 value 值，而 Kafka 的 Data Sink 则期望在流 DataFrame 中有一个名为 value 的列。

注意：如果要开发以 Kafka 为 Data Sink 的 Spark 结构化流处理程序，则必须将 Kafka 的依赖包添加到 classpath 中。可参考 7.4.4 节中的说明。

下面编写一个 Spark 结构化流应用程序作为 Kafka 的生产者，它读取 Rate 数据源，然后将数据写入 Kafka 的 rates 主题中。

在这个示例中，Spark 结构化流程序会向 Kafka 的 rates 主题发送消息（本例为读取 Rate 数据源的数据），用 Kafka 自带的消费者脚本程序订阅该主题。一旦它收到了订阅的消息，马上输出。程序的处理流程如图 7-9 所示。

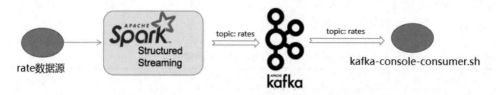

图 7-9 Spark 结构化流程序将计算结果写入 Kafka

编写 Spark 结构化流程序处理逻辑，代码如下：

```scala
//第7章/sstreaming_sink_kafka.scala

import org.apache.spark.sql.SparkSession
import org.apache.spark.sql.functions._

//以每秒10行的速度设置Rate数据源，并使用两个分区
val ratesSinkDF = spark.readStream
    .format("rate")
```

```
      .option("rowsPerSecond","10")
      .option("numPartitions","2")
      .load()

//转换 ratesSinkDF 以创建一个"key"列和"value"列
//value 列包含一个 JSON 字符串,该字符串包含两个字段:timestamp 和 value
val ratesSinkForKafkaDF = ratesSinkDF
    .select(
       col("value").cast("string") as "key",
       to_json(struct("timestamp","value")) as "value"
    )

//设置一个流查询,使用 topic "rates",将数据写到 Kafka
val query = ratesSinkForKafkaDF.writeStream
    .outputMode("append")
    .format("kafka")
    .option("kafka.Bootstrap.servers", "localhost:9092")
    .option("topic","rates")
    .option("checkpointLocation", "tmp/rates")
    .start()

//等待流程序结束
//query.awaitTermination()          //在 spark-shell 交互环境下不需要这行代码
```

要运行这个流程序,建议按以下步骤进行操作:

(1) 启动 ZooKeeper 服务。

Kafka 依赖于 Apache ZooKeeper,所以在启动 Kafka 之前,要先启动它。打开一个终端窗口,执行的命令如下:

```
$ cd ~/bigdata/kafka_2.12-2.4.1
$ ./bin/zookeeper-server-start.sh config/zookeeper.properties
```

等待 30s 左右,以便 ZooKeeper 启动。

(2) 接下来,启动 Kafka 服务器。

另打开一个终端窗口,执行的命令如下:

```
$ cd ~/bigdata/kafka_2.12-2.4.1
$ ./bin/kafka-server-start.sh config/server.properties
```

等待 30s 左右,以便 Kafka 启动。

(3) 查看和创建 Kafka 主题(如果已经有了 rates 主题,则此步略过)。

创建 rates 主题。另外打开第 3 个终端窗口,执行的命令如下:

```
$ cd ~/bigdata/kafka_2.12-2.4.1
$ ./bin/kafka-topics.sh --create --zookeeper localhost:2181 --replication-factor 1 --partitions 1 --topic rates
```

查看已有的主题,执行的命令如下:

```
$ ./bin/kafka-topics.sh --list --zookeeper localhost:2181
```

(4) 在第 3 个终端窗口,运行 Kafka 自带的消费者脚本,订阅 rates 主题消息,执行的

命令如下:

```
$ ./bin/kafka-console-consumer.sh --Bootstrap-server 192.168.190.145:9092
--topic rates
```

(5) 执行上面的流处理程序代码。

(6) 回到第 3 个终端窗口(运行消费者脚本的窗口),如果一切正常,应该可以看到在终端输出的订阅消息,部分内容如下:

```
{"timestamp":"2021-02-03T19:14:59.352+08:00","value":1960}
{"timestamp":"2021-02-03T19:14:59.552+08:00","value":1962}
{"timestamp":"2021-02-03T19:14:59.752+08:00","value":1964}
{"timestamp":"2021-02-03T19:14:59.952+08:00","value":1966}
{"timestamp":"2021-02-03T19:15:00.152+08:00","value":1968}
{"timestamp":"2021-02-03T19:14:59.452+08:00","value":1961}
{"timestamp":"2021-02-03T19:14:59.652+08:00","value":1963}
{"timestamp":"2021-02-03T19:14:59.852+08:00","value":1965}
...
```

7.6.3　使用 Foreach Data Sink

与结构化流提供的其他内置数据接收器相比,Foreach 是一个很有意思的数据接收器,因为它根据数据应该如何被写出、何时写出数据及将数据写入何处,具有完整的灵活性和可扩展性,但这种灵活性和可扩展性是有要求的,即由用户自己来负责在使用这个数据接收器时写出数据的逻辑。

要使用 Foreach 接收器,必须实现 ForeachWriter 接口(注意,该接口只支持 Scala 和 Java 语言),它包含 3 种方法:open()、process()和 close()。只要有一个触发器的输出生成一系列的行,这些方法就会被调用。

使用 Foreach 数据接收器需要注意以下细节:

(1) ForeachWriter 抽象类实现的一个实例将在驱动程序端被创建,它将被发送到 Spark 集群中的 Executors 执行。这有两个条件。第一,ForeachWriter 的实现必须是可序列化的,否则它的实例不能通过网络发送到 Executors。第二,如果在创建过程中有任何初始化都将发生在驱动程序端,因此,如果想打开一个数据库连接或套接字连接,那就不应该在类初始化期间发生,而是在其他地方。

(2) 在流 DataFrame 中分区的数量决定了有多少个 ForeachWriter 实现的实例被创建。这类似于 Dataset.foreachPartition()方法。

(3) 在 ForeachWriter 抽象类中定义的 3 种方法将在 Executors 上被调用。

(4) 执行初始化(例如打开数据库连接或套接字连接)的最佳位置是在 open()方法中,然而,每当有数据被写出来时,就会调用 open()方法,因此,这种逻辑必须是智能和高效的。

(5) open()方法签名有两个输入参数:分区 ID 和版本。返回类型是布尔型。这两个参数的组合唯一地表示一组需要被写出来的行。这个版本的值是一个单调递增的 ID,随着每个触发器的增加而增加。根据分区 ID 的值和版本参数，open()方法需要决定它是否需要写出行序列，并将适当的布尔值返回结构流引擎。

（6）如果 open()方法的返回值为 true，则对于触发器输出的每行就会调用 process()方法。

（7）无论何时调用 open()方法，不管它返回什么值，close()方法也都会被调用。如果在调用 process()方法时出现错误，则该错误将被传递到 close()方法中。调用 close()方法的目的是给用户一个机会来清理在 open()或 process()方法调用期间创建的任何必要状态。只有当 Executor 的 JVM 崩溃或者 open()方法抛出一个 Throwable 异常时，才不会调用 close()方法。

简而言之，Foreach 数据接收器为用户提供了一个在写出流 DataFrame 的数据时足够的灵活性。下面通过一个示例来演示如何使用这种类型的 Data Sink。

编写 Spark 结构化流应用程序，将 Rate 数据源中的数据写入控制台，此程序包含了一个 ForeachWriter 抽象类的简单实现。

首先导入依赖的类，代码如下：

```scala
//第7章/sstreaming_sink_foreach.scala

import org.apache.spark.sql.{ForeachWriter, Row, SparkSession}
```

接下来定义一个 ForeachWriter 抽象类的实现，并实现它的 3 种方法：open()、process()和 close()，代码如下：

```scala
//自定义一个ConsoleWriter,继承自ForeachWriter
class ConsoleWriter(private var pId:Long = 0, private var ver:Long = 0)
            extends ForeachWriter[Row] {

    //初始化方法
    def open(partitionId: Long, version: Long): Boolean = {
        pId = partitionId       //分区 id
        ver = version           //版本号
        println(s"open => ($partitionId, $version)")
        true
    }

    //业务处理方法
    def process(row: Row): Unit = {
        println(s"writing => $row")
    }

    //做一些清理性的工作
    def close(errorOrNull: Throwable): Unit = {
        println(s"close => ($pId, $ver)")
    }
}
```

然后，在流数据写出时，指定 foreach()方法并使用上面自定义的 ConsoleWriter，代码如下：

```
//以每秒10行的速度设置Rate数据源，并使用两个分区
val ratesSourceDF = spark.readStream
    .format("rate")
    .option("rowsPerSecond","10")
    .option("numPartitions","2")
    .load()

//设置Foreach Data Sink
val query = ratesSourceDF.writeStream
    .foreach(new ConsoleWriter)
    .start()

//等待流程序结束
query.awaitTermination()              //在spark-shell交互方式下不需要这一行代码
```

当开始执行时，可以看到控制台的输出，内容如下：

```
open => (0, 1)
writing => [2021-02-03 19:52:53.194,0]
writing => [2021-02-03 19:52:53.394,2]
writing => [2021-02-03 19:52:53.594,4]
writing => [2021-02-03 19:52:53.794,6]
writing => [2021-02-03 19:52:53.994,8]
close => (0, 1)
open => (1, 1)
writing => [2021-02-03 19:52:53.294,1]
writing => [2021-02-03 19:52:53.494,3]
writing => [2021-02-03 19:52:53.694,5]
writing => [2021-02-03 19:52:53.894,7]
writing => [2021-02-03 19:52:54.094,9]
close => (1, 1)
open => (0, 2)
writing => [2021-02-03 19:52:54.194,10]
writing => [2021-02-03 19:52:54.394,12]
writing => [2021-02-03 19:52:54.594,14]
writing => [2021-02-03 19:52:54.794,16]
writing => [2021-02-03 19:52:54.994,18]
close => (0, 2)
open => (1, 2)
writing => [2021-02-03 19:52:54.294,11]
writing => [2021-02-03 19:52:54.494,13]
writing => [2021-02-03 19:52:54.694,15]
writing => [2021-02-03 19:52:54.894,17]
writing => [2021-02-03 19:52:55.094,19]
close => (1, 2)
...
```

7.6.4 使用Console Data Sink

Console Data Sink非常简单，但它不是一个容错的数据接收器。它主要用于学习和测

试，不能在生产环境下使用。它只有两种配置选项：要显示的行数，以及输出太长时是否截断。这些选项都有一个默认值，见表 7-9。

表 7-9 Console Data Sink 配置选项

Option	默 认 值	描　　述
numRows	20	在控制台输出的行的数量
truncate	true	当每行的内容超过 20 个字符时，是否截断显示

Console Data Sink 的底层实现使用了与 DataFrame.show() 方法相同的逻辑来显示流 DataFrame 中的数据。下面通过一个示例来了解这些 option 参数的用法。

编写 Spark 结构化流程序，读取 Rate 数据源流数据，并将流数据的处理结果写到控制台，每次输出不超过 30 行，代码如下：

```scala
//第7章/sstreaming_sink_console.scala

//Rate 数据源读出数据
val rateSourceDF = spark.readStream
    .format("rate")
    .option("rowsPerSecond","10")     //每秒产生 10 条数据
    .option("numPartitions","2")      //两个分区
    .load()

//将结果 DataFrame 写到控制台
val query = rateSourceDF.writeStream
    .outputMode("append")
    .format("console")                //Console Data Sink
    .option("truncate",value = false) //不截断显示
    .option("numRows",30)             //每次输出 30 行
    .start()

//等待流程序结束
//query.awaitTermination()            //spark-shell 交互环境下不需要此行代码
```

执行上面的程序，输出的结果如下：

```
-------------------------------------------
Batch: 1
-------------------------------------------
+-----------------------+-----+
|timestamp              |value|
+-----------------------+-----+
|2021-02-03 20:07:36.987|0    |
|2021-02-03 20:07:37.187|2    |
|2021-02-03 20:07:37.387|4    |
|2021-02-03 20:07:37.587|6    |
|2021-02-03 20:07:37.787|8    |
|2021-02-03 20:07:37.087|1    |
|2021-02-03 20:07:37.287|3    |
```

```
|2021-02-03 20:07:37.487|5    |
|2021-02-03 20:07:37.687|7    |
|2021-02-03 20:07:37.887|9    |
+-----------------------+-----+
...
```

7.6.5 使用 Memory Data Sink

与 Console 数据接收器类似，这个数据接收器也很容易理解和使用。事实上，它非常简单，不需要任何配置。它也不是一个容错的数据接收器，主要用于学习和测试，不要在生产环境中使用。它收集的数据被发送给 Driver，并作为内存中的表存储在 Driver 中。换句话说，可以发送到 Memory 数据接收器的数据量是由 Driver 中 JVM 拥有的内存大小决定的。在设置这个数据接收器时，可以指定一个查询名称作为 DataStreamWriter.queryName 函数参数，然后就可以对内存中的表发出 SQL 查询。与 Console 数据接收器不同的是，一旦数据被发送到内存中的表，就可以使用绝大多数在 Spark SQL 组件中可用的特性进一步分析或处理数据。如果数据量很大，并且不适合内存，则最好的选择就是使用 File Data Sink 以 Parquet 格式来写出数据。

下面通过一个示例来了解 Memory 数据接收器的用法。编写 Spark 结构化流程序，读取 Rate 数据源流数据，并将流数据的处理结果写到内存表中，然后将该内存表发出。

实现代码如下：

```scala
//第 7 章/sstreaming_sink_memory.scala

//读取 Rate 数据源数据
val ratesDF = spark.readStream
    .format("rate")
    .option("rowsPerSecond","5")
    .option("numPartitions","1")
    .load()

//将数据写到 Memory Data Sink，内存表名为"rates"
val query = ratesDF.writeStream
    .outputMode("append")
    .format("memory")                       //指定 Memory Data Sink
    .queryName("rates")                     //指定查询名称
    .option("truncate", value = false)
    .start()

//针对 rates 内存表发出 SQL 查询
spark.sql("select * from rates").show(10)

//统计 rates 内存表中的行数
spark.sql("select count(*) from rates").show()

query.awaitTermination()          //在 spark-shell 交互环境下，不需要这一行代码
```

需要注意的一点是，即使在流查询 ratesSQ 停止之后，内存中的 rates 仍然会存在，然而，一旦一个新的流查询以相同的名称开始，那么来自内存中的数据就会被截断。

7.6.6 Data Sink 与输出模式

在了解了有哪些 Data Sink 之后，还有很重要的一点是要了解每种类型的 Data Sink 支持哪些输出。关于 Data Sink 和所支持的输出模式，见表 7-10。

表 7-10 Data Sink 与输出模式

Data Sink	支持的输出模式	备 注
File	Append	只支持写出新行，没有更新
Kafka	Append、Update、Complete	
Foreach	Append、Update、Complete	依赖于 ForeachWriter 实现
Console	Append、Update、Complete	
Memory	Append、Complete	不支持 in-place 更新

7.7 深入研究输出模式

在 Spark 结构化流中，输出模式可以是 complete、update 或 append，其中 complete 输出模式意味着每次都要写入全部的结果表，update 输出模式写入从上批处理中已更改的行，而 append 输出模式仅写入新行。

一般来讲，有以下两种类型的流查询：

（1）第 1 种类型称为"无状态类型"，它只对流入的流数据进行基本的转换，然后将数据写到一个数据接收器上。

（2）第 2 种类型称为"有状态类型"，它需要保持一定数量的状态，不管它是隐式还是显式地完成。

有状态类型通常执行某种聚合，或者使用像 mapGroupsWithState() 或 flatMapGroupsWithState() 这样的结构化流 API，可以维护特定用例所需的任意状态，例如，维护用户会话数据。

7.7.1 无状态流查询

无状态流查询的典型应用场景是实时流 ETL，它可以连续读取实时流数据，例如在线服务连续生成的 PV 事件，以捕获哪些页面正在被哪些用户浏览。在这种用例中，它通常执行以下操作：

（1）过滤、转换和清洗。实际生成并获取的数据是混乱的脏数据，而且这种结构可能不太适合重复分析。

（2）转换为更有效的存储格式。像 CVS 和 JSON 这样的文本文件格式易读性虽然好，但对于重复分析来讲是低效的，特别是如果数据量很大，例如几百 TB 字节。更有效的二进制格式（如 PRC、Parquet 或 Avro）通常用于减少文件大小和提高分析速度。

（3）按某些列划分数据。在将数据写到数据接收器时，可以根据常用列的值对数据进行分区，以加快组织中不同团队的重复分析。

以前的任务在将数据写到数据接收器之前不需要流查询来维护任何类型的状态，Spark 也不需要记住以前微批处理的任何数据来处理当前记录。随着新数据的出现，它被清理、转换，并可能进行重组，然后立即被写出，因此，这种无状态流类型的唯一适用的输出模式是 append。不适合使用 complete 输出模式，因为这需要结构化的流来维护所有以前的数据，这些数据可能太大而无法维护。也不适合使用 update 输出模式，因为只有新数据会被写出来，然而，当这种 update 输出模式被用于无状态流查询时，结构化流就会识别这个并将其与 append 输出模式相同对待。当不适当的输出模式用于流查询时，结构化的流引擎会抛出异常。

下面的示例展示了在使用不适当的输出模式时（使用无状态流查询时使用 complete 输出模式）会发生什么情况，代码如下：

```scala
//使用无状态流查询时使用complete输出模式
val ratesDF = spark.readStream
    .format("rate")
    .option("rowsPerSecond","10")
    .option("numPartitions","2")
    .load()

//简单转换
val ratesOddEvenDF = ratesDF.withColumn("even_odd", $"value" % 2 === 0)

//写到Console Data Sink，使用complete输出模式
val ratesSQ = ratesOddEvenDF.writeStream
    .outputMode("complete")          //注意这里
    .format("console")
    .option("truncate",false)
    .option("numRows",50)
    .start()
```

当提交以上代码执行时，会抛出异常信息，内容如下：

```
//在分析阶段来自结构化流的一个异常
org.apache.spark.sql.AnalysisException: Complete output mode not supported when there are no streaming aggregations on streaming DataFrames/Datasets;
```

7.7.2 有状态流查询

在执行诸如计数、平均、求和等聚合运算时，Spark 将跨先前的多个微批组合信息。

为此，流程序需要维护每个 Executor 上部分计数的一些元数据信息，这就是所谓的"状态"。例如，假设正在计算已经解析或过滤的记录的数量，这里的计数就是状态，每个选中的记录增加计数，代码如下：

```
df.where(col("data.type") == lit("type1")).count()
```

有状态流处理面临着以下两个挑战：

（1）确保容错。

（2）精确一次性交付语义。

结构化流在底层默认作为微批处理进行处理。当流查询运行时，状态在微批之间进行版本控制，因此，在生成一系列增量执行计划时，每个执行计划都知道它需要从哪个版本的状态中读取数据。每个微批处理会读取状态数据的前一个版本，即前一个运行计数，然后更新它并创建一个新版本。每个版本的检查点都位于在查询中提供的相同检查点位置。这样，Spark 引擎就可以从故障中恢复，并确保精确地一次交付，因为它确切地知道需要从哪里重新启动。

这些分布式状态文件存储在每个 Executor 的内存中，对这些状态文件的所有更改都通过将 WAL（Write Ahead Log，预写日志）文件写到检查点位置（如 HDFS 或 S3）来支持。

在有状态流中最简单和最常用的是流聚合。流聚合又包含多种聚合类型。

（1）按 key 进行聚合，模板代码如下：

```
eventsDF.groupBy(col("data.type")).count()
```

（2）按事件时间窗口聚合，模板代码如下：

```
eventsDF
   .groupBy(window("timestamp", "5 mins"))
   .agg(avg(col("data.clicks")))
```

（3）同时按 key 和事件时间窗口聚合，模板代码如下：

```
eventsDF
   .groupBy(col("data.type"), window("timestamp", "5 mins"))
   .agg(avg(col("data.clicks")))
```

（4）用户定义的聚合函数（UDAF）。

当一个有状态的流查询通过一个 group by 转换执行一个聚合时，这个聚合的状态是由结构化的流引擎隐式地维护的。随着更多数据的到达，新数据聚合的结果被更新到结果表中。

在每个触发点上，根据输出模式，更新后的数据或结果表中的所有数据都被写到一个数据接收器上。这意味着使用 append 输出模式是不合适的，因为这违反了输出模式的语义，该模式指定只有附加到结果表的新行将被发送到指定的输出接收器。换句话说，只有 complete 和 update 输出模式适合于有状态查询类型，并且聚合状态隐式地由结构化流引擎负责维护。使用 complete 输出模式的流查询的输出总是等于或超过使用 update 输出模式的相同流查询的输出。

下面的示例用来说明 update 和 complete 模式之间的输出差异。移动电话的开关机等事件会保存在 JSON 格式的文件中。现在编写 Spark 结构化流处理程序来读取这些事件并统计不同的 action 发生的数量。建议按以下步骤操作。

（1）准备数据。

在本示例中使用文件数据源，该数据源以 JSON 文件的格式记录了一小组移动电话动作事件。每个事件由以下 3 个字段组成。

① id：表示手机的唯一 ID。在样例数据集中，电话 ID 类似于 phone1、phone2、phone3 等这样的字符串。

② action：表示用户所采取的操作，该操作的可能值是 open 或 close。

③ ts：表示用户 action 发生时的时间戳，这是事件时间。

本例准备了两个存储移动电话事件数据的 JSON 文件：action.json 和 newaction.json。

action.json 文件的内容如下：

```
{"id":"phone1","action":"open","ts":"2018-03-02T10:02:33"}
{"id":"phone2","action":"open","ts":"2018-03-02T10:03:35"}
{"id":"phone3","action":"open","ts":"2018-03-02T10:03:50"}
{"id":"phone1","action":"close","ts":"2018-03-02T10:04:35"}
{"id":"phone3","action":"close","ts":"2018-03-02T10:07:35"}
```

newaction.json 文件的内容如下：

```
{"id":"phone4","action":"open","ts":"2018-03-02T10:07:50"}
{"id":"phone2","action":"crash","ts":"2018-03-02T11:09:13"}
{"id":"phone5","action":"swipe","ts":"2018-03-02T11:17:29"}
```

注意在这两个数据文件中，action 字段值的区别。在 action.json 文件中，包含两类 action 值，分别为 open 和 close，而在 newaction.json 文件中，包含三类 action 值，分别为 open、crash 和 swipe。

为了模拟数据流的行为，应把这两个 JSON 文件复制到 HDFS 指定目录下。

（2）编写 Spark 结构化流程序，读取文件流数据源并执行聚合操作，然后输出结果。这里使用的输入模式是 complete，代码如下：

```scala
//第7章/sstreaming_mode.scala

//导入语句
import org.apache.spark.sql.SparkSession
import org.apache.spark.sql.types._
import org.apache.spark.sql.functions._

//为手机事件数据创建一个 Schema
val mobileDataSchema = new StructType()
    .add("id", StringType, nullable = false)
    .add("action", StringType, nullable = false)
    .add("ts", TimestampType, nullable = false)

//监听的文件目录
```

```
val dataPath = "/data/spark/mobile2"

//读取指定目录下的源数据文件，一次一个
val mobileDF = spark.readStream
    .option("maxFilesPerTrigger", 1)
    .schema(mobileDataSchema)
    .json(dataPath)

//执行聚合等操作
val actionCountDF = mobileDF
    .groupBy("action")
    .count

//输出模式是complete
val query = actionCountDF.writeStream
    .format("console")
    .option("truncate", "false")
    .outputMode("complete")    //这里设为"complete"模式
    .start()

//等待流程序执行结束
//query.awaitTermination()        //在spark-Shell命令行下不需要执行这一行代码
```

执行上面的代码，流查询输出如下：

```
-------------------------------------------
Batch: 0
-------------------------------------------
+------+-----+
|action|count|
+------+-----+
|close |2    |
|open  |3    |
+------+-----+


-------------------------------------------
Batch: 1
-------------------------------------------
+------+-----+
|action|count|
+------+-----+
|close |2    |
|swipe |1    |
|crash |1    |
|open  |4    |
+------+-----+
```

观察上面的输出结果，在 Batch 1 中输出的聚合结果包含了 Batch 0 中的状态。这说明在上面的代码中，带有 complete 输出模式的流查询的输出包含了结果表中的所有 action 类型。

接下来，将上面代码中的输出模式改为 update 模式，其余代码保持不变，代码如下：

```
...
//输出模式修改为update
val query = actionCountDF.writeStream
```

```
        .format("console")
        .option("truncate", "false")
        .outputMode("update")            //这里设为 update 模式
        .start()
...
```

执行上面的代码,流查询输出如下:

```
-------------------------------------------
Batch: 0
-------------------------------------------
+------+-----+
|action|count|
+------+-----+
|close |2    |
|open  |3    |
+------+-----+


-------------------------------------------
Batch: 1
-------------------------------------------
+------+-----+
|action|count|
+------+-----+
|swipe |1    |
|crash |1    |
|open  |4    |
+------+-----+
```

观察上面的输出结果,在 Batch 1 中输出的聚合结果并不包含 Batch 0 中的状态。这说明在上面的代码中,带有 update 输出模式的流查询的输出只包含 newaction.json 文件中的 actions,这些 actions 结果(包含更新的 open 动作)以前从未出现过。

同样地,如果有状态查询类型使用了不适当的输出模式,则结构化的流引擎会抛出异常信息。例如,继续修改上面的代码,将输出模式设为 append,代码如下:

```
...
//对有状态流查询使用不适当的输出
val query = actionCountDF.writeStream
        .format("console")
        .option("truncate", "false")
        .outputMode("append")            //这里设为 append 模式
        .start()
...
```

因为执行的有 groupBy()聚合操作,所以异常,异常信息如下:

```
org.apache.spark.sql.AnalysisException: Append output mode not supported
when there are streaming aggregations on streaming DataFrames/DataSets
without watermark;
```

也有一种例外情况:如果向带有聚合的有状态的流查询提供一个水印,则所有的输出模式都是适用的。这是因为结构化的流引擎将删除旧的聚合状态数据,这些数据比指定的水印要"古老",这意味着一旦水印被逾越,新的行就可以被添加到结果表中。这时 append 输出的语义是有意义的。

7.8 深入研究触发器

触发器的设置决定了结构化的流引擎何时运行在流查询中表达的流计算逻辑,其中包括所有的转换,以及将数据写入数据接收器。换句话说,触发器的设置可控制什么时候数据被写到数据接收器上,以及使用哪种处理模式。从 Spark 2.3 开始,引入了一种称为连续的(Continuous)新处理模式。

Spark 结构化流所支持的不同类型的触发器,见表 7-11。

表 7-11 Data Sink 与输出模式

触发器类型	描述
未指定(默认)	如果没有显式地指定触发器设置,则默认情况下,查询将以微批处理模式执行,在这种模式下,上一个微批处理完成后将立即生成新的微批
固定间隔的微批	查询将以微批模式执行,其中微批将在用户指定的时间间隔启动。 如果前一个微批处理在间隔内完成,则引擎将等待直到间隔结束,然后启动下一个微批处理。 如果前一个微批需要的时间比间隔时间长(一个间隔边界丢失),则下一个微批将在前一个微批完成后立即开始(不会等待下一个间隔边界)。 如果没有新数据可用,则不会启动微批处理
一次性微批	查询将只执行一个微批处理来处理所有可用数据,然后自行停止。这在希望定期启动集群,处理自上一阶段以来可用的所有内容,然后关闭集群的场景中非常有用。在某些情况下,这可能会导致显著的成本节约
连续,具有固定的检查点间隔	查询将在新的低延迟、连续处理模式下执行

到目前为止,所有的流查询示例都没有指定触发器类型,而是使用了默认触发器类型,因为这个默认的触发器类型选择微批模式作为处理模式,而流查询中的逻辑执行不是基于时间的,而是在前一批数据完成处理后立即执行的。这意味着,在数据被写入的频率方面,可预测性会降低。

如果需要更强的可预测性,则可以指定固定的间隔触发器,示例代码如下:

```
import org.apache.spark.sql.streaming.Trigger

//默认触发器(尽可能快地运行微批处理)
df.writeStream
  .format("console")
  .start()

//2s 微批处理间隔的 ProcessingTime 触发器(按处理时间)
```

```
df.writeStream
  .format("console")
  .trigger(Trigger.ProcessingTime("2 seconds"))
  .start()

//一次性触发器
df.writeStream
  .format("console")
  .trigger(Trigger.Once())
  .start()

//具有一秒检查点间隔的 Continuous 触发器
df.writeStream
  .format("console")
  .trigger(Trigger.Continuous("1 second"))
  .start()
```

7.8.1 固定间隔触发器

固定间隔触发器可以根据用户提供的时间间隔（例如，每 3s），在特定的时间间隔内执行流查询中的逻辑。在处理模式方面，这个触发器类型使用微批处理模式。这个间隔可以用字符串格式指定，也可以作为 Scala Duration 或 Java TimeUnit 来指定。

使用固定间隔触发器的示例，代码如下：

```
//第7章/sstreaming_trigger_fixed.scala

import org.apache.spark.sql.SparkSession
import org.apache.spark.sql.streaming.Trigger

//设置每秒3行
val ratesDF = spark.readStream
    .format("rate")
    .option("rowsPerSecond","2")
    .option("numPartitions","2")
    .load()

//每3s触发一次流查询，并将其写入控制台
val query = ratesDF.writeStream
    .format("console")
    .outputMode("append")
    .option("numRows",50)
    .option("truncate",value = false)
    .trigger(Trigger.ProcessingTime("3 seconds"))   //这里指定触发器类型
    .start()

//等待流处理结束
//query.awaitTermination()           //在 spark-shell 交互环境下不需要
```

提交执行上面的代码，输出的结果如下：

```
-------------------------------------
Batch: 1
-------------------------------------
+----------------------+-----+
|timestamp             |value|
+----------------------+-----+
|2021-02-04 10:21:21.93|0    |
|2021-02-04 10:21:22.93|2    |
|2021-02-04 10:21:22.43|1    |
|2021-02-04 10:21:23.43|3    |
+----------------------+-----+

-------------------------------------
Batch: 2
-------------------------------------
+----------------------+-----+
|timestamp             |value|
+----------------------+-----+
|2021-02-04 10:21:23.93|4    |
|2021-02-04 10:21:24.93|6    |
|2021-02-04 10:21:25.93|8    |
|2021-02-04 10:21:24.43|5    |
|2021-02-04 10:21:25.43|7    |
|2021-02-04 10:21:26.43|9    |
+----------------------+-----+
...
```

因为指定 Rate 数据源每秒产生 2 行，而触发器指示每 3s 计算一批，所以可以看到每 3s 有 6 行输出。

也可以使用 Scala Duration 类型指定触发器的时间间隔，代码如下：

```
import scala.concurrent.duration._
...

//每3s 触发一次流查询，并将其写入控制台
val query = ratesDF.writeStream
    .format("console")
    .outputMode("append")
    .option("numRows",50)
    .option("truncate",value = false)
    //时间间隔使用 Scala Duration 类型指定
    .trigger(Trigger.ProcessingTime(3.seconds))
    .start()
```

固定间隔触发器并不总能保证流查询的执行会精确地在每个用户指定的时间间隔内发生，这有以下两个原因：

（1）第 1 个原因很明显，如果没有数据到达，就没有什么可处理的，因此没有任何数据被写入 Data Sink 中。

（2）第 2 个原因是，当前一批的处理时间超过指定间隔时间时，流查询的下一个执行将在处理完成后立即启动。换句话说，它不会等待下一个时间间隔边界。

7.8.2 一次性的触发器

顾名思义，一次性触发器以微批处理模式在流查询中执行逻辑，并将数据写到 Data Sink 中一次，然后处理停止。这种触发类型的存在，在开发和生产环境中都很有用。在开发阶段，通常流计算逻辑是以迭代的方式开发的，在每个迭代中，都希望测试逻辑。这个触发器类型简化了开发-测试-迭代流程。对于生产环境，这种触发器类型适合于流入流数据量较低的情况，这时只需每天运行几次数据处理逻辑。

指定这个一次性触发器类型非常简单。使用这种触发器类型，代码如下：

```scala
//第 7 章/sstreaming_trigger_once.scala

import org.apache.spark.sql.SparkSession
import org.apache.spark.sql.streaming.Trigger

//设置每秒 3 行
val ratesDF = spark.readStream
    .format("rate")
    .option("rowsPerSecond","2")
    .option("numPartitions","2")
    .load()

//每 3s 触发一次流查询，并将其写入控制台
val query = ratesDF.writeStream
    .format("console")
    .outputMode("append")
    .option("numRows",50)
    .option("truncate",value = false)
    .trigger(Trigger.Once())          //one-time 触发器
    .start()

//等待流处理结束
//query.awaitTermination()            //spark-shell 交互环境下不需要执行这一行代码
```

7.8.3 连续性的触发器

最后一个触发类型称为连续（Continuous）触发类型。这是在 Spark 2.3 中新引入的实验性的处理模式，以解决需要端到端的毫秒级延迟的情况。连续处理是 Apache Spark 的新执行引擎，每次处理事件的延迟非常低（以毫秒为单位）。在这个新的处理模式中，结构化流启动长时间运行的任务，以持续读取、处理和将数据写到一个 Data Sink。这意味着，一旦传入的数据到达数据源，它就会立即被处理并写入 Data Sink，并且端到端延迟只有几毫秒。此外，还引入了一个异步检查点机制，用于记录流查询的进度，以避免中断长时间

运行的任务,从而提供一致的毫秒级延迟。

利用这个 Continuous 触发器类型的一个比较好的案例是信用卡欺诈性交易检测。在较高的层次上,结构化流引擎根据触发器类型确定要使用哪种处理模式,如图 7-10 所示。

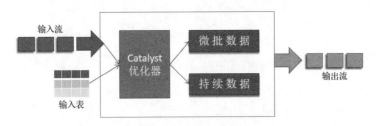

图 7-10　Spark Continuous 触发器类型

Apache Spark 一直通过微批处理提供流处理能力,这种方法的主要缺点是每个任务/微批处理必须定期收集和调度,通过这种方式 Spark 可以提供的最佳(最小)延迟大约是 1s。不存在单一事件/消息处理的概念。Spark 试图通过连续处理来克服这些限制,以提供低延迟(毫秒级响应)的流处理。

为了启用这些特性,Apache Spark 对其底层代码进行两项主要更改:

(1)创建可以连续读取消息(而不是微批处理)的新数据源和数据接收器,称为 DataSourceV2。

(2)创建一个新的执行引擎 ContinuousProcessing,它使用 ContinuousTrigger 并使用 DataSourceV2 启动长运行任务。ContinuousTrigger 内部使用 ProcessingTimeExecutor(与 ProcessingTime 触发器相同)。

在持续处理模式中只有投影和选择操作是允许的,如 select()、where()、map()、flatmap() 和 filter()。在这种处理模式下,除了聚合函数之外,所有 PySpark SQL 函数都受支持。另外需要特别注意的是,在连续处理中不支持水印,因为这涉及收集数据。

例如,在使用连续流读取 Kafka 时,其执行过程如图 7-11 所示。

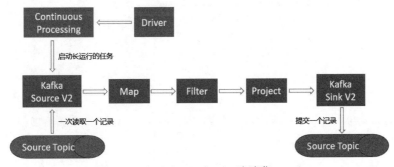

图 7-11　Spark Continuous 流消费 Kafka

要使用流查询的连续处理模式,需要指定一个连续触发器(Continuous Trigger),其中包含一个期望的检查点间隔，代码如下：

```scala
//第7章/sstreaming_trigger_continuous.scala

import org.apache.spark.sql.SparkSession
import org.apache.spark.sql.streaming.Trigger
import scala.concurrent.duration._

//设置每秒2行，分2个分区
//将数据写出到console，并使用Continuous Trigger
spark.readStream
    .format("rate")
    .option("rowsPerSecond","2")
    .option("numPartitions","2")
    .load()
    .writeStream
    .format("console")
    .outputMode("update")
    .option("truncate","false")
    .trigger(Trigger.Continuous(2.seconds))   //2s是异步检查点的时间间隔
    .start()
    .awaitTermination()
```

提交执行以上代码，可以在控制台上看到的输出内容如下：

```
-------------------------------------------
Batch: 0
-------------------------------------------
+-----------------------+-----+
|timestamp              |value|
+-----------------------+-----+
|2021-02-04 11:32:08.966|1    |
|2021-02-04 11:32:08.966|0    |
+-----------------------+-----+


-------------------------------------------
Batch: 1
-------------------------------------------
+-----------------------+-----+
|timestamp              |value|
+-----------------------+-----+
|2021-02-04 11:32:09.966|2    |
|2021-02-04 11:32:09.966|3    |
+-----------------------+-----+


-------------------------------------------
Batch: 2
-------------------------------------------
+-----------------------+-----+
|timestamp              |value|
+-----------------------+-----+
|2021-02-04 11:32:10.966|5    |
```

```
|2021-02-04 11:32:11.966|7    |
|2021-02-04 11:32:10.966|4    |
|2021-02-04 11:32:11.966|6    |
+-----------------------+-----+

-------------------------------------
Batch: 3
-------------------------------------
+-----------------------+-----+
|timestamp              |value|
+-----------------------+-----+
|2021-02-04 11:32:12.966|9    |
|2021-02-04 11:32:13.966|11   |
|2021-02-04 11:32:12.966|8    |
|2021-02-04 11:32:13.966|10   |
+-----------------------+-----+
```

在上面的代码中 ratesDF 流 DataFrame 被设置为两个分区，因此，结构化流在连续处理模式下启动了两个正在运行的任务，所以输出时所有的偶数在一起，所有的奇数出现在一起。

第 8 章 Spark 结构化流（高级）

第 7 章介绍了流处理的核心概念，Spark 结构化流处理引擎提供的特性，以及将流应用程序组合在一起的基本步骤。本章将涵盖结构化流的事件时间处理和有状态处理特性，并解释结构化流提供的支持，以帮助流应用程序对故障进行容错，并监控流应用程序的状态和进展。

8.1 事件时间和窗口聚合

基于数据创建时间处理传入的实时数据的能力是一个优秀的流处理引擎的必备功能。这一点很重要，因为要真正理解并准确地从流数据中提取见解或模式，需要能够根据数据或事件发生的时间来处理它们。

8.1.1 固定窗口聚合

一个固定的窗口（也就是一个滚动的窗口）操作本质上是根据一个固定的窗口长度将一个流入的数据流离散到非重叠的桶中。对于每一片输入的数据，根据它的事件时间将它放置到其中的一个桶中。执行聚合仅仅是遍历每个桶并在每个桶上应用聚合逻辑（例如计数或求和）。固定窗口聚合逻辑如图 8-1 所示。

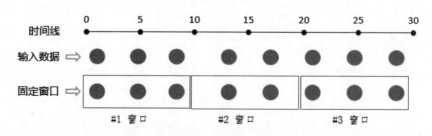

图 8-1　固定窗口聚合

下面通过一个示例程序来演示如何使用结构化流读取文件数据源。移动电话的开关机等事件会保存在 JSON 格式的文件中，现要求编写 Spark 结构化流处理程序来读取并分析这些移动电话数据，统计每 10min 内不同电话操作（如 open 或 close）发生的数量。

这实际上是在一个 10min 长的固定窗口上对移动电话操作事件的数量进行 count()聚合。建议按以下步骤操作。

(1) 准备数据。

在本示例中使用文件数据源，该数据源以 JSON 文件的格式记录了一小组移动电话动作事件。每个事件由以下 3 个字段组成，分别说明如下。

① id：表示手机的唯一 ID。在样例数据集中，电话 ID 类似于 phone1、phone2、phone3 等字符串。

② action：表示用户所采取的操作，该操作的可能值是 open 或 close。

③ ts：表示用户 action 发生时的时间戳，这是事件时间。

本示例准备了 4 个存储移动电话事件数据的 JSON 文件。

file1.json 文件的内容如下：

```
{"id":"phone1","action":"open","ts":"2018-03-02T10:02:33"}
{"id":"phone2","action":"open","ts":"2018-03-02T10:03:35"}
{"id":"phone3","action":"open","ts":"2018-03-02T10:03:50"}
{"id":"phone1","action":"close","ts":"2018-03-02T10:04:35"}
```

file2.json 文件的内容如下：

```
{"id":"phone3","action":"close","ts":"2018-03-02T10:07:35"}
{"id":"phone4","action":"open","ts":"2018-03-02T10:07:50"}
```

file3.json 文件的内容如下：

```
{"id":"phone2","action":"close","ts":"2018-03-02T10:04:50"}
{"id":"phone5","action":"open","ts":"2018-03-02T10:10:50"}
```

newaction.json 文件的内容如下：

```
{"id":"phone2","action":"crash","ts":"2018-03-02T11:09:13"}
{"id":"phone5","action":"swipe","ts":"2018-03-02T11:17:29"}
```

为了模拟数据流的行为，应把这 4 个 JSON 文件复制到 HDFS 的指定目录下。

(2) 编辑源程序，内容如下：

```scala
//第8章/sstreaming_mobile.scala

import org.apache.spark.sql.SparkSession
import org.apache.spark.sql.types._
import org.apache.spark.sql.functions._

//定义 Schema
val mobileDataSchema = new StructType()
    .add("id", StringType, nullable = false)
    .add("action", StringType, nullable = false)
    .add("ts", TimestampType, nullable = false)

//读取流数据，使用文件数据源
val dataPath = "/data/spark/mobile"
val mobileSSDF = spark.readStream
    .option("maxFilesPerTrigger", 1)
```

```
            .schema(mobileDataSchema)
            .json(dataPath)

import spark.implicits._

//在一个 10min 的窗口上执行聚合操作
val windowCountDF = mobileSSDF.groupBy(window($"ts", "10 minutes")).count()

windowCountDF.printSchema()

//执行流查询，将结果输到控制台
windowCountDF
            .select("window.start","window.end","count")
            .orderBy("start")
            .writeStream
            .format("console")
            .option("truncate", "false")
            .outputMode("complete")
            .start()
            //.awaitTermination()        //在spark-shell 交互环境下不需要执行这行代码
```

（3）执行流处理程序，输出的结果如下：

```
root
 |-- window: struct (nullable = false)
 |    |-- start: timestamp (nullable = true)
 |    |-- end: timestamp (nullable = true)
 |-- count: long (nullable = false)


+-------------------+-------------------+-----+
|start              |end                |count|
+-------------------+-------------------+-----+
|2018-03-02 10:00:00|2018-03-02 10:10:00|7    |
|2018-03-02 10:10:00|2018-03-02 10:20:00|1    |
|2018-03-02 11:00:00|2018-03-02 11:10:00|1    |
|2018-03-02 11:10:00|2018-03-02 11:20:00|1    |
+-------------------+-------------------+-----+
```

可以看出，当用窗口执行聚合时，输出窗口实际上是一个 struct 类型，它包含开始和结束时间。在上面的代码中，分别取 window 的 start 和 end 列，并按 start 窗口的开始时间排序。可以看到，每 10min 作为一个窗口进行统计。

（4）除了可以在 groupBy()转换中指定一个窗口之外，还可以从事件本身指定额外的列。对上面的例子稍做修改，使用一个窗口并在 action 列上执行聚合，实现对每个窗口和该窗口中 action 类型的 count()计数，代码如下：

```
//第 8 章/sstreaming_mobile2.scala

import org.apache.spark.sql.SparkSession
import org.apache.spark.sql.types._
import org.apache.spark.sql.functions._
```

```
//定义 Schema
val mobileDataSchema = new StructType()
    .add("id", StringType, nullable = false)
    .add("action", StringType, nullable = false)
    .add("ts", TimestampType, nullable = false)

//读取流数据,使用文件数据源
val dataPath = "/data/spark/mobile"
val mobileSSDF = spark.readStream.schema(mobileDataSchema).json(dataPath)

import spark.implicits._

//在一个 10min 的窗口上执行聚合操作
val windowCountDF = mobileSSDF
    .groupBy(window($"ts", "10 minutes"), $"action")          //修改 1
    .count()

windowCountDF.printSchema()

//执行流查询,将结果输到控制台
windowCountDF
    .select("window.start", "window.end", "action", "count")  //修改 2
    .orderBy("start")
    .writeStream
    .format("console")
    .option("truncate", "false")
    .outputMode("complete")
    .start()
    //.awaitTermination()   //在 spark-shell 交互环境下不需要执行这行代码
```

执行以上代码,输出的结果如下:

```
root
 |-- window: struct (nullable = false)
 |    |-- start: timestamp (nullable = true)
 |    |-- end: timestamp (nullable = true)
 |-- action: string (nullable = false)
 |-- count: long (nullable = false)

-------------------------------------------
Batch: 0
-------------------------------------------
+-------------------+-------------------+------+-----+
|start              |end                |action|count|
+-------------------+-------------------+------+-----+
|2018-03-02 10:00:00|2018-03-02 10:10:00|close |3    |
|2018-03-02 10:00:00|2018-03-02 10:10:00|open  |4    |
|2018-03-02 10:10:00|2018-03-02 10:20:00|open  |1    |
|2018-03-02 11:00:00|2018-03-02 11:10:00|crash |1    |
|2018-03-02 11:10:00|2018-03-02 11:20:00|swipe |1    |
+-------------------+-------------------+------+-----+
```

8.1.2 滑动窗口聚合

除了固定窗口类型之外，还有另一种称为滑动窗口（Sliding Window）的窗口类型。定义一个滑动窗口需要两个信息，即窗口长度和一个滑动间隔，滑动间隔通常比窗口的长度要小。由于聚合计算在传入的数据流上滑动，因此结果通常比固定窗口类型的结果更平滑，这种窗口类型通常用于计算移动平均。关于滑动窗口，需要注意的一点是，由于重叠的原因，一块数据可能会落入多个窗口，如图 8-2 所示。

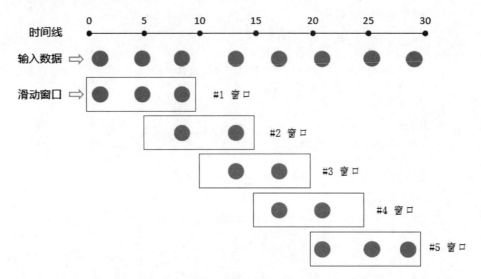

图 8-2 滑动窗口聚合

下面通过一个示例程序使用和理解滑动窗口。应用滑动窗口聚合解决一个 IoT 流数据分析需求：在一个数据中心中，按一定的时间间隔周期性地检测每个服务器机架的温度，并生成一个报告，显示每个机架在窗口长度 10min、滑动间隔 5min 的平均温度。

建议按以下步骤操作。

（1）准备数据。

在本示例中使用文件数据源，该数据源以 JSON 文件的格式记录了某数据中心两个机架的温度数据。每个事件均由以下 3 个字段组成。

① rack：表示机器的唯一 ID，字符串类型。
② temperature：表示采集到的温度值，double 类型。
③ ts：表示该事件发生时的时间戳，这是事件时间。

本示例准备了两个 IoT 事件数据的 JSON 文件。

file1.json 文件的内容如下：

```
{"rack":"rack1","temperature":99.5,"ts":"2017-06-02T08:01:01"}
{"rack":"rack1","temperature":100.5,"ts":"2017-06-02T08:06:02"}
```

```
{"rack":"rack1","temperature":101.0,"ts":"2017-06-02T08:11:03"}
{"rack":"rack1","temperature":102.0,"ts":"2017-06-02T08:16:04"}
```

file2.json 文件的内容如下:

```
{"rack":"rack2","temperature":99.5,"ts":"2017-06-02T08:01:02"}
{"rack":"rack2","temperature":105.5,"ts":"2017-06-02T08:06:04"}
{"rack":"rack2","temperature":104.0,"ts":"2017-06-02T08:11:06"}
{"rack":"rack2","temperature":108.0,"ts":"2017-06-02T08:16:08"}
```

为了模拟数据流的行为,应把这两个 JSON 文件复制到 HDFS 的指定目录下。

(2) 代码编写。

实现的流查询代码如下:

```scala
//第8章/sstreaming_iot.scala

import org.apache.spark.sql.SparkSession
import org.apache.spark.sql.types._
import org.apache.spark.sql.functions._

//定义 Schema
val iotDataSchema = new StructType()
    .add("rack", StringType, nullable = false)
    .add("temperature", DoubleType, nullable = false)
    .add("ts", TimestampType, nullable = false)

//读取流数据,使用文件数据源
val dataPath = "/data/spark/iotd"
val iotSSDF = spark.readStream
    .schema(iotDataSchema)
    .json(dataPath)

import spark.implicits._

//group by 一个滑动窗口,并在 temperature 列上求平均值
val windowAvgDF = iotSSDF
    .groupBy(window($"ts", "10 minutes", "5 minutes"))
    .agg(avg("temperature") as "avg_temp")

windowAvgDF.printSchema()

windowAvgDF
    .select("window.start", "window.end", "avg_temp")
    .orderBy("start")
    .writeStream
    .format("console")
    .option("truncate", "false")
    .outputMode("complete")
    .start()
    //.awaitTermination()        //spark-shell 交互环境下不需要这一行代码
```

在上面的代码中,首先读取温度数据,然后在 ts 列上构造一个长 10min、每 5min 进

行滑动的滑动窗口，并在这个窗口上执行 groupBy()转换。对于每个滑动窗口，avg()函数被应用于 temperature 列。

（3）执行以上代码，输出的结果如下：

```
root
 |-- window: struct (nullable = true)
 |    |-- start: timestamp (nullable = true)
 |    |-- end: timestamp (nullable = true)
 |-- avg_temp: double (nullable = true)

-------------------------------------------
Batch: 0
-------------------------------------------
+-------------------+-------------------+--------+
|start              |end                |avg_temp|
+-------------------+-------------------+--------+
|2017-06-02 07:55:00|2017-06-02 08:05:00|99.5    |
|2017-06-02 08:00:00|2017-06-02 08:10:00|101.25  |
|2017-06-02 08:05:00|2017-06-02 08:15:00|102.75  |
|2017-06-02 08:10:00|2017-06-02 08:20:00|103.75  |
|2017-06-02 08:15:00|2017-06-02 08:25:00|105.0   |
+-------------------+-------------------+--------+
```

上面的输出显示在合成数据集中有 5 个窗口。注意每个窗口的开始时间间隔为 5min，这是因为在 groupBy()转换中指定的滑动间隔的长度。

在上面的分析结果中，可以看出 avg_temp 列所代表的机架平均温度在上升。那么大家思考一下，机架平均温度的上升，是因为其中某个机架的温度升高从而导致平均温度的升高，还是所有机架的温度都在升高？

（4）为了弄清楚到底是哪些机架在不断升温，重构上面的代码，把 rack 列添加到 groupBy()转换中，代码如下（只显示不同的代码部分）：

```scala
//第8章/sstreaming_iot2.scala

import org.apache.spark.sql.SparkSession
import org.apache.spark.sql.types._
import org.apache.spark.sql.functions._

//定义 Schema
val iotDataSchema = new StructType()
    .add("rack", StringType, nullable = false)
    .add("temperature", DoubleType, nullable = false)
    .add("ts", TimestampType, nullable = false)

//读取流数据，使用文件数据源
val dataPath = "/data/spark/iotd"
val iotSSDF = spark.readStream
    .option("maxFilesPerTrigger", 1)
    .schema(iotDataSchema)
```

```
    .json(dataPath)

import spark.implicits._

//group by 一个滑动窗口和rack列，并在temperature列上求平均值
val windowAvgDF = iotSSDF
    .groupBy(window($"ts", "10 minutes", "5 minutes"), $"rack")
    .agg(avg("temperature") as "avg_temp")

windowAvgDF.printSchema()

//分别报告每个机架随时间而变化的温度
windowAvgDF
    .select("rack","window.start", "window.end", "avg_temp")
    .orderBy("rack","start")
    .writeStream
    .format("console")
    .option("truncate", "false")
    .outputMode("complete")
    .start()
    //.awaitTermination()            //spark-shell交互环境下不需要这一行代码
```

执行以上代码，输出的结果如下：

```
root
 |-- window: struct (nullable = true)
 |    |-- start: timestamp (nullable = true)
 |    |-- end: timestamp (nullable = true)
 |-- rack: string (nullable = false)
 |-- avg_temp: double (nullable = true)

-------------------------------------------
Batch: 0
-------------------------------------------
+-----+-------------------+-------------------+--------+
|rack |start              |end                |avg_temp|
+-----+-------------------+-------------------+--------+
|rack1|2017-06-02 07:55:00|2017-06-02 08:05:00|99.5    |
|rack1|2017-06-02 08:00:00|2017-06-02 08:10:00|100.0   |
|rack1|2017-06-02 08:05:00|2017-06-02 08:15:00|100.75  |
|rack1|2017-06-02 08:10:00|2017-06-02 08:20:00|101.5   |
|rack1|2017-06-02 08:15:00|2017-06-02 08:25:00|102.0   |
|rack2|2017-06-02 07:55:00|2017-06-02 08:05:00|99.5    |
|rack2|2017-06-02 08:00:00|2017-06-02 08:10:00|102.5   |
|rack2|2017-06-02 08:05:00|2017-06-02 08:15:00|104.75  |
|rack2|2017-06-02 08:10:00|2017-06-02 08:20:00|106.0   |
|rack2|2017-06-02 08:15:00|2017-06-02 08:25:00|108.0   |
+-----+-------------------+-------------------+--------+
```

从上面的输出结果表中，可以清楚地看出来，机架1的平均温度值低于101，而机架2升温的速度要远快于机架1，所以应该关注的是机架2。

8.2 水印

在流处理引擎中,水印是一种常用的技术,用于限制维护它所需的状态数量,以及处理延迟数据。

8.2.1 限制维护的聚合状态数量

通过应用于事件时间上的窗口聚合(固定窗口聚合或滑动窗口聚合),在 Spark 结构化流中可以很容易地执行常见的和复杂的流处理操作。事实上,任何时候在流查询上执行聚合时,都必须维护中间聚合状态。

首先理解"维护中间聚合状态"的含义。例如,在实时单词统计流程序中,随着每行文本的到来,流引擎会进行分词和计数处理。在分组聚合中,在用户指定的分组列中为每个唯一值(每个单词)维护聚合值(例如计数),代码如下:

```scala
//第8章/sstreaming_state.scala

import org.apache.spark.sql.functions._
import org.apache.spark.sql.SparkSession

import spark.implicits._
//创建 DataFrame,表示从连接到 localhost:9999 获得的输入行流
val lines = spark.readStream
  .format("socket")
  .option("host", "localhost")
  .option("port", 9999)
  .load()

val words = lines.as[String].flatMap(_.split(" "))  //把每行分割成单词
val wordCounts = words.groupBy("value").count()     //生成运行时单词计数

//开始运行将运行计数打印到控制台的查询
val query = wordCounts.writeStream
  .outputMode("complete")
  .format("console")
  .start()
```

处理过程和状态维护过程如图 8-3 所示。

在这个过程中,Spark 结构化流并没有具体化整个表,而只是从流数据源读取最新的可用数据,对其进行增量处理以更新结果,然后丢弃源数据。它只保留更新结果所需的最小中间状态数据,即在上面的示例中的中间计数。

这个中间状态存储在 Spark Executors 的内存中、版本化的 key-value 状态存储中(其中 key 是分组名称,value 是中间聚合值),并将其写到一个 WAL 预写日志中(该日志应该被配置为驻留在像 HDFS 这样的稳定存储系统中)。在每个触发点上,该状态都在内存

的状态存储中读取和更新，然后写入预写日志中。在失败的情况下，当 Spark 结构化的流应用程序重新启动时，状态从预写日志中还原并恢复。这种容错的状态管理显然会在结构化流引擎中产生一些资源和处理开销，因此，开销的大小与它需要维护的状态量成正比，保持状态的数量在一个可以接受大小是很重要的；换句话说，状态的规模不应该无限增长，也就是流引擎不应该无限地维护状态，必须在合适时清除旧的、不再需要的状态。

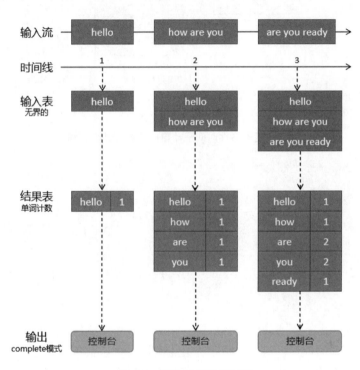

图 8-3 实时单词计数过程

对于结构化流来讲，在滑动事件时间窗口上的聚合与分组聚合非常相似。在基于窗口的聚合中，对于输入行的事件时间落入的每个窗口，该窗口的聚合值都需要被维护。用一个例子来理解这一点。假设对前面的示例进行修改，让输入流包含文本行及行生成的时间。现在想要统计在 10min 的窗口（每 5min 更新一次）中的单词计数。也就是说，统计在 12:00—12:10、12:05—12:15、12:10—12:20 等 10min 的窗口（时间间隔内）收到的单词数量。需要注意，12:00—12:10 表示在 12:00 之后和 12:10 之前到达的数据。现在，考虑 12:07 收到的一个单词。这个单词应该增加对应的两个窗口 12:00—12:10 和 12:05—12:15 的计数，因此计数将由分组键(单词)和窗口（可以从事件时间计算）两者进行索引。这个过程和结果表如图 8-4 所示。

由于这个窗口类似于分组，在代码中，可以使用 groupBy()和 window()操作来表示窗口聚合，代码如下：

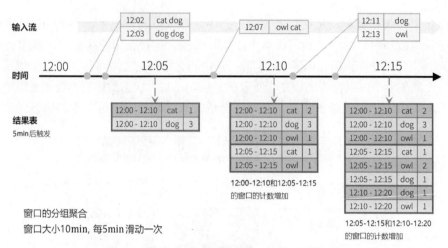

图 8-4 基于事件时间滑动窗口的单词计数过程

```
import spark.implicits._

//流 DataFrame, schema { timestamp: Timestamp, word: String }
val words = ...

//按窗口和单词对数据进行分组，并计算每组的单词计数
val windowedCounts = words.groupBy(
  window($"timestamp", "10 minutes", "5 minutes"),
  $"word"
).count()

...
```

考虑到滑动窗口的性质，窗口的数量将会无限增长。这意味着执行滑动窗口聚合会导致中间状态无限地增长，因此，必须有一种方法可以删除不再更新的旧状态。在 Spark 结构化流处理技术中，这是通过一种叫作"水印（watermark）"的技术完成的。

从 Spark 2.1 开始，在结构化流 API 中引入了水印。本质上，水印是一个时间阈值，用来指定一个时间窗口结束后最多再等待多长时间就开始计算、更新状态并删除旧的状态。可以通过简单地在查询中添加 withWatermark()运算符来启用水印，该方法定义如下：

withWatermark(eventTime: String, delayThreshold: String): Dataset[T]

这种方法需要两个参数：

（1）一个事件时间列（eventTime，必须与聚合操作的时间相同）。

（2）一个阈值（delayThreshold），用于指定延迟多长时间的数据应该被处理（以事件时间为单位，可以是秒、分钟或小时）。

聚合的状态将由 Spark 引擎进行维护，直到 max(eventTime) – delayThreshold > T，其中 max(eventTime)是引擎看到的最近事件时间，T 是窗口的结束时间。

指定水印的最大好处之一是能让结构化流引擎可以安全地删除比水印更"古老"的窗

口的聚合状态。生产环境下执行任何类型聚合的流应用程序都应该指定一个水印来避免内存不足的问题。

8.2.2 处理迟到的数据

在现实系统中，因为存在着各种可能，例如网络拥挤、网络中断或数据生成器（如移动设备等）不在线和传感器断开连接等，因此，不能保证数据会按照创建的顺序到达流处理引擎。为了容错，因此有必要处理这种乱序数据。要处理这个问题，流引擎必须保持聚合的状态。如果发生延迟事件，则当延迟事件到达时可以重新处理查询，但这意味着所有聚合的状态必须无限期地保持，这将导致内存使用量也无限地增长。这在实际生产场景中是不现实的，除非系统有无限的资源和无限的预算，因此，作为一个实时流应用程序的开发人员，必须知道要怎样处理延迟到达的数据。换句话说，数据延迟到达时间量是多少时才是可以接受的，或者说对这段时间之后迟到的数据置之不理？这取决于应用场景。

处理延迟到达的事件是流处理引擎的一个关键功能。解决这个问题的方法是使用水印。水印技术是流处理引擎处理延迟的一种有效方法。水印实际上是一个时间阈值，用来指定系统等待延迟事件的时间。如果延迟到达的事件位于水印内，则用于更新查询。否则，如果它的事件时间比水印的时间还要早，它就会被丢弃，而不会被流引擎进一步处理。

现在考虑一下前面基于事件时间滑动窗口的单词计数示例：如果某个事件延迟到达应用程序，会发生什么情况？例如，在 12:04（事件时间）生成的单词因为某些原因，在 12:11 到达流应用程序，该程序接收此迟到的事件，并使用时间 12:04 来更新窗口 12:00—12:10 中的旧计数。Spark 结构化流在基于窗口的分组中会自动这样来处理迟到的数据，因为结构化流可以长期维护部分聚合的中间状态，以便后期数据能够正确地更新旧窗口的聚合。这个过程描述如图 8-5 所示。

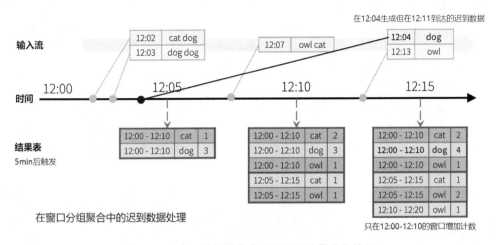

图 8-5　在窗口分组聚合中对延迟到达数据的处理

但是，要运行该查询数天，系统需要绑定它积累的中间内存状态的数量。这意味着系统需要知道何时从内存状态中删除一个旧的聚合，因为应用程序不会再接收到延迟到达的数据。为了实现这一点，需要水印技术，它允许引擎自动跟踪数据中的当前事件时间，并尝试相应地清除旧的状态。对于在时间 T 结束的特定窗口，Spark 引擎将维护其状态，并允许延迟到达的数据更新状态，直到 "max(eventTime)–延迟阈值> T"。换句话说，在阈值内到达的迟到数据将被聚合，而迟于阈值时间到达的迟到数据将会被丢弃。

要在结构化流中指定水印作为流 DataFrame 的一部分，只需使用 Watermark API 并提供两个参数：事件时间列和阈值，这些数据可以是秒、分钟或小时。使用 Watermark API 的模板代码如下：

```scala
import spark.implicits._

//流 DataFrame, schema { timestamp: Timestamp, word: String }
val words = ...

//通过窗口和单词分组数据，计算每个组的计数
val windowedCounts = words
    .withWatermark("timestamp", "10 minutes")    //水印，延迟阈值10min
    .groupBy(
        window($"timestamp", "10 minutes", "5 minutes"),
        $"word")
    .count()
```

在本例中，在 timestamp 列值上定义查询的水印，并定义 10 minutes 作为允许延迟到达数据的阈值。如果该查询在 update 输出模式下运行，引擎则会持续更新结果表中窗口的计数，直到窗口的水位超过水印，它落后 timestamp 列的当前事件时间 10min，如图 8-6 所示。

正如图 8-6 所示，引擎跟踪的最大事件时间是蓝色虚线，在每个触发器开始时设置为 (最大事件时间–10min)的水印为红线。例如，当引擎观察到数据（12:14, dog）时，它为下一次触发器将水印设置为 12:04。此水印使引擎保持中间状态额外 10min，以允许统计后期数据。例如，数据（12:09, cat）是乱序和延迟到达的，它落在窗口 12:00—12:10 和 12:05—12:15。由于它仍然在触发器中水印 12:04 之前，引擎仍然将中间计数作为状态维护，并正确地更新相关窗口的计数，然而，当水印更新到 12:11 时，窗口（12:00—12:10）的中间状态被清除，所有后续数据（例如(12:04, donkey)）被认为太迟了而因此被忽略。注意，在每个触发器之后，更新的计数(紫色行)被写入 sink 作为触发器输出，这是由 update 模式决定的。

一些接收器（例如文件）可能不支持 update 模式需要的细粒度更新。为了使用它们，Spark 结构化流还支持 append 模式。在这种模式下，只有最后的计数被写入 sink 中，如图 8-7 所示。

与前面的 update 模式类似，Spark 引擎维护每个窗口的中间计数，但是，部分计数不会更新到结果表，也不会写入接收器。引擎等待 10min 来统计延迟到达的数据，然后删除小于水印的窗口的中间状态，并将最终的计数添加到结果表/接收器。例如，窗口 12:00—12:10 的最终计数只有在水印更新为 12:11 后才会被添加到结果表中。

第8章　Spark结构化流(高级)

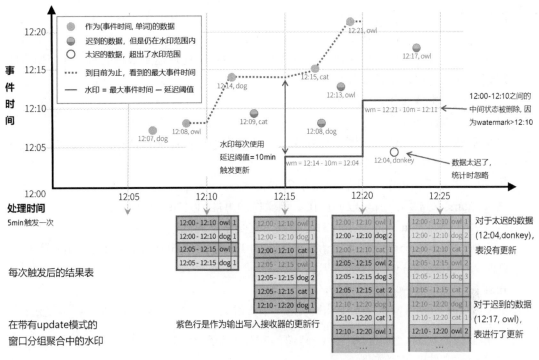

图 8-6　在窗口分组聚合中对延迟到达数据的处理

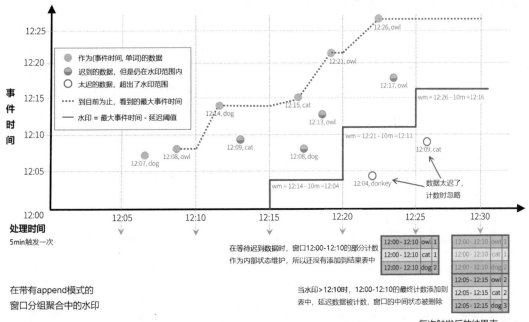

图 8-7　在 append 模式的窗口分组聚合中使用水印

注意：在非流数据集上使用 withWatermark()是无效的。由于水印不应该以任何方式影响任何批处理查询，所以 Spark 引擎将直接忽略它。

下面通过一个完整的应用示例来理解和掌握使用水印处理迟到数据的方法。使用水印来处理延迟到达的移动电话操作事件数据。

建议按以下步骤操作。

（1）准备数据。

在本示例中使用文件数据源，代表移动电话操作事件数据被存储在两个 JSON 格式的文件中。每个事件由以下 3 个字段组成。

① id：表示手机的唯一 ID，字符串类型。

② action：表示用户所采取的操作，该操作的可能值是"open"或"close"。

③ ts：表示用户 action 发生时的时间戳，这是事件时间。

本示例准备了两个存储移动电话事件数据的 JSON 文件 file1.json 和 file2.json，其中第 2 个数据文件 file2.json 中包含的是迟到的数据。

file1.json 文件的内容如下：

```
{"id":"phone1","action":"open","ts":"2018-03-02T10:15:33"}
{"id":"phone2","action":"open","ts":"2018-03-02T10:22:35"}
{"id":"phone3","action":"open","ts":"2018-03-02T10:33:50"}
```

file2.json 文件的内容如下：

```
{"id":"phone4","action":"open","ts":"2018-03-02T10:29:35"}
{"id":"phone5","action":"open","ts":"2018-03-02T10:11:35"}
```

注意观察这两个数据文件中的数据。数据以这样一种方式设置，即 file1.json 文件中的每行进入了它自己的 10min 窗口，那么 file1.json 的处理会形成 3 个窗口：10:10:00—10:20:00、10:20:00—10:30:00 和 10:30:00—10:40:00。示例中将水印指定为 10min。file2.json 文件中的数据代表迟到的数据，其中第 1 行落在 10:20:00—10:30:00 窗口中，所以即使它到达的时间较晚，它的时间戳仍然在水印的阈值范围内，因此它将被处理。file2.json 文件中的最后一行数据的时间戳在 10:10:00—10:20:00 窗口中，由于它超出了水印的阈值，所以它将被忽略，而不会被处理。

为了模拟数据流的行为，应把这两个 JSON 文件复制到 HDFS 的指定目录下。

（2）代码编写。

实现的流查询代码如下：

```scala
//第 8 章/sstreaming_late_date.scala

import org.apache.spark.sql.SparkSession
import org.apache.spark.sql.functions._
import org.apache.spark.sql.types._

//定义 Schema
val mobileDataSchema = new StructType()
```

```
        .add("id", StringType, nullable = false)
        .add("action", StringType, nullable = false)
        .add("ts", TimestampType, nullable = false)

//读取流数据,使用文件数据源
val dataPath = "/data/spark/mobile3"
val mobileSSDF = spark.readStream
    .schema(mobileDataSchema)
    .json(dataPath)

import spark.implicits._

//设置一个带有水印的 streaming DataFrame,并按 ts 和 action 列分组
val windowCountDF = mobileSSDF
    .withWatermark("ts", "10 minutes")  //水印,10min,必须先于 groupBy 调用
    .groupBy(window($"ts", "10 minutes"), $"action")
                                    //指定的窗口列必须与水印中指定的列一致
    .count()

windowCountDF.printSchema()

//输到控制台
windowCountDF
    .select("window.start", "window.end", "action", "count")
    .writeStream
    .format("console")
    .option("truncate", "false")
    .outputMode("update")
    .start()
```

(3)执行程序,输出源数据的结构,内容如下:

```
root
 |-- window: struct (nullable = false)
 |    |-- start: timestamp (nullable = true)
 |    |-- end: timestamp (nullable = true)
 |-- action: string (nullable = false)
 |-- count: long (nullable = false)
```

当它读到第 1 个流数据文件 file1.json 时,输出的结果如下:

```
-------------------------------------------
Batch: 0
-------------------------------------------
+-------------------+-------------------+------+-----+
|start              |end                |action|count|
+-------------------+-------------------+------+-----+
|2018-03-02 10:30:00|2018-03-02 10:40:00|open  |1    |
|2018-03-02 10:10:00|2018-03-02 10:20:00|open  |1    |
|2018-03-02 10:20:00|2018-03-02 10:30:00|open  |1    |
+-------------------+-------------------+------+-----+
```

正如期望的,每行都落在它自己的窗口内。

当它读取到第 2 个流数据文件 file2.json 时，输出的结果如下：

```
-------------------------------------------
Batch: 1
-------------------------------------------
+-------------------+-------------------+------+-----+
|start              |end                |action|count|
+-------------------+-------------------+------+-----+
|2018-03-02 10:20:00|2018-03-02 10:30:00|open  |2    |
+-------------------+-------------------+------+-----+
```

注意到窗口 10:20:00—10:30:00 的 count 现在被更新为 2，窗口 10:10:00—10:20:00 没有变化。如前所述，因为 file2.json 文件中的最后一行的时间戳落在 10min 的水印阈值之外，因此它不会被处理。

（4）如果删除对 Watermark API 的调用，则输出的结果如下：

```
-------------------------------------------
Batch: 1
-------------------------------------------
+-------------------+-------------------+------+-----+
|start              |end                |action|count|
+-------------------+-------------------+------+-----+
|2018-03-02 10:10:00|2018-03-02 10:20:00|open  |2    |
|2018-03-02 10:20:00|2018-03-02 10:30:00|open  |2    |
+-------------------+-------------------+------+-----+
```

可以看出，因为没有指定水印，迟到的数据也不会被删除，所以对窗口 10:10:00—10:20:00 的 count 计数被更新为 2。

最后，因为水印是一个有用的特性，因此理解聚合状态被正确清理的条件是很重要的：

（1）输出模式不能是 complete 模式，必须是 update 或 append 模式。原因是，complete 模式的语义规定必须维护所有聚合数据，并且不违反这些语义，水印不能删除任何中间状态。

（2）通过 groupBy() 转换的聚合必须直接位于 event-time 列或 event-time 列的窗口上。

（3）在 Watermark API 和 groupBy() 转换中指定的 event-time 列必须是同一个。

（4）当设置一个流 DataFrame 时，必须在调用 groupBy() 转换之前调用 Watermark API，否则它将被忽略。

8.3 任意状态处理

如前所述，按 key 或事件窗口聚合的中间状态由结构化流自动维护，然而，并不是所有的事件时间处理都可以通过简单地在一个或多个列上聚合，并且在没有窗口的情况下得到满足。例如，在 IoT 实时温度监控程序中，当看到 3 个连续的温度读数超过 100℃时，需要发出一个警报。再例如，关于维护用户会话，其中每个会话的长度不是由固定的时间决定的，而是由用户的活动和缺乏活动决定的。要解决这两个示例和类似的应用场景，就需要能够在每组数据上应用任意处理逻辑，以控制每组数据的窗口长度，并在触发器点上保持任意状态。这就需要应用结构化流的任意状态处理（有时也称为复杂状态处理）。

8.3.1 结构化流的任意状态处理

结构化流为流应用程序提供了一种回调机制来执行任意的状态处理，并且它将负责确保中间状态的维护和以容错的方式存储。这种处理方式基本上可以归结为执行以下任务之一的能力，如图8-8所示。

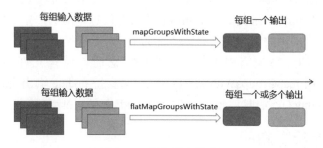

图 8-8 任意状态处理

这两个任务如下：

（1）映射一组数据，对每组数据应用任意处理，然后每组只输出一行。

（2）映射一组数据，对每组数据进行任意处理，然后每组输出任意数量的行，包括none。

对于其中每个任务（Task），结构化流都提供了一个特定的API来处理它，其中第1个API被称为mapGroupsWithState()方法，第2个API被称为flatMapGroupsWithState()方法。这些API从Spark 2.2开始引入，并且只使用Scala或Java绑定在类型化DataSet API上工作。

当使用任何类型的回调机制时，重要的是要清楚地了解何时及多长时间调用它一次及输入参数的详细信息。顺序如下：

（1）要在一个流DataFrame上执行任意的有状态处理，必须首先通过调用groupByKey()转换并提供进行分组的列来指定分组，然后它会返回KeyValueGroupedDataset类的一个实例。

（2）从KeyValueGroupedDataset类的一个实例中，可以调用mapGroupsWithState()或flatMapGroupsWithState()函数。这两个API都需要一组不同的输入参数。

（3）当调用mapGroupsWithState()函数时，需要提供超时类型和用户定义的回调函数。超时部分将在稍后解释。

（4）当调用flatMapGroupsWithState()函数时，需要提供一个输出模式、超时类型和一个用户定义的回调函数。

结构化流和用户定义的回调函数之间的关系说明如下：

（1）对于每一组输入数据，用户定义的回调函数将在每个触发器中被重复调用。对于每次调用，它都是针对每个在触发器中有数据的组。如果一个特定的组在触发器中没有任何数据，就不会对该组进行调用，因此，不应该假设这个函数对于每个组在每个触发器中都被调用。

（2）每次调用用户定义的回调函数时，都会传入以下信息：

① 该组 key 的值。

② 一个 group 的所有数据。不保证它们有任何特定的顺序。

③ 一个 group 的前一种状态，它是由先前的同一 group 调用返回的。一个 group 的状态由一个称为 GroupState 的状态持有者类管理。当需要更新一个 group 的状态时，必须使用新状态调用该类的 update 函数。每个 group 的状态信息是由用户定义的类定义的。当调用 update 函数时，所提供的用户定义状态不能为 null。

正如已了解到的，每当需要维护一个中间状态时，就只允许某些输出模式。在 Spark 2.3 中，当调用 mapGroupsWithState()方法时只支持 update 输出模式，然而，当调用 flatMapGroupsWithState()方法时，append 模式和 update 模式都支持。

8.3.2 处理状态超时

在带有水印的事件时间聚合的情况下，中间状态的超时是由结构化流内部管理的，用户无法来影响它，而结构化流任意状态处理提供了控制中间状态超时的灵活性。由于有能力维护任意状态，所以对于某些特定的用例来讲，控制中间状态超时是非常有意义的。

结构化流有状态处理提供了 3 种不同的超时类型。

第 1 种超时类型是基于处理时间的，第 2 种超时类型是基于事件时间的。超时类型是在全局级别配置的，这意味着它适用于特定的流 DataFrame 中的所有组。可以为每个单独的组配置超时值，并且可以随意更改。如果中间状态被配置了超时，则在处理回调函数中给定的值列表之前，检查它是否超时了，这是很重要的。

在某些场景中，不需要超时，第 3 种超时类型是为这个场景设计的。

超时类型是在类 GroupStateTimeout 中定义的，可以在调用 mapGroupsWithState()或 flatMapGroupsWithState()函数时指定类型。超时时间是使用分别用于处理时间超时和事件时间超时的函数 GroupState.setTimeoutDuration()或 GroupState.setTimeoutTimeStamp()函数指定的。

当一个特定 group 的中间状态超时时，结构化流服务将调用用户定义的回调函数，其中包含一个空的值列表，并将标志 GroupState.hasTimedOut 设置为 true。

在这 3 种不同的超时类型中，事件时间超时是最复杂的。事件时间超时意味着它是基于事件发生的时间，因此对于这个超时类型，必须通过 DataFrame.withWatermark()方法在流 DataFrame 中设置水印。为了控制每组的超时，需要在一个特定组处理期间向 GroupState.setTimeoutTimestamp()函数提供一个时间戳值。当水印超出所提供的时间戳时，一个组的中间状态就是超时的。在用户会话场景中，当用户与网站交互时，可以简单地根据用户的最新交互时间加上某个阈值来更新超时时间戳，从而扩展用户会话。这是为了确保只要用户与网站交互，用户会话就保持活跃，中间日期将不会超时。

处理时间超时类型的工作方式与事件时间超时类型类似，然而，不同之处在于，它是基于服务器的挂钟（Wall Clock）的，它一直向前。为了控制每组的超时，在处理一个特定组期间为 GroupState.setTimeoutDuration() 函数提供了一个持续时间。该持续时间可以是 1min、1h 或者 2d。当时钟超过了所提供的持续时间时，一个组的中间状态超时。由于这种超时类型依赖于系统时钟，所以在时区改变或有时钟偏差时考虑这种情况是很重要的。

重要的是要认识到，如果流中无数据状况存在一段时间，就不会有任何触发器，因此用户定义的回调函数将不会被调用。此外，水印不会被推进，超时函数调用也不会发生。

8.3.3 节将通过几个示例来演示如何实现任意状态处理。

8.3.3 任意状态处理实战

本节将通过两个用例来演示结构化流中的任意状态处理。

（1）第 1 个是关于从数据中心计算机机架温度数据中提取模式，并维护中间状态下每个机架的状态。每当遇到 3 个连续的 100℃ 及以上的温度时，机架状态将升级到 warning（警告）级别。这个例子将使用 mapGroupsWithState() API。

（2）第 2 个例子是关于用户会话的，它将根据用户与网站的交互来跟踪用户状态。这个例子将使用 flatMapGroupsWithState() API。

在执行任意状态处理时，无论使用哪种 API，都遵循一组通用的步骤：

（1）定义几个类来表示输入数据、中间状态和输出。

① 输入事件（I）。

② 要保持的任意状态（S）。

③ 输出（O）（如果合适，这种类型则可能与状态表示相同）。

（2）定义状态转换函数。有了上面这些类型，就可以制定实现自定义状态处理逻辑的状态转换函数。需要定义两种状态转换函数，第 1 种是回调函数，由结构化流来调用；第 2 种函数包含在每组数据上执行的任意状态处理逻辑，以及维护状态的逻辑。

（3）决定一个超时类型和一个适当的超时值。

首先实现第 1 个复杂模式提取示例：数据中心计算机机架温度数据复杂事件模式识别和处理。

在本例中，感兴趣的模式是从同一个机架上采集到连续 3 个温度读数在 100℃ 或以上，并且两个连续高温读数之间的时间差必须在 60s 内。当检测到这种模式时，该特定机架的状态将升级为 warning（警告）状态。如果下一个进入的温度读数低于 100℃ 阈值，则机架状态就会降级为正常值。

建议按以下步骤操作。

(1) 准备数据

在本示例中,使用文件数据源,事件数据存储在 JSON 格式的文件中。每个事件由以下 3 个字段组成。

① rack:表示机架的唯一 ID,字符串类型。

② temperature:表示采集到的温度值,double 类型。

③ ts:表示事件发生时的时间戳,这是事件时间。

这里提供了 3 个数据文件,其中 file1.json 的内容用于显示机架 rack1 的温度,在 100℃ 上下交替变化。文件 file2.json 用于显示 rack2 的温度,连续升温。在 file3.json 文件中,rack3 也在升温,但温度读数间隔超过 60s。

file1.json 文件的内容如下:

```
{"rack":"rack1","temperature":99.5,"ts":"2017-06-02T08:01:01"}
{"rack":"rack1","temperature":100.5,"ts":"2017-06-02T08:02:02"}
{"rack":"rack1","temperature":98.3,"ts":"2017-06-02T08:02:29"}
{"rack":"rack1","temperature":102.0,"ts":"2017-06-02T08:02:44"}
```

file2.json 文件的内容如下:

```
{"rack":"rack1","temperature":97.5,"ts":"2017-06-02T08:02:59"}
{"rack":"rack2","temperature":99.5,"ts":"2017-06-02T08:03:02"}
{"rack":"rack2","temperature":105.5,"ts":"2017-06-02T08:03:44"}
{"rack":"rack2","temperature":104.0,"ts":"2017-06-02T08:04:06"}
{"rack":"rack2","temperature":108.0,"ts":"2017-06-02T08:04:49"}
```

file3.json 文件的内容如下:

```
{"rack":"rack2","temperature":108.0,"ts":"2017-06-02T08:06:40"}
{"rack":"rack3","temperature":100.5,"ts":"2017-06-02T08:06:20"}
{"rack":"rack3","temperature":103.7,"ts":"2017-06-02T08:07:35"}
{"rack":"rack3","temperature":105.3,"ts":"2017-06-02T08:08:53"}
```

为了模拟数据流的行为,应把这 3 个 JSON 文件复制到 HDFS 的指定目录下。

(2) 首先导入项目所依赖的包,代码如下:

```scala
//第8章/sstreaming_complex_state1.scala

import org.apache.spark.sql.SparkSession
import org.apache.spark.sql.streaming.{GroupState, GroupStateTimeout, OutputMode}
import org.apache.spark.sql.types._
```

(3) 接下来,准备两个 case class。对于这个用例,机架温度输入数据由类 RackInfo 来表示(输入事件 I),中间状态和输出都由一个名为 RackState 的类表示(任意状态 S),代码如下:

```scala
//用于表示输入数据的 case class
case class RackInfo(rack:String, temperature:Double, ts:java.sql.Timestamp)

//用于表示中间状态和输出数据的 case class
//注意,构造函数参数被定义为可修改的,所以可以更新它们
//lastTS 变量用于比较以前和当前温度读数之间的时间
```

```
case class RackState(var rackId: String,
                     var highTempCount: Int,
                     var status: String,
                     var lastTS: java.sql.Timestamp)
```

(4)然后定义两个函数。第 1 个被称为 updateRackState,它包含了模式检测的核心逻辑,用来检测在 60s 内产生 3 个连续超过 100℃的高温读数,在每个 group 上执行。第 2 个函数叫作 updateAcrossAllRackStatus,它是一个回调函数,它将被传递到 mapGroupsWithState() 函数。这个函数可确保根据事件时间的顺序来处理机架温度读数,代码如下:

```
//包含用于检测上述温度模式并更新中间状态的主要逻辑
def updateRackState(rackState: RackState, rackInfo: RackInfo):RackState = {

  //设置条件来决定是否更新机架状态
  val lastTS = Option(rackState.lastTS).getOrElse(rackInfo.ts)
  val withinTimeThreshold = (rackInfo.ts.getTime - lastTS.getTime) <= 60000
  val meetCondition = if (rackState.highTempCount < 1) true else
withinTimeThreshold
  val greaterThanEqualTo100 = rackInfo.temperature >= 100.0

  //判断匹配条件
  (greaterThanEqualTo100, meetCondition)  match {
    case (true, true) => {          //如果两个条件都满足了
      rackState.highTempCount = rackState.highTempCount + 1   //发现一个高温
      rackState.status = if (rackState.highTempCount >= 3) "Warning" else
"Normal"
    }
    case _ => {                     //如果两个条件都不满足,或之一不满足
      rackState.highTempCount = 0   //则清零,重新开始统计
      rackState.status = "Normal"
    }
  }

  rackState.lastTS = rackInfo.ts   //将状态时间更新为最后一个事件的时间
  rackState
}

//回调函数,提供 mapGroupsWithState API
//GroupState[S]是一个由结构化流提供的包装器,并在内部用于跨执行管理状态 S
//在该函数中,GroupState 提供了对状态的突变访问,以及检查和设置超时的能力
def updateAcrossAllRackStatus( rackId: String,
                    inputs: Iterator[RackInfo],
                    oldState: GroupState[RackState]) : RackState = {
  //初始化 rackState,如果之前有状态存在,则使用它,否则创建一个新的状态
  var rackState = if (oldState.exists) oldState.get else RackState(rackId, 3,
"", null)

  //按时间戳对输入进行升序排序
  inputs.toList.sortBy(_.ts.getTime).foreach( input => {
    rackState = updateRackState(rackState, input)
```

```
        //在状态持有类 GroupState 中更新该 rackState 非常重要
        oldState.update(rackState)
    })

    rackState
}
```

设置步骤现在已经完成。

（5）接下来，在结构化流应用程序中将回调函数连接到 mapGroupsWithState()方法，代码如下：

```
//主程序
def main(args: Array[String]): Unit = {

    val spark = SparkSession.builder()
      .master("local")
      .appName("cep demo")
      .getOrCreate()

    //用于 IoT 数据的 Schema
    val iotDataSchema = new StructType()
      .add("rack", StringType, nullable = false)
      .add("temperature", DoubleType, nullable = false)
      .add("ts", TimestampType, nullable = false)

    //读取流数据，使用文件数据源
    val dataPath = "/data/spark/iot3"
    val iotSSDF = spark.readStream
      .option("maxFilesPerTrigger", 1)
      .schema(iotDataSchema)
      .json(dataPath)

    iotSSDF.printSchema()

    import spark.implicits._

    val iotPatternDF = iotSSDF.as[RackInfo]
      .groupByKey(_.rack)           //按机架 ID 分组
      //将给定函数应用于每组数据，同时维护用户定义的每组数据状态
      //对于一个流 DataSet，该函数将在每个触发器中为每个 group 重复调用
      //每个组状态的更新将在调用之间保存
      .mapGroupsWithState[RackState,RackState](
GroupStateTimeout.NoTimeout)(updateAcrossAllRackStatus)

    //设置输出并启动流查询
    iotPatternDF.writeStream
      .format("console")
      .outputMode("update")
      .start()
      .awaitTermination()
}
```

（6）执行以上程序。当读取到第 1 个 file1.json 数据文件时，输出的结果如下：

```
+------+-------------+------+-------------------+
|rackId|highTempCount|status|             lastTS|
+------+-------------+------+-------------------+
| rack1|            1|Normal|2017-06-02 08:02:44|
+------+-------------+------+-------------------+
```

当读取到第 2 个 file2.json 数据文件时，输出的结果如下：

```
+------+-------------+-------+-------------------+
|rackId|highTempCount| status|             lastTS|
+------+-------------+-------+-------------------+
| rack1|            0| Normal|2017-06-02 08:02:59|
| rack2|            3|Warning|2017-06-02 08:04:49|
+------+-------------+-------+-------------------+
```

当读到第 3 个 file3.json 数据文件时，输出的结果如下：

```
+------+-------------+------+-------------------+
|rackId|highTempCount|status|             lastTS|
+------+-------------+------+-------------------+
| rack3|            1|Normal|2017-06-02 08:08:53|
| rack2|            0|Normal|2017-06-02 08:06:40|
+------+-------------+------+-------------------+
```

从以上的输出结果中分以看到，rack1 有一些温度读数超过 100℃，然而，它们不是连续的，因此输出状态处于正常水平。在 file2.json 文件中 rack2 有 3 个连续的温度读数超过 100℃，而每个和前一个之间的时间间隔小于 60s，所以 rack2 的状态处于 warning（警告）级别。rack3 有 3 个连续的温度读数超过 100℃，然而，每个和前一个之间的时间间隔超过了 60s，因此，它处于正常水平。

下面是第 2 个示例，基于用户会话活动的复杂事件模式识别和处理。在这个例子中，会话处理逻辑是基于用户活动的。当用户采取 login 动作时，会创建一个会话，并且在用户采取 logout 动作时结束会话。当没有用户活动持续 30min 时，会话将自动结束。

本例需要利用前面描述的超时特性来执行此检测。就输出而言，每当会话开始或结束时，该信息将被发送到输出。输出信息由用户 ID、会话开始和结束时间及访问页面的数量组成。这个用例使用 flatMapGroupsWithState()方法执行用户会话，它支持每组输出多于一行的能力。

建议按以下步骤操作。

（1）准备数据。

在本示例中使用文件数据源，事件数据存储在 JSON 格式的文件中。每个事件由以下 4 个字段组成。

① user：表示用户的唯一 ID，字符串类型。
② action：表示用户的行为，例如 login、click、send、view 或 logout，字符串类型。
③ page：表示用户浏览的页面，字符串类型。
④ ts：表示事件发生时的时间戳，这是事件时间。

这个用例的数据由 3 个 JSON 文件组成。

file1.json 文件的内容如下：

```
{"user":"user1","action":"login","page":"page1", "ts":"2017-09-06T08:08:53"}
{"user":"user1","action":"click","page":"page2", "ts":"2017-09-06T08:10:11"}
{"user":"user1","action":"send","page":"page3", "ts":"2017-09-06T08:11:10"}
```

file2.json 文件的内容如下：

```
{"user":"user2","action":"login","page":"page1", "ts":"2017-09-06T08:44:12"}
{"user":"user2","action":"view","page":"page7", "ts":"2017-09-06T08:45:33"}
{"user":"user2","action":"view","page":"page8", "ts":"2017-09-06T08:55:58"}
{"user":"user2","action":"view","page":"page6", "ts":"2017-09-06T09:10:58"}
{"user":"user2","action":"logout","page":"page9", "ts":"2017-09-06T09:16:19"}
```

file3.json 文件的内容如下：

```
{"user":"user3","action":"login","page":"page4", "ts":"2017-09-06T09:17:11"}
```

仔细观察上面的文件内容。文件 file1.json 包含用户 user1 的行为记录，它包含一个 login 动作，但是没有 logout 操作。文件 file2.json 包含用户 user2 的所有活动，包括 login 和 logout 操作。文件 file3.json 只包含用户 user3 的 login 动作。在这 3 个文件中，用户活动的时间戳是以这样一种方式设置的，即 user1 的会话将在 file3.json 文件被处理时超时。

为了模拟数据流的行为，应把这 3 个 JSON 文件复制到 HDFS 指定目录下。

（2）首先导入项目所依赖的包，代码如下：

```scala
//第 8 章/sstreaming_complex_state2.scala

import org.apache.spark.sql.SparkSession
import org.apache.spark.sql.types._
import org.apache.spark.sql.streaming.{GroupState, GroupStateTimeout, OutputMode}
import scala.collection.mutable.ListBuffer
```

（3）接下来，准备 3 个 case class。对于这个用例，用户活动输入数据由类 UserActivity 来表示。用户会话数据的中间状态由类 UserSessionState 表示，用户会话输出由类 UserSessionInfo 表示。实现这 3 个类的代码如下：

```scala
//表示用户活动数据的 case class
case class UserActivity(
    user:String,
    action:String,
    page:String,
    ts:java.sql.Timestamp
)

//表示会话中间状态的 case class
case class UserSessionState(
    var user: String,                       //用户
    var status: String,                     //会话状态
    var startTS: java.sql.Timestamp,        //会话开始时间
    var endTS: java.sql.Timestamp,          //会话结束时间
    var lastTS: java.sql.Timestamp,         //用户最后的活动时间
```

```
        var numPage: Int
)

//表示用户会话信息的 case class。当会话结束时,end 时间戳被填充
case class UserSessionInfo(
    userId:String,                          //用户 ID
    start:java.sql.Timestamp,               //会话开始时间
    end:java.sql.Timestamp,                 //会话结束时间
    numPage:Int
)
```

(4)接下来定义两个函数。

第 1 个函数叫作 updateUserActivity,它负责根据单用户活动更新用户会话状态。它根据用户所采取的操作,适当地更新会话开始或结束时间。此外,它还更新了最新的活动时间戳,代码如下:

```
/**
 * 负责根据单用户活动更新用户会话状态。
 * 它根据用户所采取的操作,适当地更新会话开始或结束时间。
 * 此外,它还更新了最新的活动时间戳。
 *
 * @param userSessionState    用户会话状态
 * @param userActivity        用户活动信息
 * @return
 */
def updateUserActivity(userSessionState: UserSessionState,
                userActivity:UserActivity) : UserSessionState = {
userActivity.action match {
  case "login" =>
    //如果用户处于登录状态,则将状态设置为 Online,并更新开始时间戳
    userSessionState.startTS = userActivity.ts
    userSessionState.status = "Online"
  case "logout" =>
    //如果用户处于登出状态,则将状态设置为 Offline,并更新结束时间戳
    userSessionState.endTS = userActivity.ts
    userSessionState.status = "Offline"
  case _ =>
    //否则将状态设置为 Active,访问页面计数 + 1
    userSessionState.numPage += 1
    userSessionState.status = "Active"
}

//更新最新时间戳
userSessionState.lastTS = userActivity.ts
userSessionState
}
```

第 2 个函数叫作 updateAcrossAllUserActivities,是回调函数,它将被传递到 flatMapGroupsWithState()函数。这个函数有两个主要的职责。第 1 个职责是处理中间会话状态的超时,并且当出现这种情况时,它会更新用户会话结束时间。第 2 个职责是确定

何时及什么被发送到输出。所需的输出是当用户会话启动时的一行,当用户会话结束时的另一行,代码如下:

```scala
import org.apache.spark.sql.streaming.GroupState
import scala.collection.mutable.ListBuffer

def updateAcrossAllUserActivities(user:String,
    inputs:Iterator[UserActivity],
    oldState: GroupState[UserSessionState]) :Iterator[UserSessionInfo] = {

var userSessionState = if (oldState.exists) oldState.get else
UserSessionState(user, "",new java.sql.Timestamp(System.currentTimeMillis),
null, null, 0)
var output = ListBuffer[UserSessionInfo]()

inputs.toList.sortBy(_.ts.getTime).foreach( userActivity => {
userSessionState = updateUserActivity(userSessionState, userActivity)
oldState.update(userSessionState)
if (userActivity.action == "login") {
output += UserSessionInfo(user, userSessionState.startTS,
userSessionState.endTS, 0)
}
})

val sessionTimedOut = oldState.hasTimedOut
val sessionEnded = !Option(userSessionState.endTS).isEmpty
val shouldOutput = sessionTimedOut || sessionEnded

shouldOutput match {
case true => {
if (sessionTimedOut) {
userSessionState.endTS =
new java.sql.Timestamp(oldState.getCurrentWatermarkMs)
}
oldState.remove()
output += UserSessionInfo(user,userSessionState.startTS,userSessionState.endTS,
userSessionState.numPage)
}
case _ => {
//extend session
oldState.update(userSessionState)
oldState.setTimeoutTimestamp(userSessionState.lastTS.getTime,"30 minutes")
}
}

output.iterator
}
```

设置步骤现在已经完成。

(5)在结构化流应用程序中将回调函数连接到 flatMapGroupsWithState()函数。在这个例子中，将利用超时特性，因此需要设置水印和事件超时类型，代码如下：

```scala
//主程序
def main(args: Array[String]): Unit = {

  val spark = SparkSession.builder()
    .master("local")
    .appName("cep demo")
    .getOrCreate()

  val userActivitySchema = new StructType()
    .add("user", StringType, nullable = false)
    .add("action", StringType, nullable = false)
    .add("page", StringType, nullable = false)
    .add("ts", TimestampType, nullable = false)

  //读取流数据，使用文件数据源
  val dataPath = "/data/spark/session"
  val userActivityDF = spark.readStream
    .option("maxFilesPerTrigger", 1)
    .schema(userActivitySchema)
    .json(dataPath)

  import spark.implicits._

  //转换到类型为 UserActivity 的 DataSet
  val userActivityDS = userActivityDF.withWatermark("ts", "30 minutes").as[UserActivity]

  //指定 event-time timeout type 并连接到回调函数
  val userSessionDS = userActivityDS
    .groupByKey(_.user)
    .flatMapGroupsWithState[UserSessionState,UserSessionInfo](
        OutputMode.Append, GroupStateTimeout.EventTimeTimeout)(updateAcrossAllUserActivities)

  //设置输出并启动流查询
  userSessionDS.writeStream
    .format("console")
    .option("truncate",value = false)
    .outputMode("append")
    .start()
    .awaitTermination()
}
```

（6）执行以上程序。当读到第 1 个 file1.json 数据文件时，输出的结果如下：

```
+------+-------------------+----+-------+
|userId|start              |end |numPage|
+------+-------------------+----+-------+
|user1 |2017-09-06 08:08:53|null|0      |
+------+-------------------+----+-------+
```

当读到第 1 个 file2.json 数据文件时，输出的结果如下：

```
+------+-------------------+-------------------+-------+
|userId|start              |end                |numPage|
+------+-------------------+-------------------+-------+
|user2 |2017-09-06 08:44:12|null               |0      |
|user2 |2017-09-06 08:44:12|2017-09-06 09:16:19|3      |
+------+-------------------+-------------------+-------+
```

当读到第 1 个 file3.json 数据文件时，输出的结果如下：

```
+------+-------------------+-------------------+-------+
|userId|start              |end                |numPage|
+------+-------------------+-------------------+-------+
|user1 |2017-09-06 08:08:53|2017-09-06 08:46:19|2      |
|user3 |2017-09-06 09:17:11|null               |0      |
+------+-------------------+-------------------+-------+
```

在处理 file1.json 文件中的用户活动之后，可以看到输出中有一行。这是因为每当函数 updateAcrossAllUserActivities 在用户活动中看到 login 动作时，它就会将 UserSessionInfo 类的一个实例添加到输出 ListBuffer 中。在处理 file2.json 文件之后，输出中有两行。一行是 login 动作，另一行用于 logout 操作。现在 file3.json 文件只包含一个带有动作 login 的 user3 的用户活动，但是输出包含两行。user1 的一行是检测 user1 会话超时的结果，这意味着由于 user1 缺乏活动，水印已经传递了该特定会话的超时值。

通过这两个用例的演示，可以看到结构化流中的任意有状态处理特性提供了灵活而强大的方法，可以将用户定义的处理逻辑应用于每个组，并完全控制何时将何内容发送到输出，从而实现对复杂事件模式的检测和处理。

8.4 处理重复数据

当数据源多次发送相同的数据时，实时流数据中的数据就会产生重复。在流处理中，由于流数据的无界性，去除重复数据是一种非常具有挑战性的任务。

不过，Spark 结构化流使流应用程序能够轻松地执行数据去重操作，因此这些应用程序可以通过在到达时删除重复的数据来保证精确一次处理。结构化流所提供的数据去重特性可以与水印一起工作，也可以不使用水印。不过，需要注意的一点是，在执行数据去重操作时，如果没有指定水印，在流应用程序的整个生命周期中，结构化流需要维护的状态将无限增长，这可能会导致内存不足的问题。使用水印，比水印更老的数据会被自动删除，以避免重复的可能。

在结构化流中执行重复数据删除操作的 API 很简单，它只有一个输入参数，该输入参数是用来唯一标识每行的列名的列表。这些列的值将被用于执行重复检测，并且结构化流将把它们存储为中间状态。

下面通过一个示例来演示这个 API 的使用。在这个示例程序中,处理重复到达的移动电话操作事件数据。

建议按以下步骤操作。

(1) 准备数据。

在本示例中使用文件数据源,代表移动电话操作事件数据存储在两个 JSON 格式的文件中。每个事件由以下 3 个字段组成。

① id:表示手机的唯一 ID,字符串类型。

② action:表示用户所采取的操作,该操作的可能值是 open 或 close。

③ ts:表示用户 action 发生时的时间戳,这是事件时间。

本示例准备了两个存储移动电话操作事件数据的 JSON 文件 file1.json 和 file2.json,其中 file2.json 文件中包含了与 file1.json 文件中重复的记录。

file1.json 文件的内容如下:

```
{"id":"phone1","action":"open","ts":"2018-03-02T10:15:33"}
{"id":"phone2","action":"open","ts":"2018-03-02T10:22:35"}
{"id":"phone3","action":"open","ts":"2018-03-02T10:23:50"}
```

观察上面的数据,每行都是唯一的 id 和 ts 列。

file2.json 文件的内容如下:

```
{"id":"phone1","action":"open","ts":"2018-03-02T10:15:33"}
{"id":"phone2","action":"open","ts":"2018-03-02T10:22:35"}
{"id":"phone4","action":"open","ts":"2018-03-02T10:29:35"}
{"id":"phone5","action":"open","ts":"2018-03-02T10:01:35"}
```

观察 file2.json 文件中的数据,前两行是 file1.json 文件中前两行的重复,第 3 行是唯一的,第 4 行也是唯一的,但延迟到达(所以在后面的代码中应该不被处理)。

为了模拟数据流的行为,应把这两个 JSON 文件复制到 HDFS 的指定目录下。

(2) 编写流处理代码。

编写流处理代码逻辑,在代码中基于 id 列进行分组 count()聚合。将 id 列和 ts 列共同定义为 key,代码如下:

```scala
//第8章/sstreaming_duplicates.scala

import org.apache.spark.sql.SparkSession
import org.apache.spark.sql.types._

val mobileDataSchema = new StructType()
    .add("id", StringType, nullable = false)
    .add("action", StringType, nullable = false)
    .add("ts", TimestampType, nullable = false)

//读取流数据,使用文件数据源
val dataPath = "/data/spark/mobile4"
val mobileDupSSDF = spark.readStream
```

```
            .option("maxFilesPerTrigger", 1)
            .schema(mobileDataSchema)
            .json(dataPath)

val windowCountDupDF = mobileDupSSDF
        .withWatermark("ts", "10 minutes")      //水印
        .dropDuplicates("id", "ts")             //去重
        .groupBy("id")
        .count

windowCountDupDF.writeStream
        .format("console")
        .option("truncate", "false")
        .outputMode("update")
        .start()
```

（3）执行上面的程序代码。

当读到 file1.json 源数据文件时，输出的结果如下：

```
+------+-----+
|id    |count|
+------+-----+
|phone3|1    |
|phone1|1    |
|phone2|1    |
+------+-----+
```

当读到 file2.json 源数据文件时，输出的结果如下：

```
+------+-----+
|id    |count|
+------+-----+
|phone4|1    |
+------+-----+
```

如预期所料，当读到 file2.json 源数据文件时，输出结果中只有一行显示在控制台中。原因是前两行是 file1.json 文件中前两行的重复，因此它们被过滤掉了（去重）。最后一行的时间戳是 10:10:00，这被认为是迟到的数据，因为时间戳比 10min 的水印阈值更迟，因此，最后一行被删除了。

8.5 容错

当开发重要的流应用程序并将其部署到生产环境中时，最重要的考虑之一就是故障恢复。根据墨菲定律，任何可能出错的地方都会出错。机器将会有故障，软件将会有缺陷。当有故障时，结构化流提供了一种方法来重新启动或恢复流应用程序，并从停止的地方继续执行。

要利用这种恢复机制，需要配置流应用程序使用检查点和预写日志。理想情况下，检

查点的位置应该是一个可靠的、容错的文件系统的路径，例如 HDFS 或 S3。结构化流将定期将所有的进度信息保存到检查点位置，例如正在处理的数据的偏移细节和中间状态值，因此，在任何失败的情况下，它可以完全重新处理相同的数据，确保得到端到端的一次保证。这些存储已处理偏移量信息的 WAL 文件以 JSON 格式保存到 HDFS、S3 bucket 或 Azure Blob 等容错存储中，以实现前向兼容性，如图 8-9 所示。

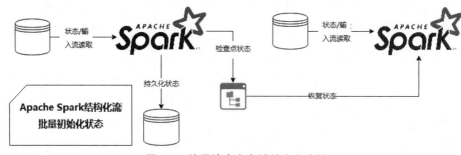

图 8-9　使用检查点存储状态和容错

向流查询添加检查点位置非常简单，只需在流查询中添加一个选项，将 checkpointLocation 作为名称和将路径作为值，代码如下：

```
//向一个流查询添加 checkpointLocation 选项
val userSessionSQ = userSessionDS.writeStream
    .format("console")
    .option("truncate",false)
    .option("checkpointLocation","/ck/location")          //设置检查点
    .outputMode("append")
    .start()
```

如果查看指定的检查点位置，则应该可以看到以下子目录：commits、metadata、offsets、sources 和 stats。这些目录中的信息是专用于特定流查询的，因此，每个流查询都必须使用不同的检查点位置，如图 8-10 所示。

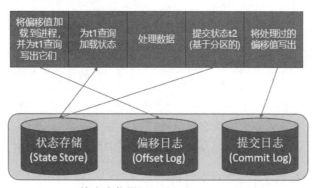

图 8-10　检查点位置

通过检查点，得到了完全的容错保证，从而确保从故障中恢复。结构化流可以结合特定的接收器、数据源和 WAL 文件提供恰好一次的交付语义。接收器和数据源必须是可重放的，就像 Kafka 和 Kinesis 等。否则它将是 at-least once（精确一次）交付语义。

就像大多数软件应用程序一样，流应用程序将随着时间的推移而不断重构，因为需要改进处理逻辑、性能或者修复 Bug。重要的是要记住，这可能会影响在检查点位置保存的信息，并知道哪些更改被认为是安全的。概括地说，有两类变化：一类是对流应用程序代码的更改，另一类是对 Spark 运行时的更改。

1. 流应用程序代码更改

检查点位置的信息被设计为对流应用程序的变化有一定的弹性。有一些变化将被认为是不相容的变化。第 1 个是通过改变 key 列、添加更多的 key 列或者删除一个现存的 key 列来改变聚合的方式。第 2 种方法是改变用于存储中间状态的类结构，例如，当一个字段被移除或者字段的类型从字符串转换为整数时。当在重新启动期间检测到不兼容的更改时，结构化流将通过一个异常进行通知。在这种情况下，必须使用新的检查点位置，或者删除先前检查点位置的内容。

2. 运行时更改

检查点格式被设计为向前兼容，这样当 Spark 跨越补丁版本或小版本的更新时（例如从 Spark 2.2.0 升级到 2.2.1 或者从 Spark 2.2.x 升级到 2.3.x），流应用程序应该能够从一个旧的检查点重新启动，甚至可以通过在流转换中做有限的更改来恢复处理（例如，添加新的过滤器子句来删除损坏的数据等），并且 WAL 文件可以无缝地工作。唯一的例外是当有严重的 Bug 修复时。不过通常不需要担心，当 Spark/PySpark 引入不兼容的变更时，它会在发行说明中清楚地进行说明。

如果由于不兼容的问题，无法启动一个使用现有检查点位置的流应用程序，则需要使用一个新的检查点位置，并且可能还需要为应用程序提供一些关于偏移量的信息来读取数据。

8.6 流查询度量指标和容错

与其他长时间运行的应用程序（如在线服务）类似，有必要了解流应用程序正在进行的进展、传入的数据速率或者中间状态所消耗的内存数量等信息。结构化流提供了一些 API 来提取关于最近执行进度的信息，以及在流应用程序中监控所有流媒体查询的异步方式。

8.6.1 流查询指标

关于流式查询的最基本的有用信息是它的当前状态。通过调用 StreamingQuery.status

属性，可以以可读的格式检索和显示这些信息。返回的对象是类型 StreamingQueryStatus，它将状态信息转换成 JSON 格式。例如，调用 StreamingQuery.status 属性，代码如下：

```
//以 JSON 格式查询状态信息
//从上面的示例中使用一个流查询
query.status
```

输出如下：

```
res11: org.apache.spark.sql.streaming.StreamingQueryStatus =
{
  "message" : "Waiting for data to arrive",
  "isDataAvailable" : false,
  "isTriggerActive" : false
}
```

很明显，在当前状态函数被调用时，前面的状态提供了关于流查询的基本信息。为了从最近的进展中获得更多的细节，例如传入的数据速率、处理速率、水印、数据源的偏移量，以及一些关于中间状态的信息，可以调用 StreamingQuery.recentProgress()函数。这个函数会返回 StreamingQueryProgress 类的实例的一个数组，它将细节转换成 JSON 格式。默认情况下，每个流查询都被配置为保持 100 个进度更新，这个数字可以通过更新 Spark 配置 spark.sql.streaming.numRecentProgressUpdates 来改变。要查看最新的流查询进度，可以调用函数 StreamingQuery.lastProgress()。关于流查询进度的示例，内容如下：

```
//流查询进度结点
{
  "id" : "9ba6691d-7612-4906-b64d-9153544d81e9",
  "runId" : "c6d79bee-a691-4d2f-9be2-c93f3a88eb0c",
  "name" : null,
  "timestamp" : "2018-04-23T17:20:12.023Z",
  "batchId" : 0,
  "numInputRows" : 3,
  "inputRowsPerSecond" : 250.0,
  "processedRowsPerSecond" : 1.728110599078341,
  "durationMs" : {
    "addBatch" : 1548,
    "getBatch" : 8,
    "getOffset" : 36,
    "queryPlanning" : 110,
    "triggerExecution" : 1736,
    "walCommit" : 26
  },
  "eventTime" : {
    "avg" : "2017-09-06T15:10:04.666Z",
    "max" : "2017-09-06T15:11:10.000Z",
    "min" : "2017-09-06T15:08:53.000Z",
    "watermark" : "1970-01-01T00:00:00.000Z"
  },
  "stateOperators" : [ {
    "numRowsTotal" : 1,
```

```
      "numRowsUpdated" : 1,
      "memoryUsedBytes" : 16127
    } ],
    "sources" : [ {
      "description" : "FileStreamSource[file:<path>/data/input]",
      "startOffset" : null,
      "endOffset" : {
        "logOffset" : 0
      },
      "numInputRows" : 3,
      "inputRowsPerSecond" : 250.0,
      "processedRowsPerSecond" : 1.728110599078341
    } ],
    "sink" : {
      "description" : "org.apache.spark.sql.execution.streaming.
ConsoleSinkProvider@37dc4031"
    }
}
```

查看上面显示的流进度的详细信息，有一些重要的关键指标值得注意。输入率表示从输入源流入流应用程序的传入数据量。处理速率代表流应用程序处理传入数据的速度有多快。在理想状态下，处理速率应该高于输入速率，如果不是这样，则需要考虑在 Spark 集群中增加节点的数量。如果流应用程序通过隐式 groupBy() 转换或显式地通过任意状态处理 APIs 来保持状态，则关注 stateOperators 部分中的指标是很重要的。

Spark UI 在 Job、Stages 和 Task 级别上提供了丰富的度量标准。流应用程序中的每个触发器都被映射到 Spark UI 中的一个 Job，在那里查询计划和任务持续时间可以很容易地检查。

8.6.2 流监控指标

结构化流提供了一种回调机制，可以在流应用程序中异步接收事件和流查询的进展。这是通过 StreamingQueryListener 接口完成的，它告诉用户什么时候启动了流查询，什么时候它已经取得了一些进展，什么时候它被终止了。这个接口的一个实现可以控制如何处理所提供的信息。一个明显的实现是将这些信息发送到 Kafka 主题或其他用于离线分析的发布-订阅系统，或者另一个流应用程序进行处理。例如，在 StreamingQueryListener 接口的一个简单实现中，将信息打印到控制台，代码如下：

```
//StreamingQueryListener 接口的一个简单实现
import org.apache.spark.sql.streaming.StreamingQueryListener
import
org.apache.spark.sql.streaming.StreamingQueryListener.QueryStartedEvent
import
org.apache.spark.sql.streaming.StreamingQueryListener.QueryProgressEvent
import
org.apache.spark.sql.streaming.StreamingQueryListener.QueryTerminatedEvent
```

```
class ConsoleStreamingQueryListener extends StreamingQueryListener {
  override def onQueryStarted(event: QueryStartedEvent): Unit = {
    println(s"streaming query started: ${event.id} - ${event.name} - ${event.runId}")
  }

  override def onQueryProgress(event: QueryProgressEvent): Unit = {
    println(s"streaming query progress: ${event.progress}")
  }

  override def onQueryTerminated(event: QueryTerminatedEvent): Unit = {
    println(s"streaming query terminated: ${event.id} - ${event.runId}")
  }
}
```

一旦有了 StreamingQueryListener 的实现，接下来就可以用 StreamQueryManager 注册，StreamQueryManager 可以处理多个监听器。注册和注销监听器的代码如下：

```
//注册和注销一个 StreamingQueryListener 实例，使用 StreamQueryManager
Val listener = new ConsoleStreamingQueryListener

//注册
spark.streams.addListener(listener)

//注销
spark.streams.removeListener(listener)
```

需要记住的一点是，每个监听器都从流应用程序中的所有流查询接收流查询事件。如果需要将特定的事件处理逻辑应用到某个流查询中，则它可以利用该流查询名称。

8.7 结构化流案例：运输公司车辆超速实时监测

想象一个物流公司车队管理解决方案，其中车队中的车辆启用了无线网络功能。每辆车定期报告其地理位置和许多操作参数，如燃油水平、速度、加速度、轴承和发动机温度等。物流公司希望利用这一遥测数据流实现一系列应用程序，以帮助他们管理业务的运营。

假设现在需要开发一个流应用程序，计算车辆每几秒的平均速度，用来检查车辆是否超速。程序整体架构如图 8-11 所示。

图 8-11　运输公司车辆超速实时监测程序架构

在本案例中，使用 Kafka 作为流数据源，将从 Kafka 的 cars 主题来读取这些事件。同时也将 Kafka 作为流的 Data Sink，检测出的超速事件将写入 Kafka 的 fastcars 主题。

为了模拟车辆向 Kafka 发送传感器数据，将创建一个 Kafka producer，它将 id、speed、acceleration 和 timestamp 写入 Kafka 的 cars 主题，代码如下：

```scala
//第 8 章/RandomCarsKafkaProducer.scala

import java.util.Properties
import org.apache.kafka.clients.producer.{KafkaProducer, ProducerRecord}
import scala.annotation.tailrec
import scala.util.{Random => random}

object RandomCarsKafkaProducer {
  def main(args: Array[String]): Unit = {
    //Kafka broker 连接属性
    val props = new Properties()
    props.put("Bootstrap.servers", "localhost:9092")
    props.put("key.serializer",
              "org.apache.kafka.common.serialization.StringSerializer")
    props.put("value.serializer",
              "org.apache.kafka.common.serialization.StringSerializer")

    //Kafka 生产者
    val producer = new KafkaProducer[String, String](props)
    val interval = 1000
    val topic = "cars"             //主题
    //val numRecsToProduce: Option[Int] = None      //None = infinite
    val numRecsToProduce: Option[Int] = Option(1000) //连续产生1000条数据

    //生成事件的方法
    @tailrec//@tailrec 确保方法是尾递归。尾递归可以保持内存需求不变
    def produceRecord(numRecToProduce: Option[Int]): Unit = {
      numRecToProduce match {
        //如果是有限数据集
        case Some(x) if x > 0 =>        //⇒
          //生成一条数据，发送一条数据
          producer.send(generateCarRecord(topic))
          Thread.sleep(interval)
          produceRecord(Some(x - 1))
        //如果是无限数据集
        case None =>
          producer.send(generateCarRecord(topic))
          Thread.sleep(interval)
          produceRecord(None)
        case _ =>
      }
    }

    produceRecord(numRecsToProduce)
  }

  //每次调用下面这种方法，将一条车辆行驶数据发送给 Kafka "cars" 主题
```

```
def generateCarRecord(topic: String): ProducerRecord[String, String] = {
    val carName = s"car${random.nextInt(10)}"       //车名
    val speed = random.nextInt(150)                  //速度
    val acc = random.nextFloat * 100                 //加速

    val value = s"$carName, $speed, $acc, ${System.currentTimeMillis()}"
    print(s"Writing $value\n")
    val d = random.nextFloat() * 100                 //模拟随机延迟时间
    if (d < 2) {
      println("抱歉！有一些网络延迟！")
      Thread.sleep((d*100).toLong)                   //产生随机延迟
    }
    //构造一条记录（事件）
    new ProducerRecord[String, String](topic, "key", value)
  }
}
```

注意：这里的时间戳是在源处生成事件（消息）的时间，即事件时间。

接下来，将原始数据解析到一个 case class 类中，这样就有了一个可以使用的结构，代码如下：

```
case class CarEvent(
    carId: String,
    speed: Option[Int],
    acceleration: Option[Double],
    timestamp: Timestamp)

object CarEvent {
    def apply(rawStr: String): CarEvent = {
      val parts = rawStr.split(",")
      CarEvent(parts(0),
          Some(Integer.parseInt(parts(1))),
          Some(java.lang.Double.parseDouble(parts(2))),
          new Timestamp(parts(3).toLong))
    }
}

val cars: Dataset[CarEvent] = df
    .selectExpr("CAST(value AS STRING)")
    .map(r => CarEvent(r.getString(0)))
```

这会产生 CarEvent 类型的 DataSet。最后完成的完整代码如下：

```
//第8章/KafkaSourceStreaming.scala

import java.sql.Timestamp
import org.apache.spark.sql.functions._
import org.apache.spark.sql.{DataFrame, SparkSession}

object KafkaSourceStreaming {

  //将聚合转换为类型化数据
```

```scala
case class CarEvent(
  carId: String,
  speed: Option[Int],
  acceleration: Option[Double],
  timestamp: Timestamp)

object CarEvent {
  def apply(rawStr: String): CarEvent = {
    val parts = rawStr.split(",")
    CarEvent(
      parts(0),
      Some(Integer.parseInt(parts(1))),
      Some(java.lang.Double.parseDouble(parts(2))),
      new Timestamp(parts(3).toLong)
    )
  }
}

def main(args: Array[String]): Unit = {

  //创建一个Spark Session，并以local模式运行
  val spark = SparkSession.builder()
    .appName("KafkaSourceStreaming")
    .master("local[*]")
    .getOrCreate()

  spark.sparkContext.setLogLevel("WARN")

  import spark.implicits._

  //读取数据源
  //不能为Kafka源设置一个模式。Kafka源有一个固定的模式(key, value)
  val df: DataFrame = spark.readStream
    .format("kafka")
    .option("kafka.Bootstrap.servers", "localhost:9092")
    .option("subscribe", "cars")
    .load()

  val cars = df
    .selectExpr("CAST(value AS STRING)")
    .map(r => CarEvent(r.getString(0)))

  //带有窗口（大小为4s的滚动窗口）和水印的聚合
  val aggregates = cars
    .withWatermark("timestamp", "3 seconds")
    .groupBy(window($"timestamp","4 seconds"), $"carId")
    .agg(avg("speed").alias("speed"))
    .where("speed > 120")        //这里定义超过120为超速

  aggregates.printSchema()
  //将结果写到控制台
  val writeToConsole = aggregates.writeStream
    .format("console")
    .option("truncate", "false")
```

```
        .queryName("cars streaming")
        .outputMode("update")
        .start()

    //将结果写到Kafka
    val writeToKafka = aggregates
        .selectExpr("CAST(carId AS STRING) AS key", "CAST(speed AS STRING) AS value")
        .writeStream
        .format("kafka")
        .option("kafka.Bootstrap.servers","localhost:9092")
        .option("topic", "fastcars")
        //.option("startingOffsets", "earliest")   //earliest，默认为 latest
        .option("checkpointLocation", "/tmp/carsck/")
        .queryName("cars streaming")
        .outputMode("update")                     //输出新的和更新的
        .start()

    spark.streams.awaitAnyTermination()           //一次运行多个流
    }
}
```

当指定非 memory 或 console 输出时，代码中必须指定检查点位置以实现容错。

要运行此程序，建议按以下步骤执行：

（1）启动 ZooKeeper，命令如下：

```
$ ./bin/zookeeper-server-start.sh config/zookeeper.properties
```

（2）启动 Kafka，命令如下：

```
$ ./bin/kafka-server-start.sh config/server.properties
```

（3）创建两个 Kafka 主题，代码如下：

```
$ ./bin/kafka-topics.sh --list --zookeeper localhost:2181
$ ./bin/kafka-topics.sh --zookeeper localhost:2181 --replication-factor 1 --partitions 1 --create --topic cars
$ ./bin/kafka-topics.sh --zookeeper localhost:2181 --replication-factor 1 --partitions 1 --create --topic fastcars
```

（4）先在一个新的终端窗口中执行消费者脚本，以此来拉取 fastcars 主题数据，代码如下：

```
$ ./bin/kafka-console-consumer.sh --Bootstrap-server localhost:9092 --topic fastcars
```

（5）执行前面编写的流计算程序。

（6）执行数据源程序，命令如下：

```
$ cd cars
$ java -jar fastcars.jar
```

（7）回到消息者脚本执行窗口（第 4 步的窗口），查看超速数据。如果一切正常，则应该看到在 fastcars 主题收到的超速数据，内容如下（部分）：

```
...
{"carId":"car2","avgSpeed":144.0,"start":"2021-08-27 12:20:48.0","end":"2021-08-27 12:20:53.0"}
```

```
{"carId":"car7","avgSpeed":130.0,"start":"2021-08-27 12:20:48.0","end":"2021-08-27 12:20:53.0"}
{"carId":"car7","avgSpeed":130.0,"start":"2021-08-27 12:20:50.0","end":"2021-08-27 12:20:55.0"}
{"carId":"car4","avgSpeed":148.0,"start":"2021-08-27 12:20:56.0","end":"2021-08-27 12:21:01.0"}
{"carId":"car4","avgSpeed":148.0,"start":"2021-08-27 12:20:58.0","end":"2021-08-27 12:21:03.0"}
{"carId":"car4","avgSpeed":148.0,"start":"2021-08-27 12:21:00.0","end":"2021-08-27 12:21:05.0"}
{"carId":"car1","avgSpeed":126.5,"start":"2021-08-27 12:21:00.0","end":"2021-08-27 12:21:05.0"}
{"carId":"car1","avgSpeed":134.0,"start":"2021-08-27 12:21:02.0","end":"2021-08-27 12:21:07.0"}
{"carId":"car1","avgSpeed":134.0,"start":"2021-08-27 12:21:04.0","end":"2021-08-27 12:21:09.0"}
{"carId":"car7","avgSpeed":149.0,"start":"2021-08-27 12:21:12.0","end":"2021-08-27 12:21:17.0"}
{"carId":"car7","avgSpeed":149.0,"start":"2021-08-27 12:21:14.0","end":"2021-08-27 12:21:19.0"}
{"carId":"car7","avgSpeed":149.0,"start":"2021-08-27 12:21:16.0","end":"2021-08-27 12:21:21.0"}
{"carId":"car6","avgSpeed":125.0,"start":"2021-08-27 12:21:16.0","end":"2021-08-27 12:21:21.0"}
{"carId":"car1","avgSpeed":139.0,"start":"2021-08-27 12:21:20.0","end":"2021-08-27 12:21:25.0"}
{"carId":"car1","avgSpeed":139.0,"start":"2021-08-27 12:21:22.0","end":"2021-08-27 12:21:27.0"}
{"carId":"car0","avgSpeed":144.0,"start":"2021-08-27 12:21:40.0","end":"2021-08-27 12:21:45.0"}
{"carId":"car0","avgSpeed":144.0,"start":"2021-08-27 12:21:42.0","end":"2021-08-27 12:21:47.0"}
{"carId":"car0","avgSpeed":144.0,"start":"2021-08-27 12:21:44.0","end":"2021-08-27 12:21:49.0"}
{"carId":"car4","avgSpeed":130.0,"start":"2021-08-27 12:21:46.0","end":"2021-08-27 12:21:51.0"}
{"carId":"car4","avgSpeed":130.0,"start":"2021-08-27 12:21:48.0","end":"2021-08-27 12:21:53.0"}
{"carId":"car4","avgSpeed":130.0,"start":"2021-08-27 12:21:50.0","end":"2021-08-27 12:21:55.0"}
...
```

8.8 结构化流案例：实时订单分析

本节使用 Spark 结构化流重构第 6 章的实时股票交易分析（参考 6.2.2 节）。在这个示例中，数据源是证券公司实时发布的市场交易数据，例如证券、股票等的买卖。我们现在需要构建一个实时仪表板应用程序，用来回答以下问题：

(1) 计算每十秒的销售和购买订单数量。
(2) 根据购买或出售的总金额来统计前 5 个客户。
(3) 找出过去一小时内前 5 个交易量最多的股票。

8.8.1 数据集说明和数据源

本案例准备了一个文件 orders.txt,包含 500 000 行数据,每行代表一个买/卖订单。每行包含以下用逗号分隔的元素。

(1) Order 时间戳:格式为 yyyy-mm-dd hh:MM:ss。
(2) Order ID:连续递增的整数。
(3) Client ID:从 1 到 100 的整数。
(4) Stock 代码:共 80 个股票代码。
(5) 股票买卖的数量:从 1 到 1000。
(6) 股票买卖的价格:从 1 到 100。
(7) 字符 B 或 S:代表一个订单的买(B)/卖(S)事件。

查看文件的前 5 行内容,命令如下:

```
$ head -n 5 ~/spark/streaming/orders.txt
```

前 5 行的内容如下:

```
2016-03-22 20:25:28,1,80,EPE,710,51.00,B
2016-03-22 20:25:28,2,70,NFLX,158,8.00,B
2016-03-22 20:25:28,3,53,VALE,284,5.00,B
2016-03-22 20:25:28,4,14,SRPT,183,34.00,B
2016-03-22 20:25:28,5,62,BP,241,36.00,S
```

以上这些格式的数据就是本案例要处理的流程序的数据。

为了模拟实时交易数据,本案例中使用一个 Linux Shell 脚本将股票交易数据发送给 Kafka orders 主题。Spark 结构化流处理程序将从这个主题读取交易数据并将计算出的结果写到另一个 Kafka metrics 主题,然后使用 Kafka 的 kafka-console-consumer.sh 脚本来接收并显示结果。处理流程如图 8-12 所示。

图 8-12 股票交易数据分析流程

编写的 Shell 脚本 streamOrders.sh 会调用 Kafka 自带的生产者脚本,依次读取 orders.txt 文件中的每条记录并发送给 Kafka 的 order 主题,间隔时间为 100ms,脚本代码如下:

```
#!/bin/bash
BROKER=$1
```

```
if [ -z "$1" ]; then
        BROKER="localhost:9092"
fi

cat orders.txt | while read line; do
        echo "$line"
        sleep 0.1
done | ~/bigdata/kafka_2.11-2.4.1/bin/kafka-console-producer.sh --broker-list
$BROKER --topic orders
```

暂时不需要启动该脚本,稍后在执行阶段会用到它。

8.8.2 计算每 10s 的销售和购买订单数量

第 1 个任务是实时计算每 10s 的销售和购买订单数量。建议按以下步骤进行操作:
(1) 启动 ZooKeeper 服务。

Kafka 依赖于 Apache ZooKeeper,所以在启动 Kafka 之前,要先启动它。打开一个终端窗口,执行的命令如下:

```
$ cd ~/bigdata/kafka_2.12-2.4.1
$ ./bin/zookeeper-server-start.sh config/zookeeper.properties &
```

这将在 2181 端口启动 ZooKeeper 进程,并让 ZooKeeper 在后台工作。
(2) 接下来,启动 Kafka 服务器。

另打开一个终端窗口,执行的命令如下:

```
$ cd ~/bigdata/kafka_2.12-2.4.1
$ ./bin/kafka-server-start.sh config/server.properties &
```

(3) 最后,需要创建用于发送股票买卖数据和计算结果数据的主题。

分别创建 orders 主题和 metrics 主题。另外打开第 3 个终端窗口,执行的命令如下:

```
$ ./bin/kafka-topics.sh --create --Bootstrap-server localhost:9092
--replication-factor 1 --partitions 1 --topic orders
$ ./bin/kafka-topics.sh --create --Bootstrap-server localhost:9092
--replication-factor 1 --partitions 1 --topic metrics
```

查看已有的主题,使用的命令如下:

```
$ ./bin/kafka-topics.sh --list --Bootstrap-server localhost:9092
```

(4) 重新启动 spark-shell。

在启动时将 Kafka 库和 Spark Kafka 连接器库添加到类路径。需要使用--package 参数让 Spark 自动下载这些文件,命令如下:

```
$ spark-shell --master spark://localhost:7077 --packages
org.apache.spark:spark-sql-kafka-0-10_2.12:3.1.2
```

(5) 在 spark-shell 中交互式地执行结构化流计算程序代码。完整的结构化流处理代码如下:

```
//第 8 章/sstream_stock_orders1.scala

import java.sql.Timestamp
```

```scala
import java.text.SimpleDateFormat

import org.apache.spark.sql.SparkSession
import org.apache.spark.sql.functions._
import org.apache.spark.sql.types._

//case class 定义 Schema
case class Order(time: java.sql.Timestamp,
                 orderId:Long,
                 clientId:Long,
                 symbol:String,
                 amount:Int,
                 price:Double,
                 buy:Boolean)

def main(args: Array[String]): Unit = {
    val spark = SparkSession.builder()
      .master("local")
      .appName("orders streaming")
      .getOrCreate()

    spark.sparkContext.setLogLevel("WARN")         //设置日志级别

    //创建一个流来监听 test topic 的消息
    val dataDF = spark.readStream
        .format("kafka")
        .option("kafka.Bootstrap.servers", " localhost:9092")
        .option("subscribe", "orders")
        .option("startingOffsets", "earliest")
        .load()

    //获得 data 这个 DataFrame 的 Schema
    dataDF.printSchema()

    //将消息内容转换为 String 类型，再从 DataFrame 转换为 DataSet[String]
    import spark.implicits._      //将 DataFrame 转换为 DataSet，必须引入这个包
    val ordersDF = dataDF.selectExpr("CAST(value AS STRING) as v").as[String]

    //解析从 Kafka 读取的数据
    //数据格式: 2016-03-22 20:25:28,1,80,EPE,710,51.00,B
    val orders = ordersDF.flatMap(record => {
      val dateFormat = new SimpleDateFormat("yyyy-MM-dd hh:mm:ss")
      val s = record.split(",")
      try {
        assert(s(6) == "B" || s(6) == "S")
        List(Order(new Timestamp(dateFormat.parse(s(0)).getTime),
              s(1).toLong,
              s(2).toLong,
              s(3),
              s(4).toInt,
```

```
                    s(5).toDouble,
                    s(6) == "B"))
        }catch {
            case e : Throwable => println("错误的行格式 (" + e + "): " + record)
                List()           //如果无法正确解析,就返回空列表
            }
        })

    //1. 统计股票买/卖总数量
    val bool2str = udf((flag:Boolean) => if (flag) "buys" else "sells")
    val numPerType = orders.groupBy(window($"ts", "10 seconds"), $"buy")
                    .agg(count("*") as "total")
                    .select(bool2str($"buy") as "type", $"total")

    /* 将该流写到控制台
    val query = numPerType.writeStream
        .outputMode("complete")
        .format("console")
        .option("truncate","false")
        .start()
    */

    //将要写入 Kafka 的结果重新组织,放在名为"value"的列中(Kafka 的要求)
    import org.apache.spark.sql.types._
    val numPerTypeValues = numPerType.select(concat_ws(",",col("type"),
col("total")).cast(StringType).as("value"))

    //执行
    val query = numPerType.writeStream
        .outputMode("complete")         //写出模式
        .format("kafka")                //格式
        .option("kafka.Bootstrap.servers","localhost:9092")//kafka broker
        .option("topic","metrics")      //要写入的主题
        .option("checkpointLocation", "tmp/ck")    //指定检查点
        .start()

    println("starting......")
    query.awaitTermination()
}
```

(6) 运行。

另外再打开一个终端窗口(第 4 个终端窗口),执行脚本 streamOrders.sh。这个脚本会从 orders.txt 文件中流式地读取行并发送给 Kafka 的 orders 主题(需要注意,orders.txt 文件要和 streamOrders.sh 脚本文件在同一目录下,另外 Kafka 的 bin 目录要位于 Path 中,因为脚本中要调用 Kafka 自带的 kafka-console-producer.sh 脚本文件),命令如下:

```
$ chmod +x streamOrders.sh
$ ./streamOrders.sh localhost:9092
```

(7) 另外再打开一个终端窗口(第 5 个终端窗口),查看流程序的输出内容。在这个

终端窗口中，启动 Kafka 自带的 kafka-console-consumer.sh 脚本，它会消费来自 Kafka metrics 主题的消息，命令如下：

```
$ ./bin/kafka-console-consumer.sh --Bootstrap-server localhost:9092
--topic metrics --from-beginning
```

输出的结果如下：

```
sells, List(12)
buys, List(20)

sells, List(28)
buys, List(21)

sells, List(37)
buys, List(12)
```

（8）等待所有的文件都被处理完后，从 Shell 停止运算流，代码如下：

```
sinkQuery.stop()
```

当 Spark 结构化流使用 Kafka 作为数据源时，它返回的流 DataFrame 有一个固定的模式，结构如下：

```
|-- key: binary (nullable = true)
|-- value: binary (nullable = true)
|-- topic: string (nullable = true)
|-- partition: integer (nullable = true)
|-- offset: long (nullable = true)
|-- timestamp: timestamp (nullable = true)
|-- timestampType: integer (nullable = true)
```

其中 value 列包含 Kafka 中每条消息的实际内容，其类型是二进制的，如图 8-13 所示。

图 8-13　Kafka 消息格式

Kafka 并不真正关心每条信息的内容，因此它将其视为一个二进制的 blob。模式中的其余列包含每条消息的元数据。如果消息的内容在发送给 Kafka 时以某种二进制格式序列化，则需要在这些消息可以在 Spark 中处理之前使用一种方法来对其进行反序列化，可以是 Spark SQL 函数或 UDF。在上面的例子中，内容是一个字符串，所以只需将其转换为 String 类型。

8.8.3　根据购买或售出的总金额统计前 5 个客户

第 2 个任务是编程 Spark 结构化流应用程序，来根据购买或出售的总金额统计前 5 个客户。这是一个 Top N 问题。修改代码如下：

```scala
//第8章/sstream_stock_orders2.scala

import java.sql.Timestamp
import java.text.SimpleDateFormat

import org.apache.spark.sql.SparkSession
import org.apache.spark.sql.functions._

//case class 定义 Schema
case class Order(ts: java.sql.Timestamp,
                 orderId:Long,
                 clientId:Long,
                 symbol:String,
                 amount:Int,
                 price:Double,
                 buy:Boolean)

def main(args: Array[String]): Unit = {
  val spark = SparkSession.builder()
    .master("local")
    .appName("orders streaming")
    .getOrCreate()

  spark.sparkContext.setLogLevel("WARN")         //设置日志级别

  //创建一个流来监听 test topic 的消息
  val dataDF = spark.readStream
    .format("kafka")
    .option("kafka.Bootstrap.servers", "localhost:9092")
    .option("subscribe", "orders")
    .option("startingOffsets", "earliest")
    .load()

  //获得 data 这个 DataFrame 的 Schema
  dataDF.printSchema()

  //将消息内容转换为 String 类型，再从 DataFrame 转换为 DataSet[String]
  import spark.implicits._       //将 DataFrame 转换为 DataSet, 必须引入这个包
  val ordersDS = dataDF.selectExpr("CAST(value AS STRING) as v").as[String]

  //解析从 Kafka 读取的数据
  //数据格式: 2016-03-22 20:25:28,1,80,EPE,710,51.00,B
  val orders = ordersDS.flatMap(record => {
    val dateFormat = new SimpleDateFormat("yyyy-MM-dd hh:mm:ss")
    val s = record.split(",")
    try {
      assert(s(6) == "B" || s(6) == "S")
      List(Order(new Timestamp(dateFormat.parse(s(0)).getTime),
        s(1).toLong,
        s(2).toLong,
```

```
                    s(3),
                    s(4).toInt,
                    s(5).toDouble,
                    s(6) == "B"))
    }catch {
      case e : Throwable => println("错误的行格式 (" + e + "): " + record)
        List()           //如果无法正确解析，就返回空列表
    }
  })

  //2. Top 5 客户
  val amountPerClient = orders.select($"clientId",($"amount" * $"price") as
"volume")
    val top5clients = amountPerClient
      .groupBy(window($"ts", "10 seconds"), $"clientId")
      .agg(sum("volume") as "volume")
      .orderBy($"volume".desc)
      .limit(5)

  //将该流写到控制台
  /* val query = top5clients.writeStream
              .outputMode("complete")
              .format("console")
              .option("truncate","false")
              .start()
  */

  //将要写入 Kafka 的结果重新组织，放在名为"value"的列中(Kafka 的要求)
  import org.apache.spark.sql.types._
  val numPerTypeValues = top5clients.select(concat_ws(",", col("clientId"),
col("volume")).cast(StringType).as("value"))

  //执行
  val query = numPerTypeValues
    .writeStream
    .outputMode("complete")         //写出模式
    .format("kafka")                //格式
    .option("kafka.Bootstrap.servers", " localhost:9092")//kafka broker
    .option("topic","metrics")      //要写入的主题
    .option("checkpointLocation", "tmp/ck")  //指定检查点
    .start()

  println("starting......")
  query.awaitTermination()
}
```

8.8.4　找出过去一小时内前 5 个交易量最多的股票

这种类型的计算是通过窗口操作完成的。窗口的持续时间是一小时，但是滑动周期与

微批持续时间相同，均为 5s，因为想要每 5s 报告一次交易量排名前五的股票。

修改代码如下：

```scala
//第 8 章/sstream_stock_orders3.scala

import java.sql.Timestamp
import java.text.SimpleDateFormat

import org.apache.spark.sql.SparkSession
import org.apache.spark.sql.functions._

//case class 定义 Schema
case class Order(ts: java.sql.Timestamp,
                 orderId:Long,
                 clientId:Long,
                 symbol:String,
                 amount:Int,
                 price:Double,
                 buy:Boolean
                )

def main(args: Array[String]): Unit = {
    val spark = SparkSession.builder()
      .master("local")
      .appName("orders streaming")
      .getOrCreate()

    spark.sparkContext.setLogLevel("WARN")          //设置日志级别

    //创建一个流来监听 test topic 的消息
    val dataDF = spark.readStream
        .format("kafka")
        .option("kafka.Bootstrap.servers", "localhost:9092")
        .option("subscribe", "orders")
        .option("startingOffsets", "earliest")
        .load()

    //获得 data 这个 DataFrame 的 Schema
    dataDF.printSchema()

    //将消息内容转换为 String 类型，再从 DataFrame 转换为 DataSet[String]
    import spark.implicits._      //将 DataFrame 转换为 DataSet，必须引入这个包
    val ordersDS = dataDF.selectExpr("CAST(value AS STRING) as v").as[String]

    //解析从 Kafka 读取的数据
    //数据格式 2016-03-22 20:25:28,1,80,EPE,710,51.00,B
    val orders = ordersDS.flatMap(record => {
      val dateFormat = new SimpleDateFormat("yyyy-MM-dd hh:mm:ss")
      val s = record.split(",")
      try {
```

```
        assert(s(6) == "B" || s(6) == "S")
        List(Order(new Timestamp(dateFormat.parse(s(0)).getTime),
                s(1).toLong,
                s(2).toLong,
                s(3),
                s(4).toInt,
                s(5).toDouble,
                s(6) == "B"))
    }catch {
      case e : Throwable => println("错误的行格式 (" + e + "): " + record)
        List()          //如果无法正确解析,就返回空列表
    }
  })

  //3. 找出过去一小时内前 5 个交易量最多的股票
  //如果每 5s 报告一次过去一小时内前 5 个交易量最多的股票
  val topStocks = orders
    .select($"ts",$"symbol",$"amount")
    .groupBy(window($"ts", "60 minutes", "5 seconds"), $"symbol")
    .agg(sum("amount") as "amount")
    .orderBy($"amount".desc)

  //将该流写到控制台
  /* val query = topStocks.writeStream
    .outputMode("complete")
    .format("console")
    .option("truncate","false")
    .start()
*/

  //将要写入 Kafka 的结果重新组织,放在名为"value"的列中(Kafka 的要求)
  import org.apache.spark.sql.types._
  val numPerTypeValues = topStocks.select(concat_ws(",", col("symbol"),
col("amount")).cast(StringType).as("value"))

  //执行
  val query = numPerTypeValues
    .writeStream
    .outputMode("complete")             //写出模式
    .format("kafka")                    //格式
    .option("kafka.Bootstrap.servers", "localhost:9092")//kafka broker
    .option("topic","metrics")          //要写入的主题
    .option("checkpointLocation", "tmp/ck")  //指定检查点
    .start()

  println("starting......")
  query.awaitTermination()
}
```

第9章 Spark 图处理库 GraphFrame
CHAPTER 9

图提供了一种强大的方法来分析数据集中的连接。对于面向图的数据，图为处理数据提供了易于理解和直观的模型。此外，还可以使用专门的图算法来处理面向图的数据。这些算法为不同的分析任务提供了有效的工具。

Apache Spark 本身包含一个位于 Spark Core 之上的分布式图处理框架 GraphX，用于图并行和数据并行计算。它建立在一个称为"图论"的数学分支上，使在 Spark 中运行图算法成为可能，但是，GraphX API 存在一些限制。首先，GraphX 只支持 Scala 语言，所以无法使用 Python 进行大型的图计算。其次，GraphX 只能在 RDD 上工作，因此不能从 DataFrame 和 Catalyst 查询优化器提供的性能改进中受益。

GraphFrames 是一个开源的 Spark 包，创建它的目的是解决以上这两个问题。它具有以下特点：

（1）提供一组 Python API。

（2）适用于 DataFrame。

GraphFrames 扩展了 Spark GraphX 以提供 DataFrame API，使分析更容易使用、更有效，并简化了数据管道。GraphFrames 集成了 GraphX 和 DataFrame，使用户可以在不将数据移动到专门的图数据库的情况下执行图模式查询。

GraphFrames 和 GraphX 的比较见表 9-1。

表 9-1 GraphFrames 和 GraphX 的比较

比较内容	GraphFrames	GraphX
核心 API	Scala、Java、Python	仅 Scala
编程抽象	DataFrame	RDD
用例	算法、查询、Motif 查找	算法
顶点 ID	任何类型（Catalyst 支持的）	Long
顶点属性/边属性	任意数量的 DataFrame 列	任何类型（VD,ED）
返回类型	GraphFrame/DataFrame	Graph[VD,ED]

9.1 图基本概念

图作为链接对象的数学概念，由顶点（图中的对象）和连接顶点的边组成。

一旦表示为图，有些问题就变得更容易解决，它们自然地产生了图算法。例如，使用传统的数据组织方法（如关系数据库）呈现分层数据是很复杂的操作，所以可以用图来简化。除了使用它们来代表社交网络和网页之间的链接外，图算法还应用在生物学、计算机芯片设计、旅行、物理、化学等领域。

首先介绍本章中用到的基本图的相关术语。

图是一种数学结构，用来建模对象之间的关系。图是由顶点和连接它们的边组成的。顶点是对象，而边是它们之间的关系。顶点是图中的一个节点。一条边连接图中的两个顶点。一般来讲，顶点代表一个实体，边代表两个实体之间的关系。图在概念上等价于顶点和边的集合，如图9-1所示。

图可以有方向也可以无方向。

无向图是具有没有方向的边的图。无向图中的边没有源顶点或目标顶点。例如，用户张三和李四构成图的顶点，他们之间的关系构成边，如图9-2所示。

图 9-1　图是等价于顶点和边的集合　　　　图 9-2　无向图

有向图就是边具有方向的图。有向图中的边具有源顶点和目标顶点。例如，Twitter follower 就是一个有向图。用户张三可以关注用户李四，而不需要暗示用户李四也关注用户张三，如图9-3所示。

有向多图是一张有向图，它包含由两条或多条平行边连接的顶点对。有向多图中的平行边是具有相同起始和终止顶点的边。它们用于表示一对顶点之间的多个关系，如图9-4所示。

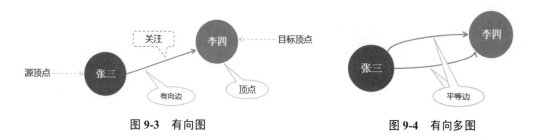

图 9-3　有向图　　　　图 9-4　有向多图

属性图是一张有向多图，它具有与顶点和边相关联的数据。属性图中的每个顶点都有一个或多个属性。类似地，每条边都有一个标签或属性，属性图如图9-5所示。

属性图为处理面向图的数据提供了丰富的抽象。它是用图建模数据的最流行的形式。例如，表示公司中的两个员工及其关系的属性图，如图9-6所示。

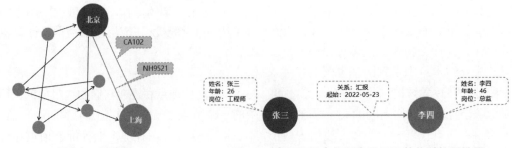

图 9-5　属性图　　　　　图 9-6　表示公司中的两个员工及其关系的属性图

属性图的另一个例子是表示微博上的社交网络的图。用户具有姓名、年龄、性别和位置等属性。此外，用户可以关注其他用户，并可能有自己的用户。在表示微博上的社交网络的图中，顶点表示用户，边表示"关注"关系。例如，一个简单的社交网络图如图 9-7 所示。

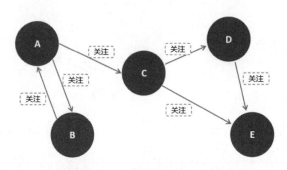

图 9-7　一个简单的社交网络图

9.2　GraphFrame 图处理库简介

Spark GraphFrames 提供了一个声明性 API，可用于大型图模型上的交互式查询和独立程序。由于 GraphFrame 是在 Spark SQL 上实现的，它支持在计算过程中进行并行处理和优化。GraphFrame 与 Spark 和 GraphX 的关系如图 9-8 所示。

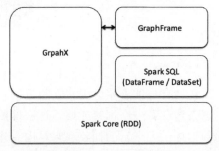

图 9-8　GraphFrame 与 Spark 和 GraphX 的关系

GraphFrame API 中的主要编程抽象是一个 GraphFrame。从概念上讲，它由两个 DataFrame 组成，分别表示图的顶点和边。顶点和边可以有多个属性，这些属性也可以用于查询。例如，在社交网络中，顶点可以包含名称、年龄、位置和其他属性，而边可以表示节点（社交网络中的个人）之间的关系。因为 GraphFrame 模型可以支持用户定义的每个顶点属性和边属性，因此它等价于属性图模型。此外，可以使用模式来定义视图，以匹配网络中各种形状的子图。

Spark GraphFrame 支持分布式属性图的图计算。使用 GraphFrames，顶点和边均表示为 DataFrame，这可以充分利用 Spark SQL 查询的优点，并支持 DataFrame 数据源，如 Parquet、JSON、CSV 等。

GraphFrame 优化了计算的关系和图部分的执行。可以使用关系运算符、模式和对算法的调用指定这些计算。

9.3 GraphFrame 的基本使用

在本节中，我们将使用 Spark GraphFrames 来建立图模型并进行研究。图的顶点和边被存储为 DataFrame，并且支持基于 Spark SQL 和 DataFrame 的查询对它们进行操作。由于 DataFrame 可以支持各种数据源，因此可以从关系表、文件（JSON、Parquet、Avro 和 CSV）等读取顶点和边信息。

顶点 DataFrame 必须包含一个叫作 id 的列，它为每个顶点指定唯一的 ID。类似地，边 DataFrame 必须包含名为 src（源顶点的 ID）和 dst（目标顶点的 ID）的两列。顶点 DataFrame 和边 DataFrame 都可以包含额外的属性列。

GraphFrames 公开了一个简洁的语言集成 API，它统一了图分析和关系查询，集成了图算法、模式匹配和查询。机器学习代码、外部数据源和 UDF 可以与 GraphFrames 集成来构建更复杂的应用程序。

9.3.1 添加 GraphFrame 依赖

因为 GraphFrames 并不存在于 Spark 的发行包中，因此首先需要添加依赖包。当前 GraphFrames 的最新版本是 0.8.1。需要指定兼容的 Spark 版本，目前直接支持的是 Spark 3.0，对应的 Scala 版本为 2.12。根据开发方式的不同，使用不同的依赖添加方式，说明如下：

（1）如果是通过 IDE 进行开发（如 IntelliJ IDEA），则可以在 Maven 项目的 pom.xml 配置文件中添加如下的依赖：

```xml
<dependency>
    <groupId>graphframes</groupId>
    <artifactId>graphframes</artifactId>
    <version>0.8.1-spark3.0-s_2.12</version>
</dependency>
```

（2）或者在 SBT 项目的 build.sbt 配置文件中添加的依赖如下：

```
libraryDependencies += "graphframes" % "graphframes" % "0.8.1-spark3.0-s_2.12"
```

（3）如果使用 spark-shell，则可以在启动 spark-shell 时，通过--packages 选项参数指定依赖，命令如下：

```
$ spark-shell --master spark://localhost:7077 --packages graphframes:graphframes:0.8.1-spark3.0-s_2.12
```

然后 spark-shell 会自动下载 GraphFrames JAR 包并缓存。

（4）如果使用 Zeppelin Notebook 进行 Spark 图计算开发，则最简单的方式是先下载 graphframes-0.8.1-spark3.0-s_2.12.jar 包，并将其复制到$SPARK_HOME/jars/目录下和$ZEPPELIN_HOME/lib/目录下，然后重启 Spark 和 Zeppelin（如果已经启动了）。

9.3.2 构造图模型

假设现在要分析一个社交网络，如图 9-9 所示。

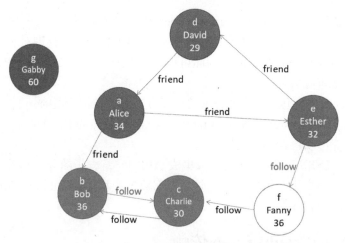

图 9-9 社交网络图

首先需要导入相应的依赖包，代码如下：

```
import org.apache.spark.sql.SparkSession
import org.graphframes._
```

可以从顶点和边 DataFrames 创建 GraphFrame 图对象。顶点 DataFrame 应该包含一个名为 id 的特殊列，它为图中的每个顶点指定唯一的 id。边 DataFrame 应该包含两个特殊列：src（边的源顶点 ID）和 dst（边的目标顶点 ID）。

这两个 DataFrame 都可以有任意其他列，这些列可以表示顶点和边属性，代码如下：

```
//顶点 DataFrame
val v = spark.createDataFrame(List(
```

```
        ("a", "Alice", 34, 234, "Apples"),
        ("b", "Bob", 36, 23232323, "Bananas"),
        ("c", "Charlie", 30, 2123, "Grapefruit"),
        ("d", "David", 29, 2321111, "Bananas"),
        ("e", "Esther", 32, 1, "Watermelon"),
        ("f", "Fanny", 36, 333, "Apples" ),
        ("g", "Gabby", 60, 23433, "Oranges")
)).toDF("id", "name", "age", "cash", "fruit")

//边 DataFrame
val e = spark.createDataFrame(List(
        ("a", "b", "friend"),
        ("b", "c", "follow"),
        ("c", "b", "follow"),
        ("f", "c", "follow"),
        ("e", "f", "follow"),
        ("e", "d", "friend"),
        ("d", "a", "friend"),
        ("a", "e", "friend")
    )).toDF("src", "dst", "relationship")

v.show()
e.show()

//使用这些顶点和这些边构建图模型
val g = GraphFrame(v, e)
```

这个示例图还可来自 GraphFrame 的 examples 包，代码如下：

```
import org.graphframes.examples.Graphs

val g2 = examples.Graphs.friends
```

9.3.3 简单图查询

GraphFrame 提供简单的图查询，如节点的入度、出度和度数。此外，由于 GraphFrame 将图表示为顶点 DataFrame 和边 DataFrame，因此很容易直接对顶点 DataFrame 和边 DataFrame 进行查询。例如，查看图的顶点和边信息，代码如下：

```
//查看顶点和边
g.vertices.show()
g.edges.show()
```

输出的结果如下：

```
+---+-------+---+--------+----------+
| id|   name|age|    cash|     fruit|
+---+-------+---+--------+----------+
|  a|  Alice| 34|     234|    Apples|
|  b|    Bob| 36|23232323|   Bananas|
|  c|Charlie| 30|    2123|Grapefruit|
```

```
|  d|  David| 29|2321111|   Bananas|
|  e| Esther| 32|      1|Watermelon|
|  f|  Fanny| 36|    333|    Apples|
|  g|  Gabby| 60|  23433|   Oranges|
+---+-------+---+-------+----------+

+---+---+------------+
|src|dst|relationship|
+---+---+------------+
|  a|  b|      friend|
|  b|  c|      follow|
|  c|  b|      follow|
|  f|  c|      follow|
|  e|  f|      follow|
|  e|  d|      friend|
|  d|  a|      friend|
|  a|  e|      friend|
+---+---+------------+
```

查看所有节点的入度，代码如下：

```
//查看所有节点的入度
g.inDegrees.show()
g.inDegrees.sort("id").show()
```

输出的结果如下：

```
+---+--------+
| id|inDegree|
+---+--------+
|  f|       1|
|  e|       1|
|  d|       1|
|  c|       2|
|  b|       2|
|  a|       1|
+---+--------+

+---+--------+
| id|inDegree|
+---+--------+
|  a|       1|
|  b|       2|
|  c|       2|
|  d|       1|
|  e|       1|
|  f|       1|
+---+--------+
```

查看所有节点的出度，代码如下：

```
//查看所有节点的出度
g.outDegrees.show()
g.outDegrees.sort("id").show()
```

输出的结果如下：

```
+---+---------+
| id|outDegree|
+---+---------+
|  f|        1|
|  e|        2|
|  d|        1|
|  c|        1|
|  b|        1|
|  a|        2|
+---+---------+

+---+---------+
| id|outDegree|
+---+---------+
|  a|        2|
|  b|        1|
|  c|        1|
|  d|        1|
|  e|        2|
|  f|        1|
+---+---------+
```

查看所有节点的总度数（出入度之和），代码如下：

```
//查看所有节点的总度数(出入度之和)
g.degrees.show()
g.degrees.sort("id").show()
```

输出的结果如下：

```
+---+------+
| id|degree|
+---+------+
|  f|     2|
|  e|     3|
|  d|     2|
|  c|     3|
|  b|     3|
|  a|     3|
+---+------+

+---+------+
| id|degree|
+---+------+
|  a|     3|
|  b|     3|
|  c|     3|
|  d|     2|
|  e|     3|
|  f|     2|
+---+------+
```

可以直接在顶点 DataFrame 上运行查询。例如，找到图中最年轻的人的年龄，代码如下：

```
//找到图中最年轻的人的年龄
val youngest = g.vertices.groupBy().min("age")
youngest.show()
```

输出的结果如下：

```
+--------+
|min(age)|
+--------+
|      29|
+--------+
```

找出年龄小于 30 岁的人，代码如下：

```
//找出年龄小于30岁的人
val lt30 = g.vertices.filter("age < 30")
lt30.show()
```

输出的结果如下：

```
+---+-----+---+
| id| name|age|
+---+-----+---+
|  d|David| 29|
+---+-----+---+
```

同样，可以在边 DataFrame 上运行查询。例如，统计一下图中 follow 关系的数量，代码如下：

```
//统计一下图中follow关系的数量
g.edges.filter("relationship = 'follow'").count()      //4
```

可以看到输出的结果为 4。

如果想知道 c 的用户有多少，分别是谁呢？GraphFrame 对象有一个 triplets 属性，它是一个 DataFrame，具有 src、edge、dst 三列，分别代表源顶点、边和目标顶点，代码如下：

```
g.triplets.printSchema()
```

输出的结果如下：

```
root
 |-- src: struct (nullable = false)
 |    |-- id: string (nullable = true)
 |    |-- name: string (nullable = true)
 |    |-- age: integer (nullable = false)
 |-- edge: struct (nullable = false)
 |    |-- src: string (nullable = true)
 |    |-- dst: string (nullable = true)
 |    |-- relationship: string (nullable = true)
 |-- dst: struct (nullable = false)
 |    |-- id: string (nullable = true)
 |    |-- name: string (nullable = true)
 |    |-- age: integer (nullable = false)
```

查看 triplets 的内容，代码如下：

```
g.triplets.show()
```

输出的结果如下：

```
+---------------+---------------+---------------+
|            src|           edge|            dst|
+---------------+---------------+---------------+
|  [a, Alice, 34]|  [a, b, friend]|    [b, Bob, 36]|
|    [b, Bob, 36]|  [b, c, follow]|[c, Charlie, 30]|
|[c, Charlie, 30]|  [c, b, follow]|    [b, Bob, 36]|
|  [f, Fanny, 36]|  [f, c, follow]|[c, Charlie, 30]|
| [e, Esther, 32]|  [e, f, follow]|  [f, Fanny, 36]|
| [e, Esther, 32]|  [e, d, friend]|  [d, David, 29]|
|  [d, David, 29]|  [d, a, friend]|  [a, Alice, 34]|
|  [a, Alice, 34]|  [a, e, friend]| [e, Esther, 32]|
+---------------+---------------+---------------+
```

从中找出关系为 follow 的数据，代码如下：

```
g.triplets.filter("edge.relationship = 'follow'").show()
```

输出的结果如下：

```
+---------------+---------------+---------------+
|            src|           edge|            dst|
+---------------+---------------+---------------+
|    [b, Bob, 36]|  [b, c, follow]|[c, Charlie, 30]|
|[c, Charlie, 30]|  [c, b, follow]|    [b, Bob, 36]|
| [e, Esther, 32]|  [e, f, follow]|  [f, Fanny, 36]|
|  [f, Fanny, 36]|  [f, c, follow]|[c, Charlie, 30]|
+---------------+---------------+---------------+
```

进一步找出与 c 有 follow 关系的数据，代码如下：

```
g.triplets
  .filter("edge.relationship = 'follow'")
  .filter("edge.dst = 'c'").show()
```

输出的结果如下：

```
+---------------+---------------+---------------+
|            src|           edge|            dst|
+---------------+---------------+---------------+
|    [b, Bob, 36]|  [b, c, follow]|[c, Charlie, 30]|
|  [f, Fanny, 36]|  [f, c, follow]|[c, Charlie, 30]|
+---------------+---------------+---------------+
```

那么，c 的用户有多少？用户都有谁？代码如下：

```
val followers = g
    .triplets
    .filter("edge.relationship = 'follow'")
    .filter("edge.dst = 'c'")
    .count
println("c 的用户数: " + followers)

g.triplets
  .filter("edge.relationship = 'follow'")
  .filter("edge.dst = 'c'")
  .select("src")
  .show
```

输出的结果如下:

```
c 的用户数:2
+--------------+
|           src|
+--------------+
|  [b, Bob, 36]|
|[f, Fanny, 36]|
+--------------+
```

9.3.4 示例:简单航班数据分析

本节通过一个简单的示例,分析 3 条航线的信息,每条航线中所有列消息见表 9-2。

表 9-2 列消息

出发机场	目的机场	距离/英里
SFO	ORD	1800
ORD	DFW	800
DFW	SFO	1400

用网络图表示,如图 9-10 所示。

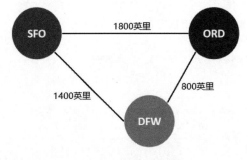

图 9-10 航线网络图

在构建的图模型中,将机场表示为顶点,将航线表示为边。图 9-10 中有 3 个顶点,每个顶点代表一个机场。每个顶点都有机场代码作为 ID,将机场所在城市名称作为属性。表示机场的顶点见表 9-3。

表 9-3 机场的顶点

id	city
SFO	旧金山(San Francisco)
ORD	芝加哥(Chicago)
DFW	达拉斯(Dallas)

边具有源 ID、目标 ID 和作为属性的距离及延误时间。表示航线的边见表 9-4。

表 9-4 航线的边

src	dst	distance	delay
SFO	ORD	1800	40
ORD	DFW	800	0
DFW	SFO	1400	10

接下来使用 GraphFrames 进行分析。建议按以下步骤操作。

（1）首先，导入相关的依赖包，代码如下：

```
import org.apache.spark.sql.SparkSession
import org.apache.spark.sql.functions._
import org.graphframes.GraphFrame
```

（2）定义两个 case class，分别用来表示顶点和边的 Schema，代码如下：

```
//case class 定义顶点和边的 Schema
case class Airport(id: String, city: String) extends Serializable

case class Flight(id: String, src: String,dst: String, dist: Double, delay: Double) extends Serializable
```

（3）定义顶点。

将机场定义为顶点。顶点 DataFrame 必须有一个 ID 列，并且可能有多个属性列。在本例中，每个机场顶点由(顶点 id→id，顶点属性→city)组成，代码如下：

```
//创建顶点 DataFrame
val vertices = spark.createDataFrame(
    Array(
        Airport("SFO","San Francisco"),
        Airport("ORD","Chicago"),
        Airport("DFW","Dallas Fort Worth")
    )
)

//查看顶点
vertices.show()
```

执行以上代码，输出内容如下：

```
+---+-----------------+
| id|             city|
+---+-----------------+
|SFO|    San Francisco|
|ORD|          Chicago|
|DFW|Dallas Fort Worth|
+---+-----------------+
```

（4）定义边。

边是机场之间的航班。一条边 DataFrame 必须有 src 和 dst 列，并且可能有多个关系列。

在本例中，边包括以下内容：
① 源 ID → src。
② 目标 ID → dst。
③ 边属性：距离 → dist。
④ 边属性：延误 → delay。

定义边的代码如下：

```
val edges = spark.createDataFrame(
    Array(
        Flight("SFO_ORD_2017-01-01_AA","SFO","ORD",1800, 40),
        Flight("ORD_DFW_2017-01-01_UA","ORD","DFW",800, 0),
        Flight("DFW_SFO_2017-01-01_DL","DFW","SFO",1400, 10)
    )
)
edges.show(false)
```

执行代码，输出的结果如下：

```
+---------------------+---+---+------+-----+
|id                   |src|dst|dist  |delay|
+---------------------+---+---+------+-----+
|SFO_ORD_2017-01-01_AA|SFO|ORD|1800.0|40.0 |
|ORD_DFW_2017-01-01_UA|ORD|DFW|800.0 |0.0  |
|DFW_SFO_2017-01-01_DL|DFW|SFO|1400.0|10.0 |
+---------------------+---+---+------+-----+
```

（5）构建 GraphFrame。

通过提供顶点 DataFrame 和边 DataFrame 来创建一个 GraphFrame。也可以只使用一条边 DataFrame 创建一个 GraphFrame，然后从边 DataFrame 的 src 和 dst 列获得顶点，代码如下：

```
//定义图
val graph = GraphFrame(vertices, edges)

//显示图的顶点
graph.vertices.show()

//显示图的边
graph.edges.show(false)
```

（6）现在可以查询 GraphFrame 来回答以下问题：

```
//有多少个机场
println("\n有多少个机场？" + graph.vertices.count())

//机场之间有多少条航线
println("\n机场之间有多少条航线？" + graph.edges.count())

//哪条航线的距离大于 1000 英里
println("\n哪些航线的距离大于 1000 英里？")
graph.edges.filter("dist > 1000").show(false)
```

```
//距离最长的航线
println("\n距离最长的航线是？")
graph.edges
    .groupBy("src", "dst")
    .agg(max("dist").as("longest"))
    .sort(desc("longest"))
    .show(false)
```

执行上面的代码，输出内容如下：

有多少个机场？3

机场之间有多少条航线？3

哪些航线的距离大于 1000 英里？
```
+----------------------+---+---+------+-----+
|id                    |src|dst|dist  |delay|
+----------------------+---+---+------+-----+
|SFO_ORD_2017-01-01_AA |SFO|ORD|1800.0|40.0 |
|DFW_SFO_2017-01-01_DL |DFW|SFO|1400.0|10.0 |
+----------------------+---+---+------+-----+
```

距离最长的航线是？
```
+---+---+-------+
|src|dst|longest|
+---+---+-------+
|SFO|ORD|1800.0 |
|DFW|SFO|1400.0 |
|ORD|DFW|800.0  |
+---+---+-------+
```

9.4　应用 motif 模式查询

10min

图分析有两种形式：图算法和图模式查询。

GraphFrame 集成了图算法和图查询，支持跨图和 Spark SQL 查询的优化，而不需要将数据移动到专门的图数据库，如图 9-11 所示。

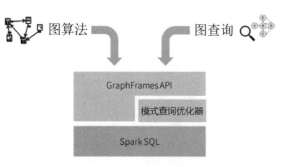

图 9-11　GraphFrame 集成了图算法和图查询

9.4.1 简单 motif 查询

Graph motif 是在图中重复出现的子图或模式,表示顶点之间的交互或关系。图查询在图中搜索符合 motif 模式的结构,找到 motif 可以帮助用户执行查询来发现图中的结构模式。例如,用户可以使用 motif 来分析用户所购买产品的网络关系图,根据表示产品的图的结构属性及其属性和它们之间的关系洞察用户行为(找出经常同时购买的商品)。这些信息可用于推荐和/或广告引擎。

例如,可以搜索这样的模式:A 关注了 B,B 关注了 C,但是 A 并不关注 C。找到这样的结果后,就可以把 C 推荐给 A,如图 9-12 所示。

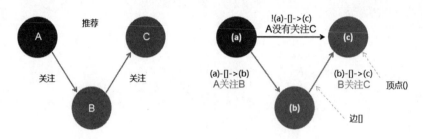

图 9-12　motif 模式表示

motif 的语法形式如下:

```
g.find("(start)-[pass]->(end)")
```

其中 g 为图对象,start 为起点,pass 为经过的边,end 为目标点,顶点用括号表示,边用方括号[]表示。命名,指定 GraphFrames Motif 查询模式,用来找到从 A 到 B 并且从 B 到 C,但没有从 A 到 C 的边的结构,代码如下:

```
graph.find("(a)-[]->(b); (b)-[]->(c);!(a)-[]->(c)")
```

继续 9.3.3 节中的示例,使用 motif 来构建包含边和顶点的更复杂的关系。

例如,找到在两个方向上都有边的顶点对,结果是一个 DataFrame,其中的列名是 motif 键,代码如下:

```
val motifs = g.find("(a)-[e]->(b); (b)-[e2]->(a)")
motifs.show(false)
```

执行以上的代码,输出的结果如下:

```
+--------------------------------+-------------+--------------------------------+-------------+
|a                               |e            |b                               |e2           |
+--------------------------------+-------------+--------------------------------+-------------+
|[b, Bob, 36, 23232323, Bananas] |[b, c, follow]|[c, Charlie, 30, 2123, Grapefruit]|[c, b, follow]|
|[c, Charlie, 30, 2123, Grapefruit]|[c, b, follow]|[b, Bob, 36, 23232323, Bananas] |[b, c, follow]|
+--------------------------------+-------------+--------------------------------+-------------+
```

注意:查询条件中,a 和 b 可以指代相同的顶点。如果需要限制为不同的顶点,则需要在返回结果中使用过滤器。

由于执行结果是一个 DataFrame，所以可以在 motif 上构建更复杂的查询。例如，找出 30 岁以上的所有相互关系，代码如下：

```
val filtered = motifs.filter("b.age > 30")
filtered.show(false)
```

执行以上代码，输出的结果如下：

```
+-------------------------------+---------------+-------------------------+---------------+
|a                              |e              |b                        |e2             |
+-------------------------------+---------------+-------------------------+---------------+
|[c, Charlie, 30, 2123, Grapefruit]|[c, b, follow]|[b, Bob, 36, 23232323, Bananas]|[b, c, follow]|
+-------------------------------+---------------+-------------------------+---------------+
```

当不需要返回路径中的元素时，可以使用匿名顶点和边，代码如下：

```
val motif = g.find("(start)-[]->()")
motif.show(false)
```

执行以上代码，输出的结果如下：

```
+-------------------------------+
|start                          |
+-------------------------------+
|[a, Alice, 34, 234, Apples]    |
|[b, Bob, 36, 23232323, Bananas]|
|[c, Charlie, 30, 2123, Grapefruit]|
|[f, Fanny, 36, 333, Apples]    |
|[e, Esther, 32, 1, Watermelon] |
|[e, Esther, 32, 1, Watermelon] |
|[d, David, 29, 2321111, Bananas]|
|[a, Alice, 34, 234, Apples]    |
+-------------------------------+
```

如果只查询单向路径，则可使用下面的模式：

```
val motif = g.find("(a)-[]->(b); !(b)-[]->(a)")
motif.show(false)
```

执行以上代码，输出的结果如下：

```
+-------------------------------+-------------------------------+
|a                              |b                              |
+-------------------------------+-------------------------------+
|[f, Fanny, 36, 333, Apples]    |[c, Charlie, 30, 2123, Grapefruit]|
|[a, Alice, 34, 234, Apples]    |[e, Esther, 32, 1, Watermelon] |
|[e, Esther, 32, 1, Watermelon] |[f, Fanny, 36, 333, Apples]    |
|[d, David, 29, 2321111, Bananas]|[a, Alice, 34, 234, Apples]   |
|[e, Esther, 32, 1, Watermelon] |[d, David, 29, 2321111, Bananas]|
|[a, Alice, 34, 234, Apples]    |[b, Bob, 36, 23232323, Bananas]|
+-------------------------------+-------------------------------+
```

执行有条件的路径搜索，代码如下：

```
val paths = g.find("(a)-[e]->(b)")
        .filter("e.relationship = 'follow'")
        .filter("a.age < b.age")
paths.show
```

执行以上代码，输出的结果如下：

```
+--------------------------------+---------------+-----------------------------+
|a                               |e              |b                            |
+--------------------------------+---------------+-----------------------------+
|[c, Charlie, 30, 2123, Grapefruit]|[c, b, follow]|[b, Bob, 36, 23232323, Bananas]|
|[e, Esther, 32, 1, Watermelon]  |[e, f, follow]|[f, Fanny, 36, 333, Apples]  |
+--------------------------------+---------------+-----------------------------+
```

进一步选择关系数据集中的列，代码如下：

```
val e2 = paths.select("e.src", "e.dst", "e.relationship")
//val e2 = paths.select("e.*")

e2.show()
```

执行以上代码，输出的结果如下：

```
+---+---+------------+
|src|dst|relationship|
+---+---+------------+
|  c|  b|      follow|
|  e|  f|      follow|
+---+---+------------+
```

9.4.2 状态查询

无状态查询，没有指定任务限制条件，虽易于表达，但只能在查询完后再进行过滤，会返回一个较大的数据集。大多数 motif 查询是无状态的，如 9.4.1 节中的示例。

motif 支持更复杂的查询，这些查询沿着 motif 中的路径携带状态。通过将 GraphFrame motif 查找与结果集上的过滤器相结合来表达这些查询，过滤器使用序列操作来构造一系列 DataFrame 列。

例如，要在图 g 中查找 4 个顶点（人）构成的关系链，要求其中至少有两个是 friend（朋友）关系。在本例中，要维护的状态是属性为 friend 的边的计数，代码如下：

```
//查找 4 个顶点构成的关系链，要求其中至少有两个是 friend（朋友）关系
val chain4 = g.find("(a)-[ab]->(b); (b)-[bc]->(c); (c)-[cd]->(d)")
chain4.printSchema()

//序列查询，带状态(cnt)
//(a) 定义每个顶点更新状态的方法：如果关系为 friend，则 cnt+1
def sumFriends(cnt: Column, relationship: Column): Column = {
    when(relationship === "friend", cnt + 1).otherwise(cnt)
}

//(b) 将更新方法应用到整个链，链上每有一个关系是 friend 就加 1，链上共 3 个关系(3 条边)
val numFriends = Seq("ab", "bc", "cd")
    .foldLeft(lit(0))((cnt, e) => sumFriends(cnt, col(e)("relationship")))

//(c) 传入限制条件：对 DataFrame 应用过滤器
val chainWith2Friends = chain4.where(numFriends >= 2)

chainWith2Friends.show()
```

执行以上代码，输出的结果如下：

```
+------------------+------------+-----------------+------------+-----------------+------------+-----------------+
|                 a|          ab|                b|          bc|                c|          cd|                d|
+------------------+------------+-----------------+------------+-----------------+------------+-----------------+
|[e, Esther, 32, 1...|[e, d, friend]|[d, David, 29, 23...|[d, a, friend]|[a, Alice, 34, 23...|[a, e, friend]|[e, Esther, 32, 1...|
|[e, Esther, 32, 1...|[e, d, friend]|[d, David, 29, 23...|[d, a, friend]|[a, Alice, 34, 23...|[a, b, friend]|[b, Bob, 36, 2323...|
|[d, David, 29, 23...|[d, a, friend]|[a, Alice, 34, 23...|[a, e, friend]|[e, Esther, 32, 1...|[e, d, friend]|[d, David, 29, 23...|
|[d, David, 29, 23...|[d, a, friend]|[a, Alice, 34, 23...|[a, e, friend]|[e, Esther, 32, 1...|[e, f, follow]|[f, Fanny, 36, 33...|
|[d, David, 29, 23...|[d, a, friend]|[a, Alice, 34, 23...|[a, b, friend]|[b, Bob, 36, 2323...|[b, c, follow]|[c, Charlie, 30, ...|
|[a, Alice, 34, 23...|[a, e, friend]|[e, Esther, 32, 1...|[e, d, friend]|[d, David, 29, 23...|[d, a, friend]|[a, Alice, 34, 23...|
+------------------+------------+-----------------+------------+-----------------+------------+-----------------+
```

如果想进一步统计每个查找出的关系链中的朋友数量，则可以对上面的代码进行修改，修改后的代码如下：

```
val chainWith2Friends = chain4
    .select("ab","bc","cd")
    .withColumn("num_friends", numFriends)
    .where("num_friends>=2")
chainWith2Friends.show()
```

执行以上代码，输出的结果如下：

```
+-------------+-------------+-------------+-----------+
|           ab|           bc|           cd|num_friends|
+-------------+-------------+-------------+-----------+
|[e, d, friend]|[d, a, friend]|[a, e, friend]|          3|
|[e, d, friend]|[d, a, friend]|[a, b, friend]|          3|
|[d, a, friend]|[a, e, friend]|[e, d, friend]|          3|
|[d, a, friend]|[a, e, friend]|[e, f, follow]|          2|
|[d, a, friend]|[a, b, friend]|[b, c, follow]|          2|
|[a, e, friend]|[e, d, friend]|[d, a, friend]|          3|
+-------------+-------------+-------------+-----------+
```

在上面这个示例中，查找由 4 个顶点组成的关系链。在这 4 个顶点 a→b→c→d 的关系链中，找出匹配这个过滤条件的子集，其实现过程如下：

（1）初始化路径上的状态。
（2）基于顶点 a 更新状态。
（3）基于顶点 b 更新状态。
（4）基于顶点 c 更新状态。
（5）基于顶点 d 更新状态。
（6）对最终状态进行过滤。如果最终状态匹配某个条件，则该关系链被过滤器接受。

9.5　构建子图

GraphFrames 提供了通过对边和顶点进行过滤来构建子图的 API。这些过滤器可以组合在一起使用。例如，构建只包含年龄在 30 岁以上的朋友的子图，代码如下：

```
//选择用户年龄超过30岁，并且边的类型是friend的子图
val g2 = g
    .filterEdges("relationship = 'friend'")
    .filterVertices("age > 30")
    .dropIsolatedVertices()

g2.vertices.show()
g2.edges.show()
```

输出的结果如下：

```
+---+------+---+--------+---------+
| id|  name|age|    cash|    fruit|
+---+------+---+--------+---------+
|  a| Alice| 34|     234|   Apples|
|  b|   Bob| 36|23232323|  Bananas|
|  e|Esther| 32|       1|Watermelon|
+---+------+---+--------+---------+

+---+---+------------+
|src|dst|relationship|
+---+---+------------+
|  a|  b|      friend|
|  a|  e|      friend|
+---+---+------------+
```

在这个例子中，分别使用 filterVertices()方法和 filterEdges()方法过滤顶点和边，用来创建子图。最后对子图调用 dropIsolatedVertices()方法，删除孤立的没有连接的点。

基于 motif 查找和 DataFrame 过滤器的组合，GraphFrames 提供了一种强大的方式来选择子图。例如，基于在边及其 src 和 dst 顶点上操作的三重过滤器来选择子图，代码如下：

```
//根据类型follow的边e来选择子图
//从年轻的用户a指向较年长的用户b
val paths = g.find("(a)-[e]->(b)")
    .filter("e.relationship = 'follow'")
    .filter("a.age < b.age")

//paths 包含顶点信息，提取边
val e2 = paths.select("e.src", "e.dst", "e.relationship")
//在 Spark 1.5+，用户可以简化这个调用
//val e2 = paths.select("e.*")

//构造子图
val g2 = GraphFrame(g.vertices, e2)

g2.vertices.show()
g2.edges.show()
```

执行上面的代码，输出的结果如下：

```
+---+-------+---+--------+----------+
| id|   name|age|    cash|     fruit|
+---+-------+---+--------+----------+
|  a|  Alice| 34|     234|    Apples|
|  b|    Bob| 36|23232323|   Bananas|
|  c|Charlie| 30|    2123|Grapefruit|
|  d|  David| 29| 2321111|   Bananas|
|  e| Esther| 32|       1|Watermelon|
|  f|  Fanny| 36|     333|    Apples|
|  g|  Gabby| 60|   23433|   Oranges|
+---+-------+---+--------+----------+

+---+---+------------+
|src|dst|relationship|
+---+---+------------+
|  c|  b|      follow|
|  e|  f|      follow|
+---+---+------------+
```

上述示例的执行过程如下：

（1）首先用(a)–[e]–>(b)过滤出所有连接的顶点，得到顶点数据集。

（2）使用条件过滤指定关系 follow 和节点年龄的大小关系，得到一条边的数据集。

（3）将顶点和边的数据集传入 GraphFrame 可以得到一张子图。

通过更复杂的 motif 将这个例子扩展到三重之外是很简单的。

9.6　内置图算法

GraphFrames 实现了一些标准的图算法：

（1）广度优先搜索（BFS）算法。

（2）连通分量算法。

（3）强连通分量算法。

（4）标签传播算法。

（5）PageRank 算法。

（6）最短路径算法。

（7）三角计数算法。

下面学习如何使用 GraphFrames 自带的这些图算法。

9.6.1　广度优先搜索（BFS）算法

广度优先搜索算法（Breadth-First-Search，BFS）是一种图搜索算法。简单地说，BFS 是从根节点开始的，沿着树（图）的宽度遍历树（图）的节点。如果所有节点均被访问，则算法终止。BFS 属于盲目搜索。

算法步骤如下：

（1）首先将根节点放入队列中。

（2）从队列中取出第1个节点，并检验它是否为目标。如果找到目标，则结束搜寻并回传结果。否则将它所有尚未检验过的直接子节点加入队列中。

（3）若队列为空，表示整张图都检查过了，即图中没有欲搜寻的目标。结束搜寻并回传"找不到目标"。

（4）重复步骤（2）。

广度优先搜索算法的遍历顺序为 1→2→3→4→5→6→7→8→9→10→11→12，如图9-13所示。

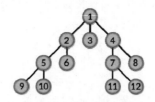

图9-13　广度优先搜索算法的遍历顺序

例如，要从g图对象中查找从源顶点为Esther、目标顶点（人）年龄小于32岁的用户，代码如下：

```
val paths = g.bfs.fromExpr("name = 'Esther'").toExpr("age < 32").run()
paths.show(false)
```

执行以上代码，输出的结果如下：

```
+--------------------------+-------------+----------------------------+
|from                      |e0           |to                          |
+--------------------------+-------------+----------------------------+
|[e, Esther, 32, 1, Watermelon]|[e, d, friend]|[d, David, 29, 2321111, Bananas]|
+--------------------------+-------------+----------------------------+
```

搜索还可以使用边过滤器和最大路径长度进行限制，代码如下：

```
val filteredPaths = g.bfs
    .fromExpr("name = 'Esther'").toExpr("age < 32")
    .edgeFilter("relationship != 'friend'")
    .maxPathLength(3)
    .run()
filteredPaths.show()
```

执行以上代码，输出的结果如下：

```
+-------------------+-------------+-----------------+-------------+------------------+
|               from|           e0|               v1|           e1|                to|
+-------------------+-------------+-----------------+-------------+------------------+
|[e, Esther, 32, 1...|[e, f, follow]|[f, Fanny, 36, 33...|[f, c, follow]|[c, Charlie, 30, ...|
```

例如，寻找从"现金超过20 000元的人"到"年龄不到50岁"的"非朋友"的连接，

代码如下:

```
val f = g.bfs
    .fromExpr("cash > 20000").toExpr("age < 50")
    .edgeFilter("relationship = 'follow'")
    .maxPathLength(3)
    .run()
f.show(false)
//f.filter("from.id != to.id").show(false)
```

执行以上代码,输出的结果如下:

```
+--------------------------------+--------------------------------+
|from                            |to                              |
+--------------------------------+--------------------------------+
|[b, Bob, 36, 23232323, Bananas] |[b, Bob, 36, 23232323, Bananas] |
|[d, David, 29, 2321111, Bananas]|[d, David, 29, 2321111, Bananas]|
+--------------------------------+--------------------------------+
```

9.6.2 连通分量算法

在图论中,无向图的一个组件(有时称为连通组件)是一张子图,其中任意两个顶点通过路径相互连接,并且在超图中不与任何其他的顶点连接。有 3 个组件的图如图 9-14 所示。

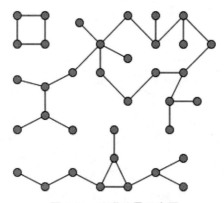

图 9-14 连通分量示意图

没有关联边的顶点本身就是一个组件。一个自身连通的图只有一个分量,由整个图组成。

连通分量算法寻找孤立的集群或孤立的子图。这些集群是图中的连接顶点的集合,其中每个顶点都可以从同一集合中的任何其他顶点到达。该算法返回一个包含每个顶点和该顶点所在连通组件的 GraphFrame,代码如下:

```
//设置检查点
spark.sparkContext.setCheckpointDir("tmp/gfcc")
val result = g.connectedComponents.run()
result.show()
```

注意：对于 GraphFrames 0.3.0 及之后的版本，默认的连通分量算法需要设置一个 Spark 检查点目录。

执行以上代码，输出的结果如下：

```
+---+-------+---+--------+----------+------------+
| id|   name|age|    cash|     fruit|   component|
+---+-------+---+--------+----------+------------+
|  a|  Alice| 34|     234|    Apples|412316860416|
|  b|    Bob| 36|23232323|   Bananas|412316860416|
|  c|Charlie| 30|    2123|Grapefruit|412316860416|
|  d|  David| 29| 2321111|   Bananas|412316860416|
|  e| Esther| 32|       1|Watermelon|412316860416|
|  f|  Fanny| 36|     333|    Apples|412316860416|
|  g|  Gabby| 60|   23433|   Oranges|146028888064|
+---+-------+---+--------+----------+------------+
```

9.6.3 强连通分量算法

在有向图的数学理论中，如果一张图的每个顶点都能从其他顶点到达，则这张图就被称为强连通图。任意有向图的强连通分量构成子图的划分，子图本身是强连通的。在线性时间内，可以测试图的强连通性，或者找到它的强连通部分。

如果有向图的每对顶点之间在每个方向上都有一条路径，则称为强连通图。也就是说，从第 1 个顶点到第 2 个顶点存在一条路径，从第 2 个顶点到第 1 个顶点存在另一条路径。强连通图的概念如图 9-15 所示。

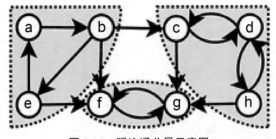

图 9-15 强连通分量示意图

注意：如果每个强连通分量都收缩到一个顶点，则由此产生的图就是一个有向无环图（DAG）。例如，图 9-16 中外层大有向无环图是内层小有向图的浓缩。它是通过将内层小有向图的每个强连通部分缩并成一个顶点而形成的。

强连通分量算法用于计算每个顶点的强连通分量（SCC），并返回一张图，其中每个顶点分配给包含该顶点的 SCC，代码如下：

```
val result = g.stronglyConnectedComponents.maxIter(10).run()
result.orderBy("component").show()
```

执行以上代码，输出的结果如下：

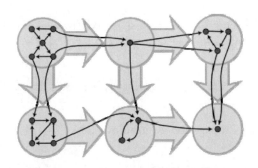

图 9-16 有向无环图(DAG)原理

```
+---+-------+---+--------+----------+------------+
| id|   name|age|    cash|     fruit|   component|
+---+-------+---+--------+----------+------------+
|  g|  Gabby| 60|   23433|   Oranges| 146028888064|
|  f|  Fanny| 36|     333|    Apples| 412316860416|
|  e| Esther| 32|       1|Watermelon| 670014898176|
|  d|  David| 29| 2321111|   Bananas| 670014898176|
|  a|  Alice| 34|     234|    Apples| 670014898176|
|  c|Charlie| 30|    2123|Grapefruit|1047972020224|
|  b|    Bob| 36|23232323|   Bananas|1382979469312|
+---+-------+---+--------+----------+------------+
```

9.6.4 标签传播算法

标签传播是一种半监督机器学习算法,它将标签分配给之前未标记的数据点。在算法开始时,数据点的一个(通常很小的)子集有标签(或分类)。这些标签将在整个算法过程中传播到未标记的点。

注意:在复杂的网络中,真实的网络往往具有社区结构。标签传播是一种查找社区的算法。与其他算法相比,标签传播在运行时间和网络结构所需的先验信息量(不需要预先知道参数)方面具有优势。缺点是它产生的不是唯一的解决方案,而是许多解决方案的集合。

在初始条件下,节点携带一个标签,该标签表示它们所属的社区。社区中的成员关系会随着邻近节点所拥有的标签而变化。此更改取决于节点一度内标签的最大数量。每个节点都用一个唯一的标签初始化,然后这些标签在网络中扩散,因此,紧密相连的组很快就会达到一个共同的标签。当许多这样密集(一致)的团体在整个网络中被创建时,它们继续向外扩展,直到不可能这样做。

该过程有以下4个步骤:

(1)初始化网络中所有节点上的标签。对于给定的节点 x,$Cx(0) = x$。

(2)设 $t = 1$。

(3)将网络中的节点按随机顺序排列,并设为 X。

（4）对于以该顺序选取的每个 $x \in X$，令 $Cx(t) = f(Cxi1(t), …, Cxim(t), Cxi(m+1)(t-1), …, Cxik(t-1)$。这里返回相邻区域中出现频率最高的标签。如果有多个最高频率的标签，则随机选择一个标签。

如果每个节点都有其相邻节点的最大数量的标签，则停止该算法。否则设 $t = t + 1$，转到步骤（3）重复执行这个过程。

标签传播算法（Label Propagation Algorithm，LPA）运行静态标签传播算法来检测网络中的社区。

网络中的每个节点最初都被分配给它自己的社区。在每个 superstep 中，节点将它们的社区从属关系发送给所有邻居，并将它们的状态更新为传入消息的模式社区从属关系。

LPA 是一种标准的图社区检测算法。它的计算成本非常低，尽管它有以下不足：(1) 不能保证收敛；(2) 但最终可以得到简单的解决方案（所有节点都被标识为单个社区），代码如下：

```
val result = g.labelPropagation.maxIter(5).run()
result.orderBy("label").show()
```

执行以上代码，输出的结果如下：

```
+---+-------+---+--------+----------+-------------+
| id|   name|age|    cash|     fruit|        label|
+---+-------+---+--------+----------+-------------+
|  g|  Gabby| 60|   23433|   Oranges| 146028888064|
|  d|  David| 29| 2321111|   Bananas| 670014898176|
|  e| Esther| 32|       1|Watermelon| 670014898176|
|  c|Charlie| 30|    2123|Grapefruit|1047972020224|
|  b|    Bob| 36|23232323|   Bananas|1382979469312|
|  f|  Fanny| 36|     333|    Apples|1460288880640|
|  a|  Alice| 34|     234|    Apples|1460288880640|
+---+-------+---+--------+----------+-------------+
```

9.6.5　PageRank 算法

PageRank 是谷歌搜索引擎使用的一种算法，用于在搜索结果中对网页进行排名。它用节点代表网页，用边代表网页间的链接，并按链接页面的数量和排名加上每个链接的页面的数量和排名计算页面重要性。根据谷歌的描述：

"PageRank 通过计算链接的数量和质量来粗略估计网站的重要性。潜在的假设是，更重要的网站有可能收到更多来自其他网站的链接。"

PageRank 是以谷歌的创始人之一 Larry Page 的名字命名的。PageRank 是衡量网站页面重要性的一种方法。PageRank 算法在测量图中某个顶点的重要性时非常有用。

在 GraphFrames 中，PageRank 算法有以下两种实现：

（1）第 1 种实现使用 org.apache.spark.graphx.graph 接口与 aggregateMessages() 方法，并运行 PageRank 为固定的迭代次数。这可以通过设置 maxIter 参数来执行。

（2）第 2 种实现使用 org.apache.spark.graphx.Pregel，并运行 PageRank，直到收敛，这可以通过设置 tol 参数来运行。

这两种实现都支持非个性化和个性化 PageRank，其中设置 sourceId 对该顶点的结果进行个性化处理。

注意：个性化 PageRank 算法的目标是要计算所有节点相对于用户 u 的相关度。个性化的 PageRank 与传统 PageRank 不同的是，每次重新遍历时，总是从用户 u 节点开始。另外，每个节点权重初始化时，个性化的 PageRank 是这样子的，假如对用户 u 推荐，则将用户 u 节点初始化为 1，其他节点都初始化为 0。

例如，运行 PageRank，识别图中的重要顶点，代码如下：

```
import org.apache.spark.sql.functions._

//resetProbability 和 tol 是收敛参数
val results = g.pageRank.resetProbability(0.15).tol(0.01).run()

results.vertices.orderBy(desc("pagerank")).show()
results.edges.orderBy(desc("weight")).show()
```

执行以上代码，输出的结果如下：

```
+---+-------+---+--------+----------+-------------------+
| id|   name|age|    cash|     fruit|           pagerank|
+---+-------+---+--------+----------+-------------------+
|  c|Charlie| 30|    2123|Grapefruit|  2.6878300011606218|
|  b|    Bob| 36|23232323|   Bananas|  2.655507832863289|
|  a|  Alice| 34|     234|    Apples| 0.44910633706538744|
|  e| Esther| 32|       1|Watermelon|  0.37085233187676075|
|  d|  David| 29| 2321111|   Bananas|  0.3283606792049851|
|  f|  Fanny| 36|     333|    Apples|  0.3283606792049851|
|  g|  Gabby| 60|   23433|   Oranges|  0.17998213862397111|
+---+-------+---+--------+----------+-------------------+

+---+---+------------+------+
|src|dst|relationship|weight|
+---+---+------------+------+
|  c|  b|      follow|   1.0|
|  b|  c|      follow|   1.0|
|  d|  a|      friend|   1.0|
|  f|  c|      follow|   1.0|
|  a|  b|      friend|   0.5|
|  a|  e|      friend|   0.5|
|  e|  f|      follow|   0.5|
|  e|  d|      friend|   0.5|
+---+---+------------+------+
```

对固定次数的迭代运行 PageRank，代码如下：

```
val results2 = g.pageRank.resetProbability(0.15).maxIter(10).run()

results2.vertices.orderBy(desc("pagerank")).show()
results2.edges.orderBy(desc("weight")).show()
```

其中，

（1）maxIter()：要运行的页面排名的迭代次数，推荐值为 20，如果推荐值太小，则会降低质量，如果推荐值太大，则会降低性能。

（2）resetProbability()：随机重置概率(alpha)，它越低，分数差距就越大，有效范围为 0~1。通常，0.15 是一个不错的分数。

执行以上代码，输出的结果如下：

```
+---+-------+---+--------+----------+-------------------+
| id|   name|age|    cash|     fruit|           pagerank|
+---+-------+---+--------+----------+-------------------+
|  b|    Bob| 36|23232323|   Bananas| 2.7025217677349773|
|  c|Charlie| 30|    2123|Grapefruit| 2.6667877057849627|
|  a|  Alice| 34|     234|    Apples| 0.4485115093698443|
|  e| Esther| 32|       1|Watermelon| 0.3613490987992571|
|  f|  Fanny| 36|     333|    Apples|0.32504910549694244|
|  d|  David| 29| 2321111|   Bananas|0.32504910549694244|
|  g|  Gabby| 60|   23433|   Oranges| 0.17073170731707318|
+---+-------+---+--------+----------+-------------------+

+---+---+------------+------+
|src|dst|relationship|weight|
+---+---+------------+------+
|  c|  b|      follow|   1.0|
|  b|  c|      follow|   1.0|
|  d|  a|      friend|   1.0|
|  f|  c|      follow|   1.0|
|  a|  b|      friend|   0.5|
|  e|  d|      friend|   0.5|
|  a|  e|      friend|   0.5|
|  e|  f|      follow|   0.5|
+---+---+------------+------+
```

为顶点 a 运行个性化的 PageRank，代码如下：

```
val results3 = g
    .pageRank
    .resetProbability(0.15)
    .maxIter(10)
    .sourceId("a")
    .run()
results3.vertices.show()
```

执行以上代码，输出的结果如下：

```
+---+------+---+--------+----------+-------------------+
| id|  name|age|    cash|     fruit|           pagerank|
+---+------+---+--------+----------+-------------------+
|  b|   Bob| 36|23232323|   Bananas| 0.3366143039702568|
|  e|Esther| 32|       1|Watermelon|0.07657840357273027|
|  a| Alice| 34|     234|    Apples| 0.17710831642683564|
|  f| Fanny| 36|     333|    Apples|0.03189213697274781|
```

```
|  g|  Gabby| 60|    23433|   Oranges|                   0.0|
|  d|  David| 29|  2321111|   Bananas|0.03189213697274781|
|  c|Charlie| 30|     2123|Grapefruit| 0.3459147020846817|
+---+-------+---+---------+----------+-------------------+
```

对顶点["a", "b", "c", "d"]并行运行个性化的 PageRank，代码如下：

```
val ids = Array("a", "b", "c", "d").asInstanceOf[Array[Any]]
val results3 = g
    .parallelPersonalizedPageRank
    .resetProbability(0.15)
    .maxIter(10)
    .sourceIds(ids)
    .run()
results3.vertices.show()
```

执行以上代码，输出的结果如下：

```
+---+-------+---+---------+----------+--------------------+
| id|   name|age|     cash|     fruit|           pageranks|
+---+-------+---+---------+----------+--------------------+
|  b|    Bob| 36| 23232323|   Bananas|(4,[0,1,2,3],[0.3...|
|  e| Esther| 32|        1|Watermelon|(4,[0,1,2,3],[0.0...|
|  a|  Alice| 34|      234|    Apples|(4,[0,1,2,3],[0.1...|
|  f|  Fanny| 36|      333|    Apples|(4,[0,1,2,3],[0.0...|
|  g|  Gabby| 60|    23433|   Oranges|(4,[0,1,2,3],[0.0...|
|  d|  David| 29|  2321111|   Bananas|(4,[0,1,2,3],[0.0...|
|  c|Charlie| 30|     2123|Grapefruit|(4,[0,1,2,3],[0.3...|
+---+-------+---+---------+----------+--------------------+
```

9.6.6 最短路径算法

在图论中，最短路径问题是在图中两个顶点之间找到一条路径，使其组成边的权值之和最小。对于有向图，路径的定义要求连续的顶点由合适的有向边连接。例如，一个加权有向图，图 9-17 中顶点 A 和 F 之间的最短路径是（A，C，E，D，F），如图 9-17 所示。

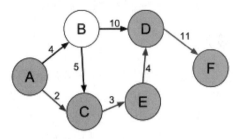

图 9-17　有向无环图（DAG）原理

GraphFrames 中最短路径算法通过 shortestPaths 实现，它计算从每个顶点到给定的地标顶点集的最短路径（但是不考虑权值），地标由顶点 ID 指定。注意，这考虑了边的方向。

例如，计算图中每个顶点到地标顶点 a 和 d 之间的最短路径，代码如下：

```
val paths = g.shortestPaths.landmarks(Seq("a", "d")).run()
paths.show(false)
```

执行以上代码，输出的结果如下：

```
+---+-------+---+--------+----------+----------------+
|id |name   |age|cash    |fruit     |distances       |
+---+-------+---+--------+----------+----------------+
|b  |Bob    |36 |23232323|Bananas   |[]              |
|e  |Esther |32 |1       |Watermelon|[d -> 1, a -> 2]|
|a  |Alice  |34 |234     |Apples    |[a -> 0, d -> 2]|
|f  |Fanny  |36 |333     |Apples    |[]              |
|g  |Gabby  |60 |23433   |Oranges   |[]              |
|d  |David  |29 |2321111 |Bananas   |[d -> 0, a -> 1]|
|c  |Charlie|30 |2123    |Grapefruit|[]              |
+---+-------+---+--------+----------+----------------+
```

进一步进行条件过滤，代码如下：

```
paths.filter(col("distances")("d") > 0).show(false)
paths.filter(size(map_keys(col("distances")))=!=0).show(false)
```

执行以上代码，输出的结果如下：

```
+---+------+---+----+----------+----------------+
|id |name  |age|cash|fruit     |distances       |
+---+------+---+----+----------+----------------+
|e  |Esther|32 |1   |Watermelon|[d -> 1, a -> 2]|
|a  |Alice |34 |234 |Apples    |[a -> 0, d -> 2]|
+---+------+---+----+----------+----------------+

+---+------+---+-------+----------+----------------+
|id |name  |age|cash   |fruit     |distances       |
+---+------+---+-------+----------+----------------+
|e  |Esther|32 |1      |Watermelon|[d -> 1, a -> 2]|
|a  |Alice |34 |234    |Apples    |[a -> 0, d -> 2]|
|d  |David |29 |2321111|Bananas   |[d -> 0, a -> 1]|
+---+------+---+-------+----------+----------------+
```

9.6.7 三角计数算法

顶点三角形计数算法是一种社区检测图算法，用于计算图中每个顶点所属的三角形的数量。三角形被定义为由三条边（a-b、b-c、c-a）连接的 3 个节点，其中每个顶点都与三角形中的其他两个顶点有关系。在一张图中，如果一条边的两个点有共同邻居点，则这 3 个点就构成了三角形结构。例如，一个顶点三角形结构示例如图 9-18 所示。

在图 9-18 中，左边的图中可找出两个三角形结构，分别是 1-2-4 和 1-3-4。

三角计数在社交网络分析中很受欢迎，它被用来检测社区并测量这些社区的凝聚力。它也可以用来确定一张图的稳定性，并经常被用作计算网络指标的一部分，如聚类系数。采用三角形计数算法计算局部聚类系数。

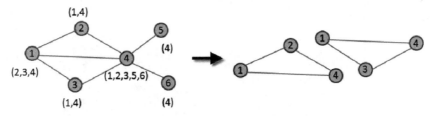

图 9-18　顶点三角形结构

注意：聚类系数有以下两种类型。

（1）局部聚类系数。一个节点的局部聚类系数是它的邻居也被连接的可能性。这个分数的计算涉及三角形计数。

（2）全局聚类系数。全局聚类系数是局部聚类系数的归一化和。

在 GraphFrames 中，通过 triangleCount 实现了该算法，它返回一个 GraphFrame，其中包含通过每个顶点的三角形数量。

例如，计算图 g 中通过每个顶点的三角形的数量，代码如下：

```
val results = g.triangleCount.run()
results.select("id", "count").show()
```

执行以上代码，输出的结果如下：

```
+---+-----+
| id|count|
+---+-----+
|  g|    0|
|  f|    0|
|  e|    1|
|  d|    1|
|  c|    0|
|  b|    0|
|  a|    1|
+---+-----+
```

9.7　保存和加载 GraphFrame

由于 GraphFrame 是在 DataFrame API 上构建的，因此它们支持保存和加载各种数据源。在下面的代码中，展示了如何将顶点和边保存到 HDFS 上的 Parquet 文件中，然后从持久化存储中重新创建顶点和边 DataFrame，并创建图模型。

首先定义两个 case class，代表顶点的边的结构，代码如下：

```
//case class 定义顶点和边的 Schema
case class Airport(id: String, city: String) extends Serializable

case class Flight(id: String, src: String,dst: String, dist: Double, delay:
Double) extends Serializable
```

然后创建边 DataFrame、顶点 DataFrame 并构造图模型,代码如下:

```
//创建顶点 DataFrame
val vertices = spark.createDataFrame(
    Array(
      Airport("SFO","San Francisco"),
      Airport("ORD","Chicago"),
      Airport("DFW","Dallas Fort Worth")
    )
  )

//创建边 DataFrame
val edges = spark.createDataFrame(
    Array(
      Flight("SFO_ORD_2017-01-01_AA","SFO","ORD",1800, 40),
      Flight("ORD_DFW_2017-01-01_UA","ORD","DFW",800, 0),
      Flight("DFW_SFO_2017-01-01_DL","DFW","SFO",1400, 10)
    )
  )

//定义图
val graph = GraphFrame(vertices, edges)
```

分别保存顶点和边,代码如下:

```
//用于保存和加载图模型
graph.vertices.write.mode("overwrite").parquet("tmp/gf/vertices")
graph.edges.write.mode("overwrite").parquet("tmp/gf/edges")
```

然后可以从持久化存储中重新创建顶点和边 DataFrame,并重新创建图,代码如下:

```
val v = spark.read.parquet("tmp/gf/vertices")
val e = spark.read.parquet("tmp/gf/edges")

//创建图
val g = GraphFrame(v, e)
g.triplets.show(false)
```

执行上面的代码,输出内容如下:

```
+------------------------+-------------------------------------------+------------------------+
|src                     |edge                                       |dst                     |
+------------------------+-------------------------------------------+------------------------+
|[SFO, San Francisco]    |[SFO_ORD_2017-01-01_AA, SFO, ORD, 1800.0, 40.0]|[ORD, Chicago]       |
|[ORD, Chicago]          |[ORD_DFW_2017-01-01_UA, ORD, DFW, 800.0, 0.0]  |[DFW, Dallas Fort Worth]|
|[DFW, Dallas Fort Worth]|[DFW_SFO_2017-01-01_DL, DFW, SFO, 1400.0, 10.0]|[SFO, San Francisco] |
+------------------------+-------------------------------------------+------------------------+
```

9.8 深入理解 GraphFrame

下面将简要介绍 GraphFrame 的内部结构及其执行计划和分区。

由于 GraphFrame 是基于 Spark SQL DataFrame 构建的,因此可以通过查看物理计划来理解图操作的执行，代码如下：
```
//查看 GraphFrame 物理执行计划
g.edges.filter("salerank < 100").explain(true)
```
Spark 将数据分割为多个分区，并在这些分区上并行执行计算。可以调整分区的级别，以提高 Spark 的计算效率。

在下面的示例中，将检查一个 GraphFrame 重新分区的结果。可以根据顶点 DataFrame 的列值对 GraphFrame 进行分区。在这里，使用 group 列中的值按组或产品类型进行分区，并将通过比较记录的前后分布来展示重新分区的结果。

首先，创建两个 GraphFrame。由于在 group 列中有 null，所以应将它们替换为一个值 unknown，代码如下：
```
//理解 GraphFrame 中的分区
val v1 = g.vertices.select("id", "group").na.fill("unknown")
v1.show()
v1.groupBy("group").count().show
val g1 = GraphFrame(v1, g.edges)
```
接下来，在对原始 GraphFrame 进行重新分区之后创建第 2 个 GraphFrame。在这里，使用组的数量作为初始分区数量，代码如下：
```
val v2 = g.vertices.select("id", "group").na.fill("unknown")
val g2t1 = GraphFrame(v2, g.edges)
val g2t2 = g2t1.vertices.repartition(11, $"group")
val g2 = GraphFrame(g2t2, g.edges)

//g2 中的记录是按组聚在一起的
g1.vertices.show()
g2.vertices.show()

//g1 中的默认分区数为 9，g2 中指定的分区数为 11
g1.vertices.rdd.partitions.size
g2.vertices.rdd.partitions.size

//也可以将分区的内容写在文件中，以便查看它们的内容
g1
  .vertices
  .write
  .mode("overwrite")
  .csv("/data/spark_demo/graphx/g1/partitions")
g2
  .vertices
  .write
  .mode("overwrite")
  .csv("/data/spark_demo/graphx/g2/partitions")
```
如果查看保存的文件内容，则会注意到大多数记录位于 5 个主要的产品组中。用户可能希望减少分区的总数，这可以使用 coalesce() 操作实现，代码如下：
```
val g2c = g2.vertices.coalesce(5)
g2c.rdd.partitions.size
```

9.9 案例：亚马逊产品联购分析

本节通过一个案例，演示使用 GraphFrame 分析亚马逊产品 2003 年 6 月 1 日联购网络。本案例所使用的数据集由爬取亚马逊网站收集。它是基于亚马逊网站的"购买了该商品的顾客同时也购买了"功能。如果产品 i 经常与产品 j 共同购买，则图中包含从 i 到 j 的有向边。

数据集统计如图 9-19 所示。

DataSet statistics	
Nodes	403394
Edges	3387388
Nodes in largest WCC	403364 (1.000)
Edges in largest WCC	3387224 (1.000)
Nodes in largest SCC	395234 (0.980)
Edges in largest SCC	3301092 (0.975)
Average clustering coefficient	0.4177
Number of triangles	3986507
Fraction of closed triangles	0.06206
Diameter (longest shortest path)	21
90-percentile effective diameter	7.6

图 9-19 亚马逊产品联购数据集

首先导入依赖包，代码如下：

```
import org.apache.spark.sql.types._
import org.apache.spark.sql.functions._
import org.apache.spark.sql.Row
import org.graphframes._
import spark.implicits._
```

加载联购数据，先把文件中的注释过滤掉，然后创建边 DataFrame，再从边 DataFrame 中找出所有的顶点，创建顶点 DataFrame。最后，由顶点 DataFrame 和边 DataFrame 创建图模型，代码如下：

```
//1. 构造 GraphFrame
//将数据加载到 RDD
val filePath = "/data/amazon/Amazon0601.txt"
val edgesRDDRow = spark.sparkContext.textFile(filePath)

//一定要对数据进行过滤，把注释过滤掉
val edgesRDD = edgesRDDRow.filter(line => !line.startsWith("#"))

//schema
val schemaString = "src dst"
```

```
val fields = schemaString
    .split(" ")
    .map(fieldName => StructField(fieldName, StringType, nullable = false))
val edgesSchema = new StructType(fields)

//边 DataFrame 和顶点 DataFrame
val rowRDD = edgesRDD
    .map(_.split("\t"))
    .map(attr => Row(attr(0).trim, attr(1).trim))
val edgesDF = spark.createDataFrame(rowRDD, edgesSchema)

//从边 RDD 中计算出顶点 RDD
val srcVerticesDF = edgesDF.select(col("src")).distinct()
val destVerticesDF = edgesDF.select(col("dst")).distinct()
val verticesDF = srcVerticesDF.union(destVerticesDF).distinct.select(col("src").
alias("id"))

println("边的数量: " + edgesDF.count())
println("顶点的数量: " + verticesDF.count())

//构造图模型
val g = GraphFrame(verticesDF, edgesDF)
```

执行以上代码,输出以下内容:

```
边的数量: 3387388
顶点的数量: 403394
```

9.9.1 基本图查询和操作

接下来,介绍图结构上的简单的图查询和操作,包括顶点、边及顶点入度和出度的显示,代码如下:

```
//2.基本的图查询操作
//顶点
g.vertices.show(5)

//边
g.edges.show(5)

//入度
g.inDegrees.show(5)

//出度
g.outDegrees.show(5)

//对边和顶点及其属性应用过滤器
g.edges.filter("src == 2").count()
g.edges.filter("src == 2").show()
g.edges.filter("dst == 2").show()
g.inDegrees.filter("inDegree >= 10").show()
```

```
//对入度和出度使用 groupBy 和 sort 操作
g.inDegrees.groupBy("inDegree").count().sort(desc("inDegree")).show(5)
g.outDegrees.groupBy("outDegree").count().sort(desc("outDegree")).show(5)
```

9.9.2 联购商品分析

motif 可用于产品联购图中，根据表示产品的图的结构属性及其属性和它们之间的关系洞察用户行为。这些信息可用于推荐和/或广告引擎。

例如，指定 motif 模式表示一个用例，其中购买了产品 a 的客户还购买了另外两种产品 b 和 c，如图 9-20 所示。

在下面这个简单查询中，搜索一系列产品，其中对产品 a 的购买也意味着对产品 b 的购买，反之亦然。这里的 find 操作将搜索由两个方向的边连接的顶点对，代码如下：

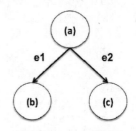

图 9-20　motif 模式 1

```
//定义 motif
val motifs = g.find("(a)-[e]->(b); (b)-[e2]->(a)")
motifs.show(5)
```

执行上面的代码，输出内容如下：

```
+--------+----------------+--------+----------------+
|       a|               e|       b|              e2|
+--------+----------------+--------+----------------+
|  [85609]|  [85609, 100018]|[100018]|  [100018, 85609]|
|  [86839]|  [86839, 100042]|[100042]|  [100042, 86839]|
|  [55528]|  [55528, 100087]|[100087]|  [100087, 55528]|
| [178970]| [178970, 100124]|[100124]| [100124, 178970]|
| [100124]| [100124, 100125]|[100125]| [100125, 100124]|
```

还可以对结果应用过滤器。例如，在下面的过滤器中将顶点 b 的 id 值指定为 2，这会找出所有与 id 为 2 的商品一齐被购买的其他商品，代码如下：

```
motifs.filter("b.id == 2").show()
```

执行上面的代码，输出内容如下：

```
+---+------+---+------+
|  a|     e|  b|    e2|
+---+------+---+------+
|[3]|[3, 2]|[2]|[2, 3]|
|[0]|[0, 2]|[2]|[2, 0]|
|[1]|[1, 2]|[2]|[2, 1]|
+---+------+---+------+
```

下面这种模式通常代表了这样一种情况：当顾客购买一种产品 a 时，他也会购买 b 和 c 其中之一，或都购买，如图 9-21 所示。

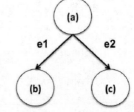

图 9-21　motif 模式 2

在该模式中，指定相同的顶点 a 是边 e1 和边 e2 的公共源，代码如下：

```
//定义motif
val motifs3 = g.find("(a)-[e1]->(b); (a)-[e2]->(c)").filter("(b != c)")
motifs3.show(5)
```

执行以上代码，输出内容如下：

```
+--------+----------------+--------+----------------+--------+
|       a|              e1|       b|              e2|       c|
+--------+----------------+--------+----------------+--------+
|[109254]|  [109254, 8742]|  [8742]|[109254, 100010]|[100010]|
|[109254]|  [109254, 8741]|  [8741]|[109254, 100010]|[100010]|
|[109254]| [109254, 59782]| [59782]|[109254, 100010]|[100010]|
|[109254]|[109254, 115349]|[115349]|[109254, 100010]|[100010]|
|[109254]| [109254, 53996]| [53996]|[109254, 100010]|[100010]|
+--------+----------------+--------+----------------+--------+
```

由于边列包含冗余信息，当不需要顶点或边时，可以省略它们的名称。例如，在模式 (a)–[]->(b) 中，[] 表示顶点 a 和 b 之间的任意边。结果中没有用于边的列。同样，(a)–[e]->() 表示顶点 a 的向外边，但不指定目标顶点。指定模式的代码如下：

```
val motifs3 = g.find("(a)-[]->(b); (a)-[]->(c)").filter("(b != c)")
motifs3.show(10)
println(motifs3.count())
```

执行以上代码，输出内容如下：

```
+--------+--------+--------+
|       a|       b|       c|
+--------+--------+--------+
|[109254]|  [8742]|[100010]|
|[109254]|  [8741]|[100010]|
|[109254]| [59782]|[100010]|
|[109254]|[115349]|[100010]|
|[109254]| [53996]|[100010]|
|[109254]|[109257]|[100010]|
|[109254]| [62046]|[100010]|
|[109254]| [94411]|[100010]|
|[109254]|[115348]|[100010]|
|[117041]| [73722]|[100010]|
+--------+--------+--------+
only showing Top 10 rows

28196586
```

下面这种模式通常代表的情况是有一个往复 a 和 b 之间的关系（一个强连通分量表明这两种产品有相似之处），如图 9-22 所示。

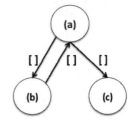

图 9-22　motif 模式 3

指定这个 motif 模式，代码如下：

```
val motifs3 = g
    .find("(a)-[]->(b); (a)-[]->(c); (b)-[]->(a)")
```

```
      .filter("(b != c)")
motifs3.show(10)
println(motifs3.count())
```

执行以上代码,输出的结果如下:

```
+-------+--------+-------+
|     a|       b|      c|
+-------+--------+-------+
|[85609]|[100018]| [85611]|
|[85609]|[100018]| [85610]|
|[85609]|[100018]| [85752]|
|[85609]|[100018]| [28286]|
|[85609]|[100018]| [93910]|
|[85609]|[100018]| [85753]|
|[85609]|[100018]| [60945]|
|[85609]|[100018]| [47246]|
|[85609]|[100018]| [85614]|
|[86839]|[100042]|[100040]|
+-------+--------+-------+
only showing top 20 rows
```

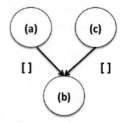

图 9-23　motif 模式 4

下面这种模式代表了这样一种情况:当客户购买了不相关的产品(a 和 c)时,他们也会购买 b。这是一个趋同的 motif,例如,企业可以使用这些信息,将这些产品囤积在一起,如图 9-23 所示。

指定这个 motif 模式,代码如下:

```
val motifs3 = g.find("(a)-[]->(b); (c)-[]->(b)").filter("(a != c)")
motifs3.show(10)
println(motifs3.count())
```

执行以上代码,输出的结果如下:

```
+--------+------+--------+
|       a|     b|       c|
+--------+------+--------+
|[365079]|[8742]|[100010]|
|[241393]|[8742]|[100010]|
| [33284]|[8742]|[100010]|
|[198072]|[8742]|[100010]|
|[203728]|[8742]|[100010]|
|[109254]|[8742]|[100010]|
|[256286]|[8742]|[100010]|
| [45415]|[8742]|[100010]|
|[372540]|[8742]|[100010]|
| [96588]|[8742]|[100010]|
+--------+------+--------+
only showing Top 10 rows
```

在下面的模式中,指定了从 a 到 b 及从 b 到 c 的边,还有一条从 c 到 b 的边。这种模式通常表示当客户购买产品 a 时,他也可能购买 b,然后继续购买 c。这可以表示对所购买的物品有一定的优先次序。此外，motif 中的强连通组件表明 b 和 c 之间的密切关系，如图 9-24 所示。

指定这个 motif 模式，代码如下：

```
val motifs3 = g.find("(a)-[]->(b); (b)-[]->(c); (c)-[]->(b)")
motifs3.show(5)
println(motifs3.count())
```

执行以上代码，输出的结果如下：

```
+--------+------+--------+
|       a|     b|       c|
+--------+------+--------+
|[188454]|[85609]|[100018]|
| [85611]|[85609]|[100018]|
| [98017]|[85609]|[100018]|
|[142029]|[85609]|[100018]|
| [64516]|[85609]|[100018]|
| [85610]|[85609]|[100018]|
|[106584]|[85609]|[100018]|
|[261049]|[85609]|[100018]|
|[331868]|[85609]|[100018]|
| [98015]|[85609]|[100018]|
+--------+------+--------+
only showing Top 10 rows
```

在下面这种模式中，指定了一个 4 节点的 motif。该模式表示客户购买 b 的可能性较高的情况，如图 9-25 所示。

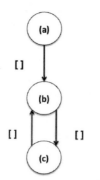

图 9-24　motif 模式 5

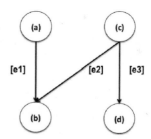

图 9-25　motif 模式 6

注意：4 节点 motif 示例非常消耗资源，需要超过 100GB 的磁盘空间和 14GB 的 RAM。另外，可以参考 9.9.3 节创建一个更小的子图来运行这个示例。

指定这个 motif 模式，代码如下：

```
//确保磁盘空间>100GB 并且 RAM≥14GB
val motifs4 = g
    .find("(a)-[e1]->(b); (c)-[e2]->(b); (c)-[e3]->(d)")
    .filter("(a != c) AND (d != b) AND (d != a)")
motifs4.show(5)
motifs4.count()
```

9.9.3 处理子图

基于 motif 查找和 DataFrame 过滤器的组合，GraphFrames 提供了一种强大的方式来选择子图。例如，选择一个基于顶点和边过滤器的子图，代码如下：

```
//选择子图
val v2 = g.vertices.filter("id < 10")
val e2 = g.edges.filter("src < 10")
val g2 = GraphFrame(v2, e2)

g2.edges.groupBy("src").count().show()
```

执行以上代码，输出的结果如下：

```
+---+-----+
|src|count|
+---+-----+
|  7|   10|
|  3|   10|
|  8|   10|
|  0|   10|
|  5|   10|
|  6|   10|
|  9|   10|
|  1|   10|
|  4|   10|
|  2|   10|
+---+-----+
```

然后基于子图进行查找和过滤，代码如下：

```
val paths = g.find("(a)-[e]->(b)").filter("e.src < e.dst")
paths.select("e.*").show(10)
```

执行以上代码，输出的结果如下：

```
+------+------+
|   src|   dst|
+------+------+
|100008|100010|
|100226|100227|
|100225|100227|
|100224|100227|
|100223|100227|
|100262|100263|
|100551|100553|
|100552|100553|
| 10095| 10096|
|100963|100964|
+------+------+
only showing Top 10 rows
```

9.9.4 应用图算法进行分析

首先，计算每个顶点的强连通分量（Strongly Connected Component，SCC），并返回一张图，其中每个顶点都分配给包含该顶点的 SCC。在 SCC 中显示节点数，代码如下：

```
//设置检查点
spark.sparkContext.setCheckpointDir("hdfs://master:8020/cp")

//强连通分量
val result = g.stronglyConnectedComponents.maxIter(10).run()
result.select("id", "component")
    .groupBy("component")
    .count()
    .sort(col("count").desc)
    .show(10)
```

接下来,计算通过每个顶点的三角形的数量,代码如下:

```
val results = g.triangleCount.run()
results.select("id", "count").show(10)
```

执行结果如下:

```
+------+-----+
|    id|count|
+------+-----+
|100010|   73|
|100140|   15|
|100227|  332|
|100263|    9|
|100320|    8|
|100553|   41|
|100704|    3|
|100735|   13|
|100768|   37|
| 10096|   58|
+------+-----+
only showing Top 10 rows
```

在下面的示例中,应用 PageRank 算法来确定产品的重要性。潜在的假设是,更受欢迎的产品很可能从其他产品节点接收到更多的链接,代码如下:

```
val results = g.pageRank.resetProbability(0.15).tol(0.01).run()
val prank = results.vertices.sort(desc("pagerank"))
prank.show(10)
```

在下面的示例中,使用标签传播算法(Label Propagation Algorithm)来查找图中产品的社区,代码如下:

```
val results = g.labelPropagation.maxIter(10).run()
results.select("id", "label").show(10)
results.select("id", "label")
    .groupBy("label")
    .count()
    .sort(desc("count"))
    .show(10)
```

下面使用最短路径算法来寻找图中两个顶点之间的路径,从而使组成边的数量最小化,代码如下:

```
val results = g.shortestPaths.landmarks(Seq("1110", "352")).run()
results.select("id", "distances").take(5).foreach(println)
```

第 10 章 Delta Lake 数据湖

随着数据量增长到新的、前所未有的水平，新的工具和技术正在出现，以处理这种规模的数据，其中一个发展的领域是数据湖。

10.1 从数据仓库到数据湖

传统上，数据仓库工具用于从数据中驱动商业智能（BI）。业界随后认识到，数据仓库通过强制 Schema on Write 限制了智能的潜力。显然，在收集数据时不能考虑到所收集数据集的所有方面，但是可能有许多其他有价值的数据可从未处理的、自然状态的数据中获得。这意味着，强制执行一个模式或删除一些看起来无用的项，从长远来看可能会损害商业智能。此外，数据仓库技术无法跟上数据增长的步伐。由于数据仓库通常基于数据库和结构化数据格式，因此它不足以应对当前数据驱动世界所面临的挑战。

因为在人工智能、ML 和大数据的时代，数据仓库对于实时决策并不是很有用，因为处理时间长，以及其他一些缺点。这导致了数据湖的出现，这些数据湖针对非结构化和半结构化数据进行了优化，可以轻松地扩展到 PB 级，并允许更好地集成各种工具，以帮助企业最大限度地利用数据。数据湖提供了一个完整和权威的数据存储，可以为数据分析、商业智能和机器学习提供动力。

数据湖是保存大量原始数据的中心位置。数据湖是企业数据的中央存储池，它接收各种数据源的输入信息。这些源可能以非结构化、半结构化和结构化的格式包括从数据库到原始声频和视频片段的任何内容。无论格式如何，数据湖都可以广泛接收新数据。这明显不同于传统关系数据库中充满规则、高度结构化的存储（在数据仓库只存放结构化数据）。

数据仓库和数据湖在处理输入数据的方式上是不同的，如图 10-1 所示。

数据湖被划分为一个或多个数据区域，具有不同程度的转换和清洁度，其中原始区域是所有其他湖泊区域数据构建的基础。数据湖倡导的是"store now, find value later"，即先存储数据，后寻找数据的价值。

与将数据存储在文件或文件夹中的分层数据仓库相比，数据湖使用扁平架构和对象存储来存储数据。对象存储使用元数据标签和唯一标识符存储数据，便于跨分区查找和检索数据，提高性能。数据湖的建立向用户隐藏了底层数据结构和物理数据存储的复杂性。数据湖和数据仓库的一般特性对比，见表 10-1。

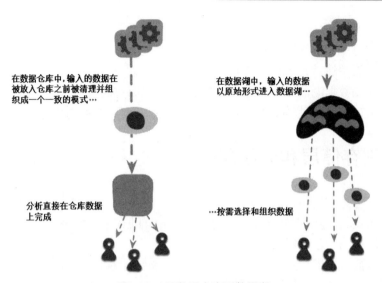

图 10-1 从数据仓库到数据湖

表 10-1 数据湖和数据仓库特性对比

维　度	数　据　湖	数　据　仓　库
数据主要类型	所有类型：结构化数据、半结构化数据和非结构化数据（原始数据）	仅结构化数据
模式 Schema	Schema on Read 数据存储之后才需要定义 Schema 提供敏捷、简单的数据集成 数据的价值尚未明确	Schema on Write 数据存储之前需要定义 Schema 数据集成之前完成大量工作 数据的价值应提前明确
成本	￥	￥￥￥
可扩展性	可扩展以低成本保存任意数量的数据，而不管数据类型如何	由于供应商成本的原因，扩展的成本将呈指数级增长
复杂性	复杂的大数据处理	复杂的 SQL 连接
目标用户	数据分析师、数据科学家	数据分析师
厂商锁定	非	是
优点	低成本，灵活性，可扩展性，允许存储机器学习所需的原始数据	用户接口是传统数据库用户所熟悉的
缺点	如果没有组织和编目数据的工具，则探索大量的原始数据可能会很困难	昂贵的、always-on 架构、专有软件，无法保存机器学习所需的非结构化（原始）数据

数据湖作为一个术语，通常在不同的上下文中有不同的含义，但是在所有数据湖定义中有几个重要的属性是一致的，即它们都支持以下特性：

（1）支持非结构化和半结构化数据。

（2）可扩展到 PB 及以上级规模。

(3) 使用类 SQL 的接口与存储的数据交互。
(4) 能够尽可能无缝地连接各种分析工具。
(5) 现代数据湖通常是解耦存储和分析工具的组合。

从数据仓库到数据湖的过渡具有将业务逻辑与存储分离的优点,以及能够独立扩展计算和存储的能力,但在这种转变中,人们忽略了确保数据可靠性的问题。

10.2 解耦存储层和分析层

在大数据流行之初,Hadoop 作为事实上的大数据平台,HDFS 充当存储层,而 Hive 充当分析层。Hadoop 能够存储几百 TB 的数据,并可以对半结构化数据进行 SQL 之类的查询,而且速度在当时已足够快了。

后来,随着数据量增长到新的规模,企业的需求变得更加雄心勃勃,即用户现在期望更短的查询时间,更好的可扩展性,易于管理等。这时,Hive 和 HDFS 开始为新的更好的技术平台让路。

为了解决 Hadoop 的复杂性和扩展性方面的挑战,业界现在正转向一种分解的架构,使用 REST API 将存储层和分析层松耦合起来。这使每层在扩展和管理方面都更加独立,并允许为每项工作使用更合适的工具。例如,在这个分解的模型中,用户可以选择对批处理工作负载使用 Spark 进行分析,而对 SQL 繁重的工作负载使用 Presto,Spark 和 Presto 都使用相同的后端存储平台。

这种方法现在正迅速成为标准。常用的存储平台是对象存储平台,如 AWS S3、Azure Blob 存储、GCS、Ceph、MinIO 等。分析平台从简单的基于 Python 和 R 的 Notebook 到 TensorFlow、Spark、Presto、Splunk、Vertica 等多种多样。

新的分解模型通常在可扩展性和易于管理方面更好。这种方法在架构上也非常适合,但仍有数据一致性和管理方面的挑战需要解决。

(1) Files 或 Tables:分解模型是指存储系统将数据视为对象或文件的集合,但是终用户端对数据的物理排列不感兴趣,他们想要看到数据的更合乎逻辑的视图。RDBMS 数据库在实现这种抽象方面做得很好。随着大数据平台逐渐成为未来的数据平台,现在人们期望这些系统能够以一种用户友好的方式行事,即不强制用户了解物理存储的任何信息。

(2) SQL 接口:如上所述,用户不再愿意考虑低效率的底层平台。例如,现在数据湖是 ACID 兼容的,这样终用户端就不会有额外的开销来确保与数据相关的保证。

(3) 变更管理:在这种规模下管理数据的另一个非常重要的方面是,能够回滚并查看何时发生了什么变化,以及检查特定的细节。目前,使用对象存储的版本管理可以实现这一点,但正如我们前面所看到的,这是在较低的物理细节层,在较高的逻辑级别上可能没有用处。现在用户期望在逻辑层看到版本控制。

可用性挑战和数据一致性需求导致了一个新的软件项目类别。这些项目位于存储平台和分析平台之间，在以本地方式处理对象存储平台的同时，为最终用户提供强 ACID 保证，其中比较引人注目的一些项目，包括 Delta Lake、Apache Iceberg、Apache Hudi 和 Apache Hive。

1. Delta Lake

Delta Lake 是一种基于文件的开源存储格式，它提供了 ACID 事务、可扩展的元数据处理，并统一了流数据和批数据处理。Delta Lake 是由 Databricks 开发的，它与 Spark 进行了深度的读写集成，可运行在现有的存储平台（S3、HDFS、Azure）上，并且完全兼容 Apache Spark API。特别地，它提供了如下特性。

（1）Spark 上的 ACID 事务：可序列化的隔离级别确保用户读取数据时永远不会看到不一致的数据。Delta Lake 允许在现有的数据湖上添加事务层。既然在它之上有了事务，就可以确保拥有可靠、高质量的数据，并且可以对其进行各种计算。

（2）可扩展的元数据处理：利用 Spark 的分布式处理能力，轻松处理具有数十亿个文件的 PB 级规模的表的所有元数据。

（3）流和批处理的统一：Delta Lake 中的一张表是一个批处理表，同时也是一个流源和 Data Sink。流数据摄入、批量历史回填、交互式查询都是开箱即用的。

（4）模式演变：自动处理模式演变，以防止在摄入期间插入坏记录。

（5）时间旅行：数据版本化功能，使用户能够关注特定的时间点，支持回滚、完整的历史审计跟踪和可重复的机器学习实验。

（6）upsert 和 delete：支持 merge、update 和 delete 操作，以支持复杂的使用，如 Change-Data-Capture（CDC）、Slowly-Changing-Dimension（SCD）操作、流更新等。

2. Apache Iceberg

Apache Iceberg 最初由 Netflix 发布，旨在解决在 S3 上存储大型 Hive-Partitioned 数据集时出现的性能、可伸缩性和可管理性挑战。

Apache Iceberg 是一种用于大型分析数据集的开放表格格式。Iceberg 向 Presto 和 Spark 添加了高性能格式的表，其工作方式与 SQL 表类似。Iceberg 的重点是避免不愉快的意外，帮助模式演化和避免无意的数据删除，用户不需要了解分区就可以快速地查询。

Apache Iceberg 具有以下特性：

（1）模式演化支持添加、删除、更新或重命名，并且没有副作用。

（2）隐藏分区可以防止用户错误导致不正确的结果或非常慢的查询。

（3）分区布局演化可以在数据卷或查询模式更改时更新表的布局。

（4）时间旅行支持使用完全相同的表快照的可重复查询，或者让用户轻松地检查更改。

（5）版本回滚允许用户通过将表重新设置为良好状态来快速纠正问题。

Apache Iceberg 支持 Apache Spark 的读和写，包括 Spark 的结构化流。Trino（PrestoSQL）也支持读取，但对删除的支持有限。Apache Flink 支持读写。最后，Iceberg

为 Apache Hive 提供读取支持。

Apache Iceberg 的优势在于包含大量分区的表的读性能。通过维护将对象映射到分区的清单文件并保持列级统计信息，Iceberg 避免了昂贵的对象存储目录清单或从 Hive 获取分区数据的需要。

此外，Iceberg 的清单允许将单个文件同时分配给多个分区。这使 Iceberg 表在分区修剪方面非常高效，并改善了高选择性查询的延迟。

3. Apache Hudi

同样重要的还有 Apache Hudi。它比 Iceberg 或 Delta Lake 更面向流，更像一个额外的处理层，而不仅是数据湖中使用的表格格式。Hudi 的主要特点是引入了增量架构。

Hudi 最初由 Uber 开源，旨在支持柱状数据格式的增量更新。它支持从多个来源获取数据，主要是 Apache Spark 和 Apache Flink。它还提供了一个基于 Spark 的实用程序来读取外部资源，例如 Apache Kafka。

支持从 Apache Hive、Apache Impala 和 PrestoDB 中读取数据。还有一个专门的工具将 Hudi 表模式同步到 Hive Metastore。

4. Apache Hive

Apache Hive 数据仓库软件已经出现一段时间了。随着新挑战的出现，Hive 现在正试图解决一致性和可用性问题。它方便了使用 SQL 读取、写入和管理分布存储中的大型数据集。结构可以投射到存储中已经存在的数据上。使用 ACID 语义的事务已经被添加到 Hive 中，以解决以下应用场景中的问题。

（1）数据流摄取：使用 Apache Flume、Apache Storm 或 Apache Kafka 等工具将数据流传输到 Hadoop 集群中。虽然这些工具可以以每秒数百行或更多行的速度写入数据，但 Hive 只能每 15min～1h 添加一次分区，而添加分区通常会很快导致表中分区数量过多。这些工具也可以将数据流传输到现有的分区中，但这会导致读取器进行脏读（在开始查询之后会看到写入的数据），并在目录中留下许多小文件，从而对 NameNode 造成压力。Hive 现在支持这样的场景，即允许读取器获得一致的数据视图，并避免过多的文件。

（2）缓慢变化的维度（SCD）：在典型的星型模式数据仓库中，维度表会随着时间缓慢地变化。这些变化会导致插入单个记录或更新记录（取决于所选择的策略）。Hive 从 0.14 开始就能够支持这一功能。

（3）数据重申：有时收集的数据被发现是不正确的，需要纠正。从 Hive 0.14 开始，可以通过 INSERT、UPDATE 和 DELETE 来支持这些用例。

（4）使用 SQL MERGE 语句进行批量更新。

在构建数据湖时，可能没有比存储数据的格式更重要的决策了。其结果将对其性能、可用性和兼容性产生直接影响。Delta Lake、Iceberg 和 Hive 的对比见表 10-2。

表 10-2 Delta Lake、Iceberg 和 Hive 对比

平台	支持的分析工具	文件格式	回滚	压缩和清理	时间旅行
Delta Lake	Apache Spark	Parquet	支持	支持，手动处理	支持
Iceberg	Apache Spark、Presto	Parquet、ORC	支持	不支持	支持
Hive	Hive、Apache Spark	ORC	不支持	手动 & 自动处理	不支持

随着 Apache Hudi、Delta Lake 和 Iceberg 等项目的成熟，表格式逐渐在大数据领域占据了主导地位，取代了世界上大多数公司使用的经过验证的传统数据湖技术。

10.3 Delta Lake 介绍

Apache Spark 没有提供数据处理系统最基本的特性，例如 ACID 事务、其他数据管理和数据治理功能。为了克服这个不足，Databricks（由 Apache Spark 的第一批工程师建立的组织）发布了 Delta Lake，这是一个开放格式的数据管理和治理层，它结合了数据湖和数据仓库的优点，可为数据湖提供可靠性、安全性和高性能，并且可用于流和批处理操作。

在 Delta Lake 中，将所有数据置于一个公共存储层，即数据湖，然后 Delta Lake 会让不同的引擎使用这些数据，如图 10-2 所示。

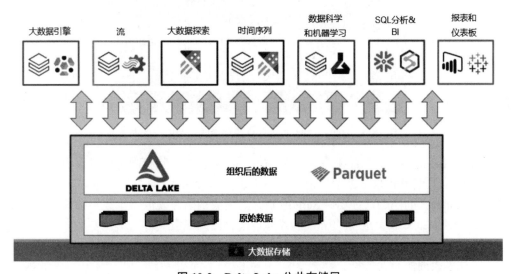

图 10-2 Delta Lake 公共存储层

可以将 Delta Lake 理解为一个数据管理框架，它为数据湖带来了可靠性、安全性和高性能。本质上，Delta Lake 是一个文件系统，它在对象存储上存储批和流数据，以及用于表结构和模式实施的增量元数据。将数据放入湖中是使用 Delta ACID API 完成的，而将数

据从湖中取出是使用 Delta JDBC 连接器完成的。Delta 数据不能被其他 SQL 查询引擎查询，如图 10-3 所示。

图 10-3　Delta Lake 管理元数据

用户的程序不直接与存储层交互，而是通过 Delta Lake API 与数据湖通信来读写数据。Delta Lake 提供了一个统一的平台，在单一平台上支持批处理和流处理工作负载。它充当计算和存储层之间的中间服务，如图 10-4 所示。

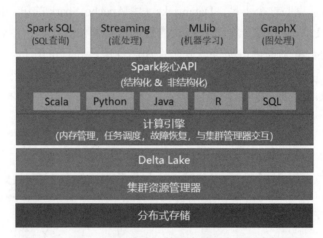

图 10-4　Delta Lake 充当计算和存储层之间的中间服务

Delta Lake 提供了 ACID 事务、快照隔离、数据版本控制和回滚，以及模式强制，以更好地处理模式变更和数据类型转换。具体来讲，Delta Lake 具有以下特性。

（1）Spark 上的 ACID 事务：Serializable 隔离级别确保读取器永远不会看到不一致的数据。在 Delta Lake 表上进行的所有事务都直接存储到磁盘。

（2）可扩展的元数据处理：利用 Spark 分布式处理能力，轻松处理具有数十亿个文件的 PB 级规模表的所有元数据。

（3）流和批处理的统一：Delta Lake 中的表既是批处理表，又是流源和流汇。流式数据摄取，批量历史回填，以及交互式查询都是开箱即用的。

（4）模式演变：自动处理模式变化，以防止在摄取期间插入错误的记录。通过模式强制，以确保上传到表中的数据与它的模式相匹配。

（5）时间旅行：数据版本化控制支持回滚、完整的历史审计跟踪和可复制的机器学习实验。

（6）upsert 和 delete：支持合并（merge）、更新（update）和删除（delete）操作，以支持复杂的用例，如变更数据捕获、渐变维度操作、流式 upsert 等。

Delta Lake 支持两个隔离级别：Serializable 和 WriteSerializable。WriteSerializable 比快照隔离更强大，它提供了可用性和性能的最佳组合，并且是默认设置。最强级别的 Serializable 隔离可确保串行序列与表的历史记录完全匹配。

Delta Lake 通过管理提交的并发性来负责并发的读写访问。这是通过乐观锁实现的。这意味着：

（1）当一个提交执行开始时，线程快照当前 DeltaLog。

（2）提交 actions 完成后，线程会检查 DeltaLog 是否被另一个线程同时更新。如果没有，则它将在 DeltaLog 中记录提交。否则它将更新其 DeltaTable 视图，并在重新处理步骤之后（如果需要）再次尝试注册提交。

这确保了隔离属性。

在 Delta Lake 表上进行的所有事务都直接存储到磁盘上。这个过程满足 ACID 持久性的特性，这意味着即使在系统出现故障时，它也将持续存在。

10.4 Delta Lake 架构

Delta Lake 的架构如图 10-5 所示。

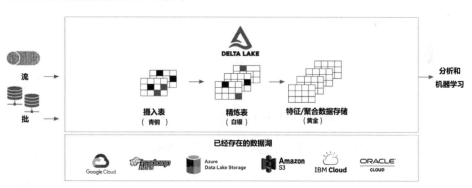

图 10-5　Delta Lake 架构图

在 Delta Lake 中，将数据表分为三类：铜表（Bronze）、银表（Silver）和金表（Gold）。这三类表分别代表数据在处理的不同阶段，说明如下。

1. 铜表

（1）因为数据来自不同的数据源，这可能是脏的数据，因此，它是原始数据的垃圾场。

(2）通常保留时间很长（例如，保留一年以上）。

(3）避免容易出错的解析。

2. 银表

(1）由中间数据组成，并应用了一些清理操作。

(2）数据是可查询的，便于调试。

3. 金表

(1）由准备使用的干净数据组成。

(2）可以使用 Spark 或 Presto 读取数据。

Delta Lake 的真正优势是操作本身提供了 ACID 事务保护。要理解 Delta Lake 是如何提供这些保护的，读者首先需要解包并理解事务日志。

Delta Lake 的核心是事务日志，这是一个中央存储库，用于跟踪用户所做的所有更改，它保证了客户端启动的事务的原子性、一致性、隔离性和持久性。事务日志以 JSON 文件的形式记录每个变更，并按照更改的顺序进行记录。如果有人做了更改，但是删除了它，则仍然会有一个记录来简化审计。它作为真实的单一来源，使用户能够访问 Delta Table 状态的最后一个版本。它提供了序列化能力，这是隔离级别的最强级别。

例如，_delta_log 目录下就是存储事务日志的目录，如图 10-6 所示。

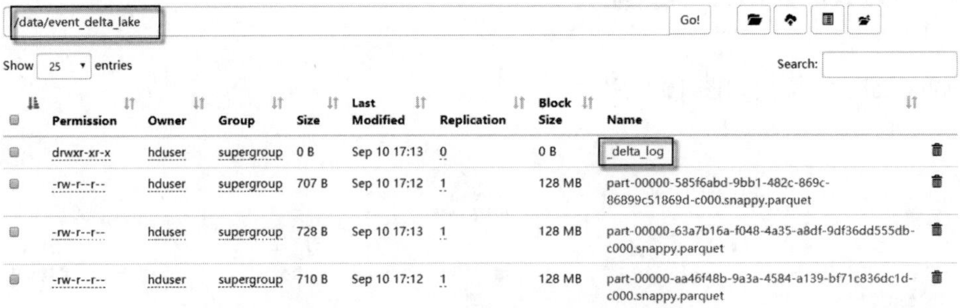

图 10-6 Delta Lake 的核心是事务日志

进入该事务日志目录，可以看到事务日志存储在 JSON 文件中，如图 10-7 所示。

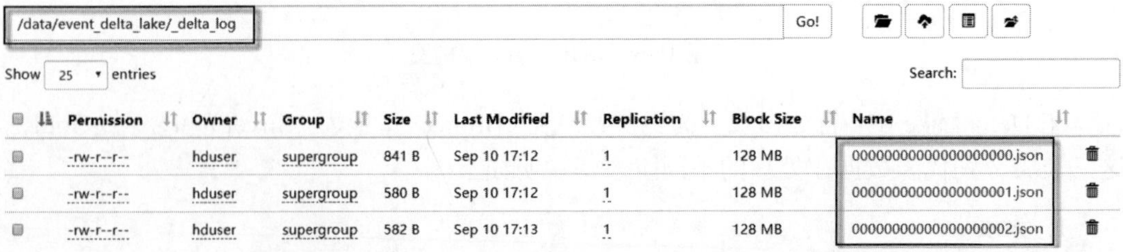

图 10-7 Delta Lake 事务日志以 JSON 文件形式记录

因此，在 Delta Lake 中，读取一张表也会重放这张表的历史记录，例如表的重命名、修改 Schema 等操作。

Delta Lake 将客户端执行的每个活动分隔为原子提交，每个原子提交由 actions 组成。成功完成提交的所有 actions 可以确保 Delta Log 记录提交的操作。对于任何失败的作业，commit 不会记录在 Delta Log 中。

更详细来讲，在 Delta Lake 中的每个 JSON 文件都是一次 commit，这个 commit 是原子性的，保存了与事务相关的详细记录。另外，Delta Lake 还可以保证多个用户同时 commit 而不会产生冲突，它用的是一种基于乐观锁处理的方式。这种解决冲突的方案适用于"写比较少、读取比较多"的场景。

假设用户要处理一个非常大的表，有百万级别的文件，那么如何高效地处理元数据呢？Delta Lake 用 Spark 来读取事务日志，然后 Delta Lake 隔一段时间对 commit 做一次合并，之后可以从 Checkpoint 开始应用后续的 commit。

10.5 Delta Lake 使用

本节将介绍 Delta Lake 的使用，包括模式强制和模式演化、批处理和流工作负载之间的互操作性、时间旅行及 Delete 和 Merge 等 DML 命令，并展示如何从交互式、批处理和流查询读取和写入增量表。

Delta Lake 被很好地集成在 Apache Spark 中，因此，最简单的方法是将 Delta Lake 与 Apache Spark 一起使用。2021 年 5 月 25 日，Delta 发布了 1.0.0 版本，支持 Spark 3.1.x，并于 2022 年 2 月 11 日发布了 1.0.1 版本，修复了 Delta Lake 1.1.0 到 Delta Lake 1.0.0 的 back-ports Bug。2021 年 12 月 3 日，Delta 发布了 1.1.0 版本，支持 Spark 3.2.x。Delta Lake 版本和 Apache Spark 的兼容版本见表 10-3。

表 10-3 Delta Lake 和 Apache Spark 的兼容版本

Delta Lake 版本	Apache Spark 版本
1.1.x	3.2.x
1.0.x	3.1.x
0.7.x 和 0.8.x	3.0.x
低于 0.7.0	2.4.2～2.4<latest>

10.5.1 安装 Delta Lake

兼容 Spark 3.1.x 的 Delta Lake 最新版本是 1.0.1（如果使用的是 Spark 2.x.x，则应使用 Delta Lake 0.6.1 版本）。下面介绍在不同场景下安装 Delta Lake 的方式。

1. 交互式安装

当使用 Apache Spark Shell 时，Delta Lake 包与 --packages 选项一起应用，并指定两个

其他配置,命令如下:

```
$ bin/spark-shell --packages io.delta:delta-core_2.12:1.0.1
    --conf "spark.sql.extensions=io.delta.sql.DeltaSparkSessionExtension"
    --conf "spark.sql.catalog.spark_catalog=org.apache.spark.sql.delta.catalog.DeltaCatalog"
```

2. 项目安装

如果使用 IntelliJ IDEA 进行项目开发,则根据是开发 Maven 项目还是 SBT 项目,进行不同的依赖配置。

Maven 中的 pom.xml 文件的配置内容如下:

```xml
<dependency>
    <groupId>io.delta</groupId>
    <artifactId>delta-core_2.12</artifactId>
    <version>1.0.1</version>
</dependency>
```

SBT 中的 build.sbt 文件的配置内容如下:

```
libraryDependencies += "io.delta" %% "delta-core" % "1.0.1"
```

3. Zeppelin 解释器配置

如果使用 Zeppelin Notebook 进行交互式开发,则需要将 delta-core_2.12-1.0.1.jar 包复制到$SPARK_HOME/jars/目录下,然后,在 Zeppelin 的 Spark 解释器中添加相关属性。需要添加的属性见表 10-4。

表 10-4　Zeppelin 解释器中的 Delta Lake 配置

属 性 名	属 性 值	备　注
spark.home	/home/hduser/bigdata/spark-3.1.2	应修改为自己的 Spark 安装目录
spark.sql.extensions	io.delta.sql.DeltaSparkSessionExtension	
spark.sql.catalog.spark_catalog	org.apache.spark.sql.delta.catalog.DeltaCatalog	

以上配置的 Spark 解释器是全局的。如果只想局部配置,则需要在每个 Notebook 前指定配置,内容如下:

```
%spark.conf
spark.sql.extensions io.delta.sql.DeltaSparkSessionExtension
spark.sql.catalog.spark_catalog org.apache.spark.sql.delta.catalog.DeltaCatalog
spark.databricks.delta.retentionDurationCheck.enabled false
```

10.5.2　表批处理读写

Delta Lake 支持 Apache Spark DataFrame 读写 API 提供的大部分选项,用于对表执行批量读写。

对于许多 Delta Lake 操作，可以在创建新的 SparkSession 时通过设置相应配置来启用与 Apache Spark DataSourceV2 和 Catalog API（自 3.0 以来）的集成。设置配置的代码如下：

```
import org.apache.spark.sql.SparkSession

val spark = SparkSession
  .builder()
  .appName("...")
  .master("...")
  .config("spark.sql.extensions","io.delta.sql.DeltaSparkSessionExtension")
  .config("spark.sql.catalog.spark_catalog","org.apache.spark.sql.delta
.catalog.DeltaCatalog")
  .getOrCreate()
```

可以在使用 spark-submit 提交 Spark 应用程序时添加配置，或者在启动 spark-shell 或 pyspark 时指定它们作为命令行参数，命令如下：

```
$ spark-submit
--conf "spark.sql.extensions=io.delta.sql.DeltaSparkSessionExtension"
--conf "spark.sql.catalog.spark_catalog=org.apache.spark.sql.delta.catalog.
DeltaCatalog"
...
```

1. 创建表

Delta Lake 支持创建两种类型的表：在 Metastore 中定义的表和由路径定义的表。

要使用 Metastore 定义的表，必须在创建新的 SparkSession 时通过设置配置来启用与 Apache Spark DataSourceV2 和 Catalog API 的集成。

要创建 Delta 表，可以先创建一张表并定义模式，或者像数据湖一样，简单地编写一个 Spark DataFrame 以 Delta 格式存储。

1）使用 SQL DDL 命令

可以使用 Apache Spark 支持的标准 SQL DDL 命令（例如 CREATE TABLE 和 REPLACE TABLE）来创建 Delta 表，代码如下：

```
CREATE TABLE IF NOT EXISTS default.people10m (
  id INT,
  firstName STRING,
  middleName STRING,
  lastName STRING,
  gender STRING,
  birthDate TIMESTAMP,
  ssn STRING,
  salary INT
) USING DELTA

CREATE OR REPLACE TABLE default.people10m (
  id INT,
  firstName STRING,
  middleName STRING,
```

```
    lastName STRING,
    gender STRING,
    birthDate TIMESTAMP,
    ssn STRING,
    salary INT
) USING DELTA
```

SQL 还支持在一个路径上创建一张表，而不需要在 Hive Metastore 中创建一个条目，代码如下：

```
CREATE OR REPLACE TABLE delta.`/delta/people10m` (
    id INT,
    firstName STRING,
    middleName STRING,
    lastName STRING,
    gender STRING,
    birthDate TIMESTAMP,
    ssn STRING,
    salary INT
) USING DELTA
```

2）使用 DataFrameWriter API

如果想创建一张表并同时从 Spark DataFrame 或 DataSet 中插入数据，则可以使用 DataFrameWriter API，代码如下：

```
//使用 DataFrame 的模式在 Metastore 中创建表并写入数据
df.write.format("delta").saveAsTable("default.people10m")

//使用 DataFrame 的模式创建带有路径的表，并向其写入数据
df.write.format("delta").mode("overwrite").save("/delta/people10m")
```

3）使用 DeltaTableBuilder API

也可以使用 Delta Lake 的 DeltaTableBuilder API 来创建表。与 DataFrameWriter API 相比，这个 API 更容易指定额外的信息，例如列注释、表属性和生成的列，代码如下：

```
//在 Metastore 中创建表
DeltaTable.createOrReplace(spark)
  .tableName("default.people10m")
  .addColumn("id", "INT")
  .addColumn("firstName", "STRING")
  .addColumn("middleName", "STRING")
  .addColumn(
    DeltaTable.columnBuilder("lastName")
      .dataType("STRING")
      .comment("surname")
      .build())
  .addColumn("lastName", "STRING", comment = "surname")
  .addColumn("gender", "STRING")
  .addColumn("birthDate", "TIMESTAMP")
  .addColumn("ssn", "STRING")
  .addColumn("salary", "INT")
  .execute()
```

```
//使用路径创建表,并添加属性
DeltaTable.createOrReplace(spark)
  .addColumn("id", "INT")
  .addColumn("firstName", "STRING")
  .addColumn("middleName", "STRING")
  .addColumn(
    DeltaTable.columnBuilder("lastName")
      .dataType("STRING")
      .comment("surname")
      .build())
  .addColumn("lastName", "STRING", comment = "surname")
  .addColumn("gender", "STRING")
  .addColumn("birthDate", "TIMESTAMP")
  .addColumn("ssn", "STRING")
  .addColumn("salary", "INT")
  .property("description", "table with people data")
  .location("/delta/people10m")
  .execute()
```

当创建一个 Delta 表时,实际上是对文件进行存储(例如文件系统,云对象存储)。所有文件一起存储在特定结构的目录中组成数据表,因此,当创建一个 Delta 表时,实际上是在将文件写入某个存储位置。例如,下面的 Spark 代码片段接受一个现有的 Spark DataFrame,并使用 Spark DataFrame API 以 Apache Parquet 存储格式将其写入文件夹/data 中,代码如下:

```
dataframe.write.format("parquet").save("/data")
```

只需对上面的代码进行简单修改,就可以创建 Delta 表,代码如下:

```
dataframe.write.format("delta").save("/data")
```

在下面的代码示例中,将首先创建 Spark DataFrame 数据,然后使用 write()方法将表保存到存储中,代码如下:

```
//创建数据 DataFrame
val data = spark.range(0, 5)

//将数据 DataFrame 写入/delta 位置
data.write.format("delta").save("/data_lake/delta_lake/delta")
```

务必注意,在大多数生产环境中,当处理大量数据时,对数据进行分区非常重要。例如,按日期字段 date 对数据进行分区,代码如下:

```
//将 Spark DataFrame 写入 Delta 表中,按 date 列分区
data.write.partitionBy("date").format("delta").save("/data_lake/delta_lake/delta")
```

如果已经有一个现有的 Delta 表,并且想要向该表追加或覆盖数据,则可在语句中包含 mode 方法,代码如下:

```
//向 Delta 表追加新数据
data.write.format("delta").mode("append").save("/data_lake/delta_lake/delta")
```

```
//覆盖 Delta 表
data.write.format("delta").mode("overwrite").save("/data_lake/delta_lake/
delta")
```

2. 读取表

与写 Delta 表类似,可以使用 DataFrame API 从 Delta 表读取相同的文件。可以通过指定表名或路径来将 Delta 表加载为 DataFrame,代码如下:

```
spark.table("default.people10m")                            //Metastore 中的查询表

spark.read.format("delta").load("/delta/people10m")         //按路径创建的表

import io.delta.implicits._
spark.read.delta("/delta/people10m")                        //简捷写法
```

例如,读取前面写入的 Delta Lake 表数据,代码如下:

```
//从/delta 位置将数据读到 DataFrame
spark.read.format("delta").load("/data_lake/delta_lake/delta").show()
```

返回的 DataFrame 为任何查询自动读取表的最新快照。当查询中有适用的谓词时,Delta Lake 会自动使用分区和统计来读取最小数量的数据。

还可以使用 SQL 读取表,方法是在指定 delta 之后指定文件位置,代码如下:

```
%sql
SELECT * FROM delta.`/data_lake/delta_lake/delta`
```

前面是直接从文件系统读取了 Delta 表,但是怎么读取一个 Metastore 中定义的 Delta 表呢? 为此,首先需要使用 saveAsTable()方法或 CREATE TABLE 语句在 Metastore 中定义表,代码如下:

```
//将数据 DataFrame 写入 Metastore,定义为 myTable
data.write.format("delta").saveAsTable("myTable")
```

注意,当使用 saveAsTable()方法时,将把 Delta 表文件保存到 Metastore 管理的位置(例如/user/hive/warehouse/myTable)。如果想使用 SQL 或控制 Delta 表的位置,则可首先使用 save 方法在指定的地方保存它(例如/delta),然后使用 SQL 语句创建表,代码如下:

```
%sql
-- 在 Metastore 中创建表
CREATE TABLE myTable (
  id INTEGER
)
USING DELTA
LOCATION "/delta"
```

如前所述,对大型表进行分区非常重要。例如,按 date 列分区,代码如下:

```
%sql
CREATE TABLE id (
  date DATE
  id INTEGER
)
```

```
USING DELTA
PARTITION BY date
LOCATION "/delta"
```

注意 LOCATION 属性指向构成 Delta 表的底层文件。

注意：什么是 Metastore？

通常在 Hadoop 生态系统中，通过 SQL（Spark SQL、Hive、Presto、Impala 等）引用数据集，需要在 Metastore（通常是 Hive Metastore）中定义表。这个表定义是一个元数据条目，它将向 Apache Spark 等数据处理框架描述数据位置、存储格式、表模式及其他属性。

3. 写入表

写入表时，可以指定不同的模式。

1）append 模式

要自动向现有的 Delta 表添加新数据，需要使用 append 模式，代码如下：

```
df.write.format("delta").mode("append").save("/delta/people10m")
df.write.format("delta").mode("append").saveAsTable("default.people10m")

import io.delta.implicits._
df.write.mode("append").delta("/delta/people10m")
```

或者使用如下的 SQL 语句：

```
INSERT INTO default.people10m SELECT * FROM morePeople
```

2）overwrite 模式

如果需要自动替换表中的所有数据，则可使用 overwrite 模式，代码如下：

```
df.write.format("delta").mode("overwrite").save("/delta/people10m")
df.write.format("delta").mode("overwrite").saveAsTable("default.people10m")

import io.delta.implicits._
df.write.mode("overwrite").delta("/delta/people10m")
```

或者使用如下的 SQL 语句：

```
INSERT OVERWRITE TABLE default.people10m SELECT * FROM morePeople
```

使用 DataFrame，还可以选择性地只覆盖与任意表达式匹配的数据。此功能在 Delta Lake 1.1.0 及以上版本中可用。例如，使用 df 中的数据自动替换目标表中按 start_date 分区的一月份的事件，代码如下：

```
df.write
  .format("delta")
  .mode("overwrite")
  .option("replaceWhere", "start_date >= '2017-01-01' AND end_date <= '2017-01-31'")
  .save("/delta/events")
```

此示例代码会写出 df 中的数据，验证数据是否与谓词匹配，并执行原子替换。如果想写出不完全匹配谓词的数据，替换目标表中匹配的行，则可以通过相应的设置来禁用约束检查，代码如下：

```
spark.conf.set("spark.databricks.delta.replaceWhere.constraintCheck.enabled",
false)
```

在 Delta Lake 1.0.0 及以下版本中，replaceWhere 选项的意思是只覆盖分区列上匹配谓词的数据。例如，使用 df 中的数据自动替换目标表中按 birthDate 分区的一月份的数据，代码如下：

```
df.write
  .format("delta")
  .mode("overwrite")
  .option("replaceWhere", "birthDate >= '2017-01-01' AND birthDate <=
'2017-01-31'")
  .save("/delta/people10m")
```

在 Delta Lake 1.1.0 及以上版本中，如果想恢复到以前的行为，则可以禁用 spark.databricks.delta.replaceWhere.dataColumns.enabled 标志，代码如下：

```
spark.conf.set("spark.databricks.delta.replaceWhere.dataColumns.enabled",
false)
```

注意：与 Apache Spark 中的文件 API 不同，Delta Lake 会记住并强制表的模式。这意味着在默认情况下，覆盖不会替换现有表的模式。

可以设置用户自定义提交元数据，指定用户定义的字符串作为这些操作提交中的元数据，这可以通过 DataFrameWriter 选项 userMetadata 或 SparkSession 配置 spark.databricks.delta.commitInfo.userMetadata 实现。如果两者都已指定，则选项 option 优先。用户定义的元数据在 history 操作中是可读的，代码如下：

```
df.write.format("delta")
  .mode("overwrite")
  .option("userMetadata", "overwritten-for-fixing-incorrect-data")
  .save("/delta/people10m")
```

10.5.3 表流处理读写

Delta Lake 通过 readStream 和 writeStream 与 Spark 结构化流深度集成。Delta Lake 克服了许多通常与流系统和文件相关的限制：

（1）使用多个流（或并发批处理作业）维护 exactly-once 的处理，即使有其他流或批查询并发地运行在该表上，Delta Lake 事务日志也能保证精确一次性处理。

（2）当使用文件作为流的源时，可有效地发现哪些文件是新的。

对于表上的许多 Delta Lake 操作，可以在创建新的 SparkSession 时通过设置配置来支持与 Apache Spark DataSourceV2 和 Catalog API（自 3.0 以来）的集成。

1. Delta 表作为数据源

当将 Delta 表加载为流源并在流查询中使用它时，查询将处理表中出现的所有数据及在流启动后到达的任何新数据，示例代码如下：
```
spark.readStream.format("delta").load("/delta/events")

import io.delta.implicits._
spark.readStream.delta("/delta/events")            //简写
```

1）限制输入的速度

以下选项可用于控制微批。

（1）maxFilesPerTrigger：每个微批处理中需要考虑多少个新文件，默认值为1000。

（2）maxBytesPerTrigger：每个微批处理的数据量。此选项设置了一个"软最大值"，意味着一个微批大约处理 maxBytesperTrigger 这个参数所指定的数量的数据，并可能处理超过这个限制。如果为流指定 Trigger.Once，则该选项将被忽略。这在默认情况下没有设置。

如果将 maxBytesPerTrigger 与 maxFilesPerTrigger 结合使用，则微批处理数据直到达到 maxBytesPerTrigger 或 maxFilesPerTrigger 限制为止。

2）忽略 update 和 delete

结构化流不处理非追加的输入，并在用作源的表上发生任何修改时抛出异常。处理不能自动向下游传播的更改有以下两种主要策略：

（1）可以删除输出和检查点，然后从头重新启动流。

（2）可以设置这两个选项。ignoreDeletes：忽略在分区边界上删除数据的事务。ignoreChanges：如果由于数据更改操作（如 UPDATE、MERGE INTO、DELETE（分区内）或 OVERWRITE）而不得不重写源表中的文件，则重新处理更新。未更改的行仍然可能被发出，因此下游消费者应该能够处理重复的副本。删除不会向下游传播。ignoreChanges 包容 ignoreDeletes，因此，如果使用 ignoreChanges，则流将不会被对源表的删除或更新所中断。

例如，假设有一张表 user_events，其中包含 date、user_email 和 action 列，按 date 分区。将 user_events 表作为流源，由于 GDPR 的关系，需要从该表中删除数据。

当在分区边界删除时（WHERE 位于分区列上），文件已经按值分段，因此删除只是从元数据中删除这些文件，因此，如果只是想从一些分区中删除数据，则可执行的代码如下：
```
spark.readStream.format("delta")
  .option("ignoreDeletes", "true")
  .load("/delta/user_events")
```

然而，如果必须删除基于 user_email 的数据，则代码如下：
```
spark.readStream.format("delta")
  .option("ignoreChanges", "true")
  .load("/delta/user_events")
```

如果使用 update 语句更新 user_email，则包含问题 user_email 的文件将被重写。当使用 ignoreChanges 时，新记录将与同一文件中所有其他未更改的记录一起向下传播。计算逻辑应该能够处理这些传入的重复记录。

3）指定初始位置

可以使用以下选项指定 Delta Lake 流源的起始点，而无须处理整个表。

（1）startingVersion：开始的 Delta Lake 版本。从这个版本（包括在内）开始的所有表更改都将被流源读取。提交的版本号可以从 DESCRIBE HISTORY 命令输出的 version 列中获取。若要只返回最新的更改，则可将该选项值指定为 latest。

（2）startingTimestamp：开始的时间戳。在这段时间戳（包括在内）或之后提交的所有表更改都将被流源读取。其值可以是以下两种类型之一：一个时间戳字符串，例如，"2019-01-01T00:00:00.000Z"；一个日期字符串，例如，"2019-01-01"。

不能同时设置这两个选项，只能使用其中一种。它们只在启动一个新的流查询时生效。如果一个流查询已经启动，并且进度已经记录在检查点中，则这些选项将被忽略。

需要注意的是，虽然可以从指定的版本或时间戳启动流源，但流源的模式始终是 Delta 表的最新模式。必须确保在指定的版本或时间戳之后对 Delta 表没有不兼容的模式更改。否则当使用不正确的模式读取数据时，流源可能会返回不正确的结果。

例如，假设有一张表 user_events。如果想读取自版本 5 以来的更改，则可执行的代码如下：

```
spark.readStream.format("delta")
  .option("startingVersion", "5")
  .load("/delta/user_events")
```

如果想读取 2018-10-18 以来的更改，则可执行的代码如下：

```
spark.readStream.format("delta")
  .option("startingTimestamp", "2018-10-18")
  .load("/delta/user_events")
```

2. Delta 表作为数据接收器

还可以使用结构化流将数据写入 Delta 表。事务日志使 Delta Lake 能够保证 exactly-once 处理，即使有其他流或批查询同时对表运行，示例代码如下：

```
val streamingDf = spark
    .readStream
    .format("rate")
    .option("rowsPerSecond","10")
    .load()

val stream = streamingDf
    .select($"value" as "number")
    .writeStream.format("delta")
    .option("checkpointLocation", "/tmp/checkpoint")
    .start("file://home/hduser/data/delta-table/demo")

stream.awaitTermination()
stream.stop()
```

将流数据写入 Delta 表时有两种模式。

1）append 模式

默认情况下，流以追加（append）模式运行，该模式将新记录添加到表中，代码如下：

```
//path 方法
events.writeStream
  .format("delta")
  .outputMode("append")
  .option("checkpointLocation","/delta/events/_checkpoints/etl-from-json")
  .start("/delta/events")

#简写
import io.delta.implicits._
events.writeStream
  .outputMode("append")
  .option("checkpointLocation","/delta/events/_checkpoints/etl-from-json")
  .delta("/delta/events")

//表方法
events.writeStream
  .outputMode("append")
  .option("checkpointLocation","/delta/events/_checkpoints/etl-from-json")
  .table("events")
```

2）complete 模式

也可以使用结构化流来替换每批处理的整个表。一个应用场景是使用聚合计算汇总信息，代码如下：

```
spark.readStream
  .format("delta")
  .load("/delta/events")
  .groupBy("customerId")
  .count()
  .writeStream
  .format("delta")
  .outputMode("complete")
  .option("checkpointLocation","/delta/eventsByCustomer/_checkpoints/streaming-agg")
  .start("/delta/eventsByCustomer")
```

上面的示例不断地更新一张表，该表包含按客户聚合的事件数。

对于延迟要求更宽松的应用程序，可以使用一次性触发器节省计算资源。使用它们根据给定的时间表更新汇总聚合表，只处理自上次更新以来到达的新数据。

10.5.4 文件移除

Delta Lake 进行了版本控制，因此可以轻松地恢复或访问数据的旧版本，因此，在某些情况下，Delta Lake 需要存储多个版本的数据才能实现回滚功能。

存储相同数据的多个版本可能成本很高，因此 Delta Lake 包含了一个 vacuum 命令，可以删除数据的旧版本（任何比指定的保留期限早的数据）。下面解释如何使用 vacuum 命令及其适用场景。

假设要使用下面的 people.csv 文件来构造一个 Delta Lake 表，文件的内容如下：

```
first_name,last_name,country
miguel,cordoba,colombia
luisa,gomez,colombia
li,li,china
wang,wei,china
hans,meyer,germany
mia,schmidt,germany
```

假设将该数据文件存储到 HDFS 的/data/data_lake/delta_lake/目录下。

首先将 people.csv 文件内容加载到 DataFrame 中，代码如下：

```
val path = "/data/data_lake/delta_lake/people.csv"
val df = spark
  .read
  .option("header", "true")
  .option("charset", "UTF8")
  .csv(path)

df.printSchema()
df.show()
```

执行以上代码，输出内容如下：

```
root
 |-- first_name: string (nullable = true)
 |-- last_name: string (nullable = true)
 |-- country: string (nullable = true)

+----------+---------+--------+
|first_name|last_name| country|
+----------+---------+--------+
|    miguel|  cordoba|colombia|
|     luisa|    gomez|colombia|
|        li|       li|   china|
|      wang|      wei|   china|
|      hans|    meyer| germany|
|       mia|  schmidt| germany|
+----------+---------+--------+
```

然后将上面的 DataFrame 写入一个 Delta Lake 表中，代码如下：

```
val outputPath = "/data_lake/delta_lake/vacuum_example"
df
  .repartition(1)
  .write
  .format("delta")
  .save(outputPath)
```

查看该 Delta Lake 的内容，代码如下：

```
val path = "/data_lake/delta_lake/vacuum_example/"
val df = spark.read.format("delta").load(path)
df.show()
```

执行以上代码，输出的结果如下：

```
+----------+---------+--------+
|first_name|last_name| country|
+----------+---------+--------+
|    miguel|  cordoba|colombia|
|     luisa|    gomez|colombia|
|        li|       li|   china|
|      wang|      wei|   china|
|      hans|    meyer| germany|
|       mia|  schmidt| germany|
+----------+---------+--------+
```

下面用另一个 new_people.csv 文件覆盖 Delta Lake 中的数据。new_people.csv 文件的内容如下：

```
first_name,last_name,country
li,li,china
aa,li,china
zhang,zhang,china
```

覆盖 Delta Lake 表的代码如下：

```
val path = "/data/data_lake/delta_lake/new_people.csv"
val outputPath = "/data_lake/delta_lake/vacuum_example/"

spark
  .read
  .option("header", "true")
  .option("charset", "UTF8")
  .csv(path)
.repartition(1)
  .write
  .format("delta")
  .mode("overwrite")
  .save(outputPath)
```

然后查看覆盖后的数据，代码如下：

```
spark.read.format("delta").load(outputPath).show()
```

执行以上代码，输出覆盖后的数据，内容如下：

```
+----------+---------+-------+
|first_name|last_name|country|
+----------+---------+-------+
|        li|       li|  china|
|        aa|       li|  china|
|     zhang|    zhang|  china|
+----------+---------+-------+
```

在终端窗口中，执行命令来查看当前的文件系统，命令如下：

```
$ hdfs dfs -ls - /data_lake/delta_lake/vacuum_example
```

图 10-8 是第 2 次写入后 HDFS 上显示的文件系统的目录结构和内容。

```
/data_lake/delta_lake/vacuum_example/_delta_log
/data_lake/delta_lake/vacuum_example/_delta_log/00000000000000000000.json
/data_lake/delta_lake/vacuum_example/_delta_log/00000000000000000001.json
/data_lake/delta_lake/vacuum_example/part-00000-36e8dca8-1378-4de5-97f4-4d4304de8dcb-c000.snappy.parquet
/data_lake/delta_lake/vacuum_example/part-00000-9d6a938e-f928-41a6-a129-482c5e36fe48-c000.snappy.parquet
```

图 10-8　Delta Lake 第 2 次写入后 HDFS 上显示的文件系统的目录结构和内容

可以看到，虽然第 1 次写入的数据不再读到 DataFrame 中，但它仍然存储在文件系统中，所以可以回滚到数据的旧版本。

例如，要显示版本 0 时的 Delta Lake 内容，代码如下：

```
val df = spark.read
    .format("delta")
    .option("versionAsOf", 0)
    .load(outputPath)
df.show()
```

执行以上代码，输出内容如下：

```
+----------+---------+--------+
|first_name|last_name| country|
+----------+---------+--------+
|    miguel|  cordoba|colombia|
|     luisa|    gomez|colombia|
|        li|       li|   china|
|      wang|      wei|   china|
|      hans|    meyer| germany|
|       mia|  schmidt| germany|
+----------+---------+--------+
```

运行 vacuum 命令并验证文件系统中是否删除了文件，代码如下：

```
import io.delta.tables._

val deltaTable = DeltaTable.forPath(spark, outputPath)
deltaTable.vacuum(0.000001)
```

将保留时间设置为 0.000001h，这样就可以立即运行这个 vacuum 命令了。

注意：当设置保留的时间小于 168h 时，需要设置 spark.databricks.delta.retentionDurationCheck.enabled = false。

运行 vacuum() 命令后，在终端窗口中执行命令来查看当前的文件系统，命令如下：

```
$ hdfs dfs -ls -R /data_lake/delta_lake/vacuum_example
```

这时文件系统的结构和内容如图 10-9 所示。

可以看到，有一个 Parquet 数据文件（旧版本的数据文件）已经被删除了。可以通过

```
/data_lake/delta_lake/vacuum_example/_delta_log
/data_lake/delta_lake/vacuum_example/_delta_log/00000000000000000000.json
/data_lake/delta_lake/vacuum_example/_delta_log/00000000000000000001.json
/data_lake/delta_lake/vacuum_example/part-00000-36e8dca8-1378-4de5-97f4-4d4304de8dcb-c000.snappy.parquet
```

图 10-9 运行 vacuum 命令后的文件系统

查看 00000000000000000001.json 文件,以了解 Delta 如何知道要删除哪些文件。查看文件的命令如下:

```
$ hdfs dfs -cat
/data_lake/delta_lake/vacuum_example/_delta_log/00000000000000000001.json
```

可以得到 00000000000000000001.json 文件的内容如下(下面对显示格式进行了美化,以方便查看):

```
{
  "commitInfo":{
    "timestamp":1646118315485,
    "operation":"WRITE",
    "operationParameters":{
      "mode":"Overwrite",
      "partitionBy":"[]"
    },
    "readVersion":0,
    "isBlindAppend":false,
    "operationMetrics":{
      "numFiles":"1",
      "numOutputBytes":"925",
      "numOutputRows":"3"
    }
  }
}
{
  "add":{
    "path":"part-00000-36e8dca8-1378-4de5-97f4-4d4304de8dcb-c000.snappy.parquet",
    "partitionValues":{},
    "size":925,
    "modificationTime":1646118314217,
    "dataChange":true
  }
}
{
  "remove":{
    "path":"part-00000-9d6a938e-f928-41a6-a129-482c5e36fe48-c000.snappy.parquet",
    "deletionTimestamp":1646118315484,
    "dataChange":true,
    "extendedFileMetadata":true,
    "partitionValues":{},
    "size":973
  }
}
```

上面 JSON 文件的 remove 部分表明 part-00000-9d6a938e--…fe48-c000.snappy.parquet 可以在执行 vacuum()命令时被删除。在执行 vacuum()命令后，就无法访问 Delta Lake 的 0 版本了。如果强行访问 versionAsOf 为 0 的版本，则会抛出异常。例如，读取前面 outputPath 指定的路径下的文件，用 versionAsOf 指定 0 版本，代码如下：

```
val df = spark.read
    .format("delta")
    .option("versionAsOf", 0)
    .load(outputPath)

df.show()
```

这时访问 0 版本，会抛出以下错误信息：

```
...
java.io.FileNotFoundException: File does not exist: hdfs://xueai8:8020/data_lake/delta_lake/vacuum_example/part-00000-9d6a938e-f928-41a6-a129-482c5e36fe48-c000.snappy.parquet
It is possible the underlying files have been updated. You can explicitly invalidate the cache in Spark by running 'REFRESH TABLE tableName' command in SQL or by recreating the Dataset/DataFrame involved.
...
```

对于旧数据，系统默认的保存期为 7 天，因此执行 deltable .vacuum()不会马上生效，除非等待 7 天（168 小时）才能运行该命令。需要设置特殊的命令来调用保留时间小于 7 天的 vacuum()方法，否则会抛出错误信息。需要修改 Spark 配置，以允许小于 168 小时的保留时间。可以静态也可以动态地将 spark.databricks.delta.retentionDurationCheck.enabled 配置参数的值设置为 false，代码如下：

```
val spark: SparkSession = {
  SparkSession
    .builder()
    .master("local")
    .appName("spark session")
    .config("spark.databricks.delta.retentionDurationCheck.enabled", "false")
    .getOrCreate()
}
```

如果使用 Zeppelin Notebook，则可以在 Spark 解释器中添加的配置如下：

```
spark.databricks.delta.retentionDurationCheck.enabled       false
```

需要注意，只有事务日志中有 remove 项时，执行 vacuum()方法才会执行删除操作。如果事务日志（JSON 文件中）就没有 remove 项，则即使调用了 vacuum()方法，也什么都不会做。

在下面这个示例中，仅将数据添加到 Delta Lake，这样，运行 vacuum()命令将什么也不做。

dogs1.csv 文件的内容如下：

```
first_name,breed
```

```
fido,lab
spot,bulldog
dogs2.csv:
first_name,breed
fido,beagle
lou,pug
```

将 dog1.csv 和 dogs2.csv 写入 Delta Lake 文件,代码如下:

```
val df = spark
  .read
  .option("header", "true")
  .option("charset", "UTF8")
  .csv("/dog_data/dogs1.csv")

df
  .repartition(1)
  .write
  .format("delta")
  .save("/data_lake/delta_lake/vacuum_example2")

val df2 = spark
  .read
  .option("header", "true")
  .option("charset", "UTF8")
  .csv("/dog_data/dogs2.csv")

df2
  .repartition(1)
  .write
  .format("delta")
  .mode(SaveMode.Append)
  .save("/data_lake/delta_lake/vacuum_example2")
```

在这两个文件被写入后,Delta Lake 所包含的内容如下:

```
+----------+-------+
|first_name|  breed|
+----------+-------+
|      fido|    lab|
|      spot|bulldog|
|      fido| beagle|
|       lou|    pug|
+----------+-------+
```

查看对应的文件系统,目录结构如下:

```
vacuum_example2/
  _delta_log/
    00000000000000000000.json
    00000000000000000001.json
  part-00000-57db2297-9aaf-44b6-b940-48c504c510d1-c000.snappy.parquet
  part-00000-6574b35c-677b-4423-95ae-993638f222cf-c000.snappy.parquet
```

其中 0000000000000000.json 文件的内容如下：

```
{
  "add":{
     "path":"part-00000-57db2297-9aaf-44b6-b940-48c504c510d1-c000.snappy.parquet",
     "partitionValues":{},
     "size":606,
     "modificationTime":1568685380000,
     "dataChange":true
  }
}
```

下面是 0000000000000001.json 文件的内容：

```
{
  "add":{
     "path":"part-00000-6574b35c-677b-4423-95ae-993638f222cf-c000.snappy.parquet",
     "partitionValues":{},
     "size":600,
     "modificationTime":1568685386000,
     "dataChange":true
  }
}
```

所有 JSON 文件都不包含任何 remove 行，因此 vacuum()方法不会删除任何文件，所以当执行下面这段代码后没有改变任何东西：

```
import io.delta.tables._

val path = "/delta_lake_data/vacuum_example2"
val deltaTable = DeltaTable.forPath(spark, path)
deltaTable.vacuum(0.000001)
```

由上面的内容可知，如果想节省数据存储成本，则可以使用 vacuum()方法从 Delta Lake 中删除文件。

在运行 overwrite 操作后，经常会有重复的文件。在_delta_log/下的 JSON 文件中，任何比指定的保留时间更早且被标记为 remove 的文件都将在执行 vacuum()方法时被删除。

10.5.5 压缩小文件

数据湖可以积累很多小文件，特别是当执行增量更新时。文件小，读取速度就会变慢。可以用 Spark 压缩 Delta Lake 中的小文件。通过压缩将小文件合并到大文件中是保持快速读取的一种重要的数据湖维护技术。

下面创建一个包含 1000 个文件的 Delta 数据湖，然后对文件夹进行压缩，压缩后只包含 10 个文件。首先构造一个 DataFrame，将其重分区为 1000 个分区，这样在存入 Delta Lake 时就会生成 1000 个小的存储文件，代码如下：

```
    //先构造一个 DataFrame
    val df = spark.range(0, 10000).toDF("number")

    //创建具有 1000 个文件的 Delta Lake
df.repartition(1000)
  .write
  .format("delta")
  .save("/data_lake/delta_lake/compact")
```

查看 _delta_log 下相应的日志文件 00000000000000000000.json，命令如下：

```
$ hdfs dfs -cat
/data_lake/delta_lake/compact/_delta_log/00000000000000000000.json
```

会看到其中包含 1000 行类似下面这样的 JSON 行（为了便于阅读，这里格式化显示其中一行），内容如下：

```
{
  "add":{
    "path":"part-00000-05e5bda3-f20f-4452-b10a-c675f7557b1d-c000.snappy.parquet",
    "partitionValues":{},
    "size":525,
    "modificationTime":1622099342864,
    "dataChange":true
  }
}
```

从 Delta Lake 中把这些数据读取出来，然后重分区为 10 个分区，再使用 overwrite 模式写入相同的 delta 表，这样便可把数据压缩到只包含 10 个文件，代码如下：

```
val data = spark.read.format("delta").load("/delta_data_lake/compact")

data.repartition(10)
    .write
    .format("delta")
    .mode("overwrite")
    .save("/data_lake/delta_lake/compact")
```

现在 /data_lake/delta_lake/compact 目录下包含 1010 个文件，其中 1000 个原始未压缩的文件及 10 个压缩过的文件。

这时再查看文件 /data_lake/delta_lake/compact/_delta_log/00000000000000000001.json，会看到其中包含 10 行类似下面这样的 JSON 行：

```
{
  "add":{
    "path":"part-00000-9f8f50a7-b31b-49f9-bdfc-6e6072002c2e-c000.snappy.parquet",
    "partitionValues":{},
    "size":4526,
    "modificationTime":1622100154614,
    "dataChange":true
  }
}
```

另外还包含 1000 行类似下面这样的 JSON 行：

```
{
  "remove":{
    "path":"part-00097-47fcbc63-44e5-42f3-924d-47eefad90a0f-c000.snappy
.parquet",
    "deletionTimestamp":1622100157546,
    "dataChange":true
  }
}
```

该 JSON 文件中的 remove 部分表明当运行 vacuum() 命令时，part-00097-……-c000.snappy.parquet 文件会被删除。

Delta Lake 包括一个删除旧版本数据的 vacuum() 命令。可以运行 vacuum() 命令删除旧的数据文件，这样就不必存储未压缩的数据了，代码如下：

```
import io.delta.tables._
import org.apache.spark.sql.functions._

val deltaTable = DeltaTable.forPath("/data_lake/delta_lake/compact")
deltaTable.vacuum(0.000001)
```

这时再查看/data_lake/delta_lake/compact 目录，可以看到只剩下 10 个 Parquet 文件了，如图 10-10 所示。

```
/data_lake/delta_lake/compact/_delta_log
/data_lake/delta_lake/compact/part-00000-226f2435-4f2f-4789-97a3-daa7aab17b6c-c000.snappy.parquet
/data_lake/delta_lake/compact/part-00001-4fcb20f7-e7e3-497f-871a-9fc8e3736453-c000.snappy.parquet
/data_lake/delta_lake/compact/part-00002-ca4277b6-8033-4fef-8c1a-075fdf365e93-c000.snappy.parquet
/data_lake/delta_lake/compact/part-00003-763e3387-a81b-4541-a3da-fd9228bd48b5-c000.snappy.parquet
/data_lake/delta_lake/compact/part-00004-92a7ee65-8ac0-4caf-85f6-ce65026d8203-c000.snappy.parquet
/data_lake/delta_lake/compact/part-00005-910b4279-e548-4856-a001-1b4460c0708e-c000.snappy.parquet
/data_lake/delta_lake/compact/part-00006-ec733df4-1909-4a66-8e18-0fb0bd46e067-c000.snappy.parquet
/data_lake/delta_lake/compact/part-00007-7aec00ed-523e-4bde-a78c-fdbcd812ee0d-c000.snappy.parquet
/data_lake/delta_lake/compact/part-00008-642a9e4e-508d-4deb-a313-f9710f11854a-c000.snappy.parquet
/data_lake/delta_lake/compact/part-00009-96405006-f4ad-4491-9246-53664bb5156b-c000.snappy.parquet
```

图 10-10　压缩小文件后

当设置保留的时间小于 168 h 时，需要如下设置：

```
spark.databricks.delta.retentionDurationCheck.enabled = false
```

这里将保留时间设置为 0.000001h，这样就可以立即运行这个 vacuum() 命令了。在执行 vacuum() 命令后，就无法访问 Delta Lake 的 0 版本了，否则会出现错误。例如，访问 Delta Lake 的 0 版本，代码如下：

```
val df = spark.read
    .format("delta")
    .option("versionAsOf", 0)
    .load("/data_lake/delta_lake/compact")

df.show()
```

运行上面的代码，会出现错误信息，内容如下：

```
...: java.io.FileNotFoundException: File does not exist: hdfs://cda:8020/
delta_data_lake/compact/part-00766-e857aad6-839d-4042-b3ed-7c81d7ef72bb
-c000.snappy.parquet
It is possible the underlying files have been updated. You can explicitly
invalidate the cache in Spark by running 'REFRESH TABLE tableName' command
in SQL or by recreating the Dataset/DataFrame involved.
...
```

最后，还有两个小问题需要理解。

1. 关于 dataChange=false

Delta Lake 事务协议能够将事务日志中的项标记为 dataChange=false，表示它们只重新安排已经是表的一部分数据。这非常强大，因为它允许执行压缩和其他读性能优化，而不会破坏使用 Delta 表作为流源的能力。用户应该在用于重写的 DataFrame 写入器中使用选项配置它。

Delta Lake 目前在数据压缩时设置 dataChange=true，这对下游流消费者来讲是一个重大改变。当文件被压缩时，用户可以选择设置 dataChange=false，这时 Delta Lake 将被更新，这样压缩对于下游流客户来讲不是一个中断操作。

2. 合并分区的 Delta Lake

假设数据存储在/some/path/data 文件夹中，并按 year 字段进行分区。进一步假设 2019 这个目录中包含 5 个文件，现在希望将其压缩为一个文件，代码如下：

```
val table = "/some/path/data"
val partition = "year = '2019'"
val numFiles = 1
spark.read
  .format("delta")
  .load(table)
  .where(partition)
  .repartition(numFiles)
  .write
  .format("delta")
  .mode("overwrite")
  .option("replaceWhere", partition)
  .save(table)
```

10.5.6　增量更新与时间旅行

Delta Lake 支持增量更新。Delta Lake 使用事务日志存储数据湖元数据，通过事务日志，用户可以在给定的时间点进行时间旅行和数据探索。下面通过一个示例来演示 Delta Lake 中的时间旅行特性。

首先，从下面这个 CSV 文件创建一个 Delta Lake。

people.csv 文件的内容如下：

```
first_name,last_name,country
miguel,cordoba,colombia
luisa,gomez,colombia
li,li,china
wang,wei,china
hans,meyer,germany
mia,schmidt,germany
```

接下来，使用 Spark SQL API 将这个 CSV 文件读到一个 DataFrame 中，然后将该 DataFrame 作为一个 Delta 数据湖写出，代码如下：

```scala
val path = "/data/data_lake/delta_lake/people.csv"
val outputPath = "/data_lake/delta_lake/person_delta_lake"

spark
  .read
  .option("header", "true")
  .option("charset", "UTF8")
  .csv(path)
  .repartition(1)
  .write
  .format("delta")
  .mode(SaveMode.Overwrite)
  .save(outputPath)
```

执行上面的代码后，再执行 HDFS Shell 命令查看，命令如下：

```
$ hdfs dfs -ls -R /data_lake/delta_lake/person_delta_lake
```

得到 /data_lake/delta_lake/person_delta_lake 目录内容，如图 10-11 所示。

```
/data_lake/delta_lake/person_delta_lake/_delta_log
/data_lake/delta_lake/person_delta_lake/_delta_log/00000000000000000000.json
/data_lake/delta_lake/person_delta_lake/part-00000-162114ff-17cd-4d46-bd9a-7071270b0bc7-c000.snappy.parquet
```

图 10-11　person_delta_lake 目录内容

数据以 Parquet 格式保存在文件中，元数据以 _delta_log/00000000000000000000.json 格式保存在文件中。JSON 文件包含关于写事务、数据模式及添加了什么文件的信息。打开这个 JSON 文件，其内容如下（为了便于阅读，这里对一行 JSON 进行了格式化展示）：

```json
{
  "commitInfo":{
    "timestamp":1565119301357,
    "operation":"WRITE",
    "operationParameters":{
      "mode":"Overwrite",
      "partitionBy":"[]"
    }
  }
}
```

```json
{
  "protocol":{
    "minReaderVersion":1,
    "minWriterVersion":2
  }
}
{
  "metaData":{
    "id":"a3ca108e-3ba1-49dc-99a0-c9d29c8f1aec",
    "format":{
       "provider":"parquet",
       "options":{}
    },
"schemaString":"{\"type\":\"struct\",\"fields\":[{\"name\":\"first_name\",\"type\":\"string\",\"nullable\":true,\"metadata\":{}},{\"name\":\"last_name\",\"type\":\"string\",\"nullable\":true,\"metadata\":{}},{\"name\":\"country\",\"type\":\"string\",\"nullable\":true,\"metadata\":{}}]}",
    "partitionColumns":[],
    "configuration":{},
    "createdTime":1565119298882
  }
}
{
  "add":{
    "path":"part-00000-78f9c583-ea60-4962-af99-895f453dce23-c000.snappy.parquet",
    "partitionValues":{},
    "size":939,
    "modificationTime":1565119299000,
    "dataChange":true
  }
}
```

1. 增量更新

下面使用一些纽约市出租车数据来构建并增量更新一个 Delta 数据湖。首先构建一个初始的 Delta 数据湖，代码如下：

```
import org.apache.spark.sql.SaveMode

val outputPath = "/data_lake/delta_lake/incremental_delta_lake/"
val p1 = "/data/data_lake/delta_lake/taxi_data/taxi1.csv"

//将文件读到DataFrame中，然后写到Delta Lake中
spark
  .read
  .option("header", "true")
  .option("charset", "UTF8")
  .csv(p1)
  .repartition(1)
```

```
  .write
  .format("delta")
  .mode(SaveMode.Overwrite)
  .save(outputPath)
```

这段代码会创建一个 Parquet 文件和一个 _delta_log/00000000000000000000.json 文件。查看文件系统结构,命令如下:

```
$ hdfs dfs -ls -R /data_lake/delta_lake/incremental_delta_lake/
```

可以看到这时的文件目录,如图 10-12 所示。

```
/data_lake/delta_lake/incremental_delta_lake/_delta_log
/data_lake/delta_lake/incremental_delta_lake/_delta_log/00000000000000000000.json
/data_lake/delta_lake/incremental_delta_lake/part-00000-fddd1008-bb1f-4a3c-8dc8-8dc910fc2b63-c000.snappy.parquet
```

图 10-12 incremental_delta_lake 目录内容

接下来检查 Delta 数据湖的内容,代码如下:

```
spark
  .read
  .format("delta")
  .load(outputPath)
  .select("passenger_count", "fare_amount")
  .show()
```

执行以上代码,输出的结果如下:

```
+---------------+-----------+
|passenger_count|fare_amount|
+---------------+-----------+
|              2|         83|
|              1|       14.5|
|              1|          8|
|              1|          6|
|              1|          6|
+---------------+-----------+
```

可以看到,在第 1 次加载后该 Delta Lake 包含 5 行数据。

接下来,将另一个文件加载到同一个 Delta Lake 中,使用 SaveMode.Append 模式,代码如下:

```
val p2 = "/data/data_lake/delta_lake/taxi_data/taxi2.csv"

//将数据读到 DataFrame 中,再写入(覆盖)Delta Lake
spark
  .read
  .option("header", "true")
  .option("charset", "UTF8")
  .csv(p2)
  .repartition(1)
  .write
  .format("delta")
```

```
  .mode(SaveMode.Append)
  .save(outputPath)
```

这段代码创建了一个 Parquet 文件和一个 _delta_log/ 00000000000000001.json 文件。文件系统目录 incremental_data_lake 现在包含的文件如图 10-13 所示。

```
/data_lake/delta_lake/incremental_delta_lake/_delta_log
/data_lake/delta_lake/incremental_delta_lake/_delta_log/00000000000000000000.json
/data_lake/delta_lake/incremental_delta_lake/_delta_log/00000000000000000001.json
/data_lake/delta_lake/incremental_delta_lake/part-00000-a1acfa24-7ec8-4b0c-84db-ece34eeed6c3-c000.snappy.parquet
/data_lake/delta_lake/incremental_delta_lake/part-00000-fddd1008-bb1f-4a3c-8dc8-8dc910fc2b63-c000.snappy.parquet
```

图 10-13　incremental_delta_lake 增量更新后的目录内容

再一次加载该 Delta Lake。在加载文件后，现在 Delta Lake 包含 10 行数据，代码如下：

```
spark
  .read
  .format("delta")
  .load(outputPath)
  .select("passenger_count", "fare_amount")
  .show()
```

执行以上代码，得到的输出内容如下：

```
+---------------+-----------+
|passenger_count|fare_amount|
+---------------+-----------+
|              2|         83|
|              1|       14.5|
|              1|          8|
|              1|          6|
|              1|          6|
|              1|          7|
|              1|          8|
|              1|        5.5|
|              1|         52|
|              6|        7.5|
+---------------+-----------+
```

2. 时间旅行

Delta 支持进行时间旅行。下面编写一个查询来检查第 1 次数据加载之后（第 2 次数据写入之前）的 Delta 数据湖，代码如下：

```
spark
  .read
  .format("delta")
  .option("versionAsOf", 0)          //指定要查询的版本
  .load(outputPath)
  .select("passenger_count", "fare_amount")
  .show()
```

执行以上代码，得到的输出内容如下：

```
+---------------+-----------+
|passenger_count|fare_amount|
+---------------+-----------+
|              2|         83|
|              1|       14.5|
|              1|          8|
|              1|          6|
|              1|          6|
+---------------+-----------+
```

在上面的代码中，使用了 option("versionAsOf",0)选项配置，用来告诉 Delta 只抓取_delta_log/00000000000000000000.json 文件中的文件，而忽略_delta_log/00000000000000000001.json 中的文件。

假设用户正在数据湖上训练一个机器学习模型，并希望在实验时保持数据不变。Delta Lake 的时间旅行特性可以让用户在训练模型时很容易使用单一版本的数据。

也可以轻松访问 Delta Lake 事务日志的完整历史，代码如下：

```
import io.delta.tables._

val lakePath = outputPath
val deltaTable = DeltaTable.forPath(spark, lakePath)
val fullHistoryDF = deltaTable.history()

fullHistoryDF.show(false)
```

执行以上代码，得到的输出内容如图 10-14 所示。

```
+-------+---------------------+------+--------+---------+-------------------+---+--------+---------+-----------+--------------+------------+-------------------+------------+
|version|            timestamp|userId|userName|operation|operationParameters|job|notebook|clusterId|readVersion|isolationLevel|isBlindAppend|   operationMetrics|userMetadata|
+-------+---------------------+------+--------+---------+-------------------+---+--------+---------+-----------+--------------+------------+-------------------+------------+
|      1|2022-03-03 14:58:...|  null|    null|    WRITE|{mode -> Append, ...|null|    null|     null|          0|          null|        true|{numFiles -> 1, n...|        null|
|      0|2022-03-03 14:52:...|  null|    null|    WRITE|{mode -> Overwrit...|null|    null|     null|       null|          null|       false|{numFiles -> 1, n...|        null|
+-------+---------------------+------+--------+---------+-------------------+---+--------+---------+-----------+--------------+------------+-------------------+------------+
```

图 10-14 访问 Delta Lake 事务日志的完整历史

为了更清楚地看明白版本的变化，只选取 version、timestamp 和 operationParameters 共 3 个字段，代码如下：

```
fullHistoryDF
    .select("version","timestamp","operationParameters")
    .show(false)
```

执行以上代码，得到的输出内容如下：

```
+-------+-----------------------+-------------------------------------+
|version|timestamp              |operationParameters                  |
+-------+-----------------------+-------------------------------------+
|1      |2022-03-03 14:58:24.588|{mode -> Append, partitionBy -> []}  |
|0      |2022-03-03 14:52:32.843|{mode -> Overwrite, partitionBy -> []}|
+-------+-----------------------+-------------------------------------+
```

查看 Delta 历史表的模式，代码如下：

```
fullHistoryDF.printSchema()
```

得到的输出内容如下：

```
root
 |-- version: long (nullable = true)
 |-- timestamp: timestamp (nullable = true)
 |-- userId: string (nullable = true)
 |-- userName: string (nullable = true)
 |-- operation: string (nullable = true)
 |-- operationParameters: map (nullable = true)
 |    |-- key: string
 |    |-- value: string (valueContainsNull = true)
 |-- job: struct (nullable = true)
 |    |-- jobId: string (nullable = true)
 |    |-- jobName: string (nullable = true)
 |    |-- runId: string (nullable = true)
 |    |-- jobOwnerId: string (nullable = true)
 |    |-- triggerType: string (nullable = true)
 |-- Notebook: struct (nullable = true)
 |    |-- NotebookId: string (nullable = true)
 |-- clusterId: string (nullable = true)
 |-- readVersion: long (nullable = true)
 |-- isolationLevel: string (nullable = true)
 |-- isBlindAppend: boolean (nullable = true)
```

因此，除了可以通过版本号获取 Delta 表的特定版本，还可以通过时间戳进行时间旅行，代码如下：

```
spark
  .read
  .format("delta")
  .option("timestampAsOf", "2022-03-03 14:53:00")  //应修改为自己的版本时间
  .load(lakePath)
  .select("passenger_count", "fare_amount")
  .show()
```

这与抓取的 Delta 表的第 1 个版本（version 等于 0）是一样的，内容如下：

```
+---------------+-----------+
|passenger_count|fare_amount|
+---------------+-----------+
|              2|         83|
|              1|       14.5|
|              1|          8|
|              1|          6|
|              1|          6|
+---------------+-----------+
```

10.5.7　合并更新（upsert）

本节将解释如何更新表列，以及如何使用 merge 命令执行 upsert 操作。

Delta Lake 存储使用 Parquet 文件，而 Parquet 文件是不可变的，不支持更新。Delta Lake 提供了 merge 语句，以此提供类似于更新的接口，但实际上，这些并不是真正的更新，不

会改变底层文件。Delta Lake 只是在重写整个 Parquet 文件。这将使大型数据集上的 upsert 或 update 列语句在执行时变得非常慢。

1. merge 示例

下面通过一个示例来理解和掌握使用 merge 命令的方法。

首先准备一个样本文件 events_data.csv，其内容如下：

```
eventType,websitePage
click,homepage
clck,about page
mouseOver,logo
```

读取该样本文件，并写到 Delta Lake 中，代码如下：

```
//数据源文件
val path = "/data/data_lake/delta_lake/event_data.csv"

//Delta 路径（HDFS）
val outputPath = "/data_lake/delta_lake/events/"

spark
  .read
  .option("header", "true")
  .option("charset", "UTF8")
  .csv(path)
  .repartition(1)
  .write
  .format("delta")
  .save(outputPath)
```

看一下 _delta_log/ 00000000000000000000.json 事务日志文件中存储了什么内容。执行的 HDFS Shell 命令如下：

```
$ hdfs dfs -cat
/data_lake/delta_lake/events/_delta_log/00000000000000000000.json
```

可以看到事务日志文件中有以下一行 JSON 文本形式的事务日志（这里为了便于阅读，进行了格式化展示）：

```
...
{
  "add":{
    "path":"part-00000-f960ca7c-eff0-40d0-b753-1f99ea4ffb9f-c000.snappy.parquet",
    "partitionValues":{},
    "size":673,
    "modificationTime":1569079218000,
    "dataChange":true
  }
}
```

下面查看数据湖的内容，代码如下：

```
//查看 Delta Lake 的内容
spark.read.format("delta").load(outputPath).show()
```

```
df.show()
```
执行以上代码，得到的输出结果如下：
```
+---------+-----------+
|eventType|websitePage|
+---------+-----------+
|    click|   homepage|
|     clck| about page|
|mouseOver|       logo|
+---------+-----------+
```

注意：第 2 行数据在 eventType 字段中有一个输入错误。它显示的是 clck 而不是 click。那么怎么来修改这个错误呢？修改错误的代码如下：

```
val deltaTable = DeltaTable.forPath(spark, outputPath)

deltaTable.updateExpr(
  "eventType = 'clck'",
  Map("eventType" -> "'click'")
)
```

或者使用另一种形式，代码如下：

```
deltaTable.update(
  col("eventType") === "clck",
  Map("eventType" -> lit("click"))
)
```

执行上面的代码，然后可以再次检查 Delta Lake 的内容，确认拼写错误是否已经修正，代码如下：

```
spark.read.format("delta").load(outputPath).show()
```

执行以上代码，得到的输出结果如下：

```
+---------+-----------+
|eventType|websitePage|
+---------+-----------+
|    click|   homepage|
|    click| about page|
|mouseOver|       logo|
+---------+-----------+
```

可以看到，错误的拼写已经得到修正。

查看此时文件系统中的所有文件，命令如下：

```
$ hdfs dfs -ls -R /data_lake/delta_lake/events/
```

输出内容目录如图 10-15 所示。

```
2022-03-01 19:35 /data_lake/delta_lake/events/_delta_log
2022-03-01 18:47 /data_lake/delta_lake/events/_delta_log/00000000000000000000.json
2022-03-01 19:35 /data_lake/delta_lake/events/_delta_log/00000000000000000001.json
2022-03-01 18:47 /data_lake/delta_lake/events/part-00000-267cec2a-a90f-432b-98fd-0c67eee4a666-c000.snappy.parquet
2022-03-01 19:35 /data_lake/delta_lake/events/part-00000-7493fe0e-1469-46f6-aac9-6ca37be2139e-c000.snappy.parquet
```

图 10-15　执行更新后的目录内容

如前所述，Parquet 文件是不可变的，那么 Delta 是怎样做到修复拼写错误的呢？查看一下 Delta 中的日志文件_delta_log/ 00000000000000000001.json，发现其内容如下：

```json
{
  "remove":{
    "path":"part-00000-267cec2a-a90f-432b-98fd-0c67eee4a666-c000.snappy.parquet",
    "deletionTimestamp":1569079467662,
    "dataChange":true
  }
}
{
  "add":{
    "path":"part-00000-7493fe0e-1469-46f6-aac9-6ca37be2139e-c000.snappy.parquet",
    "partitionValues":{},
    "size":694,
    "modificationTime":1569079467000,
    "dataChange":true
  }
}
```

可以看出，合并命令是将一个新文件写入了文件系统，对旧文件执行了 remove 操作。可以分别检查一下 Delta 中两个 Parquet 文件的内容。先查看日期最近的 Parquet 文件，代码如下：

```
val path = "/data_lake/delta_lake/events/part-00000-7493fe0e-1469-46f6-aac9-6ca37be2139e-c000.snappy.parquet"
spark.read.parquet(path).show()
```

执行上面的代码，输出内容如下：

```
+---------+-----------+
|eventType|websitePage|
+---------+-----------+
|    click|   homepage|
|    click| about page|
|mouseOver|       logo|
+---------+-----------+
```

可以看到，其输出内容是修改过后的数据。再看日期较早的 Parquet 文件，代码如下：

```
val path = "/data_lake/delta_lake/events/part-00000-267cec2a-a90f-432b-98fd-0c67eee4a666-c000.snappy.parquet"
spark.read.parquet(path).show()
```

执行上面的代码，输出内容如下：

```
+---------+-----------+
|eventType|websitePage|
+---------+-----------+
|    click|   homepage|
```

```
|      clck| about page|
|mouseOver|       logo|
+---------+-----------+
```

因此，合并命令将所有数据写入一个全新的文件中，但是，以这种方式写出所有数据将使 merge 运行要慢得多。

2. upsert 示例

首先准备一个原始数据集 original_data.csv，建立另一个小 Delta Lake。

original_data.csv 文件的内容如下：

```
date,eventId,data
2019-01-01,4,take nap
2019-02-05,8,play smash brothers
2019-04-24,9,speak at spark summit
```

将上面的数据构建为 Delta Lake，代码如下：

```
val path = "/data/data_lake/delta_lake/original_data.csv"

//delta lake path (hdfs)
val outputPath = "/data_lake/delta_lake/upsert_event/"

spark
  .read
  .option("header", "true")
  .option("charset", "UTF8")
  .csv(path)
  .repartition(1)
  .write
  .format("delta")
  .save(outputPath)
```

看一看这个 Delta Lake 的初始状态，代码如下：

```
spark.read.format("delta").load(outputPath).show(false)
```

执行上面这行代码，得到的输出内容如下：

```
+----------+-------+---------------------+
|date      |eventId|data                 |
+----------+-------+---------------------+
|2019-01-01|4      |take nap             |
|2019-02-05|8      |play smash brothers  |
|2019-04-24|9      |speak at spark summit|
+----------+-------+---------------------+
```

接下来用另一种更"友好"的表达来更新 Delta Lake。首先准备这个新的数据文件 friendly_data.csv。

friendly_data.csv 文件的内容如下：

```
date,eventId,data
2019-01-01,4,set goals
2019-02-05,8,bond with nephew
2019-08-10,66,think about my mommy
```

在上面的新数据集中,对第 4 个和第 8 个事件的描述进行了修改,而第 66 个事件是新增加的。下面执行 upsert 操作,代码如下:

```
val updatesPath = "/data/data_lake/delta_lake/friendly_data.csv"

//新数据(包含修改的数据和新增的数据)
val updatesDF = spark.read
   .option("header", "true")
   .option("charset", "UTF8")
   .csv(updatesPath)

import io.delta.tables._

DeltaTable.forPath(spark, outputPath) //outputPath 是原始的 Delta Lake path
  .as("events")
  .merge(updatesDF.as("updates"), "events.eventId = updates.eventId")
  .whenMatched.updateExpr(Map("data" -> "updates.data"))
  .whenNotMatched.insertExpr(Map(
      "date" -> "updates.date",
      "eventId" -> "updates.eventId",
      "data" -> "updates.data"
    )
  )
  .execute()
```

执行上面的代码,然后查看 upsert 之后 Delta Lake 的内容,代码如下:

```
spark.read.format("delta").load(outputPath).show(false)
```

执行上面这行代码,输出内容如下:

```
+----------+-------+----------------------+
|date      |eventId|data                  |
+----------+-------+----------------------+
|2019-08-10|66     |think about my mommy  |
|2019-04-24|9      |speak at spark summit |
|2019-02-05|8      |bond with nephew      |
|2019-01-01|4      |set goals             |
+----------+-------+----------------------+
```

从结果可以看出,将新数据合并到了旧数据中,其中第 4 个和第 8 个事件的 data 值被修改了,而第 66 个事件是新增加的事件,第 9 个事件没有变化。

3. upsert 的事务日志

在_delta_log /00000000000000000000.json 文件中包含单个项,用于添加单个 Parquet 文件。查看该文件,内容如下:

```
{
  "add":{
    "path":"part-00000-7eaa0d54-4dba-456a-ab80-b17f9aa7b583-c000.snappy.parquet",
    "partitionValues":{
```

```
      },
      "size":900,
      "modificationTime":1569177685000,
      "dataChange":true
    }
  }
```

由 _delta_log /00000000000000000001.json 文件显示，upsert 在事务日志中添加了大量记录。查看该文件，内容如下：

```
{
  "remove":{
    "path":"part-00000-7eaa0d54-4dba-456a-ab80-b17f9aa7b583-c000.snappy.parquet",
    "deletionTimestamp":1569177701037,
    "dataChange":true
  }
}
{
  "add":{
    "path":"part-00000-36aafda3-530d-4bd7-a29b-9c1716f18389-c000.snappy.parquet",
    "partitionValues":{
    },
    "size":433,
    "modificationTime":1569177698000,
    "dataChange":true
  }
}
{
  "add":{
    "path":"part-00026-fcb37eb4-165f-4402-beb3-82d3d56bfe0c-c000.snappy.parquet",
    "partitionValues":{
    },
    "size":968,
    "modificationTime":1569177700000,
    "dataChange":true
  }
}
{
  "add":{
    "path":"part-00139-eab3854f-4ed4-4856-8268-c89f0efe977c-c000.snappy.parquet",
    "partitionValues":{
    },
    "size":1013,
    "modificationTime":1569177700000,
    "dataChange":true
  }
}
```

```
{
  "add":{
    "path":"part-00166-0e9cddc8-9104-4c11-8b7f-44a6441a95fb-c000.snappy.parquet",
    "partitionValues":{
    },
    "size":905,
    "modificationTime":1569177700000,
    "dataChange":true
  }
}
{
  "add":{
    "path":"part-00178-147c78fa-dad2-4a1c-a4c5-65a1a647a41e-c000.snappy.parquet",
    "partitionValues":{
    },
    "size":1013,
    "modificationTime":1569177701000,
    "dataChange":true
  }
}
```

下面创建一个小辅助方法,以便轻松地检查这些 Parquet 文件的内容,代码如下:

```
def displayEventParquetFile(filename: String): Unit = {
  val path = "/data_lake/delta_lake/upsert_event/$filename.snappy.parquet"
  val df = spark.read.parquet(path)
  df.show(false)
}
```

然后依次查看各个 Parquet 文件的内容。

查看第 1 个 Parquet 文件的内容,代码如下:

```
displayEventParquetFile("part-00000-36aafda3-530d-4bd7-a29b-9c1716f18389-c000")
```

得到的内容如下:

```
+----+-------+----+
|date|eventId|data|
+----+-------+----+
+----+-------+----+
```

查看第 2 个 Parquet 文件的内容,代码如下:

```
displayEventParquetFile("part-00026-fcb37eb4-165f-4402-beb3-82d3d56bfe0c-c000")
```

得到的内容如下:

```
+----------+-------+----------------+
|date      |eventId|data            |
+----------+-------+----------------+
|2019-02-05|8      |bond with nephew|
+----------+-------+----------------+
```

查看第 3 个 Parquet 文件的内容，代码如下：

```
displayEventParquetFile("part-00139-eab3854f-4ed4-4856-8268-c89f0efe977c-c000")
```

得到的内容如下：

```
+----------+-------+-------------------+
|date      |eventId|data               |
+----------+-------+-------------------+
|2019-04-24|9      |speak at spark summit|
+----------+-------+-------------------+
```

查看第 4 个 Parquet 文件的内容，代码如下：

```
displayEventParquetFile("part-00166-0e9cddc8-9104-4c11-8b7f-44a6441a95fb-c000")
```

得到的内容如下：

```
+----------+-------+---------+
|date      |eventId|data     |
+----------+-------+---------+
|2019-01-01|4      |set goals|
+----------+-------+---------+
```

查看第 5 个 Parquet 文件的内容，代码如下：

```
displayEventParquetFile("part-00178-147c78fa-dad2-4a1c-a4c5-65a1a647a41e-c000")
```

得到的内容如下：

```
+----------+-------+-------------------+
|date      |eventId|data               |
+----------+-------+-------------------+
|2019-08-10|66     |think about my mommy|
+----------+-------+-------------------+
```

这个更新代码创建了数量惊人的 Parquet 文件。

第 11 章 Iceberg 数据湖
CHAPTER 11

Apache Iceberg 是一种相对较新的开放表格式，专为巨大的 PB 级表设计。表格式的功能是决定如何管理、组织和跟踪所有组成表格的文件。可以把它看作物理数据文件（用 Parquet 或 ORC 等写的）和它们如何被构造成表之间的抽象层。

使用保存在每个表上的大量元数据，Iceberg 提供了其他表格式通常无法提供的功能。这包括模式演化、分区演化和表版本回滚，所有这些都可以在不需要代价高昂的表重写或表迁移的情况下实现。

11.1 Apache Iceberg 简介

Apache Iceberg 最初是由 Netflix 公司开发的，其目的是解决他们在使用巨大的 PB 级表时长期存在的问题。它于 2018 年作为 Apache 孵化器项目开源，并于 2020 年 5 月 19 日从孵化器中毕业。

Apache Iceberg 设计的一个关键考虑是解决各种数据一致性和性能问题，这些问题是 Hive 在使用大数据时所面临的问题。Hive 在"文件夹"级（不是文件级）跟踪数据，因此在处理表中的数据时需要执行文件列表操作。当需要执行其中的许多操作时，这可能会导致性能问题。当在最终一致的对象存储（如 S3）上执行文件列表操作时，还可能出现数据丢失的情况。Apache Iceberg 通过持久树结构跟踪表中所有文件的完整列表来避免这种情况。对表的更改使用原子对象/文件级提交将路径更新到一个新的元数据文件，该元数据文件包含所有单独数据文件的位置。

Apache Iceberg 具有以下特色功能。

（1）模式进化（Schema Evolution）：支持添加、删除、更新或重命名，并且没有副作用。

（2）隐藏分区（Hidden Partitioning）：防止因用户失误导致的无提示错误结果或极慢的查询。

（3）分区布局演变（Partition Layout Evolution）：可以随着数据量或查询模式的变化而更新表的布局。

（4）时间旅行（Time Travel）：支持使用完全相同的表快照进行重复查询，或让用户轻松地检查并更改信息。

（5）版本回退（Version Rollback）：允许用户通过将表重置为之前某一时刻的状态来快速纠正问题。

关于可靠性与性能方面，Apache Iceberg 适用于查询大型表格，其中单个表可以包含数十 PB 的数据，甚至可以在没有分布式 SQL 引擎的情况下读取它们，并且旨在解决最终一致的云对象存储中的正确性问题。

（1）快速的扫描计划：不需要分布式 SQL 引擎来读取表或查找文件。

（2）高级过滤：使用表元数据、分区和列级统计信息修剪数据文件。

（3）避免监听和重命名，适用于任何云存储并在 HDFS 中减少 NameNode 拥塞。

（4）可序列化隔离：表更改是原子的，读取时不会看到部分或未提交的更改。

（5）利用乐观锁控制并发写入，并会重试以确保兼容更新成功。

有关存储格式方面，Apache Iceberg 中的一些概念列举如下：

（1）分区规范（Partition Spec）：定义了如何从数据文件推断分区信息。

（2）快照（Snapshot）：表在某个时刻的状态，由清单列表定义。每个快照都列出了在快照时构成表内容的所有数据文件。

（3）清单列表（Manifest List）：列出组成表快照的清单的元数据文件，这是 Avro 格式文件。清单列表中的每个清单文件都存储了关于其内容的信息，例如分区值范围，以加快元数据操作。

（4）清单文件（Manifest）：列出组成快照的数据文件子集的元数据文件，这也是 Avro 格式文件。清单中的每个数据文件都存储着分区元组、列级统计信息和摘要信息，这些信息用于在扫描规划期间修剪分裂。

（5）数据文件（Data File）：真实存储数据的文件。

在 Apache Spark 3 中可以以多种方式使用 Iceberg。下面列出了其中常用的几种方法。

（1）方法一，在启动 spark-shell 时指定，命令如下：

```
$ spark-shell --packages org.apache.iceberg:iceberg-spark-runtime-3.1_
2.12:0.13.1
```

（2）方法二，将 iceberg-spark-runtime-3.1_2.12:0.13.1.jar 包添加到 Spark 的 jars 目录下。

（3）方法三，如果使用 Maven 开发，则在项目的 pom.xml 文件中添加依赖，内容如下：

```xml
<dependency>
    <groupId>org.apache.iceberg</groupId>
    <artifactId>iceberg-core</artifactId>
    <version>0.13.1</version>
</dependency>
```

（4）方法四，如果使用 SBT 开发，则在项目的 build.sbt 文件中添加依赖，内容如下：

```
libraryDependencies += "org.apache.iceberg" % "iceberg-core" % "0.13.1"
```

11.2 配置和使用 Catalog

要在 Apache Spark 中使用 Iceberg，首先需要配置 Catalog（目录）。那么，Catalog 是什么？

Catalog 是 Iceberg 对表进行管理（create、drop、rename 等）的一个组件。通过 Iceberg 附带的 Catalog 目录，可以使用 SQL 命令管理表并按名称加载表。通过设置 spark.sql.catalog 下的 Spark 属性来配置 Spark Catalogs。Catalog 名称在 SQL 查询中用于标识一张表。

11.2.1 配置 Catalog

目前 Iceberg 支持两种 Catalog，分别是 HiveCatalog 和 HadoopCatalog，其中 HiveCatalog 将当前表的元数据文件路径存储在 Hive Metastore，因为这个表元数据文件是所有读写 Iceberg 表的入口，所以每次读写 Iceberg 表都需要先从 Hive Metastore 中取出对应的表元数据文件路径，然后解析这个元数据文件以进行接下来的操作，而 HadoopCatalog 是将当前表元数据文件路径记录在一个文件目录下，因此不需要连接 Hive Metastore。

Catalog（目录）的创建和命名是通过添加 Spark 属性 spark.sql.catalog.(catalog-name) 实现的，其值为实现类。目前 Iceberg 支持以下两种实现。

（1）org.apache.iceberg.spark.SparkCatalog：支持 Hive Metastore 或 Hadoop 目录作为 Catalog。

（2）org.apache.iceberg.spark.SparkSessionCatalog：将对 Iceberg 表的支持添加到 Spark 的内置 Catalog 中，并将非 Iceberg 表委托给内置 Catalog。

例如，创建一个名为 hive_prod 的 Iceberg Catalog 目录，并指定表元数据文件存储在 Hive Metastore 中（采用 HiveCatalog 类型的 Catalog），这是通过 type=hive 配置的，代码如下：

```
spark.sql.catalog.hive_prod = org.apache.iceberg.spark.SparkCatalog
spark.sql.catalog.hive_prod.type = hive
spark.sql.catalog.hive_prod.uri = thrift://metastore-host:port
```

注意：基于 Hive 的 Catalog 只加载 Iceberg 表。要加载同一 Hive Metastore 中的非 Iceberg 表，将实现切换为 SparkSessionCatalog。

而下面的配置将创建一个名为 hadoop_prod 的 Iceberg Catalog 目录，并指定表元数据文件存储在 HDFS 的指定目录中（采用 HadoopCatalog 类型的 Catalog），这是通过 type=hadoop 配置的：

```
spark.sql.catalog.hadoop_prod = org.apache.iceberg.spark.SparkCatalog
spark.sql.catalog.hadoop_prod.type = hadoop
spark.sql.catalog.hadoop_prod.warehouse = hdfs://localhost:8020/warehouse/path
```

需要注意的是，HiveCatalog 和 HadoopCatalog 不能混合使用。也就是说，用 HiveCatalog 创建的表不能用 HadoopCatalog 正确加载，反之亦然。

那么，在哪里配置 Spark Catalog？

如果使用的是 Spark Shell，则可以在执行 spark-shell 命令时进行配置，命令如下：

```
$ spark-shell --packages org.apache.iceberg:iceberg-spark-runtime-3.1_2.12:0.13.1 \
    --conf spark.sql.extensions=org.apache.iceberg.spark.extensions.IcebergSparkSessionExtensions \
```

```
    --conf spark.sql.catalog.hadoop_prod = org.apache.iceberg.spark.SparkCatalog \
    --conf spark.sql.catalog.hadoop_prod.type=hadoop \
    --conf spark.sql.catalog.hadoop_prod.warehouse=hdfs://localhost:8020/
data_lake/iceberg
```

上面这个命令创建了一个名为 hadoop_prod 的 Catalog，它是基于 Hadoop 路径的目录。在该 Catalog 下创建的所有数据库和表都将位于该目录下。

如果是在 Zeppelin Notebook 中，则可以使用%spark.conf 进行配置。例如，配置一个名为 hive_prod 的 Catalog，它是基于 Hive Metastore 的类型。配置命令如下：

```
%spark.conf
spark.jars.packages                      iceberg-spark-runtime-
3.1_2.12:0.13.1
spark.sql.extensions
    org.apache.iceberg.spark.extensions.IcebergSparkSessionExtensions
spark.sql.catalog.hive_prod
    org.apache.iceberg.spark.SparkCatalog
spark.sql.catalog.hive_prod.type         hive
spark.sql.catalog.hive_prod.uri          thrift://localhost:9083
```

11.2.2　使用 Catalog

在 SQL 查询中使用 Catalog 名称来标识表。例如，在 11.2.1 节配置的 hadoop_prod 目录中有一个名为 my_db 的数据库，在 my_db 数据库中有一个名为 my_tb 的表。那么要查询该表中的数据，可以使用 hadoop_prod 作为前缀来标识，代码如下：

```
spark.sql("SELECT * FROM hadoop_prod.my_db.my_tb")
```

Spark 3 会跟踪当前的 Catalog 和数据库，所以在当前 Catalog 和数据库中引用表时，可以省略 Catalog 和数据库名，代码如下：

```
spark.sql("USE hadoop_prod.my_db")
spark.sql("SELECT * FROM my_tb")
```

要查看当前的 Catalog 和数据库，代码如下：

```
spark.sql("SHOW CURRENT NAMESPACE")
```

11.2.3　替换 Session Catalog

要向 Apache Spark 的内置目录添加 Iceberg 表支持，以及配置 Catalog 使用 Iceberg 的 SparkSessionCatalog，配置内容如下：

```
spark.sql.catalog.hive_prod = org.apache.iceberg.spark.SparkSessionCatalog
spark.sql.catalog.hive_prod.type = hive
```

Apache Spark 内置的 Catalog 支持在 Hive Metastore 中跟踪现有的 v1 和 v2 表。这将配置 Spark 使用 Iceberg 的 SparkSessionCatalog 作为该 Session Catalog 的包装器。当表不是 Iceberg 表时，将使用内置 Catalog 来加载它。

这个配置可以对 Iceberg 表和非 Iceberg 表使用相同的 Hive Metastore。

11.2.4 运行时配置

可以在运行时对 Spark 3 读写 Iceberg 进行动态配置。

1. 读选项配置

Spark 读选项是在配置 DataFrameReader 时被传递的，代码如下：

```
//时间旅行
spark.read
    .option("snapshot-id", 10963874102873L)
    .table("hadoop_catalog.my_db.my_table")
```

可用的读配置选项见表 11-1。

表 11-1　可用的读配置选项

Spark option 选项	默认值	描述
snapshot-id	（latest）	要读取的表快照的 ID
as-of-timestamp	（latest）	以毫秒为单位的时间戳；所使用的快照将是此时的当前快照
split-size		重写这个表的 read.split.target-size 和 read.split.metadata-target-size
lookback		重写这个表的 read.split.planning-lookback
file-open-cost		重写这个表的 read.split.open-file-cost
vectorization-enabled		重写这个表的 read.parquet.vectorization.enabled
batch-size		重写这个表的 read.parquet.vectorization.batch-size

2. 写选项配置

Spark 写选项在配置 DataFrameWriter 时被传递，代码如下：

```
//用 Avro 而不是 Parquet 写
df.write
    .option("write-format", "avro")
    .option("snapshot-property.key", "value")
    .insertInto("hadoop_catalog.my_db.my_table")
```

可用的写配置选项见表 11-2。

表 11-2　可用的写配置选项

Spark option 选项	默认值	描述
write-format	write.format.default	写操作使用的文件格式：Parquet、Avro 或 ORC
target-file-size-Bytes	As per table property	覆盖这个表的 write.target-file-size-Bytes
check-nullability	true	对字段设置 nullable 检查
snapshot-property.custom-key	null	在快照摘要中添加具有自定义 key 和相应 value 的条目
fanout-enabled	false	覆盖这个表的 write.spark.fanout.enabled
check-ordering	true	检查输入模式和表模式是否相同

11.3 管理 Catalog 中的数据库

每个 Catalog 中都有一个叫 default 的默认数据库。在 Iceberg 中,数据库与命名空间(namespace)等价。也可以明确地创建一个数据库(或命名空间)并命名。

如果要查看当前的 Catalog 和命名空间(数据库),则可执行的代码如下:

```
spark.sql("SHOW CURRENT NAMESPACE").show()
```

查看当前有哪些命名空间,代码如下:

```
spark.sql("SHOW NAMESPACES").show()
//或者
spark.sql("SHOW DATABASES").show()
```

可以使用下面的语句创建新的命名空间,代码如下:

```
spark.sql("CREATE NAMESPACES TEST_DB")
//spark.sql("CREATE DATABASE TEST_DB")      //等价
```

也可以查看指定 Catalog 中包含哪些数据库/命名空间。例如,查看名为 hadoop_catalog 的目录下有哪些数据库/命名空间,代码如下:

```
spark.sql("show namespaces in hadoop_prod").show()
spark.sql("show databases in hadoop_prod").show()
```

如果要查看指定命名空间中有哪些表,则可执行的代码如下:

```
spark.sql("show tables in hadoop_prod.default").show()
```

如果要删除某个命名空间,则可用 drop database 或 drop namespace。不过需要注意的是,只可以删除空的命名空间(不存在表的命名空间),代码如下:

```
spark.sql("drop database hadoop_prod.jd_db")
//spark.sql("drop namespace hadoop_prod.jd_db")
```

11.4 管理 Iceberg 表

Apache Iceberg 作为分析数据集的开放表格式,提供了很好的用户易用性。Iceberg 支持以 SQL 或 DataFrame API 的方式来读写数据湖表。

11.4.1 基本的 CRUD 操作

要在 Apache Spark 中创建 Iceberg 表,可使用 spark-shell 或 spark.sql(…)来运行 CREATE TABLE 命令。Apache Spark 3.0 可以使用 USING Iceberg 子句在任何 Iceberg Catalog 中创建表,Iceberg 将把 Spark 中的列类型转换为相应的 Iceberg 类型。

要创建 Iceberg 表,有以下两种 Catalog 目录选项。

(1) HadoopCatalog:支持存储在 HDFS 或本地文件系统中的表。

(2) HiveCatalog:使用 Hive Metastore 通过存储对最新元数据文件的引用来跟踪 Iceberg 表。

在 11.2.1 节，已经创建了 HadoopCatalog 类型的 hadoop_catalog 和 HiveCatalog 类型的 hive_prod。接下来，在 hadoop_prod 目录中创建数据库和表，代码如下：

```
//hadoop_prod 是前面定义的基于路径的 Catalog
spark.sql("""
   CREATE TABLE hadoop_prod.my_db.my_tb1 (
     id bigint,
     data string
   )
   USING iceberg
""")
```

查看表是否创建成功，代码如下：

```
spark.sql("SHOW TABLES IN hadoop_prod.my_db").show()
```

一旦表被创建，就可以使用 INSERT INTO 来插入数据了，代码如下：

```
spark.sql("""
   INSERT INTO hadoop_prod.my_db.my_tb1
   VALUES
   (1, 'a'),
   (2, 'b'),
   (3, 'c')
""")
```

查询 Iceberg 表中插入的数据，代码如下：

```
spark.sql("select * from hadoop_prod.my_db.my_tb1").show()
```

执行以上代码，可以看到查询结果如下：

```
+---+----+
| id|data|
+---+----+
|  1|   a|
|  2|   b|
|  3|   c|
+---+----+
```

Apache Spark 3.1 增加了对 UPDATE 查询的支持，该查询可以更新表中匹配的行。UPDATE 查询接受过滤器来匹配要更新的行。例如，将 my_tb1 表中 id 为 1 这条记录的 data 字段值修改为 b，代码如下：

```
spark.sql("""
   update hadoop_prod.my_db.my_tb1
   set data="b"
   where id=1
""")
```

然后查询更新后的数据，代码如下：

```
spark.sql("select * from hadoop_prod.my_db.my_tb1").show()
```

执行以上代码，查询结果如下：

```
+---+----+
| id|data|
+---+----+
|  1|   b|
|  2|   b|
|  3|   c|
+---+----+
```

也可以在 Iceberg 表上执行聚合计算。例如，按 data 字段分组计数，代码如下：

```
spark.sql("""
    select data, count(1) as cnt
    from hadoop_prod.my_db.my_tb1
    group by data
    order by cnt desc
""").show()
```

执行以上代码，输出的结果如下：

```
+----+---+
|data|cnt|
+----+---+
|   b|  2|
|   c|  1|
+----+---+
```

Iceberg 支持使用新的 v2 DataFrameAPI 来读写 DataFrame。在下面的代码中，有一个新的数据集 newData，将其内容增量写入上面的 hadoop_prod.my_db.my_tb1 表中，然后一次查询表数据，代码如下：

```
//有一个新的 DataFrame
val newData = Seq((4,"d"), (5,"e")).toDF("id", "data")
newData.show()

//将其数据追加到 hadoop_prod.my_db.my_tb1 表中
newData.writeTo("hadoop_prod.my_db.my_tb1").append()

//查询此时的表数据
spark.table("hadoop_prod.my_db.my_tb1").show()
```

执行以上代码，输出的查询结果如下：

```
+---+----+
| id|data|
+---+----+
|  1|   b|
|  2|   b|
|  3|   c|
|  4|   d|
|  5|   e|
+---+----+
```

Spark 3 增加了对 DELETE FROM 查询的支持，可以从表中删除数据。DELETE 查询接受一个过滤器来匹配要删除的行。例如，从 my_tb1 表中删除 id 为 1 的数据，代码如下：

```
spark.sql("DELETE FROM hadoop_prod.my_db.my_tb1 WHERE id=1")

//查询此时的表数据
spark.table("hadoop_prod.my_db.my_tb1").show()
```

执行以上代码，输出的查询结果如下：

```
+---+----+
| id|data|
+---+----+
|  2|   b|
|  3|   c|
|  4|   d|
|  5|   e|
+---+----+
```

1. 多 Catalog 支持

从上面的代码中可以看到，当查询 Iceberg 表时，既可以使用 spark.sql()，也可以使用 spark.table()。实际上，在 Spark 3 中，Iceberg 0.13.1 都为 DataFrameReader 增加了对多 Catalog 的支持。

可以通过 Spark 的 DataFrameReader 接口加载路径 path 和表名。表的加载方式取决于标识符的指定方式。当使用 spark.read.format("iceberg").path(table) 或 spark.table(table) 时，table 变量可以有以下几种形式：

（1）file:/path/to/table：加载一个给定路径的 HadoopTable。

（2）tablename：加载 currentCatalog.currentNamespace.tablename。

（3）catalog.tablename：从指定的 Catalog 加载 tablename。

（4）namespace.tablename：从当前 Catalog 加载 namespace.tablename。

（5）catalog.namespace.tablename：从指定的 Catalog 加载 namespace.tablename。

（6）namespace1.namespace2.tablename：从当前 Catalog 加载 namespace1.namespace2.tablename。

上面的列表是按先后顺序排列的。例如匹配的 Catalog 将优先于任何命名空间解析。

2. 类型适配

Spark 和 Iceberg 支持不同的类型集。Iceberg 会自动进行类型转换，但不是针对所有组合，因此在设计表中的列类型之前，可能希望了解 Iceberg 中的类型转换。

1）从 Spark 类型到 Iceberg 类型

此类型转换表描述了如何将 Spark 类型转换为 Iceberg 类型。这个转换既适用于创建 Iceberg 表，也适用于通过 Spark 写入 Iceberg 表，见表 11-3。

2）Iceberg 类型到 Spark 类型

此类型转换表描述了如何将 Iceberg 类型转换为 Spark 类型。转换应用于通过 Spark 从 Iceberg 表读取数据，见表 11-4。

表 11-3 从 Spark 类型到 Iceberg 类型转换

Spark 类型	Iceberg 类型
boolean	boolean
short	integer
Byte	integer
integer	integer
long	long
float	float
double	double
date	date
timestamp	timestamp with timezone
char	string
varchar	string
string	string
binary	binary
decimal	decimal
struct	struct
array	list
map	map

表 11-4 从 Iceberg 类型到 Spark 类型转换

Iceberg 类型	Spark 类型
boolean	boolean
integer	integer
long	long
float	float
double	double
date	date
time	不支持
timestamp with timezone	timestamp
timestamp without timezone	不支持
string	string
uuid	string
fixed	binary
binary	binary
decimal	decimal
struct	struct
list	array
map	map

11.4.2 创建和删除表

11.4.1 节使用 CREATE TABLE 语句创建了 Iceberg 表。除此之外，Iceberg 也支持以 CTAS 和 RTAS 方法来创建表。下面分别来了解这两种方法。

1. CTAS（Create Table … As Select）

在使用 SparkCatalog 时，Iceberg 支持 CTAS 作为原子操作。在使用 SparkSessionCatalog 时 CTAS 虽然是受支持的，但不是原子的。

下面应用 CTAS 创建一个新的 Iceberg 表 my_tb2，其内容来自 my_tb1 中所有 id 为偶数的行，代码如下：

```
//CTAS
spark.sql("""
    CREATE TABLE hadoop_prod.my_db.my_tb2
    USING iceberg
    AS select *
        from hadoop_prod.my_db.my_tb1
        where id%2==0
""")

//查询新表的数据
spark.table("hadoop_prod.my_db.my_tb2").show()
```

执行上面的代码，查询结果如下：

```
+---+----+
| id|data|
+---+----+
|  2|   b|
|  4|   d|
+---+----+
```

2. RTAS（Replace Table … As Select）

当使用 SparkCatalog 时，Iceberg 支持 RTAS 作为原子操作。在使用 SparkSessionCatalog 时 RTAS 虽然是受支持的，但不是原子的。原子表替换使用 SELECT 查询的结果创建一个新的快照，但保留表历史。

下面应用 RTAS 使用新的内容来替换 my_tb2 表中原有的内容，代码如下：

```
//RTAS
spark.sql("""
    REPLACE TABLE hadoop_prod.my_db.my_tb2
    USING iceberg
    AS select *
        from hadoop_prod.my_db.my_tb1
        where id%2==1
""")

//查询结果
spark.table("hadoop_prod.my_db.my_tb2").show()
```

执行上面的代码，查询结果如下：

```
+---+----+
| id|data|
+---+----+
|  5|   e|
|  3|   c|
+---+----+
```

下面的代码应用 RTAS 来创建（如果目标表不存在）或替换（如果目标表存在）my_tb3 表中原有的内容，其内容来自 my_tb1 中所有 id 为奇数的行，代码如下：

```
//create or replace table ... as select
spark.sql("""
    CREATE OR REPLACE TABLE hadoop_prod.my_db.my_tb3
    USING iceberg
    AS select *
        from hadoop_prod.my_db.my_tb1
        where id%2==1
""")

//查询结果
spark.table("hadoop_prod.my_db.my_tb3").show()
```

执行上面的代码，查询结果如下：

```
+---+----+
| id|data|
+---+----+
|  3|   c|
|  5|   e|
+---+----+
```

需要注意，使用 RTAS，模式和分区规范将被替换。为了避免修改表的模式和分区，应使用 INSERT OVERWRITE 而不要使用 REPLACE TABLE。REPLACE TABLE 命令中的新表属性将与任何现有的表属性合并。如果更改了现有的表属性，则会更新它们，否则它们就会被保存下来。

表创建命令，包括 CTAS 和 RTAS，支持所有 Spark create 子句。

（1）PARTITION BY (partition-expressions)：配置分区。

（2）LOCATION '(fully-qualified-uri)'：设置表位置。

（3）COMMENT 'table documentation'：设置一张表的描述信息。

（4）TBLPROPERTIES ('key'='value', …)：设置表配置信息。

表创建命令也可以使用 USING 子句设置默认格式。这只支持 SparkCatalog，因为 Spark 对内置 Catalog 的 USING 子句处理方式不同。

3. 删除 Iceberg 表

如果想要删除一张 Iceberg 表，则可以使用 drop table 命令，代码如下：

```
//删除表my_tb3
spark.sql("drop table hadoop_prod.my_db.my_tb3")

//查看my_db中有哪些表
spark.sql("show tables in hadoop_prod.my_db").show()
```

执行上面的代码，查询结果如下：

```
+---------+---------+
|namespace|tableName|
+---------+---------+
|    my_db|   my_tb1|
|    my_db|   my_tb2|
+---------+---------+
```

11.4.3 使用分区表和分桶表

分区是一种优化技术，用于根据某些属性将表划分为若干部分。

通常，分区只是将特定类型或值的项分组以便更快地访问的一种方法。这样做的好处是，对于只访问部分数据的查询，读取和加载时间会更快。例如，跟踪日志事件、消息和事件时间的日志表可能有数百万个条目，跨度长达数月。将这些条目按天进行分区，可以更快地查询某一天发生的日志事件。

Iceberg 通过实现隐藏分区为用户简化了分区。Iceberg 没有强制用户在查询时提供单独的分区过滤器，而是在底层处理分区和查询的所有细节。用户不需要维护分区列，甚至不需要理解物理表布局就可以得到准确的查询结果。

Iceberg 有几个分区选项。用户可以按年、月、日和小时划分时间戳。Iceberg 可以跟踪列值与其分区之间的关系，而不需要额外的列。例如，当在按日(day)分区、带有"YYYY-MM-DD hh:mm:ss"时间戳格式的数据上查询时，查询语句中不需要包含"hh:mm:ss"部分。Iceberg 还可以通过标识、散列桶或截断来划分分类列值。

下面通过一个示例来了解如何使用 Iceberg 分区表，以及 Iceberg 分区表的一些特性。本示例需要一个数据集文件 jd-formated.parquet，该文件包含 2017 年 02 月 16 日到 2022 年 02 月 15 日共 5 年的京东股票交易数据。首先将该数据集加载到 DataFrame 中，代码如下：

```
//将数据集加载到 DataFrame
val file = "/data/spark/jd/jd-formated.parquet"
val jdDF = spark.read.load(file)

//查看 Schema 和部分数据
jdDF.printSchema()
jdDF.show()
```

执行以上代码，输出内容如下：

```
root
 |-- Date: date (nullable = true)
 |-- Close: double (nullable = true)
```

```
 |-- Volume: long (nullable = true)
 |-- Open: double (nullable = true)
 |-- High: double (nullable = true)
 |-- Low: double (nullable = true)

+----------+-----+-------+------+-----+-----+
|      Date|Close| Volume|  Open| High|  Low|
+----------+-----+-------+------+-----+-----+
|2022-02-15|76.13|6766205| 75.35|76.35| 74.8|
|2022-02-14|74.45|5244967| 73.94|74.62|73.01|
|2022-02-11|73.98|6673354| 75.97|76.55|73.55|
|2022-02-10| 76.4|6432184|75.955|78.39|75.24|
|2022-02-09|78.29|7061571| 76.83|78.67|76.61|
+----------+-----+-------+------+-----+-----+
only showing top 5 rows
```

接下来，将 jdDF 写入 Iceberg 数据湖中。Spark 3 引入了新的 DataFrameWriterV2 API，用于使用 DataFrame 写入表。推荐使用 v2 API 有以下几个原因：

（1）支持 CTAS、RTAS 和过滤器覆盖。

（2）所有操作都一致地按名称将列写入表。

（3）partitionedBy 支持隐藏分区表达式。

（4）覆盖行为是显式的，可以是动态的，也可以是用户提供的过滤器。

（5）每个操作的行为都对应于 SQL 语句。

① df.writeTo(t).create() 等价于 CREATE TABLE AS SELECT。

② df.writeTo(t).replace() 等价于 REPLACE TABLE AS SELECT。

③ df.writeTo(t).append() 等价于 INSERT INTO。

④ df.writeTo(t).overwritePartitions() 等价于 dynamic INSERT OVERWRITE。

（6）创建表。

要运行一个 CTAS 或 RTAS，可以使用 create、replace 或 createOrReplace 操作，代码如下：

```
val data: DataFrame = ...
data.writeTo("prod.db.table").create()
```

另外，create、replace 或 createOrReplace 操作还支持表配置方法，如 partitionedBy() 和 tableProperty()，代码如下：

```
data.writeTo("prod.db.table")
    .tableProperty("write.format.default", "orc")
    .partitionBy($"level", days($"ts"))
    .createOrReplace()
```

（7）覆盖数据。

要动态地覆盖分区，可以使用 overwritePartitions()，代码如下：

```
val data: DataFrame = ...
data.writeTo("prod.db.table").overwritePartitions()
```

要显式地覆盖分区，使用 overwrite 来提供一个过滤器，代码如下：

```
data.writeTo("prod.db.table").overwrite($"level" === "INFO")
```

注意：虽然 Spark 3 仍然支持 v1 DataFrame write API，但不推荐使用。

在 Spark 3 中使用 v1 DataFrame API 写数据时，可以使用 saveAsTable 或 insertInto 来加载带有 Catalog 目录的表。可以使用 format("iceberg")加载一个独立的表引用，该引用不会自动刷新查询使用的表。

继续使用前面几节定义的 hadoop_prod catalog 和 my_db 数据库，代码如下：

```
//将数据写到分区表
jdDF
    .sortWithinPartitions("Date")                    //要先排序
    .writeTo("hadoop_prod.jd_db.stock_part")         //写入 Iceberg 表
    .partitionedBy(months($"Date"))                  //按天分区存储，这里使用了转换函数 months
    .create()
```

在对分区表进行写操作之前，Iceberg 要求根据每个任务（Spark 分区）的分区规范对数据进行排序。这既适用于 SQL 写入，也适用于 DataFrame 写入。如果使用 DataFrame 插入数据，则可以使用 orderBy()/sort()来触发全局排序，或使用 sortWithinPartitions()来触发局部排序。全局排序（orderBy()/sort()）和本地排序（sortWithinPartitions()）都可以满足需求。

在上面的代码中，对 Date 字段应用了日期转换函数 months()，以指定按 Date 字段中的月份来分区存储。Iceberg 支持的转换函数如下。

（1）years(ts)：按年分区。

（2）months(ts)：按月分区。

（3）days(ts)或 date(ts)：按天分区。

（4）hours(ts)或 date_hour(ts)：按天和小时分区。

（5）bucket(N, col)：按哈希值对 N 个分桶取模来分区。

（6）truncate(L, col)：以截断为 L 的值进行分区。字符串会被截断到给定的长度。Integer 和 Long 会被截断到 bins：truncate(10,i)，以便产生分区 0、10、20、30、…。

执行上面的代码之后，可以查看相应的物理存储目录。这既可以通过 HDFS Shell 命令来查看，也可以通过 HDFS 的 Web UI 来查看。通过 Web UI 查看部分分区存储目录，它们位于表 stock_part 的 data 目录下，如图 11-1 所示。

下面执行条件查询，查询出 2022 年 2 月份的京东股票数据，代码如下：

```
//查询
spark.sql("""
    select *
    from hadoop_prod.jd_db.stock_part
    where date>'2022-01-31'
""").show()
```

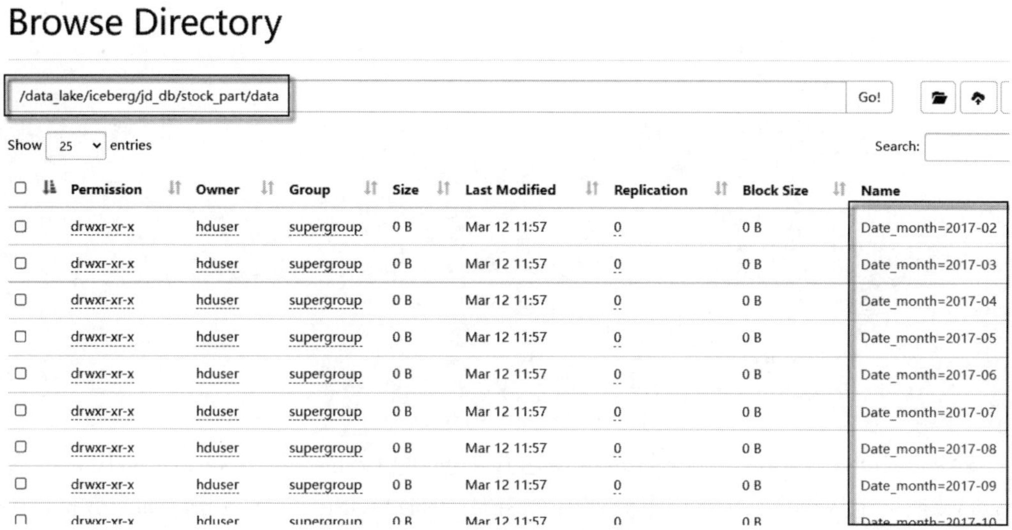

Browse Directory

	Permission	Owner	Group	Size	Last Modified	Replication	Block Size	Name
☐	drwxr-xr-x	hduser	supergroup	0 B	Mar 12 11:57	0	0 B	Date_month=2017-02
☐	drwxr-xr-x	hduser	supergroup	0 B	Mar 12 11:57	0	0 B	Date_month=2017-03
☐	drwxr-xr-x	hduser	supergroup	0 B	Mar 12 11:57	0	0 B	Date_month=2017-04
☐	drwxr-xr-x	hduser	supergroup	0 B	Mar 12 11:57	0	0 B	Date_month=2017-05
☐	drwxr-xr-x	hduser	supergroup	0 B	Mar 12 11:57	0	0 B	Date_month=2017-06
☐	drwxr-xr-x	hduser	supergroup	0 B	Mar 12 11:57	0	0 B	Date_month=2017-07
☐	drwxr-xr-x	hduser	supergroup	0 B	Mar 12 11:57	0	0 B	Date_month=2017-08
☐	drwxr-xr-x	hduser	supergroup	0 B	Mar 12 11:57	0	0 B	Date_month=2017-09
☐	drwxr-xr-x	hduser	supergroup	0 B	Mar 12 11:57	0	0 B	Date_month=2017-10

路径：/data_lake/iceberg/jd_db/stock_part/data

图 11-1　Iceberg 分区存储目录

执行以上代码，查询结果如下：

```
+----------+-----+--------+------+------+------+
|      Date|Close|  Volume|  Open|  High|   Low|
+----------+-----+--------+------+------+------+
|2022-02-15|76.13| 6766205| 75.35| 76.35|  74.8|
|2022-02-14|74.45| 5244967| 73.94| 74.62| 73.01|
|2022-02-11|73.98| 6673354| 75.97| 76.55| 73.55|
|2022-02-10| 76.4| 6432184|75.955| 78.39| 75.24|
|2022-02-09|78.29| 7061571| 76.83| 78.67| 76.61|
|2022-02-08|75.36| 7903249| 73.12| 76.07| 72.05|
|2022-02-07|73.15| 6135832| 74.09| 74.99| 72.81|
|2022-02-04|73.77| 6082889| 71.94| 74.95| 71.86|
|2022-02-03|71.85| 7493688| 72.08|  73.3| 71.33|
|2022-02-02|73.21| 5887066| 75.58| 75.71| 72.41|
|2022-02-01|75.08| 7408338| 74.26| 75.89| 73.84|
|2022-01-31|74.88| 9331051| 71.84| 74.95| 71.79|
|2022-01-28|69.34| 9026547|  67.6| 69.42|  66.2|
|2022-01-27|66.68|13137080| 70.75|  70.8|65.785|
|2022-01-26|71.11| 9546074|  72.9| 73.49| 70.51|
|2022-01-25|72.15| 7291178| 70.28| 72.97|  70.2|
|2022-01-24|71.74|15503400| 72.52| 72.52| 68.11|
|2022-01-21|73.46|14440500|77.602| 77.68| 72.66|
|2022-01-20| 77.0|26218260| 77.25| 81.24| 76.77|
|2022-01-19| 72.3| 6318328| 73.18| 74.22| 72.03|
+----------+-----+--------+------+------+------+
only showing top 20 rows
```

在上面的例子中，使用了 DataFrameWriterV2 API 写入分区表数据。Iceberg 也支持使用 SQL 创建分区表和写入数据。例如，使用 partitioned by 创建一个分区表，应用 SQL 语句，代码如下：

```
//先创建表
spark.sql("""
    create table hadoop_prod.jd_db.stock_part2(
        Date date,
        Close double,
        Volume bigint,
        Open double,
        High double,
        Low double
    )
    using iceberg
    partitioned by(years(Date))        //按年分区
""")
```

PARTITIONED BY 子句支持转换表达式来创建隐藏的分区，在上面的代码中通过 years()转换指定按年进行分区。下面使用 INSERT INTO … SELECT … 语句插入数据，代码如下：

```
//写入分区表数据
spark.sql("""
    INSERT INTO hadoop_prod.jd_db.stock_part2
    SELECT Date,Close,Volume,Open,High,Low
    FROM hadoop_prod.jd_db.stock_part
    ORDER BY Date
""")

//查看写入是否成功
spark.table("hadoop_prod.jd_db.stock_part2").show(5)
```

执行以上代码，输出的结果如下：

```
+----------+-----+-------+-----+------+-----+
|      Date|Close| Volume| Open|  High|  Low|
+----------+-----+-------+-----+------+-----+
|2017-02-16|30.23|7694667|30.32| 30.57|30.03|
|2017-02-17|29.85|7080140|29.57| 30.27|29.51|
|2017-02-21|30.23|5131289| 30.0|30.285|29.81|
|2017-02-22|30.47|6604136|30.27| 30.67| 30.1|
|2017-02-23|30.61|9921265|30.75| 30.88|30.23|
+----------+-----+-------+-----+------+-----+
only showing top 5 rows
```

然后通过 Web UI 查看物理分区布局，如图 11-2 所示。

1. RTAS（Replace Table … As Select）

当使用 SparkCatalog 时，Iceberg 支持 RTAS 作为原子操作。也可以使用 RTAS 方法根据已有的表来替换并导入一个新的分区表。在下面的示例中创建了分区表 stock_part2，指定按股票交易日期中的月份来分区，其数据来源于对 stock_part 表的查询，代码如下：

Permission	Owner	Group	Size	Last Modified	Replication	Block Size	Name
drwxr-xr-x	hduser	supergroup	0 B	Mar 12 12:30	0	0 B	Date_year=2017
drwxr-xr-x	hduser	supergroup	0 B	Mar 12 12:30	0	0 B	Date_year=2018
drwxr-xr-x	hduser	supergroup	0 B	Mar 12 12:30	0	0 B	Date_year=2019
drwxr-xr-x	hduser	supergroup	0 B	Mar 12 12:30	0	0 B	Date_year=2020
drwxr-xr-x	hduser	supergroup	0 B	Mar 12 12:30	0	0 B	Date_year=2021
drwxr-xr-x	hduser	supergroup	0 B	Mar 12 12:30	0	0 B	Date_year=2022

路径：/data_lake/iceberg/jd_db/stock_part2/data

图 11-2　Iceberg 表插入数据以后的物理分区布局

```
//覆盖 stock_part2 表，改为按月分区
spark.sql("""
    REPLACE TABLE hadoop_prod.jd_db.stock_part2
    USING iceberg
    PARTITIONED BY (months(Date))
    AS SELECT Date,Close,Volume,Open,High,Low
      FROM hadoop_prod.jd_db.stock_part
      ORDER BY Date
""")
```

执行以上代码后，查看 stock_part2 表的物理布局，可以看到每个月份的交易数据被单独存储在一个分区中，结构如图 11-3 所示。

Browse Directory

路径：/data_lake/iceberg/jd_db/stock_part2/data/Date_year=null

Permission	Owner	Group	Size	Last Modified	Replication	Block Size	Name
drwxr-xr-x	hduser	supergroup	0 B	Mar 12 12:45	0	0 B	Date_month=2017-02
drwxr-xr-x	hduser	supergroup	0 B	Mar 12 12:45	0	0 B	Date_month=2017-03
drwxr-xr-x	hduser	supergroup	0 B	Mar 12 12:45	0	0 B	Date_month=2017-04
drwxr-xr-x	hduser	supergroup	0 B	Mar 12 12:45	0	0 B	Date_month=2017-05
drwxr-xr-x	hduser	supergroup	0 B	Mar 12 12:45	0	0 B	Date_month=2017-06
drwxr-xr-x	hduser	supergroup	0 B	Mar 12 12:45	0	0 B	Date_month=2017-07

图 11-3　执行 RTAS 后 Iceberg 表的物理布局

2. CTAS（Create Table … As Select）

在使用 SparkCatalog 时，Iceberg 支持 CTAS 作为原子操作。也可以使用 CTAS 方法根据已有的表来创建并导入一个新的分区表。下面的代码创建了分区表 stock_part3，指定按

股票交易日期中的月份来分区，其数据来源于对 stock_part 表中所有 2021 年交易数据的查询，代码如下：

```
//将2021年的交易记录查询出来，然后单独写到分区表stock_part3中，并按月分区存储
spark.sql("""
    CREATE TABLE hadoop_prod.jd_db.stock_part3
    USING iceberg
    PARTITIONED BY (months(Date))
    AS SELECT Date,Close,Volume,Open,High,Low
        FROM hadoop_prod.jd_db.stock_part
        WHERE year(Date)="2021"
        ORDER BY Date
""")
```

执行以上代码后，查看 stock_part3 表的物理布局，可以看到 2021 年每个月份的交易数据被单独存储在一个分区中，结构如图 11-4 所示。

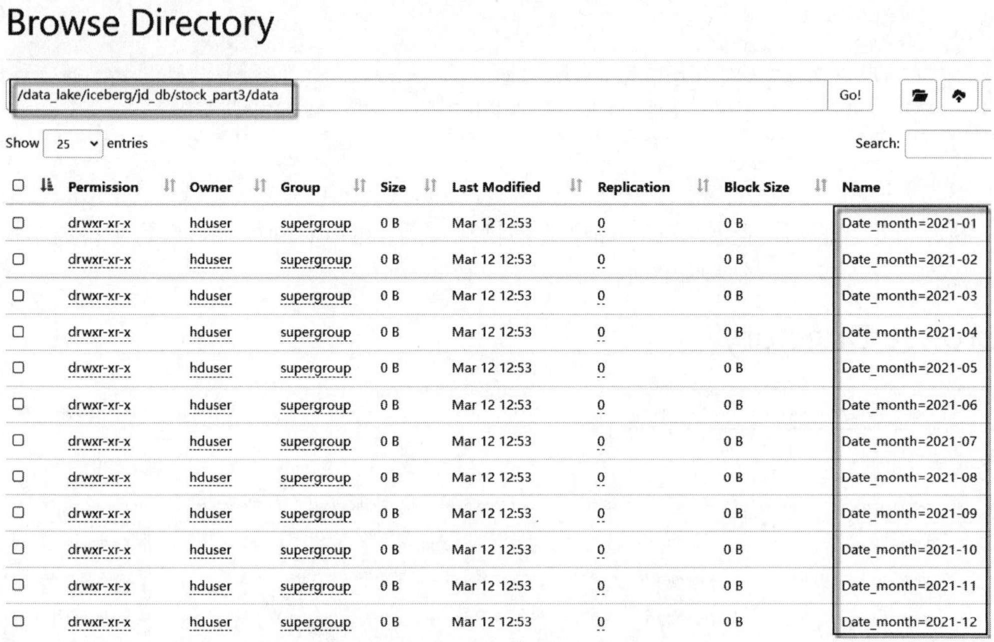

图 11-4 执行 CTAS 后 Iceberg 表的物理布局

3. 桶表

对于大多数分区转换，可以简单地将原始列添加到排序条件中，但 bucket() 除外。对于 bucket() 分区转换，需要在 Spark 中注册 Iceberg 转换函数，以便在排序过程中指定它。

下面是另一个有桶分区的示例。首先创建一个桶表，按 Date 字段分，代码如下：

```
spark.sql("""
    create table hadoop_prod.jd_db.stock_part4(
        Date date,
```

```
        Close double,
        Volume bigint,
        Open double,
        High double,
        Low double
    )
    using iceberg
    partitioned by(bucket(5, Date))
""")
```

需要注册函数来处理 bucket，注册代码如下：

```
import org.apache.iceberg.spark.IcebergSpark
import org.apache.spark.sql.types.DataTypes

IcebergSpark.registerBucketUDF(spark, "stock_bucket5", DataTypes.DateType, 5)
```

这里将 bucket 函数注册为 stock_bucket5，它可以在 sort 子句中使用。

如果用 SQL 语句插入数据，则可以像下面这样使用该函数，代码如下：

```
spark.sql("""
    INSERT INTO hadoop_prod.jd_db.stock_part4
    SELECT Date,Close,Volume,Open,High,Low
    FROM hadoop_prod.jd_db.stock_part
    ORDER BY stock_bucket5(Date)
""")
```

如果用 DataFrame 插入数据，则可以像下面这样使用该函数，代码如下：

```
spark
    .table("hadoop_prod.jd_db.stock_part")
    .sortWithinPartitions(expr("stock_bucket5(Date)"))
    .writeTo("hadoop_prod.jd_db.stock_part4")
    .append()
```

查看 stock_part4 表的物理布局，可以看到有 5 个与桶对应的文件夹，如图 11-5 所示。

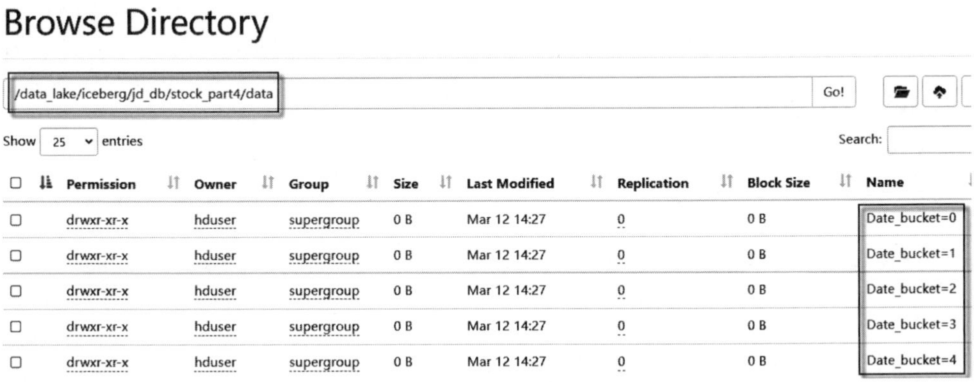

图 11-5 分桶存储后 Iceberg 表的物理布局

还支持使用 CTAS 创建分桶表，代码如下：

```
spark
    .table("hadoop_prod.jd_db.stock_part")
    .sortWithinPartitions(expr("stock_bucket5(Date)"))
    .writeTo("hadoop_prod.jd_db.stock_part5")
    .partitionedBy(bucket(5, $"Date"))
    .create()
```

查看 stock_part5 表的物理布局,同样可以看到有 5 个与桶对应的文件夹,如图 11-6 所示。

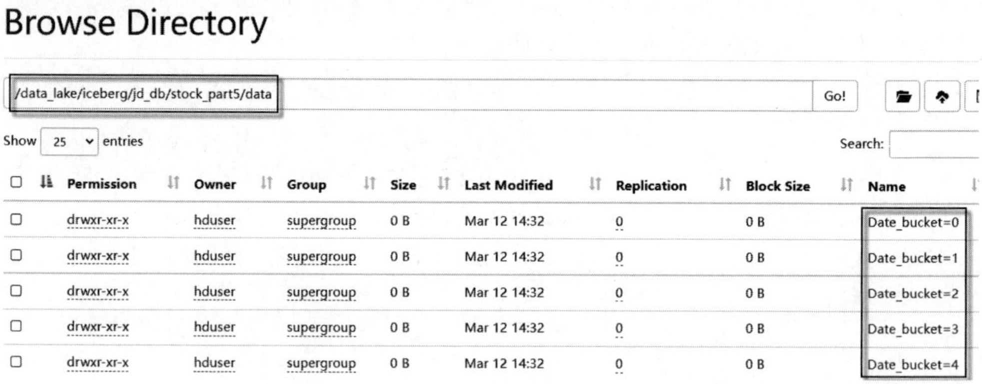

图 11-6 使用 CTAS 创建分桶表后 Iceberg 表的物理布局

也可以混合使用分桶和分区。例如,先把数据分为 5 个桶,每个桶下面再按月份分区(这里使用 CTAS 方法,创建了表 stock_part6),代码如下:

```
//先分桶,再在每个桶下面分区
spark
    .table("hadoop_prod.jd_db.stock_part")
    .sortWithinPartitions(expr("stock_bucket5(Date)"))
    .writeTo("hadoop_prod.jd_db.stock_part6")
    .partitionedBy(bucket(5, $"Date"), months($"Date"))
    .create()
```

查看 stock_part6 表的物理布局,可以看到有 5 个与桶对应的文件夹,如图 11-7 所示。

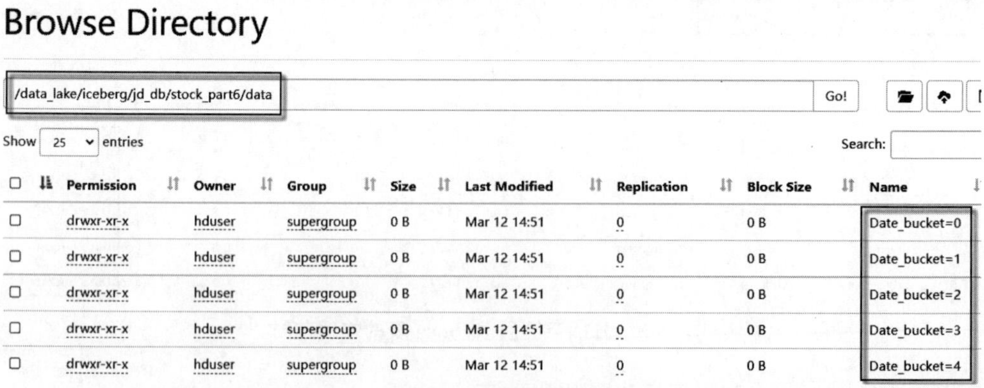

图 11-7 先分桶

导航到 stock_part6 表的每个桶文件夹中，可以看到每个桶中按月进一步进行了分区，如图 11-8 所示。

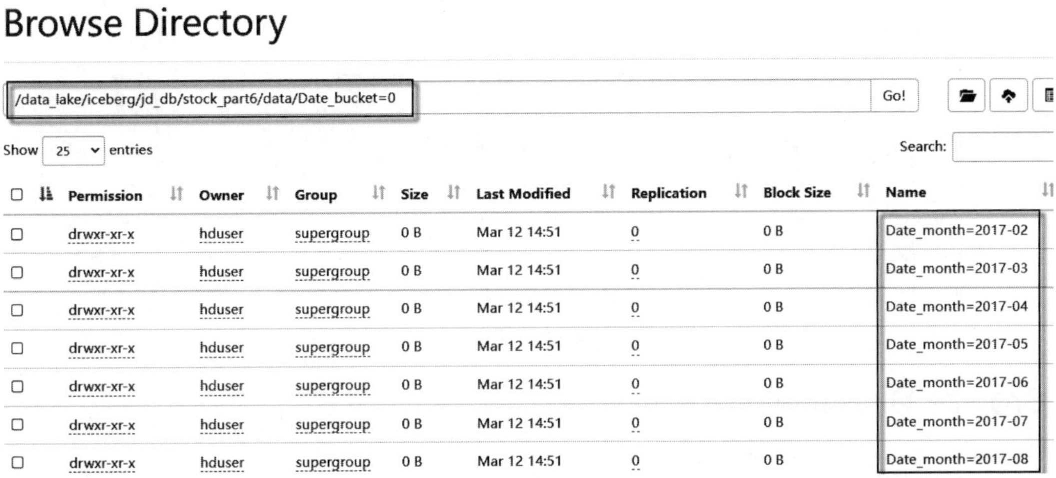
Browse Directory
/data_lake/iceberg/jd_db/stock_part6/data/Date_bucket=0
Go!
Show 25 entries
Search:
Permission Owner Group Size Last Modified Replication Block Size Name
drwxr-xr-x hduser supergroup 0 B Mar 12 14:51 0 0 B Date_month=2017-02
drwxr-xr-x hduser supergroup 0 B Mar 12 14:51 0 0 B Date_month=2017-03
drwxr-xr-x hduser supergroup 0 B Mar 12 14:51 0 0 B Date_month=2017-04
drwxr-xr-x hduser supergroup 0 B Mar 12 14:51 0 0 B Date_month=2017-05
drwxr-xr-x hduser supergroup 0 B Mar 12 14:51 0 0 B Date_month=2017-06
drwxr-xr-x hduser supergroup 0 B Mar 12 14:51 0 0 B Date_month=2017-07
drwxr-xr-x hduser supergroup 0 B Mar 12 14:51 0 0 B Date_month=2017-08

图 11-8　每个桶下再分区

也可以先分区，每个分区下面再分桶。在下面的代码中，先把数据按年份进行分区，每个分区下面再按 Volumn 列分为 6 个桶（纯粹出于演示的目的选择 Volumn 列）。为此，需要先将 bucket 函数注册为 stock_bucket6，再在 sort 子句中使用，代码如下：

```
//需要注册函数来处理 bucket，如下所示
import org.apache.iceberg.spark.IcebergSpark
import org.apache.spark.sql.types.DataTypes

IcebergSpark.registerBucketUDF(spark, "stock_bucket6", DataTypes.LongType, 6)
```

然后使用 CTAS 创建一个新的表 stock_part7 并插入数据，代码如下：

```
spark
    .table("hadoop_prod.jd_db.stock_part")
    .sortWithinPartitions(expr("stock_bucket6(Volume)"))
    .writeTo("hadoop_prod.jd_db.stock_part7")
    .partitionedBy(years($"Date"), bucket(6, $"Volume"))
    .create()
```

查看 stock_part7 表的物理布局，可以看到每年有一个对应的文件夹，如图 11-9 所示。

导航到 stock_part7 表的每个分区文件夹中，可以看到每个分区中进一步被划分为 6 个桶，如图 11-10 所示。

4. DELETE FROM

Spark 3 增加了对 DELETE FROM 查询的支持，可以从表中删除数据。

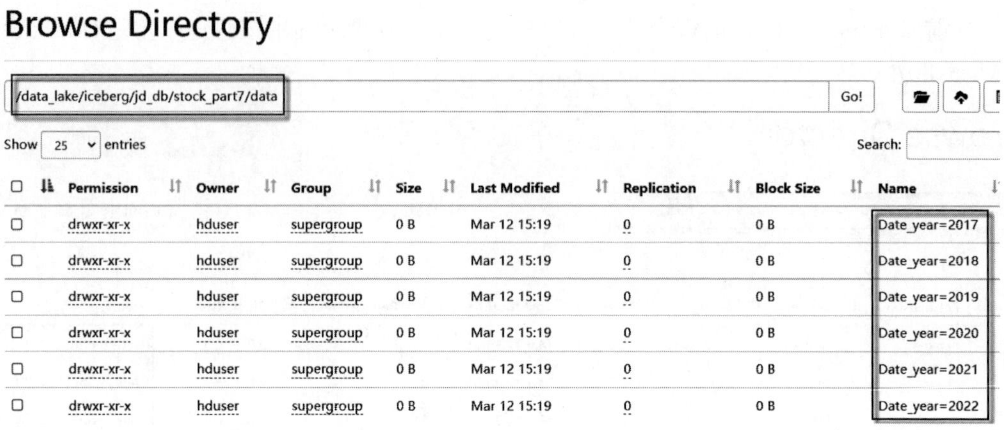

图 11-9　先分区

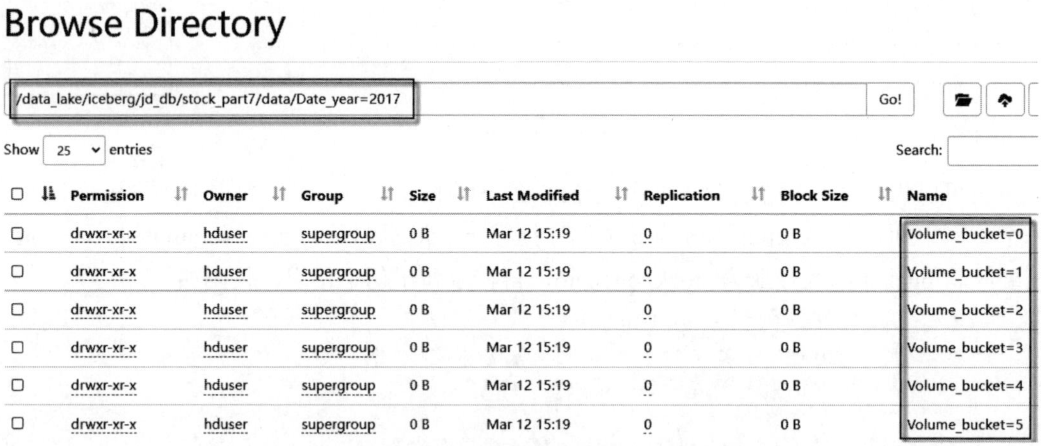

图 11-10　每个分区下再分桶

Delete 查询接受一个过滤器来匹配要删除的行，代码如下：

```
spark.sql("""
  DELETE FROM hadoop_prod.jd_db.stock_part
  WHERE Date >= '2020-05-01 00:00:00' and Date < '2020-06-01 00:00:00'
""")
```

如果删除过滤器匹配表的整个分区，Iceberg 则将执行一个只删除元数据的操作。如果过滤器匹配表中的各个行，则 Iceberg 将只重写受影响的数据文件。

11.4.4　数据覆盖

INSERT OVERWRITE 可以用查询的结果替换表中的数据。OVERWRITE 是 Iceberg 表的原子操作。要被 INSERT OVERWRITE 替换的分区取决于 Spark 的分区覆盖模式和表的分区。建议使用 MERGE INTO 而不是 INSERT OVERWRITE，因为 Iceberg 可以仅重写受影响的数据文件，并且具有更容易理解的行为，而且如果表的分区发生变化，则被动态重写覆盖的数据可能会发生变化。

Spark 的默认覆盖模式是静态的（static），但是在写 Iceberg 表时建议使用动态覆盖模式。静态覆盖模式通过将 PARTITION 子句转换为过滤器来确定要覆盖表中的哪些分区，但是 PARTITION 子句只能引用表的列。

动态覆盖模式是通过设置 spark.sql.sources.partitionOverwriteMode=dynamic 来配置的。

为了演示动态和静态覆盖的行为，创建一个由 DDL 定义的 logs 表，代码如下：

```
CREATE TABLE hadoop_prod.my_app.logs (
    uuid string NOT NULL,
    level string NOT NULL,
    ts timestamp NOT NULL,
    message string)
USING iceberg
PARTITIONED BY (level, hours(ts))
```

1. 动态覆盖

当 Spark 的覆盖模式是动态时，包含 SELECT 查询产生的行的分区将被替换。

例如，查询从示例的 logs 表中删除重复的日志事件，代码如下：

```
INSERT OVERWRITE hadoop_prod.my_app.logs
SELECT uuid, first(level), first(ts), first(message)
FROM prod.my_app.logs
WHERE cast(ts as date) = '2020-07-01'
GROUP BY uuid
```

在 dynamic 模式下，这将用 SELECT 结果中的行替换任何分区。因为所有行的日期都限制在 7 月 1 日，所以只替换当天的 hours。

2. 静态覆盖

当 Spark 的覆盖模式为静态时，PARTITION 子句将被转换为用于从表中删除的过滤器。如果省略了 PARTITION 子句，则所有的分区都将被替换。

由于在上面的查询中没有 PARTITION 子句，所以当以静态模式运行时，它将删除表中所有现有的行，但只写入 7 月 1 日的日志。

要覆盖已加载的分区，需要添加一个与 SELECT 查询过滤器对齐的 PARTITION 子句，代码如下：

```
INSERT OVERWRITE hadoop_prod.my_app.logs
PARTITION (level = 'INFO')
SELECT uuid, first(level), first(ts), first(message)
FROM prod.my_app.logs
WHERE level = 'INFO'
GROUP BY uuid
```

注意：这种模式不能像动态示例查询那样替换小时分区，因为 PARTITION 子句只能引用表中的列，而不能引用隐藏分区。

11.4.5 修改表结构

Iceberg 支持使用 ALTER TABLE 命令对表结构进行修改。在 Apache Spark 3 中，Iceberg 有完整的 ALTER TABLE 支持：

（1）重命名表。
（2）设置或删除表属性。
（3）添加、删除和重命名列。
（4）添加、删除和重命名嵌套字段。
（5）重新排序顶级列和嵌套的 struct 字段。
（6）扩大 int、float 和 decimal 字段的类型。
（7）使必须列变为可选列。

此外，可以使用 SQL 扩展来添加对分区演化和设置表的写顺序的支持。不过需要特别注意的是，ALTER TABLE 不能修改 Hadoop 表，也就是说，在 HadoopCatalog 类型的 Catalog 中创建的表，不能使用 ALTER TABLE 命令，只有在 HiveCatalog 类型的 Catalog 中创建的表才支持 ALTER TABLE。

为了演示 ALTER TABLE 命令的使用，先配置一个名为 hive_prod 的 HiveCatalog 类型的 Catalog。在 Zeppelin Notebook 中指定配置（如果使用 spark-shell，则可参考 11.2.1 节），代码如下：

```
%spark.conf
spark.jars.packages                    iceberg-spark-runtime-3.1_2.12:0.13.1
spark.sql.extensions
   org.apache.iceberg.spark.extensions.IcebergSparkSessionExtensions
spark.sql.catalog.hive_prod
   org.apache.iceberg.spark.SparkCatalog
spark.sql.catalog.hive_prod.type       hive
spark.sql.catalog.hive_prod.uri        thrift://localhost:9083
```

接下来在 hive_prod 中创建一个名为 my_db 的命名空间（数据库），代码如下：

```
//创建命名空间
spark.sql("CREATE NAMESPACE hive_prod.my_db")

//查看是否创建成功
spark.sql("show namespaces in hive_prod").show()
```

执行上面的代码，在输出结果中应当能够看到创建的 my_db 命名空间。

然后，在 hive_prod.my_db 命名空间中创建一个名为 my_tb 的 Iceberg 表，代码如下：

```
//创建 Iceberg 表
spark.sql("""
   CREATE TABLE hive_prod.my_db.my_tb (
     id bigint,
     data string
   )
   USING iceberg
""")
```

```
//向 Iceberg 表中插入数据
spark.sql("""
    INSERT INTO hive_prod.my_db.my_tb
    VALUES
    (1, 'a'),
    (2, 'b'),
    (3, 'c')
""")

//查看创建和插入是否成功
spark.sql("select * from hive_prod.my_db.my_tb").show()
```

执行上面的代码，应该可以看到输出内容如下：

```
+---+----+
| id|data|
+---+----+
|  1|   a|
|  2|   b|
|  3|   c|
+---+----+
```

接下来，学习一些常用的 ALTER TABLE 命令。

1. 重命名表

语法：ALTER TABLE … RENAME TO

下面将表 my_tb 重命名为 new_tb，代码如下：

```
spark.sql("ALTER TABLE hive_prod.my_db.my_tb RENAME TO hive_prod.my_db.new_tb")
```

执行以上代码，然后查看 my_db 中的表名，代码如下：

```
//查询 hadoop_catalog.my_db 数据库中有哪些表
spark.sql("SHOW TABLES IN hive_prod.my_db").show()
```

执行以上代码，输出的结果如下：

```
+---------+---------+
|namespace|tableName|
+---------+---------+
|    my_db|   new_tb|
+---------+---------+
```

2. 设置表属性

语法：ALTER TABLE … SET TBLPROPERTIES

例如，将 new_tb 表的 read.split.target-size 的属性值设置为 256M，代码如下：

```
//设置表属性
spark.sql("""
    ALTER TABLE hive_prod.my_db.new_tb SET TBLPROPERTIES (
        'read.split.target-size'='268435456'
    )
```

```
//查看表的描述
spark.sql("desc formatted hive_prod.my_db.new_tb").show(false)
```

Iceberg 使用表属性来控制表行为。执行上面的代码，输出内容如图 11-11 所示。

```
+----------------------+------------------------------------------------------------------+---------+
|col_name              |data_type                                                         |comment  |
+----------------------+------------------------------------------------------------------+---------+
|id                    |bigint                                                            |         |
|data                  |string                                                            |         |
|                      |                                                                  |         |
|# Partitioning        |                                                                  |         |
|Not partitioned       |                                                                  |         |
|                      |                                                                  |         |
|# Detailed Table Information|                                                            |         |
|Name                  |hive_prod.my_db.new_tb                                            |         |
|Location              |hdfs://xueai8:8020/user/hive/warehouse/my_db.db/my_tb             |         |
|Provider              |iceberg                                                           |         |
|Owner                 |hduser                                                            |         |
|Table Properties      |[current-snapshot-id=5948140814430359531,format=iceberg/parquet,read.split.target-size=268435456]||
+----------------------+------------------------------------------------------------------+---------+
```

图 11-11　设置表属性

可以看出，表属性设置成功。可以用 UNSET 删除指定的属性，代码如下：

```
//删除属性
spark.sql("""
    ALTER TABLE hive_prod.my_db.new_tb UNSET TBLPROPERTIES ('read.split.target-size')
""")

//查看表的描述
spark.sql("desc formatted hive_prod.my_db.new_tb").show(false)
```

再一次查看表的属性，可以看到属性删除成功，如图 11-12 所示。

```
+----------------------+------------------------------------------------------------------+---------+
|col_name              |data_type                                                         |comment  |
+----------------------+------------------------------------------------------------------+---------+
|id                    |bigint                                                            |         |
|data                  |string                                                            |         |
|                      |                                                                  |         |
|# Partitioning        |                                                                  |         |
|Not partitioned       |                                                                  |         |
|                      |                                                                  |         |
|# Detailed Table Information|                                                            |         |
|Name                  |hive_prod.my_db.new_tb                                            |         |
|Location              |hdfs://xueai8:8020/user/hive/warehouse/my_db.db/my_tb             |         |
|Provider              |iceberg                                                           |         |
|Owner                 |hduser                                                            |         |
|Table Properties      |[current-snapshot-id=5948140814430359531,format=iceberg/parquet]  |         |
+----------------------+------------------------------------------------------------------+---------+
```

图 11-12　删除表属性

3. 增加新的属性列

语法：ALTER TABLE … ADD COLUMN

要向 Iceberg 添加一列，可以使用 ALTER TABLE 中的 ADD COLUMNS 子句，代码如下：

```
//添加列
spark.sql("""
    ALTER TABLE hive_prod.my_db.new_tb
    ADD COLUMNS (
        new_column string comment 'new_column docs'
    )
""")

//查看表的描述
spark.sql("desc formatted hive_prod.my_db.new_tb").show(false)
```

可以同时添加多个列，以逗号分隔。执行以上代码，然后查看表的属性，可以看到表多了一个新的属性列，如图 11-13 所示。

```
+-------------------------+----------------------------------------------------------------+----------------+
|col_name                 |data_type                                                       |comment         |
+-------------------------+----------------------------------------------------------------+----------------+
|id                       |bigint                                                          |                |
|data                     |string                                                          |                |
|new_column               |string                                                          |new_column docs |
|                         |                                                                |                |
|# Partitioning           |                                                                |                |
|Not partitioned          |                                                                |                |
|                         |                                                                |                |
|# Detailed Table Information|                                                             |                |
|Name                     |hive_prod.my_db.new_tb                                          |                |
|Location                 |hdfs://xueai8:8020/user/hive/warehouse/my_db.db/my_tb           |                |
|Provider                 |iceberg                                                         |                |
|Owner                    |hduser                                                          |                |
|Table Properties         |[current-snapshot-id=5948140814430359531,format=iceberg/parquet]|                |
+-------------------------+----------------------------------------------------------------+----------------+
```

图 11-13 添加新的属性列

还可以添加嵌套列。例如，为表 new_tb 添加一个 struct 类型的列 point，代码如下：

```
//创建一个 struct 列
spark.sql("""
    ALTER TABLE hive_prod.my_db.new_tb
    ADD COLUMN point struct<x: double, y: double>
""")

//查看表的描述
spark.sql("desc formatted hive_prod.my_db.new_tb").show(false)
```

查看表的属性，如图 11-14 所示。

对于嵌套的列，在引用时应该使用完整的列名来标识。例如，为 point 添加一个嵌套的子列 z，代码如下：

```
//向该 struct 添加一个字段
spark.sql("ALTER TABLE hive_prod.my_db.new_tb ADD COLUMN point.z double")
```

```
+--------------------------+--------------------------+------------------------------------+
|col_name                  |data_type                 |comment                             |
+--------------------------+--------------------------+------------------------------------+
|id                        |bigint                    |                                    |
|data                      |string                    |                                    |
|new_column                |string                    |new_column docs                     |
|point                     |struct<x:double,y:double> |                                    |
|                          |                          |                                    |
|# Partitioning            |                          |                                    |
|Not partitioned           |                          |                                    |
|                          |                          |                                    |
|# Detailed Table Information|                        |                                    |
|Name                      |hive_prod.my_db.new_tb    |                                    |
|Location                  |hdfs://xueai8:8020/user/hive/warehouse/my_db.db/my_tb|                      |
|Provider                  |iceberg                   |                                    |
|Owner                     |hduser                    |                                    |
|Table Properties          |[current-snapshot-id=5948140814430359531,format=iceberg/parquet]|         |
+--------------------------+--------------------------+------------------------------------+
```

图 11-14　添加新的嵌套属性列

查看表的属性，如图 11-15 所示。

```
+--------------------------+----------------------------------+------------------------------+
|col_name                  |data_type                         |comment                       |
+--------------------------+----------------------------------+------------------------------+
|id                        |bigint                            |                              |
|data                      |string                            |                              |
|new_column                |string                            |new_column docs               |
|point                     |struct<x:double,y:double,z:double>|                              |
|                          |                                  |                              |
|# Partitioning            |                                  |                              |
|Not partitioned           |                                  |                              |
|                          |                                  |                              |
|# Detailed Table Information|                                |                              |
|Name                      |hive_prod.my_db.new_tb            |                              |
|Location                  |hdfs://xueai8:8020/user/hive/warehouse/my_db.db/my_tb|            |
|Provider                  |iceberg                           |                              |
|Owner                     |hduser                            |                              |
|Table Properties          |[current-snapshot-id=5948140814430359531,format=iceberg/parquet]|  |
+--------------------------+----------------------------------+------------------------------+
```

图 11-15　添加嵌套子列

为表 new_tb 添加新的数组列、map 列及如何修改这些嵌套列，代码如下：

```
//创建一个嵌套的 struct 数组列
spark.sql("""
    ALTER TABLE hive_prod.my_db.new_tb
    ADD COLUMN points2 array<struct<x: double, y: double>>
""")

//为数组列中嵌套的 struct 添加字段。使用关键字"element"访问数组的元素列
spark.sql("""
    ALTER TABLE hive_prod.my_db.new_tb
    ADD COLUMN points2.element.z double
```

```
    """)
    //创建一个 map 列，key 和 value 都是 struct
    spark.sql("""
        ALTER TABLE hive_prod.my_db.new_tb
        ADD COLUMN points3 map<struct<x: int>, struct<a: int>>
    """)

    //为 map 的 struct 值增加一个字段。使用关键字"value"访问 map 的值列
    spark.sql("""
        ALTER TABLE hive_prod.my_db.new_tb
        ADD COLUMN points3.value.b int
    """)

    //查看表的描述信息
    spark.sql("desc formatted hive_prod.my_db.new_tb").show(false)
```

可以看到修改以后表的结构和属性如图 11-16 所示。

```
+--------------------------+----------------------------------------------------------+-----------------+
|col_name                  |data_type                                                 |comment          |
+--------------------------+----------------------------------------------------------+-----------------+
|id                        |bigint                                                    |                 |
|data                      |string                                                    |                 |
|new_column                |string                                                    |new_column docs  |
|point                     |struct<x:double,y:double,z:double>                        |                 |
|points2                   |array<struct<x:double,y:double,z:double>>                 |                 |
|points3                   |map<struct<x:int>,struct<a:int,b:int>>                    |                 |
|                          |                                                          |                 |
|# Partitioning            |                                                          |                 |
|Not partitioned           |                                                          |                 |
|                          |                                                          |                 |
|# Detailed Table Information|                                                        |                 |
|Name                      |hive_prod.my_db.new_tb                                    |                 |
|Location                  |hdfs://xueai8:8020/user/hive/warehouse/my_db.db/my_tb     |                 |
|Provider                  |iceberg                                                   |                 |
|Owner                     |hduser                                                    |                 |
|Table Properties          |[current-snapshot-id=5948140814430359531,format=iceberg/parquet]|           |
+--------------------------+----------------------------------------------------------+-----------------+
```

图 11-16　修改后的表结构和属性

在 Spark 2.4.4 及以后的版本中，可以通过添加 FIRST 或 AFTER 子句在任何位置添加列，但要改变列的位置，有一个隐含的要求，即所有位置改变的字段类型要兼容，代码如下：

```
    //在 new_column 后增加一列 other_column
    spark.sql("""
        ALTER TABLE hive_prod.my_db.new_tb
        ADD COLUMN other_column string AFTER new_column
    """)
```

上面这句代码在执行时会失败。它在 new_column 后增加 other_column，会造成 other_column 和后续的 point、points2、points3 不兼容。

如果添加的字段类型与之前的类型兼容，则一切正常，代码如下：

```
//在嵌套的point列中添加一个子列q，放在第1个位置
spark.sql("""
    ALTER TABLE hive_prod.my_db.new_tb
    ADD COLUMN point.w double FIRST
""")

//查看表结构描述信息
spark.sql("desc hive_prod.my_db.new_tb").show(false)
```

执行以后，可以看到修改以后表的结构和属性如图11-17所示。

```
+---------------+------------------------------------------------+-----------------+
|col_name       |data_type                                       |comment          |
+---------------+------------------------------------------------+-----------------+
|id             |bigint                                          |                 |
|data           |string                                          |                 |
|new_column     |string                                          |new_column docs  |
|point          |struct<w:double,x:double,y:double,z:double>     |                 |
|points2        |array<struct<x:double,y:double,z:double>>       |                 |
|points3        |map<struct<x:int>,struct<a:int,b:int>>          |                 |
|               |                                                |                 |
|# Partitioning |                                                |                 |
|Not partitioned|                                                |                 |
+---------------+------------------------------------------------+-----------------+
```

图 11-17　在任意位置添加列

4. 重命名列

语法：ALTER TABLE … RENAME COLUMN

Iceberg 允许对任何字段进行重命名。要重命名一个字段，需要使用 RENAME COLUMN。例如，对字段和嵌套字段进行重命名，代码如下：

```
spark.sql("""
    ALTER TABLE hive_prod.my_db.new_tb RENAME COLUMN data TO payload
""")

spark.sql("""
    ALTER TABLE hive_prod.my_db.new_tb RENAME COLUMN point.z TO zip
""")

//查看表结构描述信息
spark.sql("desc hive_prod.my_db.new_tb").show(false)
```

执行以上代码以后，可以看到修改后表的结构和属性如图11-18所示。

注意：嵌套的重命名命令只重命名叶子字段。上面的命令将 location.lat 重命名为 location.latitude。

```
+---------------+-------------------------------------------+-----------------+
|col_name       |data_type                                  |comment          |
+---------------+-------------------------------------------+-----------------+
|id             |bigint                                     |                 |
|payload        |string                                     |                 |
|new_column     |string                                     |new_column docs  |
|point          |struct<w:double,x:double,y:double,zip:double>|               |
|points2        |array<struct<x:double,y:double,z:double>>  |                 |
|points3        |map<struct<x:int>,struct<a:int,b:int>>     |                 |
|               |                                           |                 |
|# Partitioning |                                           |                 |
|Not partitioned|                                           |                 |
+---------------+-------------------------------------------+-----------------+
```

图 11-18　在任意位置添加列

5. 修改列定义

语法：ALTER TABLE … ALTER COLUMN

Alter column 可用于扩展类型、使字段可选、设置注释和重新排序字段。

如果修改是安全的，则 Iceberg 允许改变列类型。安全的列类型修改，代码如下：

```
int => bigint
float => double
decimal(P,S) => decimal(P2,S)//当 P2 > P（规模不能改变）
```

例如，为 payload 字段添加设置注释，代码如下：

```
spark.sql("""
    ALTER TABLE hive_prod.my_db.new_tb ALTER COLUMN payload comment '原字段名为 data'
""")

//查看表结构描述信息
spark.sql("desc hive_prod.my_db.new_tb").show(false)
```

执行以上代码以后，可以看到修改后表的结构和属性如图 11-19 所示。

```
+---------------+-------------------------------------------+-----------------+
|col_name       |data_type                                  |comment          |
+---------------+-------------------------------------------+-----------------+
|id             |bigint                                     |                 |
|payload        |string                                     |原字段名为 data  |
|new_column     |string                                     |new_column docs  |
|point          |struct<w:double,x:double,y:double,zip:double>|               |
|points2        |array<struct<x:double,y:double,z:double>>  |                 |
|points3        |map<struct<x:int>,struct<a:int,b:int>>     |                 |
|               |                                           |                 |
|# Partitioning |                                           |                 |
|Not partitioned|                                           |                 |
+---------------+-------------------------------------------+-----------------+
```

图 11-19　在任意位置添加列

在 Iceberg 中，允许使用 FIRST 和 AFTER 子句对结构中的顶级列或列进行重新排序，代码如下：

```
spark.sql("""
    ALTER TABLE hive_prod.my_db.new_tb ALTER COLUMN payload AFTER new_column
""")

spark.sql("""
    ALTER TABLE hive_prod.my_db.new_tb ALTER COLUMN point.w AFTER x
""")

//查看表结构描述信息
spark.sql("desc hive_prod.my_db.new_tb").show(false)
```

执行以后，可以看到字段顺序进行了调整，如图 11-20 所示。

```
+---------------+-------------------------------------------+-------------------+
|col_name       |data_type                                  |comment            |
+---------------+-------------------------------------------+-------------------+
|id             |bigint                                     |                   |
|new_column     |string                                     |new_column docs    |
|payload        |string                                     |原字段名为data     |
|point          |struct<x:double,w:double,y:double,zip:double>|                 |
|points2        |array<struct<x:double,y:double,z:double>>  |                   |
|points3        |map<struct<x:int>,struct<a:int,b:int>>     |                   |
|               |                                           |                   |
|# Partitioning |                                           |                   |
|Not partitioned|                                           |                   |
+---------------+-------------------------------------------+-------------------+
```

图 11-20　调整字段顺序

注意：改变列的位置操作，有一个隐含的要求，即所有位置改变的字段类型要兼容，否则执行时会失败；ALTER COLUMN 不用于更新结构(struct)类型。可以使用 ADD COLUMN 和 DROP COLUMN 来添加或删除结构字段。

可以使用 DROP NOT NULL 将不允许为空的列更改为允许为空。例如，设置 id 列允许为空，代码如下：

```
spark.sql("""
    ALTER TABLE hive_prod.my_db.new_tb ALTER COLUMN id DROP NOT NULL
""")
```

注意：目前 Iceberg 还不支持将允许为空的列更改为不允许为空。

6. 删除列

语法：ALTER TABLE … DROP COLUMN

要删除列，使用 ALTER TABLE … DROP COLUMN。例如，删除 my_tb 表中嵌套的 point.w 列，代码如下：

```
spark.sql("ALTER TABLE hive_prod.my_db.new_tb DROP COLUMN point.w")
```

执行以上代码，可以看到表结构和属性如图 11-21 所示。

```
+--------------+---------------------------------------------+-----------------+
|col_name      |data_type                                    |comment          |
+--------------+---------------------------------------------+-----------------+
|id            |bigint                                       |                 |
|new_column    |string                                       |new_column docs  |
|payload       |string                                       |原字段名为data    |
|point         |struct<x:double,y:double,zip:double>         |                 |
|points2       |array<struct<x:double,y:double,z:double>>    |                 |
|points3       |map<struct<x:int>,struct<a:int,b:int>>       |                 |
|other_column  |string                                       |                 |
|              |                                             |                 |
|# Partitioning|                                             |                 |
|Not partitioned|                                            |                 |
+--------------+---------------------------------------------+-----------------+
```

图 11-21 删除列后的表属性

注意：同样地，删除某个列时，要保证其后面的列在各自的位置上与现有的列兼容。如果不兼容，则会导致删除失败。

11.5 探索 Iceberg 表

为了检查表的历史、快照和其他元数据，Iceberg 支持元数据表。

元数据表通过在原始表名之后添加元数据表名来标识。例如，db.table 的历史记录是通过 db.table.history 表读取的。也就是说，元数据表，如 history 和 snapshots，可以使用 Iceberg 表名作为命名空间。

例如，从 prod.db.table 的 files 元数据表中读取，运行以下语句：

```
SELECT * FROM prod.db.table.files
```

注意：从 Spark 3.0 开始，检查的表名格式（catalog.database.table.metadata）与 Spark 的默认 catalog(spark_catalog)不兼容。如果已经替换了默认目录，则可能需要使用 DataFrameReader API 来检查表。

11.5.1 History 历史表

要显示表历史，代码如下：

```
spark.sql("SELECT * FROM prod.db.table.history").show()
```

执行以上代码，输出结果如图 11-22 所示。

```
+-------------------------+-------------------+-------------------+---------------------+
| made_current_at         | snapshot_id       | parent_id         | is_current_ancestor |
+-------------------------+-------------------+-------------------+---------------------+
| 2019-02-08 03:29:51.215 | 5781947118336215154| NULL              | true                |
| 2019-02-08 03:47:55.948 | 5179299526185056830| 5781947118336215154| true                |
| 2019-02-09 16:24:30.13  | 296410040247533544 | 5179299526185056830| false               |
| 2019-02-09 16:32:47.336 | 2999875608062437330| 5179299526185056830| true                |
| 2019-02-09 19:42:03.919 | 8924558786060583479| 2999875608062437330| true                |
| 2019-02-09 19:49:16.343 | 6536733823181975045| 8924558786060583479| true                |
+-------------------------+-------------------+-------------------+---------------------+
```

图 11-22 查询 Iceberg 表的历史表信息

这显示了一个回滚的提交。该示例有两个具有相同父节点的快照，其中一个不是当前表状态的祖先。

11.5.2 Snapshots 快照表

要显示一张表的有效快照，代码如下：

```
spark.sql("SELECT * FROM prod.db.table.snapshots").show()
```

执行以上代码，输出结果如图 11-23 所示。

```
+-------------------------+------------------+-----------+-----------+----------------------------------------------------+----------------------------------------------------------------+
| committed_at            | snapshot_id      | parent_id | operation | manifest_list                                      | summary                                                        |
+-------------------------+------------------+-----------+-----------+----------------------------------------------------+----------------------------------------------------------------+
| 2019-02-08 03:29:51.215 | 57897183625154   | null      | append    | s3://.../table/metadata/snap-57897183625154-1.avro | { added-records -> 2478404, total-records -> 2478404,          |
|                         |                  |           |           |                                                    |   added-data-files -> 438, total-data-files -> 438,            |
|                         |                  |           |           |                                                    |   spark.app.id -> application_1520379288616_155055 }           |
| ...                     | ...              | ...       | ...       | ...                                                |                                                                |
+-------------------------+------------------+-----------+-----------+----------------------------------------------------+----------------------------------------------------------------+
```

图 11-23 查询 Iceberg 表的快照表信息

还可以将快照连接到表历史记录。例如，查询显示表历史，以及写每个快照的应用程序 ID，代码如下：

```
spark.sql("""
   select
      h.made_current_at,
      s.operation,
      h.snapshot_id,
      h.is_current_ancestor,
      s.summary['spark.app.id']
   from prod.db.table.history h
   join prod.db.table.snapshots s
     on h.snapshot_id = s.snapshot_id
   order by made_current_at
""")
```

执行以上代码，输出结果如图 11-24 所示。

```
+-------------------------+-----------+-----------------+---------------------+---------------------------------+
| made_current_at         | operation | snapshot_id     | is_current_ancestor | summary[spark.app.id]           |
+-------------------------+-----------+-----------------+---------------------+---------------------------------+
| 2019-02-08 03:29:51.215 | append    | 57897183625154  | true                | application_1520379288616_155055 |
| 2019-02-09 16:24:30.13  | delete    | 296410040247531 | false               | application_1520379288616_151109 |
| 2019-02-09 16:32:47.336 | append    | 57897183625154  | true                | application_1520379288616_155055 |
| 2019-02-08 03:47:55.948 | overwrite | 51792995261850  | true                | application_1520379288616_152431 |
+-------------------------+-----------+-----------------+---------------------+---------------------------------+
```

图 11-24 对历史表和快照表执行 join 连接的结果

11.5.3 Files 数据文件表

要显示一张表的数据文件和每个文件的元数据，代码如下：

```
spark.sql("SELECT * FROM prod.db.table.files").show()
```

输出结果如图 11-25 所示。

```
+--------------------------------------------------------------+-------------+--------------+-------------------+--------------------+-------------------+----------------+-------------------+--------------+---------------+
| file_path                                                    | file_format | record_count | file_size_in_bytes| column_sizes       | value_counts      | null_value_counts | lower_bounds   | upper_bounds   | key_metadata | split_offsets |
+--------------------------------------------------------------+-------------+--------------+-------------------+--------------------+-------------------+----------------+-------------------+--------------+---------------+
| s3://.../table/data/00000-3-8d6d0e8-d427-4809-bcf0-f5d45a4aad96.parquet | PARQUET | 1 | 597 | [1 -> 90, 2 -> 62] | [1 -> 1, 2 -> 1] | [1 -> 0, 2 -> 0] | [1 -> , 2 -> c] | [1 -> , 2 -> c] | null | [4] |
| s3://.../table/data/00001-4-8d6d0e8-d427-4809-bcf0-f5d45a4aad96.parquet | PARQUET | 1 | 597 | [1 -> 90, 2 -> 62] | [1 -> 1, 2 -> 1] | [1 -> 0, 2 -> 0] | [1 -> , 2 -> b] | [1 -> , 2 -> b] | null | [4] |
| s3://.../table/data/00002-5-8d6d0e8-d427-4809-bcf0-f5d45a4aad96.parquet | PARQUET | 1 | 597 | [1 -> 90, 2 -> 62] | [1 -> 1, 2 -> 1] | [1 -> 0, 2 -> 0] | [1 -> , 2 -> a] | [1 -> , 2 -> a] | null | [4] |
+--------------------------------------------------------------+-------------+--------------+-------------------+--------------------+-------------------+----------------+-------------------+--------------+---------------+
```

图 11-25　查询 Iceberg 表的数据文件和每个文件的元数据

11.5.4　Manifests 文件清单表

要显示一张表的文件清单（Manifests）和每个文件的元数据，代码如下：

```
spark.sql("SELECT * FROM prod.db.table.manifests").show()
```

输出结果如图 11-26 所示。

```
+-------------------------------------------------------------------+--------+------------------+---------------------+-------------------------+--------------------------+-------------------------+---------------------------------+
| path                                                              | length | partition_spec_id| added_snapshot_id   | added_data_files_count  | existing_data_files_count| deleted_data_files_count| partitions                      |
+-------------------------------------------------------------------+--------+------------------+---------------------+-------------------------+--------------------------+-------------------------+---------------------------------+
| s3://.../table/metadata/45b5290b-ee61-4788-b324-b1e2735c0e10-m0.avro | 4479 | 0              | 6668963634911763636 | 8                       | 0                        | 0                       | [[false,2019-05-13,2019-05-15]] |
+-------------------------------------------------------------------+--------+------------------+---------------------+-------------------------+--------------------------+-------------------------+---------------------------------+
```

图 11-26　查询 Iceberg 表的文件清单和每个文件的元数据

在 Spark 2.4 或 Spark 3 中可以通过 DataFrameReader API 加载元数据表，代码如下：

```
//命名的元数据表
spark.read
    .format("iceberg")
    .load("prod.db.table.files")
    .show(truncate = false)


//Hadoop path 表
spark.read
    .format("iceberg")
    .load("hdfs://localhost:8020/path/to/table#files")
    .show(truncate = false)
```

Apache Iceberg 提供了在给定时间点加载任何快照或数据的灵活性。例如，要检查版本历史，代码如下：

```
spark.read.table("local.db.table.history").show(10, false)
```

按照快照 ID 读取，代码如下：

```
Spark
    .read
    .option("snapshot-id", 26265153310899177788L)
    .table("local.db.table")
```

或者在任何给定的时间点读取，代码如下：

```
Spark
    .read
    .option("as-of-timestamp", 1637879675001L)
    .table("local.db.table")
```

11.6 Apache Iceberg 架构

有了前面对 Apache Iceberg 的初步了解，接下来深入了解它的架构体系。

Iceberg 表的结构有 3 层，分别如下：

（1） Iceberg Catalog。

（2） Metadata Layer（元数据层），包含元数据文件（Metadata File）、清单列表（Manifest List）、清单文件（Manifest File）。

（3） Data Layer（数据层）。

下面是一张 Iceberg 表结构的结构图，如图 11-27 所示。

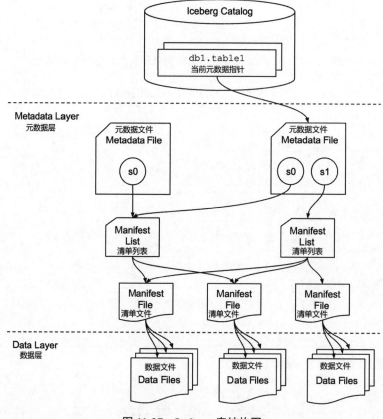

图 11-27　Iceberg 表结构图

11.6.1　Iceberg Catalog

读取 Iceberg 表的第 1 步是找到该表当前元数据指针的位置。在这个可以找到当前元数据指针的当前位置的中心位置，就是 Iceberg Catalog。

Iceberg Catalog 的主要需求是它必须支持更新当前元数据指针的原子操作（例如，HDFS、Hive Metastore、Nessie）。这就是允许 Iceberg 表上的事务是原子的并提供正确性保证的原因。

在 Catalog 目录中，每个表都有一个指向该表当前元数据文件的引用或指针。例如，在图 11-28 中，有两个元数据文件。目录中表的当前元数据指针的值是右侧元数据文件的位置。

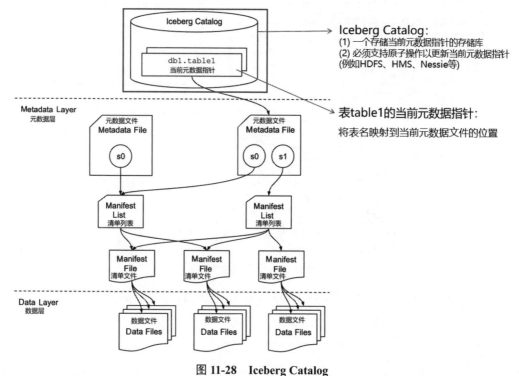

图 11-28　Iceberg Catalog

这些数据的形式取决于使用的是什么 Iceberg 目录。

（1）使用 HDFS 作为 Catalog 目录，在表的元数据文件夹中有一个名为 version-hint.text 的文件，其内容是当前元数据文件的版本号。

（2）使用 Hive Metastore 作为 Catalog 目录，Metastore 中的表条目有一张表属性，用来存储当前元数据文件的位置。

（3）以 Nessie 为 Catalog 目录，Nessie 存储了该表当前元数据文件的位置。

因此，当一个 SELECT 查询读取一个 Iceberg 表时，查询引擎首先进入 Iceberg Catalog，然后检索它要读取的表的当前元数据文件所在位置的条目，然后打开该文件。

11.6.2　元数据文件

Iceberg 使用 JSON 文件跟踪表元数据。对表的每次更改都会生成一个新的元数据文件

(Metadata File),以提供原子性。默认情况下,旧的元数据文件将作为历史记录保存,如图 11-29 所示。

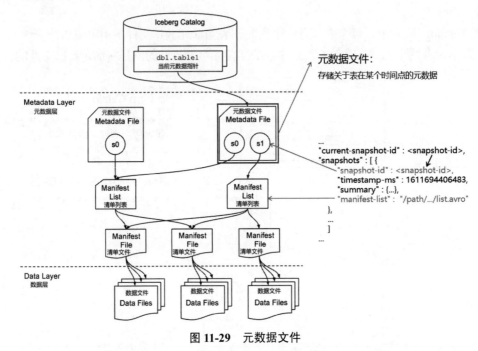

图 11-29　元数据文件

顾名思义,元数据文件存储着关于表的元数据。这包括关于表的模式、分区信息、快照及哪个快照是当前快照的信息。图 11-29 中的例子是为了演示而进行了删减,一个元数据文件完整内容的例子,内容如下(v3.metadata.json):

```
{
    "format-version" : 1,
    "table-uuid" : "4b96b6e8-9838-48df-a111-ec1ff6422816",
    "location" : "/home/hadoop/warehouse/db2/part_table2",
    "last-updated-ms" : 1611694436618,
    "last-column-id" : 3,
    "schema" : {
        "type" : "struct",
        "fields" : [ {
            "id" : 1,
            "name" : "id",
            "required" : true,
            "type" : "int"
        }, {
            "id" : 2,
            "name" : "ts",
            "required" : false,
            "type" : "timestamptz"
        }, {
```

```
          "id" : 3,
          "name" : "message",
          "required" : false,
          "type" : "string"
      } ]
    },
    "partition-spec" : [ {
      "name" : "ts_hour",
      "transform" : "hour",
      "source-id" : 2,
      "field-id" : 1000
    } ],
    "default-spec-id" : 0,
    "partition-specs" : [ {
      "spec-id" : 0,
      "fields" : [ {
          "name" : "ts_hour",
          "transform" : "hour",
          "source-id" : 2,
          "field-id" : 1000
      } ]
    } ],
    "default-sort-order-id" : 0,
    "sort-orders" : [ {
      "order-id" : 0,
      "fields" : [ ]
    } ],
    "properties" : {
      "owner" : "hadoop"
    },
    "current-snapshot-id" : 1257424822184505371,
    "snapshots" : [ {
      "snapshot-id" : 8271497753230544300,
      "timestamp-ms" : 1611694406483,
      "summary" : {
        "operation" : "append",
        "spark.app.id" : "application_1611687743277_0002",
        "added-data-files" : "1",
        "added-records" : "1",
        "added-files-size" : "960",
        "changed-partition-count" : "1",
        "total-records" : "1",
        "total-data-files" : "1",
        "total-delete-files" : "0",
        "total-position-deletes" : "0",
        "total-equality-deletes" : "0"
      },
      "manifest-list" : "/home/hadoop/warehouse/db2/part_table2/metadata/snap-8271497753230544300-1-d8a778f9-ad19-4e9c-88ff-28f49ec939fa.avro"
    },
```

```
    {
        "snapshot-id" : 1257424822184505371,
        "parent-snapshot-id" : 8271497753230544300,
        "timestamp-ms" : 1611694436618,
        "summary" : {
            "operation" : "append",
            "spark.app.id" : "application_1611687743277_0002",
            "added-data-files" : "1",
            "added-records" : "1",
            "added-files-size" : "973",
            "changed-partition-count" : "1",
            "total-records" : "2",
            "total-data-files" : "2",
            "total-delete-files" : "0",
            "total-position-deletes" : "0",
            "total-equality-deletes" : "0"
        },
        "manifest-list" : "/home/hadoop/warehouse/db2/part_table2/metadata/snap-1257424822184505371-1-eab8490b-8d16-4eb1-ba9e-0dede788ff08.avro"
    } ],
    "snapshot-log" : [ {
        "timestamp-ms" : 1611694406483,
        "snapshot-id" : 8271497753230544300
    }, {
        "timestamp-ms" : 1611694436618,
        "snapshot-id" : 1257424822184505371
    } ],
    "metadata-log" : [ {
        "timestamp-ms" : 1611694097253,
        "metadata-file" : "/home/hadoop/warehouse/db2/part_table2/metadata/v1.metadata.json"
    }, {
        "timestamp-ms" : 1611694406483,
        "metadata-file" : "/home/hadoop/warehouse/db2/part_table2/metadata/v2.metadata.json"
    } ]
}
```

当一个SELECT查询读取一个Iceberg表并在从Catalog目录中的表条目获取其位置后打开其当前元数据文件时，查询引擎会读取current-snapshot-id的值，然后，它使用这个值在snapshots数组中查找快照的条目，最后检索该快照的manifest-list条目的值，并打开位置指向的清单列表（Manifest List）。

11.6.3 清单列表（Manifest List）

清单列表（Manifest List）是清单文件（Manifest Files）的列表。清单列表包含了组成该快照的每个清单文件的信息，例如清单文件的位置、作为其中一部分添加的快照、关于它所属分区的信息及它所跟踪的数据文件的分区列的下界和上界，如图 11-30 所示。

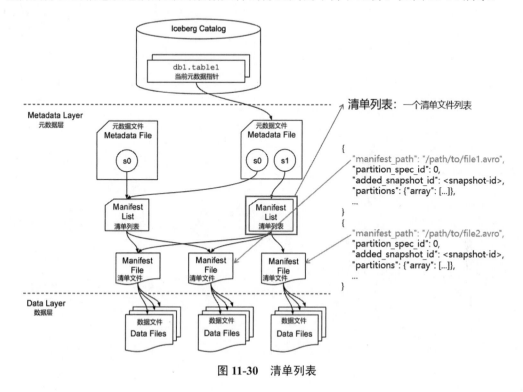

图 11-30 清单列表

一个清单列表文件的完整内容如下（为了友好显示，将 Avro 格式的文件转换为 JSON 格式显示，snap-1257424822184505371-1-eab8490b-8d16-4eb1-ba9e-0dede788ff08.avro）：

```
{
    "manifest_path": "/home/hadoop/warehouse/db2/part_table2/metadata/eab8490b-8d16-4eb1-ba9e-0dede788ff08-m0.avro",
    "manifest_length": 4884,
    "partition_spec_id": 0,
    "added_snapshot_id": {
        "long": 1257424822184505300
    },
    "added_data_files_count": {
        "int": 1
    },
    "existing_data_files_count": {
        "int": 0
    },
    "deleted_data_files_count": {
        "int": 0
    },
    "partitions": {
```

```
            "array": [ {
                "contains_null": false,
                "lower_bound": {
                    "Bytes": "¹Ô\\u0006\\u0000"
                },
                "upper_bound": {
                    "Bytes": "¹Ô\\u0006\\u0000"
                }
            } ]
        },
        "added_rows_count": {
            "long": 1
        },
        "existing_rows_count": {
            "long": 0
        },
        "deleted_rows_count": {
            "long": 0
        }
    }
    {
        "manifest_path": "/home/hadoop/warehouse/db2/part_table2/metadata/d8a778f9-ad19-4e9c-88ff-28f49ec939fa-m0.avro",
        "manifest_length": 4884,
        "partition_spec_id": 0,
        "added_snapshot_id": {
            "long": 8271497753230544000
        },
        "added_data_files_count": {
            "int": 1
        },
        "existing_data_files_count": {
            "int": 0
        },
        "deleted_data_files_count": {
            "int": 0
        },
        "partitions": {
            "array": [ {
                "contains_null": false,
                "lower_bound": {
                    "Bytes": ",Ô\\u0006\\u0000"
                },
                "upper_bound": {
                    "Bytes": ",Ô\\u0006\\u0000"
                }
            } ]
        },
        "added_rows_count": {
            "long": 1
```

```
    },
    "existing_rows_count": {
        "long": 0
    },
    "deleted_rows_count": {
        "long": 0
    }
}
```

当 SELECT 查询读取 Iceberg 表并在从元数据文件中获取快照的位置后为快照打开清单列表时，查询引擎会读取 manifest-path 条目的值，并打开清单文件。它还可以在这个阶段进行一些优化，例如使用行计数或使用分区信息过滤数据。

11.6.4 清单文件（Manifest File）

清单文件跟踪数据文件及关于每个文件的附加细节和统计信息，从而实现文件级跟踪数据。每个清单文件都是跟踪数据文件的一个子集，以实现并行性和大规模重用效率，如图 11-31 所示。

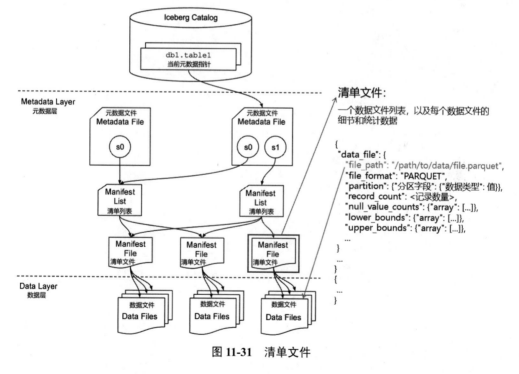

图 11-31　清单文件

清单文件包含许多有用的信息，用于在从这些数据文件中读取数据时提高效率和性能，例如关于分区成员、记录计数及列的下限和上限的详细信息。这些统计数据是在写入操作期间为每个清单的数据文件子集写入的，因此比 Hive 中的统计数据更可能存在、更准确、更及时。Iceberg 是文件格式不可知的，因此清单文件还指定了数据文件的文件格式，例如 Parquet、ORC 或 Avro。

一个清单文件的完整内容如下（为了提高可读性，转换为 JSON 格式显示）：

```
{
    "status": 1,
    "snapshot_id": {
        "long": 1257424822184505300
    },
    "data_file": {
        "file_path": "/home/hadoop/warehouse/db2/part_table2/data/ts_hour=2021-01-26-01/00000-6-7c6cf3c0-8090-4f15-a4cc-3a3a562eed7b-00001.parquet",
        "file_format": "PARQUET",
        "partition": {
            "ts_hour": {
                "int": 447673
            }
        },
        "record_count": 1,
        "file_size_in_Bytes": 973,
        "block_size_in_Bytes": 67108864,
        "column_sizes": {
            "array": [ {
                "key": 1,
                "value": 47
            },
            {
                "key": 2,
                "value": 57
            },
            {
                "key": 3,
                "value": 60
            } ]
        },
        "value_counts": {
            "array": [ {
                "key": 1,
                "value": 1
            },
            {
                "key": 2,
                "value": 1
            },
            {
                "key": 3,
                "value": 1
            } ]
        },
        "null_value_counts": {
            "array": [ {
                "key": 1,
```

```
                "value": 0
            },
            {
                "key": 2,
                "value": 0
            },
            {
                "key": 3,
                "value": 0
            } ]
        },
        "lower_bounds": {
            "array": [ {
                "key": 1,
                "value": "\\u0002\\u0000\\u0000\\u0000"
            },
            {
                "key": 2,
                "value": "\\u0000„ ‚Ã¹\\u0005\\u0000"
            },
            {
                "key": 3,
                "value": "test message 2"
            } ]
        },
        "upper_bounds": {
            "array": [ {
                "key": 1,
                "value": "\\u0002\\u0000\\u0000\\u0000"
            },
            {
                "key": 2,
                "value": "\\u0000„ ‚Ã¹\\u0005\\u0000"
            },
            {
                "key": 3,
                "value": "test message 2"
            } ]
        },
        "key_metadata": null,
        "split_offsets": {
            "array": [
                4
            ]
        }
    }
}
```

当 SELECT 查询读取 Iceberg 表并在从清单列表中获取清单文件的位置后打开一个清单文件时，查询引擎会读取每个 data-file 对象的 file-path 条目的值，并打开该数据文件。它还可以在这个阶段进行一些优化，例如使用行计数，以及使用分区或列统计信息过滤数据。

11.7 CRUD 操作的底层实现

了解了 Iceberg 表的不同组件及访问 Iceberg 表中的数据的任何引擎或工具所采取的路径之后，接下来更深入地了解在 Iceberg 表上执行 CRUD 操作时，底层会发生什么。

1. CREATE TABLE

首先，在 Iceberg Catalog 的 db1 数据库（Iceberg 中称为 namespace）中创建一个分区表，代码如下：

```
CREATE TABLE db1.table1 (
    order_id BIGINT,
    customer_id BIGINT,
    order_amount DECIMAL(10, 2),
    order_ts TIMESTAMP
)
USING iceberg
PARTITIONED BY ( HOUR(order_ts) );
```

上面在数据库 db1 中创建了一个名为 table1 的表。该表有 4 列，并按 order_ts 时间戳列的小时粒度进行分区。

当执行上面的查询时，在元数据层中创建一个带有快照 s0 的元数据文件（快照 s0 没有指向任何清单列表，因为表中还没有数据），然后，db1.table1 的当前元数据指针的 Catalog 条目被更新为指向这个新元数据文件的路径。在这个语句被执行之后，环境看起来如图 11-32 所示。

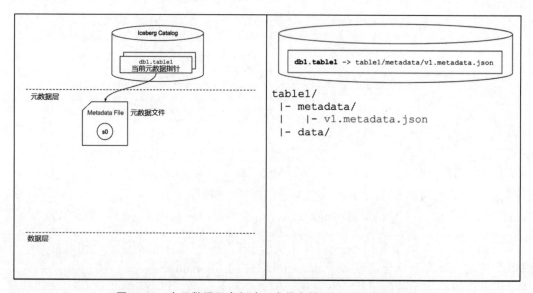

图 11-32　在元数据层中创建一个带有快照 s0 的元数据文件

2. INSERT

现在，向表 table1 中添加一些数据，代码如下：

```
INSERT INTO table1 VALUES (
    123,
    456,
    36.17,
    '2021-01-26 08:10:23'
);
```

整个创建过程如图 11-33 所示。

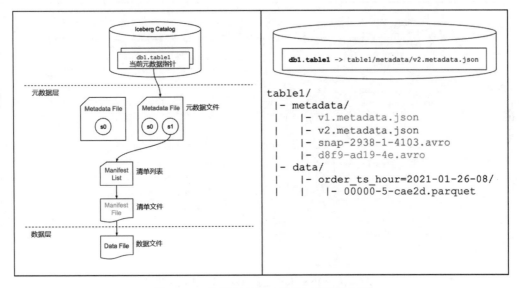

图 11-33　执行 INSERT 操作时的创建过程

当执行这个 INSERT 语句时，执行过程如下。

（1）创建 Parquet 文件形式的数据：table1/data/order_ts_hour = 2021-01-26-08/00000-5-cae2d.parquet。

（2）创建一个指向该数据文件的 Manifest 清单文件（包括额外的详细信息和统计信息）- table1/metadata/d8f9-ad19-4e.avro，每个数据文件都是 Manifest 清单文件中的一条记录。

（3）创建一个指向这个清单文件的清单列表（Manifests 文件，包括额外的详细信息和统计信息）- table1/metadata/snap-2938-1-4103.avro。Manifests 文件中的每条记录都是所有 Manifest 文件统计信息的当前集合。每个 Manifest 清单文件都是 Manifests 文件中的一个记录。

（4）基于以前的当前元数据文件创建一个新的元数据文件（快照文件），快照文件中记录与表 Schema 信息、分区的路径及 Manifests 文件的路径相对应的快照，其中包含一个新的快照 s1，并跟踪以前的快照 s0，指向这个清单列表（包括额外的详细信息和统计信息）- table1/metadata/v2.metadata.json。

(5) 用于 db1.table1 的当前元数据指针的值在 Catalog 目录中被自动更新，现在指向这个新的元数据文件。

在所有这些步骤中，读取表的任何人都将继续读取第 1 个元数据文件，直到原子步骤（5）完成，这意味着使用数据的任何人都不会看到表的状态和内容不一致的视图（整个提交过程是一个事务执行，即 ACID 保证）。

3. SELECT

再次查看 SELECT 路径，但这次是在正在处理的 Iceberg 表上，如图 11-34 所示。

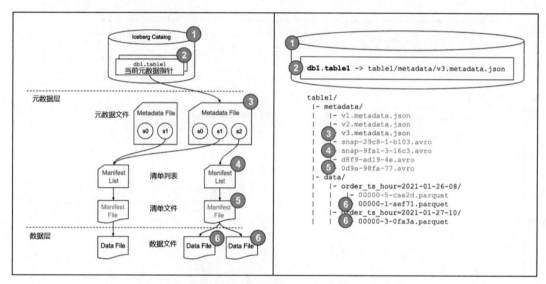

图 11-34　执行 INSERT 之后再次执行 SELECT 查询

当这个 SELECT 语句被执行时，执行过程如下。
（1）查询引擎转到 Iceberg Catalog 目录。
（2）检索 db1.table1 的当前元数据文件位置条目。
（3）打开这个元数据文件，并检索当前快照 s2 的清单列表位置的条目。
（4）打开这个清单列表，检索唯一的清单文件的位置。
（5）打开这个清单文件，检索两个数据文件的位置。
（6）读取这些数据文件，因为它是一个 SELECT*，所以将数据返回客户端。

4. 压缩

Iceberg 设计的另一个关键功能是压缩，这有助于平衡写侧和读侧。

在 Iceberg 中，压缩是一个异步后台进程，它将一组小文件压缩成更少的大文件。因为它是异步的，并且是在后台运行的，所以它不会对用户产生负面影响。实际上，它基本上是一种普通的 Iceberg 写作业，输入和输出具有相同的记录，但在写作业提交其事务后，文件大小和属性对分析有了很大的改善。

任何时候在处理数据时，都需要对所要实现的目标进行权衡，一般来讲，写侧和读侧的动机是相反的。

（1）在写操作方面，通常需要低延迟：使数据尽可能快地可用，这意味着希望在获得记录后立即写入数据，甚至可能不需要将其转换为柱状格式，但是，如果对每条记录都这样做，则最终每个文件会形成一个记录（小文件问题的最极端形式）。

（2）在读取端，通常需要高吞吐量：在一个文件中以柱状格式包含许多记录，这样与数据相关的可变成本（读取数据）将超过固定成本（保存记录、打开每个文件等的开销）。通常还需要最新的数据，但需要在读取操作中付出额外的开销。

压缩有助于平衡写侧和读侧，可以立即写入数据，在最极端的情况下，每个文件只有一行记录，读取器可以马上看到并使用，而后台压缩过程则周期性地将所有这些小文件合并成更少、更大、列状格式的文件。

通过压缩，读取器可以持续地以所需的高吞吐量形式获得 99% 的数据，但仍然可以以低延迟、低吞吐量的形式看到最近 1% 的数据。

对于这个用例，还需要注意的是，压缩作业的输入文件格式和输出文件格式可以是不同的文件类型。例如，从流写入中写入 Avro，这些写入被压缩成更大的 Parquet 文件用于分析。

另一个重要的注意事项是，由于 Iceberg 不是引擎或工具，因此调度/触发和实际压缩工作是由与 Iceberg 集成的其他工具和引擎完成的。

11.8 增量更新与合并更新

Iceberg 不但支持增量更新，还支持合并更新。Apache Spark 3 增加了对 MERGE INTO 查询的支持，可以表示行级更新。通过重写包含在 overwrite 提交中需要更新的行的数据文件，Iceberg 支持 MERGE INTO。

本节通过一个示例来演示 Iceberg 表中的增量更新实现和合并更新实现。

1. 配置 Catalog

要支持增量更新和合并更新，需要配置扩展 spark.sql.extensions，并指定 Catalog。下面以 Zeppelin Notebook 为例，配置它并配置基于 Hadoop 路径的 Catalog，代码如下：

```
%spark.conf
spark.sql.extensions
    org.apache.iceberg.spark.extensions.IcebergSparkSessionExtensions
spark.sql.catalog.hadoop_prod
    org.apache.iceberg.spark.SparkCatalog
spark.sql.catalog.hadoop_prod.type         hadoop
spark.sql.catalog.hadoop_prod.warehouse
    hdfs://xueai8:8020/data_lake/iceberg
```

2. 初始数据湖表

首先将一个原始商品数据集写入 Iceberg 数据湖中,创建一个原始的 Iceberg 表。这个原始商品数据集文件 phones.csv 的内容如下:

```
id,price,stock
小米,1299.00,100
苹果,3299.00,300
华为,2299.00,200
```

将这个 CSV 文件读到 DataFrame,然后写入一个 Iceberg 表中,代码如下:

```
//使用 Spark SQL API 将这个 CSV 文件读到一个 DataFrame 中
val path = "/data/data_lake/iceberg/phones.csv"
val df = spark
  .read
  .option("header", "true")
  .option("charset", "UTF8")
  .csv(path)

//查看模式和内容
df.printSchema()
df.show
```

执行以上代码,输出内容如下:

```
root
 |-- id: string (nullable = true)
 |-- price: string (nullable = true)
 |-- stock: string (nullable = true)

+----+-------+-----+
|  id|  price|stock|
+----+-------+-----+
|小米|1299.00|  100|
|苹果|3299.00|  300|
|华为|2299.00|  200|
+----+-------+-----+
```

然后将该 DataFrame 作为一个 Iceberg 数据湖写出,代码如下:

```
df
  .repartition(1)
  .writeTo("hadoop_prod.my_db.phones")
  .create()          //创建 Iceberg 表

//查看写入是否成功
spark.table("hadoop_prod.my_db.phones").show()
```

执行以上代码,应该可以看到输出的表数据,内容如下:

```
+----+-------+-----+
|  id|  price|stock|
+----+-------+-----+
|小米|1299.00|  100|
```

```
| 苹果|3299.00|  300|
| 华为|2299.00|  200|
+----+-------+-----+
```

查看 hadoop_prod.my_db.phones 表的物理存储结构，在终端命令行下执行的命令如下：

```
$ hdfs dfs -ls -R /data_lake/iceberg/my_db/phones
```

可以看到目录结构如图 11-35 所示。

```
/data_lake/iceberg/my_db/phones/data
/data_lake/iceberg/my_db/phones/data/00000-55-98bc1a36-e36e-43f3-bed7-f399b7e423f6-00001.parquet
/data_lake/iceberg/my_db/phones/metadata
/data_lake/iceberg/my_db/phones/metadata/46d1d556-f331-4e23-ae8d-469a95efc304-m0.avro
/data_lake/iceberg/my_db/phones/metadata/snap-4093447139497540820-1-46d1d556-f331-4e23-ae8d-469a95efc304.avro
/data_lake/iceberg/my_db/phones/metadata/v1.metadata.json
/data_lake/iceberg/my_db/phones/metadata/version-hint.text
```

图 11-35　hadoop_prod.my_db.phones 表的物理存储结构

3. 增量更新

Apache Iceberg 支持增量更新。接下来向 hadoop_prod.my_db.phones 表插入一条新的数据记录，代码如下：

```
//增加一条记录
spark.sql("insert into hadoop_prod.my_db.phones values('Nokia',699.00,350)")

//查看这时表中的数据
spark.table("hadoop_prod.my_db.phones").show()
```

执行以上代码，输出的结果如下：

```
+------+-------+-----+
|    id|  price|stock|
+------+-------+-----+
|  小米|1299.00|  100|
|  苹果|3299.00|  300|
|  华为|2299.00|  200|
| Nokia| 699.00|  350|
+------+-------+-----+
```

也可以批量增加数据。例如有一个新的数据文件 append_phones.csv，该数据文件的内容如下（注意其包含的数据与 phones.csv 文件中的数据并无任何重复）：

```
id,price,stock
oppo,1688.00,230
noov,499.00,310
```

将数据文件追加到 hadoop_prod.my_db.phones 表中，代码如下：

```
//增量数据文件
val appendData = "/data/data_lake/iceberg/append_phones.csv"

//追加到表中
spark
```

```
    .read
    .option("header", "true")
    .option("charset", "UTF8")
    .csv(appendData)
    .writeTo("hadoop_prod.my_db.phones")
    .append()

//查看此时表中的数据
spark.table("hadoop_prod.my_db.phones").show()
```

执行以上代码，输出的结果如下：

```
+-----+-------+-----+
|   id|  price|stock|
+-----+-------+-----+
| oppo|1688.00|  230|
| noov| 499.00|  310|
|Nokia| 699.00|  350|
|   小米|1299.00|  100|
|   苹果|3299.00|  300|
|   华为|2299.00|  200|
+-----+-------+-----+
```

4. 合并更新（MERGE INTO / UPSERT）

随着 Apache Spark 3.0 的发布，Apache Iceberg 通过 MERGE INTO 查询支持 upsert。它们使用直接的"写时复制（copy-on-write）"方法，即需要更新记录的文件会立即被重写。

MERGE INTO 使用一组更新数据（称为 source）来更新一张表（称为 target 表）。使用类似于连接条件的 ON 子句可以找到 target 表中某一行的更新。MERGE INTO 的语法如下：

```
MERGE INTO prod.db.target t           -- target 表
USING (SELECT ...) s                  -- source 更新
ON t.id = s.id                        -- 用于查找目标行更新的条件
WHEN ...                              -- 更新
```

对目标表中行的更新使用 WHEN MATCHED…THEN…列出。多个 MATCHED 子句可以添加条件，以确定何时应用每个匹配。使用第 1 个匹配的表达式，模板代码如下：

```
WHEN MATCHED AND s.op = 'delete' THEN DELETE
WHEN MATCHED AND t.count IS NULL AND s.op = 'increment' THEN UPDATE SET t.count = 0
WHEN MATCHED AND s.op = 'increment' THEN UPDATE SET t.count = t.count + 1
```

不匹配的 source 行(更新)可以插入，模板代码如下：

```
WHEN NOT MATCHED THEN INSERT *
```

插入还支持其他条件，模板代码如下：

```
WHEN NOT MATCHED AND s.event_time > still_valid_threshold THEN INSERT (id, count) VALUES (s.id, 1)
```

源数据（source）中只有一条记录可以更新目标表中的任何给定行，否则将抛出错误。

假设现在有一个包含一组更新数据的文件 merge_phones.csv，其中既包含对表 my_db.phones 中部分数据的修改（例如修改价格 price、库存 stock 等），也包含表 my_db.phones 中未曾存储过的数据（在新增加的数据），内容如下：

```
id,price,stock
小米,1199.00,100
苹果,2799.00,150
vivo,2199.00,120
```

现在希望将这个 merge_phones.csv 文件中的数据与 my_db.phones 表中原有数据进行合并：如果商品 ID 已经在表中，则更新该商品价格和库存数量；如果表中还没有那种商品的记录，就将这个新商品的记录插到表中。

首先将这个新数据文件加载到 DataFrame 中，再写入 Iceberg 数据湖的 new_phones 表中，代码如下：

```
//新的数据文件
val mergeData = "/data/data_lake/iceberg/merge_phones.csv"

//读到 DataFrame 中
val mergeDF = spark
  .read
  .option("header", "true")
  .option("charset", "UTF8")
  .csv(mergeData)

//mergeDF.show()

//写入 Iceberg 中
mergeDF
  .repartition(1)
  .writeTo("hadoop_prod.my_db.new_phones")
  .create()

//查看写入是否成功
spark.table("hadoop_prod.my_db.new_phones").show()
```

执行以上代码，输出的结果如下：

```
+----+-------+-----+
|  id|  price|stock|
+----+-------+-----+
|小米|1199.00|  100|
|苹果|2799.00|  150|
|vivo|2199.00|  120|
+----+-------+-----+
```

从输出结果可以看出，因为小米和苹果产品已在源表中，所以对这两个产品的价格和库存数量做了修改，而 vivo 在源表中是没有的，所以执行插入操作。

现在有了一个源表 new_phones 和一个目标表 phones。接下来，将源表合并到目标表

中,执行 merge into 操作,代码如下:

```
//merge into / upsert
spark.sql( """
    merge into hadoop_prod.my_db.phones p
    using (select * from hadoop_prod.my_db.new_phones) n
    on p.id=n.id
    when matched then update set p.price=n.price,p.stock=n.stock
    when not matched then insert *
""")
//查看合并结果
spark.table("hadoop_prod.my_db.phones").show()
```

执行以上合并更新代码后,再次查询表中数据,得到的输出内容如下:

```
+-----+-------+-----+
|   id|  price|stock|
+-----+-------+-----+
| oppo|1688.00|  230|
| noov| 499.00|  310|
| vivo|2199.00|  120|
|   华为|2299.00|  200|
|   小米|1199.00|  100|
|   苹果|2799.00|  150|
|Nokia| 699.00|  350|
+-----+-------+-----+
```

从上面的输出结果可以看出,vivo 所在行是新增的商品记录,而小米和苹果所在行的两种商品价格已得到调整。

5. 增量读取

要增量地读取追加的数据,可以使用以下两种方式。

(1) start-snapshot-id:用于增量扫描的起始快照 ID(不包含)。

(2) end-snapshot-id:用于增量扫描的结束快照 ID(包含)。这是可选的,如果忽略它,则将默认为当前快照。

首先查看所有的快照信息,代码如下:

```
spark.sql("select * from hadoop_prod.my_db.phones.snapshots")
    .select("committed_at","snapshot_id","parent_id","operation")
    .show(false)
```

执行以上代码,输出的结果如下:

```
+-----------------------+-------------------+-------------------+---------+
|committed_at           |snapshot_id        |parent_id          |operation|
+-----------------------+-------------------+-------------------+---------+
|2022-03-14 19:54:35.235|4093447139497540820|null               |append   |
|2022-03-14 20:02:29.811|3260015930261212891|4093447139497540820|append   |
|2022-03-14 20:02:36.021|3128063198822559538|3260015930261212891|append   |
|2022-03-14 20:08:16.768|4514244045809687626|3128063198822559538|append   |
+-----------------------+-------------------+-------------------+---------+
```

下面获取 start-snapshot-id（3128063198822559538L）之后到 end-snapshot-id（4514244045809687626L）的增量数据，代码如下：

```
spark.read
  .format("iceberg")
  .option("start-snapshot-id", "3128063198822559538")
  .option("end-snapshot-id", "4514244045809687626")
  .load("hadoop_prod.my_db.phones")
  .show()
```

执行以上代码，可以得到的输出内容如下：

```
+----+-------+-----+
|  id| price |stock|
+----+-------+-----+
|oppo|1688.00|  230|
|noov| 499.00|  310|
+----+-------+-----+
```

注意：当前仅从 append 追加操作获取数据。不支持 replace、overwrite、delete 操作。Spark 的 SQL 语法不支持增量读。

11.9　时间旅行

Iceberg 表格式支持的另一个关键功能是所谓的"时间旅行"。Iceberg 提供了开箱即用的功能，可以查看一张表在过去不同时间点的样子。

为了选择特定的表快照或某个时间点的快照，Iceberg 支持以下两个 Spark 读选项。

（1）snapshot-id：选择一个特定的表快照。

（2）as-of-timestamp：选择某个时间戳（以毫秒为单位）的当前快照。

用户可以查看某张表的 snapshots 快照元数据表来查看该表的快照信息。例如，要查看 hadoop_prod.my_db.phones 表的所有快照，代码如下：

```
//查看所有的快照
spark.sql("select * from hadoop_prod.my_db.phones.snapshots").show()
```

可以得到的输出信息如图 11-36 所示。

```
+-------------------+-------------------+-------------------+---------+-------------------+-------------------+
|       committed_at|        snapshot_id|          parent_id|operation|      manifest_list|            summary|
+-------------------+-------------------+-------------------+---------+-------------------+-------------------+
|2022-03-14 19:54:..|4093447139497540820|               null|   append|hdfs://xueai8:802..|{spark.app.id -> ..|
|2022-03-14 20:02:..|3260015930261212891|4093447139497540820|   append|hdfs://xueai8:802..|{spark.app.id -> ..|
|2022-03-14 20:02:..|3128063198822559538|3260015930261212891|   append|hdfs://xueai8:802..|{spark.app.id -> ..|
|2022-03-14 20:08:..|4514244045809687626|3128063198822559538|   append|hdfs://xueai8:802..|{spark.app.id -> ..|
+-------------------+-------------------+-------------------+---------+-------------------+-------------------+
```

图 11-36　查看 hadoop_prod.my_db.phones 表的所有快照

可以只选取部分字段以查看完整的提交时间，代码如下：

```
spark.sql("select * from hadoop_prod.my_db.phones.snapshots")
    .select("committed_at","snapshot_id","parent_id","operation")
    .show(false)
```

执行上面的代码，输出内容如图 11-37 所示。

```
+-----------------------+--------------------+-------------------+---------+
|committed_at           |snapshot_id         |parent_id          |operation|
+-----------------------+--------------------+-------------------+---------+
|2022-03-14 18:49:58.408|6656301523862263880 |null               |append   |
|2022-03-14 18:50:20.936|6959645350842185769 |6656301523862263880|append   |
|2022-03-14 18:50:29.341|7048516951834814451 |6959645350842185769|append   |
|2022-03-14 18:59:32.779|2071183445152070492 |7048516951834814451|overwrite|
+-----------------------+--------------------+-------------------+---------+
```

图 11-37　查看 hadoop_prod.my_db.phones 表的所有快照的完整提交时间

其中 snapshot_id 字段代表每个快照的版本编号，committed_at 字段代表该版本快照的提交时间。

如果想要查看指定版本或时间点的表内容(所谓时间旅行)，则可以使用 as-of-timestamp 选项。例如，指定查看 2022-03-14 19:00:25 时刻的表的内容，可以执行时间旅行，代码如下：

```
//时间旅行到 2022-03-14 19:00:25
import java.sql.Timestamp

spark.read
    .option("as-of-timestamp", Timestamp.valueOf("2022-03-14 19:00:25").getTime())
    .format("iceberg")
    .load("hadoop_prod.my_db.phones")
    .show()
```

执行以上代码，可以得到的输出内容如下：

```
+-----+-------+-----+
|   id|  price|stock|
+-----+-------+-----+
| oppo|1688.00|  230|
| noov| 499.00|  310|
|Nokia| 699.00|  350|
| vivo|2199.00|  120|
| 华为|2299.00|  200|
| 小米|1199.00|  100|
| 苹果|2799.00|  150|
+-----+-------+-----+
```

如果再旅行到时间点 2022-03-14 18:52:25，查看这个时刻表中的内容，则可执行的代码如下：

```
//时间旅行到 2022-03-14 18:52:25
import java.sql.Timestamp

spark.read
    .option("as-of-timestamp", Timestamp.valueOf("2022-03-14 18:52:25").
```

```
   getTime())
      .format("iceberg")
      .load("hadoop_prod.my_db.phones")
      .show()
```

可以得到的输出内容如下：

```
+-----+-------+-----+
|   id|  price|stock|
+-----+-------+-----+
| oppo|1688.00|  230|
| noov| 499.00|  310|
| 小米|1299.00|  100|
| 苹果|3299.00|  300|
| 华为|2299.00|  200|
|Nokia| 699.00|  350|
+-----+-------+-----+
```

也可以通过 snapshot-id 选项查看指定快照版本对应的表内容。例如，要查看 snapshot-id 为 7625222922103268949 的快照对应的表内容，代码如下：

```
//时间旅行到 ID 为 7625222922103268949 的快照
spark.read
   .option("snapshot-id", 6959645350842185769L)
   .format("iceberg")
   .load("hadoop_prod.my_db.phones")
   .show()
```

执行上面的代码，可以得到的输出内容如下：

```
+-----+-------+-----+
|   id|  price|stock|
+-----+-------+-----+
| 小米|1299.00|  100|
| 苹果|3299.00|  300|
| 华为|2299.00|  200|
|Nokia| 699.00|  350|
+-----+-------+-----+
```

如果将时间旅行到 ID 为 8262259194646368185 的快照，查看此时的表内容，则可执行的代码如下：

```
//将时间旅行到 ID 为 8262259194646368185 的快照
spark.read
   .option("snapshot-id", 7048516951834814451L)
   .format("iceberg")
   .load("hadoop_prod.my_db.phones")
   .show()
```

执行以上代码，可以得到的表内容如下：

```
+-----+-------+-----+
|   id|  price|stock|
+-----+-------+-----+
| oppo|1688.00|  230|
```

```
| noov  | 499.00 | 310 |
| 小米  |1299.00 | 100 |
| 苹果  |3299.00 | 300 |
| 华为  |2299.00 | 200 |
|Nokia  | 699.00 | 350 |
+-------+--------+-----+
```

注意：Spark 目前不支持在 DataFrameReader 命令中使用带有 option 的 table()方法，所有 option 选项将被无声地忽略。当尝试时间旅行或使用其他选项时，不要使用 table。Options 选项将在 Spark 3.1 - Spark-32592 中支持 table；Spark 的 SQL 语法还不支持时间旅行。

11.10 隐藏分区和分区演变

与其他数据湖产品相比，Iceberg 提供了对隐藏分区和分区演变的支持。在深入了解隐藏分区和分区演变之前，先来重温一下分区的概念。

11.10.1 分区概念

分区是一种编写时通过将相似的行分组在一起来加快查询速度的方法。例如，对 logs 日志表中的日志项的查询通常包含一个时间范围，例如对上午 10 点到 12 点的日志的查询，可以使用的 SQL 语句如下：

```
SELECT level, message FROM logs
WHERE event_time BETWEEN '2018-12-01 10:00:00' AND '2018-12-01 12:00:00'
```

将 logs 表配置为按 event_time 日期分区将会把日志事件分组到具有相同事件日期的文件中。Iceberg 会跟踪这个日期，并使用它来跳过没有有用数据的其他日期的文件。

Iceberg 可以按年、月、日和小时粒度对时间戳分区。它还可以使用分类列（如 logs 表中的 level 列，代表日志级别）将行存储在一起，从而加快查询速度。

与其他支持分区的表格式（如 Hive）相比，Iceberg 还支持隐藏分区。什么是隐藏分区？

（1）Iceberg 负责处理为表中的行生成分区值的烦琐且容易出错的任务。

（2）Iceberg 避免自动读取不必要的分区。使用者不需要知道表是如何分区的，也不需要在查询中添加额外的过滤器。

（3）Iceberg 分区布局可以根据需要发展演化。

为了进一步理解 Iceberg 支持的隐藏分区，考虑一下 Hive 是如何实现和管理分区的。

1. Hive 分区

在 Hive 中，分区是显式的，以列的形式出现，所以 logs 分区表中会有一个名为 event_date 的列。在每次写表时都必须提供分区，所以当写入时，insert 需要为 event_date 列提供数据，代码如下：

```
INSERT INTO logs PARTITION (event_date)
  SELECT level, message, event_time, format_time(event_time, 'YYYY-MM-dd')
    FROM unstructured_log_source
```

类似地，Hive 中的查询还必须为分区列显式地提供一个过滤器。例如在 logs 分区表中执行查询操作，除了需要提供 event_time 过滤器外，还必须具有 event_date 过滤器，代码如下：

```
SELECT level, count(1) as count FROM logs
WHERE event_time BETWEEN '2018-12-01 10:00:00' AND '2018-12-01 12:00:00'
  AND event_date = '2018-12-01'
```

如果缺少 event_date 过滤器，Hive 则会执行全表扫描（扫描表中的每个文件），因为它不知道 event_time 列与 event_date 列相关，所以 Hive 分区的问题就在于，Hive 必须被赋予分区值。在 logs 日志表中，它并不知道 event_time 和 event_date 之间的关系。

这导致了以下几个问题：

（1）Hive 不能验证分区值，要生成正确的值取决于写入器 writer。

（2）如果使用错误的格式（例如 2018-12-01 而不是 20181201），则将产生静默的错误结果，而不是查询失败。

（3）使用错误的源列（如 processing_time 或时区）也会导致不正确的结果，而不是失败。

（4）正确地编写查询取决于用户。

（5）使用错误的格式也会导致静默的错误结果。

（6）不理解表的物理布局的用户会得到不必要的慢查询，Hive 不能自动翻译过滤器。

（7）工作查询被绑定到表的分区模式（Schema），因此不能在不破坏查询的情况下更改分区配置。

2. Iceberg 隐藏分区

Iceberg 通过获取列值并可选地转换它来生成分区值。Iceberg 负责将 event_time 转换为 event_date，并跟踪它们之间的关系。表分区是使用这些关系配置的。日志表 logs 将按照日期（event_time）和级别（level）进行分区。

因为 Iceberg 不需要用户维护分区列，所以它可以隐藏分区。Iceberg 通过实现隐藏分区为用户简化了分区。每次查询都会正确地生成分区值，并且总是在可能的情况下用于加速查询。Iceberg 没有强制用户在查询时提供单独的分区过滤器，而是在底层处理分区和查询所有细节。用户不需要维护分区列，甚至不需要理解物理表布局就可以得到准确的查询结果。

隐藏分区最重要的好处是，用户在查询 Iceberg 表时不需要提供分区布局信息，查询不再依赖于表的物理布局。通过物理和逻辑的分离，Iceberg 表可以随着数据量的变化而演化分区模式。不需要进行昂贵的迁移就可以修复配置错误的表。这不仅使 Iceberg 分区非常友好，而且还允许在不破坏预先编写的查询的情况下随时间改变分区布局。在演化分区规范时，更改之前表中的数据不受影响，其元数据也不受影响。只有在演化之后写入表的数据才会用新规范进行分区，并且这个新数据集的元数据会单独保存。在查询时，每个分区布局的各自元数据被用来标识它需要访问的文件，这被称为拆分规划（split-planning）。拆分规划是 Iceberg 的众多特性之一，这些特性是由于表元数据而得以实现的，它创建了物理和逻辑之间的分离。

3. 分区演变

Hive 表不支持就地分区演变。在 Hive 中，要更改分区，必须使用新的分区列完全重写整个表。这对于大型表来讲代价很高，并可能导致数据准确性问题。此外，依赖于表分区的查询现在必须为新表重写，而 Iceberg 以一种不会导致这些问题的方式实现分区。

Iceberg 支持就地表进化，可以在数据量改变时改变分区布局。Iceberg 不需要昂贵的迁移，例如重写表数据或迁移到新表。例如，因为 Hive 表分区不能改变，所以从按 day 分区布局到按 hour 分区布局就需要一个新表，而由于查询依赖于分区，因此必须为新表重写查询。

因为在 Iceberg 表中查询不直接引用分区值，所以分区可以在现有表中更新。当演变一个分区规范时，使用早期规范编写的旧数据保持不变。在新的布局中使用新规范写入新数据。每个分区版本的元数据是分开保存的。正因为如此，当用户开始编写查询时，就得到了拆分计划。在拆分计划中，每个分区布局使用它为特定分区布局派生的过滤器分别规划文件。这个过程如图 11-38 所示。

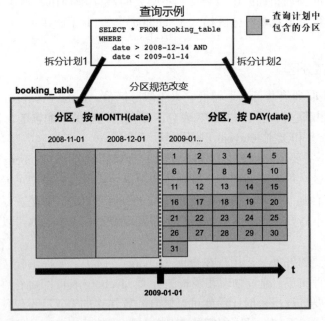

图 11-38 分区演变过程

在图 11-38 中，2008 年的数据按月分区，从 2009 年开始对表进行了更新，这样数据就可以按天分区了。这两种分区布局可以在同一张表中共存。

由于 Iceberg 使用了隐藏的分区，因此不需要为特定的分区布局编写查询来提高速度。相反，可以编写选择所需数据的查询，Iceberg 会自动删除不包含匹配数据的文件。

分区演化是一种元数据操作，不会立即重写文件。Spark 支持通过 ALTER TABLE SQL 语句更新分区规范。

11.10.2 分区演变示例

本节通过一个应用示例来深入理解 Iceberg 的分区演变及其底层实现过程。该应用示例模拟了如下的业务场景：

> 某公司（暂且称为 X 公司）从 2008 年开始开发软件业务，拓展客户和市场。其日志事件数据是按月分区存储在 Iceberg 数据湖中。随着业务的发展，日志事件信息发生得越来越频繁。公司决定，从 2009 年开始，将日志事件按天进行分区存储。为此，开发如下代码来完成这个任务。

为了简单起见，本示例使用一个简单的数据集 logdata.csv，它模仿 X 公司开发的某些软件产品的日志表，包含 3 个字段列，分别为 ts、log_id 和 log_msg。注意，数据中的 ts 列显示为与 UNIX 时间戳（以秒为单位）对应的 long 数据类型。

数据集 logdata.csv 文件的内容如下：

```
1225526400,1,a
1225699200,2,b
1225785600,3,c
1226476800,4,d
1226908800,5,e
1226995200,6,f
1227513600,7,g
1227772800,8,h
1228032000,9,i
1228118400,10,j
1228377600,11,k
1228809600,12,l
1228982400,13,m
1229673600,14,n
1230019200,15,o
1230278400,16,p
1230451200,17,q
1230624000,18,r
1230710400,19,s
1230796800,20,t
1230969600,21,u
1231747200,22,v
1232352000,23,w
1232784000,24,x
1233216000,25,y
1233302400,26,z
```

继续使用 HadoopCatalog 类型的 Catalog，基于 Hadoop 路径。名为 hadoop_prod 的 Catalog 配置如下：

```
%spark.conf
spark.sql.extensions
   org.apache.iceberg.spark.extensions.IcebergSparkSessionExtensions
spark.sql.catalog.hadoop_prod
   org.apache.iceberg.spark.SparkCatalog
spark.sql.catalog.hadoop_prod.type          hadoop
spark.sql.catalog.hadoop_prod.warehouse
   hdfs://localhost:8020/data_lake/iceberg
```

建议按以下步骤操作。

(1) 在 my_db 命名空间中创建一个名为 logtable 的 Iceberg 表，这个表最初是按事件月份进行分区的，代码如下：

```
//在名为 my_db 的命名空间中创建表(logtable)
spark.sql("""
   CREATE TABLE hadoop_prod.my_db.logtable (
      ts         timestamp,
      log_id     bigint,
      log_msg    string
   )
   USING iceberg
   PARTITIONED BY (months(ts))
""")
```

如果没有 my_db 命名空间，则先创建它，代码如下：

```
spark.sql("crete namespace hadoop_prod.my_db")
```

执行以上代码，然后在终端窗口中查看 logtable 对应的物理结构，Shell 命令如下：

```
$ hdfs dfs -ls -R /data_lake/iceberg/my_db/logtable
```

可以看到 hadoop_prod.my_db.logtable 对应的物理存储结构如图 11-39 所示。

```
/data_lake/iceberg/my_db/logtable/metadata
/data_lake/iceberg/my_db/logtable/metadata/v1.metadata.json
/data_lake/iceberg/my_db/logtable/metadata/version-hint.text
```

图 11-39　分区演变过程

在每个 Iceberg 表文件夹中，都有一个元数据文件夹（metadata）和一个数据文件夹（data）。元数据文件夹（metadata）包含关于分区规范、它们的唯一 ID 的信息，以及使用适当的分区规范 ID 连接各个数据文件的清单。数据文件夹（data）包含构成整个 Iceberg 表的所有表数据文件。在图 11-39 中我们只看到 metadata 元数据目录，而没有 data 目录，这是因为刚创建的空表中还没有任何数据。另外，v1.metadata.json 是元数据文件，而 version-hint.text 文件中则标记了当前的元数据版本号。可以查看当前的版本信息，代码如下：

```
//查看当前的版本信息
spark
   .read
```

```
    .text("/data_lake/iceberg/my_db/logtable/metadata/version-hint.text")
    .show()
```

执行上面的代码,输出内容包含下一个快照的 ID,内容如下:

```
+-----+
|value|
+-----+
|    1|
+-----+
```

再查看 v1.metadata.json 文件的内容,代码如下:

```
spark
    .read
    .text("/data_lake/iceberg/my_db/logtable/metadata/v1.metadata.json")
    .show(100,false)
```

下面显示了其中的部分内容:

```
...
 "partition-spec" : [ {
    "name" : "ts_month",
    "transform" : "month",
    "source-id" : 1,
    "field-id" : 1000
  } ],
  "default-spec-id" : 0,
  "partition-specs" : [ {
    "spec-id" : 0,
    "fields" : [ {
      "name" : "ts_month",
      "transform" : "month",
      "source-id" : 1,
      "field-id" : 1000
    } ]
  } ],
  ...
  "current-snapshot-id" : -1,
  "snapshots" : [ ],
  "snapshot-log" : [ ],
  "metadata-log" : [ ]
```

可以看出,当前的分区规范要求通过 month 转换按月进行分区。另外当前的快照 ID 是–1。

(2)将数据添加到表中。在下面的命令中,只添加时间戳为 2009 年 1 月 1 日之前的数据,模拟了示例中的场景。当写入表时,确保在分区列上排序数据,代码如下:

```
//首先将数据源读到 DataFrame 中
import org.apache.spark.sql.types._

//数据文件路径
val filePath = "/data/data_lake/iceberg/logdata.csv"

//指定 Schema 中的字段及类型
```

```
    val fields = Array(
              StructField("ts", LongType, true),
              StructField("log_id", IntegerType, true),
              StructField("log_msg", StringType, true)
            )
//创建 Schema
val schema = StructType(fields)

//将数据加载到 DataFrame 中
val data = spark
  .read
  .option("delimiter", ",")
  .option("header", "false")
  .option("charset", "UTF8")
  .schema(schema)
  .csv(filePath)

//data.printSchema()
//data.show()

//将 ts 字段转换为时间戳,并选择2009年1月1日之前的数据,写入 Iceberg 表中
data.where(col("ts").lt(1230768000L))
  .select(
      col("ts").cast(DataTypes.TimestampType),
      col("log_id"),
      col("log_msg"))
  .sort(col("ts"))
  .repartition(1)
  .write
  .format("iceberg")
  .mode("overwrite")
  .save("hadoop_prod.my_db.logtable")
```

以上命令执行完毕后,在终端窗口中执行以下 Shell 命令查看 logtable 对应的物理结构:

```
$ hdfs dfs -ls -R  /data_lake/iceberg/my_db/logtable
```

可以看到现在 hadoop_prod.my_db.logtable 对应的物理存储结构如图 11-40 所示。

从图 11-40 可以看出,现在多了一个 data 文件夹,其中包含两个分区文件夹 ts_month=2008-11 和 ts_month=2008-12,在这两个分区文件夹下是新增加的 Parquet 格式的数据文件。在 metadata 元数据文件夹下,多了几个新的文件,其中/metadata/v2.metadata.json 文件是当前的元数据文件,它包含了对/metadata/snap-*.avro 文件的引用(该文件是清单列表文件),而/metadata/61224930-*.avro 是清单文件。

注意:在每个 Iceberg 表文件夹中,都有一个元数据文件夹(metadata)和一个数据文件夹(data)。元数据文件夹包含关于分区规范、它们的唯一 ID 的信息,以及使用适当的分区规范 ID 连接各个数据文件的清单。数据文件夹包含构成整个 Iceberg 表的所有表数据文件。当向带有分区的表写入数据时,Iceberg 会在数据文件夹中创建多个文件夹。每个分区都用分区描述和值命名。例如,一个以 time 为标题并以 month 为分区的列将有文件夹

time_month=2008-11、time_month=2008-12 等。我们将在下面的例子中看到这一点。在多个列上分区的数据创建多个分层的文件夹，每个顶级文件夹包含一个或多个用于每个二级分区值的子文件夹。

```
/data_lake/iceberg/my_db/logtable/data
/data_lake/iceberg/my_db/logtable/data/ts_month=2008-11
/data_lake/iceberg/my_db/logtable/data/ts_month=2008-11/00000-59-6c97ce6d-e6bf-48cb-aeaa-07f07b7558f2-00001.parquet
/data_lake/iceberg/my_db/logtable/data/ts_month=2008-11/00001-60-607c28d1-631f-411e-a543-2666f8ef0758-00001.parquet
/data_lake/iceberg/my_db/logtable/data/ts_month=2008-11/00002-61-7195b35b-3231-4b2c-aac5-30ff3db02513-00001.parquet
/data_lake/iceberg/my_db/logtable/data/ts_month=2008-11/00003-62-c21e4442-0764-4979-a1a6-dd23eb05e8c6-00001.parquet
/data_lake/iceberg/my_db/logtable/data/ts_month=2008-11/00004-63-dc281250-49be-4f9a-a727-6447beae5b00-00001.parquet
/data_lake/iceberg/my_db/logtable/data/ts_month=2008-11/00005-64-8afb885f-b111-4c31-a928-93fa537af2f1-00001.parquet
/data_lake/iceberg/my_db/logtable/data/ts_month=2008-11/00006-65-0e91c1fa-6124-4104-ba94-0a78dfce22e9-00001.parquet
/data_lake/iceberg/my_db/logtable/data/ts_month=2008-11/00007-66-9df2de50-89fa-4545-b9bb-ff4ecbdf6759-00001.parquet
/data_lake/iceberg/my_db/logtable/data/ts_month=2008-11/00008-67-75624e68-e0a3-4af1-8c40-b5741ee038ee-00001.parquet
/data_lake/iceberg/my_db/logtable/data/ts_month=2008-12
/data_lake/iceberg/my_db/logtable/data/ts_month=2008-12/00009-68-090a9d96-e084-4696-997b-48a1489bcac6-00001.parquet
/data_lake/iceberg/my_db/logtable/data/ts_month=2008-12/00010-69-49cd18e7-1479-4009-8a18-300ef6afa43d-00001.parquet
/data_lake/iceberg/my_db/logtable/data/ts_month=2008-12/00011-70-6f2ff456-d33c-401a-adc3-fb6e3b2b76f7-00001.parquet
/data_lake/iceberg/my_db/logtable/data/ts_month=2008-12/00012-71-a2b9da6d-e089-48c0-aec1-6a1ba478a442-00001.parquet
/data_lake/iceberg/my_db/logtable/data/ts_month=2008-12/00013-72-f3a9d6e2-4b4d-471f-93d3-78d1126d1448-00001.parquet
/data_lake/iceberg/my_db/logtable/data/ts_month=2008-12/00014-73-6aa425dc-10a5-4ea6-8213-525ec0fafbc0-00001.parquet
/data_lake/iceberg/my_db/logtable/data/ts_month=2008-12/00015-74-89bc93c5-1bd9-4fc8-8ccc-64a2450758b9-00001.parquet
/data_lake/iceberg/my_db/logtable/data/ts_month=2008-12/00016-75-30b3d8bc-ecfe-4534-a150-e9055e5a43fd-00001.parquet
/data_lake/iceberg/my_db/logtable/data/ts_month=2008-12/00017-76-4b20c2eb-a176-4e46-91f3-4d8bae012749-00001.parquet
/data_lake/iceberg/my_db/logtable/data/ts_month=2008-12/00018-77-1f9ef4e2-3bed-42d6-ac2a-f4da411e392b-00001.parquet
/data_lake/iceberg/my_db/logtable/metadata
/data_lake/iceberg/my_db/logtable/metadata/61224930-1e9e-424f-b089-4f0ddbdf0cd7-m0.avro
/data_lake/iceberg/my_db/logtable/metadata/snap-6579228024748445391-1-61224930-1e9e-424f-b089-4f0ddbdf0cd7.avro
/data_lake/iceberg/my_db/logtable/metadata/v1.metadata.json
/data_lake/iceberg/my_db/logtable/metadata/v2.metadata.json
/data_lake/iceberg/my_db/logtable/metadata/version-hint.text
```

图 11-40　分区演变过程

查看元数据文件 v2.metadata.json，其中部分内容如下：

```
...
  "current-snapshot-id" : 6579228024748445391,
  "snapshots" : [ {
    "snapshot-id" : 6579228024748445391,
    "timestamp-ms" : 1647324680804,
    "summary" : {
      "operation" : "append",
      "spark.app.id" : "app-20220315124150-0001",
      ...
    },
    "manifest-list" : "hdfs://192.168.190.133:8020/data_lake/iceberg/my_db/logtable/metadata/snap-6579228024748445391-1-61224930-1e9e-424f-b089-4f0ddbdf0cd7.avro",
    "schema-id" : 0
  } ],
  "snapshot-log" : [ {
    "timestamp-ms" : 1647324680804,
    "snapshot-id" : 6579228024748445391
  } ],
```

```
    "metadata-log" : {
      "timestamp-ms" : 1647324555320,
      "metadata-file" : "hdfs://192.168.190.133:8020/data_lake/iceberg/my_db/
logtable/metadata/v1.metadata.json"
    }
```

这时查看表中的数据,代码如下:

```
spark.table("hadoop_prod.my_db.logtable").show()
```

可以看到数据表中仅包含 2008 年的数据,内容如下:

```
+-------------------+------+-------+
|                 ts|log_id|log_msg|
+-------------------+------+-------+
|2008-11-01 16:00:00|     1|      a|
|2008-11-03 16:00:00|     2|      b|
|2008-11-04 16:00:00|     3|      c|
|2008-11-12 16:00:00|     4|      d|
|2008-11-17 16:00:00|     5|      e|
|2008-11-18 16:00:00|     6|      f|
|2008-11-24 16:00:00|     7|      g|
|2008-11-27 16:00:00|     8|      h|
|2008-11-30 16:00:00|     9|      i|
|2008-12-01 16:00:00|    10|      j|
|2008-12-04 16:00:00|    11|      k|
|2008-12-09 16:00:00|    12|      l|
|2008-12-11 16:00:00|    13|      m|
|2008-12-19 16:00:00|    14|      n|
|2008-12-23 16:00:00|    15|      o|
|2008-12-26 16:00:00|    16|      p|
|2008-12-28 16:00:00|    17|      q|
|2008-12-30 16:00:00|    18|      r|
|2008-12-31 16:00:00|    19|      s|
+-------------------+------+-------+
```

下面应用条件查询,只查询 2008 年 12 月份的数据(注意,between…and…包含的范围),代码如下:

```
val sql_select = """
    select ts, log_id, log_msg
    from hadoop_prod.my_db.logtable
    where ts between '2008-12-01' and '2009-01-01'
    order by ts
  """
spark.sql(sql_select).show()
```

执行以上代码,输出的内容如下:

```
+-------------------+------+-------+
|                 ts|log_id|log_msg|
+-------------------+------+-------+
|2008-12-01 16:00:00|    10|      j|
|2008-12-04 16:00:00|    11|      k|
|2008-12-09 16:00:00|    12|      l|
|2008-12-11 16:00:00|    13|      m|
|2008-12-19 16:00:00|    14|      n|
```

```
|2008-12-23 16:00:00|     15|        o|
|2008-12-26 16:00:00|     16|        p|
|2008-12-28 16:00:00|     17|        q|
|2008-12-30 16:00:00|     18|        r|
|2008-12-31 16:00:00|     19|        s|
+-------------------+-------+---------+
```

(3)修改表的分区规范,将按月分区修改为按天分区,代码如下:

```
spark.sql("ALTER TABLE hadoop_prod.my_db.logtable ADD PARTITION FIELD
days(ts)")
```

(4)手动向表中添加新的日志记录。在这个写操作中,只添加在 2009 年 1 月 1 日或之后发生的日志事件数据,代码如下:

```
data.where(col("ts").gt(1230768000L))
    .select(
        col("ts").cast(DataTypes.TimestampType),
        col("log_id"),
        col("log_msg")
    )
    .sort(col("ts"))
    .writeTo("hadoop_prod.my_db.logtable")
    .overwritePartitions()
```

(5)正如在代码中看到的,在 Iceberg 中进行分区演变后,不需要重写整个表。这时如果查看 logtable 的 data 文件夹,将看到 Iceberg 已经根据分区值对数据文件进行了组织。2009 年 1 月 1 日之前的时间戳是按 month 组织的;该日期和之后的时间戳是按 day 进行组织的,命令如下:

```
$ hdfs dfs -ls -R /data_lake/iceberg/my_db/logtable/data
```

结果如图 11-41 所示。

```
/data_lake/iceberg/my_db/logtable/data/ts_month=2008-12
/data_lake/iceberg/my_db/logtable/data/ts_month=2008-12/00009-68-090a9d96-e084-4696-997b-48a1489bcac6-00001.parquet
/data_lake/iceberg/my_db/logtable/data/ts_month=2008-12/00010-69-49cd18e7-1479-4009-8a18-300ef6afa43d-00001.parquet
/data_lake/iceberg/my_db/logtable/data/ts_month=2008-12/00011-70-6f2ff456-d33c-401a-adc3-fb6e3b2b76f7-00001.parquet
/data_lake/iceberg/my_db/logtable/data/ts_month=2008-12/00012-71-a2b9da6d-e089-48c0-aec1-6a1ba478a442-00001.parquet
/data_lake/iceberg/my_db/logtable/data/ts_month=2008-12/00013-72-f3a9d6e2-4b4d-471f-93d3-78d1126d1448-00001.parquet
/data_lake/iceberg/my_db/logtable/data/ts_month=2008-12/00014-73-6aa425dc-10a5-4ea6-8213-525ec0fafbc0-00001.parquet
/data_lake/iceberg/my_db/logtable/data/ts_month=2008-12/00015-74-89bc93c5-1bd9-4fc8-8ccc-64a2450758b9-00001.parquet
/data_lake/iceberg/my_db/logtable/data/ts_month=2008-12/00016-75-30b3d8bc-ecfe-4534-a150-e9055e5a43fd-00001.parquet
/data_lake/iceberg/my_db/logtable/data/ts_month=2008-12/00017-76-4b20c2eb-a176-4e46-91f3-4d8bae012749-00001.parquet
/data_lake/iceberg/my_db/logtable/data/ts_month=2008-12/00018-77-1f9ef4e2-3bed-42d6-ac2a-f4da411e392b-00001.parquet
/data_lake/iceberg/my_db/logtable/data/ts_month=2009-01
/data_lake/iceberg/my_db/logtable/data/ts_month=2009-01/ts_day=2009-01-01
/data_lake/iceberg/my_db/logtable/data/ts_month=2009-01/ts_day=2009-01-01/00000-93-06a5f0bb-c105-45a6-9a03-c6a94607dca7-00001.parquet
/data_lake/iceberg/my_db/logtable/data/ts_month=2009-01/ts_day=2009-01-03
/data_lake/iceberg/my_db/logtable/data/ts_month=2009-01/ts_day=2009-01-03/00001-94-1fdbbe01-c1ba-4e29-b560-4d7e83df33dd-00001.parquet
/data_lake/iceberg/my_db/logtable/data/ts_month=2009-01/ts_day=2009-01-12
/data_lake/iceberg/my_db/logtable/data/ts_month=2009-01/ts_day=2009-01-12/00002-95-55d61f3a-90b1-4b45-84d0-44f09718bd12-00001.parquet
/data_lake/iceberg/my_db/logtable/data/ts_month=2009-01/ts_day=2009-01-19
/data_lake/iceberg/my_db/logtable/data/ts_month=2009-01/ts_day=2009-01-19/00003-96-bcbb7f15-9b75-4e1d-a1ab-e23050ee4c3a-00001.parquet
/data_lake/iceberg/my_db/logtable/data/ts_month=2009-01/ts_day=2009-01-24
/data_lake/iceberg/my_db/logtable/data/ts_month=2009-01/ts_day=2009-01-24/00004-97-ed4d08bf-9a86-4a0c-8611-6c0d51751d3d-00001.parquet
/data_lake/iceberg/my_db/logtable/data/ts_month=2009-01/ts_day=2009-01-29
/data_lake/iceberg/my_db/logtable/data/ts_month=2009-01/ts_day=2009-01-29/00005-98-f90eff92-341a-4af2-b016-dd2aa0e38c5a-00001.parquet
/data_lake/iceberg/my_db/logtable/data/ts_month=2009-01/ts_day=2009-01-30
/data_lake/iceberg/my_db/logtable/data/ts_month=2009-01/ts_day=2009-01-30/00006-99-d5f8a694-fc52-400f-936e-33bdb1349ddc-00001.parquet
```

图 11-41 分区演变过程

该公司现在想要查询在跨 2008 年 12 月和 2009 年 1 月期间发生的所有日志事件。查询将跨越多个分区布局,但仍然无缝地工作,而不需要用户指定任何额外的信息或知道任何关于表的分区,代码如下:

```
spark.sql("""
    SELECT *
    FROM hadoop_prod.my_db.logtable
    WHERE ts > '2008-12-14' AND ts < '2009-1-14'
""").show()
```

查询的结果如下:

```
+-------------------+------+-------+
|                 ts|log_id|log_msg|
+-------------------+------+-------+
|2008-12-19 16:00:00|    14|      n|
|2008-12-23 16:00:00|    15|      o|
|2008-12-26 16:00:00|    16|      p|
|2008-12-28 16:00:00|    17|      q|
|2008-12-30 16:00:00|    18|      r|
|2008-12-31 16:00:00|    19|      s|
|2009-01-01 16:00:00|    20|      t|
|2009-01-03 16:00:00|    21|      u|
|2009-01-12 16:00:00|    22|      v|
+-------------------+------+-------+
```

总体来讲,Iceberg 为分区和分区演变提供了大量的功能。所执行的大多数分区测试的工作情况与 Iceberg 文档中所声明的完全一样。

11.11 使用存储过程维护表

要在 Spark 中使用 Iceberg,首先需要配置 Spark Catalog。只有在 Spark 3.x 中使用 Iceberg SQL 扩展时,存储过程才可用。

1. 用法

可以通过 CALL 从任何配置的 Iceberg Catalog 中使用存储过程。所有存储过程都位于命名空间 system 中。CALL 支持通过名称或位置传递参数,推荐通过名称传递参数。不支持混合位置参数和命名参数。

1)命名参数

所有过程参数都被命名。当通过名称传递参数时,参数可以按任意顺序传递,任何可选参数都可以省略,代码如下:

```
CALL catalog_name.system.procedure_name(arg_name_2 => arg_2, arg_name_1 =>
arg_1)
```

2)位置参数

当按位置传递参数时,如果结尾参数是可选的,则只能省略它们,代码如下:

```
CALL catalog_name.system.procedure_name(arg_1, arg_2, ... arg_n)
```

2. 快照管理

1）rollback_to_snapshot()

将表回滚到指定的快照 ID。

要回滚到特定的时间，需要使用 rollback_to_timestamp() 方法。此过程将使所有引用受影响表的缓存 Spark 计划失效。

例如，将表 db.sample 回滚到快照 ID 1，执行语句如下：

```
CALL catalog_name.system.rollback_to_snapshot('db.sample', 1)
```

2）rollback_to_timestamp()

将表回滚到某个时间点的当前快照。此过程将使所有引用受影响表的缓存 Spark 计划失效。例如，将 db.sample 回滚到一天前，执行语句如下：

```
CALL catalog_name.system.rollback_to_timestamp('db.sample', date_sub(current_date(), 1))
```

3）set_current_snapshot()

设置表的当前快照 ID。与 rollback 不同，快照不需要是当前表状态的祖先。此过程将使所有引用受影响表的缓存 Spark 计划失效。

例如，为 db.sample 将当前快照设置为 1，代码如下：

```
CALL catalog_name.system.set_current_snapshot('db.sample', 1)
```

4）cherrypick_snapshot()

cherry-pick 从快照更改到当前表状态。在不修改或删除原始快照的情况下，cherry-picking 从现有快照创建一个新的快照。

只有 append 和动态覆盖快照可以被 cherry-picked。此过程将使所有引用受影响表的缓存 Spark 计划失效。例如，cherry-pick 快照 1，代码如下：

```
CALL catalog_name.system.cherrypick_snapshot('my_table', 1)
```

使用命名参数 cherry-pick 快照 1，代码如下：

```
CALL catalog_name.system.cherrypick_snapshot(snapshot_id => 1, table => 'my_table' )
```

3. 元数据管理

可以使用 Iceberg 存储过程执行许多维护操作。

1）expire_snapshots()

在 Iceberg 中，每次写/更新/删除/upsert/压缩都会生成一个新快照，同时保留旧数据和元数据，以便快照隔离和时间旅行。expire_snapshots() 过程可以用来删除不再需要的旧快照及其文件。

此过程将删除仅旧快照所需的旧快照和数据文件。这意味着 expire_snapshots() 将不会删除未过期快照所需要的文件。

例如，删除超过 10 天的快照，但保留最近的 100 个快照，代码如下：

```
CALL hive_prod.system.expire_snapshots('db.sample', date_sub(current_date(),
10), 100)
```

删除所有比当前时间戳老的快照，但保留最后 5 个快照，代码如下：

```
CALL hive_prod.system.expire_snapshots(table => 'db.sample', older_than =>
now(), retain_last => 5)
```

2）remove_orphan_files()

用于删除 Iceberg 表的任何元数据文件中没有引用的文件，因此可以认为是"孤立的"文件。

例如，通过在这个表上执行 remove_orphan_files 命令，列出所有可能需要删除的文件，而不是真正删除它们，代码如下：

```
CALL catalog_name.system.remove_orphan_files(table => 'db.sample', dry_run =>
true)
```

删除表 db.sample 所不知道的 tablelocation/data 文件夹中的任何文件，代码如下：

```
CALL catalog_name.system.remove_orphan_files(table => 'db.sample',
location => 'tablelocation/data')
```

3）rewrite_manifests()

重写表清单（Manifest）以优化扫描计划。

清单中的数据文件是按照分区规范中的字段进行排序的。此过程使用 Spark 作业并行运行，并且将使所有引用受影响表的缓存 Spark 计划失效。

例如，重写表 db.sample 中的清单文件并将其与表分区对齐，代码如下：

```
CALL catalog_name.system.rewrite_manifests('db.sample')
```

重写表 db.sample 中的清单，并禁用 Spark 缓存的使用。这样做可以避免执行器的内存问题，代码如下：

```
CALL catalog_name.system.rewrite_manifests('db.sample', false)
```

4. 表迁移

snapshot()和 migrate()过程有助于测试和将现有的 Hive 或 Spark 表迁移到 Iceberg。

1）snapshot()

在不更改源表的情况下，创建用于测试的表的轻量级临时副本。

可以对新创建的表进行更改或写入，而不会影响源表，但是快照使用原始表的数据文件。当在快照上运行插入或覆盖操作时，新文件被放置在快照表的位置，而不是原始表的位置。

测试完快照表后，运行 DROP TABLE 清理快照表。

例如，在 Catalog 的 default 位置为 db.snap 创建一个孤立的 Iceberg 表，它引用表 db.sample，命名为 db.snap，代码如下：

```
CALL catalog_name.system.snapshot('db.sample', 'db.snap')
```

在一个手动指定的位置/tmp/temptable/，迁移一个引用表 db.sample 的孤立的 Iceberg 表 db.snap，代码如下：

```
CALL catalog_name.system.snapshot('db.sample', 'db.snap', '/tmp/temptable/')
```

2）migrate()

用装载源数据文件的 Iceberg 表替换表。表模式、分区、属性和位置将从源表复制。

如果任何表分区使用不支持的格式，则迁移将失败。支持的格式有 Avro、Parquet 和 ORC。现有数据文件被添加到 Iceberg 表的元数据中，可以使用从原始表模式创建的 name-to-id 映射来读取。

为了在测试时保持原始表不变，可以使用 snapshot 创建共享源数据文件和模式的新临时表。例如，将 Spark 的默认目录中的 db.sample 表迁移到一个 Iceberg 表，并添加一个属性 foo，设置为 bar，代码如下：

```
CALL catalog_name.system.migrate('spark_catalog.db.sample', map('foo', 'bar'))
```

将当前 Catalog 中的 db.sample 迁移到一个 Iceberg 表，而不添加任何其他属性，代码如下：

```
CALL catalog_name.system.migrate('db.sample')
```

11.12 整合 Spark 结构化流

Apache Iceberg 使用 Apache Spark 的 DataSourceV2 API 实现数据源和 Catalog 目录。Spark DataSourceV2 是一个不断发展的 API，在 Spark 版本中提供了不同级别的支持。

从 Spark 3.0 开始，支持 DataFrame 读写。

11.12.1 流读取

可以使用快照历史记录来支持从 Iceberg 表中作为流源读取数据。Iceberg 支持处理 Spark 结构化流作业中的增量数据，这些作业从历史时间戳开始，模板代码如下：

```
val df = spark.readStream
  .format("iceberg")
  .option("stream-from-timestamp", Long.toString(streamStartTimestamp))
  .load("database.table_name")
```

Iceberg 仅支持从 append 快照读取数据，而 overwrite 快照不能被处理，会引起异常。类似地，默认情况下 delete 快照会导致异常，但是通过设置 stream-skip-delete-snapshot=true 可以忽略 delete。

11.12.2 流写入

使用 Iceberg 表作为 Spark 结构化流的 Data Sink（数据接收器）。Spark 为所有的 sink

提供了 queryId 和 epochId，必须保证所有的写操作都是幂等的。Spark 可能会尝试多次提交同一个批处理，因此，用户需要知道每个查询最新提交的 epochId。一种方法是在快照摘要中持久化 queryId 和 epochId。在写操作上，可以简单地遍历快照并检查给定查询最新提交的 epochId，以使写操作幂等。

要将流查询的值写入 Iceberg 表，需要使用 DataStreamWriter，模板代码如下：

```
val tableIdentifier: String = ...
data.writeStream
    .format("iceberg")
    .outputMode("append")
    .trigger(Trigger.ProcessingTime(1, TimeUnit.MINUTES))
    .option("path", tableIdentifier)
    .option("checkpointLocation", checkpointPath)
    .start()
```

其中 tableIdentifier 可以是：

（1）HDFS 表的完全限定路径，例如 hdfs://nn:8020/path/to/table。

（2）如果表由 catalog 跟踪，则为表名，如 database.table_name。

Iceberg 不支持 continuous processing，因为它不提供 commit 输出的接口。

Iceberg 支持 append 和 complete 的输出模式。

（1）append：将每个微批的行追加到表中。

（2）complete：每个微批替换表内容。

Iceberg 表应该在开始流查询之前创建。

1. 对分区表进行写入

在对分区表进行写操作之前，Iceberg 要求根据每个任务（Spark 分区）的分区规范对数据进行排序。对于批处理查询，鼓励执行显式排序来满足需求，但是这种方法会带来额外的延迟，因为重分区和排序被认为是流工作负载的繁重操作。为了避免额外的延迟，可以启用 Fanout Writer（扇出写入器）来消除这一要求。模板代码如下：

```
val tableIdentifier: String = ...
data.writeStream
    .format("iceberg")
    .outputMode("append")
    .trigger(Trigger.ProcessingTime(1, TimeUnit.MINUTES))
    .option("path", tableIdentifier)
    .option("fanout-enabled", "true")
    .option("checkpointLocation", checkpointPath)
    .start()
```

Fanout Writer 按每个分区值打开文件，直到写任务完成才关闭这些文件。不鼓励批处理查询使用此功能，因为对输出行进行显式排序对于批处理工作负载来讲并不昂贵。

2. 读写表示例

本示例是从上游 Kafka 中读取数据，写入 Iceberg 表，打包放到 PBLP 集群上通过

spark-submit 提交执行。

（1）通过 Kafka 脚本创建供测试使用的主题并准备测试数据。

使用 SSH 方式登录到 Kafka 集群，并切换到 Kafka 的安装目录，命令如下：

```
$ cd ~/bigdata/kafka-4.1.2
```

创建名为 iceberg_test 的 Kafka 主题，命令如下：

```
$ ./bin/kafka-topics.sh --zookeeper localhost:2181 --topic iceberg_test --partitions 1 --replication-factor 1 --create
```

生成测试数据，命令如下：

```
$ kafka-console-producer.sh --broker-list localhost:9092 --topic iceberg_test
```

（2）通过 Spark SQL 创建测试使用的数据库 iceberg_db 和表 iceberg_table。

（3）编写 Spark 处理逻辑，代码如下：

```scala
def main(args: Array[String]): Unit = {

  //配置使用数据湖元数据
  val sparkConf = new SparkConf()
  sparkConf.set("spark.sql.extensions", "org.apache.iceberg.spark.extensions.IcebergSparkSessionExtensions")
  sparkConf.set("spark.sql.catalog.dlf_catalog", "org.apache.iceberg.spark.SparkCatalog")
  sparkConf.set("spark.sql.catalog.dlf_catalog.type", "org.apache.iceberg.spark.SparkCatalog")
  sparkConf.set("spark.sql.catalog.dlf_catalog.warehouse", "<yourOSSWarehousePath>")

  val spark = SparkSession
    .builder()
    .config(sparkConf)
    .appName("StructuredSinkIceberg")
    .getOrCreate()

  val checkpointPath = "/tmp/iceberg_table_checkpoint"
  val BootstrapServers = "localhost:9092"
  val topic = "iceberg_test"

  //从上游 Kafka 读取数据
  val df = spark.readStream
    .format("kafka")
    .option("kafka.Bootstrap.servers", BootstrapServers)
    .option("subscribe", topic)
    .load()

  import spark.implicits._
  val resDF = df.selectExpr("CAST(key AS STRING)", "CAST(value AS STRING)")
    .as[(String, String)].toDF("id", "data")

  //流式写入 Iceberg 表
```

```
    val query = resDF.writeStream
      .format("iceberg")
      .outputMode("append")
      .trigger(Trigger.ProcessingTime(1, TimeUnit.MINUTES))
      .option("path", "dlf_catalog.iceberg_db.iceberg_table")
      .option("checkpointLocation", checkpointPath)
      .start()

  query.awaitTermination()
}
```

11.12.3　维护流表

流查询可以快速创建新的表版本,这将创建大量的表元数据来跟踪这些版本。强烈建议通过调优提交速率、旧快照过期和自动清理元数据文件来维护元数据。

1. 调优提交速率

高提交率会产生大量的数据文件、清单和快照,这会导致表难以维护。鼓励至少 1min 的触发间隔,并在需要时增大间隔,代码如下:

```
import org.apache.spark.sql.streaming.Trigger

//默认触发器 (尽可能快地运行微批处理)
df.writeStream
  .format("console")
  .start()

//ProcessingTime Trigger, 使用处理时间, 指定 2S 的微批间隔
df.writeStream
  .format("console")
  .trigger(Trigger.ProcessingTime("2 seconds"))
  .start()

//一次性触发器
df.writeStream
  .format("console")
  .trigger(Trigger.Once())
  .start()
```

2. 过期旧快照

每个写入表的微批都会生成一个新的快照,在表元数据中跟踪该快照直到过期,从而删除不再需要的元数据和任何数据文件。快照随着频繁提交而迅速累积,因此强烈建议定期维护由流查询写的表。

维护操作需要 Table 实例,使用 Java API。例如,将时间超过 1 天的快照设为过期,代码如下:

```
Table table = ...
long tsToExpire = System.currentTimeMillis() - (1000*60*60*24); //1天
table.expireSnapshots()
    .expireOlderThan(tsToExpire)
    .commit();
```

还有一个 Spark Action 可以并行运行大表的表过期，模板代码如下：

```
Table table = ...
SparkActions
  .get()
  .expireSnapshots(table)
  .expireOlderThan(tsToExpire)
  .execute();
```

过期的快照会从元数据中删除，因此它们不可以再用于时间旅行查询。

3. 清理元数据文件

Iceberg 使用 JSON 文件跟踪表元数据。对表的每次更改都会生成一个新的元数据文件，以提供原子性。默认情况下，旧的元数据文件将作为历史记录保存。频繁提交的表，例如由流作业编写的表，可能需要定期清理元数据文件。

如果需要自动清理元数据文件，则可在表属性中设置 write.metadata.delete-after-commit.enabled = true。这将保留一些元数据文件（最高可达 write.metadata.previous-versions-max），并在每个新创建的元数据文件后删除最旧的元数据文件，包括以下这些属性。

（1）write.metadata.delete-after-commit.enabled 属性：每次表提交后是否删除旧的元数据文件。赋值为 true 表示自动清理元数据文件。

（2）write.metadata.previous-versions-max 属性：要保存的旧元数据文件的数量。

4. 压缩数据文件

Iceberg 在一张表中跟踪每个数据文件。数据文件越多会导致在清单文件中存储的元数据越多，而较小的数据文件会导致不必要的元数据量和文件打开成本的低效率查询。

微批处理中写入的数据量通常很小，这可能导致表元数据跟踪大量小文件。将小文件压缩为大文件可以减少表所需的元数据，并提高查询效率。Iceberg 可以使用 Spark 和 rewriteDataFiles() 操作并行压缩数据文件。这将把小文件组合成更大的文件，以减少元数据开销和运行时文件打开成本。模板代码如下：

```
Table table = ...
SparkActions
  .get()
  .rewriteDataFiles(table)
  .filter(Expressions.equal("date", "2020-08-18"))
  .option("target-file-size-Bytes", Long.toString(500 * 1024 * 1024))   //500 MB
  .execute();
```

可以通过 files 元数据表来检查数据文件大小和确定何时压缩分区。

5. 重写清单

Iceberg 在其清单列表和清单文件中使用元数据来加快查询规划，并删除不必要的数据文件。元数据树的作用是作为表数据的索引。

元数据树中的清单将按照添加顺序自动压缩，这使在写入模式与读取过滤器对齐时查询速度更快。例如，在数据到达时写入按小时分区的数据是与时间范围查询过滤器对齐的。

当表的写模式与查询模式不一致时，可以重写元数据，使用 rewriteManifests()或rewriteManfests()动作（用于使用 Spark 进行并行重写）将数据文件重新组合到清单中。

为了优化流工作负载上的写延迟，Iceberg 可以使用不自动压缩清单的 fast append 来写新快照，这可能导致大量的小清单文件，可以重写清单以优化查询和压缩。Spark Action 使用 rewriteManifestsAction()进行并行重写元数据，以将数据文件重新组合到清单中。

下面重写了小清单文件，并根据第 1 个分区字段对数据文件进行分组，代码如下：

```
Table table = ...
SparkActions
    .get()
    .rewriteManifests(table)
    .rewriteIf(file -> file.length() < 10 * 1024 * 1024) //10 MB
    .execute();
```

第 12 章 Hudi 数据湖

Apache Hudi 是下一代流数据湖平台。Apache Hudi 直接为数据湖带来了核心数据仓库和数据库功能。Hudi 提供表、事务、高效的 upserts/delete、高级索引、流吸收服务、数据集群/压缩优化和并发，同时保持数据为开源文件格式。

Apache Hudi 不仅非常适合流式工作负载，还允许创建高效的增量批处理管道。Apache Hudi 可以轻松地在任何云存储平台上使用。Hudi 的高级性能优化，使任何流行的查询引擎（包括 Apache Spark、Flink、Presto、Trino、Hive 等）的分析工作更快。

当前 Hudi 的最新版本是 0.10.1。

12.1 Apache Hudi 特性

下面将讨论一些关键概念和术语，这些概念和术语对于理解和有效使用这些原语非常重要。

12.1.1 Hudi Timeline

Apache Hudi 使用 Timeline 来管理事务和其他表服务。

Hudi 的核心是维护表上在不同时间 instant（瞬态）执行的所有操作的 Timeline（时间线），这有助于提供表的瞬时视图，同时还有效地支持按到达顺序检索数据。一个 Hudi Instant 包括以下组成部分。

（1）Instant Action：对表执行的操作类型。

（2）Instant Time：瞬态时间通常是一个时间戳（例如 20190117010349），它按照 action 的开始时间顺序单调地增加。

（3）State：瞬态的当前状态。

Hudi 保证在时间轴上执行的操作是原子的、基于瞬态时间的时间轴是一致的。执行的关键操作如下。

（1）COMMITS：一个 commit 表示将一批记录原子地写到一张表中。

（2）CLEANS：清除表中不再需要的旧版本文件的后台活动。

（3）DELTA_COMMIT：增量提交指的是将一批记录原子地写入 MergeOnRead 类型表，其中部分/所有数据都可以写入增量日志。

（4）COMPACTION：协调 Hudi 内部不同数据结构的后台活动，例如将更新从基于行的日志文件移动到柱状格式。在内部，compaction 显示为时间轴上的一个特殊提交。

（5）ROLLBACK：表示提交/增量提交不成功&回滚，删除写入过程中产生的所有部分文件。

（6）SAVEPOINT：将某些文件组标记为"已保存"，这样 cleaner 就不会删除它们。在灾难/数据恢复场景中，它有助于将表恢复到时间轴上的某个点。

任何给定的瞬态都可能处于下列状态之一。

（1）REQUESTED：表示某个 action 已安排，但尚未启动。

（2）INFLIGHT：表示当前正在执行的 action。

（3）COMPLETED：表示时间轴上 action 的完成。

例如，下面的例子显示了 10:00 到 10:20 发生在 Hudi 表上的 upserts 事件，大约每 5min 发生一次，在 Hudi 时间轴上留下提交元数据，以及其他后台清理/压缩，如图 12-1 所示。

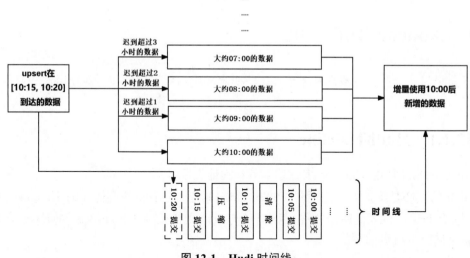

图 12-1　Hudi 时间线

需要注意，提交时间表示数据到达的时间（10:20AM），而实际的数据组织反映的是实际时间或事件时间，这些数据是为从 07:00 开始的每小时存储桶准备的。在考虑延迟和数据完整性之间的权衡时，这是两个关键的概念。

当有延迟到达的数据时（预计 9:00 到达的数据，在 10:20 到达，迟到超过了 1h），可以看到 upsert 将新数据生成到更旧的时间桶/文件夹中。在时间轴的帮助下，尝试获取 10:00 以后成功提交的所有新数据的增量查询能够非常有效地只使用更改的文件，而不必扫描所有超过 07:00 的时间桶。

12.1.2　Hudi 文件布局

Apache Hudi 的一般文件布局结构描述如下：

（1）Hudi 将数据表组织到分布式文件系统的基路径（Base Path）下的目录结构中。

（2）表被分解成分区。

（3）在每个分区内，文件被组织到文件组中，由一个文件 ID 唯一地标识。

（4）每个文件组包含几个文件片。

（5）每个文件片包含在某个提交/压缩瞬态生成的一个基文件（.parquet），以及一组日志文件（.log.*），这些日志文件包含自基本文件生成以来对基文件的插入/更新。

Hudi 采用 Multiversion Concurrency Control（MVCC），其中压缩操作将日志和基文件合并生成新的文件片，清理操作将删除未使用的/旧的文件片，以回收文件系统上的空间。

Apache Hudi 的一般文件布局结构如图 12-2 所示。

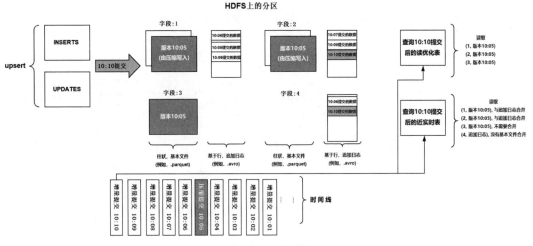

图 12-2　Hudi 文件布局

12.1.3　Hudi 表类型

Hudi 表类型定义了如何在 DFS 上索引和布局数据，以及如何在这样的组织上实现上述原语和时间轴活动（如何写入数据）。

Hudi 支持的表类型如下。

（1）Copy On Write：仅使用柱状文件格式存储数据（例如 Parquet）。通过在写入过程中执行同步合并，简单地更新版本和重写文件。

（2）Merge On Read：使用柱状（如 Parquet）+ 基于行（如 Avro）的文件格式组合存储数据。更新被记录到增量文件中，随后压缩，以同步或异步的方式生成柱状文件的新版本。

这两种表类型之间的比较见表 12-1。

表 12-1　Copy On Write 和 Merge On Read 的比较

比较指标	Copy On Write	Merge On Read
数据延迟	较高	较低
查询延迟	较低	较高
更新成本（I/O）	较高（重写整个 Parquet）	较低（追加到增量日志文件）
Parquet 文件大小	较小（高更新（I/O）成本）	较高（低更新（I/O）成本）
写入放大	较高	较低（取决于压缩策略）

1. Copy On Write 表

Copy On Write 表中的文件片只包含 base/columnar 文件（基本文件和柱状文件），每次提交都会生成基本文件的新版本。换句话说，我们隐式地压缩每个提交，这样只存在列数据，因此，写放大（输入数据的 1 字节所写入的字节数）要高得多，其中读放大为 0。这是分析工作负载非常需要的属性，因为分析工作负载主要是读密集型的。

当数据写入 Copy On Write 表并在其上运行两个查询时，它在概念上的工作原理如图 12-3 所示。

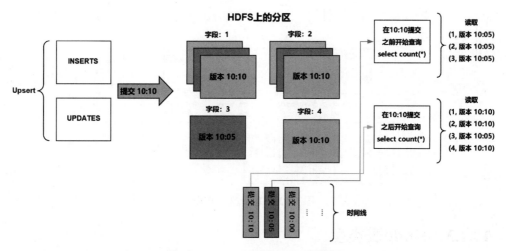

图 12-3　Copy On Write 工作原理

当写入数据时，对现有文件组的更新会为该文件组生成一个新片，并在该片上标记提交瞬态时间（Instant Time），而 insert 则分配一个新的文件组，并为该文件组写入它的第 1 个片。这些文件切片和它们的提交瞬态时间在图 12-3 中是用颜色编码的。对这样一张表运行的 SQL 查询（例如 select count() 计算该分区中的总记录），首先检查最近提交的时间轴，并过滤每个文件组的所有文件片（除了最新的文件片），如图 12-3 所示，一个旧的查询不会看到当前正在提交的文件，但是一个新的查询会在提交后开始获得新数据，因此，查询不受任何写失败/部分写的影响，只在已提交的数据上运行。

Copy On Write 表的目的是从根本上改善当前对表的管理方式：
（1）第一类支持在文件级自动更新数据，而不是重写整个表/分区。
（2）能够增量地使用变更，而不是使用比较耗费性能的扫描或摸索启发式。
（3）严格控制文件大小以保持优异的查询性能（小文件会严重影响查询性能）。

2. Merge On Read 表

Merge On Read 表是 Copy On Write 表的超集，在某种意义上，它仍然支持读优化查询，只暴露最新的文件片中的 base/columnar 文件。此外，它将每个文件组的传入 upsert 存储到基于行的增量日志上，以便在查询期间动态地将增量日志应用到每个文件 ID 的最新版本上，从而支持快照查询。因此，这种表类型试图智能地平衡读和写放大，以提供近实时的数据。这里最重要的变化是压缩器，它现在小心地选择哪些增量日志文件需要压缩到它们的柱状基文件中，以检查查询性能（更大的增量日志文件将导致在查询侧合并数据更长的合并时间）。

Merge On Read 表的工作原理，以及两种查询类型（快照查询和读取优化查询）的展示，如图 12-4 所示。

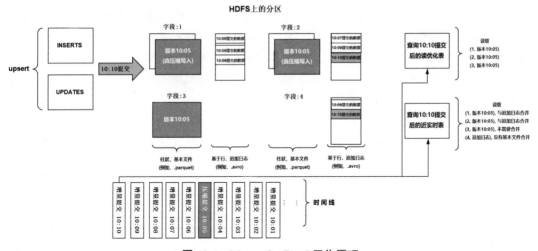

图 12-4 Merge On Read 工作原理

在这个例子中发生了许多有趣的事情，这些事情揭示了这种方法的微妙之处：
（1）现在大约每分钟提交一次，这是其他表类型无法做到的。
（2）在每个文件 ID 组中，现在都有一个增量日志文件，它保存对基本柱状文件中的记录的传入更新。在这个示例中，增量日志文件保存了从 10:05 到 10:10 的所有数据。与前面一样，基本柱状文件仍然使用提交进行版本控制，因此，如果只看基文件，则表布局看起来就像写表（Copy On Write）。
（3）定期压缩过程协调来自增量日志的这些更改，并生成一个新版本的基文件，就像示例中的 10:05 所发生的那样。

(4) 查询同一个底层表有两种方式：读取优化查询和快照查询，这取决于选择的是查询数据的性能还是新鲜度。

(5) 对于读优化查询，提交的数据何时可用于查询的语义发生了微妙的变化。注意，这样一个在 10:10 运行的查询不会看到 10:05 之后的数据，而快照查询总会看到最新的数据。

(6) 什么时候触发压缩及压缩什么是解决这些难题的关键。通过实施压缩策略，我们积极地压缩最新的分区，而不是旧的分区，可以确保读取优化查询在 X 分钟内以一致的方式看到发布的数据。

在读表上进行合并的目的是直接在 DFS 上支持近实时的处理，而不是将数据复制到可能无法处理数据量的专门系统。这个表还有一些次要方面的好处，例如通过避免同步合并数据来减少写放大（也就是一批中每字节的数据写入的数据量）。

12.1.4 Hudi 查询类型

查询类型定义了如何查询公开底层数据（如何读取数据）。Hudi 支持的查询类型如下。

(1) 快照查询：查询可以查看给定提交或压缩操作时表的最新快照。在对读表（Merge On Read）进行合并的情况下，它会动态地合并最新文件片的基本文件和增量文件，从而公开近乎实时的数据（几分钟）。对于写表（Copy On Write）上的复制，它提供了对现有 Parquet 表的 drop-in 替换，同时提供了 upsert/delete 和其他写侧功能。

(2) 增量查询：由于给定的 commit/compaction，查询只能看到写入表的新数据。这有效地提供了变更流，以启用增量数据管道。

(3) 读优化查询：查询可以查看一个给定的提交/压缩操作时的表的最新快照。仅在最新的文件切片中显示 base/columnar 文件，并保证与非 Hudi 柱状表相比具有相同的柱状查询性能。

不同查询类型之间的比较见表 12-2。

表 12-2　快照查询和读优化查询的比较

比较指标	快 照 查 询	读优化查询
数据延迟	较低	较高
查询延迟	较高（合并基/柱状文件+基于行的增量/日志文件）	较低（原始基/柱状文件性能）

不同类型的表所支持的查询类型见表 12-3。

表 12-3　不同类型的表所支持的查询类型

表 类 型	支持的查询类型
Copy On Write（COW）	快照查询+增量查询
Merger On Read（MOR）	快照查询+增量查询+读优化的查询

12.2 在 Spark 3 中使用 Hudi

使用 Spark 数据源,可以插入和更新具有默认表类型(Copy On Write)的 Hudi 表。在每次写操作之后,还将展示如何同时读取快照数据和增量数据。

Hudi 当前最新版本是 0.10.1,支持 Spark 3.0.3 和 Spark 3.1.2 版本。

12.2.1 配置 Hudi

分别运行支持 Hudi 的 spark-shell、PySpark 和 Spark SQL。

使用 spark-shell 时的命令如下:

```
//spark-shell for Spark 3.1
$ spark-shell \
  --packages \ org.apache.hudi:hudi-spark3.1.2-bundle_2.12:0.10.1,org.apache.spark:spark-avro_2.12:3.1.2 \
  --conf 'spark.serializer=org.apache.spark.serializer.KryoSerializer'
```

通过 HoodieSparkSessionExtension SQL 扩展,Hudi 支持使用 Spark SQL 写入和读取数据。从提取的目录运行 Spark SQL 与 Hudi,命令如下:

```
#Spark SQL for Spark 3.1
$ spark-sql
--packages org.apache.hudi:hudi-spark3.1.2-bundle_2.12:0.10.1,org.apache.spark:spark-avro_2.12:3.1.2 \
--conf 'spark.serializer=org.apache.spark.serializer.KryoSerializer' \
--conf 'spark.sql.extensions=org.apache.spark.sql.hudi.HoodieSparkSessionExtension'
```

以上命令会自动下载需要的 JAR 包。注意以下几点:

(1) spark-avro 模块需要在--packages 中指定,因为 spark-shell 默认不包含它。

(2) spark-avro 和 Spark 版本必须匹配(本书使用的是 Spark 3.1.2)。

(3) 本书中使用了为 Scala 2.12 构建的 hudi-spark-bundle,因为所使用的 spark-avro 模块也依赖于 2.12 版本。

也可以先下载对应的 JAR 包,上传到$SPARK_HOME/jars 目录下自动加载。

如果使用 Maven 进行项目开发,则要在项目的 pom.xml 文件中添加依赖,内容如下:

```xml
//for Spark 3
<dependency>
    <groupId>org.apache.hudi</groupId>
    <artifactId>hudi-spark3.1.2-bundle_2.12</artifactId>
    <version>0.10.1</version>
</dependency>
<dependency>
    <groupId>org.apache.spark</groupId>
    <artifactId>spark-avro_2.12</artifactId>
    <version>3.1.2</version>
</dependency>
```

如果使用 Zeppelin 执行交互式开发，则需要执行的配置如下。

（1）把 spark-avro_2.12-3.1.2.jar 复制到$ZEPPELIN_HOME/lib/目录下。

（2）在运行所有的 Spark 代码之前，先执行的配置如下：

```
%spark.conf
spark.jars.packages org.apache.hudi:hudi-spark3.1.2-bundle_2.12:0.10.1,
org.apache.spark:spark-avro_2.12:3.1.2
spark.serializer       org.apache.spark.serializer.KryoSerializer
```

12.2.2 初始设置

设置表名、基本路径和数据生成器，以生成示例所需要的记录，代码如下：

```
//spark-shell
import org.apache.hudi.QuickstartUtils._
import scala.collection.javaConversions._
import org.apache.spark.sql.SaveMode._
import org.apache.hudi.DataSourceReadOptions._
import org.apache.hudi.DataSourceWriteOptions._
import org.apache.hudi.config.HoodieWriteConfig._

val tableName = "hudi_trips_cow"
val basePath = "hdfs://xueai8:8020/hudi/hudi_trips_cow"
val dataGen = new DataGenerator

//测试生成的 JSON 数据集
convertToStringList(dataGen.generateInserts(2)).foreach(println)
```

执行以上代码，可以看到 Hudi 的数据生成器生成了两条 JSON 数据，数据格式如下：

```
{"ts": 1647196090688, "uuid": "421b6078-2d2c-4e23-a6f5-b64713bdf81d",
"rider": "rider-284", "driver": "driver-284", "begin_lat": 0.7340133901254792,
"begin_lon": 0.5142184937933181, "end_lat": 0.7814655558162802, "end_lon":
0.6592596683641996, "fare": 49.527694252432056, "partitionpath": "asia/
india/chennai"}
{"ts": 1646976254550, "uuid": "30d0b36f-ca4b-43ad-abb0-71de287ae259",
"rider": "rider-284", "driver": "driver-284", "begin_lat": 0.1593867607188556,
"begin_lon": 0.010872312870502165, "end_lat": 0.9808530350038475, "end_lon":
0.7963756520507014, "fare": 29.47661370147079, "partitionpath": "americas/
united_states/san_francisco"}
//DataGenerator 生成的示例数据包含的字段类型如下
{"name": "ts","type": "long"},
{"name": "uuid", "type": "string"},
{"name": "rider", "type": "string"},
{"name": "driver", "type": "string"},
{"name": "begin_lat", "type": "double"},
{"name": "begin_lon", "type": "double"},
{"name": "end_lat", "type": "double"},
{"name": "end_lon", "type": "double"},
{"name": "fare", "type": "double"}
```

Hudi 支持通过 Spark 对 Hudi 数据集的数据进行插入、更新和删除。

12.2.3 插入数据

Hudi 的 hudi-spark 模块提供了 DataSource API 来将一个 Spark DataFrame 写入（和读取）一个 Hudi 表。读写时可以指定多个选项。

（1）HoodieWriteConfig：TABLE_NAME，必需项。

（2）DataSourceWriteOptions：数据源写入选项。

每次向 Hudi 数据集写入 DataFrame 时，必须指定 DataSourceWriteOptions。这些选项在写操作之间可能是相同的。DataSourceWriteOptions 中各个字段选项的含义见表 12-4。

表 12-4　DataSourceWriteOptions 中各个字段选项的含义

选　项	描　述
TABLE_NAME	要在其下注册数据集的表名。
TABLE_TYPE_OPT_KEY	可选的。要写入的表的类型。指定数据集是创建为 COPY_ON_WRITE 还是 MERGE_ON_READ。默认值为 COPY_ON_WRITE。 可用值有 COW_TABLE_TYPE_OPT_VAL (default) 和 MOR_TABLE_TYPE_OPT_VAL。 注意，在初次创建表之后，当使用 Spark SaveMode.Append 模式写入（更新）表时，这个值必须保持一致
RECORDKEY_FIELD_OPT_KEY	必选项。主键字段，其值将用作 HoodieKey 的 recordKey 组件。记录键唯一地标识每个分区中的记录/行。记录键可以是单个列，也可以引用多个列。KEYGENERATOR_CLASS_OPT_KEY 属性应该根据它是简单键还是复杂键进行相应的设置。例如 col1 表示简单字段，col1,col2,col3，等表示复杂字段。嵌套字段可以使用点符号来指定，例如 a.b.c。实际值将通过对字段值调用 .tostring() 来获得。默认值为 uuid
PARTITIONPATH_FIELD_OPT_KEY	必选项。用于对表进行分区的列，其值将用作 HoodieKey 的 partitionPath 组件。实际的值将通过对字段值调用 .tostring() 来获得。为了防止分区，提供空字符串作为值，例如""。使用 KEYGENERATOR_CLASS_OPT_KEY 指定分区/不分区。如果需要对分区路径进行 url 编码，可以设置 URL_ENCODE_PARTITIONING_OPT_KEY。如果同步到 Hive，也要使用 HIVE_PARTITION_EXTRACTOR_CLASS_OPT_KEY 指定。默认值为 partitionpath
PRECOMBINE_FIELD_OPT_KEY	必选项。在实际写入之前用于预组合的字段。当同一批内的两条记录具有相同的键值时，将选择指定字段中值最大的记录（由 Object.compareTo(..) 确定）。默认值为 ts
OPERATION_OPT_KEY	可选的。要使用的写操作。可用的值有 UPSERT_OPERATION_OPT_VAL (default)、BULK_INSERT_OPERATION_OPT_VAL、INSERT_OPERATION_OPT_VAL 和 DELETE_OPERATION_OPT_VAL
KEYGENERATOR_CLASS_OPT_KEY	可选项。 用于指定写入 Hudi 的分区字段生成器。默认值为 org.apache.huoli.keygen.SimpleKeyGenerator
HIVE_PARTITION_EXTRACTOR_CLASS_OPT_KEY	如果使用 Hive，则应指定表是否应该分区。可用值有 classOf[SlashEncodedDayPartitionValueExtractor].getCanonicalName (default)、classOf[MultiPartKeysValueExtractor].getCanonicalName、classOf[TimestampBasedKeyGenerator].getCanonicalName、classOf[NonPartitionedExtractor].getCanonicalName 和 classOf[GlobalDeleteKeyGenerator].getCanonicalName（当 OPERATION_OPT_KEY 设置为 DELETE_OPERATION_OPT_VAL 时使用）

下面生成一些新的行程数据,将它们加载到 DataFrame 中,然后将 DataFrame 写入 Hudi 表中,代码如下:

```
//将数据加载到 DataFrame
val inserts = convertToStringList(dataGen.generateInserts(10))
val df = spark.read.json(spark.sparkContext.parallelize(inserts, 2))
df.printSchema()
df.show(5)

//然后将该 DataFrame 写入 Hudi 表中
df.write.format("hudi")
  .options(getQuickstartWriteConfigs)
  .option(PRECOMBINE_FIELD_OPT_KEY, "ts")
  .option(RECORDKEY_FIELD_OPT_KEY, "uuid")
  .option(PARTITIONPATH_FIELD_OPT_KEY, "partitionpath")
  .option(TABLE_NAME, tableName)
  .mode(SaveMode.Overwrite)
  .save(basePath)
```

注意上面将写入模式指定为 SaveMode.Overwrite,含义是如果表已经存在,则会覆盖并重新创建表。这里提供了一个记录键(模式中的 uuid)、分区字段(region/country/city)和组合逻辑(模式中的 ts),以确保每个分区中的行程记录是唯一的。这里使用默认的写操作:upsert。如果工作负载没有更新,则可以执行 insert 或 bulk_insert 操作,这样会更快。

12.2.4 查询数据

Hudi 支持以下查询类型。

(1)快照查询:查看给定提交或压缩操作时表的最新快照。在对 Merge On Read 表进行合并的情况下,它会动态地合并最新文件片的基本文件和增量文件,从而公开近乎实时的数据(几分钟)。对于 Copy On Write 表,它提供了对现有 Parquet 表的就地替换,同时提供了 upsert/delete 和其他写侧功能。

(2)增量查询:查询只看到从给定的提交/压缩开始写入表的新数据。这有效地提供了变更流,以启用增量数据管道。

(3)读优化查询:查询可以看到表自提交/压缩操作开始的最新快照。仅在最近的文件片中显示基文件/柱状文件并保证与非 Hudi 柱状表相比具有相同的柱状查询性能。

不同查询类型之间的比较见表 12-5。

表 12-5 不同查询类型比较

比较指标	快 照 查 询	读优化查询
数据延迟	低	高
查询延迟	高(合并 base/columnar 文件+基于行的 delta/log 文件)	低(原始的 base/columnar 文件性能)

将数据文件加载到一个 DataFrame 中，代码如下：

```
//从 Hudi 表中将数据读到 DataFrame 中
val basePath = "hdfs://xueai8:8020/hudi/hudi_trips_cow"

val tripsSnapshotDF = spark
  .read
  .format("hudi")
  .load(basePath)

tripsSnapshotDF.printSchema()
tripsSnapshotDF.show(5)
```

在上面的代码中，load(basePath)使用/partitionKey=partitionValue 文件夹结构，用于 Spark 自动分区发现。

在下面的查询中提供对摄入数据的快照查询，代码如下：

```
//将从 Hudi 表中查询的 DataFrame 注册到临时表中
tripsSnapshotDF.createOrReplaceTempView("hudi_trips_snapshot")

//执行条件查询与字段投影
spark.sql("select fare, begin_lon, begin_lat, ts from  hudi_trips_snapshot where fare > 20.0").show()

//查询快照信息
spark.sql("""
    select _hoodie_commit_time,
           _hoodie_record_key,
           _hoodie_partition_path,
           rider,
           driver,
           fare
    from hudi_trips_snapshot""")
.show()
```

自 0.9.0 版本以来，Hudi 支持时间旅行查询。目前支持 3 种查询时间格式，代码如下：

```
//查询快照信息
spark.read.
  format("hudi").
  option("as.of.instant", "20220317123157975").
  load(basePath).
  show()

spark.read.
  format("hudi").
  option("as.of.instant", "2022-03-17 12:31:57.975").
  load(basePath).
  show()

//等价于 "as.of.instant = 2022-03-18 00:00:00"
spark.read.
```

```
format("hudi").
option("as.of.instant", "2022-03-18").
load(basePath).
show()
```

从 0.9.0 版本开始，Hudi 支持一个内置的 Hudi FileIndex: HoodieFileIndex 来查询 Hudi 表，它支持分区修剪和元表查询。这将有助于提高查询性能。它还支持非全局查询路径，这意味着用户可以通过基路径查询表，而无须在查询路径中指定"*"。默认情况下，该特性已为非全局查询路径启用。对于全局查询路径，Hudi 使用旧的查询路径。

12.2.5 更新数据

这类似于插入新数据。使用数据生成器生成对现有行程的更新，加载到 DataFrame 中，并将 DataFrame 写入 Hudi 表，代码如下：

```
//生成对现有行程进行更新的数据
val updates = convertToStringList(dataGen.generateUpdates(10))
val df = spark.read.json(spark.sparkContext.parallelize(updates, 2))

//以 Append 模式将新的数据写入 Hudi 表中
df.write.format("hudi")
  .options(getQuickstartWriteConfigs)
  .option(PRECOMBINE_FIELD_OPT_KEY, "ts")
  .option(RECORDKEY_FIELD_OPT_KEY, "uuid")
  .option(PARTITIONPATH_FIELD_OPT_KEY, "partitionpath")
  .option(TABLE_NAME, tableName)
  .mode(SaveMode.Append)         //注意这里，Append
  .save(basePath)
```

注意，保存模式现在是 SaveMode.Append。一般来讲，除非第 1 次创建表，否则总是使用 Append 模式。再次查询数据将显示更新的行程。每个写操作都会生成一个由时间戳表示的新提交，代码如下：

```
//再次从 Hudi 表中将数据读到 DataFrame 中
val tripsSnapshotDF2 = spark
  .read
  .format("hudi")
  .load(basePath)

tripsSnapshotDF2.show(5)

println(tripsSnapshotDF.count)           //10
println(tripsSnapshotDF2.count)          //17
```

12.2.6 增量查询

Hudi 还提供了获取自给定提交时间戳以来更改的记录流的功能。这可以通过 Hudi 的增量查询实现，并提供一个开始时间，更改需要从该时间流开始。如果想要在提交之后进行的所有更改（这是常见的情况），则不需要指定 endTime，代码如下：

```
//spark-shell
//重新加载数据
spark.
  read.
  format("hudi").
  load(basePath).
  createOrReplaceTempView("hudi_trips_snapshot")

val commits = spark.sql("""
    select distinct(_hoodie_commit_time) as commitTime
    from  hudi_trips_snapshot
    order by commitTime
""").map(k => k.getString(0)).take(50)

//commits.foreach(println)

val beginTime = commits(commits.length - 2)     //感兴趣的提交时间

//增量查询数据
val tripsIncrementalDF = spark.read.format("hudi").
  option(QUERY_TYPE_OPT_KEY, QUERY_TYPE_INCREMENTAL_OPT_VAL).
  option(BEGIN_INSTANTTIME_OPT_KEY, beginTime).
  load(basePath)

//注册到临时表中
tripsIncrementalDF.createOrReplaceTempView("hudi_trips_incremental")

//查询
spark.sql("""
    select `_hoodie_commit_time`, fare, begin_lon, begin_lat, ts
    from hudi_trips_incremental
    where fare > 20.0
""").show()
```

这将使用 fare > 20.0 的过滤器给出在 beginTime 提交后发生的所有更改。查询结果如图 12-5 所示。

```
+------------------+------------------+-------------------+-------------------+-------------+
|_hoodie_commit_time|              fare|          begin_lon|          begin_lat|           ts|
+------------------+------------------+-------------------+-------------------+-------------+
|  20220317135008436| 89.45841313717807| 0.22991770617403628| 0.6923616674358241|1647077319121|
|  20220317135008436| 51.42305232303094| 0.7071871604905721|  0.876334576190389|1647173963333|
|  20220317135008436| 71.08018349571618| 0.8150991077375751|0.019252379188893319|1647268308603|
|  20220317135008436| 58.42042255207710| 0.8849896596590882| 0.07076797401073076|1647279514764|
|  20220317135008436|26.636532270940915| 0.12314538318119372| 0.35527775182006427|1647236775801|
|  20220317135008436|44.596839246210095| 0.38697902072535484|  0.904518901778190 2|1446977812971|
|  20220317135008436|59.071923248697225| 0.06105928762642976|  0.508361582050114|1647455162466|
+------------------+------------------+-------------------+-------------------+-------------+
```

图 12-5　增量查询结果

这个特性的独特之处在于，它现在允许在批处理数据上编写流管道。

12.2.7 时间点查询

接下来学习如何查询特定时间的数据。具体的时间可以通过将 endTime 指向特定的提交时间，将 beginTime 指向 000（表示尽可能早的提交时间）来表示，代码如下：

```
val beginTime = "000"                           //表示所有的提交大于这段时间
val endTime = commits(commits.length - 2)       //感兴趣的提交时间

//增量查询数据
val tripsPointInTimeDF = spark.read.format("hudi").
  option(QUERY_TYPE_OPT_KEY, QUERY_TYPE_INCREMENTAL_OPT_VAL).
  option(BEGIN_INSTANTTIME_OPT_KEY, beginTime).
  option(END_INSTANTTIME_OPT_KEY, endTime).
  load(basePath)
tripsPointInTimeDF.createOrReplaceTempView("hudi_trips_point_in_time")

spark.sql("""
    select `_hoodie_commit_time`, fare, begin_lon, begin_lat, ts
    from hudi_trips_point_in_time
    where fare > 20.0
""").show()
```

可以看到，查询出来的是追加操作前的数据，如图 12-6 所示。

```
+-------------------+------------------+--------------------+-------------------+-------------+
|_hoodie_commit_time|              fare|           begin_lon|          begin_lat|           ts|
+-------------------+------------------+--------------------+-------------------+-------------+
|    20220317123157975| 29.47661370147079|0.010872312870502165| 0.1593867607188556|1646970998794|
|    20220317123157975| 91.99515909032544|  0.2783086084578943| 0.2110206104048945|1647129242333|
|    20220317123157975| 90.25710109008239|  0.4006983139989222| 0.08528650347654165|1647307166364|
|    20220317123157975| 86.75932789048282|  0.1375535486249935| 0.7180196467760873|1647448789987|
|    20220317123157975| 98.3428192817987 |  0.3349917833248327| 0.4777395067707303|1647209231061|
|    20220317123157975| 90.90538095331540|  0.1994932322922063| 0.18294079059016366|1647006459641|
|    20220317123157975|49.527694252432056|  0.5142184937933181| 0.7340133901254792|1647310094520|
|    20220317123157975| 63.72504913279929|  0.888493603696927 | 0.6570857443423376|1646943125563|
+-------------------+------------------+--------------------+-------------------+-------------+
```

图 12-6 时间点查询结果

12.2.8 删除数据

Hudi 支持对存储在 Hudi 表中的数据实现两种类型的删除，允许用户指定不同的记录负载实现。

（1）软删除：保留记录键，并将所有其他字段的值清空。这可以通过确保适当的字段在表模式中为空，并在将这些字段设置为空后简单地 upsert 表实现。

（2）硬删除：更强的删除形式是从表中物理地删除记录的任何痕迹。这可以通过 3 种不同的方式实现。

① 使用数据源，设置 OPERATION_OPT_KEY 为 DELETE_OPERATION_OPT_VAL。这将删除被提交的数据集中的所有记录。

② 使用数据源，将 PAYLOAD_CLASS_OPT_KEY 的值设置为 org.apache.hudi.EmptyHoodieRecordPayload。这将删除被提交的数据集中的所有记录。

③ 使用数据源或 DeltaStreamer，将名为 _hoodie_is_deleted 的列添加到数据集。对于所有要删除的记录，该列的值必须设置为 true，对于任何要被 upsert 的记录，该列的值要么被设置为 false，要么被设置为 null。

对于其中硬删除的第 3 种方式，假设有原始模式内容如下：

```
{
  "type":"record",
  "name":"example_tbl",
  "fields":[{
    "name": "uuid",
    "type": "String"
  }, {
    "name": "ts",
    "type": "string"
  }, {
    "name": "partitionPath",
    "type": "string"
  }, {
    "name": "rank",
    "type": "long"
  }
]}
```

确保添加了 _hoodie_is_deleted 列，内容如下：

```
{
  "type":"record",
  "name":"example_tbl",
  "fields":[{
    "name": "uuid",
    "type": "String"
  }, {
    "name": "ts",
    "type": "string"
  }, {
    "name": "partitionPath",
    "type": "string"
  }, {
    "name": "rank",
    "type": "long"
  }, {
    "name" : "_hoodie_is_deleted",
    "type" : "boolean",
    "default" : false
  }
]}
```

然后对于任何想要删除的记录，都可以将_hoodie_is_deleted标记为true，内容如下：

```
{
"ts": 0.0,
"uuid": "19tdb048-c93e-4532-adf9-f61ce6afe10",
"rank": 1045,
"partitionpath": "americas/brazil/sao_paulo",
"_hoodie_is_deleted" : true
}
```

要删除传入的HoodieKeys记录，代码如下：

```
//获取总记录计数
spark.sql("select uuid, partitionpath from hudi_trips_snapshot").count()

//获取两条要删除的记录
val ds = spark.sql("select uuid, partitionpath from hudi_trips_snapshot").limit(2)
//val ds = spark.sql("select uuid, partitionPath from hudi_ro_table where rider = 'rider-213'")

//删除问题数据
val deletes = dataGen.generateDeletes(ds.collectAsList())
val df = spark.read.json(spark.sparkContext.parallelize(deletes, 2))

df.write.format("hudi").
  options(getQuickstartWriteConfigs).
  option(OPERATION_OPT_KEY,"delete").
  option(PRECOMBINE_FIELD_OPT_KEY, "ts").
  option(RECORDKEY_FIELD_OPT_KEY, "uuid").
  option(PARTITIONPATH_FIELD_OPT_KEY, "partitionpath").
  option(TABLE_NAME, tableName).
  mode(Append).
  save(basePath)

//重新加载表并验证记录已被删除
val roAfterDeleteViewDF = spark.
  read.
  format("hudi").
  load(basePath)

roAfterDeleteViewDF.registerTempTable("hudi_trips_snapshot")

//fetch应该返回(total-2)条记录
spark.sql("select uuid, partitionpath from hudi_trips_snapshot").count()
```

注意：删除操作只支持Append模式。

12.2.9 插入覆盖

生成一些新的行程数据，覆盖输入中出现的所有分区。对于批处理ETL作业，此操作可以比upsert更快，后者是一次性重新计算整个目标分区（而不是增量地更新目标表）。这是因为，能够完全绕过 upsert 写路径中的索引、预组合和其他重新分区步骤，代码如下：

```
spark
  .read.format("hudi")
  .load(basePath)
  .select("uuid","partitionpath")
  .sort("partitionpath","uuid")
  .show(100, false)

//生成要插入的新数据
val inserts = convertToStringList(dataGen.generateInserts(10))

//将这些数据构造为 DataFrame
val df = spark.
    read.json(spark.sparkContext.parallelize(inserts, 2)).
    filter("partitionpath = 'americas/united_states/san_francisco'")
                                            //挑出特定的分区数据

//将新的数据插入并覆盖表数据
df.write.format("hudi").
  options(getQuickstartWriteConfigs).
  option(OPERATION.key(),"insert_overwrite").
  option(PRECOMBINE_FIELD.key(), "ts").
  option(RECORDKEY_FIELD.key(), "uuid").
  option(PARTITIONPATH_FIELD.key(), "partitionpath").
  option(TBL_NAME.key(), tableName).
  mode(SaveMode.Append).
  save(basePath)

//与之前的查询相比，现在 San Francisco 应该有不同的 key spark.
  read.format("hudi").
  load(basePath).
  select("uuid","partitionpath").
  sort("partitionpath","uuid").
  show(100, false)
```

执行以上代码，输出内容如图 12-7 所示。

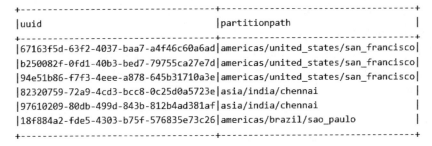

图 12-7　插入覆盖结果

第 13 章 Spark 大数据处理综合案例

CHAPTER 13

本章的综合案例涉及数据的采集（使用爬虫程序）、数据集成、数据预处理、大数据存储、Hive 数据仓库应用、大数据 ETL 实现和大数据结果展现等全流程所涉及的各种典型操作，涵盖 Linux、MySQL、Hadoop 3、Flume、Spark 3.x.x、Hive、Spring MVC 框架、ECharts 组件和 IntelliJ IDEA、Zeppelin Noebook 等系统和软件的使用方法。通过本项目，将有助于读者综合运用主流大数据技术及各种工具软件，掌握大数据离线批处理的全流程操作。

13.1 项目需求说明

就业是最大的民生，是社会稳定的主要支撑。在经历新冠肺炎疫情的两年多里，就业压力越发明显，毋庸置疑，2022 年就业形势会更严峻。

从新增数据看，2022 年，全国大学生毕业人数，将首次突破 1000 万人，达到惊人的 1020 万人，其中高学历毕业生人数，如 985、211 高校毕业生达到 75 万人。其他一本毕业生人数超过 100 万人，二本毕业生人数高达 370 万人。专科大学生毕业人数近 460 万人。2022 年，全国各大高校将近有 130 万人研究生毕业。每年约 85% 的应届毕业生会选择直接工作。

据 2020 年年底教育部统计，2019 年度我国出国留学人员总数为 70.35 万人，较上一年度增长 6.25%；各类留学回国人员总数为 58.03 万人。基于该数据推测，大概会有 100 万国外留学人员和其他人员回国求职。

综合测算，新增就业大学生人数将超过 1100 万人。

面对如此严峻的就业形势，求职乃至职业选择，对于求职者来讲就显得至关重要，所以招聘需求分析应成为每个职场人和将要走入职场者关注的焦点，要不断地思考和回答：在特定的发展阶段、在特定的文化背景下，面对变动的市场环境和弹性的岗位要求，企业到底需要什么样的求职者？不同城市、不同行业、不同岗位对求职者的要求及给出的薪酬待遇是什么样的？国企和民企、内资和外资企业该怎么选？

因此，广大求职者及整个社会都需要及时地分析行业企业招聘信息，为求学、求职提供指导性建议。本案例综合运用大数据分析和可视技术，对使用爬虫程序从互联网上采集到的某头部招聘网站招聘岗位数据进行多维度分析，并可视化展示分析结果，以供有需要的用户参考。

13.2 项目架构设计

本项目涵盖了大数据处理的完整流程，包括数据采集（爬虫）、数据集成、数据 ETL、Hive 数据仓库、数据清洗和预处理、数据分析、数据分析结果可视化等步骤，涉及的技术包括 Java 爬虫程序开发、Flume 数据采集技术、Spark SQL 实现 ETL 数据抽取、Hive 数据仓库技术、Spring MVC Web 开发及 Apache ECharts 图表应用技术等。通过本项目，读者可以完整地掌握一个大数据项目的处理流程。

本项目架构流程如图 13-1 所示。

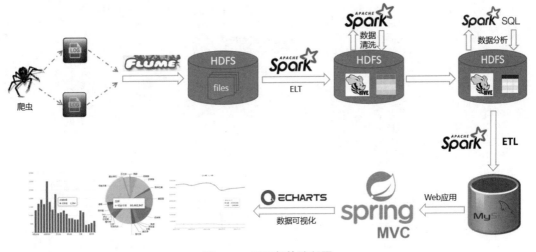

图 13-1　项目架构流程图

项目开发步骤如下。

（1）数据采集：编写爬虫程序，爬取 51job 网站的 5 个热门城市（北京、上海、广州、深圳、杭州）的热门岗位（大数据、数据分析、Java、Python）的招聘数据，并保存到文件中。

（2）数据集成：使用 Flume 自动监测并将采集到的数据文件导入 HDFS 中存储。

（3）数据 ETL：使用 Spark 建立 ETL 管道，将集成的数据文件导入 Hive 数据仓库中的 ODS 层。

（4）数据清洗：使用 Spark + Hive 进行数据清洗和整理。

（5）数据分析：使用 Spark SQL + Hive 进行数据多维度分析。

（6）数据导出：使用 Spark 建立 ETL 管道，将分析结果导到 MySQL 数据库。

（7）数据可视化：使用 Spring MVC 框架+ECharts 实现分析结果网页可视化展示。

13.3 项目实现：数据采集

编写 Java 爬虫程序，从 51job 招聘网站上爬取 5 个热门城市（北京、上海、广州、深圳、杭州）的热门岗位（大数据、数据分析、Java、Python）的招聘信息。

注意：如果读者想要了解该爬虫程序的 Python 版本实现，可以参考笔者另一本书《PySpark 原理深入与编程实战》的 8.3 节，可以将它看作本书的 Python 版本。

爬虫程序实际上是一个网络客户端程序，可以使用它发送各种 HTTP 请求，然后解析请求回来的数据内容。首先编写一个 HTTP 请求处理的客户端工具，代码如下：

```java
//第13章/HttpUtil.java
package com.xueai8.spider;

import org.apache.http.HttpEntity;
import org.apache.http.HttpHeaders;
import org.apache.http.client.methods.CloseableHttpResponse;
import org.apache.http.client.methods.HttpGet;
import org.apache.http.impl.client.CloseableHttpClient;
import org.apache.http.impl.client.HttpClients;
import org.apache.http.util.EntityUtils;

import java.io.IOException;

/*
  自定义网络请求工具类，定义 doGetRequest() 请求方法
*/
public class HttpUtil {

    //请求指定的 url
    public String doGetRequest(String url,String encode){
        //定义变量，保存请求回来的内容
        String content = "";

        //创建 httpClient 对象
        HttpGet httpGet = new HttpGet(url);

        //在 HTTP 接受报头中设置 API 媒体类型
        httpGet.addHeader(HttpHeaders.USER_AGENT,"Mozilla/5.0");
        httpGet.addHeader(HttpHeaders.ACCEPT,"application/json, text/javascript, */*; q=0.01");

        //try-with-resources
        try (CloseableHttpClient httpClient = HttpClients.createDefault();
            //发送请求，它会立即在 HttpResponse 对象中返回响应
            CloseableHttpResponse response = httpClient.execute(httpGet)) {

            //Get HttpResponse Status
            //System.out.println(response.getProtocolVersion());
```

```
            //System.out.println(response.getStatusLine().getStatusCode());
            //System.out.println(response.getStatusLine().getReasonPhrase());
            //System.out.println(response.getStatusLine().toString());

            //首先验证状态码
            int statusCode = response.getStatusLine().getStatusCode();
            if (statusCode != 200){
                throw new RuntimeException("失败。HTTP 错误的状态码是: " + statusCode);
            }

            //现在拉回响应对象
            HttpEntity entity = response.getEntity();
            //判断实体 Entity 是否为空，如果不为空，则可以使用 EntityUtils
            if (entity != null) {
                content = EntityUtils.toString(entity, encode);
                return content;
            }

        }catch (IOException e) {
            e.printStackTrace();
        }

        return content;
    }
}
```

接下来编写爬虫程序，爬取 51job 网站最新的招聘信息，并将爬取到的招聘信息保存到 TSV 格式的文件中。招聘信息限定为热门城市（北京、上海、广州、深圳和杭州）的热门岗位（大数据、数据分析、Java、Python）。

实现代码如下：

```
//第13章/JobSpider.java
package com.xueai8.spider;

import com.google.gson.Gson;
import com.google.gson.JsonArray;
import com.google.gson.JsonObject;

import java.io.FileOutputStream;
import java.io.OutputStreamWriter;
import java.util.Arrays;

public class JobSpider {

    private static HttpUtil httpUtil = new HttpUtil();
    private static Gson gson = new Gson();            //Google gson 对象

    public static void main(String[] args) throws Exception {
        //输出流，将爬取的数据写入文件中
```

```java
            OutputStreamWriter os = new OutputStreamWriter(
                new FileOutputStream("51job招聘信息.csv"),"utf-8");
        //指定存储文件
        String result_file = "zhaopin.tsv";
        //先写入标题
        String[] file_header = {
            "jobid","job_name","job_title","job_href",
            "company_name","company_href", "providesalary_text",
            "workarea","workarea_text","companytype_text",
            "degreefrom","workyear", "issuedate","jobwelf",
            "jobwelf_list", "attribute_text","companysize_text",
            "companyind_text"};
        os.write(String.join("\t", Arrays.asList(file_header)));

        //岗位关键字列表
        String[] job_list = {"Java","Python","数据分析","大数据"};

        for(String job_name : job_list){
            String job_url = "https://search.51job.com/list/010000%252c020000%252c030200%252c040000%252c080200,000000,0000,00,9,99," + job_name + ",2,1.html";
            String content = httpUtil.doGetRequest(job_url, "gbk");

            //目标1：获取总页数
            //将请求返回的JSON字符串转换为JSON对象
            JsonObject jsonObject = gson.fromJson(content, JsonObject.class);
            int total_page = Integer
                .parseInt(jsonObject.get("total_page")
                .getAsString());
            System.out.println(job_name + "岗位的总页数为" + total_page);

            //分页搜索
            int page_index = 1;
            while(page_index <= total_page){
                System.out.println("#########岗位: " + job_name + ", 页码: " + page_index + " #########");

                //向服务器发起请求，并保存返回的响应对象
                String page_url = "https://search.51job.com/list/010000%252c020000%252c030200%252c040000%252c080200,000000,0000,00,9,99," + job_name + ", 2," + page_index + ".html";
                String response_page = httpUtil.doGetRequest(page_url, "gbk");

                //将请求返回的JSON字符串转换为JSON对象
                JsonObject jsonObjectOfPage =
                    gson.fromJson(response_page, JsonObject.class);

                //提取搜索引擎搜索的结果
                JsonArray engine_search_result_of_page =
                    jsonObjectOfPage.getAsJsonArray("engine_search_result");
```

```
        //遍历一页中的每个岗位招聘信息
        for(int i=0;i<engine_search_result_of_page.size();i++){
          try{
            JsonObject job_one =
              engine_search_result_of_page.get(i).getAsJsonObject();
            os.write(job_one.get("jobid").getAsString() + "\t" +
                job_one.get("job_name").getAsString() + "\t" +
                job_one.get("job_title").getAsString() + "\t" +
                job_one.get("job_href").getAsString() + "\t" +
                job_one.get("company_name").getAsString() + "\t" +
                job_one.get("company_href").getAsString() + "\t" +
                job_one.get("providesalary_text").getAsString()+"\t"+
                job_one.get("workarea").getAsString() + "\t" +
                job_one.get("workarea_text").getAsString() + "\t" +
                job_one.get("companytype_text").getAsString()+"\t"+
                job_one.get("degreefrom").getAsString() + "\t" +
                job_one.get("workyear").getAsString() + "\t" +
                job_one.get("issuedate").getAsString() + "\t" +
                job_one.get("jobwelf").getAsString() + "\t" +
                job_one.get("jobwelf_list")
                    .getAsJsonArray().toString() + "\t" +
                job_one. get("attribute_text")
                    .getAsJsonArray().toString() + "\t" +
                job_one.get("companysize_text").getAsString()+"\t"+
                job_one.get("companyind_text").getAsString() + "\n"
            );
          }catch(Exception e){
            e.printStackTrace();
          }
        }
        page_index++;
      } //end while
    }
    os.close();
  }
}
```

运行以上的爬虫程序，可以获得一个名为 zhaopin.tsv 的招聘数据存储文件。

13.4　项目实现：数据集成

本步骤使用 Flume 自动监测并将采集到的数据文件导入 HDFS 中存储。这一步是可选的，用户根据自己的要求决定是否采用。如果没有要求，则可以直接采用 HDFS Shell 命令将采集到的数据上传到 HDFS 上存储。

13.4.1　Flume 简介

Flume 是一个高可用的、高可靠的、分布式的海量日志采集、聚合和传输的系统，Flume

支持在日志系统中定制各类数据发送方，用于收集数据；同时，Flume 可对数据进行简单处理，并写到各种数据接受方（可定制）。

Flume 主要由以下 3 个重要的组件构成。

（1）Source：完成对日志数据的收集，分成 Transition 和 Event 打入 Channel 中。

（2）Channel：主要提供一个队列功能，对 Source 提供的数据进行简单缓存。

（3）Sink：取出 Channel 中的数据，保存到相应的存储文件系统或数据库，或者提交到远程服务器。

Flume 架构如图 13-2 所示。

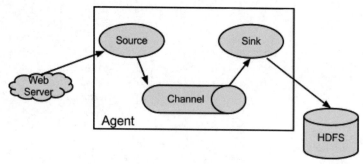

图 13-2　Flume 架构

13.4.2　安装和配置 Flume

首先从 Flume 的官网下载并安装包（本书使用的是 apache-flume-1.9.0-bin.tar.gz），并将安装包上传到 Linux 系统中，然后按以下步骤安装和配置 Flume。

（1）切换到安装包所在目录，解压下载的安装包，命令如下：

```
$ tar -zxvf apache-flume-1.9.0-bin.tar.gz -C /home/hduser/bigdata
```

上面将 apache-flume-1.9.0-bin.tar.gz 解压到了 /home/hduser/bigdata 目录下，然后重命名文件夹，以简化引用，命令如下：

```
$ cd /home/hduser/bigdata/
$ mv apache-flume-1.9.0-bin flume-1.9.0
```

（2）现在编辑 /etc/profile 文件（或 .bashrc 文件）以更新 Apache Flume 的环境变量，以便可以从任何目录访问它。例如，在 nano 编辑器中打开 profile 文件，命令如下：

```
$ nano /etc/profile
```

在打开的文件中，添加如下几行内容：

```
export FLUME_HOME=/home/hduser/bigdata/flume-1.9.0
export FLUME_CONF_DIR=$FLUME_HOME/conf
export PATH=$PATH:$FLUME_HOME/bin
```

然后按下快捷键 Ctrl + O 保存修改，按下快捷键 Ctrl + X 退出 nano 编辑器，并执行 source 命令使环境变量生效，命令如下：

```
$ source /etc/profile
```
（3）修改 flume-env.sh 配置文件。

Flume 安装包中默认没有 flume-env.sh 文件，用户需要复制 flume-env.sh.template 文件并去掉后缀.template，命令如下：
```
$ cd /home/hduser/bigdata/flume-1.9.0/conf
$ cp ./flume-env.sh.template ./flume-env.sh
$ nano ./flume-env.sh
```
使用编辑器打开 flume-env.sh 文件后，在文件的最后增加下面这行内容，用于设置 JAVA_HOME 变量：
```
export JAVA_HOME=/opt/java/jdk1.8.0_281
```
注意：这里应修改为读者自己的 JDK 安装目录。

然后按下快捷键 Ctrl + O 保存修改，按下快捷键 Ctrl + X 退出 nano 编辑器。

（4）验证 Flume 的安装是否成功。在命令行执行的命令如下：
```
$ flume-ng version
```
如果安装成功，则应该出现 Flume 的版本信息，类似下面这样的内容：
```
Flume 1.9.0
Source code repository: https://git-wip-us.apache.org/repos/asf/flume.git
Revision: d4fcab4f501d41597bc616921329a4339f73585e
Compiled by fszabo on Mon Dec 17 20:45:25 CET 2018
From source with checksum 35db629a3bda49d23e9b3690c80737f9
```
如果看到 Flume 的版本信息，则说明 Apache Flume 已经安装成功了。

13.4.3 实现数据集成

接下来，将 Flume 集成到项目中，使用 Flume 监视招聘数据文件所在目录，并将新增加的招聘数据文件自动上传到指定的 HDFS 目录中。

编写 Flume 配置文件，将 source 类型指定为 spoolDir，将 sink 类型指定为 hdfs。

首先，在$FLUME_HOME/conf/目录下创建一个名为 zhaopin.conf 的纯文本文件，并编辑内容如下：
```
#第13章/zhaopin.conf

#列出Agent包含的sources、sinks和channels
LogAgent.sources = mysource
LogAgent.channels = mychannel
LogAgent.sinks = mysink

#为source设置channel
LogAgent.sources.mysource.channels = mychannel
#配置名为"mysource"的source，采用spooldir类型，从本地目录~/job51获取数据
LogAgent.sources.mysource.type = spooldir
LogAgent.sources.mysource.spoolDir = /home/hduser/job51
```

```
#为sink设置channel
LogAgent.sinks.mysink.channel = mychannel
#配置名为"mysink"的sink,将结果写入HDFS中,每个文件10000行数据
LogAgent.sinks.mysink.type = hdfs
LogAgent.sinks.mysink.hdfs.fileType = DataStream
LogAgent.sinks.mysink.hdfs.path = hdfs://localhost:8020/job51
LogAgent.sinks.mysink.hdfs.batchSize = 1000
LogAgent.sinks.mysink.hdfs.rollSize = 0
LogAgent.sinks.mysink.hdfs.rollCount = 10000
LogAgent.sinks.mysink.hdfs.useLocalTimeStamp = true

#配置名为"mychannel"的channel,采用memory类型
LogAgent.channels.mychannel.type = memory
LogAgent.channels.mychannel.capacity = 10000
```

上面配置的含义是,将/home/hduser/job51/目录下新增的数据文件,自动上传到HDFS文件系统的/job51/目录下。

然后启动flume agent。将目录切换到Flume的安装目录,在命令行执行的命令如下:

```
$ ./bin/flume-ng agent -c ./conf -f ./conf/zhaopin.conf -n LogAgent
-Dflume.root.logger=INFO,console
```

最后,将爬虫程序爬取到的zhaopin.tsv文件复制到/home/hduser/job51/目录下,然后Flume会自动将其复制/上传到hdfs://localhost:8020/job51/目录下(Flume Sink指定的位置)。

13.5 项目实现:数据 ETL

到此为止,已经爬取了热门城市的热门岗位数据,并保存到了HDFS分布式文件系统中。接下来使用Spark编写批处理程序,建立ETL数据处理管道,将集成到HDFS中的数据文件导入Hive数据仓库中的ODS层。

注意:ETL(Extract-Transform-Load),用来描述将数据从来源端经过抽取(Extract)、转换(Transform)、加载(Load)至目的端的过程。

Spark SQL内置提供了对众多常用数据源的支持,可以很容易地使用它来构建ETL管道。编写Spark批处理程序,建立ETL数据处理管道,将集成到HDFS中的数据文件导入Hive数据仓库中的ODS层。

为了便于理解数据ETL的过程,这一阶段的开发工具选择使用Zeppelin Notebook,以交互式的方法实现数据ETL。关于Zeppelin Notebook的使用,可参考2.4节的内容。

首先定义一个用于数据ETL的工具类,代码如下:

```
//第13章/定义ETL工具_spark.zpln

import org.apache.spark.sql._
import org.apache.spark.sql.types._

//ETL工具类
```

```
object EltUtil {

  //extract from file
  /**
   * @param spark         SparkSession 实例
   * @param fileMap       要加载的文件参数
   *
   * @return              返回一个 DataFrame（Dataset[Row]）
   *
   */
  def extractFileToDataFrame(spark: SparkSession, fileMap: Map[String,String]):
Dataset[Row] = {
    //读取文件数据源，创建 DataFrame
    val df = spark.read
      .option("header",fileMap("header"))
      .option("inferSchema",true)
      .option("sep",fileMap("sep"))
      .csv(fileMap("filePath"))

    //返回
    df
  }

  //extract from mysql
  /**
   * @param spark         SparkSession 实例
   * @param jdbcMap       要加载的 JDBC 配置项
   *
   * @return              返回一个 DataFrame（Dataset[Row]）
   *
   */
  def extractFromJdbc(spark: SparkSession, jdbcMap:Map[String,String]):
Dataset[Row] = {
    //读取 JDBC 数据源，创建 DataFrame
    val df = spark
      .read
      .format("jdbc")
      .options(jdbcMap)
      .load()

    //返回
    df
  }

  //load to hive
  /**
   * @param spark
   * @param df        要装载到 Hive 中的 DataFrame
   * @param db        要装载到的 Hive 数据库
   * @param tb        要装载到的 Hive ODS 表
   * @param partitionColumn 指定分区列
```

```scala
 *
 * @return unit
 */
def loadToHive(spark: SparkSession, df:Dataset[Row], hiveMap: Map[String,
String]):Unit = {
   val database = hiveMap("db")              //要写入的数据库
   val table = hiveMap("tb")                 //要写入的数据表
   val partitionColumn = hiveMap.get("partitionColumn")   //分区列

   spark.sql(s"use $database")               //打开指定数据库,这里使用了字符串插值

   //有的表需要分区,有的不需要。这里使用模式匹配来分别处理
   partitionColumn match{
     case Some(column) =>
       df.write
         .format("parquet")
         .mode("overwrite")         //覆盖
         .partitionBy(column)       //指定分区
         .saveAsTable(table)
     case None =>
       df.write
         .format("parquet")
         .mode("overwrite")         //覆盖
         .saveAsTable(table)        //会将 DataFrame 数据保存到 Hive 表中
   }
}

//定义一个 ETL 方法,包含 extract + load
def eltFromFileToHive(spark: SparkSession,
                fileMap: Map[String,String],
                hiveMap: Map[String,String]) = {
  //extract from file
  //调用抽取文件数据源的方法,返回一个 DataFrame
  val df = extractFileToDataFrame(spark, fileMap)

  //load to hive
  //调用装载数据仓库的方法
  loadToHive(spark, df, hiveMap)
}

//定义一个 ETL 方法,包含 extract + load
def eltFromJdbcToHive(spark: SparkSession,
                jdbcMap:Map[String,String],
                hiveMap: Map[String,String]) = {
  //extract from JDBC
  val df = extractFromJdbc(spark, jdbcMap)

  //load to Hive
  loadToHive(spark, df, hiveMap)
}

}
```

然后定义主业务逻辑,调用上面的工具方法实现 ETL 任务,代码如下:

```
//第13章/01_数据ETL_spark.zpln

//加载外部数据源，构造DataFrame
//这里为了测试，使用本地文件（生产环境下应使用HDFS文件路径）
val file = "file://home/hduser/data/job51/zhaopin.tsv"
val df1 = spark.read
    .option("header",true)
    .option("infeSchema",true)
    .option("sep","\t")
    .csv(file)

//缓存数据集
df1.cache()

/* 执行ETL - 将数据从文件装载到Hive ODS层 */
//指定要抽取的文件数据源
val filePath = "file://home/hduser/data/job51/zhaopin.tsv"
val fileOptions = Map(
    "filePath" -> filePath,
    "header" -> "true",
    "sep" -> "\t"
)

//定义要装载的Hive配置项
val hiveOptions = Map(
    "db" -> "job51_db",
    "tb" -> "ods_job",
    "partitionColumn" -> "label"
)

//执行ETL
EltUtil.eltFromFileToHive(spark, fileOptions, hiveOptions)

//测试ETL是否成功
spark.table("job51_db.ods_job").show(false)
```

执行以上代码，将把 zhaopin.tsv 文件中的数据抽取到 Hive 的 job51_db.ods_job 表中并分区存储。在运行这个任务时，有两点需要注意。

（1）要确保已经启动了 Hive Metastore 服务。启动该服务的命令如下：

```
$ hive --service metastore
```

（2）提前在 Hive 中创建数据库 job51_db。

13.6 项目实现：数据清洗与整理

现在招聘数据已经抽取到了 Hive 的 ODS 层，但是还不能直接对数据进行分析，因为原始爬取的数据总是存在着各种各样的问题，如有可能包含重复的招聘信息、有可能包含错误的招聘信息。另外，类似薪资这样的数据，还必须先提取出来，才能进一步分析，因此有必要对数据进行清洗和整理。

编写 Spark 批处理程序，对抽取到 Hive 数据仓库中的招聘数据进行清洗，包括去重、错误数据处理、空值处理、属性转换、属性提取等数据预处理任务。

在项目的这一步骤中，需要完成以下几项预处理任务：

（1）薪资数据的抽取和转换。在爬虫程序爬取的原始招聘数据中，薪资的表示形式和单位都不统一，有日薪、月薪、年薪，有单位万元或单位千元，有的是固定薪资，有的是薪资范围，因此，要将这些不统一的薪资表示形式进行抽取和转换，统一到以元为单位，包含最低月薪和最高月薪。

（2）数据去重。在爬虫程序爬取的原始招聘数据中，有的招聘信息是重复的，为此需要对数据进行去重处理。这里判断是否重复的依据是 jobid 字段。

（3）工作城市抽取。在有的招聘信息中，工作地点细分为区，例如北京市海淀区、北京市朝阳区等，需要统一到城市维度即可，所以对这个字段（workarea_city）只取城市名。

（4）删除无意义的数据。在对招聘信息进行分析时，薪资是非常重要的一个分析指标，但是在爬虫程序爬取的原始招聘数据中，有的招聘信息并没有给出薪资信息。这里认为这些无薪资信息的招聘信息为无效数据，予以删除处理。

（5）删除非关注城市的招聘信息。本案例只分析一线热门城市"北京""上海""广州""深圳""杭州"的招聘信息，但是在爬虫程序爬取的原始招聘数据中，有的招聘信息的招聘地点不属于这 5 个热门一线城市，因此需要予以排除。

（6）最后，是属性选择。根据项目的需求，只选择保留其中对于后续分析有意义的 11 个属性，分别是 jobid、job_name、company_name、providesalary_text、workarea_city、companytype_text、jobwelf、attribute_text、companysize_text、companyind_text、label 字段。

在这些预处理任务中，最复杂的是薪资的抽取和转换，因此，在下面的代码中，将这个处理封装到一个单独的方法 processSalaryText(wage) 中，并将其注册为一个 UDF 函数。

为了便于理解数据清洗和整理的过程，这一阶段的开发工具选择使用 Zeppelin Notebook，以交互式的方法实现数据清洗与整理。关于 Zeppelin Notebook 的使用，可参考 2.4 节的内容。

最终实现的代码如下：

```
//第13章/02_数据清洗_spark.zpln

//将 Hive ODS 层数据加载到 DataFrame
val df1 = spark.table("job51_db.ods_job")

//缓存数据集
df1.cache

//1. 数据去重
//每个招聘信息，有一个唯一的 jobid。因为下载时可能不同的关键字对应的是同一个 jobid
```

```
//所以有可能有重复的信息。根据 jobid 去重
val df2 = df1.dropDuplicates("jobid")

//2. 预处理：过滤掉错误（或无意义）的数据
/*
    错误/无意义数据认定标准：
    (1) 不提供薪资信息。
    (2) 非北、上、广、深、杭这 5 个一线城市的招聘数据。
*/
//将无薪资的招聘信息过滤掉
val df3 = df2.filter(col("providesalary_text").isNotNull)
println("过滤掉无薪资的招聘信息后，还剩下的招聘信息数：" + df3.count)

//处理 workarea_text 字段，只取城市名
val df4 = df3.withColumn("workarea_city",
            split(col("workarea_text"),"-").getItem(0))
df4.select("workarea_city", "workarea_text").show()

//过滤一线城市的招聘信息
val hot_cities = Array("北京","上海","广州","深圳","杭州")
val df5 = df4.filter(col("workarea_city").isin(hot_cities:_*))
println(df5.count)
df5.show()

//3. 预处理：选择属性
/*
    属性选择
    (1) job_name 和 job_title 重复，保留一个。
    (2) job_href 和 company_href 不需要，删除；degreefrom、workyear、issuedate
也不需要，删除。
    (3) workarea、workarea_text 字段删除，保留 workarea_city。
    (4) jobwelf 保留，jobwelf_list 字段删除。
*/
val cols = List("jobid","job_name","company_name","providesalary_text",
            "workarea_city","companytype_text","jobwelf",
            "attribute_text","companysize_text","companyind_text",
            "label")
val df6 = df5.select(cols.head, cols.tail:_*)    //必须有第 1 个参数
df6.printSchema()

//4. 预处理：提取最低薪资和最高薪资
//自定义薪资转换函数（UDF），将薪资字段转换为最低月薪和最高月薪
def processSalaryText(wage:String): (Double,Double) = {
    var min_wage = 0.0      //最低薪资
    var max_wage = 0.0      //最高薪资
    if (!"".equals(wage) && wage.contains("-")){
        //如果是月薪
        if (wage.endsWith("月")){
            if (wage.contains("万")){
```

```
                min_wage = wage.substring(0,wage.indexOf("-")).toDouble * 10000
                max_wage = wage.substring(wage.indexOf("-")+1,wage.indexOf("万")).
toDouble * 10000
            }else{
                min_wage = wage.substring(0,wage.indexOf("-")).toDouble * 1000
                max_wage = wage.substring(wage.indexOf("-")+1,wage.indexOf("千")).
toDouble * 1000
            }
        }
        //如果是年薪
        else{
            if (wage.contains("万")){
                min_wage = wage.substring(0,wage.indexOf("-")).toDouble * 10000/12
                max_wage = wage.substring(wage.indexOf("-")+1,wage.indexOf("万")).
toDouble * 10000/12
            }else{
                min_wage = wage.substring(0,wage.indexOf("-")).toDouble * 1000/12
                max_wage = wage.substring(wage.indexOf("-")+1,wage.indexOf("千")).
toDouble * 1000/12
            }
        }
    }
    else if(!"".equals(wage) && !wage.contains("-")){
        //如果是月薪
        if (wage.endsWith("月")){
            if (wage.contains("万")){
                min_wage = wage.substring(0,wage.indexOf("万")).toDouble * 10000
            }else{
                min_wage = wage.substring(0,wage.indexOf("千")).toDouble * 1000
            }
            max_wage = min_wage
        }
        //如果是年薪
        else if (wage.endsWith("年")){
            if (wage.contains("万")){
                min_wage = wage.substring(0,wage.indexOf("万")).toDouble * 10000/12
            }else{
                min_wage = wage.substring(0,wage.indexOf("千")).toDouble * 1000/12
            }
            max_wage = min_wage
        }
        //如果是日薪
        else if (wage.endsWith("天")){
            min_wage = wage.substring(0,wage.indexOf("元")).toDouble * 30
            max_wage = min_wage
        }//如果是时薪
        else if (wage.endsWith("小时")){
            min_wage = wage.substring(0,wage.indexOf("元")).toDouble * 8 * 30
            max_wage = min_wage
```

```
            }
        }
        (min_wage, max_wage)
}

//注册 UDF
//注册为一个 UDF（在 DSL API 中使用时的注册方法）
val convertSalaryTextUDF =
            udf(processSalaryText(_:String):(Double,Double))

val df7 = df6
    .withColumn("salary_min",
            convertSalaryTextUDF(col("providesalary_text"))("_1"))
    .withColumn("salary_max",
            convertSalaryTextUDF(col("providesalary_text"))("_2"))
    .drop("providesalary_text")
df7.printSchema()

//5. 将整理后的数据存储起来：写入数仓
//定义要装载的 Hive 配置项
val hiveOptions = map(
    "db": "job51_db",
    "tb": "f_job_cleaned",
    "partitionColumn": "label"
)

//将清洗过后的数据加载到 Hive 数据仓库中
EltUtil.loadToHive(spark, df7, hiveOptions)

//重新加载，确定保存成功
val data = spark.table("job51_db.f_job_cleaned")
println(data.count)
data.printSchema()
data.show()
```

执行以上代码，它将从 Hive 中将 job51_db.ods_job 表数据加载到内存中进行处理，并把清洗后的分区数据存储到 Hive 的 job51_db.f_job_cleaned 表中。在运行这个任务时，要确保已经启动了 Hive Metastore 服务。启动该服务的命令如下：

```
$ hive --service metastore
```

13.7 项目实现：数据分析

经过上一步清洗和整理，现在数据已经准备好了，可以用于业务分析了。

使用 Spark SQL + Hive 从多个维度对整理后的数据集进行分析，并将分析结果存入数据集，其中部分分析维度如下：

(1) 按岗位类别（大数据、数据分析、Java、Python）统计招聘岗位数。
(2) 按城市（北京、上海、广州、深圳、杭州）统计招聘岗位数。
(3) 按城市和岗位类别统计招聘岗位数。
(4) 按公司类型（民营公司、上市公司等）统计招聘岗位数。
(5) 按公司类型和岗位类别统计招聘岗位数。
(6) 统计每个城市各种类型公司的数量。
(7) 查看公司规模及划分情况。
(8) 统计数据分析相关岗位在不同行业的招聘数量。
(9) 统计大数据相关岗位在不同行业的招聘数量。
(10) 统计 Python 相关岗位在不同行业的招聘数量。
(11) 统计 Java 相关岗位在不同行业的招聘数量。
(12) 以各种方式统计平均薪资，如按城市、按岗位类别、按城市和岗位类别、按公司类型、按行业类型等。

编写 Spark 批处理程序，从多个维度对整理后的数据集进行分析，并将分析结果存入数据集。首先将 13.6 节清洗后并存储的数据加载到 DataFrame 中，代码如下：

```
//第13章/03_数据分析_spark.zpln

//加载清洗后的数据 (hive --service metastore)
data = spark.table("job51_db.f_job_cleaned")

//缓存
data.cache
```

1. 探索性分析

首先查看数据量和数据模式，代码如下：

```
println("数据量: " + data.count)

//查看数据模式
data.printSchema()
```

执行以上代码，输出内容如下：

```
数据量: 176092

root
 |-- jobid: string (nullable = true)
 |-- job_name: string (nullable = true)
 |-- job_title: string (nullable = true)
 |-- job_href: string (nullable = true)
 |-- company_name: string (nullable = true)
 |-- company_href: string (nullable = true)
 |-- providesalary_text: string (nullable = true)
 |-- workarea: string (nullable = true)
```

```
 |-- workarea_text: string (nullable = true)
 |-- companytype_text: string (nullable = true)
 |-- degreefrom: string (nullable = true)
 |-- workyear: string (nullable = true)
 |-- issuedate: string (nullable = true)
 |-- jobwelf: string (nullable = true)
 |-- jobwelf_list: string (nullable = true)
 |-- attribute_text: string (nullable = true)
 |-- companysize_text: string (nullable = true)
 |-- companyind_text: string (nullable = true)
 |-- label: string (nullable = true)
```

可以看到，可用于分析的数据共有 176 092 条。

下面在 data 数据集上调用 describe()方法，执行描述性统计操作，查看薪资相关的基本信息，代码如下：

```
data.describe("salary_min", "salary_max").show(false)
```

执行以上代码，输出内容如下：

```
+-------+------------------+------------------+
|summary| salary_min       | salary_max       |
+-------+------------------+------------------+
|count  | 171714           | 171714           |
|mean   | 12018.094389508102| 18989.19857049134 |
|stddev | 110623.25154958254| 163562.43467731785|
|min    | 833.3333333333334 | 1000.0           |
|max    | 3.5E7            | 4.5E7            |
+-------+------------------+------------------+
```

从输出结果可以看出，薪资的均值范围为 12 000～19 000 元（月薪）。

接下来，看一下在采集的数据中，不同公司规模的划分情况，代码如下：

```
data.groupBy("companysize_text").count.show
```

执行以上代码，输出内容如下：

```
+----------------+-----+
|companysize_text|count|
+----------------+-----+
|            null| 4466|
|         50-150人|37138|
|         少于50人|18757|
|     5000-10000人| 6673|
|        150-500人|37557|
|       1000-5000人|26566|
|        500-1000人|23635|
|        10000人以上|16922|
+----------------+-----+
```

从以上输出结果来看，提供最多就业岗位的公司规模主要集中在小型公司，通常人数在 50～500 人，其次是 500～5000 人的中型公司，超过 5000 人的大型公司在招聘网站公开进行社招的数量很有限。

为了便于进一步分析，将 data 数据集注册到一个名为 job_tb 的临时视图中，代码如下：

```
//注册临时表
data.createOrReplaceTempView("job_tb")
```

然后以可视化的形式查询招聘公司规模分布情况，代码如下：

```
%sql
select companysize_text,count(*) as cnt
from job_tb
group by companysize_text
```

执行以上代码，输出结果如图 13-3 所示。

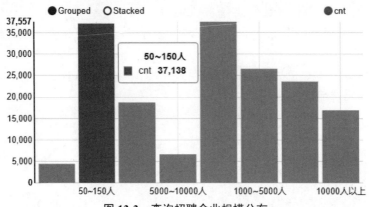

图 13-3　查询招聘企业规模分布

接下来，使用 Spark SQL 从多个维度对招聘数据集进行分析。

2. 按岗位类别分析

按岗位类别（大数据、数据分析、Java、Python）统计招聘岗位数，代码如下：

```
%sql
select label as post_type, count(*) as post_number
from job_tb
```

执行以上代码，输出结果如图 13-4 所示。

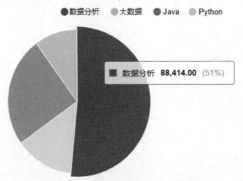

图 13-4　不同招聘岗位占比分析

从上面的输出结果可以得知,数据分析的招聘岗位最多,超过 50%,其次是 Java 岗位的招聘,约占总招聘数的 25%。

3. 按城市进行分析

分析不同城市(北京、上海、广州、深圳、杭州)的招聘岗位数量,代码如下:

```sql
%sql
select workarea_city,count(*) as post_number
from job_tb
group by workarea_city;
```

执行以上代码,输出结果如图 13-5 所示。

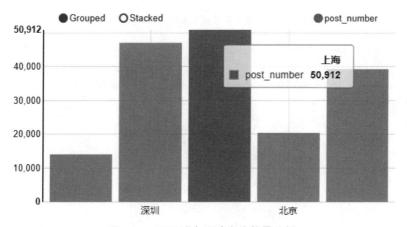

图 13-5 不同城市招聘岗位数量分析

从图 13-5 可以看出,经济发达程度和就业程度是成正比的,越发达的城市(例如上海、深圳),招聘岗位(人才需求)越旺盛。

4. 按城市和岗位两个维度进行分析

分析每个城市各个岗位的招聘岗位数量,代码如下:

```sql
%sql
select workarea_city, label as post_type, count(*) as post_number
from job_tb
group by workarea_city,label;
```

执行以上代码,输出结果如图 13-6 所示。

5. 按公司类型进行分析

按公司类型(民营企业、上市公司等)分别统计招聘岗位数,代码如下:

```sql
%sql
select companytype_text, count(*) as post_number
from job_tb
group by companytype_text;
```

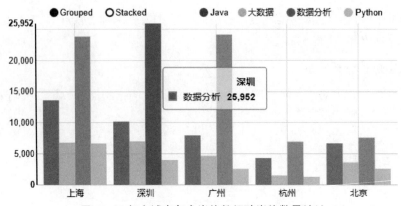

图 13-6　每个城市各个岗位的招聘岗位数量统计

执行以上代码,输出结果如图 13-7 所示。

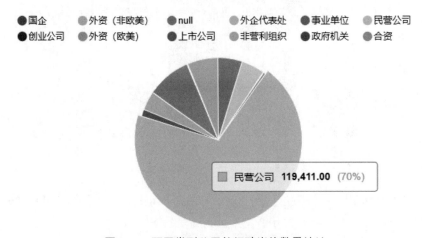

图 13-7　不同类型公司的招聘岗位数量统计

从图 13-7 可以看出,民营企业在吸纳人才就业方面是绝对的主力。

6. 按公司类型和岗位类别进行分析

从公司类型和岗位类别两个维度,进一步细粒度地分析每个城市对不同岗位的招聘需求量,代码如下:

```
%sql
select companytype_text, label as post_type, count(*) as post_number
from job_tb
group by companytype_text,label;
```

执行以上代码,输出结果如图 13-8 所示。

从图 13-8 可以看出,所有类型企业中,民营企业提供的岗位最多,其中又以数据分析岗位为最,遥遥领先其他岗位的招聘数量。

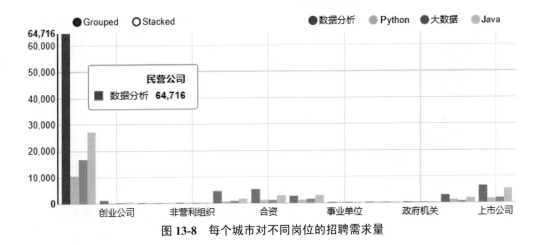

图 13-8　每个城市对不同岗位的招聘需求量

7. 按城市和公司类型进行分析

从城市和公司类型两个维度，统计各个城市提供招聘岗位的不同类型企业的数量，代码如下：

```sql
%sql
select workarea_city,companytype_text,count(*) as company_number
from job_tb
group by workarea_city,companytype_text;
```

执行以上代码，输出结果如图 13-9 所示。

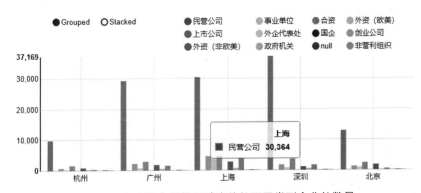

图 13-9　各个城市提供招聘岗位的不同类型企业的数量

从图 13-9 可以看出，深圳、上海、广州这 3 个城市的民营公司提供了大多数的社招岗位，即使在北京和杭州，也是民营企业提供的社招岗位数量最多。

8. 按行业进行分析

统计不同行业的招聘岗位数量，显示其中 Top 10 的行业，代码如下：

```
%sql
select companyind_text, count(*) as cnt
```

```
    from job_tb
    group by companyind_text
    order by cnt desc
    limit 10;
```

执行以上代码，输出结果如图13-10所示。

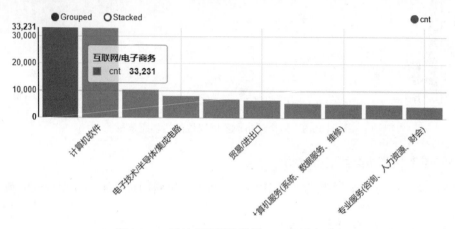

图13-10　统计招聘岗位数量Top 10的行业

从图13-10可以看出，互联网/电子商务、计算机软件这两个行业对大数据、数据分析、Java、Python岗位的需求量最大。排在第3位的是房地产，第4位是电子技术/半导体/集成电路。

9. 将分析结果保存到数据集

通常需要将分析结果存放到Hive的数据集层，以便企业的业务部门从数据集层访问业务相关数据。因为这些分析任务很相似，为了简单清晰，这里只提供了两个实现：

（1）按岗位类别（大数据、数据分析、Java、Python）统计招聘岗位数，并将分析结果分别写入Hive的job51_db.m_job_post_tb表中，代码如下：

```
//执行SQL查询
val sql = """
    select label as post_type, count(*) as post_number
    from job_tb
    group by label
"""
val postByTypeDF = spark.sql(sql)

//定义要装载的Hive配置项
val hiveOptions = map(
    "db" -> "job51_db",
    "tb" -> "m_post_tb"
)

//将分析后的数据加载到Hive数据仓库中
EltUtil.loadToHive(spark, postByTypeDF, hiveOptions)
```

(2)所有类型招聘的行业分布数据,并将分析结果写入 Hive 的 job51_db.m_post_by_companyind 表中,代码如下:

```
//执行 SQL 查询
val sql = """
    select label as post_type, companyind_text as companyind, count(*) as post_num
    from job_tb
    group by label, companyind_text;
"""

val postByCompanyindDF = spark.sql(sql)

//定义要装载的 Hive 配置项
val hiveOptions = map(
    "db" -> "job51_db",
    "tb" -> "m_post_by_companyind"
)

//将清洗过后的数据加载到 Hive 数据仓库中
EltUtil.loadToHive(spark, postByCompanyindDF, hiveOptions)
```

其余的分析任务,读者可自行参考实现。

在运行这个任务时,要确保已经启动了 Hive Metastore 服务。启动该服务的命令如下:

```
$ hive --service metastore
```

13.8　项目实现:分析结果导出

作为分析的最后一步,同样需要使用 Spark 建立 ETL 管道,将 Hive 数据集中的分析结果导出到 MySQL 数据库中,以便前端使用数据进行可视化展示。

Spark SQL 内置提供了对众多常用数据源的支持,可以很容易地使用它来构建 ETL 管道。编写 Spark 批处理程序,建立 ETL 管道,将 Hive 数据集中的分析结果导出到 MySQL 数据库中。

1. 导出岗位类别统计数据

从 Hive 的 job51_db.m_job_post_tb 表中将数据导出到 MySQL 的 post_tb 表中,代码如下:

```
//第 13 章/04_数据导出_spark.zpln

import java.util.Properties

//加载分析结果数据集
val postByTypeDF = spark.table("job51_db.m_post_tb")

//下面创建一个 prop 变量,用来保存 JDBC 连接参数
val props = new Properties()
props.put("user", "root")                        //表示用户名是 root
```

```
    props.put("password", "123456")              //表示密码是123456
    props.put("driver","com.mysql.jdbc.Driver")  //表示JDBC 驱动程序

//下面就可以连接数据库了,采用append 模式,表示将记录追加到数据库Spark 的students 表中
//数据库名为job51
val DB_URL= "jdbc:mysql://localhost:3306/job51?useSSL=false"
postByTypeDF.write
    .mode("append")
    .jdbc(DB_URL, "post_tb", props)
```

2. 导出招聘的行业分布数据

从 Hive 的 job51_db.m_post_by_companyind 表中将数据导出到 MySQL 的 post_of_companyind 表中,代码如下:

```
//第13章/04_数据导出_spark.zpln

import java.util.Properties

//加载分析结果数据集
val postByCompanyindDF = spark.table("job51_db.m_post_by_companyind")

//下面创建一个prop 变量,用来保存JDBC 连接参数
val props = new Properties()
props.put("user", "root")                     //表示用户名是root
props.put("password", "123456")               //表示密码是123456
props.put("driver","com.mysql.jdbc.Driver")   //表示JDBC 驱动程序

val table = "post_of_companyind"

//下面就可以连接数据库了,采用append 模式,表示将记录追加到数据库Spark 的students 表中
//数据库名为job51
val DB_URL= "jdbc:mysql://localhost:3306/job51?useSSL=false"
postByCompanyindDF.write
    .mode("append")
    .jdbc(DB_URL, "post_of_companyind", props)
```

在运行这个任务时,需要注意两点:

(1) 要确保已经启动了 Hive Metastore 服务。启动该服务的命令如下:

```
$ hive --service metastore
```

(2) 执行导出前,首先需要在 MySQL 数据库中创建名为 job51 的数据库,SQL 语句如下:

```
mysql> create database job51;
```

13.9 项目实现:数据可视化

项目的最后一步,是将分析结果进行可视化展示。数据可视化的技术有很多,这里选择 Java 技术栈中的 Spring MVC 框架,通过开发 Spring MVC Web 项目,使用 ECharts 作

为可视化组件,以此来展示前面的分析结果。

注意:这里选择 Spring MVC + ECharts 组合,是为了 Java 技术栈的项目完整性。实际上,读者可以任意选择自己喜欢的可视化工具和方法。

13.9.1 Spring MVC 框架简介

Spring MVC 是一种基于 Java 语言、实现了 Web MVC 设计模式、请求驱动类型的轻量级 Web 框架,即使用了 MVC 架构模式的思想,将 Web 层进行职责解耦。基于请求驱动指的就是使用请求-响应模型,Spring MVC 的目的就是帮助用户简化日常 Web 开发。

1. MVC 设计模式

MVC 设计模式的任务是将包含业务数据的模块与显示模块的视图解耦。MVC 在模型和视图之间引入重定向层可以解决问题。此重定向层是控制器,控制器将接受请求,执行更新模型的操作,然后通知视图关于模型更改的消息。

2. Spring MVC 核心组件

Spring MVC 的核心组件共有 5 个,分别如下。

(1)DispatcherServlet:前端控制器,用于接受所有请求。
(2)HandlerMapping:处理器映射器,用于配置请求路径与 Controller 组件的对应关系。
(3)Controller:控制器,具体处理请求的组件。
(4)ModelAndView:Controller 组件处理完请求后得到的结果,由数据与视图名称组成。
(5)ViewResolver:视图解析器,可根据视图名称确定需要使用的视图组件。

Spring MVC 的核心执行流程如图 13-11 所示。

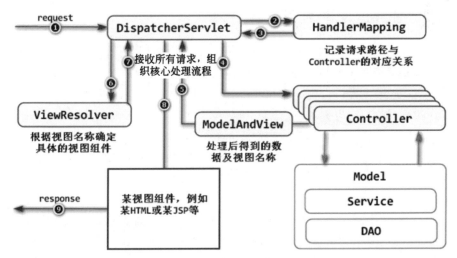

图 13-11　Spring MVC 的核心执行流程

3. Spring MVC 的执行流程

Spring MVC 的执行流程如下。

第 1 步：用户将 request 请求发送到前端控制器 DispatcherServlet。

第 2 步：前端控制器 DispatcherServlet 请求处理映射器 HandlerMapping 查找**Handler **（可以根据 xml 配置、注解进行查找）。

第 3 步：处理映射器 HandlerMapping 向前端控制器 DispatcherServlet 返回执行链处理对象 HandlerExecutionChain。

第 4 步：前端控制器调用处理器适配器 HandlerAdapter 去执行 Handler。

第 5 步：处理器适配器在 Handler 处理器中执行 Handler，Handler 处理器就是平常所讲的 Controller 控制器，是具体处理请求的组件。

第 6 步：Handler 执行完成后，由控制器 Controller 给适配器 HandlerAdapter 返回 ModelAndView（数据模型和视图）。

第 7 步：处理器适配器向前端控制器返回 ModelAndView（ModelAndView 是 Spring MVC 框架的一个底层对象，包括 Model 模型和 View 视图）。

第 8 步：前端控制器请求视图解析器 ViewResolver 去进行视图解析（根据逻辑视图名解析成真正的视图（JSP 页面）），通过这种策略可以很容易地更换其他视图技术，只需更改视图解析器。

第 9 步：视图解析器向前端控制器返回 View。

第 10 步：前端控制器进行视图渲染 （视图渲染将模型数据（在 ModelAndView 对象中）填充到 request 域）。

第 11 步：前端控制器向用户响应结果（response 对象）。

13.9.2　ECharts 图表库介绍

ECharts 是一个开源的纯 JavaScript 图表库，可以流畅地运行在 PC 和移动设备上，兼容当前绝大部分浏览器（IE 8/9/10/11、Chrome、Firefox、Safari 等），底层依赖轻量级的向量图形库 ZRender，提供直观、交互丰富、可高度个性化定制的数据可视化图表，是一款非常优秀的可视化前端框架。

ECharts 提供了常规的折线图、柱状图、散点图、饼图、K 线图，用于统计的盒形图，用于地理数据可视化的地图、热力图、线图，用于关系数据可视化的关系图、旭日图，多维数据可视化的平行坐标，还有用于 BI 的漏斗图，仪表盘，并且支持图与图之间的混搭。

ECharts 最初由百度团队开源，并于 2018 年初捐赠给 Apache 基金会，成为 ASF 孵化级项目。2021 年 1 月 26 日晚，Apache 基金会官方宣布 ECharts 项目正式毕业。

Apache ECharts 图表库的使用非常简单，下面是来自 ECharts 官网的一个简单示例，在 HTML 页面中显示柱状图，代码如下：

```html
//第13章/echarts_demo.html

<!DOCTYPE html>
<html>
<head>
    <meta charset="utf-8">
    <title>简单 ECharts 实例</title>
    <!-- 引入 echarts.js -->
    <script src="echarts.js"></script>
</head>
<body>
    <!-- 为 ECharts 准备一个具备大小（宽和高）的 DOM -->
    <div id="main" style="width: 600px;height:400px;"></div>
    <script type="text/javascript">
        //基于准备好的 DOM，初始化 ECharts 实例
        var myChart = echarts.init(document.getElementById('main'));

        //指定图表的配置项和数据
        var option = {
            title: {
                text: '简单 ECharts 实例'
            },
            tooltip: {},
            legend: {
                data:['销量']
            },
            xAxis: {
                data: ["衬衫","羊毛衫","雪纺衫","裤子","高跟鞋","袜子"]
            },
            yAxis: {},
            series: [{
                name: '销量',
                type: 'bar',
                data: [5, 20, 36, 10, 10, 20]
            }]
        };

        //使用刚指定的配置项和数据显示图表
        myChart.setOption(option);
    </script>
</body>
</html>
```

在浏览器中打开这个网页，可以看到生成的柱状图如图 13-12 所示。

13.9.3　Spring MVC Web 程序开发

对于 Spring MVC Web 程序开发，本案例选择使用 IntelliJ IDEA + Maven 的方式进行开发。

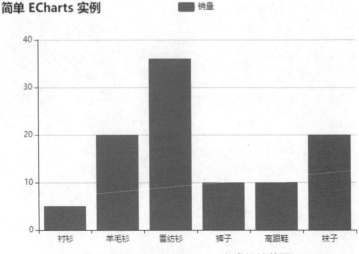

图 13-12　Apache ECharts 生成的柱状图

1. 创建 Maven 项目

首先在 IntelliJ IDEA 中创建一个 Maven 项目，建议按以下步骤操作。

（1）在 IntelliJ IDEA 中，选择菜单项 File→New Project，打开新建项目向导窗口，在窗口左侧选择 Maven，在右侧勾选复选框 Create from archetype，然后在右侧正文选择项目模板 org.apache.maven.archetypes:maven-archetype-webapp，如图 13-13 所示。

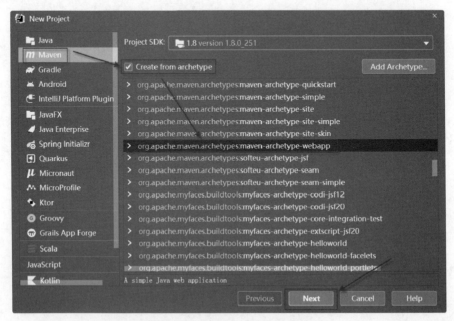

图 13-13　新建 Maven 项目向导

（2）在图 13-13 中单击 Next 按钮，进入项目创建向导的下个窗口，在其中输入项目的名称，选择项目文件的存放位置，然后单击 Next 按钮继续，如图 13-14 所示。

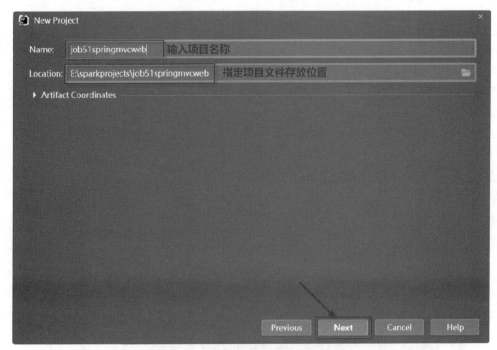

图 13-14　为新项目指定名称和存放目录

这样就新建了一个全新的 Maven 项目。此时的项目结构如图 13-15 所示。

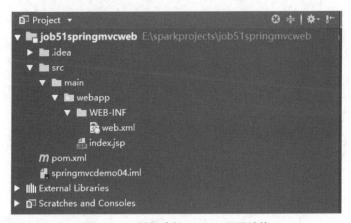

图 13-15　新创建的 Maven 项目结构

当然到这里只是新建了一个 Maven 项目，还没有添加 Spring MVC 相关的依赖和配置，接下来还需要添加对 Spring MVC 的支持。

2. 添加 Spring MVC 支持

为项目添加 Spring MVC 支持，建议按以下步骤操作。

（1）在项目名称上右击，在弹出的环境菜单中选择第 2 项 Add Framework Support...菜单项，如图 13-16 所示。

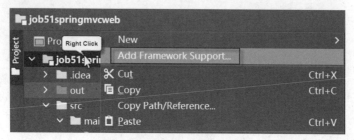

图 13-16　在项目名称上右击

（2）在打开的 Add Frameworks Support 窗口中，在左侧选择 Spring 和 Spring MVC，然后单击 OK 按钮确认，如图 13-17 所示。

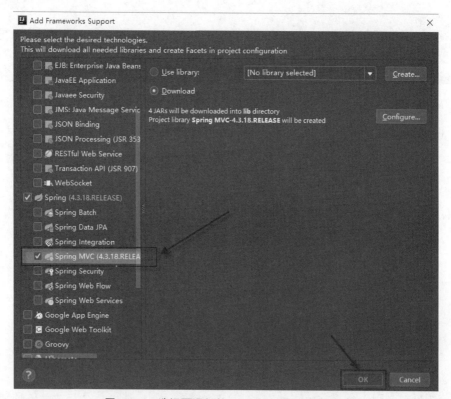

图 13-17　选择要添加的 Spring 和 Spring MVC

（3）接下来它会下载一些对 Spring MVC 支持的相关库，这可能需要几分钟时间，如图 13-18 所示。

图 13-18　下载 Spring MVC 支持库

（4）下载完毕后，会发现项目结构有了一些变化，如图 13-19 所示。

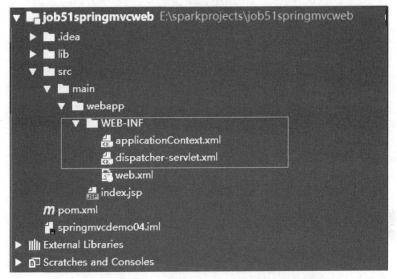

图 13-19　添加 Spring MVC 支持后的项目结构

如图 13-19 所示，在 WEB-INF 目录下，新建了两个配置文件 applicationContext.xml 和 dispatcher-servlet.xml。

（5）接下来对项目结构进行完善。在 src/main/目录下，新建两个目录 java 和 resources，并分别将 java 目录设为源代码根目录，将 resources 目录设为资源根目录。最终完成的项目结构如图 13-20 所示。

3. 项目 Web 配置和数据源配置

Spring MVC 具有非常棒的依赖注入特性，可以自动管理项目的请求映射转发、组件依赖注入和数据源管理。不过在此之前，需要进行一些配置。

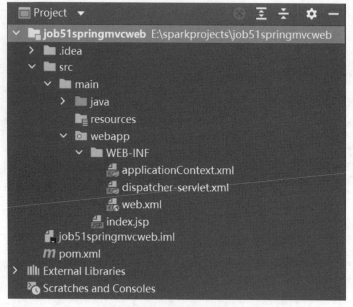

图 13-20 最终的项目结构

（1）首先配置 WEB-INF/web.xml 文件，内容如下：

```xml
//第13章/job51springmvcweb/src/main/webapp/WEB-INF/web.xml

<?xml version="1.0" encoding="UTF-8"?>
<web-app xmlns="http://xmlns.jcp.org/xml/ns/javaee"
     xmlns:xsi="http://www.w3.org/2001/XMLSchema-instance"
     xsi:schemaLocation="http://xmlns.jcp.org/xml/ns/javaee
http://xmlns.jcp.org/xml/ns/javaee/web-app_3_1.xsd"
     version="3.1">

  <display-name>SpringMVCExample</display-name>
  <welcome-file-list>
    <welcome-file>index.html</welcome-file>
    <welcome-file>index.jsp</welcome-file>
  </welcome-file-list>
  <context-param>
    <param-name>contextConfigLocation</param-name>
    <param-value>/WEB-INF/dispatcher-servlet.xml</param-value>
  </context-param>
  <listener>
    <listener-class>org.springframework.web.context.ContextLoaderListener</listener-class>
  </listener>
  <!-- 编码过滤器配置 -->
  <filter>
    <filter-name>characterEncodingFilter</filter-name>
```

```xml
    <filter-class>org.springframework.web.filter.CharacterEncodingFilter
</filter-class>
    <init-param>
      <param-name>encoding</param-name>
      <param-value>UTF-8</param-value>
    </init-param>
  </filter>
  <filter-mapping>
    <filter-name>characterEncodingFilter</filter-name>
    <url-pattern>/*</url-pattern>
  </filter-mapping>

  <servlet>
    <servlet-name>dispatcher</servlet-name>
    <servlet-class>org.springframework.web.servlet.DispatcherServlet
</servlet-class>
    <load-on-startup>1</load-on-startup>
  </servlet>
  <servlet-mapping>
    <servlet-name>dispatcher</servlet-name>
    <url-pattern>/</url-pattern>
  </servlet-mapping>
</web-app>
```

（2）配置 WEB-INF/dispatcher-servlet.xml 文件，内容如下：

```xml
//第13章/job51springmvcweb/src/main/webapp/WEB-INF/dispatcher-servlet.xml

<?xml version="1.0" encoding="UTF-8"?>
<beans xmlns="http://www.springframework.org/schema/beans"
       xmlns:xsi="http://www.w3.org/2001/XMLSchema-instance"
       xmlns:context="http://www.springframework.org/schema/context"
       xmlns:mvc="http://www.springframework.org/schema/mvc"
       xsi:schemaLocation="http://www.springframework.org/schema/beans
       http://www.springframework.org/schema/beans/spring-beans.xsd
       http://www.springframework.org/schema/context
       http://www.springframework.org/schema/context/spring-context.xsd
       http://www.springframework.org/schema/mvc
       http://www.springframework.org/schema/mvc/spring-mvc.xsd">

    <!-- 开启注解驱动 -->
    <mvc:annotation-driven />

    <!--静态资源请求检查-->
    <mvc:default-servlet-handler default-servlet-name="default"/>

    <mvc:resources location="/js/" mapping="/js/**"/>
    <mvc:resources location="/css/" mapping="/css/**"/>
    <mvc:resources location="/images/" mapping="/images/**"/>

    <!-- 提供组件扫描支持 -->
```

```xml
            <context:component-scan base-package="com.xueai8.job51"/>

            <!-- view resolver: -->
            <!-- InternalResourceViewResolver, UrlBasedViewResolver -->
            <bean class="org.springframework.web.servlet.view.InternalResourceViewResolver">
                <!-- viewClass: 解析视图的类型，默认为 JstlView 类型 -->
                <!-- 如果视图是带有 JSTL 标记的 JSP 模板，则可以忽略 viewClass 属性 -->
                <property name="viewClass" value="org.springframework.web.servlet.view.JstlView" />
                <property name="prefix" value="/WEB-INF/jsp/" />
                <property name="suffix" value=".jsp" />
            </bean>

            <!--数据源配置-->
            <bean id="dataSource" class="org.springframework.jdbc.datasource.DriverManagerDataSource">
                <!--MySQL 5.x-->
                <property name="driverClassName" value="com.mysql.jdbc.Driver" />
                <property name="url" value="jdbc:mysql://localhost:3306/job51?useSSL=false&amp;useUnicode=true&amp;characterEncoding=UTF-8" />
                <property name="username" value="root" />
                <property name="password" value="123456" />
            </bean>
</beans>
```

需要注意的是，上面的数据源配置中使用的 MySQL 5.x 的配置属性，如果读者使用的是 MySQL 8.x 及以上版本，则应修改配置属性，修改后的配置属性如下：

```xml
<!--MySQL 8-->
<property name="driverClassName" value="com.mysql.cj.jdbc.Driver" />
<property name="url" value="jdbc:mysql://localhost:3306/job51?useSSL=false&amp;useUnicode=true&amp;characterEncoding=UTF-8&amp;serverTimezone=UTC" />
```

另外需要注意将数据库的 username 属性（账号）和 password 属性（密码）修改为读者自己使用的数据账号和密码。

4. 编写数据模型类 Model

本项目需要用到两个数据模型类，分别封装从 MySQL 中查询出来的招聘岗位统计数据和招聘企业行业分布数据。在 src/main/java/目录下创建包 com.xueai8.job51.model，然后在该包下创建两个模型类：JobtypeNum 和 PostOfCompanyind。

JobtypeNum.java 源文件的内容如下：

```java
//第13章/job51springmvcweb/src/main/java/com/xueai8/job51/model/JobtypeNum.java

package com.xueai8.job51.model;

/**
 * Model
 * 岗位：招聘数量
 */
```

```java
public class JobtypeNum {
    private String postType;
    private int postNum;

    public JobtypeNum() {
    }

    public JobtypeNum(String postType, int postNum) {
        this.postType = postType;
        this.postNum = postNum;
    }

    public String getPostType() {
        return postType;
    }

    public void setPostType(String postType) {
        this.postType = postType;
    }

    public int getPostNum() {
        return postNum;
    }

    public void setPostNum(int postNum) {
        this.postNum = postNum;
    }

    @Override
    public String toString() {
        return "JobtypeNum{" +
                "postType='" + postType + '\'' +
                ", postNum=" + postNum +
                '}';
    }
}
```

PostOfCompanyind.java 源文件的内容如下：

```java
//第13章/job51springmvcweb/src/main/java/com/xueai8/job51/model/PostOfCompanyind.java
package com.xueai8.job51.model;

/**
 * model
 * 不同行业的岗位招聘数量
 */
public class PostOfCompanyind {
    private String postType;       //岗位类型
    private String companyind;     //行业
    private int postNum;           //招聘数量
```

```java
    public PostOfCompanyind() {
    }

    public PostOfCompanyind(String postType, String companyind, int postNum) {
        this.postType = postType;
        this.companyind = companyind;
        this.postNum = postNum;
    }

    public String getPostType() {
        return postType;
    }

    public void setPostType(String postType) {
        this.postType = postType;
    }

    public String getCompanyind() {
        return companyind;
    }

    public void setCompanyind(String companyind) {
        this.companyind = companyind;
    }

    public int getPostNum() {
        return postNum;
    }

    public void setPostNum(int postNum) {
        this.postNum = postNum;
    }

    @Override
    public String toString() {
        return "PostOfCompanyind{" +
                "postType='" + postType + '\'' +
                ", companyind='" + companyind + '\'' +
                ", postNum=" + postNum +
                '}';
    }
}
```

5. 编写数据访问接口 DAO

在 src/main/java/ 目录下创建包 com.xueai8.job51.dao，然后在该包下创建 DAO 数据访问接口文件 Job51Dao.java，代码如下：

```java
//第13章/job51springmvcweb/src/main/java/com/xueai8/job51/dao/Job51Dao.java

package com.xueai8.job51.dao;

import com.xueai8.job51.model.JobtypeNum;
import com.xueai8.job51.model.PostOfCompanyind;
```

```java
import java.util.List;

/**
 * DAO 接口
 */
public interface Job51Dao {

    //查看"岗位 - 招聘总数"
    public List<JobtypeNum> find();

    //查看 "岗位 - 行业 - 招聘总数"前 Top 10，根据岗位类型查找
    public List<PostOfCompanyind> Top 10(String postType);
}
```

在同一位置，创建该接口的实现类，代码如下：

```java
//第13章/job51springmvcweb/src/main/java/com/xueai8/job51/dao/Job51DaoImpl.java

package com.xueai8.job51.dao;

import com.xueai8.job51.model.JobtypeNum;
import com.xueai8.job51.model.PostOfCompanyind;
import org.springframework.beans.factory.annotation.Autowired;
import org.springframework.jdbc.core.JdbcTemplate;
import org.springframework.jdbc.core.RowMapper;
import org.springframework.stereotype.Component;

import javax.sql.DataSource;
import java.sql.ResultSet;
import java.sql.SQLException;
import java.util.List;

/**
 * DAO 实现类
 */
@Component
public class Job51DaoImpl implements Job51Dao{

    private JdbcTemplate jdbcTemplate;

    @Autowired
    public Job51DaoImpl(DataSource dataSoruce) {
        jdbcTemplate = new JdbcTemplate(dataSource);
    }

    @Override
    public List<JobtypeNum> find() {
        String sql = "select * from post_tb";
        List<JobtypeNum> jobList = jdbcTemplate.query(
                sql,
                new RowMapper<JobtypeNum>() {
                    @Override
                    public JobtypeNum mapRow(ResultSet rs, int rowNum) throws SQLException {
                        JobtypeNum job = new JobtypeNum();
```

```
                        job.setPostType(rs.getString("post_type"));
                        job.setPostNum(rs.getInt("post_number"));
                        return job;
                    }
                });

        return jobList;
    }

    @Override
    public List<PostOfCompanyind> Top 10(String postType) {
        String sql = "select * from post_of_companyind where post_type=? order by post_num limit 10";
        List<PostOfCompanyind> jobList = jdbcTemplate.query (
                sql,
                new Object[] { postType },
                new RowMapper<PostOfCompanyind>() {
                    @Override
                    public PostOfCompanyind mapRow(ResultSet rs, int rowNum) throws SQLException {
                        PostOfCompanyind job = new PostOfCompanyind();
                        job.setPostType(rs.getString("post_type"));
                        job.setCompanyind(rs.getString("companyind"));
                        job.setPostNum(rs.getInt("post_num"));
                        return job;
                    }
                });

        return jobList;
    }
}
```

DAO 接口的主要作用是访问 MySQL 数据源，查询数据并返回数据模型（或数据模型集合）。

6. 编写控制器 Controller

在 src/main/java/目录下创建包 com.xueai8.job51.controller，然后在该包下创建控制器类 Job51Controller.java，代码如下：

```
//第13章/job51springmvcweb/src/main/java/com/xueai8/job51/dao/Job51Controller.java

package com.xueai8.job51.controller;

import com.xueai8.job51.dao.Job51Dao;
import com.xueai8.job51.model.JobtypeNum;
import com.xueai8.job51.model.PostOfCompanyind;
import org.springframework.beans.factory.annotation.Autowired;
import org.springframework.stereotype.Controller;
import org.springframework.ui.Model;
import org.springframework.web.bind.annotation.RequestMapping;
import org.springframework.web.bind.annotation.RequestMethod;
import org.springframework.web.bind.annotation.RequestParam;
```

```java
import java.util.List;

/**
 * 控制器类
 */
@Controller
public class Job51Controller {

    //依赖注入DAO
    @Autowired
    private Job51Dao job51Dao;

    //动态数据柱状图 bar2.jsp
    @RequestMapping(value="/bar2", method=RequestMethod.GET)
    public String bar2(Model model){
        //查询数据，封装到模型对象中
        List<JobtypeNum> listJob = job51Dao.find();
        model.addAttribute("listJob", listJob);

        //返回要重定向的视图页面
        return "bar2";
    }

    //动态数据柱状图 bar2.jsp
    @RequestMapping(value="/pie2", method=RequestMethod.GET)
    public String pie2(Model model, @RequestParam("pt") String postType){
        //查询数据，封装到模型对象中
        List<PostOfCompanyind> listJob = job51Dao.Top 10(postType);
        model.addAttribute("listJob", listJob);

        //返回要重定向的视图页面
        return "pie2";
    }
}
```

在上面这个程序文件中，配置路径/bar2 请求映射到对 bar2()方法的调用，它将通过 DAO 数据访问类加载数据库中的数据，并将数据封装到模型 model 中，然后与视图页面一起返回，数据模型会在视图页面中进行渲染。同样，配置路径/pie2 请求映射到对 pie2() 方法的调用，它将通过 DAO 数据访问类加载数据库中的数据，并将数据封装到模型 model 中，然后与视图页面一起返回，数据模型会在视图页面中进行渲染，其中参数 postType 会被映射到客户端请求参数 pt，代表要查询的岗位类型，其值可以是"大数据""数据分析""Java"或"Python"这 4 个值之一。

13.9.4 前端 ECharts 组件开发

在 13.9.3 节中，通过 DAO 访问类加载数据库中的数据，会在视图页面中进行渲染。所谓渲染，指的是将数据按模板语法填充到模板文件内的 ECharts 图表组件中。

在 src/main/webapp/WEB-INF/目录下创建 jsp 文件夹,用来存放 jsp 模板文件,然后在 jsp 文件夹下,分别创建 bar2.jsp 文件和 pie2.jsp 文件,如图 13-21 所示。

图 13-21 项目架构图

1. 可视化按类别统计的岗位数量

编辑 bar2.jsp 模板文件,将按类别统计的岗位数量用 ECharts 中的柱状图组件可视化显示,代码如下:

```jsp
//第13章/job51springmvcweb/src/main/webapp/WEB-INF/jsp/bar2.jsp

<%@ page contentType="text/html;charset=UTF-8" language="java" %>
<%@ taglib uri="http://java.sun.com/jsp/jstl/core" prefix="c"%>
<c:set var="baseurl" value="${pageContext.request.contextPath }"></c:set>
<html>
    <head>
        <title>柱状图</title>
        <meta charset="UTF-8">
        <!-- 引入 ECharts 文件 -->
        <script src="/static/js/echarts.min.js"></script>
    </head>
    <body class="bg-light">
    <section id="content">
        <div class="wrapper doc">
            <!-- 正文内容 -->
            <article>
                <h3>柱状图</h3>

                <!-- 在绘图前需要为 ECharts 准备一个具备大小(宽和高)的 DOM 容器 -->
                <div id="main" style="width: 100%;height:550px;"></div>

                <script type="text/javascript">
                    //基于准备好的 DOM,初始化 ECharts 实例
                    var dom1 = document.getElementById('main');
                    var myChart = echarts.init(dom1);
```

```
//指定图表的配置项和数据
var option = {
    title: {
        text: '各岗位招聘数量统计',
        subtext: "一线城市",
        textStyle: {//主标题的属性
            color: '#C28D21',           //颜色
            fontSize: 28,                //大小
            fontStyle: 'oblique',        //斜体
            fontWeight: '700',           //粗细
            fontFamily: 'monospace'      //字体
        },
        subtextStyle: {//副标题的属性
            color: '#25664A',
            fontSize: 20                 //大小
        }
    },
    tooltip: {},
    legend: {
        data:['岗位招聘数量统计'],
        align: "right",
        top: "5%", //bottom:"20%"    //组件离容器的距离
        right: "10%", //left:"10%"   //组件离容器的距离
        width: "auto", //图例组件的宽度
        height: "auto", //图例组件的高度
    },
    //x轴
    xAxis: {
        data: [
            <c:forEach var="job" items="${listJob}">
            "${job.postType}",
            </c:forEach>
        ]
    },
    //y轴
    yAxis: {},
    //在ECharts里，series是指一组数值及它们映射成的图
    series: [{
        name: '招聘数量',
        type: 'bar',
        data: [
            <c:forEach var="job" items="${listJob}">
            ${job.postNum},
            </c:forEach>
        ]
    }],
    grid: {
        left: 300,
        top: 60
```

```
                    }
                };
                //使用刚指定的配置项和数据显示图表
                myChart.setOption(option);
        </script>
            </article>
        </div>
    </section>

    </body>
</html>
```

注意：为了便于阅读，这里只显示了最主要的代码。完整文件内容可参考配套的源码。

上面的柱状图在浏览器中执行时展示的结果如图13-22所示。

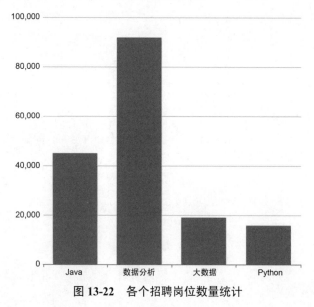

图13-22 各个招聘岗位数量统计

2. 可视化按指定招聘岗位统计的 Top 10 行业

在 pie2.jsp 模板文件中渲染的是按指定招聘岗位统计的 Top 10 行业，例如，招聘"大数据"岗位的 Top 10 行业。可以根据 url 参数来选择查询哪个岗位（大数据、数据分析、Java、Python 之一）的 Top 10 招聘企业。在 pie2.jsp 页面中使用 ECharts 的饼状图进行可视化展示，代码如下：

```
//第 13 章/job51springmvcweb/src/main/webapp/WEB-INF/jsp/pie2.jsp

<%@ page contentType="text/html;charset=UTF-8" language="java" %>
<%@ taglib uri="http://java.sun.com/jsp/jstl/core" prefix="c"%>
<!DOCTYPE html>
```

```html
<html>
    <head>
        <title>饼状图</title>
        <meta charset="UTF-8">
        <!-- 引入 ECharts 文件 -->
        <script src="/static/js/echarts.min.js"></script>
    </head>

    <body class="bg-light">

    <section id="content">
        <div class="wrapper doc">
            <!-- 正文内容 -->
            <article>
                <h3>按招聘企业所属行业分析</h3>
                <!-- 在绘图前需要为 ECharts 准备一个具备大小（宽和高）的 DOM 容器 -->
                <div id="main" style="width: 100%;height:550px;"></div>

                <script type="text/javascript">
                    //基于准备好的 DOM，初始化 ECharts 实例
                    var dom1 = document.getElementById('main');
                    var myChart = echarts.init(dom1);

                    //指定图表的配置项和数据
                    var option = {
                        title: {
                            text: '51job 招聘行业排名 Top 10',
                            subtext: '${listJob[0].postType}',
                            left: 'left'
                        },
                        tooltip: {
                            trigger: 'item'
                        },
                        legend: {
                            orient: 'vertical',
                            left: 'right',
                        },
                        series: [
                            {
                                name: '行业',
                                type: 'pie',
                                radius: '50%',
                                data: [
                                    <c:forEach var="job" items="${listJob}">
                                        {value:${job.postNum}, name:'${job.companyind}'},
                                    </c:forEach>
                                ],
                                emphasis: {
                                    itemStyle: {
                                        shadowBlur: 10,
                                        shadowOffsetX: 0,
                                        shadowColor: 'rgba(0, 0, 0, 0.5)'
```

```
                                }
                            }
                        }
                    ]
                };
                //使用刚指定的配置项和数据显示图表
                myChart.setOption(option);
            </script>
        </article>
    </div>
</section>

    </body>
</html>
```

注意：为了便于阅读，这里只显示了最主要的代码。完整文件内容可参考配套的源码。

在浏览器中访问这个页面，展示结果如图 13-23 所示。

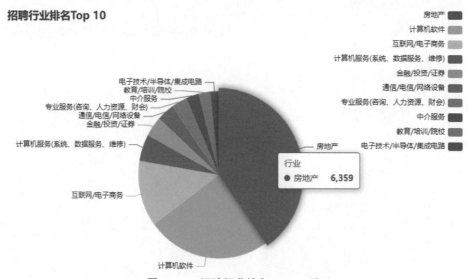

图 13-23　招聘行业排名 Top 10 统计

13.10　项目部署和测试

要部署 Spring MVC Web 项目，需要为项目添加一个 Tomcat 服务器，然后将 Web 项目打为 WAR 包，并部署到 Tomcat 服务器上运行。

1. 为项目添加 Tomcat 服务器

在 IntelliJ IDEA 中，选择菜单 File→Settings...，打开 Settings 设置面板，然后在面板左侧依次选择 Build,Execution,Deployment→Application Servers，然后在右侧单击加号按

钮（+），通过选择正确的 Tomcat 安装目录，向项目添加一个 Tomcat 服务器，然后单击 OK 按钮确认。过程如图 13-24 所示。

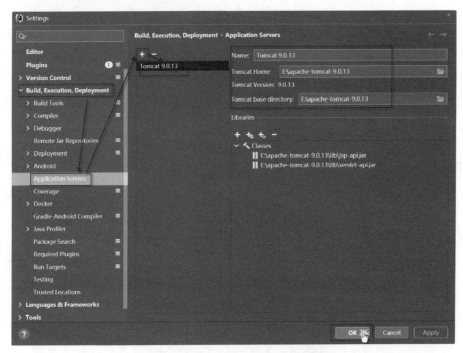

图 13-24　为项目添加 Tomcat 服务器

2. 部署 Spring Web 项目

在 IntelliJ IDEA 中的右上方工具栏，单击下拉列表中的下三角按钮，在打开的菜单中选择 Edit Configurations 菜单项，进行运行时配置，如图 13-25 所示。

图 13-25　选择运行时配置编辑菜单

在打开的 Run/Debug Configurations 面板中，添加一个新的运行时配置，然后切换到 Deployment 选项卡，单击加号（+）按钮，选择要部署的 Web 程序，如图 13-26 所示。

接下来，切换到 Server 选项卡，将应用程序服务器选为 Tomcat，选择默认打开网页的浏览器，并检查其他默认值为否正确，如图 13-27 所示。

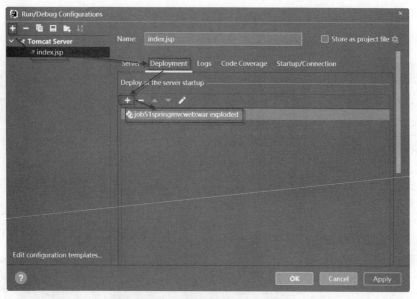

图 13-26 选择要部署的 Web 程序

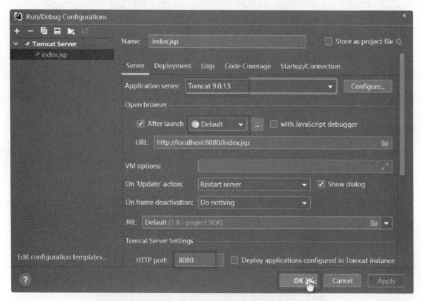

图 13-27 指定 Web 程序要部署的服务器等选项

一切准备就绪，然后单击 OK 按钮确认，这时会启动 Tomcat 服务器并部署 WAR 包，最后会自动弹出浏览器。因为项目中配置的两个访问路径分别是/bar2 和 pie2，所以可以在浏览器中分别访问的两个地址如下：

（1）http://localhost:8080/bar2，这时会打开可视化岗位统计数据的柱状图。

（2）http://localhost:8080/pie2，这时会打开可视化 Top 10 行业分布的饼状图。

图 书 推 荐

书名	作者
Flink 原理深入与编程实战（Scala + Java）	辛立伟
HarmonyOS 应用开发实战（JavaScript 版）	徐礼文
HarmonyOS 原子化服务卡片原理与实战	李洋
鸿蒙操作系统开发入门经典	徐礼文
鸿蒙应用程序开发	董昱
鸿蒙操作系统应用开发实践	陈美汝、郑森文、武延军、吴敬征
HarmonyOS 移动应用开发	刘安战、余雨萍、李勇军 等
HarmonyOS App 开发从 0 到 1	张诏添、李凯杰
HarmonyOS 从入门到精通 40 例	戈帅
JavaScript 基础语法详解	张旭乾
华为方舟编译器之美——基于开源代码的架构分析与实现	史宁宁
Android Runtime 源码解析	史宁宁
鲲鹏架构入门与实战	张磊
鲲鹏开发套件应用快速入门	张磊
华为 HCIA 路由与交换技术实战	江礼教
深度探索 Go 语言——对象模型与 runtime 的原理、特性及应用	封幼林
深入理解 Go 语言	刘丹冰
深度探索 Flutter——企业应用开发实战	赵龙
Flutter 组件精讲与实战	赵龙
Flutter 组件详解与实战	[加]王浩然（Bradley Wang）
Flutter 跨平台移动开发实战	董运成
Dart 语言实战——基于 Flutter 框架的程序开发（第 2 版）	亢少军
Dart 语言实战——基于 Angular 框架的 Web 开发	刘仕文
IntelliJ IDEA 软件开发与应用	乔国辉
深度探索 Vue.js——原理剖析与实战应用	张云鹏
Vue+Spring Boot 前后端分离开发实战	贾志杰
Vue.js 快速入门与深入实战	杨世文
Vue.js 企业开发实战	千锋教育高教产品研发部
Python 从入门到全栈开发	钱超
Python 全栈开发——基础入门	夏正东
Python 全栈开发——高阶编程	夏正东
Python 全栈开发——数据分析	夏正东
Python 游戏编程项目开发实战	李志远
Python 人工智能——原理、实践及应用	杨博雄 主编,于营、肖衡、潘玉霞、高华玲、梁志勇 副主编
Python 深度学习	王志立
Python 预测分析与机器学习	王沁晨
Python 异步编程实战——基于 AIO 的全栈开发技术	陈少佳
Python 数据分析实战——从 Excel 轻松入门 Pandas	曾贤志
Python 数据分析从 0 到 1	邓立文、俞心宇、牛瑶
Python Web 数据分析可视化——基于 Django 框架的开发实战	韩伟、赵盼

续表

书名	作者
FFmpeg 入门详解——音视频原理及应用	梅会东
Python 玩转数学问题——轻松学习 NumPy、SciPy 和 Matplotlib	张骞
Pandas 通关实战	黄福星
深入浅出 Power Query M 语言	黄福星
云原生开发实践	高尚衡
云计算管理配置与实战	杨昌家
虚拟化 KVM 极速入门	陈涛
虚拟化 KVM 进阶实践	陈涛
边缘计算	方娟、陆帅冰
物联网——嵌入式开发实战	连志安
动手学推荐系统——基于 PyTorch 的算法实现（微课视频版）	於方仁
人工智能算法——原理、技巧及应用	韩龙、张娜、汝洪芳
跟我一起学机器学习	王成、黄晓辉
TensorFlow 计算机视觉原理与实战	欧阳鹏程、任浩然
分布式机器学习实战	陈敬雷
计算机视觉——基于 OpenCV 与 TensorFlow 的深度学习方法	余海林、翟中华
深度学习——理论、方法与 PyTorch 实践	翟中华、孟翔宇
深度学习原理与 PyTorch 实战	张伟振
AR Foundation 增强现实开发实战（ARCore 版）	汪祥春
ARKit 原生开发入门精粹——RealityKit + Swift + SwiftUI	汪祥春
HoloLens 2 开发入门精要——基于 Unity 和 MRTK	汪祥春
巧学易用单片机——从零基础入门到项目实战	王良升
Altium Designer 20 PCB 设计实战（视频微课版）	白军杰
Cadence 高速 PCB 设计——基于手机高阶板的案例分析与实现	李卫国、张彬、林超文
Octave 程序设计	于红博
ANSYS 19.0 实例详解	李大勇、周宝
ANSYS Workbench 结构有限元分析详解	汤晖
AutoCAD 2022 快速入门、进阶与精通	邵为龙
SolidWorks 2020 快速入门与深入实战	邵为龙
SolidWorks 2021 快速入门与深入实战	邵为龙
UG NX 1926 快速入门与深入实战	邵为龙
西门子 S7-200 SMART PLC 编程及应用（视频微课版）	徐宁、赵丽君
三菱 FX3U PLC 编程及应用（视频微课版）	吴文灵
全栈 UI 自动化测试实战	胡胜强、单镜石、李睿
pytest 框架与自动化测试应用	房荔枝、梁丽丽
敏捷测试从零开始	陈霁、王富、武夏
深入理解微电子电路设计——电子元器件原理及应用（原书第 5 版）	[美]理查德•C.耶格（Richard C. Jaeger）、[美]特拉维斯•N.布莱洛克（Travis N. Blalock）著；宋廷强 译
深入理解微电子电路设计——数字电子技术及应用（原书第 5 版）	[美]理查德•C.耶格（Richard C.Jaeger）、[美]特拉维斯•N.布莱洛克（Travis N.Blalock）著；宋廷强 译
深入理解微电子电路设计——模拟电子技术及应用（原书第 5 版）	[美]理查德•C.耶格（Richard C.Jaeger）、[美]特拉维斯•N.布莱洛克（Travis N.Blalock）著；宋廷强 译